Texts and
Monographs
in Physics

Hartmut M. Pilkuhn

# Relativistic Particle Physics

Springer-Verlag
New York Heidelberg Berlin

Hartmut M. Pilkuhn

Institut für Kernphysik
Universität Karlsruhe
D-7500 Karlsruhe 1, Postfach 6380
Federal Republic of Germany

*Editors:*

Wolf Beiglböck

Institut für Angewandte Mathematik
Universität Heidelberg
Im Neuenheimer Feld 5
D-6900 Heidelberg 1
Federal Republic of Germany

Maurice Goldhaber

Department of Physics
Brookhaven National Laboratory
Associated Universities, Inc.
Upton, NY 11973
USA

Elliott H. Lieb

Department of Physics
Joseph Henry Laboratories
Princeton University
P. O. Box 708
Princeton, NJ 08540
USA

Walter Thirring

Institut für Theoretische Physik
der Universität Wien
Boltzmanngasse 5
A-1090 Wien
Austria

With 85 Figures

ISBN 0-387-09348-6 Springer-Verlag New York
ISBN 3-540-09348-6 Springer-Verlag Berlin Heidelberg

**Library of Congress Cataloging in Publication Data**

Pilkuhn, Hartmut.
Relativistic particle physics.

(Texts and monographs in physics)
Bibliography: p.
Includes index.
1. Particles (Nuclear physics) 2. Quantum field theory. I. Title.
QC793.2.P54 530.1'2 79-10666



Printed in the United States of America.



9 8 7 6 5 4 3 2 1

# Preface

Why study relativistic particle physics? Because of deeper understanding, curiosity and applications.

Consider first deeper understanding. Physics forms the basis of many other sciences, and relativistic particle physics forms the basis of physics. Starting from nonrelativistic point mechanics, there are three major steps: first to classical (unquantized) relativistic electrodynamics, then to nonrelativistic quantum mechanics and finally to relativistic quantum physics. This book describes the third step. Relativistic particle problems which are mainly classical (such as synchrotron radiation) are largely omitted (see for example Jackson 1975).

I have divided the subject into several smaller steps. The step from the Schrödinger equation to the Klein–Gordon and Dirac equations (chapter 1) is easy, apart from logical inconsistencies in limiting cases. Chapter 2 deals mainly with two-particle problems. From two-particle unitarity (sect. 2-5) and a symmetric treatment of projectile and target in the Born approximation to scattering (sect. 2-7), one is able to deduce recoil corrections to the relativistic one-particle equations (mainly the reduced mass, sect. 2-9). The final formulas provide a rather firm basis for atomic physics.

Quantum electrodynamics (QED) is presented in chapter 3. Clearly, many things must be omitted if one allots one chapter to the subject of whole books (Jauch and Rohrlich 1976, Källén 1958, Akhiezer and Berestetskii 1965, Bjorken and Drell 1965, Landau and Lifshitz 1971, 1975, and others). I have kept the formalism to a minimum, but I have tried to avoid sketchy derivations. There is even some relatively new material, for example the derivation (and limitation) of the Dirac equation of chapters 1 and 2 from QED (sect. 3-14). As in solid state physics and nuclear physics, second quantization is derived from the desired particle properties of the field operator, rather than by means of the "canonical formalism". Higher order perturbation theory is developed from unitarity and analyticity, mainly because this method is mandatory for the strong interactions (chapter 6). In practical calculations, the method is sometimes more elegant than the direct use of Feynman rules, and renormalization also looks somewhat nicer. However, I have not gone far in this direction. Mass renormalization is only mentioned in connection with the $\pi^+ - \pi^0$ mass difference (sect. 4-15), and the general proof of renormalizability is omitted. A possible unification of weak and electromagnetic interactions is discussed in sects. 5-10 and 5-11. The deeper understanding ends here, at least as far as the electromagnetic interaction is concerned.

Chapter 4 contains the particle zoo, particularly the many strongly interacting particles and resonances (hadrons). Among these, the "nonstrange" hadrons (nucleons, pions and the $\rho$, $\omega$, $\Delta$ . . . resonances) are important for the theory of nuclear forces. The strange, charmed and other "new particles" are presented for completeness. All these particles are classified by $SU(2)$-symmetry (=isospin), $SU(3)$-symmetry and by the quark model. Two-particle spin states, $P$, $C$, $T$, 3-particle states, the theory of unstable particle production and decay angular correlations are also included in chapter 4.

Chapter 5 is on weak interactions, about half of it on $\beta$-decay. The effects of the nuclear Coulomb field on electron spectra are discussed in some detail, in view of applications.

Chapter 6 contains that part of the theory of strong interactions which may be regarded as well established. The presently much discussed "quantum chromodynamics" (QCD) is still in the speculative stage and is barely mentioned. Some of the material of chapter 6 is more concerned with phenomenological parametrization (for example the Regge pole model).

Chapter 7 discusses particular hadronic processes. Much of this material is necessary for understanding nuclear physics. In fact, $\pi\pi$ scattering is necessary for understanding pion-nucleon ($\pi N$) scattering, and $\pi N$ scattering is necessary for understanding not only pion-nucleus scattering but also $NN$ scattering and nuclear physics as a whole.

Nuclear potentials are constructed in sect. 7-5. However, potential theory can hardly be the final basis of nuclear theory. The theoretical basis for a Schrödinger or Bethe-Salpeter equation for hadronic interactions is still quite weak. (The fact that we are better off in the electromagnetic case is connected with gauge invariance.) I have therefore indicated in several places (beginning with sect. 6-3 and ending with sect. 7-8) how the methods of analyticity and unitarity can be applied in nuclear physics more directly. The successes of these methods are still modest, but this could be partly due to the fact that they are unfamiliar to many nuclear theorists.

Chapter 8 is brief but has been included in order to give an idea of some of the most important applications of QED. I have also in several places mentioned effects which play a role for muonic and pionic atoms but hardly for normal atoms.

This book has grown over a period of >5 years. I have used much of its material in my lectures at Karlsruhe and also at Oslo. Suggestions for improvements and error corrections are welcome. Among the many people who have helped me, I should particularly mention Dr. E. Borie. Finally, I must thank our secretary E. Xanke for her patient and perfect typing of all the different versions of the manuscript.

I dedicate this book to my wife Aud, who delayed her own career and raised our 4 common indices.

Karlsruhe — HARTMUT PILKUHN
*May, 1979*

## Some Non-standard Definitions Used in This Book:

1. $d_L^3 p \equiv (2\pi)^{-3}\, d^3p/2E$ (see 1-2.19); $dLips$ is defined in (2-1.8).
2. $\sigma^{\mu\nu} = \frac{1}{2}(\gamma^\mu\gamma^\nu - \gamma^\nu\gamma^\mu)$ without extra factor $i$ (compare 1-5.13)
3. $S = 1 + iT$ instead of $S = 1 - iT$ (compare 1-3.3, 2-1.4).
4. Clebsch–Gordon sign conventions are used for all particle states which form an isospin (or $U$-spin) multiplet. This is inconvenient in the quark model (compare section 4-10).
5. Matrix elements are written as $T_{if} = \langle f | T | i \rangle$. Some authors have begun to write $T_{fi}$ instead, but there is little point in continuing in that direction until English is written from right to left. Perhaps mathematicians will start writing operators from left to right.
6. Read the lines between eqs. (1-0.1) and (1-0.3)!
7. Unit vectors have a circumflex, $\hat{\mathbf{p}} = \vec{\mathbf{p}}/p$.
8. Many matrix elements (particularly those of weak interactions) contain 2-component spinors $\chi = \chi(\vec{\mathbf{p}})$ which exist in right- and lefthanded versions, $\chi_r$ and $\chi_l$. The two versions coincide for $\vec{\mathbf{p}} = \vec{0}$ (Pauli spinors). From section 1-6 onwards, the notation $\chi_0$ is used for $\chi(\vec{0})$.

# Contents

## Appendices

CHAPTER ONE

# One-Particle Problems

## 1-0 Lorentz Invariance

Lorentz discovered that Maxwell's equations maintain their form under certain linear transformations of the time $t$ and coordinates $\mathbf{x}$. The transformations were further studied by Poincaré and generalized to massive particles of arbitrary interactions by Einstein. The homogeneous Maxwell equations are

$$\nabla \mathbf{H} = 0, \qquad \nabla \times \mathbf{E} + c^{-1}\dot{\mathbf{H}} = 0. \tag{0.1}$$

The form of the inhomogeneous equations depends on the measure of the electric charge $e$. Most books on relativistic quantum theory use "rationalized" (= Heavyside-Lorentz) units, in which the equations contain no factor $4\pi$:

$$\nabla \mathbf{E} = e\rho, \qquad \nabla \times \mathbf{H} - c^{-1}\dot{\mathbf{E}} = c^{-1}e\mathbf{j}. \tag{0.2}$$

Here $e$ is the elementary charge unit (proton charge). The static potential $\phi$ due to a proton at the origin is $(4\pi r)^{-1}e$ and the Coulomb potential between an electron (charge $-e$) and a nucleus of charge $Z_{\mathcal{N}}e$ is

$$V(r) = -e\phi = -\hbar c Z_{\mathcal{N}} \alpha r^{-1}, \qquad \alpha = e^2(4\pi\hbar c)^{-1} = 137^{-1}. \tag{0.3}$$

The term $\dot{\mathbf{E}}/c$ in the second equation of (2) was invented by Maxwell in order to save the continuity equation for the current $\mathbf{j}$:

$$\nabla \mathbf{j} + \dot{\rho} = 0. \tag{0.4}$$

If the six functions $\mathbf{E}(t, \mathbf{x})$, $H(t, \mathbf{x})$ are expressed in terms of four new ones (the scalar potential $\phi(t, \mathbf{x})$ and the vector potential $\mathbf{A}(t, x)$)

$$\mathbf{H} = \nabla \times \mathbf{A}, \qquad \mathbf{E} = -\nabla\phi - \dot{\mathbf{A}}/c, \tag{0.5}$$

then (1) is automatically satisfied.

The further discussion of these equations is simplified by grouping the time $t$ and the space coordinates $\mathbf{x}$ into one 4-vector:

$$x^\mu \equiv (ct, \mathbf{x}) \equiv (x^0, x^1, x^2, x^3) = (ct, x, y, z), \qquad \partial_\nu \equiv \partial/\partial x^\nu. \tag{0.6}$$

Similarly, $c\rho$ and $\mathbf{j}$ as well as $\phi$ and $\mathbf{A}$ are grouped into 4-vectors

$$j^\mu \equiv (c\rho, \mathbf{j}), \qquad A^\mu \equiv (\phi, \mathbf{A}), \tag{0.7}$$

by which we mean that they transform as $x^\mu$, as we shall see. The continuity equation (4) is then written as

$$\partial_\mu j^\mu \equiv \partial_0 j^0 + \sum_{i=1}^{3} \partial_i j^i = 0. \tag{0.8}$$

We shall generally need 4-vectors with a lower index:

$$A_\mu \equiv (A^0, -\mathbf{A}) = g_{\mu\nu} A^\nu, \qquad g_{\mu\nu} \equiv \begin{pmatrix} 1 & 0 & 0 & 0 \\ 0 & -1 & 0 & 0 \\ 0 & 0 & -1 & 0 \\ 0 & 0 & 0 & -1 \end{pmatrix} = g^{\mu\nu}, \tag{0.9}$$

and define the scalar product of two 4-vectors $A$ and $B$ as follows

$$A \cdot B \equiv A^0 B^0 - \mathbf{AB} = A_\mu B^\mu = A^\mu B_\mu = A^\mu g_{\mu\nu} B^\nu. \tag{0.10}$$

Repetition of a greek index implies summation over the products of the components as in (8). The tensor $g$ which transforms upper-index (= contravariant) 4-vectors into lower-index (= covariant) 4-vectors is called the metric tensor. It is then seen that $\mathbf{E}$ and $\mathbf{H}$ are components of the antisymmetric 4-tensor

$$F^{\mu\nu} = \partial^\mu A^\nu - \partial^\nu A^\mu = \begin{pmatrix} 0 & -E^1 & -E^2 & -E^3 \\ E^1 & 0 & -H^3 & H^2 \\ E^2 & H^3 & 0 & -H^1 \\ E^3 & -H^2 & H^1 & 0 \end{pmatrix}, \qquad \partial^\mu = (\partial^0, -\nabla), \tag{0.11}$$

and that the four equations (2) form one 4-vector equation:

$$-\partial_\nu F^{\mu\nu} = \partial_\nu \partial^\nu A^\mu - \partial^\mu(\partial_\nu A^\nu) = e j^\mu / c. \tag{0.12}$$

Since all these equations are differential ones, they are invariant against translations in time and space:

$$x'^\mu = x^\mu + a^\mu. \tag{0.13}$$

Maxwell's equations are also form-invariant under Lorentz transformations, which are those homogeneous transformations of 4-vectors which do not change the scalar product (10)

$$x'^{\mu} = \Lambda^{\mu}_{\rho} x^{\rho}, \qquad g_{\mu\nu} \Lambda^{\mu}_{\rho} \Lambda^{\nu}_{\sigma} = g_{\rho\sigma}. \tag{0.14}$$

Together with the inhomogeneous transformations (13), they form the inhomogeneous Lorentz group or "Poincaré" group.

Already in nonrelativistic mechanics, invariance of the equations of motion under (13) for a closed system results in conservation of energy $E$ and momentum $\mathbf{p}$ of that system. Einstein thus thought it natural to combine $E/c$ and $\mathbf{p}$ into one 4-momentum $P^{\mu}$. From here on, we put $c = 1$ (see appendix C-1):

$$P^{\mu} \equiv (P^{0}, \mathbf{p}) = (E, \mathbf{p}). \tag{0.15}$$

Since the system may interact via electromagnetic forces, it was then necessary to have $P^2 = P \cdot P = P_{\mu} P^{\mu}$ Lorentz invariant. For a single free particle of mass $m$, one must have

$$P^2 = E^2 - p^2 = m^2, \tag{0.16}$$

in order to regain for $E$ the nonrelativistic ($nr$) expression $p^2/2m$ in the limit $p \ll m$:

$$E = (m^2 + p^2)^{\frac{1}{2}} \xrightarrow[p \ll m]{} m + E_{nr}, \qquad E_{nr} = p^2/2m. \tag{0.17}$$

Thus in Einstein's theory the mass $m$ of a nonrelativistic particle appears as an additional energy $mc^2$.

A classical particle moves on a line $\mathbf{x} = \mathbf{x}(t)$ in 4-space, with velocity $\mathbf{v}(t) = d\mathbf{x}/dt$. Since $d\mathbf{x}$ and $dt$ form a 4-vector, $\mathbf{v}$ does not represent the spatial components of a 4-vector. On the other hand

$$d\tau^2 = dt^2 - d\mathbf{x}^2 = dt^2(1 - v^2) \tag{0.18}$$

is Lorentz-invariant. $\tau$ is called the proper time. The only possible 4-vector generalization of $\mathbf{v}(t)$ for a classical particle is therefore the 4-velocity

$$u^{\mu} = dx^{\mu}/d\tau, \qquad u_{\mu} u^{\mu} = 1. \tag{0.19}$$

A nonrelativistic particle of charge $Z$ moves according to $m d\mathbf{v}/dt = Ze\mathbf{E}$ in an electric field $\mathbf{E}(t, \mathbf{x})$.

Since $\mathbf{E}$ is part of the tensor $F^{\mu\nu}$, the only possible "covariant" generalization of this equation of motion is

$$du^{\mu}(t)/d\tau = ZeF^{\mu\nu} u_{\nu}(t)/m. \tag{0.20}$$

Insertion of (11) gives the Lorentz force law

$$m d\mathbf{u}/dt = Ze(\mathbf{E} + \mathbf{v} \times \mathbf{H}), \tag{0.21}$$

which is thus relativistically correct!

All these laws have been tested experimentally. As it is difficult to accelerate macroscopic bodies to relativistic velocities, the most accurate tests involve elementary particles. For example, the best relativistic clocks are provided by the decay rates of relativistic particles. The exponential decay laws must be relativistically invariant, i.e., for a beam of $N$ unstable particles moving with fixed velocity $\mathbf{v}$, we have

$$-dN(t) = d\tau N(t) \cdot \Gamma, \qquad -dN/dt = N\Gamma d\tau/dt = N\Gamma/\gamma \tag{0.22}$$

where $\Gamma$ is the decay rate of particles at rest ($\Gamma^{-1}$ is the "lifetime"). The ratio $\gamma = dt/d\tau = (1 - v^2)^{-\frac{1}{2}}$ is called the Lorentz factor. According to (22), it reduces the decay rate of moving particles. This "time dilatation" has been verified for decaying muons of $\gamma = 12$ with 1% accuracy (Bailey et al., 1972).

The great merit of Einstein was that he postulated Lorentz invariance (or rather Poincaré-invariance, which includes energy-momentum conservation) for all interactions in closed systems (i.e., excepting external forces). Consider for example the decay of a pion into a muon and an "antineutrino," $\pi^- \to \mu^- \bar{\nu}$. (The muon is like a heavy electron, and the antineutrino is massless and chargeless but has spin-$\frac{1}{2}$ just as $e^-$ and $\mu^-$). Before the decay, the total 4-momentum of the system is $P_\pi$. The decay is induced by a nonelectromagnetic interaction where $\mu$ and $\bar{\nu}$ appear instantly when the pion disappears. The total 4-momentum in the final state is $P_\mu + P_{\bar{\nu}}$ (the 4-vector index is omitted to avoid confusion with the particles $\mu$ and $\bar{\nu}$), and 4-momentum conservation requires

$$P_\pi = P_\mu + P_{\bar{\nu}}: \qquad E_\pi = E_\mu + E_{\bar{\nu}}, \qquad \mathbf{p}_\pi = \mathbf{p}_\mu + \mathbf{p}_{\bar{\nu}}. \tag{0.23}$$

In particular, if the pion is at rest ($\mathbf{p}_\pi = 0$, $E_\pi = m_\pi$), we have $\mathbf{p}_\mu = -\mathbf{p}_{\bar{\nu}}$ in the final state. With $m_{\bar{\nu}} = 0$, we have $E_{\bar{\nu}} = p_{\bar{\nu}}$ according to (16), and find

$$m_\pi = E_{\bar{\nu}} + (m_\mu^2 + p_\mu^2)^{\frac{1}{2}} = p_\mu + (m_\mu^2 + p_\mu^2)^{\frac{1}{2}}. \tag{0.24}$$

Thus, if $m_\mu$ is known and $\mathbf{p}_\mu$ is measured in $\pi$-decay, one can calculate $m_\pi$. The calculated value agrees well with the pion mass obtained from the energy levels of pionic atoms, where the pion is bound in a Bohr orbit (Daum et al. 1978).

Since Maxwell's equation as well as the equations of motion (20) for particles contain the $A^\mu$ only in the combination $F^{\mu\nu}$ of (11), they are invariant under "gauge transformations"

$$A'^\mu(t, \mathbf{x}) = A^\mu(t, \mathbf{x}) + \partial^\mu \Lambda(t, \mathbf{x}), \tag{0.25}$$

where $\Lambda(t, x)$ is an arbitrary scalar field. This arbitrariness in $A^\mu$ was used by Lorentz to decouple the $A^\nu$ in (12). He simply took $\Lambda$ such that $\partial_\nu \partial^\nu \Lambda + \partial_\nu A^\nu = 0$, obtaining

$$\partial_\nu \partial^\nu A'^\mu = ej^\mu, \qquad \partial_\mu A'^\mu = 0. \qquad (\partial_\nu \partial^\nu = \partial_0^2 - \nabla^2 \equiv \Box = \text{"quabla".}) \tag{0.26}$$

The solutions of these equations in vacuum ($j^\mu = 0$) can be decomposed into plane waves of the type

$$A^\mu(x) = \varepsilon^\mu e^{-iK \cdot x}, \qquad K \cdot x = K^0 t - \mathbf{kx}, \qquad K^{02} - k^2 = 0. \tag{0.27}$$

The last constraint in (27) guarantees $\Box A^\mu = 0$. The vector $\mathbf{k}$ is called the wave number vector and gives the direction of propagation of the wave, $K^0 = \omega$ is the frequency, and $\lambda = 2\pi/k$ is the wavelength. The wave has velocity $\lambda\omega/2\pi = 1$, i.e., the constant $c$ in (1) which we have put $= 1$ is the velocity of light in vacuum. In fact, $p/E \approx 1 = c$ is the velocity of any free particle of negligible mass.

From (10) we see that the only purpose of covariant 4-vectors and of the metric tensor $g$ is to hide the minus sign which occurs in the scalar product. This can also be achieved by defining

$$A^4 = iA^0, \qquad A \cdot B = -\sum_{\nu=1}^{4} A^\nu B^\nu. \tag{0.28}$$

Unfortunately, authors who use this convention drop the common minus sign on the right-hand side of (28). For them, a particle of 4-momentum $P$ has $P^2 = -m^2$. The method (28) becomes confusing for complex 4-vectors.

## 1-1 The Klein–Gordon Equation

Nonrelativistically, the Schrödinger equation is obtained from the classical relation between energy $E$ and momentum $\mathbf{p}$ of a particle by substituting

$$E \to i\hbar\, \partial_t, \qquad \mathbf{p} \to -i\hbar\nabla, \tag{1.1}$$

and by applying the resulting operators to the Schrödinger wave function $\psi_S$. Equations (0.11) and (0.15) tell us that (1) can be rewritten in terms of 4-vectors as follows:

$$P^\mu \to i\hbar\, \partial^\mu = i\hbar(\partial^0, -\nabla). \tag{1.2}$$

In this way, the free-particle equation $P_\nu P^\nu = m^2$ gives rise to the free Klein-Gordon equation (in the following, we take $\hbar = 1$ (see C-1)):

$$-\partial_\nu\, \partial^\nu \psi = m^2 \psi. \tag{1.3}$$

We already know that the Schrödinger equation contains the scalar potential in the combination $i\,\partial_t - V = i\,\partial_t - Ze\phi$, where $Ze$ now denotes the charge of the particle described by the wave function $\psi$. This combination must be generalized to a 4-vector and inserted into (3):

$$\pi_\nu \pi^\nu \psi = m^2 \psi, \qquad \pi^\mu \equiv i\,\partial^\mu - ZeA^\mu. \tag{1.4}$$

Equation (4) applies only to spinless particles (pions, for example), as we shall see. One can formally take the square root of the operator, which gives

$\pi^0\psi = \pm(\pi^2 + m^2)^{\frac{1}{2}}\psi$ and shows that the sign of the total energy is undetermined. In the nonrelativistic limit $\psi \to \psi_S$, however, the positive sign must be taken:

$$(i\,\partial_t - Ze\phi)\psi_S = (m^2 + \pi^2)^{\frac{1}{2}}\psi_S \approx (m + \pi^2/2m - \pi^4/8m^3)\psi_S. \tag{1.5}$$

The first term on the right-hand side is a constant which is normally subtracted in the energy scale (compare 0.17), and the last term is a relativistic correction.

In reality, the classical equations are, of course, consequences of the quantum equations and not vice versa. The classical limit of (4) is obtained by putting

$$\psi(t, \mathbf{x}) = B \exp\{iS\}, \qquad P^\mu(t, \mathbf{x}) = -\partial^\mu S \tag{1.6}$$

($B$ and $S$ are real functions) and by neglecting $\partial_\mu\,\partial^\mu B$ (see Section 1-11).

We now show that a theory based on (4) is invariant under the gauge transformation (0.25). We admit that $\psi$ transforms into another wave function $\psi'$, but only such that

$$\pi'^\mu\psi' = e^{iZ\chi(\Lambda(x))}\pi^\mu\psi, \tag{1.7}$$

because the additional phase of (7) will disappear in all matrix elements $\langle f|$ operator $(\pi^\mu)$ $|i\rangle$ even if the final state $|f\rangle$ contains different particles (as in $\pi$-decay), provided the total charge is conserved, $\sum_f Z_f = \sum_i Z_i$. Rewriting now

$$e^{iZ\chi}(i\,\partial^\mu - ZeA^\mu)\psi = [i\,\partial^\mu - ZeA^\mu + Z(\partial^\mu\chi)]e^{iZ\chi}\psi, \tag{1.8}$$

we see that (7) is satisfied if and only if

$$\chi = -e\Lambda, \qquad \psi' = \psi \exp(-iZe\Lambda). \tag{1.9}$$

Next, consider the current density $\mathbf{j}(t, \mathbf{x})$ and probability density $\rho(t, \mathbf{x})$. Nonrelativistically,

$$\mathbf{j} = \frac{1}{2m}(\psi^*\boldsymbol{\pi}\psi + \psi\boldsymbol{\pi}^*\psi^*) = \frac{1}{2mi}(\psi^*\nabla\psi - \psi\nabla\psi^* - 2iZe\mathbf{A}\psi^*\psi), \tag{1.10}$$

and $\rho = |\psi|^2$. Relativistically, this is inconsistent: the probability density must be defined such that the norm $\int \rho d^3x$ is time-independent, which means that it must satisfy the continuity equation (0.8). If (10) remains valid relativistically, then one needs

$$\rho = j^0 = \frac{1}{2m}(\psi^*\pi^0\psi + \psi\pi^{0*}\psi^*) = \frac{i}{2m}(\psi^*\,\partial_t\psi - \psi\,\partial_t\psi^* + 2iZe\phi\psi^*\psi). \tag{1.11}$$

It is now easily checked from (3) and its complex conjugate

$$(\pi_\nu^*\pi^{\nu*} + m^2)\psi^* = 0, \tag{1.12}$$

that (10) and (11) do in fact satisfy $\partial_\mu j^\mu = 0$. For stationary solutions, we have

$$\psi(x) = e^{-iEt}\psi(\mathbf{x}), \qquad \rho = m^{-1}(E - Ze\phi)|\psi|^2, \tag{1.13}$$

which shows that $\rho \to |\psi|^2$ for $E \to m$ and $Ze\phi \ll m$. In addition to the norm, the scalar product of two states $\psi_1$ and $\psi_2$ must be time-independent, which forces us to define

$$(\psi_2, \psi_1) = \int d^3x \, [\psi_2^*(i\,\partial_t - Ze\phi)\psi_1 - \psi_1(i\,\partial_t + Ze\phi)\psi_2^*]. \tag{1.14}$$

A factor $(2m)^{-1}$ is omitted, i.e., the norm is redefined as $2m \int \rho d^3x$. For free particles ($A^\mu = 0$ or $Z = 0$), the Klein-Gordon equation (3) has plane-wave solutions:

$$\psi_\mathbf{p}(x) = e^{-iEt + i\mathbf{p}\mathbf{x}}, \qquad (\psi_{\mathbf{p}'}, \psi_\mathbf{p}) \equiv \langle \mathbf{p}' | \mathbf{p} \rangle = (2\pi)^3 \, \delta_3(\mathbf{p} - \mathbf{p}')2E. \tag{1.15}$$

The reader might worry that with $\hbar = c = 1$ and a redefinition of the norm, it would be difficult to do dimensional checks of later results. We hope that the simplification of formulas over-compensates this disadvantage. A particular bonus of (15) is its applicability to massless particles. In fact, the free-field Maxwell equations in the Lorentz gauge, $\partial_\nu \partial^\nu A^\mu = 0$ are identical with Klein-Gordon equations (one for each value of $\mu$) for massless particles (photons). Their plane-wave solutions (0.27) are also normalized by (15) (except for a factor $\varepsilon_{2\mu}^* \varepsilon_1^\mu$ (see Section 2-8)).

Since (11) may become negative, it cannot really represent a probability density. Instead, $Z\rho$ and $Z\mathbf{j}$ may be interpreted as charge and current densities, in accordance with the original definitions (0.2). In fact, gauge invariance makes it absolutely necessary to have charge conservation in a local form. For probability conservation, such a strong statement cannot be made.

Our modified interpretation is supported by invariance of (4) under "charge conjugation" ($C$). It transforms $\phi$, $\mathbf{A}$ and $\psi$ as follows:

$$\phi_c = -\phi, \qquad \mathbf{A}_c = -\mathbf{A}, \qquad \psi_c = \eta_c \psi^* \tag{1.16}$$

where $\eta_c$ is a phase which will be fixed in section 4-5. The operation transforms (4) into (12) and amounts to a reversal of the particle's charge, $Ze \to -Ze$. Thus, if $\psi$ is a solution of the Klein-Gordon equation for a $\pi^+$, then $\psi^*$ is a solution of the Klein-Gordon equation for $\pi^-$. The charge density $e\rho$ changes sign under this operation, as it should.

Although we have avoided an obvious inconsistency at this point, it must be admitted that relativistic equations for states containing a fixed number of particles (one pion for example) are only approximately valid, except for free particles. The correct interpretation of the Klein-Gordon equation will be given in chapter 3. Limits of validity of single-particle equations will be mentioned in 3-14.

One may wonder whether the modified density will not give additional relativistic corrections to a standard nonrelativistic Schrödinger treatment of (5). To find these, we must transform $\psi$ into another function $\psi_S$ for which the probability density is $|\psi_S|^2$. This is done by considering the pair of functions (Feshbach and Villars 1958)

$$\psi_g = \tfrac{1}{2}[\psi + m^{-1}(i\,\partial_t - Ze\phi)\psi], \qquad \psi_k = \tfrac{1}{2}[\psi - m^{-1}(i\,\partial_t - Ze\phi)\psi], \tag{1.17}$$

which satisfy coupled differential equations of first order in $t$:

$$(i\,\partial_t - Ze\phi \mp m)\psi_{g,k} = \pm(2m)^{-1}\pi^2(\psi_g + \psi_k). \tag{1.18}$$

For $E \sim m$, $\psi_g$ is much larger than $\psi_k$, and the time dependence of $\psi_k$ is approximately $\exp(-imt)$. To lowest nonvanishing order in $m$, we thus obtain from (18)

$$\psi_k \approx -(4m^2)^{-1}\pi^2\psi_g. \tag{1.19}$$

The density $\rho$ of (11) is expressed in terms of $\psi_k$ and $\psi_g$ as follows:

$$\begin{aligned}\rho &= (\psi_g^* + \psi_k^*)\tfrac{1}{2}(\psi_g - \psi_k) - \tfrac{1}{2}(\psi_g + \psi_k)(\psi_k^* - \psi_g^*)\\ &= \psi_g^*\psi_g - \psi_k^*\psi_k.\end{aligned} \tag{1.20}$$

This shows that we can use (5) with $\psi_S = \psi_g$ up to terms of order $m^{-4}$. The method of getting higher-order corrections will be discussed in section 1-7 for spin-$\frac{1}{2}$ particles. It is not particularly useful for spinless particles such as $\pi^{\pm}$ because (4) gets modified by additional "strong interactions."

For stationary solutions (13), the Klein-Gordon equation becomes

$$[(E - Ze\phi)^2 - m^2 + (\nabla - iZe\mathbf{A})^2]\psi = 0, \tag{1.21}$$

which is brought into the form of the Schrödinger equation

$$[(\nabla - iZe\mathbf{A})^2 + k^2 - U(\mathbf{r})]\psi = 0, \tag{1.22}$$

by defining the wave number $k$ and the quantity $U$ as follows:

$$k = (E^2 - m^2)^{\frac{1}{2}}, \qquad U = V(2E - V), \qquad V = Ze\phi. \tag{1.23}$$

Note that $U$ is now energy-dependent. For $V \ll m$, $E \sim m$ we get $U = 2mV$, as in the nonrelativistic case.

## 1-2 Lorentz Transformations and Density of States

The transformations (0.14) can be written in matrix form $x' = \Lambda x$ with $x = (x^\rho)$ as a column and $\Lambda = (\Lambda^{\mu\rho}) = (\Lambda^\mu_\rho)$ as a $4 \times 4$ matrix. The scalar product (0.10) becomes $A_{\text{tr}}\,gB$, where $A_{\text{tr}}$ is the transpose of $A$, i.e., a row, and Lorentz invariance of the scalar product requires

$$\Lambda_{\text{tr}}\,g\Lambda = g. \tag{2.1}$$

The product of two Lorentz transformations $\Lambda$ and $\Lambda'$ is again a Lorentz transformation, i.e., if both $\Lambda$ and $\Lambda'$ satisfy (1), then $\Lambda'\Lambda$ also satisfies (1) (remember $(\Lambda'\Lambda)_{tr} = \Lambda_{tr}\Lambda'_{tr}$). $\Lambda^{-1}$ also satisfies (1), because of $g^{-1} = g$. Thus Lorentz transformations form a group. A familiar subgroup is the "3-dimensional orthogonal group" $O_3$, which does not transform the time $t = x^0$

$$\Lambda(O_3) = \begin{pmatrix} 1 & 0 & 0 & 0 \\ 0 & & & \\ 0 & & O_3^{ik} & \\ 0 & & & \end{pmatrix}, \qquad O_{3\text{tr}} O_3 = 1. \tag{2.2}$$

Since the determinant of a product is the product of the determinants, (1) requires $(\det \Lambda)^2 = 1$, i.e., $\det \Lambda = \pm 1$. The transformations with $\det \Lambda = 1$ form a subgroup. In the case of $O_3$, these are called rotations ($R$). $O_3$-matrices with $\det O_3 = -1$ can be decomposed as $S \cdot R$, where $S$ is the space reflection

$$S = \begin{pmatrix} -1 & 0 & 0 \\ 0 & -1 & 0 \\ 0 & 0 & -1 \end{pmatrix}, \qquad \Lambda(S) = \begin{pmatrix} 1 & 0 \\ 0 & S \end{pmatrix} = g. \tag{2.3}$$

$O_3$ has $O_2$ as a subgroup, which leaves one of the coordinates invariant. Let $z = x^3$ be that coordinate:

$$O_3(O_{2(z)}) = \begin{pmatrix} O_{2(z)}^{ik} & & 0 \\ & & 0 \\ 0 & 0 & 1 \end{pmatrix}, \qquad \begin{aligned} &(0^{11})^2 + (0^{21})^2 = (0^{12})^2 + (0^{22})^2 = 1, \\ &0^{11}0^{12} + 0^{21}0^{22} = 0. \end{aligned} \tag{2.4}$$

The 3 conditions on $O_{2(z)}^{ik}$ allow us to express the matrix $O_{2(z)}$ in terms of one single parameter $\omega$ as follows

$$O_{2(z)}(\omega) = \begin{pmatrix} \cos\omega & \sin\omega \\ -\sin\omega & \cos\omega \end{pmatrix}. \tag{2.5}$$

This parametrization has the following property: If $O''_{2(z)} = O'_{2(z)} \cdot O_{2(z)}$, then $\omega'' = \omega + \omega'$, as can be seen by rewriting $\sin\omega\cos\omega' + \cos\omega\sin\omega' = \sin(\omega + \omega')$, etc. Clearly, $\omega$ is the familiar "angle of rotation around the $z$ axis."

We now apply the same method to Lorentz transformations which do affect the $x^0$-component of $x^\mu$. There is first of all the time reversal $\mathscr{T}$:

$$x' = (-x^0, \mathbf{x}), \qquad \Lambda(\mathscr{T}) = -g, \tag{2.6}$$

which has det $\Lambda = -1$. The remaining transformations of det $\Lambda = +1$ are called "special" Lorentz transformations. We immediately jump to transformations $L_2$ which leave two of the space coordinates invariant. To be definite, let $y$ and $z$ be the invariant coordinates. The transformations $L_{2(y,z)}$ are called "transformations along the $x$-axis"

$$\Lambda(L_{2(y,z)}) = \begin{pmatrix} L_{2(y,z)} & & 0 & 0 \\ & & 0 & 0 \\ 0 & 0 & 1 & 0 \\ 0 & 0 & 0 & 1 \end{pmatrix}, \qquad \begin{aligned} &(L^{11})^2 - (L^{21})^2 = (L^{12})^2 - (L^{22})^2 = 1, \\ &L^{11}L^{12} - L^{21}L^{22} = 0. \end{aligned} \tag{2.7}$$

The 3 conditions again follow from (1) and allow us to express $L_{2(y,z)}$ in terms of one single parameter $\eta$ as follows

$$L_{2(y,z)}(\eta) = \begin{pmatrix} \cosh \eta & \sinh \eta \\ \sinh \eta & \cosh \eta \end{pmatrix}. \tag{2.8}$$

$\eta$ is called the "rapidity" and is related to the "velocity" $\beta = v/c$ as follows

$$\cosh \eta = (1 - \beta^2)^{-\frac{1}{2}}, \qquad \sinh \eta = \beta(1 - \beta^2)^{-\frac{1}{2}}, \qquad \beta = \tanh \eta. \tag{2.9}$$

Clearly, Lorentz transformations along the same axis are additive in the rapidity ($\eta'' = \eta + \eta'$) but not in the velocity ($\beta'' \neq \beta + \beta'$). For this reason, the parametrization in terms of rapidity is the most convenient one. For small $\eta$, we have of course $\beta \approx \eta$.

A general rotation has 3 real parameters and can be decomposed into rotations around given axes, as we shall see later. A general "special" Lorentz transformation $\Lambda(\boldsymbol{\eta})$ (along an arbitrary direction $\hat{\boldsymbol{\eta}}$) is

$$L^0_0 = \cosh \eta,\ L^i_0 = L^0_i = \hat{\eta}_i \sinh \eta,\ L^i_j = \delta_{ij} + \hat{\eta}_i\hat{\eta}_j(\cosh \eta - 1). \tag{2.10}$$

The action of $S$ and $\mathcal{T}$ on quantum mechanical states is not yet determined by (3) and (6), due to the existence of an additional transformation, namely charge conjugation C(1.16). The study of weak interactions (for example of the decays $\pi^- \to \mu^- \bar{\nu}$ and $\pi^+ \to \mu^+ \nu$) shows that these interactions are not parity-invariant, where the parity operator $P$ for spinless particles such as $\pi^\pm$ merely reverses the argument $\mathbf{r}$ of the wave function (for a plane wave (1.15), this is equivalent to replacing $\mathbf{p}$ by $-\mathbf{p}$). They are, however, invariant under $C\mathcal{P}$:

$$C\mathcal{P}\,|\pi^+(\mathbf{p})\rangle = -\,|\pi^-(-\mathbf{p})\rangle. \tag{2.11}$$

Thus, if we insist on the Lorentz invariance of all interactions as postulated by Einstein, we must identify $S = C\mathcal{P}$.

Unfortunately, there exists a "superweak" interaction in kaon decays which is not invariant under either $S$ or $\mathcal{T}$. There is no way of redefining the corresponding quantum mechanical operators, and we are forced to conclude that the interactions of particles are not strictly Lorentz invariant.

Fortunately, the damage appears to be limited to Lorentz transformations having det $\Lambda = -1$. In particular, kaon decays appear to be invariant under the joint transformation $S\mathscr{T} = C\mathscr{P}\mathscr{T}$, the corresponding transformations matrix

$$\Lambda(S)\Lambda(\mathscr{T}) = -1 \tag{2.12}$$

having det $\Lambda = +1$. Since any matrix of det $-1$ can be decomposed into a product of a matrix of det $+1$ and the matrix $g$, all one needs to exclude is $\pm g$. The remaining Lorentz transformations with det $\Lambda = +1$ form the subgroup of "proper Lorentz transformations." For these, Einstein's conjecture appears to be exact.

The general proper Lorentz transformations consist again of two pieces, one of which constitutes a subgroup. Putting $\rho = \sigma = 0$ in (0.14), we find

$$(\Lambda_0^0)^2 = 1 + \sum_{k=1}^{3} (\Lambda_0^k)^2 \geq 1, \tag{2.13}$$

which shows that $\Lambda_0^0$ cannot lie between $-1$ and $+1$. Thus the transformations with $\Lambda_0^0 \geq 1$ form a subgroup (the orthochronous group). The full proper group is obtained by multiplying a member of the orthochronous group by the $C\mathscr{P}\mathscr{T}$ transformation (12). Thus, in the following, we can restrict ourselves to proper orthochronous transformations.

We must also consider the Lorentz transformations of space-time differentials. The Jacobi determinant is just det $\Lambda$, which gives us

$$d^4x' = \det \Lambda \, d^4x = d^4x = dt\, dx\, dy\, dz \tag{2.14}$$

for proper Lorentz transformations. On the other hand, $d^3x$ is invariant only under rotations. One can of course also verify that (8) has the Jacobi determinant 1. Now it is also evident that the norm $\int \rho \, d^3x$ is Lorentz-invariant only if $\rho$ is the 0-component of a 4-vector as in (1.11).

In momentum space, $d^4P = dP_0 \, d^3p$ is Lorentz invariant, with a fourth differential $dP_0$ in addition to $d^3p$. It is then convenient to redefine the so-called phase space differential, which is the number of plane waves having momenta between $\mathbf{p}$ and $\mathbf{p} + d^3p$ (the density of states), divided by the normalization of these states. This is the quantity which enters "Fermi's golden rule" for transition rates. Nonrelativistically, one encloses the physical system in a cube of length $L$ and requires periodicity of the wave function at the boundary, in order to keep the momentum operator Hermitean. This allows only those discrete values of $p_x$, $p_y$, $p_z$ which satisfy

$$Lp_i = 2\pi n_i, \qquad n_i = \text{integer}. \tag{2.15}$$

The nonrelativistic normalization of the plane waves is $L^3$. One then takes $L \to \infty$ and gets the nonrelativistic phase space differential

$$L^{-3}\, d^3n = (2\pi)^{-3}\, d^3p. \tag{2.16}$$

We now define the "Lorentz invariant phase space differential" $d_L^3 p$ as that invariant quantity which reduces to (16) in the nonrelativistic limit, apart from a constant. Obviously we must start from $d^4P$ and ensure that the integration over $P_0$ in the transition rate picks only the value $P_0 = E \equiv (p^2 + m^2)^{\frac{1}{2}}$. This is accomplished by the function $\delta(P^2 - m^2)\theta(P_0)$, where $\theta$ is the step function

$$\theta(P_0) \equiv \tfrac{1}{2}(1 + \text{sign}\,(P_0)) = 0 \quad \text{for } P_0 < 0, \qquad 1 \quad \text{for } P_0 > 0. \quad (2.17)$$

The $\delta$-function depends only on $P^2 = P_0^2 - p^2$ and is obviously Lorentz-invariant. It can be rewritten as follows:

$$\begin{aligned}\delta(P^2 - m^2) &= \delta(P_0^2 - p^2 - m^2) = \delta[(P_0 - E)(P_0 + E)] \\ &= (2E)^{-1}[\delta(P_0 - E) + \delta(P_0 + E)], \qquad E \equiv (p^2 + m^2)^{\frac{1}{2}}. \quad (2.18)\end{aligned}$$

The step function kills the unwanted part $\delta(P_0 + E)$. The delicate point here is that sign $(P_0)$ is Lorentz-invariant only for $P^2 = m^2 > 0$, which follows from $\Lambda_0^0 \geq 1$. Thus "tachyons" (particles with $m^2 < 0$) are excluded, and zero-mass particles (photons and neutrinos) must be treated as limits of massive particles. Thus we obtain

$$d_L^3 p = (2\pi)^{-3}\, d^4P\, \delta(P^2 - m^2)\theta(P_0) = (2\pi)^{-3}(2E)^{-1}\, d^3p. \quad (2.19)$$

In the nonrelativistic limit $E \to m$, (19) reduces to (16) apart from the factor $(2m)^{-1}$. Equation (19) conforms with the state normalization (1.15) in the sense of the completeness relation

$$|\mathbf{p}\rangle = \int |\mathbf{p}'\rangle\langle\mathbf{p}'|\mathbf{p}\rangle\, d_L^3 p'. \quad (2.20)$$

## 1-3 Scattering of Spinless Particles

We first review some concepts of scattering theory (see Taylor 1972 for details). Scattering is basically a time-dependent process. At a large negative time $-T$, the wave function $\psi(t)$ which is an eigenstate of the full Hamilton operator coincides with a free wave packet $\phi_i(t)$. This packet is weakly localized in a region where the potential vanishes, and moves towards the scatterer. As soon as it begins to overlap with the potential, $\psi$ starts to deviate from $\phi_i$. At the large positive time $+T$ the probability density in the potential region is again zero, such that $\psi(t = +T)$ can be expanded in terms of weakly localized free wave packets $\phi_f(t)$:

$$\psi(-T) = \phi_i(-T), \qquad \psi(+T) = \sum_f S_{if}\phi_f(T). \quad (3.1)$$

The expansion coefficients $S_{if}$ form the "scattering matrix" or "$S$-matrix"; $\frac{1}{2}E_i^{-1}|S_{if}|^2$ is the probability of finding the final state $f$ when the initial state was $i$. The factor $\frac{1}{2}E_i^{-1}$ compensates for our normalization (1.15). When $i$ and $f$ stand for the continuous momenta $\mathbf{p}_i$ and $\mathbf{p}_f$, the differential transition probability is $\frac{1}{2}E_i^{-1}|S_{if}|^2 \, d_L^3 p_f$. The differential cross section equals this probability per time interval ($2T$ in our case) and per incident particle flux (flux $= p_i/E_i$ in our case of a plane wave of unit amplitude):

$$d\sigma = \tfrac{1}{2}E_i^{-1}|S_{if}|^2 \, d_L^3 p_f (2T)^{-1} E_i/p_i = \tfrac{1}{8}(E_f p_i)^{-1}|S_{if}|^2 T^{-1}(2\pi)^{-3} p_f^2 \, dp_f \, d\Omega. \tag{3.2}$$

Here we have used (2.19) and $d^3p = p^2 \, dp \, d\Omega$, $\Omega = (\vartheta, \varphi)$ are the polar and azimuthal angles of $\mathbf{p}_f$, and $d\Omega = d \cos \vartheta \, d\varphi$. If the potential is weak, we have little scattering, $S_{if} \approx \langle \mathbf{p}_f | \mathbf{p}_i \rangle$. In the following we assume a time-independent potential, such that energy is conserved. If we neglect the small energy spread which is necessarily present in a wave packet, $S$ must have the form

$$S_{if} = \langle \mathbf{p}_f | \mathbf{p}_i \rangle + i8\pi^2 \, \delta(E_i - E_f) f(\Omega). \tag{3.3}$$

The factor $i8\pi^2$ is inserted for later convenience. In (2) we need the square of a $\delta$-function, which is an undefined quantity. Here we help ourselves by noting that in a dynamical calculation such as perturbation theory, $\delta(E_i - E_f)$ arises from the time integral over plane waves. Therefore we may write

$$2\pi \, \delta^2(E_i - E_f) \approx \int_{-T}^{T} dt \exp\{i(E_f - E_i)t\} \, \delta(E_i - E_f) = 2T \, \delta(E_i - E_f). \tag{3.4}$$

To obtain the last expression, we have put $E_i = E_f$ in the exponent on account of the second $\delta$-function. We shall return to this trick in sections 2-1 and 3-1. The differential cross section is only measured for $\mathbf{p}_i \neq \mathbf{p}_f$, where the first term of (3) vanishes. Thus we find

$$d\sigma = \delta(E_i - E_f)|f(\Omega)|^2 (E_f p_i)^{-1} p_f^2 \, dp_f \, d\Omega = |f(\Omega)|^2 \, d\Omega. \tag{3.5}$$

To obtain the last form, we have used

$$p \, dp = E \, dE \tag{3.6}$$

which follows from $E^2 = p^2 + m^2$, and cancelled $dE_f$ against $\delta(E_i - E_f)$, taking $E_f = E_i$ afterwards. We see that the $8\pi^2$ in (3) was chosen such as to lead to the familiar form (5).

The equivalent time-independent scattering formalism runs as in the non-relativistic case. The asymptotic form of the wave function appropriate to scattering is

$$\psi(\mathbf{r}) = e^{ikz} + e^{ikr} r^{-1} f(k, \Omega), \tag{3.7}$$

with the $z$-axis along $\mathbf{k}$ and $\Omega = (\vartheta, \varphi) =$ polar and azimuthal angles of $\mathbf{r}$. For $\mathbf{A} = 0$ and spherically symmetric $V$, $\psi$ is expanded into spherical harmonics; $\nabla^2$ of (1.22) can then be replaced by $r^{-2} \, \partial_r r^2 \, \partial_r - L(L+1)r^{-2}$:

$$\psi(\mathbf{r}) = \sum_{L, M} a_{LM} R_L(r) Y_M^L(\vartheta, \varphi), \tag{3.8}$$

$$[r^{-2} \, \partial_r r^2 \, \partial_r + k^2 - U - L(L+1)r^{-2}]R_L = 0. \tag{3.9}$$

This is the familiar partial wave equation. In particular, in the regions where $V$ vanishes, it becomes

$$[\rho^{-2}\,\partial_\rho \rho^2\,\partial_\rho + 1 - L(L+1)\rho^{-2}]R_L(\rho) = 0, \qquad \rho \equiv kr, \tag{3.10}$$

which has the spherical Bessel and Neuman functions $j_L(\rho)$ and $n_L(\rho)$ as solutions (see Appendix A1). The solution corresponding to the asymptotic form (7) is obtained by expanding the plane wave and the "scattering amplitude" $f(k, \Omega)$:

$$e^{ikz} = \sum_L i^L(2L+1)j_L(\rho)P_L(\cos\vartheta) = \sum_L i^L[4\pi(2L+1)]^{\frac{1}{2}}j_L(\rho)Y_L^0, \tag{3.11}$$

$$f(k, \Omega) = \sum_L (2L+1)T_L(k)P_L(\cos\vartheta), \tag{3.12}$$

$$a_{LM} = \delta_{MO}\,i^L[4\pi(2L+1)]^{\frac{1}{2}}, \qquad \psi(\mathbf{r}) = \sum_L i^L(2L+1)R_L(\rho)P_L(\cos\vartheta). \tag{3.13}$$

In the region $V = 0$, the solution corresponding to (7) can be written as

$$R_L = e^{i\,\delta_L}[j_L(\rho)\cos\delta_L - n_L(\rho)\sin\delta_L] = \tfrac{1}{2}[h_L(\rho)e^{2i\,\delta_L} + h_L^*(\rho)], \tag{3.14}$$

where $\delta_L = \delta_L(k)$ is the "phase shift" and $h_L = j_L + in_L$ is the Hankel function. Probability conservation guarantees that $\delta_L$ is real as long as $V$ remains real. The "partial wave amplitude" $T_L$ of (12) is given by $\delta_L$ as follows:

$$T_L(k) = k^{-1}e^{i\,\delta_L}\sin\delta_L. \tag{3.15}$$

In the case of an unscreened Coulomb potential, $V(r \to \infty) \sim r^{-1}$ decreases so slowly that (14) must be modified (section 1-8).

## 1-4 Pauli Equation and Spinor Rotations

Electrons, muons, neutrinos, and their antiparticles have spin-$\frac{1}{2}$, which means that their quantum mechanical state is described by two wave functions instead of just one. These functions are combined into a spinor, the nonrelativistic version of which will be called a "Pauli spinor"

$$\psi_P(t, \mathbf{x}) = \begin{pmatrix} \psi_P^{(\frac{1}{2})} \\ \psi_P^{(-\frac{1}{2})} \end{pmatrix} = \psi_P^{(\frac{1}{2})}\chi(\tfrac{1}{2}) + \psi_P^{(-\frac{1}{2})}\chi(-\tfrac{1}{2}),$$
$$\chi(\tfrac{1}{2}) = \begin{pmatrix} 1 \\ 0 \end{pmatrix}, \qquad \chi(-\tfrac{1}{2}) = \begin{pmatrix} 0 \\ 1 \end{pmatrix}. \tag{4.1}$$

The $\psi$'s depend on $t$ and $\mathbf{x}$, the $\chi$'s don't. $\psi_P$ satisfies the "Pauli equation," which differs from the Schrödinger equation by a spin-dependent interaction with a magnetic field $\mathbf{H} = \text{rot}\,\mathbf{A}$. Defining the nonrelativistic Hamiltonian $H_P$ as usual without the particle rest energy $m$, the equation reads

$$(i\,\partial_t - m - H_P)\psi_P = 0, \qquad H_P = V + (\boldsymbol{\pi}^2 - Ze\boldsymbol{\sigma}\mathbf{H})/2m, \qquad \boldsymbol{\pi} \equiv -i\nabla - Ze\mathbf{A},$$

$$\sigma^1 = \sigma_x = \begin{pmatrix} 0 & 1 \\ 1 & 0 \end{pmatrix}, \qquad \sigma^2 = \sigma_y = \begin{pmatrix} 0 & -i \\ i & 0 \end{pmatrix}, \qquad \sigma^3 = \sigma_z = \begin{pmatrix} 1 & 0 \\ 0 & -1 \end{pmatrix}. \tag{4.2}$$

The charge $Z$ is $-1$ for electrons and muons and 0 for neutrinos. The $\boldsymbol{\sigma}$'s are Pauli's spin matrices and satisfy

$$\sigma^i \sigma^j = \delta_{ij} + i\varepsilon_{ijk}\sigma^k, \qquad (\mathbf{p}\boldsymbol{\sigma})(\mathbf{q}\boldsymbol{\sigma}) = \mathbf{pq} + i\boldsymbol{\sigma}(\mathbf{p} \times \mathbf{q}). \tag{4.3}$$

When $\mathbf{p}$ and $\mathbf{q}$ contain differential operators, the second form of (3) can be misleading. Using the first form we find, for example,

$$(\boldsymbol{\pi}\boldsymbol{\sigma})^2 = \boldsymbol{\pi}^2 - Ze(\nabla \times \mathbf{A} + \mathbf{A} \times \nabla)\boldsymbol{\sigma} = \boldsymbol{\pi}^2 - Ze\boldsymbol{\sigma}\mathbf{H}, \tag{4.4}$$

because of $\nabla \times \mathbf{A} = \text{rot}\,\mathbf{A} - \mathbf{A} \times \nabla$. This shows that if the electromagnetic interaction is introduced via the substitution $i\,\partial^\mu \to \pi^\mu = i\,\partial^\mu - ZeA^\mu$ as in the spinless case, $H_P$ can depend on $\boldsymbol{\sigma}$ only in the combination $-Ze\boldsymbol{\sigma}\mathbf{H}/2m$. On the other hand, since $\mathbf{H}$ is gauge invariant by itself, it is in principle possible to have

$$H_P - V = (\boldsymbol{\pi}\boldsymbol{\sigma})^2/2m - e\kappa\boldsymbol{\sigma}\mathbf{H}/2m = \boldsymbol{\pi}^2/2m - \mu\boldsymbol{\sigma}\mathbf{H}, \tag{4.5}$$

$$\mu \equiv (Z + \kappa)e/2m \equiv Zeg/4m, \tag{4.6}$$

where $\kappa$ is an "anomalous magnetic moment," $2 + 2\kappa/Z$ is the "$g$-factor" and $\mu$ is the "magnetic moment." In fact, protons and neutrons have also spin-$\frac{1}{2}$ but have vastly different values of $\mu$ (see table C-4), due to their inner structure. Even for electrons and muons, a small value $\kappa \approx -\alpha/2\pi$ arises from the interaction with virtual photons.

The charge density is now

$$\rho = \psi_P^+ \psi_P = \psi_P^{(\frac{1}{2})*}\psi_P^{(\frac{1}{2})} + \psi_P^{(-\frac{1}{2})*}\psi_P^{(-\frac{1}{2})}, \tag{4.7}$$

where the row spinor $\psi_P^+$ is the Hermitean adjoint of the column spinor $\psi_P$. (7) expresses the fact that at each point in space, a spin-$\frac{1}{2}$ particle has two states available. The current operator can be adopted from (1.10) with the same generalization. The continuity equation $\partial_\mu j^\mu = 0$ is then fulfilled because the new term in $H_P$ drops out of $\partial_t(\psi_P^+\psi_P)$, due to the Hermiticity of $\boldsymbol{\sigma}$.

How do spinors transform under rotations? The simplest possibility would be $\psi'(t, \mathbf{x}) = \psi(t, \mathbf{x}')$ as for spinless particles. However, since $H_P$ must be rotationally invariant and $\mathbf{H}$ transforms as a vector, $H_i'(\mathbf{x}) = O_{3ik} H_k(\mathbf{x}')$, $\psi^+\boldsymbol{\sigma}\psi$ must also transform as a vector. If each component of $\psi$ transformed into itself, $\psi^+\boldsymbol{\sigma}\psi$ would be a triplet of scalars instead of one vector. On the other hand, $\psi^+\psi$ must be a scalar, i.e., the transformation matrices are unitary. A phase factor can be omitted from the matrices, since it commutes with all $\boldsymbol{\sigma}$ and disappears in $\psi^+\boldsymbol{\sigma}\psi$. According to (B-1.10), we can therefore put

$$\psi'(\mathbf{x}) = SU_2(\boldsymbol{\omega})\psi(\mathbf{x}'),\ SU_2 = \exp\left(-i\sum_{k=1}^{3}\omega^k\lambda^k\right), \tag{4.8}$$

where the $\lambda^k$ are linearly independent Hermitean traceless matrices, for example the $\sigma^k$ of (2), and the $\omega^k$ are parameters of the rotation as in (2.5). One can choose these parameters such that they form a vector whose direction $\hat{\omega}$ is the rotation axis and whose length $\omega$ is the rotation angle. Rotations around the $z$-axis have $\boldsymbol{\omega} = (0, 0, \omega)$: These must transform $\psi^+\sigma^3\psi$ into itself (apart from the substitution $\mathbf{r}\to\mathbf{r}'$):

$$\exp(i\omega\lambda^3)\sigma^3\exp(-i\omega\lambda^3) = \sigma^3.$$

This requires $\lambda^3$ to commute with $\sigma^3$, which is only possible if $\lambda^3$ is proportional to $\sigma^3$. Moreover, we must have

$$\psi'^+\sigma_x\psi' = \cos\omega\ \psi^+\sigma_x\psi + \sin\omega\ \psi^+\sigma_y\psi \tag{4.9}$$

according to the first row of (2.5), from which we find $\lambda^3 = \frac{1}{2}\sigma^3$. The generalization to arbitrary $\boldsymbol{\omega}$ is obviously

$$SU_2(\boldsymbol{\omega}) = \exp(-\tfrac{1}{2}i\boldsymbol{\omega\sigma}) = \cos\tfrac{1}{2}\omega - i\hat{\boldsymbol{\omega}}\boldsymbol{\sigma}\sin\tfrac{1}{2}\omega. \tag{4.10}$$

One may also invert the problem, i.e., determine the transformations of a 3-component function $\mathbf{H}$ such that $\psi^+\boldsymbol{\sigma}\mathbf{H}\psi$ is invariant under the transformation (10) of $\psi$. It is convenient to rewrite the scalar product

$$\boldsymbol{\sigma}\mathbf{H} = \sigma^3H^3 - \sigma^+H^- - \sigma^-H^+, \qquad H^\pm = 2^{-\frac{1}{2}}(\mp H_x - iH_y),$$

$$\sigma^\pm = 2^{-\frac{1}{2}}(\mp\sigma_x - i\sigma_y), \qquad \sigma^+ = -2^{\frac{1}{2}}\begin{pmatrix}0 & 1\\ 0 & 0\end{pmatrix}, \qquad \sigma^- = 2^{\frac{1}{2}}\begin{pmatrix}0 & 0\\ 1 & 0\end{pmatrix}, \tag{4.11}$$

because each of the combinations $\psi^+\sigma^3\psi$, $\psi^+\sigma^+\psi$ and $\psi^+\sigma^-\psi$ transforms into itself for $\omega^1 = \omega^2 = 0$. Consequently, also $H^+$ and $H^-$ must transform into themselves for this choice of parameters. A short calculation shows that $\psi^+\sigma^\pm\psi$ is multiplied by $\exp(\mp i\omega)$, and invariance of (11) requires the same to be true for $H^\pm$. Transformation back to Cartesian coordinates shows that $H_x$ and $H_y$ must be transformed according to (2.5). Thus, from the mathematical point of view, the 3-dimensional orthogonal transformations form the 3-dimensional representation of the group $SU_2$ (see B-1). The components $a^3$, $a^\pm$ of a vector $\mathbf{a}$ are called spherical harmonics or Cartan components.

Later we shall need the unitary transformation from the Pauli matrices $\boldsymbol{\sigma}$ to the matrices $-\boldsymbol{\sigma}^*$ which also satisfy (3)

$$c\boldsymbol{\sigma}^*c^+ = -\boldsymbol{\sigma}. \tag{4.12}$$

Apart from a possible phase, $c$ is a rotation by $\pi$ around the $y$-axis:

$$c = \begin{pmatrix} 0 & -1 \\ 1 & 0 \end{pmatrix} = \exp\{-\tfrac{1}{2}i\pi\sigma_y\} = -i\sigma_y. \tag{4.13}$$

Let us now investigate the matrices $SU_2$ more closely. For rotations around the $z$- and $y$-axes ($\boldsymbol{\omega} = (0, 0, \varphi)$ and $(0, \vartheta, 0)$ respectively), (10) gives the following matrix elements:

$$(\exp(-\tfrac{1}{2}i\varphi\sigma_z))_{mm'} = \delta_{mm'}e^{-im\varphi}, \tag{4.14}$$

$$(\exp(-\tfrac{1}{2}i\vartheta\sigma_y))_{mm'} = \delta_{mm'} \cos \tfrac{1}{2}\vartheta - 2m\, \delta_{m,-m'} \sin \tfrac{1}{2}\vartheta \equiv d^{\frac{1}{2}}_{mm'}(\vartheta), \tag{4.15}$$

with $m = \pm\frac{1}{2}$ as usual. For explicit calculations, it is therefore useful to parametrize a general rotation not by $\boldsymbol{\omega}$, but by the Euler angles $\alpha$, $\beta$, and $\gamma$ which rotate successively around the $z$-, $y$-, and $z$-axes, by $\gamma$, $\beta$, and $\alpha$. In particular, for plane waves of momentum $\mathbf{p}$ with polar angle $\vartheta$ and azimuth $\varphi$ we shall need

$$D(\varphi, \vartheta, -\varphi)_{mm'} = (e^{-\frac{1}{2}i\varphi\sigma_z}e^{-\frac{1}{2}i\vartheta\sigma_y}e^{\frac{1}{2}i\varphi\sigma_z})_{mm'} = e^{i\varphi(m'-m)}\, d^{\frac{1}{2}}_{mm'}(\vartheta), \tag{4.16}$$

which consists of a rotation by $-\varphi$ around the $z$-axis, $\vartheta$ around the $y$-axis and $\varphi$ around the rotated $z$-axis. (The last rotation serves to reduce $D$ to the unit matrix for $\vartheta = 0$). The transformed states are explicitly

$$\chi_m(\lambda, \vartheta, \varphi) = \sum_{m'} D_{mm'}\chi_{m'}(\lambda) = D_{m\lambda} = e^{i\varphi(\lambda-m)}\, d_{m\lambda}(\vartheta), \tag{4.17}$$

$$\chi(\tfrac{1}{2}, \vartheta, \varphi) = \begin{pmatrix} \cos \frac{1}{2}\vartheta \\ \sin \frac{1}{2}\vartheta e^{i\varphi} \end{pmatrix}, \qquad \chi(-\tfrac{1}{2}, \vartheta, \varphi) = \begin{pmatrix} -\sin \frac{1}{2}\vartheta e^{-i\varphi} \\ \cos \frac{1}{2}\vartheta \end{pmatrix}. \tag{4.18}$$

They are eigenstates of the helicity operator

$$\begin{aligned} \tfrac{1}{2}\boldsymbol{\sigma}\hat{p} &= \tfrac{1}{2}(\sigma_z \cos \vartheta + \sigma_x \sin \vartheta \cos \varphi + \sigma_y \sin \vartheta \sin \varphi) \\ &= \frac{1}{2}\begin{pmatrix} \cos \vartheta & \sin \vartheta e^{-i\varphi} \\ \sin \vartheta e^{i\varphi} & -\cos \vartheta \end{pmatrix}. \end{aligned} \tag{4.19}$$

with eigenvalues $\lambda$. They can also be expressed as linear combinations of the original states $\chi(M) \equiv |M, 0, 0\rangle$ of (1) (we call $M$ the "magnetic quantum number"):

$$|\lambda, \vartheta, \varphi\rangle = \sum_{M'} D_{M'\lambda}(\varphi, \vartheta, -\varphi)|M', 0, 0\rangle, \tag{4.20}$$

which is an alternative form of (17). For later use, we also collect the matrix elements of $\boldsymbol{\sigma}$ between $\chi(M)$ of (1) and $\chi(\lambda, \vartheta, \varphi)$:

$$\begin{aligned} \chi^+(\lambda')\chi(M) &= e^{i\varphi(M-\lambda')}\, d^{\frac{1}{2}}_{M\lambda'}(\vartheta), \\ \chi^+(\lambda')\sigma_z\chi(M) &= 2Me^{i\varphi(M-\lambda')}\, d^{\frac{1}{2}}_{M\lambda'}(\vartheta), \\ \chi^+(\lambda')\sigma_x\chi(M) &= e^{-i\varphi(M+\lambda')}\, d^{\frac{1}{2}}_{-M\lambda'}(\vartheta), \\ -i\chi^+(\lambda')\sigma_y\chi(M) &= 2Me^{-i\varphi(M+\lambda')}\, d^{\frac{1}{2}}_{-M\lambda'}(\vartheta). \end{aligned} \tag{4.21}$$

## 1-5 The Dirac Equation

Dirac (1928) wrote his famous equation for relativistic spin-$\frac{1}{2}$ particles, using 4-component spinors. Van der Waerden (1929) realized that 2-component spinors are sufficient from the point of view of Lorentz invariance (see also the review by Bade and Jehle 1953), only the parity transformation looks less elegant there. Note that both the Klein-Gordon equation and Pauli's equation are invariant under

$$\mathbf{x}^{\mathscr{P}} = -\mathbf{x}, \qquad \mathbf{A}^{\mathscr{P}}(\mathbf{x}) = -\mathbf{A}(-\mathbf{x}), \qquad \psi^{\mathscr{P}}(\mathbf{x}) = \eta\psi(-\mathbf{x}), \tag{5.1}$$

where $\eta$ is an arbitrary phase as in (1.16). It was only after the discovery of parity violation that the use of two-component spinors was fully recognized (Feynman and Gell-Mann 1958, see also Feynman 1961).

The free-particle Klein-Gordon equation (1.3) follows directly from $E^2 - p^2 = m^2$ and must apply to particles of arbitrary spins. Pauli's equation (4.2) tells us that $\nabla^2$ must be rewritten as $(\nabla\boldsymbol{\sigma})^2$ before the substitution $-i\nabla \to \boldsymbol{\pi}$ is made. For the covariant generalization of this trick we need a fourth Hermitean matrix $\sigma^0$ such that $\partial_t^2 = (\partial_t\sigma^0)^2$ and $\sigma^0\boldsymbol{\sigma} = \boldsymbol{\sigma}\sigma^0$. Obviously, $\sigma^0 = \pm 1$. It will be slightly more practical always to use $\sigma^0 = 1$ but to admit $-\boldsymbol{\sigma}$ in addition to $\boldsymbol{\sigma}$. We thus define two sets of Pauli 4-matrices:

$$\sigma^\mu \equiv (\sigma^0, \boldsymbol{\sigma}), \qquad \sigma_\mu = (\sigma^0, -\boldsymbol{\sigma}) \equiv \sigma_l^\mu, \qquad \sigma^0 = \begin{pmatrix} 1 & 0 \\ 0 & 1 \end{pmatrix}, \tag{5.2}$$

$$\partial_\mu\, \partial^\mu = \partial_\mu \sigma_l^\mu\, \partial_\nu \sigma^\nu = \partial_\mu \sigma^\mu\, \partial_\nu \sigma_l^\nu. \tag{5.3}$$

The two forms in (3) are no longer equal when $i\,\partial^\mu$ is replaced by $\pi^\mu$. Thus we arrive at two different relativistic equations

$$\pi_\mu \sigma_l^\mu \pi_\nu \sigma^\nu \psi_r = m^2 \psi_r, \qquad (\pi^\mu = i\,\partial^\mu - ZeA^\mu) \tag{5.4}$$

$$\pi_\mu \sigma^\mu \pi_\nu \sigma_l^\nu \psi_l = m^2 \psi_l. \tag{5.5}$$

The indices $r$ and $l$ stand for right- and left-handed, respectively. If we rewrite $\pi_\mu \sigma_l^\mu$ as $\pi^\mu \sigma^\mu$ and observe that the parity transformation replaces $\pi^\mu(\mathbf{x})$ by $\pi_\mu(-\mathbf{x})$, we see that (5) is the parity transform of (4):

$$\psi_r^{\mathscr{P}}(\mathbf{x}) = \eta\psi_l(-\mathbf{x}). \tag{5.6}$$

In electrodynamics, we can simply define $\psi_l$ as $m^{-1}\pi_\nu \sigma^\nu \psi_r$, in which case (4) becomes a linear relation between $\psi_l$ and $\psi_r$:

$$m\psi_l \equiv \pi_\nu \sigma^\nu \psi_r, \qquad \pi_\mu \sigma_l^\mu \psi_l = m\psi_r. \tag{5.7}$$

Multiplying the second equation by $\pi_\nu \sigma^\nu$ and exchanging summation indices on its left-hand side, we see that (5) is also satisfied, i.e., the theory is parity-invariant. The first-order differential eqs. (7) form Dirac's equation; they are equivalent to the second-order equation (4). For the Pauli equation limit, one defines large ($g$) and small ($k$) components

$$\psi_g = 2^{-\frac{1}{2}}(\psi_r + \psi_l), \qquad \psi_k = 2^{-\frac{1}{2}}(\psi_r - \psi_l). \tag{5.8}$$

Then the Dirac equation assumes the equivalent form

$$(\pi^0 - m)\psi_g = \boldsymbol{\pi\sigma}\psi_k, \qquad (\pi^0 + m)\psi_k = \boldsymbol{\pi\sigma}\psi_g. \tag{5.9}$$

(8) and (9) are the spin-$\frac{1}{2}$ analogues of (1.17) and (1.18). Approximating $\psi_k \approx (2m)^{-1}\boldsymbol{\pi\sigma}\psi_g$, we find $\psi_g \approx \psi_P$.

The two equations (7) are combined into a single equation as follows:

$$\pi_\mu \gamma^\mu_{(h)} \psi_{(h)} = m\psi_{(h)}, \qquad \psi_{(h)} \equiv \begin{pmatrix} \psi_r \\ \psi_l \end{pmatrix}, \qquad \gamma^\mu_{(h)} = \begin{pmatrix} 0 & \sigma^\mu_l \\ \sigma^\mu & 0 \end{pmatrix}. \tag{5.10}$$

The index $(h)$ stands for "high-energy representation." The transition (8) to the "low-energy" representation is unitary and therefore leaves the Dirac equation form-invariant. The explicit $\gamma_{(l)}$-matrices are different, however:

$$\pi_\mu \gamma^\mu \psi = m\psi, \qquad \psi_{(l)} \equiv \begin{pmatrix} \psi_g \\ \psi_k \end{pmatrix}, \qquad \gamma^0_{(l)} = \begin{pmatrix} \sigma^0 & 0 \\ 0 & -\sigma^0 \end{pmatrix}, \qquad \boldsymbol{\gamma}_{(l)} = \begin{pmatrix} 0 & \boldsymbol{\sigma} \\ -\boldsymbol{\sigma} & 0 \end{pmatrix}. \tag{5.11}$$

In view of such unitary transformations, it is useful to characterize the $\gamma$-matrices by their anticommutators, which are representation-independent:

$$\gamma^\mu\gamma^\nu + \gamma^\nu\gamma^\mu = 2g^{\mu\nu}. \tag{5.12}$$

When the Pauli Hamilton operator is given by (4.5), the Dirac equation must contain an additional piece proportional to $\kappa F^{\mu\nu}$. The new piece must not contain any further derivatives, and therefore $F^{\mu\nu}$ must be contracted with $\gamma^\mu$ and $\gamma^\nu$. The antisymmetry (0.11) of $F^{\mu\nu}$ excludes the symmetric combination (12), and we are left with the antisymmetric one:

$$\sigma^{\mu\nu} = \tfrac{1}{2}(\gamma^\mu\gamma^\nu - \gamma^\nu\gamma^\mu), \qquad \sigma^{0i}_{(l)} = \begin{pmatrix} 0 & \sigma^i \\ \sigma^i & 0 \end{pmatrix}, \qquad \sigma^{ij}_{(l)} = -i\varepsilon_{ijk}\begin{pmatrix} \sigma^k & 0 \\ 0 & \sigma^k \end{pmatrix}. \tag{5.13}$$

(Most authors define $\sigma^{\mu\nu}$ as $i\sigma^{\mu\nu}$ of (13), for which they are punished by explicit factors $i$ in covariant matrix elements). The Dirac equation can then be written as

$$\left(\pi_\mu\gamma^\mu + \frac{\kappa}{2mZ}\pi_\mu\pi_\nu\sigma^{\mu\nu}\right)\psi = m\psi, \qquad \frac{1}{Z}\pi_\mu\pi_\nu\sigma^{\mu\nu}_{(l)} = e\begin{pmatrix} \mathbf{H}\boldsymbol{\sigma} & -i\mathbf{E}\boldsymbol{\sigma} \\ -i\mathbf{E}\boldsymbol{\sigma} & \mathbf{H}\boldsymbol{\sigma} \end{pmatrix}, \tag{5.14}$$

where $\pi_\mu\pi_\nu\sigma^{\mu\nu}$ is only an elegant way of writing $-\frac{1}{2}iZeF_{\mu\nu}\sigma^{\mu\nu}$. Separating (14) into equations for $\psi_g$ and $\psi_k$ and eliminating $\psi_k$ as before, we obtain

$$\left(\pi^0 - m + \frac{\kappa e}{2m}\boldsymbol{\sigma}\mathbf{H}\right)\psi_g$$
$$= \left(\boldsymbol{\sigma\pi} + i\frac{\kappa e}{2m}\boldsymbol{\sigma}\mathbf{E}\right)\left(\pi^0 + m - \frac{\kappa e}{2m}\boldsymbol{\sigma}\mathbf{H}\right)^{-1}\left(\boldsymbol{\sigma\pi} - i\frac{\kappa e}{2m}\boldsymbol{\sigma}\mathbf{E}\right)\psi_g. \tag{5.15}$$

The right-hand side reduces again to $(\boldsymbol{\sigma\pi})^2/2m$ in the nonrelativistic limit, and we are in fact led to (4.5) (corrections will be calculated in section 1-7). In spite of this, we shall deal mainly with the simpler equation (11) in the following, for two reasons: (i) it can be solved exactly for a central potential ($\mathbf{A} = 0$, $\phi = \phi(r)$), (ii) in the development of quantum electrodynamics, the wave function $\psi$ is replaced by a field operator $\Psi$ which creates and annihilates particles, which satisfies (11) and *not* (14). In fact, (14) would make quantum electrodynamics "nonrenormalizable." The small value $\kappa \approx -\alpha/2\pi$ will be derived in section 3-12.

After having settled on the Dirac equation, we must find a conserved current in order to define a scalar product. As in the spinless case, we first write down the Hermitean conjugates of (7)

$$(-i\,\partial_\mu - ZeA_\mu)\psi_r^+ \sigma^\mu = m\psi_l^+, \qquad (-i\,\partial_\mu - ZeA_\mu)\psi_l^+ \sigma_l^\mu = m\psi_r^+. \tag{5.16}$$

We can now construct two 4-vectors

$$j_r^\mu = \psi_r^+ \sigma^\mu \psi_r, \qquad j_l^\mu = \psi_l^+ \sigma_l^\mu \psi_l. \tag{5.17}$$

From (7) and (16) we find

$$i\,\partial_\mu j_r^\mu = m(\psi_r^+ \psi_l - \psi_l^+ \psi_r) = -i\,\partial_\mu j_l^\mu. \tag{5.18}$$

Thus the sum of $j_r$ and $j_l$ is a conserved quantity:

$$j^\mu = j_r^\mu + j_l^\mu, \qquad \partial_\mu j^\mu = 0, \tag{5.19}$$

This conclusion remains valid for $\kappa \neq 0$. The scalar product becomes

$$(\psi_2, \psi_1) = \int d^3x(\psi_{2r}^+ \psi_{1r} + \psi_{2l}^+ \psi_{1l}) = \int d^3x(\psi_{2g}^+ \psi_{1g} + \psi_{2k}^+ \psi_{1k}). \tag{5.20}$$

The 4-component version of (16) is

$$(-i\,\partial_\mu - ZeA_\mu)\psi^+ \gamma^{+\mu} = m\psi^+. \tag{5.21}$$

The Hermitean conjugate matrices $\gamma^+$ are eliminated by defining an adjoint spinor $\bar{\psi}$:

$$\bar{\psi} = \psi^+ \gamma^0, \qquad \gamma^0 \gamma^{\mu +} \gamma^0 = \gamma^\mu \tag{5.22}$$

(compare 12 and 10). The Dirac equation and its adjoint are therefore

$$(i\,\partial_\mu - ZeA_\mu)\gamma^\mu \psi = m\psi, \qquad (-i\,\partial_\mu - ZeA_\mu)\bar{\psi}\gamma^\mu = m\bar{\psi}, \tag{5.23}$$

and the conserved current is

$$j^\mu = \bar{\psi}\gamma^\mu \psi, \qquad j^0 = \bar{\psi}\gamma^0 \psi = \psi^+ \psi. \tag{5.24}$$

## 1-6 Lorentz Transformations of Spinors

After our detailed study of spinor rotations, the special Lorentz transformations follow easily. A general complex transformation matrix $C$ can be written as the product of a Hermitean and a unitary matrix

$$C = AU, \qquad U^+ = U^{-1}, \qquad A^+ = A. \tag{6.1}$$

We are already through with unitary matrices, so we can concentrate on $A$. We want $\psi^+\sigma^\mu\psi$ to transform as a 4-vector (the index $r$ on $\psi$ is omitted), which means that $\psi^+\ \partial_\mu\sigma^\mu\psi$ must transform as a scalar:

$$\psi'^+\ \partial'_\mu\sigma^\mu\psi' = \psi^+A\ \partial'_\mu\sigma^\mu A\psi = \psi^+\ \partial_\mu\sigma^\mu\psi, \tag{6.2}$$

apart from the transformation of the $x^\mu$ in $\psi$ and $\psi^+$. Taking the determinant of the matrix between $\psi^+$ and $\psi$, we find

$$(\det A)^2(\partial_0'^2 - \nabla'^2) = \partial_0^2 - \nabla^2 \quad \rightarrow \quad \det A = \pm 1. \tag{6.3}$$

But $\det A = -1$ corresponds to a phase ($\alpha = \pi$ in B-1.9) and can be discarded. Thus $A$ must have unit determinant and consequently have the form

$$A(\boldsymbol{\eta}) = \exp(\tfrac{1}{2}\boldsymbol{\eta\sigma}) = \cosh \tfrac{1}{2}\eta + \hat{\boldsymbol{\eta}}\boldsymbol{\sigma} \sinh \tfrac{1}{2}\eta. \tag{6.4}$$

The factor $\frac{1}{2}$ is again necessary if we want to identify $\eta$ with the rapidity (2.8) for transformations along the $z$-axis. In general, $\hat{\boldsymbol{\eta}}$ is the direction of the Lorentz transformation as in (2.10), i.e., vector components perpendicular to $\hat{\boldsymbol{\eta}}$ remain unchanged.

Multiplying the defining equation of $\psi_l$ in (5.7) by $\psi_r^+$ and remembering (2), we see that $\psi_l$ must transform inversely to $\psi_r$, i.e., with $A^{-1}(\boldsymbol{\eta}) = A(-\boldsymbol{\eta})$. We therefore have

$$\psi_r'(x^\mu) = \exp(\tfrac{1}{2}\boldsymbol{\eta\sigma})\psi_r(x'^\mu), \qquad \psi_l'(x^\mu) = \exp(-\tfrac{1}{2}\boldsymbol{\eta\sigma})\psi_l(x'^\mu). \tag{6.5}$$

Van der Waerden called $\psi_l$ a "dotted" spinor, and $\psi_r$ an "undotted" one.

In the following we consider plane-wave solutions of the free-particle Dirac equation

$$\psi_{\mathbf{p}}(t, \mathbf{x}) = \exp(-iP_\mu x^\mu)m^{\frac{1}{2}}\chi(\mathbf{p}), \tag{6.6}$$

where $\chi(\mathbf{p})$ is the relativistic spin function of a free particle. It is obtained from the nonrelativistic spin function $\chi(\mathbf{0}) \equiv \chi_0$ of (4.1) or (4.18) by a Lorentz transformation with velocity $\beta = p/E$ in the direction $\hat{\mathbf{p}}$ (a "boost"). According to (2.9), the boost has

$$\cosh\eta = E/m, \qquad \sinh\eta = p/m, \qquad \hat{\boldsymbol{\eta}} = \hat{\mathbf{p}}. \tag{6.7}$$

For $A(\mathbf{p})$ given by (4), we need

$$\begin{pmatrix}\cosh\\ \sinh\end{pmatrix}\frac{\eta}{2} = [\tfrac{1}{2}(\cosh\eta \pm 1)]^{\frac{1}{2}} = (2m)^{-\frac{1}{2}}(E \pm m)^{\frac{1}{2}}, \tag{6.8}$$

$$A(\mathbf{p}) = [2m(E+m)]^{-\frac{1}{2}}(E + m + \mathbf{p}\boldsymbol{\sigma}) = m^{-\frac{1}{2}}(E + \mathbf{p}\boldsymbol{\sigma})^{\frac{1}{2}} = (m^{-1}P_\mu \sigma_l^\mu)^{\frac{1}{2}}. \tag{6.9}$$

The matrix square root is checked by squaring and using

$$m^2 = (E-p)(E+p), \qquad E + m \pm p = [2(E+m)(E \pm p)]^{\frac{1}{2}}. \tag{6.10}$$

One may also check that $\chi^+(\mathbf{p})P_\mu \sigma^\mu \chi(\mathbf{p})$ is independent of $\mathbf{p}$. From (5) and (6) we find for 4-component spinors

$$\psi_{\mathbf{p}} = \exp(-iP_\mu x^\mu)u(\mathbf{p}), \qquad u_{(h)} = \begin{pmatrix}u_r\\ u_l\end{pmatrix} = \begin{pmatrix}(P_\mu \sigma_l^\mu)^{\frac{1}{2}}\\ (P_\mu \sigma^\mu)^{\frac{1}{2}}\end{pmatrix}\chi_0. \tag{6.11}$$

The $m^{\frac{1}{2}}$ in (6) has been introduced such that (11) applies also to massless particles (neutrinos). For the low-energy representation, insertion of the first form of (9) gives

$$u_{(l)} = (m/2)^{\frac{1}{2}}\begin{pmatrix}A(\mathbf{p}) + A(-\mathbf{p})\\ A(\mathbf{p}) - A(-\mathbf{p})\end{pmatrix}\chi_0 = (E+m)^{\frac{1}{2}}\begin{pmatrix}1\\ \boldsymbol{\sigma}\mathbf{p}/(E+m)\end{pmatrix}\chi_0. \tag{6.12}$$

For $\chi_0$, we can use both the original spin states (4.1) or the helicity states (4.18). In the latter case, we have from (4.19)

$$\boldsymbol{\sigma}\mathbf{p}\chi_0(\lambda, \hat{\mathbf{p}}) = 2\lambda p \chi_0(\lambda, \hat{\mathbf{p}}), \tag{6.13}$$

$$\chi_r = m^{-\frac{1}{2}}(E + 2\lambda p)^{\frac{1}{2}}\chi_0, \qquad \chi_l = m^{-\frac{1}{2}}(E - 2\lambda p)^{\frac{1}{2}}\chi_0, \tag{6.14}$$

which shows that for $p/E \sim 1$, $\chi_r$ is nonzero only for $\lambda = \frac{1}{2}$ (right-handed helicity). Since the helicity operator $\frac{1}{2}\boldsymbol{\sigma}\hat{\mathbf{p}}$ commutes with $P_\mu \sigma_l^\mu$, $\lambda$ has the meaning of angular momentum along the momentum direction. For $\chi_0 = \chi_0(M)$, on the other hand, (12) is neither an eigenstate of $\boldsymbol{\sigma}\hat{\mathbf{p}}$ nor of $\sigma_z$, since $\sigma_z$ does not commute with $P_\mu \sigma_l^\mu$.

A general homogeneous Lorentz transformation to a system characterized by a particle momentum $\mathbf{p}'$ can be decomposed into an "antiboost" to the particle rest frame ($\mathbf{p} \to 0$), a rotation of the spin states in the rest frame to the direction of $\mathbf{p}'$ (Wigner rotation) and a boost along $\mathbf{p}'$. As the helicity is not affected by the antiboost and boost, we have

$$\psi'_{\mathbf{p}\lambda} = \exp(-iP'_\mu x'^\mu)(P'_\mu \sigma_l^\mu)^{\frac{1}{2}}\chi_0(\lambda, \hat{\mathbf{p}}) = \sum_{\lambda'} D_{\lambda'\lambda}(\varphi_R, \vartheta_R, -\varphi_R)\psi_{\mathbf{p}'\cdot\lambda'} \tag{6.15}$$

where $\varphi_R$ and $\vartheta_R$ now characterize the angles between $\mathbf{p}$ and $\mathbf{p}'$ in the particle rest frame. From two different plane wave spinors $\chi_a = \chi(\mathbf{p}_a, M_a)$ and $\chi_b = \chi(\mathbf{p}_b, M_b)$ we can form two Lorentz-invariants:

$$\chi_b^+ \chi_{al} = m_a^{-1}\chi_b^+ P_{a\mu}\sigma^\mu \chi_a, \qquad \chi_{bl}^+ \chi_a = m_b^{-1}\chi_b^+ P_{b\mu}\sigma^\mu \chi_a. \tag{6.16}$$

Instead of (16) one normally uses the scalar and pseudoscalar combinations, with $\bar{u} = u^+\gamma_0$:

$$\bar{u}_b u_a = m_a^{\frac{1}{2}} m_b^{\frac{1}{2}} (\chi_{bl}^+ \chi_a + \chi_b^+ \chi_{al}), \qquad \bar{u}_b \gamma_5 u_a = m_a^{\frac{1}{2}} m_b^{\frac{1}{2}} (\chi_{bl}^+ \chi_a - \chi_b^+ \chi_{al}) \quad (6.17)$$

The pseudoscalar $\bar{u}_b \gamma_5 u$ changes sign under the parity transformation (5.6). The matrix $\gamma_5$ is defined by (17) in the high-energy representation, but it will also be needed in the low-energy representation:

$$\gamma_{5(h)} = \begin{pmatrix} \sigma_0 & 0 \\ 0 & -\sigma_0 \end{pmatrix}, \qquad \gamma_{5(l)} = \begin{pmatrix} 0 & \sigma_0 \\ \sigma_0 & 0 \end{pmatrix}. \quad (6.18)$$

It can be defined in a representation-independent way as follows:

$$\gamma_5 = i\gamma^0\gamma^1\gamma^2\gamma^3 = -i\gamma_0\gamma_1\gamma_2\gamma_3. \quad (6.19)$$

For $a = b$, (17) gives the orthonormalization

$$\bar{u}(\mathbf{p}, M')u(\mathbf{p}, M) = 2m\,\delta_{MM'}. \quad (6.20)$$

Finally, we consider charge conjugation, which transforms (5.23) into equations for the charge conjugate spinor $\psi_c$ where $Z$ is replaced by $-Z$:

$$(i\,\partial_\mu + ZeA_\mu)\gamma^\mu\psi_c = m\psi_c, \qquad (-i\,\partial_\mu + ZeA_\mu)\bar{\psi}_c\gamma^\mu = m\bar{\psi}_c. \quad (6.21)$$

This sign reversal of $Z$ is accomplished by a complex conjugation as in (1.16). The 2-component equations (5.7) are transformed into

$$(i\,\partial_\mu + ZeA_\mu)\sigma^{*\mu}\psi_r^* = -m\psi_l^*, \qquad (i\,\partial_\mu + ZeA_\mu)\sigma_l^{\mu*}\psi_l^* = -m\psi_r^*, \quad (6.22)$$

and the star on the Pauli matrices is transformed away by the matrix $c$ (for "charge conjugation") of (4.12):

$$c\sigma^{\mu*}c^+ = \sigma_l^\mu, \qquad c\sigma_l^{\mu*}c^+ = \sigma^\mu. \quad (6.23)$$

This shows that $\psi_r$ and $\psi_l^*$ are exchanged under charge conjugation. One of the two spinors must be given an extra minus sign to compensate for the sign in front of $m$ in (22). We thus take

$$\psi_{cr} = \eta_c c\psi_l^*, \qquad \psi_{cl} = -\eta_c c\psi_r^*, \qquad \psi_c = i\eta_c\gamma^2\psi^* = i\eta_c\gamma^2\gamma^0\bar{\psi}_{\mathrm{tr}}. \quad (6.24)$$

The last equation is for 4-component spinors. The "charge parity" $\eta_c$ will be discussed in section 4-5. For electrons, $\eta_c = 1$. As in the spinless case, we expect $j^\mu$ (5.19) to change sign under charge conjugation. We eliminate its $j_l$-part in favour of $j_{cr}$ by transposing $\psi_l^+\sigma_l^\mu\psi_l$:

$$j_l^\mu = \psi_l^+\sigma_l^\mu\psi_l = \psi_{l,\mathrm{tr}}\sigma_l^{\mu*}\psi_l^* = \psi_{l,\mathrm{tr}}c^+c\sigma_l^{\mu*}c^+c\psi_l^* = \psi_{cr}^+\sigma^\mu\psi_{cr} \equiv j_{cr}^\mu \quad (6.25)$$

Hence we obtain $j = j_r + j_{cr}$, which must be wrong since it implies that electron and positron have equal charges, in contradiction with (21) which shows that they have opposite charges. This contradiction will be resolved in chapter 3, where $\psi$ is replaced by an "anticommuting field operator" $\Psi$.

The charge conjugate spin function is denoted by $v(\mathbf{p}, M)$. According to (24),

$$v(\mathbf{p}, M) = \begin{pmatrix} 0 & c \\ -c & 0 \end{pmatrix} u^*(\mathbf{p}, M) = i\gamma^2 u^*(\mathbf{p}, M) = iu(-E, -\mathbf{p}, -M). \quad (6.26)$$

The last expression follows from (11), (9) and the defining equation (4.12). The $+i$-branch of the square root is taken in $A(\mathbf{p})$ and the $-i$-branch in $A(-\mathbf{p})$, as required by

$$\chi_l(\mathbf{p}) = \chi_r(-\mathbf{p}) = A^2(-\mathbf{p})\chi_r(\mathbf{p}) = m^{-1}P_\mu \sigma^\mu \chi_r(\mathbf{p}). \quad (6.27)$$

With $v^+ = iu_{\text{tr}}\gamma^2$ we find the orthogonality relations

$$\bar{v}(\mathbf{p}, M)v(\mathbf{p}, M') = -2m\,\delta_{MM'}. \quad (6.28)$$

## 1-7 Relativistic Corrections to the Pauli Equation and Central Field Problem

We now wish to study the $m^{-2}$ corrections to the Pauli equation which arise from (5.15). For this purpose, we neglect the anomalous magnetic moment contribution in the second bracket on the right-hand side and expand $(\pi^0 + m)^{-1} = (2m)^{-1} - \frac{1}{4}m^{-2}(\pi^0 - m)$. The second term is part of our correction, and when it operates on $(\boldsymbol{\sigma\pi} - \frac{1}{2}im^{-1}\kappa e\boldsymbol{\sigma}\mathbf{E})\psi_g$, we can neglect both the $\kappa e$-term and also $\partial_t \boldsymbol{\pi}$ in comparison with $\partial_t \psi_g$. Thus we obtain from (5.15), with $i\,\partial_t^{nr} \equiv i\,\partial_t - m$:

$$(i\,\partial_t^{nr} - H_P)\psi_g = \tfrac{1}{4}m^{-2}\{i\kappa e[\boldsymbol{\sigma}\mathbf{E}, \boldsymbol{\sigma\pi}] + \boldsymbol{\sigma\pi}(Ze\phi - i\,\partial_t^{nr})\boldsymbol{\sigma\pi}\}\psi_g. \quad (7.1)$$

Here $H_P$ is already the operator (4.5). However, (1) is difficult to interpret because the normalization (5.20) in terms of $\psi$ still contains operators:

$$(\psi, \psi) = \int d^3x(\psi_g^+\psi_g + \psi_k^+\psi_k) \approx \int d^3x\psi_g^+\{1 + (\boldsymbol{\pi\sigma})^2/4m^2\}\psi_g. \quad (7.2)$$

The Pauli spinor $\psi_P$ must be introduced such that $(\psi, \psi) \sim \int d^3x\psi_P^+\psi_P$. To first order in $(\boldsymbol{\pi\sigma})^2 m^{-2}$, this is achieved by

$$\psi_g = (2m)^{\frac{1}{2}}[1 - \tfrac{1}{8}(\boldsymbol{\pi\sigma}/m)^2]\psi_P. \quad (7.3)$$

The unessential factor $(2m)^{\frac{1}{2}}$ is included because we want the factor $2E$ of (1.15) also for (2). We insert (3) into (1) and at the same time multiply (1) by $1 - \frac{1}{8}(\boldsymbol{\pi\sigma}/m)^2$ from the left. To lowest order in $m^{-2}$, the right-hand operator in (1) remains unchanged, while the left-hand operator acquires the extra terms $-\frac{1}{4}m^{-2}(\boldsymbol{\pi\sigma})^2 i\,\partial_t^{nr} + \frac{1}{8}m^{-2}((\boldsymbol{\pi\sigma})^2 H_P + H_P(\boldsymbol{\pi\sigma})^2)$, of which the first term cancels the unwanted operator $\boldsymbol{\sigma\pi} i\,\partial_t^{nr}\boldsymbol{\sigma\pi}$ on the right-hand side:

$$(i\,\partial_t^{nr} - H_P)\psi_P$$
$$= \tfrac{1}{4}m^{-2}\{i\kappa e[\boldsymbol{\sigma}\mathbf{E}, \boldsymbol{\sigma\pi}] + Ze\boldsymbol{\sigma\pi}\phi\boldsymbol{\sigma\pi} - \tfrac{1}{2}H_P(\boldsymbol{\pi\sigma})^2 - \tfrac{1}{2}(\boldsymbol{\pi\sigma})^2 H_P\}\psi_P. \quad (7.4)$$

The curly bracket is simplified by inserting $H_P \approx (\boldsymbol{\pi\sigma})^2/2m + Ze\phi$. Calling $\boldsymbol{\pi\sigma} \equiv A$, $\frac{1}{2}Ze\phi \equiv B$, we see that $B$ occurs in the combination

$$\begin{aligned} 2ABA - BA^2 - A^2B &= [[A, B], A] \\ &= \tfrac{1}{2}iZe[\mathbf{E}\boldsymbol{\sigma}, \boldsymbol{\pi\sigma}] = \tfrac{1}{2}iZe[\mathbf{E}\boldsymbol{\pi} - \boldsymbol{\pi}\mathbf{E} + i(\mathbf{E} \times \boldsymbol{\pi} - \boldsymbol{\pi} \times \mathbf{E})\boldsymbol{\sigma}]. \quad (7.5) \end{aligned}$$

The commutator is the same as that multiplying $i\kappa e$, and is evaluated using (4.3). We find

$$(i\,\partial_t^{nr} - H_P)\psi_P = -\tfrac{1}{8}m^{-2}\{(\boldsymbol{\pi\sigma})^4 m^{-1} + (Z + 2\kappa)e[2(\mathbf{E} \times \boldsymbol{\pi})\boldsymbol{\sigma} + \operatorname{div}\mathbf{E}]\}\psi_P. \quad (7.6)$$

This is an important formula for atomic physics. The method by which we have derived it is called the elimination method of Pauli (1958). More widely used is the transformation method of Foldy and Wouthuysen (1950). The equivalence of both methods has been proven by de Vries and Jonker (1968).

The first correction in (6) is a kinetic energy correction as in (1.5). The next term is a spin-orbit potential and can be rewritten for a central potential

$$Ze\phi = V(r), \qquad Ze\mathbf{E} = -\nabla V = -\hat{\mathbf{r}}\,\partial_r V, \quad (7.7)$$

$$V_{LS} = (1 + 2\kappa/Z)(4m^2r)^{-1}(\partial_r V)(\mathbf{r} \times \mathbf{p})\boldsymbol{\sigma} = \tfrac{1}{2}m^{-2}r^{-1}(1 + 2\kappa/Z)\mathbf{LS}\,\partial_r V. \quad (7.8)$$

This is essentially the magnetic energy of a moving fermion. The factor $\frac{1}{2}$ is not obvious from this point of view. It was derived classically in 1926 by Thomas (see Jackson 1975). The last term in (6) is called the Darwin term and appears only for $s$-states.

The Coulomb field problem can be solved exactly, at least for electrons and muons ($\kappa \approx 0$) and a point nucleus, $V \sim r^{-1}$. We rewrite (5.11) as

$$i\,\partial_t\psi = H\psi, \qquad H = m\gamma_0 - i\,\nabla\boldsymbol{\alpha} + V, \qquad \boldsymbol{\alpha} = \begin{pmatrix} 0 & \boldsymbol{\sigma} \\ \boldsymbol{\sigma} & 0 \end{pmatrix} = \gamma_0\boldsymbol{\gamma} = \boldsymbol{\sigma}\gamma_5. \quad (7.9)$$

Due to the spin-orbit coupling, $H$ does not commute with the orbital angular momentum $\mathbf{L}$, but only with the total angular momentum

$$\mathbf{J} = \mathbf{L} + \tfrac{1}{2}\boldsymbol{\sigma} = -i\mathbf{r} \times \nabla + \frac{1}{2}\begin{pmatrix} \boldsymbol{\sigma} & 0 \\ 0 & \boldsymbol{\sigma} \end{pmatrix}, \quad (7.10)$$

which follows from $[\mathbf{L}, \nabla\boldsymbol{\alpha}] = i\boldsymbol{\alpha} \times \nabla = -[\frac{1}{2}\boldsymbol{\alpha}, \nabla\boldsymbol{\alpha}]$. Therefore, the eigenstates $\psi_{j\lambda}$ of $H$, $J^2$ and $J_z$ will be linear combinations of states having $L = j \pm \frac{1}{2}$. The relevant two-component states are constructed from the basic spinors $\chi$ of (4.1) and the spherical harmonics $Y_{m_1}^L(\vartheta, \varphi)$ by means of the Clebsch-Gordan coefficients (B-4.2):

$$\chi_L^{j\lambda} = \sum_{m_2} C_{m_1\, m_2\, \lambda}^{L\; \frac{1}{2}\; j}\, Y_{m_1}^L(\vartheta, \varphi)\chi(m_2). \quad (7.11)$$

On the other hand, $H$ does commute with the parity operation $\psi^{\mathscr{P}} = \gamma_0 \psi(-\mathbf{x})$. Due to the factor $\gamma_0$, parity eigenstates $\psi_{j\pi\lambda}$ must contain only one value of $L$ in the large components and only the other value of $L$ (which will be called $\tilde{L}$) in the small components:

$$\psi_{j\pi\lambda} = \begin{pmatrix} g_\kappa(r)\chi_L^{j\lambda} \\ if_\kappa(r)\chi_{\tilde{L}}^{j\lambda} \end{pmatrix} \qquad \pi = (-1)^L, \qquad \kappa \equiv \begin{cases} j + \frac{1}{2} & \text{for} \quad L = j + \frac{1}{2} \\ -j - \frac{1}{2} & \text{for} \quad L = j - \frac{1}{2} \end{cases} \tag{7.12}$$

Thus, although $H$ does not commute with $\mathbf{L}$, the quantum number $L$ enters through the parity backdoor as the orbital angular momentum of the large components. The index $\kappa$ is defined such that it combines the pair of indices $j$, $\pi$. It is the eigenvalue of the operator

$$K = -\gamma_0(\boldsymbol{\sigma}\mathbf{L} + 1) = -\gamma_0(\mathbf{J}^2 - \mathbf{L}^2 + \tfrac{1}{4}). \tag{7.13}$$

The second expression follows from $\mathbf{J}^2 = \mathbf{L}^2 + \mathbf{L}\boldsymbol{\sigma} + \frac{3}{4}$. The eigenvalues of $K$ are evaluated for the large components as $-[j(j+1) - L(L+1) + \frac{1}{4}]$ and agree with (12). To obtain the radial Dirac equation, we transform $\gamma_5 \nabla\boldsymbol{\alpha}$ of (9) by inserting $1 = r^{-2}(\boldsymbol{\sigma}\mathbf{r})^2$ and using (4.3)

$$\nabla\boldsymbol{\sigma} = r^{-2}(\boldsymbol{\sigma}\mathbf{r})^2(\boldsymbol{\sigma}\nabla) = r^{-2}(\boldsymbol{\sigma}\mathbf{r})(\mathbf{r}\,\nabla - \boldsymbol{\sigma}\mathbf{L}) = \sigma_r[\partial_r + r^{-1}(1 + \gamma_0 K)] \tag{7.14}$$

The operator $\sigma_r = r^{-1}\boldsymbol{\sigma}\mathbf{r}$ has the property

$$\sigma_r \chi_L^{j\lambda} = \varepsilon \chi_{\tilde{L}}^{j}, \qquad \varepsilon^2 = 1. \tag{7.15}$$

This follows from the fact that $\chi_L$ and $\chi_{\tilde{L}}$ have the same rotation properties but opposite parities. Since $\sigma_r$ is a pseudoscalar operator with $\sigma_r^2 = 1$, it must exchange $\chi_L$ and $\chi_{\tilde{L}}$, possibly with an additional minus sign. By putting $\vartheta = 0$, $\sigma_r(0) = \sigma_z$ and $Y_L^m(0) = (4\pi)^{-\frac{1}{2}}(2L+1)^{\frac{1}{2}}\,\delta_{mo}$, application of $\sigma_r$ to (11) gives

$$(4\pi)^{\frac{1}{2}}\sigma_z \chi_L^{j\lambda} = 2\lambda C_{0\,\frac{1}{2}\,\lambda}^{L\,\frac{1}{2}\,j}(2L+1)^{\frac{1}{2}}\chi(\lambda) = \pm(j + \tfrac{1}{2})^{\frac{1}{2}} 2\lambda\chi(\lambda) \tag{7.16}$$

for $L = j \pm \frac{1}{2}$ according to (B-4.2). Therefore we have in fact $\varepsilon = -1$ in (15), and the stationary Schrödinger equation

$$H\psi = E\psi, \qquad H = m\gamma_0 + V - i\gamma_5\sigma_r\left[\partial_r + \frac{1}{r}(1 + \gamma_0 K)\right] \tag{7.17}$$

gives the following pair of radial equations:

$$(E - m - V)g + [\partial_r + r^{-1}(1 - \kappa)]f = 0, \tag{7.18}$$

$$(-E - m + V)f + [\partial_r + r^{-1}(1 + \kappa)]g = 0. \tag{7.19}$$

Although this formalism looks rather strange, its physical content is closely related to that of the Klein-Gordon equation. Here we shall only treat the region in which $V = 0$. Then (19) can be used to eliminate $f$,

$$f = (E + m)^{-1}k[\partial_\rho + \rho^{-1}(1 + \kappa)]g, \qquad \rho = kr = (E^2 - m^2)^{\frac{1}{2}}r. \quad (7.20)$$

Inserting this into (18), we arrive at

$$[\partial_\rho^2 + 2\rho^{-1}\,\partial_\rho + 1 - \rho^{-2}\kappa(\kappa + 1)]g = 0, \quad (7.21)$$

which coincides with (3.10) for spherical Bessel and Neuman functions (note that $\kappa(\kappa + 1) = L(L + 1)$ for both values of $\kappa$). Moreover, these functions obey

$$\partial_{\rho j_L} = \rho^{-1}Lj_L - j_{L+1} = \rho^{-1}(L + 1)j_L + j_{L-1} \quad (7.22)$$

both for $j_L$ and $n_L$, and insertion of these relations into (20) shows that $f$ is essentially the same function as $g$, only with $L$ replaced by $\tilde{L}$.

Because the small components $\psi_k$ can be evaluated from the large ones $\psi_g$, it is possible to develop the scattering formalism in term of $\psi_g$ alone. Here the orbital angular momentum **L** plays the same role as in the nonrelativistic limit, and so the nonrelativistic formalism can be used. For the scattering situation, we put in analogy with (3.14)

$$\psi_{j\pi\lambda,\,g} = (E + m)^{\frac{1}{2}}e^{i\,\delta_{L\pm}}[j_L \cos\delta_{L\pm} - n_L \sin\delta_{L\pm}]\chi_{L}^{j\lambda}, \quad (7.23)$$

with real $\delta_{L_\pm}$. The index $\pm$ stands for $j = L \pm \frac{1}{2}$, i.e., $L_+$ has $\kappa = -j - \frac{1}{2} = -L - 1$ and $L_-$ has $\kappa = L$. The factor $(E + m)^{\frac{1}{2}}$ has been taken out in view of (6.12). The asymptotic form of $\psi_b$ appropriate to scattering is an extension of (3.7):

$$\psi_g(k, \vartheta, \varphi, \lambda) = \{e^{ikz} + r^{-1}e^{ikr}[f(k, \vartheta) + ig(k, \vartheta)\boldsymbol{\sigma}\hat{\mathbf{n}}_s]\}(E + m)^{\frac{1}{2}}\chi_0(\lambda), \quad (7.24)$$

where $\hat{\mathbf{n}}_s = k^{-2}\mathbf{k} \times \mathbf{k}'$ is the normal to the scattering plane. This form follows from parity conservation. The scattering amplitudes $f(k, \vartheta)$ and $g(k, \vartheta)$ are called the "spin-nonflip" and "spin-flip" amplitudes and should not be confused with the radial functions $f_\kappa$ and $g_\kappa$. Their partial-wave expansion is

$$f + ig\boldsymbol{\sigma}\hat{\mathbf{n}}_s = \sum_{L=0}^{\infty} (2L + 1)(a_L + b_L\mathbf{L}\boldsymbol{\sigma})P_L\,(\cos\vartheta), \quad (7.25)$$

and observing $\mathbf{L}\boldsymbol{\sigma}P_L = i\boldsymbol{\sigma}\hat{\mathbf{n}}P_L^1$ on the one hand and $\mathbf{L}\boldsymbol{\sigma} = L$ for $j = L + \frac{1}{2}$ and $-L - 1$ for $j = L - \frac{1}{2}$ on the other, one finds

$$a_L + Lb_L = f_{L+}, \qquad a_L - (L + 1)b_L = f_{L-}, \qquad f_{L\pm} \equiv k^{-1}e^{i\,\delta_{L\pm}} \sin\delta_{L\pm}, \quad (7.26)$$

$$f(k, \vartheta) = \sum_{L=0}^{\infty} [(L + 1)f_{L+} + Lf_{L-}]P_L, \quad (7.27)$$

$$g(k, \vartheta) = \sum_{L=1}^{\infty} (f_{L+} - f_{L-})P_L^1. \quad (7.28)$$

The differential cross section for a particle of polarization **P** is given by

$$d\sigma/d\Omega = |f|^2 + |g|^2 + 2\hat{\mathbf{n}}_s \cdot \mathbf{P}\ \mathrm{Im}\,(fg^*), \tag{7.29}$$

and the polarization produced in the scattering of unpolarized particles is

$$\mathbf{P}' = \hat{\mathbf{n}}_s(|f|^2 + |g|^2)^{-1} \cdot 2\ \mathrm{Im}\,(fg^*). \tag{7.30}$$

These formulas will be essentially rederived in section 2-5.

For completeness, we also give the expansion

$$\psi(k, \vartheta, \varphi, \lambda) = \sum_{\kappa} a_{\kappa\lambda}\psi_{j\pi\lambda},$$

$$a_{\kappa\lambda} = C^{L\,\frac{1}{2}\,j}_{0\,\lambda\,\lambda}\, a_{L0} = C^{L\,\frac{1}{2}\,j}_{0\,\lambda\,\lambda}\, i^L[4\pi(2L+1)]^{\frac{1}{2}}. \tag{7.31}$$

Here we have used the plane wave decomposition (3.11) and inverted the expansion (11).

## 1-8 Coulomb Scattering of Spinless Particles

We now discuss the solutions of (3.9) for the Coulomb potential, $V = V_{C\,\mathrm{point}} = Z_a Z\alpha/r$, corresponding to electric charges $Z_a e$ and $Ze$ of projectile and target. We introduce

$$\eta = Z_a Z\alpha E/k = Z_a Z\alpha v^{-1}, \qquad \text{(Coulomb parameter)} \tag{8.1}$$

$$\gamma \equiv [(L + \tfrac{1}{2})^2 - Z^2\alpha^2]^{\frac{1}{2}} \qquad \text{for } Z_a = \pm 1, \tag{8.2}$$

and get the Klein-Gordon equation in the form

$$[\rho^{-2}\,\partial_\rho \rho^2\,\partial_\rho + 1 - 2\eta/\rho - \rho^{-2}(\gamma^2 - \tfrac{1}{4})]R_L = 0. \tag{8.3}$$

The solution which is regular at the origin (except for $L = 0$, where it diverges approximately with the power $-Z^2\alpha^2$ of $\rho$) is

$$R_L = (2\rho)^{\gamma-\frac{1}{2}} e^{-i\rho} N_\gamma F(\gamma + \tfrac{1}{2} - i\eta, 2\gamma + 1, 2i\rho), \tag{8.4}$$

where $F$ is the hypergeometric function (A-2.3), and $N_\gamma$ is a normalization constant. For large $r$, $F$ becomes (Mott and Massey 1965, Landau and Lifschitz, Vol. 3) $W_1 + W_2$,

$$W_2 = \Gamma^{-1}(a)\Gamma(b)e^x x^{a-b}\tilde{G}(1-a, b-a, x) \qquad \text{for} \quad \mathrm{Re}\,x > 0, \tag{8.5}$$

$$W_1 = \Gamma^{-1}(b-a)\Gamma(b)(-x)^{-a}\tilde{G}(a, a-b+1, -x) \qquad \text{for} \quad \mathrm{Re}\,x < 0.$$

$$\tilde{G}(\alpha, \beta, z) = 1 + \alpha\beta/z + \alpha(\alpha+1)\beta(\beta+1)/2!z^2 + \cdots \tag{8.6}$$

$\tilde{G}$ is semiconvergent. We need $R_L(r \to \infty)$ where $\tilde{G} = 1$:

$$e^{-i\rho}F(\rho \to \infty) = \Gamma(2\gamma + 1)$$

$$[e^{-i\rho}\Gamma^{-1}(\gamma + \tfrac{1}{2} + i\eta)(-2i\rho)^{-\gamma-\frac{1}{2}+i\eta} + e^{i\rho}\Gamma^{-1}(\gamma + \tfrac{1}{2} - i\eta)(2i\rho)^{-\gamma-\frac{1}{2}-i\eta}], \tag{8.7}$$

$$R_L(\rho \to \infty) = (2\rho)^{-1}N_\gamma\Gamma(2\gamma + 1)2\ \mathrm{Re}\,[e^{i\rho}\Gamma^{-1}(\gamma + \tfrac{1}{2} - i\eta)i^{-\gamma-\frac{1}{2}}(2i\rho)^{-i\eta}]. \tag{8.8}$$

The square bracket is separated into amplitude and phase as follows:

$$\Gamma^{-1}(\gamma + \tfrac{1}{2} - i\eta) = |\Gamma^{-1}(\gamma + \tfrac{1}{2} + i\eta)| \exp\{i \arg \Gamma(\gamma + \tfrac{1}{2} + i\eta)\}, \quad (8.9)$$

$$(2i\rho)^{-i\eta} = e^{\pi\eta/2} \exp\{-i\eta \ln 2\rho\}, \qquad i^{-\gamma-\frac{1}{2}} = \exp\{-\tfrac{1}{2}i\pi(\gamma + \tfrac{1}{2})\}, \quad (8.10)$$

and the amplitude of $R_L(\rho \to \infty)$ is normalized to $\rho^{-1}$ as in (A-2.6), appropriate to the boundary condition of an incident (almost) plane wave:

$$\begin{aligned} R_L(\rho \to \infty) &= \rho^{-1} \cos[\rho + \arg \Gamma(\gamma + \tfrac{1}{2} + i\eta) - \eta \ln 2\rho - \tfrac{1}{2}\pi(\gamma + \tfrac{1}{2})] \\ &= \rho^{-1} \sin[\rho - \eta \ln 2\rho - \tfrac{1}{2}L\pi + \sigma_\gamma], \qquad (8.11) \\ \sigma_\gamma &= \arg \Gamma(\gamma + \tfrac{1}{2} + i\eta) - \tfrac{1}{2}\pi(\gamma - \tfrac{1}{2} - L). \qquad (8.12) \end{aligned}$$

Arg $\Gamma$ is given in (A-2.7). The normalization constant implied by (11) is

$$N_\gamma = |\Gamma(\gamma + \tfrac{1}{2} + i\eta)| \Gamma^{-1}(2\gamma + 1) e^{-\pi\eta/2}, \quad (8.13)$$

and the Coulomb phase $\sigma_\gamma$ has been defined such that $\sigma_\gamma = 0$ for uncharged particles (compare A-1.1). The partial wave $S$-matrix element is

$$e^{2i\sigma_\gamma} = e^{-i\pi(\gamma - \frac{1}{2} - L)} \Gamma(\gamma + \tfrac{1}{2} + i\eta)/\Gamma(\gamma + \tfrac{1}{2} - i\eta). \quad (8.14)$$

This expression has poles at all nonpositive integers of the argument of the $\Gamma$-function,

$$\gamma + \tfrac{1}{2} + i\eta_B = -n_r, \qquad n_r = 0, 1, 2 \ldots \quad (8.15)$$

The poles thus occur at imaginary $\eta = \eta_B$ and correspond to bound states. This follows from (3.14) if one observes that for complex $\rho$, the second solution of the partial-wave equation (3.10) is $h(-\rho)$ instead of $h^*(\rho)$. Using eq. (A-1.6) we have

$$\begin{aligned} R_L &= \tfrac{1}{2}[h_L(\rho) e^{2i\delta_L} + (-1)^L h_L(-\rho)] \\ &= \tfrac{1}{2}[h_L(i\kappa r) e^{2i\delta_L} + (-1)^L h_L(-i\kappa r)], \qquad \rho \equiv kr \equiv i\kappa r. \qquad (8.16) \end{aligned}$$

The function $h_L$ goes asymptotically like $-(-i)^L e^{-\kappa r}/\kappa r$ and is therefore localized. Bound states occur for $|\exp(2i\,\delta_L)| = \infty$ where $h_L(-i\kappa r)$ can be neglected. Next, we expand $\gamma$ in (2) as

$$\gamma \approx L + \frac{1}{2} - \frac{Z^2\alpha^2}{2L+1}\left[1 + \left(\frac{Z\alpha}{2L+1}\right)^2\right], \quad (8.17)$$

and introduce the principal quantum number $n = n_r + L + 1$:

$$\begin{aligned} \eta_B^{-2} &= -\left(n - \frac{Z^2\alpha^2}{2L+1}[\;]\right)^{-2} \\ &= -n^{-2} - 2n^{-3}\frac{Z^2\alpha^2}{2L+1} - n^{-4}\left(\frac{Z^2\alpha^2}{2L+1}\right)^2\left(3 + \frac{2n}{2L+1}\right). \qquad (8.18) \end{aligned}$$

The bound state energies now follow from (1) as

$$E_B^2 = m^2(1 - Z^2\alpha^2\eta_B^{-2})^{-1}, \tag{8.19}$$

$$E_B - m = -\tfrac{1}{2}n^{-2}Z^2\alpha^2 m(1 + \delta_{KG}), \qquad \delta_{KG} \approx Z^2\alpha^2 n^{-2}\left(\frac{n}{L + \frac{1}{2}} - \frac{3}{4}\right). \tag{8.20}$$

## 1-9 Coulomb Scattering of Spin-$\frac{1}{2}$ Particles

The motion of spin-$\frac{1}{2}$ particles in a spherically symmetric electrostatic potential is described by (7.18) and (7.19). For the point Coulomb potential these equations can be solved exactly.

We follow the presentation of Rose (1961), putting

$$g = (E + m)^{\frac{1}{2}}(\varphi_+ + \varphi_-), \qquad f = i(E - m)^{\frac{1}{2}}(\varphi_+ - \varphi_-). \tag{9.1}$$

Introducing as before $\rho = kr = (E^2 - m^2)^{\frac{1}{2}}r$, we arrive at

$$\varphi_\mp(k^{-2}Vm \mp i\rho^{-1}\kappa) = (1 - k^{-2}VE \pm i\,\partial_\rho \pm i\rho^{-1})\varphi_\pm\,. \tag{9.2}$$

The point is that for $V = -Z\alpha/r = -Z\alpha k/\rho$, the $\rho$-dependence of the left-hand side of (2) is $\rho^{-1}$. We eliminate $\varphi_-$ from (2) and obtain for $\varphi_+$

$$[\rho - 2\eta + \rho^{-1}(Z^2\alpha^2 + 1 - \kappa^2) - i + \partial_\rho \rho\, \partial_\rho + 2\,\partial_\rho]\varphi_+ = 0, \tag{9.3}$$

where $\eta$ is the Coulomb parameter $VrE/k = -Z\alpha E/k$. Finally we define $\varphi_+ = (2\rho)^{-\frac{1}{2}}R_\kappa$ and find that

$$[\rho^{-2}\,\partial_\rho \rho^2\,\partial_\rho + 1 - \rho^{-1}(2\eta + i) - \rho^{-2}(\kappa^2 - Z^2\alpha^2 - \tfrac{1}{4})]R_\kappa = 0. \tag{9.4}$$

This equation differs from (8.3) for the spinless case only by the replacement $\eta \to \eta + i/2$ and the redefinition

$$\gamma = [\kappa^2 - Z^2\alpha^2]^{\frac{1}{2}} = [(J + \tfrac{1}{2})^2 - Z^2\alpha^2]^{\frac{1}{2}}. \tag{9.5}$$

It is therefore solved by

$$R_\kappa = (2\rho)^{\gamma - \frac{1}{2}}e^{-i\rho}N'F(\gamma + 1 - i\eta, 2\gamma + 1, 2i\rho). \tag{9.6}$$

The functions $g$ and $f$ can be chosen to be real, in which case $\varphi_+$ and $\varphi_-$ must be complex conjugates of each other. This is, in fact, possible, since the equations for $\varphi_\pm$ in (2) are exchanged under complex conjugation. We set

$$\varphi_+ = (2\rho)^{\gamma - 1}Ne^{i\xi}(\gamma - i\eta)e^{-i\rho}F(\gamma + 1 - i\eta, 2\gamma + 1, 2i\rho), \tag{9.7}$$

where the new normalization constant $N$ is real, and the phase is determined such that $\varphi_- = \varphi_+^*$. Thus we require

$$\varphi_- = \varphi_+ e^{-2i\xi}\frac{\gamma + i\eta}{\gamma - i\eta} = \frac{\rho - \eta + i\rho\,\partial_\rho + i}{\eta m/E - i\kappa}\varphi_+ \tag{9.8}$$

according to (2). We put $\rho \ll 1$ and get $\rho\,\partial_\rho \varphi_+ = (\gamma - 1)\varphi_+$ from (7), i.e.,

$$e^{2i\xi} = \frac{\gamma + i\eta}{\gamma - i\eta}\frac{\eta m/E - i\kappa}{-\eta + i\gamma} = -\frac{\kappa + i\eta m/E}{\gamma - i\eta}. \tag{9.9}$$

Hence, we have

$$\begin{pmatrix} g \\ f \end{pmatrix} = (E \pm m)^{\frac{1}{2}} N 2^\gamma \rho^{\gamma-1} \begin{matrix} \mathrm{Re} \\ -\mathrm{Im} \end{matrix} \left\{ (\gamma - i\eta) e^{i\xi - i\rho} F(\gamma + 1 - i\eta, 2\gamma + 1, 2i\rho) \right\}. \tag{9.10}$$

For $\rho \to \infty$, we use again $F = W_1 + W_2$ of (8.5), with $\tilde{G} = 1$, $a = \gamma + 1 - i\eta$, $b = 2\gamma + 1$, $b - a = \gamma + i\eta$. We see that contrary to the spinless case, $W_1$ vanishes this time by one power of $\rho$ faster than $W_2$. Using again the transcription (8.10), we find $F(\rho = \infty) = W_2 = \Gamma^{-1}(\gamma + 1 - i\eta)\Gamma(2\gamma + 1) \times (2i\rho)^{-\gamma - i\eta} e^{2i\rho}$ and

$$\begin{pmatrix} g \\ f \end{pmatrix} = (E \pm m)^{\frac{1}{2}} \frac{N}{\rho} e^{\pi\eta/2 - 1} \Gamma(2\gamma + 1) \begin{matrix} \mathrm{Re} \\ -\mathrm{Im} \end{matrix} \left\{ \frac{\gamma - i\eta}{\Gamma(\gamma + 1 - i\eta)} e^{-i\eta \ln 2\rho} e^{-\frac{1}{2} i\pi\gamma} e^{i\rho + i\xi} \right\}. \tag{9.11}$$

With $\Gamma(\gamma + 1 - i\eta) = (\gamma - i\eta)\Gamma(\gamma - i\eta)$ and separation of the phase of $\Gamma$ analogous to (8.9), we end up with

$$\begin{pmatrix} g \\ f \end{pmatrix} = \begin{matrix} \cos \\ -\sin \end{matrix} \left( -\eta \ln 2\rho + \rho - \tfrac{1}{2}\pi\gamma + \xi + \arg \Gamma(\gamma + i\eta) \right) (E \pm m)^{\frac{1}{2}} \rho^{-1} a_{\kappa\lambda}, \tag{9.12}$$

where we have put

$$N = a_{\kappa\lambda} \,|\Gamma(\gamma + i\eta)|\, \Gamma^{-1}(2\gamma + 1) e^{-\pi\eta/2} \tag{9.13}$$

in order to get the form of (7.23), (7.31). Analogous to (8.11), we now write

$$\begin{aligned} &\cos\,(\rho - \eta \ln 2\rho - \tfrac{1}{2}\pi\gamma + \xi + \arg \Gamma(\gamma + i\eta)) \\ &= \sin\,(\rho - \eta \ln 2\rho - \tfrac{1}{2}L\pi + \sigma_\kappa), \\ &\sigma_\kappa = \tfrac{1}{2}\pi(L + 1 - \gamma) + \xi + \arg \Gamma(\gamma + i\eta). \end{aligned} \tag{9.14}$$

The poles of

$$S_\kappa = e^{2i\sigma_\kappa} = -(\gamma - i\eta)^{-1}(\kappa + i\eta m/E) e^{i\pi(L + 1 - \gamma)} \Gamma(\gamma + i\eta)/\Gamma(\gamma - i\eta) \tag{9.15}$$

occur at the nonpositive integers of the argument of the $\Gamma$-function. This time we define the radial quantum number $n_r$ differently for $|\kappa| = L + 1$ $(J = L + \frac{1}{2})$ and for $\kappa = L$ $(J = L - \frac{1}{2})$:

$$\gamma + i\eta_B = \begin{cases} -n_r & (J = L + \frac{1}{2}) \\ -n_r + 1 & (J = L - \frac{1}{2}) \end{cases} \qquad n_r = 0, 1, 2, \ldots \tag{9.16}$$

We can now take over formulas (8.17) and (8.18) with $L$ replaced by $J$ according to (5). The resulting binding energies are

$$E_B - m = -Z^2\alpha^2 m\tfrac{1}{2}n^{-2}(1 + \delta_{D_i}), \qquad \delta_{D_i} \approx Z^2\alpha^2 n^{-2}[n/(J+\tfrac{1}{2}) - \tfrac{3}{4}]. \tag{9.17}$$

Note that to this order in $Z^2\alpha^2 n^{-2}$, (17) is more easily derived from a lowest order perturbation theoretical calculation of (7.6). That calculation is also more flexible in the sense that it allows additional perturbations. For example, to include the anomalous magnetic moment $\kappa_a$, we merely replace $-Z_a(=1)$ by $-(Z_a + 2\kappa_a)$ according to (7.6):

$$\delta_{Di} \approx -(Z_a + 2\kappa_a)Z^2\alpha^2 n^{-2}[n/(J+\tfrac{1}{2}) - \tfrac{3}{4}]. \tag{9.18}$$

It is also worth noting that all results of this section are independent of $L$ for given $J$. For the atomic bound states, this implies a degeneracy of the $nS_{\frac{1}{2}}$ and $nP_{\frac{1}{2}}$ states etc. We might as well have expressed our results in terms of linear combinations of such states, and eqs. (3)–(8) do in fact refer to such combinations. Comparison between (1) and (5.8) shows that $\varphi_+$ and $\varphi_-$ are closely related to $\psi_r$ and $\psi_l$. It would have been simpler to derive (3) directly from (5.4). For real atoms, the degeneracy is lifted by nuclear size effects (section 5-4), vacuum polarisation (sect. 3-10), and other radiative corrections (Lamb shift).

## 1-10 Summation of the High Partial Waves and Deviations from the Point Coulomb Potential

In the previous sections we have solved the Klein-Gordon and Dirac partial-wave equations for the scattering on a point Coulomb potential. The full scattering amplitude $f(\vartheta)$ remains to be found, including the deviations from the point Coulomb potential which occur in the low partial waves. We first treat the spinless case. Let

$$\delta_L = \sigma_\gamma + \delta_L' \tag{10.1}$$

denote the true phase shift in the $L$th partial wave, and $\sigma_\gamma$ its point Coulomb approximation. The scattering amplitude can be rewritten as follows:

$$f_C = (2ik)^{-1} \sum_{L=0}^{\infty} (2L+1)(e^{2i\sigma_\gamma + 2i\delta_L'} - 1)P_L = f_{C_{\text{point}}} + f', \tag{10.2}$$

$$f_{C_{\text{point}}} = (2ik)^{-1} \sum_{L=0}^{\infty} (2L+1)(e^{2i\sigma_\gamma} - 1)P_L, \tag{10.3}$$

$$f' = (2ik)^{-1} \sum_L (2L+1)e^{2i\sigma_\gamma}(e^{2i\delta_L'} - 1)P_L \equiv \sum_L (2L+1)e^{2i\sigma_\gamma}T_L' P_L, \tag{10.4}$$

where the number of partial waves needed in (4) is limited. The summation in (3) requires a similar trick. We write

$$\sigma_\gamma = \sigma_L + \delta\sigma_L, \qquad e^{2i\sigma_L} \equiv \Gamma(L+1+i\eta)/\Gamma(L+1-i\eta). \tag{10.5}$$

From (8.14) and (8.17) we see that $\delta\sigma_L$ depends on $Z^2\alpha^2(2L+1)^{-1}$, which is normally negligible except for the lowest values of $L$. The $\sigma_L$ are identical to the nonrelativistic Coulomb phases (Appendix A2), except for the relativistic definitions of the parameters $k$ and $\eta$. They are therefore easily summed up:

$$f_{C_{\text{point}}} = f_C^{(0)} + f_C^{(1)}, \tag{10.6}$$

$$f_C^{(0)} = \frac{1}{2ik}\sum_L (2L+1)(e^{2i\sigma_L}-1)P_L = \frac{-\eta}{2k}(\sin\tfrac{1}{2}\vartheta)^{-2-2i\eta}\Gamma(1+i\eta)/\Gamma(1-i\eta)$$

$$= -\eta(2k\sin^2\tfrac{1}{2}\vartheta)^{-1}\exp\{-i\eta\ln\sin^2\tfrac{1}{2}\vartheta + 2i\sigma_0\}. \tag{10.7}$$

The amplitude $f_C^{(1)}$ is discussed below

$$f_C^{(1)} = (2ik)^{-1}\sum_L (2L+1)e^{2i\sigma_L}(e^{2i\delta\sigma_L}-1)P_L. \tag{10.8}$$

Now to the spin-$\frac{1}{2}$ case. The total phase shift $\delta_{L\pm}$ in states of total angular momentum $j = L \pm \frac{1}{2}$ is decomposed as in (1)

$$\delta_{L\pm} = \sigma_{L\pm} + \delta'_{L\pm}, \tag{10.9}$$

with the point Coulomb phases $\sigma_{L\pm}$ given by (9.15). The decomposition (2) is now applied to the amplitudes $f$ and $g$ of (7.27), (7.28):

$$f = f_{C_p} + f', \qquad g = g_{C_p} + g', \tag{10.10}$$

$$f_{C_p} = (2ik)^{-1}\sum_L [(L+1)(e^{2i\sigma_{L+}}-1) + L(e^{2i\sigma_{L-}}-1)]P_L, \tag{10.11}$$

$$g_{C_p} = (2ik)^{-1}\sum_{L=1}^{\infty}[e^{2i\sigma_{L+}} - e^{2i\sigma_{L-}}]P_L^1,$$

$$f' = \sum_L [e^{2i\sigma_{L+}}(L+1)f'_{L+} + e^{2i\sigma_{L-}}f'_{L-}]P_L,$$

$$g' = \sum_{L=1}^{\infty}[e^{2i\sigma_{L+}}f'_{L+} - e^{2i\sigma_{L-}}f'_{L-}]P_L^1, \qquad f'_{L\pm} \equiv (2ik)^{-1}(e^{2i\delta'_{L\pm}}-1). \tag{10.12}$$

The summation over the high partial waves is again performed by writing $\sigma_{L\pm}$ in a form where all relativistic effects are contained in the parameters $\gamma$ and $\eta$. From (9.15) and the definition (7.12) of $\kappa$, we see that

$$e^{2i\sigma_{L\pm}} = -(\kappa + i\eta m/E)\,\text{sign}\,(\kappa)C_J, \qquad C_J = -\frac{\Gamma(\gamma+i\eta)}{\Gamma(\gamma+1-i\eta)}e^{i\pi(J+\frac{1}{2}-\gamma)}. \tag{10.13}$$

The advantage of working with $C_J$ is that these quantities depend only on $\eta$ and $J$. It is now customary to define the integer $J + \frac{1}{2} = l$ (but note that $L$ is either $l$ or $l - 1$). Then (11) can be rewritten as

$$kf_{C_p} = G + iF\eta m/E, \qquad kg_{C_p} = \tan\frac{\vartheta}{2}G - i\cot\frac{\vartheta}{2}F\eta m/E, \tag{10.14}$$

$$G = \tfrac{1}{2}i\sum_l l^2 C_l(P_l + P_{l-1}), \qquad F = \tfrac{1}{2}i\sum_l lC_l(P_l - P_{l-1}) \tag{10.15}$$

Here we have used the following relations in the expression for $g'$:

$$P_l^1 + P_{l-1}^1 = -\cot\frac{\vartheta}{2}l(P_l - P_{l-1}), \qquad P_l^1 - P_{l-1}^1 = \tan\frac{\vartheta}{2}l(P_l + P_{l-1}). \tag{10.16}$$

Using moreover the identity

$$(1 + \cos\vartheta)\,\partial_{\cos\vartheta}(P_l - P_{l-1}) = l(P_l + P_{l-1}), \tag{10.17}$$

we see that $G$ can be expressed in terms of $F$ as

$$G = (1 - \cos\vartheta)\,\partial_{\cos\vartheta}F = 2\sin^2\frac{\theta}{2}\,\partial_{\cos\vartheta}F. \tag{10.18}$$

This is about as far as one can go with the exact point Coulomb problem. As in the spinless case, the sum in (15) can be performed only in a power series expansion in $Z^2\alpha^2(2l+2)^{-2}$, analogous to (8.17)

$$\gamma \approx l - Z^2\alpha^2(2l+2)^{-1}[1 + (Z\alpha/(2l+2))^2]. \tag{10.19}$$

In the approximation $\gamma = l$, we are formally back at the nonrelativistic case for $F$ and $G$. For this case we know (from the Pauli equation (4.2) with $\mathbf{A} = 0$) that

$$f_{C_p}^{(0)}(E = m) = f_C^{(0)}, \qquad g_{C_p}^{(0)}(E = m) = 0, \tag{10.20}$$

where $f_C^{(0)}$ is given by the spinless case (7). We then obtain from (14)

$$G^{(0)}(E = m) = i\cot^2\frac{\vartheta}{2}\eta F^0(E = m), \qquad i\eta F^0(E = m) = \sin^2\frac{\vartheta}{2}kf_C^{(0)}. \tag{10.21}$$

But since $F^{(0)}$ and $G^{(0)}$ are in fact independent of $E$, (21) is already the relativistic solution to zeroth order in $Z^2\alpha^2$. Therefore

$$f_{C_p}^{(0)} = f_C^{(0)}\left(\cos^2\frac{\vartheta}{2} + \sin^2\frac{\vartheta}{2}m/E\right), \qquad g_{C_p}^{(0)} = \sin\frac{\vartheta}{2}\cos\frac{\vartheta}{2}f_C^{(0)}(1 - m/E). \tag{10.22}$$

We now return to the evaluation of $f_C^{(1)}$ defined by (8), (5) and (8.14), following a method of Lenz (private comm.). By means of the hypergeometric function

$${}_2F_1(a, b; c, 1) = \frac{\Gamma(c)\Gamma(c-a-b)}{\Gamma(c-a)\Gamma(c-b)} = 1 + \frac{ab}{c} + \frac{a(a+1)b(b+1)}{2!\,c(c+1)} + \cdots \tag{10.23}$$

we find from (8.14)

$$\begin{aligned} e^{2i\,\delta\sigma_{L\pm}} &\equiv e^{2i\sigma_\gamma}\Gamma(L+1-i\eta)/\Gamma(L+1+i\eta) \\ &= e^{i\pi(L+\frac{1}{2}-\gamma)}{}_2F_1(L+\tfrac{1}{2}-\gamma,\,-2i\eta;\,L+1-i\eta,\,1) \\ &\approx [1+i\pi Z^2\alpha^2(2L+1)^{-1}][1-2i\eta Z^2\alpha^2(2L+1)^{-1}(L+1-i\eta)^{-1}]. \end{aligned} \tag{10.24}$$

The second square bracket can be approximated by 1 except perhaps for the first few values of $L$, depending on the magnitude of $\eta$. Insertion of (24) into (8) then gives

$$f_C^{(1)} = (2ik)^{-1}i\pi Z^2\alpha^2 \sum_L e^{2i\sigma_L}P_L \approx \tfrac{1}{2}i\pi Z^2\alpha^2 f_C^{(0)}(\eta+\tfrac{1}{2}i). \tag{10.25}$$

The last expression is derived from (7):

$$\begin{aligned} 2ikf_C^{(0)}(\eta+\tfrac{1}{2}i) &= \sum (2L+1)[\Gamma(L+\tfrac{1}{2}+i\eta)/\Gamma(L+\tfrac{3}{2}-i\eta)-1]P_L \\ &= (-i\eta+\tfrac{1}{2})(\sin\tfrac{1}{2}\vartheta)^{-1-2i\eta}\Gamma(1+i\eta-\tfrac{1}{2})/\Gamma(1-i\eta+\tfrac{1}{2}), \end{aligned} \tag{10.26}$$

by observing that in the sum in (26), $\sum(2L+1)P_L = \delta(1-\cos\vartheta)$ can be omitted for $\vartheta \neq 0$ and that $(2L+1)/\Gamma(L+\frac{3}{2}-i\eta) \approx 2/\Gamma(L+\frac{1}{2}-i\eta)$. In the summed expression in (26), we can also abbreviate

$$(-i\eta+\tfrac{1}{2})\Gamma(\tfrac{1}{2}+i\eta)/\Gamma(\tfrac{3}{2}-i\eta) = \Gamma(\tfrac{1}{2}+i\eta)/\Gamma(\tfrac{1}{2}-i\eta) \equiv \exp\{2i\sigma_{-\frac{1}{2}}\}, \tag{10.27}$$

and obtain the final result for $f_{C_{\text{point}}}$:

$$f_{C_{\text{point}}} \approx f_C^{(0)}(1-\tfrac{1}{2}\pi Z_a Z_b \alpha v\,\sin\tfrac{1}{2}\vartheta\,\exp\{2i\sigma_{-\frac{1}{2}}-2i\sigma_0\}). \tag{10.28}$$

If one is willing to neglect terms of order $L^{-1}$, then (24) applies also to the spin-$\frac{1}{2}$ case. Therefore, for small angles, an improvement of (22) is

$$f_{C_p} \approx f_{C_{\text{point}}}(\cos^2\tfrac{1}{2}\vartheta + mE^{-1}\sin^2\tfrac{1}{2}\vartheta). \tag{10.29}$$

Better formulas for the spin-$\frac{1}{2}$ case are given by Gluckstern and Lin (1964), and by Bühring (1966). See also the book of Überall (1971).

## 1-11 Classical Equations of Motion Including Spin

Macroscopic electric and magnetic fields are normally so weak and slowly varying in space and time that the motion of relativistic particles in these fields can be evaluated classically using (0.20). We now wish to derive that equation from the Klein-Gordon equation (1.4), using the ansatz (1.6):

$$(i\,\partial_\mu - ZeA_\mu)(i\,\partial^\mu - ZeA^\mu)\psi = m^2\psi, \qquad \psi = Be^{iS}, \tag{11.1}$$

where $B$ and $S$ are real functions of $x^\mu$. The left-hand side of (1) is explicitly

$$\begin{aligned} &(i\,\partial_\mu - ZeA_\mu)e^{iS}(i\,\partial^\mu B - ZeA^\mu B - B\,\partial^\mu S) \\ &= e^{iS}\{(P_\mu - ZeA_\mu)[i\,\partial^\mu B + B(P^\mu - ZeA^\mu)] - \partial_\mu\,\partial^\mu B \\ &\quad + i(P^\mu - ZeA^\mu)\,\partial_\mu B + iB\,\partial_\mu(P^\mu - ZeA^\mu)\}, \qquad P^\mu \equiv -\partial^\mu S. \end{aligned} \tag{11.2}$$

The classical limit applies if and only if $\partial_\mu \partial^\mu B$ can be neglected. We insert the remaining terms of (2) into (1), divide by exp $(iS)$ and find for the real and imaginary parts

$$(P_\mu - ZeA_\mu)(P^\mu - ZeA^\mu) = m^2, \tag{11.3}$$

$$2(P_\mu - ZeA_\mu)\,\partial^\mu B + B\,\partial_\mu(P^\mu - ZeA^\mu) = 0. \tag{11.4}$$

We multiply (4) by $B/m$, whereby it assumes the form of the continuity equation

$$\partial_\mu j^\mu = 0, \qquad j^\mu = (P^\mu - ZeA^\mu)B^2/m. \tag{11.5}$$

As in the nonrelativistic case (see for example Messiah 1966), the classical limit does not correspond to a single 4-velocity trajectory $u^\mu(\mathbf{x}(t))$, but to a 4-velocity field $u^\mu(t, \mathbf{x})$ of noninteracting particles. The time differentiation along a trajectory is $d/dt = \partial_t + \mathbf{v}\nabla$, the Lorentz-invariant generalization being $d/d\tau = u_\nu\,\partial^\nu$. We can now identify

$$u^\mu = (P^\mu - ZeA^\mu)/m = j^\mu/B^2 = -\,(\partial^\mu S + ZeA^\mu). \tag{11.6}$$

For a particle in a macroscopic field, $u^\mu$ must be proportional to $j^\mu$. The proportionality function follows from the condition (0.19) $u_\mu u^\mu = 1$ and (3). Using $0 = \partial^\mu(u^\nu u_\nu) = 2u^\nu\,\partial^\mu u_\nu$, we find

$$du^\mu/d\tau = u_\nu\,\partial^\nu u^\mu = u_\nu(\partial^\nu u^\mu - \partial^\mu u^\nu) = -u_\nu Ze(\partial^\nu A^\mu - \partial^\mu A^\nu)/m, \tag{11.7}$$

which coincides with (0.20) which we wanted to derive. A similar derivation holds for spinor particles, in which case $B$ is a spinor, and $B^2$ is replaced by $\bar{B}B$.

Since an accelerated charged particle emits radiation, (7) cannot be exact. An equation which approximately includes the energy loss due to radiation is the Lorentz-Dirac equation (see Rohrlich 1965)

$$m\dot{u}^\mu = ZeF^{\mu\nu}u_\nu + \tfrac{2}{3}\alpha(\ddot{u}^\mu - \dot{u}^2 u^\mu), \qquad \dot{u}^\mu \equiv du^\mu/d\tau. \tag{11.8}$$

A somewhat different equation has recently been proposed by Mo and Papas (1971) and has been discussed and criticized by Stöckel (1976).

When (7) applies, one can also derive a classical equation of motion for the spin operator $\mathbf{S}(\mathbf{x}, t)$ of a particle in a macroscopic field $F^{\mu\nu}$. Its nonrelativistic limit follows from the equations of motion for operators,

$$\dot{\mathbf{S}} = i[H, \mathbf{S}] = 2\mu\mathbf{S} \times \mathbf{H}. \tag{11.9}$$

In the second form we have used (4.5) with $\mathbf{S} = \frac{1}{2}\boldsymbol{\sigma}$. A covariant generalization of $\mathbf{S}$ is given by a 4-vector $S^\mu$ satisfying

$$S^\mu u_\mu = 0, \tag{11.10}$$

which reduces to $S^0 = 0$ in the particle rest frame. Note that $2S^\mu$ is different from $\sigma^\mu$ of (5.6). From (7) and (9), we find

$$u_\mu \dot{S}^\mu = \;-S^\mu \dot{u}_\mu = -ZeF^{\mu\nu}u_\nu S_\mu/m = ZeF^{\mu\nu}u_\mu S_\nu/m. \tag{11.11}$$

Since $\dot{S}^\mu$ must be proportional both to $S^\nu$ and $F^{\lambda\sigma}$ with coefficients which involve only $u^\nu$ (terms proportional to $\dot{u}^\nu$ are forbidden to lowest order in $\alpha/m$), the relationship can only contain the products $F^{\mu\nu} S_\nu$ and $u^\mu F^{\lambda\nu} u_\lambda S_\nu$. The coefficients of these products follow from (11) and (9)

$$\dot{S}^\mu = m^{-1} e[(Z + \kappa)F^{\mu\nu}S_\nu - \kappa u^\mu F^{\lambda\nu} u_\lambda S_\nu]. \tag{11.12}$$

The consistency of (12) with (11) is checked by multiplying (9) by $mu_\mu$ and observing $u_\mu u^\mu = 1$. Similarly, (9) is checked by observing that the second term of (12) vanishes in the nonrelativistic limit. Equation (12) is known as the Bargmann-Michel-Telegdi equation (1959). It is based on the spin-independence of the classical equation of motion and on the smallness of the fine-structure constant $\alpha$.

Synchrotron radiation also induces a very slow self-polarization of the accelerated particle, which can become as high as 90% in electron synchrotrons. It can be evaluated semiclassically on the basis of (12). See Jackson (1976) for a review.

In view of (10), the spin operator $S^\mu$ has only 3 independent components, with

$$S_0 = \mathbf{S}\mathbf{u}/u_0 = S_l v, \tag{11.13}$$

where $S_l$ is the longitudinal component of $\mathbf{S}$. For spin-$\frac{1}{2}$ particles, $\mathbf{S}$ can be related to the polarization vector $\mathbf{P}$ of (7.29), which is the expectation value of $\frac{1}{2}\boldsymbol{\sigma}$ for a moving particle (see 2-3.8 below):

$$S^2 = S_0^2 - S_l^2 - S_t^2 = -P^2, \qquad \mathbf{S}_t = \mathbf{P}_t. \tag{11.14}$$

Here $t$ stands for the transverse components, $\mathbf{S}_t \cdot \mathbf{p} = \mathbf{P}_t \cdot \mathbf{p} = 0$. Decomposing also

$$P^2 = P_t^2 + P_l^2, \qquad P_l = \text{helicity} = h, \tag{11.15}$$

and adding $S_t^2 = P_t^2$ to the first of eqs. (14), we find that

$$S_0^2 - S_l^2 = -h^2, \qquad S_l = hE/m, \tag{11.16}$$

where the second equation is obtained from the first one by elimination of $S_0^2$ according to (13). This then defines the Lorentz transformation properties of $\mathbf{P}$.

CHAPTER TWO

# Two-Particle Problems

## 2-1 Reaction Rate Density and $S$-Matrix

The theory of relativistic particles describes elastic and inelastic collisions, particle production, and particle decays. The corresponding transition probability amplitudes $S_{if}$ form an $S$-matrix similar to the one mentioned in section 1-3, but the meaning of the indices $i$ and $f$ is more complicated.

We consider a simultaneous collision of $n_i$ initial particles of momenta $\mathbf{p}_j$, leading to the production of $n_f$ final particles of momenta $\mathbf{p}_n$. The appropriate initial and final states are products of (almost) plane waves:

$$|i\rangle = \prod_{j=1}^{n_i} |\mathbf{p}_j, M_j\rangle, \qquad |f\rangle = \prod_{n=1}^{n_f} |\mathbf{p}_n, M_n\rangle. \tag{1.1}$$

The $M_j$ and $M_n$ are possible magnetic quantum numbers. To begin with, we only consider reaction rates of spinless particles. Each type of particle initially present in a macroscopic system is then characterized by a classical phase space distribution $F_j(\mathbf{p}, \mathbf{R}, T)$, which may vary over macroscopic distances $\mathbf{R}$ and times $T$. The particle density $\rho_j$ and total particle number $N_j$ are obtained from $F_j$ as follows:

$$\int d^3pF_j(\mathbf{p}, \mathbf{R}, T) = \rho_j(\mathbf{R}, T), \qquad \int d^3R\rho_j(\mathbf{R}, T) = N_j(T). \tag{1.2}$$

We can, at most, measure the differential reaction rate density, which is the differential reaction rate per macroscopic volume element $d^3R$

$$dr_{if} = \prod_{j=1}^{n_i} d^3p_j(2E_j)^{-1}F_j(\mathbf{p}_j, \mathbf{R}, T)(dT\, d^3R)^{-1} \prod_{n=1}^{n_f} d_L^3p_n|S_{if}|^2 \tag{1.3}$$

Here we have assumed that the simultaneous collision of the $n_i$ particles happens in the common volume element $d^3R$. Normally we consider only 2-particle collisions, $n_i = 2$. The factors $(2E_j)^{-1}$ are again due to our state normalization (1-1.15). For example, the transition probability for the collision of two particles $a$ and $b$ of energies $E_a$ and $E_b$ is $(4E_a E_b)^{-1} |S_{if}|^2$.

$S_{if}$ will be zero unless energy and momentum are conserved in the reaction (this will be "derived" in 3-4.12). Also, the case where the final state is identical to the initial state must be considered separately, because it includes the probability that the particles do not interact at all. We therefore put

$$S_{if} \equiv \langle f | i \rangle + i(2\pi)^4 \, \delta_4(P_i - P_f) T_{if},$$
$$P_i \equiv \sum_{j=1}^{n_i} P_j, \qquad P_f \equiv \sum_{n=1}^{n_f} P_n, \qquad \langle f | i \rangle = \prod_{j=1}^{n_i} \langle \mathbf{p}_{jf}, M_{jf} | \mathbf{p}_j, M_j \rangle. \tag{1.4}$$

States containing different types or multiplicities of particles are said to belong to different "channels" $c$. We define $\langle f | i \rangle = 0$ for $c_i \neq c_f$. $S_{if}$ as given by (4) applies to multiparticle systems without external forces ("closed" systems), whereas our previous formula (1-3.3) applied to one particle in an external potential. The problem of squaring the energy $\delta$-function in $|S_{if}|^2$ was mentioned already in section 1-3. For the $\delta_3(\mathbf{p}_i - \mathbf{p}_f)$-function, we can return to our normalization cube of section 1-2. In the construction of the $S$-matrix in perturbation theory, one encounters an integral of the type

$$\int_0^L d^3x \exp\{i\mathbf{x}(\mathbf{p}_i - \mathbf{p}_f)\} = \delta_{\mathbf{p}_i, \mathbf{p}_f} L^3, \tag{1.5}$$

where the $\delta$ is a Kronecker-$\delta$. From the density of states $d^3n = (2\pi)^{-3} \, d^3pL^3$, one factor $L^3$ is left over which is not cancelled by a normalization integral, but instead by the macroscopic $d^3R$ in (3). With this result in mind, we can put

$$(2\pi)^4 \, \delta_4^2(P_i - P_f) = dT \, d^3R \, \delta_4(P_i - P_f). \tag{1.6}$$

Note that the interval $2T$ of (1-3.4) has been replaced by the macroscopic differential $dT$ in (6).

From (3), (4) and (6) we find that all measurable quantities contain the combination

$$(dT \, d^3R)^{-1} |S_{if}|^2 \prod_{n=1}^{n_f} d_L^3 p_n = |T_{if}|^2 \, d \text{ Lips } (P_i^2; f), \tag{1.7}$$

$$\begin{aligned} d \text{ Lips } (P_i^2, f) &\equiv (2\pi)^4 \, \delta_4(P_i - P_f) \prod_{n=1}^{n_f} d_L^3 p_n \\ &= (2\pi)^{4-3n_f} \, \delta_4(P_i - P_f) \prod d^3p_n / 2E_n . \end{aligned} \tag{1.8}$$

Equation (8) defines the "restricted Lorentz invariant phase space differential" $d$ Lips, which will be evaluated in the next section. Insertion of (7) into (3) gives

$$dr_{if} = \prod_{j=1}^{n_i} d^3p_j(2E_j)^{-1}F_j(\mathbf{p}_j, \mathbf{R}, T)|T_{if}|^2 \, d \text{ Lips } (P_i^2; f). \tag{1.9}$$

The case $n_i = 1$ has $d$ Lips $(P_i^2; f) = 0$ unless the initial particle is unstable. For unstable particles, (9) gives the decay rate density

$$dr_{if} = d^3p(2E)^{-1}F(\mathbf{p}, \mathbf{R}, T)|T_{if}|^2 \, d \text{ Lips } (P^2; f). \tag{1.10}$$

Consider now a beam of unstable particles, all of which have the same momentum $\mathbf{p}_a$. The corresponding $F$ is

$$F(\mathbf{p}, \mathbf{R}, T) = \delta_3(\mathbf{p} - \mathbf{p}_a)\rho(\mathbf{R}, T). \tag{1.11}$$

The $d^3p$ in (10) can be cancelled against the $\delta_3$-function:

$$dr_{if} = (2E_a)^{-1}\rho|T_{if}|^2 \, d \text{ Lips } (P_a^2; f). \tag{1.12}$$

More familiar than the differential decay rate density is the differential decay rate or decay width $d\Gamma_f$ (the index $i$ is dropped), which differs from (12) only by the omission of the particle density $\rho$. We shall in addition understand $p_a = 0$ in $d\Gamma_f$, i.e., the differential decay rate is evaluated in the rest frame of the unstable particle. Since $|T_{if}|^2 \, d$ Lips is Lorentz invariant, we simply have

$$dr_{if} = \rho E_a^{-1}m \, d\Gamma_f, \qquad d\Gamma_f = (2m)^{-1}|T_{if}|^2 \, d \text{ Lips } (P^2; f). \tag{1.13}$$

This quantity will be discussed in detail in the following sections. Here we merely mention that $E_a^{-1}m$ in (13) is just the inverse of the Lorentz factor $\gamma$ (compare 1-0.22) and is responsible for the time dilatation in the decays of particles in flight.

For the collisions between two types of particles $a$ and $b$, (9) gives

$$dr_{ab} = d^3p_1 \, d^3p_2(4E_1E_2)^{-1}F_a(\mathbf{p}_1, \mathbf{R}, T)F_b(\mathbf{p}_2, \mathbf{R}, T)|T_{if}|^2 \, d \text{ Lips}. \tag{1.14}$$

In normal collision experiments, all particles $a$ have velocity $\mathbf{p}_a$ and all particles $b$ have velocity $\mathbf{p}_b$ (with the exception of colliding beam experiments, one has in fact $p_b = 0$), i.e., both $F_a$ and $F_b$ are of the special form (11). This gives

$$dr_{if} = \rho_a\rho_b(4E_aE_b)^{-1}|T_{if}|^2 \, d \text{ Lips } (P_i^2; f) \equiv \rho_a\rho_b v_{ab} \, d\sigma, \tag{1.15}$$

with $P_i = P_a + P_b$. The quantities $v_{ab}$ and $d\sigma$ are called the relative velocity and differential cross section and will be discussed in section 2-3. For the calculation of thermonuclear reaction rates, one needs

$$F = \rho(R, T)C(\tau, \mu)[e^{(E-\mu)/\tau} \mp 1]^{-1}, \qquad \tau \equiv k_B T_K \tag{1.16}$$

where $T_K$ = absolute temperature, $k_B$ = Boltzmann's constant, $\mu$ = chemical potential (opposite for particles and antiparticles, $\mu = 0$ for photons), and the two signs apply for bosons and fermions, respectively. For the nondegenerate and nonrelativistic ($E = m + p^2/2m$) gas, one obtains Maxwell's distribution $F/\rho = (2\pi m\tau)^{-\frac{3}{2}} \exp(-p^2/2m\tau)$. Relativistic cases are discussed by Chiu (1968). As an example of $n_i = 3$, we mention the reaction $e^- pp \to \nu d$, which may occur in dense stars (Bahcall and May 1969).

## 2-2 Phase Space and Kinematics of Decays into Two Particles

The phase space element $d$ Lips of (1.8) is now calculated for two particles in the final state. We put $P_i^2 \equiv s$ and $P_f = P_1 + P_2$

$$d \text{ Lips } (s; P_1, P_2) = (4\pi^2 4E_1 E_2)^{-1}\, d^3p_1\, d^3p_2\, \delta_4(P_i - P_1 - P_2). \quad (2.1)$$

As $d$ Lips is Lorentz-invariant, it can be evaluated in various systems. We begin with the rest system of $i$, i.e., that system in which $\mathbf{p}_i = 0$. Inserting $P_i = (s^{\frac{1}{2}}, \mathbf{0})$ into (1), we get

$$\begin{aligned} d \text{ Lips } &(s; P_1, P_2) \\ &= (4\pi^2 4E_1E_2)^{-1}\, d^3p_1\, d^3p_2\, \delta(s^{\frac{1}{2}} - E_1 - E_2)\, \delta_3(\mathbf{p}_1 + \mathbf{p}_2). \end{aligned} \quad (2.2)$$

The last $\delta$-function tells us that $d$ Lips vanishes except for

$$\mathbf{p}_1 + \mathbf{p}_2 = 0. \quad (2.3)$$

This equation defines the cms (centre-of-momentum, centre-of-mass, or barycentric system) of particles 1 and 2. $\delta_3(\mathbf{p}_1 + \mathbf{p}_2)$ can be cancelled against $d^3p_2$; in the remainder we must replace $\mathbf{p}_2$ by $-\mathbf{p}_1$ $(E_2^2 = m_2^2 + p_1^2)$. Next, we rewrite $d^3p_1$ in spherical coordinates:

$$d \text{ Lips } (s; P_1 P_2) = (4\pi^2 4E_1 E_2)^{-1} p^2\, dp\, d\Omega\, \delta(s^{\frac{1}{2}} - E_1 - E_2). \quad (2.4)$$

Here we have dropped the index 1 on $\Omega$ and $\mathbf{p}$. For $p$, the index is unnecessary in the cms, as $|\mathbf{p}_1| = |\mathbf{p}_2| \equiv p$. To eliminate the last $\delta$-function, we introduce the total final state energy as a new variable

$$E = E_1 + E_2 = (m_1^2 + p^2)^{\frac{1}{2}} + (m_2^2 + p^2)^{\frac{1}{2}}, \quad (2.5)$$

$$dE = (E_1^{-1} + E_2^{-1})p\, dp = E_1^{-1}E_2^{-1}Ep\, dp. \quad (2.6)$$

Taking from this equation $dp$ into (4), we find

$$d \text{ Lips } (s; P_1, P_2) = (16\pi^2 E)^{-1} p\, dE\, \delta(s^{\frac{1}{2}} - E)\, d\Omega = (4\pi)^{-2} p s^{-\frac{1}{2}}\, d\Omega. \quad (2.7)$$

In (7), all 3 quantities $p$, $E_1$ and $E_2$ are now functions of $s$, $m_1^2$ and $m_2^2$ only. We first find the cms energies $E_1$ and $E_2$ by noting

$$\begin{aligned} E_1 + E_2 = s^{\frac{1}{2}} \quad \text{and} \quad m_1^2 - m_2^2 &= E_1^2 - E_2^2 \\ &= (E_1 - E_2)(E_1 + E_2) = (E_1 - E_2)s^{\frac{1}{2}}. \end{aligned}$$

Thus, knowing the sum and difference of $E_1$ and $E_2$, we find

$$E_1 = \tfrac{1}{2}s^{-\frac{1}{2}}(s + m_1^2 - m_2^2), \qquad E_2 = \tfrac{1}{2}s^{-\frac{1}{2}}(s - m_1^2 + m_2^2). \tag{2.8}$$

Finally, we obtain $p^2$ from $E_1^2$ (or $E_2^2$) as

$$p^2 = \frac{1}{4s}[(s + m_1^2 - m_2^2)^2 - 4sm_1^2] \equiv \frac{1}{4s}\lambda(s, m_1^2, m_2^2). \tag{2.9}$$

The function $\lambda$ is called the triangle function (see the discussion following 6-3.29), and is actually symmetric in all three variables:

$$\lambda(a, b, c) = a^2 + b^2 + c^2 - 2ab - 2ac - 2bc. \tag{2.10}$$

It can also be written in product forms, for example

$$\lambda = [a - (b^{\frac{1}{2}} + c^{\frac{1}{2}})^2][a - (b^{\frac{1}{2}} - c^{\frac{1}{2}})^2]. \tag{2.11}$$

We now evaluate $d$ Lips for $\mathbf{p}_i \neq 0$. The energies and momenta get an index $L$ (for "lab system"). The $z$-axis is taken along the momentum of the initial particle or state, $P_{iL} = (E_L, 0, 0, p_L)$, see fig. 2-2. From the relation $m_2^2 = (P_i - P_1)^2 = s + m_1^2 - 2P_iP_1$, we evaluate $P_iP_1$ in the lab system:

$$P_iP_1 = E_L E_{1L} - p_L p_{1L} \cos\theta_{1L} = \tfrac{1}{2}(s + m_1^2 - m_2^2) = s^{\frac{1}{2}}E_1. \tag{2.12}$$

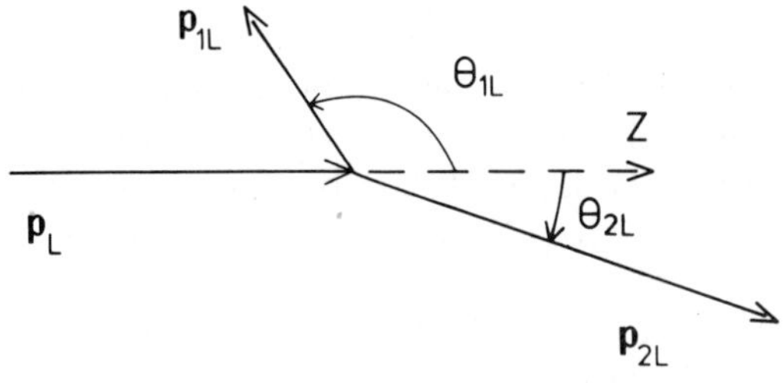


**Figure 2-2** The decay $i \to 12$ in the lab system.

Energies and momenta without subscript $L$ refer to the cms. The transverse momentum of particle 1 (or 2) has the same value in the lab system and in the cms:

$$p_{1L}\sin\theta_{1L} = p\sin\theta = p_{2L}\sin\theta_{2L}, \tag{2.13}$$

which gives

$$\begin{aligned} p_{1L}^2\cos^2\theta_{1L} &= p_{1L}^2(1 - \sin^2\theta_{1L}) = p_{1L}^2 - p^2\sin^2\theta \\ &= E_{1L}^2 - E_1^2 + p^2\cos^2\theta. \end{aligned} \tag{2.14}$$

Multiplying (14) with $p_L^2 = s - E_L^2$ and comparing with $p_L^2 p_{1L}^2 \cos^2\theta_{1L}$ from (12), we find

$$s^{\frac{1}{2}}E_{1L} = E_L E_1 + p_L p \cos\theta, \tag{2.15}$$

i.e., the lab energy of the final particle 1 is linearly related to the cosine of its cms angle. With $d\Omega = d\phi \, d \cos \theta$ and $d \cos \theta = s^{\frac{1}{2}} p_L^{-1} p^{-1} \, dE_{1L}$ according to (15),

$$d \text{ Lips } (s; P_1, P_2) = (4\pi)^{-2} p_L^{-1} \, dE_{1L} \, d\phi, \tag{2.16}$$

which is the desired expression. The maximal and minimal values of $E_{1L}$ occur at $\cos \theta = \pm 1$:

$$s^{\frac{1}{2}} E_{1L} \binom{\max}{\min} = E_L E_1 \pm p_L p. \tag{2.17}$$

Equation (16) tells us that $d \text{ Lips}/dE_{1L}$ is independent of $E_{1L}$ within its physical range (17).

As an example, we evaluate (17) for the muon lab energy $E_{1L}$ in the decay of a pion into a muon and a neutrino, $\pi \to \mu\nu$. As the neutrino has zero mass, it has cms momentum

$$p = E_2 = \tfrac{1}{2}(m_\pi - m_1^2/m_\pi) \tag{2.18}$$

according to (8), with $s^{\frac{1}{2}} = m_\pi$. We then get the maximal and minimal muon lab energies as

$$E_{1L} \binom{\max}{\min} = \tfrac{1}{2}(E_L \pm p_L) + \tfrac{1}{2} m_\pi^{-2} m_1^2 (E_L \mp p_L). \tag{2.19}$$

This formula applies also to the photon lab energy in the decay $\pi^0 \to \gamma\gamma$, if we put $m_1 = 0$:

$$E_{\gamma L} \binom{\max}{\min} = \tfrac{1}{2}(E_L \pm p_L).$$

One can also express the lab angle $\theta_{1L}$ in terms of cms quantities. Inserting $E_{1L}$ from (15) into (12), we find

$$s^{\frac{1}{2}} p_{1L} \cos \theta_{1L} = p_L E_1 + E_L p \cos \theta \tag{2.20}$$

and combining this with (13) gives

$$\tan \theta_{1L} = s^{\frac{1}{2}} \sin \theta \, (E_1 p_L p^{-1} + E_L \cos \theta)^{-1}. \tag{2.21}$$

## 2-3 Decay Rates and Cross Sections

The quantum systems of relativistic particle physics are normally simple: one or two particles in the initial states, and a few particles in the final states. It is then possible to keep track of over-all probability conservation. The probability of finding the system in any final state must be equal to the initial state probability,

$$\sum_f \int |S_{if}|^2 \prod_{n=1}^{n_f} d_L^3 p_n = \langle i | i \rangle, \tag{3.1}$$

where the summation extends over all channels, including the channel $i$. Together with the superposition principle, this is in fact an essential fundament of relativistic collision theory, as we shall see in sections 2-5, 3-11, 6-1, and 6-4. In the decays of unstable particles, one can measure the decrease of the quantum mechanical initial state probability as a function of time. If the unstable particles are all at rest and no new unstable particles are produced, their density $\rho$ decreases as follows:

$$d\rho/dT = -\Gamma(E)\rho = -E^{-1}m\Gamma\rho, \tag{3.2}$$

where $\Gamma$ is the total decay rate

$$\Gamma = \sum_{f\neq i}\Gamma_f, \qquad \Gamma_f = (2m)^{-1}\int |T_{if}|^2\, d\text{ Lips } (s_d; f), \qquad P^2 \equiv s_d. \tag{3.3}$$

The partial decay rate $\Gamma_f$ into channel $f$ is the integral over the corresponding differential decay rate $d\Gamma_f$ of (1.13). The quantity $P^2 = s_d$ is very close to $m^2$ but is not identical to it, because an unstable particle cannot be an eigenstate of the full Hamilton operator (the distinction makes little sense here, but we include it in view of the problems to be discussed in section 4-8).

Frequently one has a beam of unstable particles of well-defined velocity $\boldsymbol{\beta} = \mathbf{p}/E$. Then $\rho$ decreases along the direction of motion as

$$\frac{d\rho}{dR} = \frac{d\rho}{dT}\left(\frac{dR}{dT}\right)^{-1} = -\frac{m}{E}\Gamma\frac{\rho}{\beta} = -\Gamma\rho m/p. \tag{3.4}$$

For the case of decays into 2 particles, $d$ Lips is given by (2.7) in the cms and by (2.16) in the lab system. If the unstable particle is spinless, then $|T_{if}|^2$ is independent of $\Omega$ (there is no preferred direction in the cms), and the differential decay rate is isotropic in the cms and flat in $E_{1L}$ between $E_{1L}(\text{min})$ and $E_{1L}(\text{max})$ and zero outside this interval (rectangular distribution, see fig. 2-3). If the decaying particles have spin $S$, they are described by $(2S + 1)^2$ phase space functions $F_{MM'}$, where $M$ and $M'$ are magnetic quantum numbers. For a full understanding of this complication, one must consider the production of these particles as part of the quantum system; this will be done in section 4-8. Here we only need the Hermiticity property

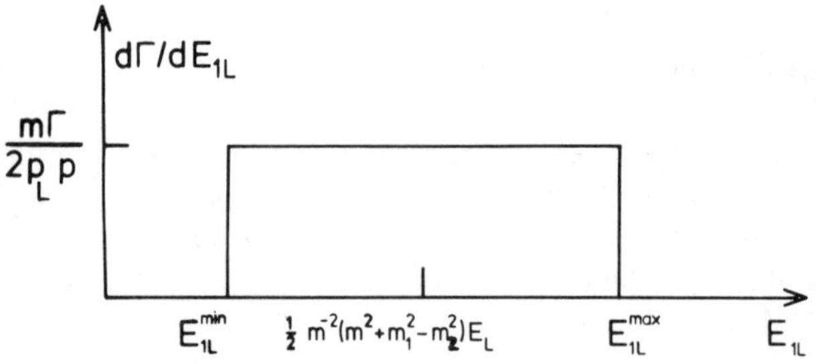


**Figure 2-3** The decay rate of a two-particle decay of a spinless particle as a function of the lab energy of one of the particles. $E_L$ and $p_L$ denote lab energy and momentum of the decaying particle, and the limits of $E_{1L}$ are given by (2.17), with $s^{1/2} = m$.

$F_{MM'} = F^*_{M'M}$. It is convenient to decompose $F$ into its trace and a normalized density matrix $\rho_{MM'}$ (not to be confused with the density $\rho$ of (2)):

$$F_{MM'} = F\rho_{MM'}, \qquad F \equiv \sum_M F_{MM}, \qquad \sum_M \rho_{MM} = 1. \tag{3.5}$$

In the differential decay rate (1.13), $|T_{if}|^2$ must now be replaced by

$$\overline{|T_{if}|^2} \equiv \sum_{MM'} \rho_{MM'} \sum_{M_f} T_{if}(M, M_f) T^*_{if}(M', M_f), \tag{3.6}$$

where $\sum_{M_f}$ indicates a summation over possible final spin states. For unpolarized particles, $\rho$ is a multiple of the unit matrix

$$\rho_{MM'} = \delta_{MM'}(2S+1)^{-1}, \qquad \overline{|T_{if}|^2} = (2S+1)^{-1} \sum_{M, M_f} |T_{if}(M, M_f)|^2. \tag{3.7}$$

The last expression is obvious: in an unpolarized beam, every spin state occurs with equal probability, namely $(2S+1)^{-1}$. Thus $|T_{if}|^2$ is averaged over initial spin states and summed over final spin states.

For spin-$\frac{1}{2}$ particles, one normally expresses $\rho$ in terms of the polarization vector $\mathbf{P}$ (which was already mentioned at the end of section 1-7) by means of the Pauli spin matrices (1-4.2):

$$\rho = \tfrac{1}{2}(1 + \mathbf{P}\boldsymbol{\sigma}) = \frac{1}{2}\begin{pmatrix} 1 + P_z & P_x - iP_y \\ P_x + iP_y & 1 - P_z \end{pmatrix}. \tag{3.8}$$

For unpolarized particles, the differential decay rate is still isotropic. Moreover, the partial decay rate $\Gamma_f$ of (3) is independent of the polarization state, because $\Gamma_f(M')$ is obtained from $\Gamma_f(M)$ by a rotation. Thus the appropriate generalization of (3) is

$$\Gamma_f = (2m)^{-1} \int d \text{ Lips } (s_a, f) \sum_{M_f} |T_{if}(M, M_f)|^2. \tag{3.9}$$

We now turn to collisions between two particles $a$ and $b$. The corresponding density matrices are denoted by $\rho^a$ and $\rho^b$. The $|T_{if}|^2$ in the reaction rate density (1.15) is replaced by

$$\overline{|T_{if}|^2} = \sum_{M_a \cdots M_{b'}} \rho^a_{M_a M_{a'}} \rho^b_{M_b M_{b'}} \sum_{M_f} T_{if}(M_a, M_b, M_f) T^*_{if}(M'_a, M'_b, M_f). \tag{3.10}$$

When particle $b$ (the "target particle") is at rest (we call this system the "lab system," although in colliding beam experiments it has nothing to do with the laboratory) we have

$$s = (P_a + P_b)^2 = m_a^2 + m_b^2 + 2E_{aL} m_b, \tag{3.11}$$

$$p_{aL} = (E_{aL}^2 - m_a^2)^{\frac{1}{2}} = (2m_b)^{-1} \lambda^{\frac{1}{2}}(s, m_a^2, m_b^2), \tag{3.12}$$

$$v_{aL} = p_{aL}/E_{aL} = \lambda^{\frac{1}{2}}/2E_{aL} m_b. \tag{3.13}$$

From (11) we can insert $E_{aL} m_b = P_a P_b$ into (13), which shows that $v_{aL}$ is Lorentz-invariant. Moreover, it is symmetric in $a$ and $b$. It is therefore not only the lab velocity, but also the "antilab" velocity $v_{b\bar{L}}$, which is the velocity of $b$ in the Lorentz system where the projectile $a$ is at rest

$$v_{aL} = v_{b\bar{L}} = (2P_a P_b)^{-1} \lambda^{\frac{1}{2}}(s, m_a^2, m_b^2) \equiv v_L. \tag{3.14}$$

For a two-particle system of arbitrary momenta $P_a$, $P_b$, we now define the relative velocity $v_{ab}$ as

$$v_{ab} \equiv 2\lambda^{\frac{1}{2}}(s, m_a^2, m_b^2)(4E_a E_b)^{-1}. \tag{3.15}$$

This quantity is not Lorentz invariant but reduces to the lab velocity when one of the particles is at rest. With this definition and the definition (1.15) of the product $v_{ab}\, d\sigma$, we arrive at

$$d\sigma = \tfrac{1}{2}\lambda^{-\frac{1}{2}}(s, m_a^2, m_b^2) \overline{|T_{if}|^2}\, d \text{ Lips } (s; f). \tag{3.16}$$

In the lab system, $\rho_a \mathbf{v}_{aL} = \mathbf{f}_a$ is the flux of particles $a$, and (1.15) can be rewritten as

$$dr_{if} = f_a \rho_b\, d\sigma, \tag{3.17}$$

which shows that $d\sigma$ coincides with the differential cross section defined in section 1-3. The flux decreases along $\mathbf{v}_{aL}$ as

$$df_a/dR = -f_a \rho_b \sigma \equiv -\Lambda^{-1} f_a, \tag{3.18}$$

where $\sigma$ is the total cross section and $\Lambda$ is the mean free path. However, since $v_{ab}$ is not Lorentz-invariant, these geometrical interpretations do not hold in other frames of reference.

An interesting limiting case is the capture of loosely bound particles (electron capture in nuclear $\beta$-decay, etc.). Here both eqs. (1.12) and (1.15) apply. Let $m_B = m_a + m_b$ be the mass of the bound state and $\mu = m_a m_b / m_B$ the reduced mass. The decay matrix element of (1.12) is called $T_B$, the scattering matrix element is $T$. In the bound state rest frame we therefore have

$$dr_{if} = (2m_B)^{-1} \rho_B\, |T_B|^2\, d \text{ Lips} = (4m_a m_b)^{-1} \rho_a \rho_b\, |T|^2\, d \text{ Lips}. \tag{3.19}$$

When the bound state wave function $\psi$ is nearly constant over the range of the capture process, we can put $\rho_a \rho_b = \rho_B\, |\psi(0)|^2$ and find

$$|T_B|^2 = |\psi(0)|^2 |T|^2 (2\mu)^{-1} \approx (2\pi)^{-1} \mu^2 Z^3 \alpha^3 |T|^2. \tag{3.20}$$

The last formula is valid for a hydrogen-like wave function $\psi = (\pi a^3)^{-\frac{1}{2}} e^{-r/a}$, with $a = (Z\alpha\mu)^{-1}$. We can also directly relate $\Gamma_B$ to $\sigma_B$ using eqs. (3) and (6)

$$d\Gamma_B = d\sigma\, |\psi(0)|^2 p/\mu. \tag{3.21}$$

However, (20) is more general in the sense that it also applies to the production of bound states, for example $pp \to \pi + d$.

Our target density $\rho_b$ is the number of particles per volume. It is related to the more commonly used matter density $\rho_m$ (measured in g cm$^{-3}$) by

$$\rho = \rho_m N_A / A, \qquad N_A = 6.0220 \times 10^{23}/\text{mole}, \tag{3.22}$$

where $N_A$ is Avogadro's number and $A$ the atomic weight, in units g/mole. Since our cross sections depend only on the atomic nucleus and not on the state of matter (solid, fluid or gas), one frequently rewrites (18) as

$$df_a/dx = -\Lambda^{-1}f_a, \qquad \Lambda^{-1} = \sigma N_A/A \tag{3.23}$$

where $x$ is the amount of material transversed, in units g cm$^{-2}$, and $\Lambda$ has now the dimension g cm$^{-2}$, too. $\Lambda^{-1}$ is called the mass absorption coefficient. Examples are given in table 2-3.

**Table 2-3** Mean free paths of 20 GeV neutrons in various materials (from Particle Data Group 1978). The mean free paths of protons and neutrons are roughly equal and energy independent above 1 GeV in these materials. Also included are radiation lengths $X_0$, which will be explained in section 8-3.

| Target | $\rho_m$[g cm$^{-3}$] | $\Lambda$[cm] | $\Lambda$[g cm$^{-2}$] | $X_0$[cm] | $X_0$[g cm$^{-2}$] |
|---|---|---|---|---|---|
| Water | 1.00 | 58.3 | 58.3 | 36.1 | 36.1 |
| Air 20°C | 0.0012 | $5 \times 10^4$ | 60.2 | 30050 | 36.2 |
| Al | 2.70 | 25.5 | 68.9 | 8.9 | 24.0 |
| Fe | 7.87 | 10.2 | 79.9 | 1.76 | 13.84 |
| Pb | 11.35 | 111.7 | 9.8 | 0.56 | 6.37 |

## 2-4 Kinematics of Two-Body Reactions

We now consider two-particle reactions $ab \to cd$. Energy-momentum conservation requires

$$P_a + P_b = P_c + P_d. \tag{4.1}$$

Thus, we have 3 independent 4-vectors, from which we can construct 6 Lorentz invariants, e.g., 3 squares and 3 scalar products. However, as $P_a^2 = m_a^2, \ldots P_d^2 = m_d^2$ are all fixed, only two Lorentz-invariant variables remain. One of them is $s$, the square of the cms energy. Following Mandelstam, we introduce two more variables

$$t = (P_c - P_a)^2 = (P_b - P_d)^2, \tag{4.2}$$

$$u = (P_a - P_d)^2 = (P_c - P_b)^2, \tag{4.3}$$

which are linearly related through

$$s + t + u = m_a^2 + m_b^2 + m_c^2 + m_d^2. \tag{4.4}$$

In the lab system, $t$ has a form similar to (3.11):

$$t = m_b^2 + m_d^2 - 2m_b E_{dL}. \tag{4.5}$$

Differences between 4-momenta of the initial and final particles are called 4-momentum transfers. Both $t$ and $u$ are squares of 4-momentum transfers.

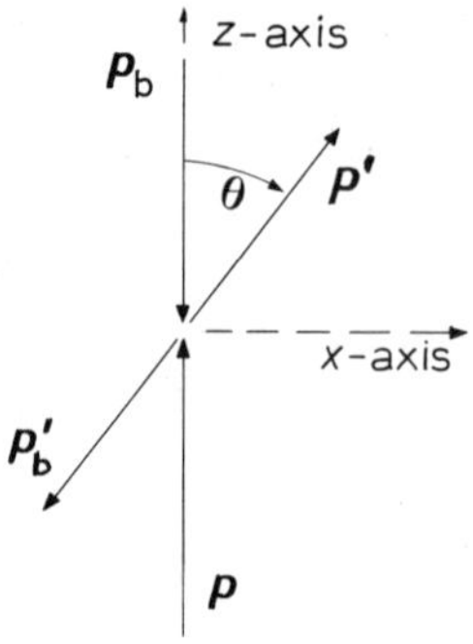


**Figure 2-4** The standard coordinate system of scattering in the cms. If only one particle carries spin, this particle is taken as particle $a$, with $\mathbf{p} = \mathbf{p}_a$ along the $z$-axis.

For the cms momenta in the initial and final states, we shall mainly use the letters $p$ and $p'$:

$$p = (4s)^{-\frac{1}{2}}\lambda^{\frac{1}{2}}(s, m_a^2, m_b^2), \qquad p' = (4s)^{-\frac{1}{2}}\lambda^{\frac{1}{2}}(s, m_c^2, m_d^2). \tag{4.6}$$

We also introduce the cms scattering angle $\theta$ between $a$ and $c$ (see fig. 2-4) and express $t$ in terms of the cms quantities as follows:

$$t = m_a^2 + m_c^2 - 2E_aE_c + 2pp' \cos\theta = m_b^2 + m_d^2 - 2E_bE_d + 2pp' \cos\theta. \tag{4.7}$$

We now insert the two-particle phase space (2.7) into (3.16) and find

$$d\sigma(ab \to cd) = \tfrac{1}{2}\lambda^{-\frac{1}{2}}\overline{|T_{if}|^2}\, d\,\text{Lips}\,(s; P_c, P_d) = (4ps)^{-1}p'\overline{|T(s, \Omega/4\pi)|^2}\, d\Omega. \tag{4.8}$$

For unpolarized particles, we can replace $d\Omega$ by $2\pi\, d\cos\theta$:

$$\begin{aligned} d\sigma(ab \to cd) &= \frac{\pi p'}{2ps}\overline{\left|\frac{T}{4\pi}\right|^2} d\cos\theta = \frac{\pi}{4p^2s}\overline{\left|\frac{T}{4\pi}\right|^2} dt = \frac{\pi}{\lambda}\overline{\left|\frac{T}{4\pi}\right|^2} dt \\ &= \frac{\pi}{\lambda}(2S_a + 1)^{-1}(2S_b + 1)^{-1} \sum_{M_aM_bM_cM_d} |T/4\pi|^2\, dt. \end{aligned} \tag{4.9}$$

We can also express $dt$ by $dE_{dL}$ (5) and regain the phase space (2.16), with index $1 = d$.

For elastic scattering, we have $m_a = m_c$, $m_b = m_d$, $p' = p$ and (7) simplifies to

$$t_{el} = -2p^2(1 - \cos\theta) = -4p^2 \sin^2\frac{\theta}{2}. \tag{4.10}$$

The upper kinematical limit of $t$ is then 0. In general, the kinematical limits of $t$ occur at $\cos\theta = \pm 1$ in (7):

$$t_{\min}^{\max} = (4s)^{-1}[(m_c^2 - m_a^2 + m_b^2 - m_d^2)^2 - \lambda - \lambda' \pm 2(\lambda\lambda')^{\frac{1}{2}}], \tag{4.11}$$

where $\lambda$ and $\lambda'$ are the triangle functions of (6).

The following trivial formulas are also useful:

$$\begin{aligned} 2P_aP_b &= s - m_a^2 - m_b^2, & 2P_cP_d &= s - m_c^2 - m_d^2, \\ 2P_aP_c &= m_a^2 + m_c^2 - t, & 2P_bP_d &= m_b^2 + m_d^2 - t, \\ 2P_aP_d &= m_a^2 + m_d^2 - u, & 2P_cP_b &= m_c^2 + m_b^2 - u. \end{aligned} \tag{4.12}$$

From (6) and (3.12) we see that initial cms momentum $p$ and lab momentum $p_{\text{lab}}$ are both proportional to $\lambda^{\frac{1}{2}}$. This has a simple origin and an important generalization: whenever we have two 4-momenta $P_i$ and $P_j$, we can define a third 4-momentum $P_k$ such that

$$P_i + P_j + P_k = 0. \tag{4.13}$$

For example, when $i = a$ and $j = b$, $P_k = -P_a - P_b = -P$, $P_k^2 = s$. In a system where one of these 4-momenta is "at rest", $P_i = ((P_i^2)^{\frac{1}{2}}, \mathbf{0})$ say, the other two 3-momenta have equal magnitudes, $p_j = p_k \equiv p(i)$. We then have

$$4\lambda(P_i^2, P_j^2, P_k^2) = p^2(k)P_k^2 = p^2(j)P_j^2 = p^2(i)P_i^2. \tag{4.14}$$

The first of these equations is merely the definition (2.9) of $\lambda$, using $(-P)^2 = P^2 = s$. The others follow from the total symmetry (2.10) of $\lambda$ in $P_i^2$, $P_j^2$, $P_k^2$.

Another useful formula is the connection between the differential cross section in the lab system $d\sigma/d\Omega_L$ and in the cms, $d\sigma/d\Omega$. As the azimuthal angle remains unchanged by the transformation between the two systems, we have

$$d\sigma/d\Omega = (d\sigma/d\Omega_L)\, d \cos \theta_{1L}/d \cos \theta. \tag{4.15}$$

Here $\theta$ is the angle between $a$ and $c$ in the cms as before, and $\theta_{1L}$ is the same angle in the lab system. We can again use fig. 2-2 and (2.21), with $E_L = E_{aL} + m_b =$ total lab energy. We take the inverse of (2.21), use $d \cot \theta_{1L} = \sin^{-3} \theta_{1L}\, d \cos \theta_{1L}$ and find

$$d \cos \theta_{1L}/d \cos \theta = \sin^{-3} \theta \sin^3 \theta_{1L} s^{-\frac{1}{2}} E_L(1 + E_L^{-1}E_1 p^{-1} p_L \cos \theta). \tag{4.16}$$

## 2-5 Unitarity and Partial-Wave Expansion

Conservation of probability led us to condition (3.1) for the $S$-matrix. The superposition principle allows us to start with any linear combination

$$\psi(-T) = c_i\phi_i(-T) + c_j\phi_j(-T), \qquad |c_i|^2 + |c_j|^2 = 1, \tag{5.1}$$

where we have returned to the time description of (1-3.1). In that case $S_{if}$ is replaced by $c_i S_{if} + c_j S_{jf}$, and conservation of probability now requires that even

$$\sum_k \int S_{ik} S_{jk}^* \prod_{n=1}^{n_k} d_L^3 P_n = \langle i | j \rangle. \tag{5.2}$$

In a symbolic notation, (2) is sometimes written as $SS^+ = 1$, i.e., $S$ is a "unitary matrix." When the decomposition (1.4) is introduced, one gets

$$-i(T_{ij} - T_{ji}^*) = \sum_k \int d \text{ Lips } (s; k) T_{ik} T_{jk}^*. \tag{5.3}$$

For $i = j$, i.e., elastic scattering at zero angle and no spin-flip, this gives the so-called optical theorem,

$$2 \text{ Im } T_{ii} = \sum_k \int d \text{ Lips } (s; k) |T_{ik}|^2 = 2\lambda^{\frac{1}{2}}(s, m_a^2, m_b^2)\sigma_i. \tag{5.4}$$

The last expression is of course only valid when the initial state contains 2 particles, $a$ and $b$. The cross section $\sigma_i = \sigma_{ab}$ is the total cross section for given magnetic quantum numbers $i = (M_a, M_b)$. For meson-nucleon scattering, however, this is the same as the spin-averaged cross section. For other reactions, one can obtain the unpolarized cross section by summing over $M_a$, $M_b$:

$$\sum_{M_a, M_b} \text{Im } T_{M_a, M_b}(s, \Omega = 0) = (2S_a + 1)(2S_b + 1)\lambda^{\frac{1}{2}}\sigma. \tag{5.5}$$

The nonlinear integral equation (3) can be reduced to a set of algebraic equations by the partial wave expansion. This reduction is complete only when three-particle channels can be neglected on the right-hand side of (3). However, it remains useful also under more general circumstances.

We go to the cms and take $\mathbf{p} = \mathbf{p}_a$ along the $z$-axis (see fig. 2-5). $\Omega = (\theta, 0)$ is the solid angle between $\mathbf{p}$ and $\mathbf{p}'$ ($\mathbf{p}'$ is the momentum of the first particle in state $j$), $\Omega' = (\theta', \phi')$ is the solid angle between $\mathbf{p}$ and $\mathbf{p}''$ ($\mathbf{p}''$ is the corresponding momentum in state $k$), and $\Omega'' = (\theta'', \phi'')$ is the solid angle between $\mathbf{p}$ and $\mathbf{p}''$. Then the unitarity equation (3) reads explicitly, with $d \text{ Lips} = d\Omega' p_k (16\pi^2 s^{\frac{1}{2}})^{-1}$

$$-i(T_{ij}(\Omega) - T_{ji}^*(\Omega)) = \sum_k p_k (16\pi^2 s^{\frac{1}{2}})^{-1} \int d\Omega' T_{ik}(\Omega') T_{jk}^*(\Omega''). \tag{5.6}$$

To get at the essential point first, we consider a situation where only spinless particles are involved, for example collisions of two $\alpha$-particles. As all angular dependence is stated explicitly, the indices $i$, $j$ and $f$ are now reduced to indices for the open channels $\alpha\alpha$, $\alpha\alpha^*$ (20.2 MeV) etc., depending on $s$. We expand $T_{ij}(\theta)$ in terms of Legendre polynomials of $\cos \theta$:

$$T_{ij}(\theta) = 8\pi s^{\frac{1}{2}} \sum_{L=0}^{\infty} (2L + 1) P_L(\cos \theta) T_{ij, L}(s) \equiv 8\pi s^{\frac{1}{2}} f_{ij}. \tag{5.7}$$

$T_{ik}(\Omega')$ is independent of $\phi'$ for spinless particles and can be expanded correspondingly in $\cos \theta'$. $T_{jk}(\Omega'')$ finally depends only on $\cos \theta''$, and from the decomposition of $P_L(\cos \theta'')$ in terms of spherical harmonics (B-5.4), we can perform the $\phi'$ integration in (6) immediately:

$$\int d\phi' P_L(\cos \theta'') = 2\pi P_L(\cos \theta) P_L(\cos \theta') \tag{5.8}$$

With $x = \cos\theta$, $x' = \cos\theta'$, (6) is now transformed into

$$-8\pi s^{\frac{1}{2}} \sum_L (2L+1)P_L(x)i(T_{ij,L} - T^*_{ji,L})$$
$$= \sum_k p_k 4s^{\frac{1}{2}} \int dx' \sum_{L'} (2L'+1)P_{L'}(x')T_{ik,L'} \sum_L (2L+1)2\pi P_L(x)P_L(x')T^*_{jk,L}$$
$$= 16\pi s^{\frac{1}{2}} \sum_k p_k \sum_{L'} (2L'+1)P_{L'}(x)T_{ik,L'} T^*_{jk,L'}. \tag{5.9}$$

The last expression is obtained using the orthogonality relation

$$\frac{1}{2}\int dx' P_{L'}(x')P_L(x')(2L+1) = \delta_{LL'}, \tag{5.10}$$

which also ensures that (9) is correct for each partial wave separately,

$$(2i)^{-1}(T_{ij,L} - T^*_{ji,L}) = \sum_k p_k T_{ik,L} T^*_{jk,L}. \tag{5.11}$$

This is the desired partial-wave unitarity. If we are only interested in elastic scattering, we may drop the indices $i$ and $j$ and obtain

$$\text{Im } T_L = \sum_k p_k |T_{k,L}|^2. \tag{5.12}$$

Occasionally, all channels except the elastic one are closed at low energies (this applies to $\alpha\alpha$-scattering below the threshold of the $\alpha\alpha^*$ (20.2 MeV) channel, but not to $\pi^+\alpha$-scattering). Then (12) implies that $T_L$ is of the form

$$T_L = p^{-1}e^{i\delta_L}\sin\delta_L = (2ip)^{-1}(e^{2i\delta_L} - 1) = p^{-1}(\cot\delta_L - i)^{-1}, \tag{5.13}$$

where $\delta_L$ is a real function of $s$. Comparing (7) and (13) with the "potential theoretical" equations (1-3.12) and (1-3.15), we find that the elastic two-body scattering formulas are obtained from the one-body formulas by replacing the lab momentum $k$ and scattering angle $\vartheta$ by the corresponding cms quantities $p$ and $\theta$. The differential cross section follows from (4.8) and (7):

$$d\sigma_{ij}/d\Omega = p^{-1}p'\left|\sum_L (2L+1)P_L(\cos\theta)T_{ij,L}\right|^2. \tag{5.14}$$

The integrated cross section is simplified by the orthogonality relations between the Legendre polynomials:

$$\sigma_{ij} = \sum_L \sigma_{ij,L}, \qquad \sigma_{ij,L} = 4\pi p^{-1}p'(2L+1)|T_{ij,L}|^2, \tag{5.15}$$

and in the "purely elastic" region where (13) applies,

$$\sigma_L = 4\pi p^{-2}(2L+1)\sin^2\delta_L.$$

We now derive the partial-wave unitarity equations for the case that only the first particle (particle $a$) carries spin (electron scattering on a spinless target), and to save some indices we assume that only the elastic channel is open. Let $\lambda = \lambda_i$ and $\lambda' = \lambda_j$ denote the helicities of particle $a$ in the initial and final states. This time $T_{ij}(\Omega)$ cannot be expanded in terms of Legendre polynomials, because it has a $\phi$-dependence from the helicity states (1-4.17), which is, in fact, valid for arbitrary spin $S$. Contrary to the scattering angle $\theta$, $\phi$ cannot be expressed in terms of the Mandelstam variables. It therefore enters $T$ via the spin states only, such that we can expand

$$\begin{aligned} T_{\lambda\lambda'}(\Omega) &= 8\pi s^{\frac{1}{2}} \sum_{J=S}^{\infty} (2J+1)\, D^{J*}_{\lambda\lambda'}(\Omega) T_J(\lambda\lambda') \\ &= 8\pi s^{\frac{1}{2}} e^{i\phi(\lambda-\lambda')} \sum_J (2J+1)\, d^J_{\lambda\lambda'}(\theta) T_J(\lambda\lambda'). \end{aligned} \tag{5.16}$$

Next, we need the expansion of $T^*_{ji}$ in (6), which involves the functions $D^J_{\lambda'\lambda}(-\Omega)$, where $-\Omega$ denotes the rotation opposite to $\Omega$. It rotates the momentum $\mathbf{p}'$ back to $\mathbf{p}$:

$$D_{\lambda'\lambda}(-\Omega) = (D^{-1}(\Omega))_{\lambda'\lambda} = D^*_{\lambda\lambda'}(\Omega), \tag{5.17}$$

due to the unitarity of the $D$-matrices, $D^{-1} = D^+$. Therefore we obtain from (6)

$$\begin{aligned} &-2\pi i \sum_J D^{J*}_{\lambda\lambda'}(\Omega)[T_J(\lambda\lambda') - T^*_J(\lambda'\lambda)] \\ &\quad = p \sum_{\lambda''} \int d\Omega' \left[\sum_{J'} (2J'+1)\, D^{J'*}_{\lambda\lambda''}(\Omega') T_{J'}(\lambda\lambda'')\right] \\ &\quad\quad \times \left[\sum_J (2J+1)\, D^J_{\lambda'\lambda''}(\Omega'') T^*_J(\lambda'\lambda'')\right]. \end{aligned} \tag{5.18}$$

On the left-hand side we have in fact $\phi = 0$ (see fig. 2-5). $\Omega''$, on the other hand, comprises three angles this time, two of which specify the direction of $\mathbf{p}''$ in a system in which $\mathbf{p}'$ is the $z$-axis (remember that the expansion (16) is only possible when the $z$-axis points along the initial momentum). Due to the group properties of the rotation matrices, we can express $D(\Omega'')$ as the

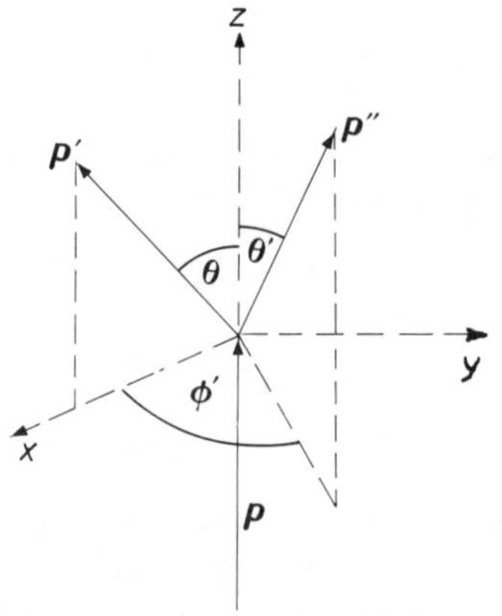


**Figure 2-5** The momenta $\mathbf{p}$, $\mathbf{p}'$ and $\mathbf{p}''$ which occur in the unitarity equation.

product of two rotation matrices, the first of which corresponds to the rotation of $\mathbf{p}'$ back to the $z$-axis while the second describes the rotation from the $z$-axis to $\mathbf{p}''$:

$$D^J_{\lambda'\lambda''}(\Omega'') = \sum_M (D^{-1}(\Omega))_{\lambda' M}\, D_{M\lambda''}(\Omega') = \sum_M D^*_{M\lambda'}(\Omega)\, D_{M\lambda''}(\Omega'). \tag{5.19}$$

The $\phi'$-dependence of the integrand of (18) consists of a factor $e^{i(\lambda-\lambda'')\phi'}$ from the first square bracket and a factor $e^{i(\lambda''-M)\phi'}$ from the second square bracket, according to (19). Therefore, the $\phi'$-integration produces a factor $2\pi\,\delta_{M\lambda}$ and the right-hand side of (18) becomes

$$2\pi p \sum_{\lambda''} \int d\cos\theta' \left[\sum_{J'} (2J'+1)\, d^{J'}_{\lambda\lambda''}(\theta') T_{J'}(\lambda\lambda'')\right] \times \left[\sum_J (2J+1)\, D^{J*}_{\lambda\lambda'}(\Omega)\, d^J_{\lambda\lambda''}(\theta') T^*_J(\lambda'\lambda'')\right] \tag{5.20}$$

$$= 4\pi p \sum_{\lambda'',J} (2J+1) T_J(\lambda\lambda'') T^*_J(\lambda'\lambda'')\, D^{J*}_{\lambda\lambda'}(\Omega) \tag{5.21}$$

The last expression is obtained by means of the orthogonality relations (B-5.1) of the $d$-functions.

Comparing (21) with the left-hand side of (18) we find the same $\phi$-dependence in both sums. From the orthogonality of the functions $d^J_{\lambda\lambda'}$ and $d^{J'}_{\lambda\lambda'}$ we conclude that the coefficients of $d^J_{\lambda\lambda'}(\theta)$ in the two expansions in $J$ must be equal term by term:

$$-i[T_J(\lambda\lambda') - T^*_J(\lambda'\lambda)] = 2\sum_{\lambda''} p T_J(\lambda\lambda'') T^*_J(\lambda'\lambda''). \tag{5.22}$$

This is the desired partial-wave unitarity involving a particle with spin. Comparison with (11) shows that the helicity enters formally like a channel index. The differential cross section for elastic scattering of unpolarized particles of spin $S$ is given by (4.8) or (4.9),

$$d\sigma/d\Omega = (4s)^{-1}(2S+1)^{-1} \sum_{\lambda\lambda'} |T(\lambda\lambda', s, \theta)/4\pi|^2$$

$$= (2S+1)^{-1} \sum_{\lambda\lambda'} \left|\sum_J (2J+1)\, d^J_{\lambda\lambda'}(\theta) T_J(\lambda\lambda', s)\right|^2, \tag{5.23}$$

and is in fact independent of $\phi$. The integrated cross section is simplified by the orthogonality relations of the $d$-functions:

$$\sigma = \sum_J \sigma_J, \qquad \sigma_J = 4\pi(2S+1)^{-1}(2J+1) \sum_{\lambda\lambda'} |T_J(\lambda\lambda')|^2. \tag{5.24}$$

These formulas can be simplified for parity-conserving scattering. The parity operator commutes with the total angular momentum operator but reverses the helicity. Thus for parity-invariant scattering

$$T_J(\lambda\lambda') = T_J(-\lambda, -\lambda'). \tag{5.25}$$

When particle $a$ carries only spin $\frac{1}{2}$, this is sufficient to reduce the unitarity eq. (22) to the spinless form. We substitute

$$T_J(\tfrac{1}{2}, \pm\tfrac{1}{2}) = \tfrac{1}{2}(f_{L+} \pm f_{(L+1)-}), \qquad L \equiv J - \tfrac{1}{2} \tag{5.26}$$

in (22) and obtain separate equations for $f_{L+}$ and $f_{L-}$:

$$\text{Im}\, f_{L\pm} = p\,|\, f_{L\pm}\,|^2. \tag{5.27}$$

The index with a minus sign has been shifted by one unit such that $f_{L-}$ has $J = L - \frac{1}{2}$. The notation $L\pm$ has the same meaning as in (1-7.23).

From (27) we conclude that $f_{L+}$ and $f_{L-}$ must be of the form $p^{-1}e^{i\delta}\sin\delta$ each, i.e., identical with the cms versions of (1-7.26):

$$f_{L+} = p^{-1}e^{i\delta_{L+}}\sin\delta_{L+}, \qquad f_{L-} = p^{-1}e^{i\delta_{L-}}\sin\delta_{L-}. \tag{5.28}$$

Clearly the scattering amplitudes $f_{L\pm}$ belong to states with parity $(-1)^L$.

The helicity states have the angular dependence (1-4.17). Therefore, with the $z$-axis along the initial momentum, $\mathbf{p}$, the general angular dependence of $T_{\lambda\lambda'}(\Omega)$ for the scattering of spin-$\frac{1}{2}$ particles is

$$\begin{aligned} T_{\lambda\lambda'}(\Omega) &= 8\pi s^{\frac{1}{2}}\chi_0^{+}(\lambda', \theta, \phi)F\chi_0(\lambda, 0, 0) = 8\pi s^{\frac{1}{2}}e^{i(\lambda-\lambda')\phi}\, d^{\frac{1}{2}}_{\lambda\lambda'}(\theta)F, \\ F &= f_1 + 4\lambda\lambda' f_2 + 2\lambda f_3 + 2\lambda' f_4, \end{aligned} \tag{5.29}$$

where $2\lambda$ and $2\lambda'$ are the eigenvalues of twice the helicity operators $\boldsymbol{\sigma}\hat{\mathbf{p}}$ and $\boldsymbol{\sigma}\hat{\mathbf{p}}'$ and the functions $f_i$ depend only on $\cos\theta$. From table B-5 we find

$$(2J+1)\, d^J_{\lambda\lambda'} = 2(P'_{J+\frac{1}{2}} - 4\lambda\lambda' P'_{J-\frac{1}{2}})\, d^{\frac{1}{2}}_{\lambda\lambda'}, \tag{5.30}$$

such that (16) can be replaced by the simpler expansion

$$F(\lambda\lambda') = 2\sum_J (P'_{J+\frac{1}{2}} - 4\lambda\lambda' P'_{J-\frac{1}{2}})T_J(\lambda\lambda'). \tag{5.31}$$

For parity-conserving interactions, we have $F(\lambda, \lambda') = F(-\lambda, -\lambda')$, $f_3 = f_4 = 0$,

$$f_1 = \sum_L (f_{L+}P'_{L+1} - f_{(L+1)-}P'_L) = f_{0+} + \sum_{L=1}^{\infty}(f_{L+}P'_{L+1} - f_{L-}P'_{L-1}), \tag{5.32}$$

$$f_2 = \sum_L (f_{(L+1)-}P'_{L+1} - f_{L+}P'_L) = \sum_{L=1}^{\infty}(f_{L-} - f_{L+})P'_L, \tag{5.33}$$

and from the orthogonality relations (B-5.12), these equations can be inverted

$$f_{L\pm} = \frac{1}{2}\int_{-1}^{1} dx[f_1 P_L(x) + f_2 P_{L\pm 1}(x)]. \tag{5.34}$$

Finally, we construct the matrix elements $T_{MM'}(\Omega)$ between states which are quantized along the $z$-axis also in the final states. In this case the eigenvalue

$2\lambda'$ must be replaced by the operator $\boldsymbol{\sigma}\hat{\mathbf{p}}'$. We put $\phi = 0$ and consider only the case of parity conservation:

$$T(M, M', \theta) = 8\pi s^{\frac{1}{2}} \chi_0^+(M')[f_1 + 2Mf_2(\sigma_z \cos\theta + \sigma_x \sin\theta)]\chi_0(M), \quad (5.35)$$

$$\begin{aligned} T(\tfrac{1}{2}, \tfrac{1}{2}, \theta) &= 8\pi s^{\frac{1}{2}} f, \qquad f \equiv f_1 + f_2 \cos\theta, \\ T(-\tfrac{1}{2}, \tfrac{1}{2}, \theta) &= 8\pi s^{\frac{1}{2}} g, \qquad g \equiv -f_2 \sin\theta. \end{aligned} \quad (5.36)$$

$f$ and $g$ are obviously the cms versions of (1-7.27) and (1-7.28).

## 2-6 Born Approximation for the Scattering by a 4-Potential

In this section we first rewrite the Klein-Gordon and Dirac equation as integral equations ("Lippmann-Schwinger equations"), and then compute from these equations the scattering from a given 4-potential $A^\mu(x)$ to first order in $eA^\mu$ ("Born approximation").

As usual, we start with the definitions of Greens functions or "propagators" which satisfy the free-particle equations with an extra $\delta$-function. For spin zero and $\frac{1}{2}$, these are denoted by $\Delta$ and $S$, respectively,

$$(\partial_\mu \partial^\mu + m^2)\,\Delta(x) = \delta_4(x), \qquad (i\,\partial^\mu \gamma_\mu - m)S(x) = \delta_4(x). \quad (6.1)$$

One sees that

$$S = -(i\,\partial_\mu \gamma^\mu + m)\,\Delta \quad (6.2)$$

reduces the equation for $S$ to that for $\Delta$. The differential equations $(i\,\partial^\mu - ZeA^\mu)^2\psi = m^2\psi$ and $(i\,\partial^\mu - ZeA^\mu)\gamma_\mu\psi = m\psi$ are then equivalent to the integral equations

$$\begin{aligned} \text{KG:}\quad \psi(x) &= \psi_0 - Ze\int d^4y\,\Delta(x-y) \\ &\qquad [iA_\mu(y)\,\partial^\mu + i\,\partial^\mu A_\mu(y) - ZeA^2(y)]\psi(y), \quad (6.3) \\ \text{Dirac:}\quad \psi(x) &= \psi_0 + Ze\int d^4y S(x-y)\gamma^\mu A_\mu(y)\psi(y), \end{aligned}$$

as can be seen by application of the operators $\Box + m^2$ and $i\,\partial_\mu\gamma^\mu - m$, respectively. For the construction of the Greens function $\Delta(x)$, we need the Fourier transforms

$$\Delta(x) = (2\pi)^{-4}\int d^4P e^{-iPx}\Phi(P_\mu), \qquad \delta_4(x) = (2\pi)^{-4}\int d^4P e^{-iPx}. \quad (6.4)$$

Then (1) reduces to

$$(-P^2 + m^2)\Phi = 1. \quad (6.5)$$

This specifies $\Phi$ except on the "mass-shell", $P^2 = m^2$. As in the nonrelativistic treatment of scattering by means of Greens functions, $\Phi(P^2 = m^2)$

must be determined by "$i\varepsilon$-rules" appropriate to the boundary conditions on scattering experiments. Taking $\psi_0$ in (3) as the incident plane wave, i.e.,

$$\psi(x) = \psi_0(x) = \psi_i(x) = \psi_{\mathbf{p}_i, M_i}(x) \qquad \text{for } x_0 \to -\infty, \tag{6.6}$$

we must use the "Feynman propagator"

$$\Phi(P_\mu) = (m^2 - P^2 - i\varepsilon)^{-1} = (m^2 - P_0^2 + p^2 - i\varepsilon)^{-1}, \qquad \varepsilon = 0^+. \tag{6.7}$$

To understand the $i\varepsilon$, it suffices to integrate over $P_0$ in (4). Mathematically, any continuous path from $-\infty$ to $\infty$ in the complex $P_0$-plane would do. With the definition (7), one keeps instead $P_0$ real and shifts $m^2 \to m^2 - i\varepsilon$. We can write $\Phi(P_0, \mathbf{p})$ as a product of poles in the complex $P_0$-plane (see fig. 2-6), with $E \equiv (m^2 + p^2)^{\frac{1}{2}}$:

$$\begin{aligned}\Phi(P_\mu) &= (E - i\varepsilon - P_0)^{-1}(E - i\varepsilon + P_0)^{-1} \\ &= \frac{1}{2E}\left(\frac{1}{E - i\varepsilon - P_0} + \frac{1}{E - i\varepsilon + P_0}\right)\end{aligned} \tag{6.8}$$

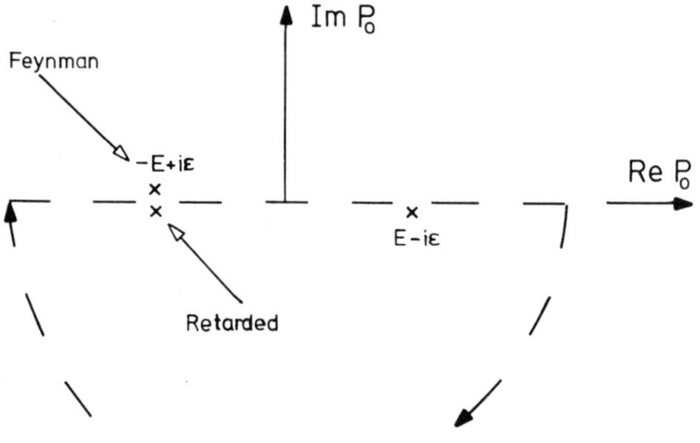


**Figure 2-6** The crosses show the positions of the poles of $\Phi_R$ and $\Phi_F$ in the complex $P_0$-plane. The dashed line is the integration contour for $x_0 > 0$.

We first integrate (4) for $x^0 > 0$. The factor $\exp(-iP_0 x^0) = \exp(-ix^0 \operatorname{Re} P_0 + x^0 \operatorname{Im} P_0)$ falls off exponentially for $\operatorname{Im} P_0 < 0$. We may therefore close the integration contour by a large circle in the lower half plane of complex $P_0$ and apply the residue method:

$$\oint (z - P_0)^{-1}\, dP_0\, f(P_0) = \begin{cases} 2\pi i f(z) & \text{for } z \text{ inside integration contour} \\ 0 & \text{otherwise,} \end{cases} \tag{6.9}$$

where the integration goes clockwise along a closed contour and $f$ is regular inside and on the contour. In our case $z = E - i\varepsilon$ lies inside the contour, and $f = (2E)^{-1} \exp(-iP_0 x^0)$, such that our Feynman propagator becomes

$$\Delta_F(x^0 > 0, \mathbf{x}) = i(2\pi)^{-3} \int \frac{d^3p}{2E}\, e^{i\mathbf{px}} e^{-iEx^0} \equiv \Delta_+(x). \tag{6.10}$$

This differs from the nonrelativistic expression only by the appearance of $d_L^3 p$ of (1-2.19). In fact, (10) can be rewritten as

$$\Delta_+(x-y) = i \int d_L^3 p \psi_{\mathbf{p}}(x) \psi_{\mathbf{p}}^*(y), \qquad \psi_{\mathbf{p}} = e^{-iEx^0 + i\mathbf{p}\mathbf{x}}. \tag{6.11}$$

Before continuing, let us take a look at the "retarded propagator"

$$\begin{aligned} \Phi_R(P_\mu) &= (m^2 - P^2 - i\varepsilon \operatorname{sign}(P_0))^{-1} \\ &= \tfrac{1}{2}E\left(\frac{1}{E - i\varepsilon - P_0} + \frac{1}{E + i\varepsilon + P_0}\right), \end{aligned} \tag{6.12}$$

which has both $P_0$-poles in the lower half plane (see fig. 2-6). The corresponding retarded Greens function $\Delta_R$ for $x^0 > 0$ also receives contributions from $P_0 = -E$, which in view of (11) would correspond to states of negative energy. Since such states do not exist in nature, $\Phi_R$ cannot be used in (3).

For $x^0 < 0$, we can evaluate $\Delta_F$ and $\Delta_R$ by closing the integration contour in the upper half plane:

$$\Delta_F(x_0 < 0) = i \int d_L^3 p e^{iEx^0 - i\mathbf{p}\mathbf{x}} \equiv \Delta_-(x) = \Delta_+(-x^0, -\mathbf{x}), \tag{6.13}$$

$$\Delta_R(x^0 < 0) = 0. \tag{6.14}$$

At a first glance, (14) looks more reasonable than (13) which seems to involve the negative-energy states and also to violate causality, in the sense that $\psi(x)$ in (3) depends on $\psi(y)$ at future times, $y^0 > x^0$. However, the theory must be invariant under the Lorentz-transformation $x^\mu \to -x^\mu$, and an alternative formulation of the scattering problem exists in which the roles of $\Delta_+$ and $\Delta_-$ are interchanged. We shall see in chapter 3 that that formulation is appropriate for the scattering of antiparticles. In the Born approximation, (13) does not enter because we shall take $x^0 \to \infty$ in (3) and $y^0 < \infty$. Collecting our results, we have

$$\Delta_F(x) = (2\pi)^{-4} \int d^4 P e^{-iP \cdot x} (m^2 - P^2 - i\varepsilon)^{-1} = \theta(x^0)\, \Delta_+ + \theta(-x^0)\, \Delta_-, \tag{6.15}$$

$$\begin{aligned} S_F(x) &= -(2\pi)^{-4} \int d^4 P e^{-iP \cdot x} (P \cdot \gamma + m)(m^2 - P^2 - i\varepsilon)^{-1} \\ &= \theta(x^0) S_+ + \theta(-x^0) S_-, \end{aligned} \tag{6.16}$$

$$\begin{aligned} S_+ &= -i \int d_L^3 p \sum_M u\bar{u} e^{-iP \cdot x} = -i \int d_L^3 p (P \cdot \gamma + m) e^{-iP \cdot x}, \\ S_- &= \phantom{-}i \int d_L^3 p \sum_M v\bar{v} e^{iP \cdot x} = -i \int d_L^3 p (-P \cdot \gamma + m) e^{iP \cdot x}. \end{aligned} \tag{6.17}$$

In (16), we have inserted (2) for $S_F$ and in (17) the spin summation from (A-4.6). Now we come to the construction of the $S$-matrix element. We take

$x^0 \to \infty$ in (3) and $\psi(y) = \psi_i(y)$ on the right-hand side, thus obtaining the once iterated wave function

$$\text{KG:} \quad \psi_1(\infty, \mathbf{x}) = \psi_i(x) - iZe \int d_L^3 p \psi_{\mathbf{p}}(x) \int d^4 y \psi_{\mathbf{p}}^* \times (iA_\mu \partial^\mu + i \partial_\mu A^\mu - ZeA^2)\psi_i, \tag{6.18}$$

$$\text{Dirac:} \quad \psi_1(\infty, \mathbf{x}) = \psi_i(x) - iZe \int d_L^3 p \sum_M \psi_{\mathbf{p}M}(x) \int d^4 y \bar{\psi}_{\mathbf{p}M} A \cdot \gamma \psi_i.$$

The coefficient of a given free state $\psi_f$ in $\psi(\infty, \mathbf{x})$ gives the $S$-matrix element:

$$S_{if} = \langle f | S | i \rangle = (\psi_f, \psi_1(\infty)) = \langle f | i \rangle - ie \int d^4 y A \cdot j_{if}, \tag{6.19}$$

where $j_{if}^\mu$ is the "matrix element of the current operator in the configuration space representation." This terminology will be explained in section 3-2. In the spin-$\frac{1}{2}$ case, we obtain (using 1-1.15 or rather 1-5.20)

$$j_{if}^\mu = Z \sum_M \int d_L^3 p \langle \mathbf{p}_f M_f | \mathbf{p} M \rangle \bar{\psi}_{\mathbf{p}M} \gamma^\mu \psi_{\mathbf{p}_i M_i} = Z \bar{\psi}_{\mathbf{p}_f M_f} \gamma^\mu \psi_{\mathbf{p}_i M_i}. \tag{6.20}$$

We can also write the $S$-matrix element for scattering of spinless particles in the form (19), if we transform the term $i \partial^\mu A_\mu \psi_i$ by partial integration

$$\int d^4 y \psi_f^* i \partial^\mu A_\mu \psi_i = -i \int d^4 y (\partial^\mu \psi_f^*) A_\mu \psi_i, \tag{6.21}$$

$$\begin{aligned} j_{if}^\mu &= Z \int d_L^3 p \langle \mathbf{p}_f | \mathbf{p} \rangle \psi_{\mathbf{p}}^* (i \overleftrightarrow{\partial^\mu} - ZeA^\mu) \psi_{\mathbf{p}_i} \\ &= Z \psi_{\mathbf{p}_f}^* (i \overleftrightarrow{\partial^\mu} - ZeA^\mu) \psi_{\mathbf{p}_i}, \end{aligned} \tag{6.22}$$

$$A \overleftrightarrow{\partial^\mu} B \equiv A \, \partial^\mu B - (\partial^\mu A) B.$$

$j_{if}^0$ is similar to the integrand of (1-1.14), but there $\psi_1$ and $\psi_2$ cannot be plane waves, and $\phi = A^0$ has an extra factor 2. In later applications, we shall mainly need the "momentum space representation" of $j^\mu$. For that purpose, we Fourier transform $A_\mu$:

$$A_\mu(y) = \int d^4 K e^{-iK \cdot y} \tilde{A}_\mu(K). \tag{6.23}$$

From here on, we work consistently to lowest order in $Ze$, neglecting the last term in (22). Then the whole $y$-dependence of $j_{if}^\mu(y)$ comes from the plane waves:

$$j_{if}^\mu(y) = J_{if}^\mu \exp\{-i(P_i - P_f) \cdot y\}. \tag{6.24}$$

The $y$-integration thus gives a factor $(2\pi)^4 \delta_4(P_i - P_j + K)$, such that the final form of (19) is

$$S_{if} = \langle f | i \rangle - i(2\pi)^4 e \tilde{A}_\mu (P_f - P_i) J_{if}^\mu. \tag{6.25}$$

The explicit forms of the "matrix elements of the current operator in momentum space" are, to lowest order in $e$,

$$\text{KG:} \quad J_{if}^{\mu} = Z(P_i + P_f)^{\mu}, \tag{6.26}$$

$$\text{Dirac:} \quad J_{if}^{\mu} = Z\bar{u}(P_f, M_f)\gamma^{\mu}u(P_i, M_i) \equiv Z\bar{u}_f\gamma^{\mu}u_i. \tag{6.27}$$

Protons and neutrons satisfy the modified Dirac equation (1-5.14), which leads to the more complicated expression

$$J_{if}^{\mu} = \bar{u}_f[Z\gamma^{\mu} + \tfrac{1}{2}m^{-1}\kappa\sigma^{\mu\nu}(P_i - P_f)_{\nu}]u_i, \tag{6.28}$$

where $\kappa$ is the anomalous magnetic moment which occurs already in the nonrelativistic limit (1-4.6). By using the free-particle Dirac equations $P_i \cdot \gamma u_i = m_i u_i$ and $\bar{u}_f P_f \cdot \gamma = m_f \bar{u}_f$, one can replace

$$\sigma^{\mu\nu}(P_i - P_f)_{\nu} \to (m_i + m_f)\gamma^{\mu} - (P_i + P_f)^{\mu} \tag{6.29}$$

for arbitrary masses $m_i$ and $m_f$. A somewhat simpler version of (28) is therefore

$$J_{if}^{\mu} = \bar{u}_f[(Z + \kappa)\gamma^{\mu} - \tfrac{1}{2}m^{-1}\kappa(P_i + P_f)^{\mu}]u_i. \tag{6.30}$$

As an application of this formalism, we compute the Born approximation for the Coulomb scattering of a light spinless particle (pion) on a spinless nucleus. In that case we have

$$\mathbf{A}(y) = 0, \qquad A^0(y) = \phi(y) = V(|\mathbf{y}|)/Ze. \tag{6.31}$$

We define a "Born amplitude" as follows, with $U \equiv 2EV$:

$$f_{\text{Born}}(q) = -(4\pi)^{-1}\int U(r)e^{i\mathbf{q}\mathbf{r}}\,d^3r = -q^{-1}\int_0^{\infty} U(r)r\,dr\,\sin qr, \tag{6.32}$$

The Fourier transform $A^0(K)$ is found by inverting (23) and inserting (31):

$$\tilde{A}^0(K) = \delta(K^0)(2\pi)^{-3}\int e^{-i\mathbf{k}\mathbf{r}}\,d^3rV/Ze = -\delta(K^0)f_{\text{Born}}(k)(4\pi^2ZeE)^{-1} \tag{6.33}$$

Inserting this and (26) into (25), we find that $S_{if}$ is of the form (1-3.3) with $f(\Omega) = f_{\text{Born}}(\mathbf{p}_f - \mathbf{p}_i)$. Thus $f_{\text{Born}}$ is the desired scattering amplitude.

In practice, $f_{\text{Born}}$ is first measured by high-energy electron scattering (see the text by Überall 1971), and the potential is found by inversion of (32):

$$V(\mathbf{r}) = -(4\pi^2E)^{-1}\int f_{\text{Born}}e^{-i\mathbf{q}\mathbf{r}}\,d^3q = -(\pi Er)^{-1}\int_0^{\infty} f_{\text{Born}}\,q\,dq\,\sin qr. \tag{6.34}$$

$V$ is then expressed in terms of the normalized nuclear charge density $\rho(r_N)$:

$$V(\mathbf{r}) = ZZ_a\alpha\int d^3r_N\,\rho(r_N)|\mathbf{r} - \mathbf{r}_N|^{-1}, \qquad \int \rho\,d^3r_N = 1. \tag{6.35}$$

Inserting this into (32) and using $\mathbf{r} - \mathbf{r}_N$ as a new integration variable, we find

$$f_{\text{Born}}(q) = -2\eta p q^{-2} F(q), \tag{6.36}$$

$$F(q) = \int d^3 r \rho(r) e^{i\mathbf{qr}} = 4\pi q^{-1} \int_0^\infty \rho(r) r \, dr \sin qr. \tag{6.37}$$

$F$ is called the nuclear "form factor." The remaining factors $-2\eta p q^{-2}$ occur also in the point Coulomb amplitude (1-10.7), which is valid to all orders in $e$, except for the correction (1-10.28). A simple improvement of (1-10.28) to order $\alpha$ is therefore

$$f_C \approx f_{C\,\text{point}} F \approx f_{\text{Born}} \exp\{-i\eta \ln \sin^2 \tfrac{1}{2}\vartheta + 2i\sigma_0\}(1 - \tfrac{1}{2}\pi \ldots) \tag{6.38}$$

which can differ drastically from $f_{C\,\text{point}}$ for larger momentum transfers $q = 2p \sin \vartheta/2$. A more accurate method is to solve the Klein-Gordon equation for each partial wave separately, using the potential (34).

Nuclear charge distributions are sometimes expressed in terms of their momenta $\langle r^{2n} \rangle$. From (37) it follows that these are related to the power series expansion of $F$ in $q^2$. To first order in $q^2$, we can put $\exp(i\mathbf{qr}) = 1 + i\mathbf{qr} - \frac{1}{2}(\mathbf{qr})^2$, the angular average of which is $1 - \frac{1}{6}q^2 r^2$. Thus, we have

$$-(dF/dq^2)|_{q^2=0} = \frac{1}{6} \int d^3 r \rho(r) r^2 = \tfrac{1}{6}\langle r^2 \rangle. \tag{6.39}$$

The quantity $\langle r^2 \rangle^{\frac{1}{2}}$ is the "mean square radius." If the target is a neutral atom, then the amplitude at very small scattering angles is reduced by the electron charge distribution (see section 8-2).

## 2-7 The Scattering of Two Particles by One-Photon Exchange

In this section we evaluate recoil and spin corrections to the scattering on nuclei in the Born approximation. These corrections are particularly important for scattering on light nuclei and at large momentum transfers. We shall interpret the scattering as a one-photon-exchange (see fig. 2-7.1 for *ep*-

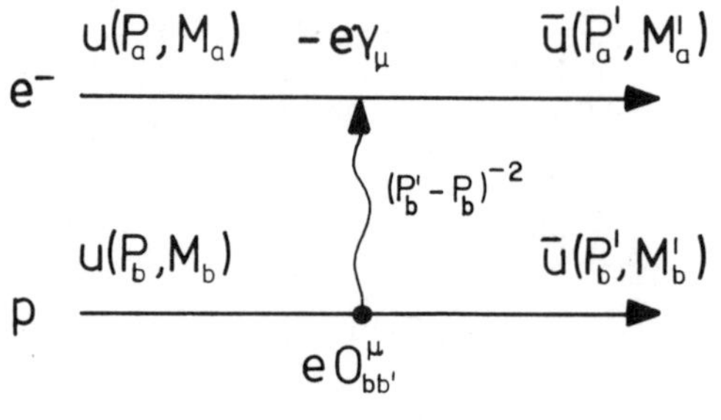


**Figure 2-7.1** The lowest order graph for $e^- p$-scattering.

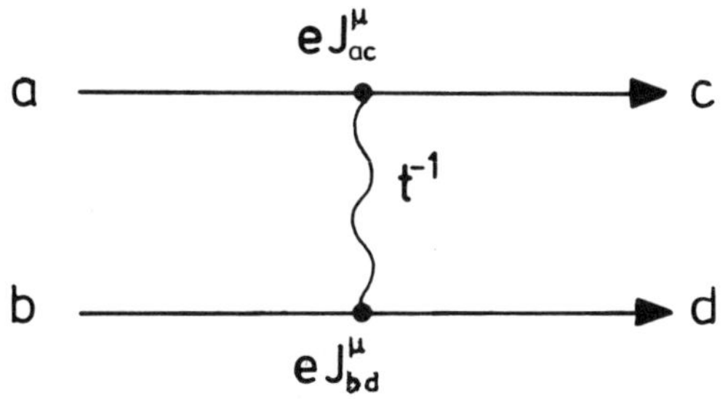


**Figure 2-7.2** The one-photon exchange graph for a general reaction $ab \to cd$. In the case of elastic scattering, $c$ and $d$ are denoted by $a'$ and $b'$, respectively.

scattering) and will be able to generalize it to arbitrary elastic ($ab \to a'b'$) and inelastic ($ab \to cd$) collisions (fig. 2-7.2).

Consider the process in the cms where projectile and target approach each other with opposite momenta $\mathbf{p}_i \equiv \mathbf{p}_a$ and $\mathbf{p}_b = -\mathbf{p}_a$. Our formalism of the previous section was covariant up to (6.30), only (6.31) cannot be used. Since the projectile sees a moving charge distribution, $A^\mu(y)$ must be calculated from the inhomogeneous Maxwell equations (1-0.12). The calculation is simplified in the Lorentz gauge (1-0.26). The differential equation $\Box A^\mu = ej^\mu$ is easily converted to an integral equation:

$$A^\mu(y) = A_0^\mu(y) + e\int d^4x\, D(y-x)j^\mu(x), \qquad \Box\, D(x) = \delta_4(x). \tag{7.1}$$

The differential equation for the Greens function $D$ is identical with (6.1) for $m = 0$, $D = \Delta(m = 0)$. From causality, we would use the retarded propagator $D_R$ in (1), and this is in fact correct when $A^\mu$ is a Hermitean field operator (see section 3-3). Presently, $A^\mu$ is only a complex function, and it will turn out that in this formulation $D_F$ must be used

$$D_F(y) = (2\pi)^{-4}\int d^4K e^{-iK\cdot y}\Phi_0(K), \qquad \Phi_0 = -(K^2 + i\varepsilon)^{-1}. \tag{7.2}$$

In this chapter, we shall always have $K^2 \equiv t < 0$. The $i\varepsilon$ can then be omitted from (2) and the distinction between $D_F$ and $D_R$ can be forgotten. We also have $A_0^\mu = 0$ in (1). The $j^\mu(x)$ is the current density due to the target. It will be denoted by $j_{\text{tar}}^\mu$ in the following to distinguish it from the projectile current density $j_{if}^\mu$ of (6.19). Insertion of (1) into (6.19) thus gives

$$S_{if} = \langle f \,|\, i\rangle - ie\int d^4x\, d^4y\, j_{if}^\mu(y) j_{\text{tar},\,\mu}(x)\, D_F(y-x). \tag{7.3}$$

$S_{if}$ is now symmetric in the projectile and target currents. Moreover, as "projectile" and "target" are reciprocal concepts (which is most evident in the cms), $j_{\text{tar}}$ must have the same structure as $j_{if}$. That means, if the target has 4-momentum $P_b$ before the collision and $P_b'$ after the collision, we must have

$$j_{\text{tar}}^\mu(x) = \exp\{-i(P_b - P_b')\cdot x\} J_{bb'}^\mu \tag{7.4}$$

as in (6.24), and if the target is a pion, muon or proton, $J^{\mu}_{bb'}$ is given by (6.26), (6.27) and (6.28) respectively, with the indices $i$ and $f$ replaced by $b$ and $b'$. To emphasize this symmetry, we call the incident particle $a$, $P_i \equiv P_a$ and $P_f \equiv P'_a$. Inserting $j^{\mu}_{aa'}$ from (6.24), $D_F$ from (2) and $j_{\text{tar}}$ from (4) into (3), we can perform the $y$-integration (which produces a factor $(2\pi)^4\, \delta_4(P_a - P'_a + K)$), the $x$-integration (which produces a factor $(2\pi)^4\, \delta_4(P_b - P'_b - K)$) and the $K$-integration (which eliminates the second $\delta_4$-function and puts $K = P_b - P'_b$ in the first. The result can be put into the form of (1.4),

$$S_{if} = \langle f\,|\,i\rangle + i(2\pi)^4\, \delta_4(P_a + P_b - P'_a - P'_b)T_{if}, \tag{7.5}$$

$$T_{if} = e^2 t^{-1} J^{\mu}_{aa'}\, g_{\mu\nu} J^{\nu}_{bb'}, \qquad t = q_0^2 - q^2 \qquad q^{\mu} = -K^{\mu} = (P_a - P'_a)^{\mu}. \tag{7.6}$$

For elastic scattering in the cms, we have $q^0 = 0$, $t = -q^2$, where $\mathbf{q}$ is the 3-momentum transfer in this system. For later use, we also quote $A^{\mu}(y)$ as given by (1), (2), and (4):

$$A^{\mu}(y) = -et^{-1} \exp\{-i(P_b - P'_b)\cdot y\}J^{\mu}_{bb'} \tag{7.7}$$

We now wish to establish some general properties of $J^{\mu}_{bb'}$. The particle $b$ need not be "elementary" but can be a nucleus, or even an atom. It is then possible that the state $b$ is excited by the collision, in which case the final target state will be denoted by $|d\rangle$, instead of $|b'\rangle$, with $P_d^2 > m_b^2$. Even in such complicated situations, $j_{\text{tar}}$ must be of the form (4), because otherwise we would not get 4-momentum conservation in (5). When (4) is inserted into the current conservation condition $\partial_{\mu}\, j^{\mu}_{\text{tar}} = 0$, one finds

$$(P_b - P_d)_{\mu} J^{\mu}_{bd} = 0, \qquad \text{or} \qquad q_{\mu} J^{\mu} = 0. \tag{7.8}$$

Moreover, $J^{\mu}$ must of course be a 4-vector. If now the free-particle states $|b\rangle$ and $|d\rangle$ are both spinless, they are completely specified by their 4-momenta $P_b$ and $P_d$, and $J^{\mu}$ can only be proportional to $P_b^{\mu}$ and $P_d^{\mu}$. For elastic scattering, we have $(P_b - P'_b)(P_b + P'_b) = P_b^2 - P_b'^2 = 0$ and $(P_b - P'_b)^2 = t$, and (8) is only satisfied if $J^{\mu}$ has the form $(P_b + P'_b)^{\mu} G(P_b^2, P_b'^2, P_b \cdot P'_b)$, where the function $G$ is still arbitrary. Inserting $P_b^2 = P_b'^2 = m_b^2$ and $P_b P'_b = \frac{1}{2}t - m_b^2$, we see that $G$ is a function of $t$ only. It must be closely related to the form factor $F(q)$ of (6.36). In the region $q^2 \ll m_b^2$, the energy transfer to the target is negligible, and we have $t = -q^2$ ($\mathbf{q} = -\mathbf{k}$). This shows that $G = Z_b F_b(t = -q^2)$,

$$J^{\mu}_{bb'} = Z_b(P_b + P'_b)^{\mu} F_b(t), \qquad F_b(0) = 1. \tag{7.9}$$

We now take (9) as the precise definition of the form factor for all $t$. Its interpretation (6.37) as the Fourier transform of the charge distribution of particle $b$ is limited to small recoil energies (the best coordinate frame for this purpose is the "Breit frame," which takes $\mathbf{p}_b = -\mathbf{p}'_b = \mathbf{q}/2$, $E_b = E'_b \approx m_b + q^2/8m_b$). Actually, the pion also has a form factor, and our previous result (6.26) is only correct for $q^2\langle r_\pi^2\rangle \ll 1$, $\langle r_\pi^2\rangle^{\frac{1}{2}}$ being the pion mean square radius (6.39).

Considering now an inelastic collision to a spinless final state $|d\rangle$ (for example the Coulomb excitation of the $0^+$ excited state of $^4$He), we find that (8) is only satisfied for

$$J_{bd}^{\mu} = [t(P_b + P_d)^{\mu} - q^{\mu}(m_d^2 - m_b^2)]F_{bd}^0, \qquad F_{bd} = tF_{bd}^0 \tag{7.10}$$

Of course, it must be remembered that these formulas are useful only at high energies where the Born approximation is reliable.

Spin-$\frac{1}{2}$ particles such as $p$, $n$, $^3$He ... have a magnetisation density in addition to a charge density. These two densities can be introduced in several equivalent ways, corresponding to equivalent forms of (6.28). The most convenient form is that of (6.30):

$$J^{\mu} = \bar{u}'[G_M(t)\gamma^{\mu} - \tfrac{1}{2}m^{-1}F_2(t)(P + P')^{\mu}]u. \tag{7.11}$$

For small $t$, this must be identical to (6.30), i.e.,

$$F_2(0) = \kappa, \qquad G_M(0) = Z + \kappa. \tag{7.12}$$

$G_M$ and $F_2$ are called "magnetic" and "Pauli" form factors, respectively. The form corresponding to (6.28) is

$$J^{\mu} = \bar{u}'0^{\mu}u, \qquad 0^{\mu} = F_1\gamma^{\mu} - \tfrac{1}{2}m^{-1}F_2\sigma^{\mu\nu}(P - P')_{\nu}, \qquad F_1 = G_M - F_2. \tag{7.13}$$

$F_1$ is called the "Dirac" form factor and has the particularly simple limit $F_1(0) = Z$ which vanishes for neutrons. There exist two more couplings which are consistent with (8), namely $\gamma_5(P + P')^{\mu}$ and $\gamma_5\sigma^{\mu\nu}(P - P')_{\nu}$. However, when these are sandwiched between $\bar{u}$ and $u$ as in (11), they form axial vectors rather than proper vectors. They are thus ruled out by parity conservation and appear only at the level of weak interactions (see chapter 5). In the nonrelativistic case, the matrix elements of the current operator between particles of equal spins $S$ contain $2S + 1$ arbitrary functions (vertex functions), which reduce to the "multipole moments" at $t = 0$. This is a consequence of rotational invariance, parity conservation, and gauge invariance, and therefore remains true in the relativistic case. In other words, as soon as we have reduced $J_{bb'}^{\mu}$ to $2S + 1$ linearly independent scalar functions, we are through (see also section 5-4). The case $S = 1$ is presented in the next section.

Whereas the $\sigma^{\mu\nu}$-part of (13) is obviously gauge invariant also for $m_d \neq m_b$, the $\gamma^{\mu}$-part is not. A conveniently symmetrized form for that case is

$$0_{bd}^{\mu} = (m_b + m_d)^{-1}[F_3(t(m_d - m_b)^{-1}\gamma^{\mu} + P_b^{\mu} - P_d^{\mu}) + F_2\sigma^{\mu\nu}(P_b - P_d)_{\nu}], \tag{7.14}$$

where condition (8) is easily checked for the first bracket (when $b$ and $d$ have opposite parities, (5-4.5) applies with $(m_b + m_d)g_1 + tg_3 = 0$).

We can also study the general implications of current conservation

$q_\mu J^\mu = 0$ by introducing the components $J''$ and $J^\perp$ of $\mathbf{J}$ which are parallel and perpendicular to $\mathbf{q}$, respectively:

$$\mathbf{qJ} = qJ'', \qquad q^0 J^0 - qJ'' = 0, \qquad J^\perp \equiv \mathbf{J} - \mathbf{q}(\mathbf{Jq})/q^2. \tag{7.15}$$

The matrix element $T_{if}$ of (5) is then rewritten as

$$T_{if} = -e^2[-q^{-2} J^0_{aa'} J^0_{bb'} + t^{-1} J^\perp_{aa'} J^\perp_{bb'}]. \tag{7.16}$$

Note that the first denominator is independent of $q^0$. Thus our old result (6.36) is better than its derivation.

The general structure of the matrix element (6) is illustrated by a "Feynman graph," which is given in fig. 2-7.1 for the case of electron-proton scattering. Incoming particles, their 4-momenta and possible plane-wave spin states are represented by arrows from the left. Arrows to the right are used for outgoing particles. The factor $t^{-1}$ is expressed by the wavy line joining the upper and lower lines. It is said to represent a "virtual photon" (although in the decomposition (16), the $q^{-2}$-part is only the Fourier transform of a point Coulomb field). The arrow on that line merely indicates a sign convention for the 4-momentum transfer: the virtual photon carries 4-momentum $P_b - P'_b$ into the upper "vertex" if the arrow points upwards, and $P'_b - P_b = P_a - P'_a$ into the lower "vertex" if the arrow points downwards. In later graphs, "inner lines" are drawn without arrows. The vertex point itself represents the matrix element of the current operator excluding the spin states, i.e., $-e\gamma_\mu$ for electrons and $e0^\mu_{bb'}$ for protons. The lower vertex is occasionally denoted by a blob to indicate the presence of form factors. The graph of a general one-photon exchange reaction $ab \to cd$ is shown more schematically in fig. 2-7.2. Feynman graphs will be discussed in more detail in sections 3-5 and 3-6.

The differential cross section for unpolarized particles is best expressed in the form (4.9),

$$d\sigma/dt = \frac{\pi}{\lambda}(2S_a + 1)^{-1}(2S_b + 1)^{-1} \sum_{M_a \cdots M_d} \left| \frac{\alpha}{t} J_{aa'\mu} J^\mu_{bd} \right|^2, \qquad \lambda = 4p^2 s. \tag{7.17}$$

If particle $a$ is spinless and has $Z_a = \pm 1$ and no form factor, insertion of (6.26) gives

$$d\sigma/dt = \lambda^{-1} t^{-2} \pi\alpha^2 (2S_b + 1)^{-1} \sum_{\lambda\lambda'} |(P_a + P'_a) J_{bd}(\lambda, \lambda')|^2. \tag{7.18}$$

Also if particle $a$ is an electron or muon, its spin summation is easily performed by means of (A-4.10), with $Q = \gamma_\mu$, $\tilde{Q} = \gamma_\nu$:

$$S_{\mu\nu} \equiv \frac{1}{2} \sum_{M_a M_{a'}} J_{aa'\mu} J^*_{aa'\nu} = \frac{1}{2} \operatorname{trace} \gamma_\mu (P_a \cdot \gamma + m_a) \gamma_\nu (P'_a \cdot \gamma + m_a) \tag{7.19}$$

$$= g_{\mu\nu} t + 2(P_{a\mu} P'_{a\nu} + P_{a\nu} P'_{a\mu}), \qquad \text{with} \quad t = 2(m_a^2 - P_a P'_a);$$

$$d\sigma/dt = \lambda^{-1} t^{-2} \pi\alpha^2 (2S_b + 1)^{-1} S_{\mu\nu} \sum_{\lambda\lambda'} J^\mu_{bd}(\lambda, \lambda') J^{\nu *}_{bd}(\lambda, \lambda'). \tag{7.20}$$

Because of $J_{aa'}(P_b - P_d) = 0$, we can put $P_d^\mu = P_b^\mu$ in $J_{bd}^\mu$. Thus, for electron scattering on a spinless nucleus, insertion of (9) into (20) gives

$$\frac{d\sigma}{dt} = \frac{4\pi}{\lambda}\left(Z_b F_b \frac{\alpha}{t}\right)^2 S_{\mu\nu} P_b^\mu P_b^\nu = \frac{4\pi}{\lambda}\left(Z_b F_b \frac{\alpha}{t}\right)^2 [tm_b^2 + 4(P_a P_b)(P'_a P_b)]. \quad (7.21)$$

This can be rewritten in terms of the Mandelstam variables $s$ and $t$ using (4.12) and the definition (2.9) of $\lambda$. However, it is more convenient to first use

$$P'_a P_b = P'_b P_a = P_a P_b + \tfrac{1}{2}t, \qquad \lambda = 4[(P_a P_b)^2 - m_a^2 m_b^2] \quad (7.22)$$

In electron scattering, $m_a^2$ is normally negligible, in which case (21) reduces to

$$d\sigma/dt = 4\pi(Z_b F_b \alpha/t)^2(1 + \tfrac{1}{4}ts(P_a P_b)^{-2}), \quad (7.23)$$

where the first term is just the Rutherford cross section. When $b$ is a nucleon, insertion of (11) into (A-4.10) gives, with $P_b'^\mu \to P_b^\mu$ as above,

$$\begin{aligned}\frac{1}{2}\sum_{\lambda\lambda'} J_{bb'}^\mu J_{bb'}^{\nu *} &= \frac{1}{2}\,\text{trace}\,(G_M \gamma^\mu - m^{-1}F_2 P_b^\mu)(P_b \cdot \gamma + m) \\ &\quad \times (G_M \gamma^\nu - m^{-1} F_2 P_b^\nu)(P'_b \cdot \gamma + m) \quad (7.24)\\ &= G_M^2 S^{\mu\nu}(b) + 2m^{-2}F_2^2 P_b^\mu P_b^\nu (P_b P'_b + m^2) - 8 G_M F_2 P_b^\mu P_b^\nu \\ &= G_M^2 g^{\mu\nu} t + 4(F_1^2 - \tfrac{1}{4}F_2^2 t/m_b^2) P_b^\mu P_b^\nu .\end{aligned}$$

In the last expression we have inserted the form corresponding to (19) for $S^{\mu\nu}(b)$ and the Dirac form factor $F_1$ of (13). From (20) we now obtain the differential cross section

$$\frac{d\sigma}{dt} = \frac{4\pi}{\lambda}\left(\frac{\alpha}{t}\right)^2 \{G_M^2 t(t + P_a P'_a) + (F_1^2 - \tfrac{1}{4}F_2^2 t/m_b^2)[tm_b^2 + 4(P_a P_b)(P'_a P_b)]\}. \quad (7.25)$$

Note that the square bracket is identical with that of (21). Observing also $P_a P'_a = m_a^2 - \frac{1}{2}t$ and neglecting $m_a^2$, we find

$$\frac{d\sigma}{dt} \approx 4\pi\left(\frac{\alpha}{t}\right)^2 \{(F_1^2 - \tfrac{1}{4}F_2^2 t/m_b^2)(1 + \tfrac{1}{4}ts(P_a P_b)^{-2}) + \tfrac{1}{8}G_M^2 t^2 (P_a P_b)^{-2}\}. \quad (7.26)$$

This formula is due to Rosenbluth (1950). Similar formulas for higher spin particles are given by Waldenström and Olsen (1971).

For the experimental determination of the nucleon form factors, it is convenient to introduce the so-called "electric form factor"

$$G_E = F_1 - \tau F_2, \qquad \tau \equiv -\tfrac{1}{4}tm_b^{-2} \quad (7.27)$$

Expressing now $F_2$ and $F_1$ in terms of $G_E$ and $G_M$ of (13)

$$F_2 = (G_M - G_E)(1 + \tau)^{-1}, \qquad F_1 = (G_E + \tau G_M)(1 + \tau)^{-1} \quad (7.28)$$

we find

$$F_1^2 + \tau F_2^2 = (G_E^2 + \tau G_M^2)(1 + \tau)^{-1} \quad (7.29)$$

in which case the Rosenbluth formula (26) depends on the form factors only through $G_E^2$ and $G_M^2$.

For elastic or inelastic scattering on arbitrary nuclear targets, the spin-averaged and -summed $J_{bd}^{\mu} J_{bd}^{\nu}$ which enters (20) can only depend on $P_b^{\mu}$, $K^{\mu}$, $P_b^{\nu}$, $K^{\nu}$ and $g^{\mu\nu}$ via the two gauge invariant combinations

$$(2S_b + 1)^{-1} \sum_{\lambda\lambda'} J_{bd}^{\mu} J_{bd}^{\nu} = m_b^{-2} W_2 (P_b^{\mu} - K^{\mu} P_b \cdot K/t)(P_b^{\nu} - K^{\nu} P_b \cdot K/t) - W_1 (g^{\mu\nu} - K^{\mu} K^{\nu}/t). \tag{7.30}$$

Here $W_1$ and $W_2$ are "structure functions" which depend on $t = K^2$ (and in the case of excitation of continuum states such as $ed \to e'pn$ also on $sp_d^2 = m_b^2 + 2P_b \cdot K + t$). In the contraction of (30) with $S_{\mu\nu}$, we can, of course, omit $K^{\mu}$ and $K^{\nu}$:

$$\frac{d\sigma}{dt} = \frac{4\pi}{\lambda} \left(\frac{\alpha}{t}\right)^2 \{-W_1(t + P_a \cdot P_a') + \tfrac{1}{4} m_b^{-2} W_2 [t m_b^2 + 4(P_a P_b)(P_a' P_b)]\}. \tag{7.31}$$

Comparison with (25) shows that in the case of elastic scattering on nucleons, $W_1 = -tG_M^2$, while $\frac{1}{4} m_b^{-2} W_2$ is given by (29).

## 2-8 Particles of Spin 1, $\frac{3}{2}$, and 2; Photons

A particle of spin $S$ is described by $2S + 1$ independent wave functions, which transform into each other under rotations. For $S > \frac{1}{2}$, a covariant description requires additional components and a corresponding number of auxiliary equations. Spin 1 can be described by a 4-vector function $A^{\mu}(t, \mathbf{x})$, the free particle equations being

$$(\partial_{\nu} \partial^{\nu} + m^2) A^{\mu}(t, \mathbf{x}) = 0, \qquad \partial_{\mu} A^{\mu} = 0. \tag{8.1}$$

Plane-wave solutions of (1) are

$$A_{\mathbf{p}M}^{\mu}(t, \mathbf{x}) = \exp(-iP \cdot x) \varepsilon^{\mu}(\mathbf{p}, M), \qquad P_{\mu} \varepsilon^{\mu} = 0. \tag{8.2}$$

In the particle rest frame $P^{\mu} = (m, \mathbf{0})$, (2) requires $\varepsilon^0 = 0$, which is just the requirement that the particle state has no spin-0 component. We put

$$\varepsilon^{\mu}(0, M) = (0, \mathbf{e}(M)), \quad \mathbf{e}(\pm 1) = 2^{-\frac{1}{2}}(\mp 1, -i, 0), \quad \mathbf{e}(0) = (0, 0, 1), \tag{8.3}$$

in agreement with (1-4.11). The spin-1 matrices are

$$S_x = \begin{pmatrix} 0 & 0 & 0 \\ 0 & 0 & -i \\ 0 & i & 0 \end{pmatrix}, \quad S_y = \begin{pmatrix} 0 & 0 & i \\ 0 & 0 & 0 \\ -i & 0 & 0 \end{pmatrix}, \quad S_z = \begin{pmatrix} 0 & -i & 0 \\ i & 0 & 0 \\ 0 & 0 & 0 \end{pmatrix} \tag{8.4}$$

Generally, $\varepsilon^{\mu}(\mathbf{p}, M)$ is given by application of (1-2.10) to (3). With $\hat{\boldsymbol{\eta}} = \hat{\mathbf{p}}$, $\hat{\boldsymbol{\eta}} \sinh \eta = \mathbf{p}/m$ and $\cosh \eta = E/m$, we find

$$\varepsilon^{\mu}(\mathbf{p}, M) = (\mathbf{ep}/m, \mathbf{e} + \mathbf{p}(\mathbf{ep}) m^{-1}(E + m)^{-1}). \tag{8.5}$$

For $\hat{p}_x = \sin\vartheta > 0$, $\hat{p}_y = 0$, $\hat{p}_z = \cos\vartheta$, helicity states $\varepsilon^\mu(\mathbf{p}, \lambda)$ are given by

$$\begin{aligned}\varepsilon^\mu(\pm 1) &= 2^{-\frac{1}{2}}(0, \ \mp\cos\vartheta, \ -i, \ \pm\sin\vartheta),\\ \varepsilon^\mu(0) &= m^{-1}(p, E\sin\vartheta, 0, E\cos\vartheta).\end{aligned} \tag{8.6}$$

States with $|\lambda| = 1$ will also be needed for arbitrary $\hat{\mathbf{p}} = (\sin\vartheta\cos\phi, \sin\vartheta \sin\phi, \cos\vartheta)$:

$$\varepsilon^\mu(\pm 1) = 2^{-\frac{1}{2}}(0, \mp\cos\vartheta\cos\phi + i\sin\phi, \ -i\cos\phi \mp \cos\vartheta\sin\phi, \ \pm\sin\vartheta). \tag{8.7}$$

The orthonormality and completeness relations are

$$\varepsilon_\mu^*(\mathbf{p}, M')\varepsilon^\mu(\mathbf{p}, M) = \varepsilon^{0*}(\mathbf{p}, M')\varepsilon^0(\mathbf{p}, M) - \boldsymbol{\varepsilon}^*(\mathbf{p}, M')\boldsymbol{\varepsilon}(\mathbf{p}, M) = -\delta_{MM'}, \tag{8.8}$$

$$\sum_M \varepsilon^\mu(\mathbf{p}, M)\varepsilon^\nu(\mathbf{p}, M) = -g^{\mu\nu} + P^\mu P^\nu/m^2. \tag{8.9}$$

In the absence of an electromagnetic field, the conserved current is analogous to the spinless case,

$$j^\mu = i[A_\nu^* \, \partial^\mu A^\nu - A^\nu \, \partial^\mu A_\nu^*], \tag{8.10}$$

with the resulting scalar product

$$\langle B | A \rangle = i \int d^3x (B_\nu^* \, \partial_t A^\nu - A^\nu \, \partial_t B_\nu^*). \tag{8.11}$$

Relativistic particles of spin $S$ may be described by $(2S + 1)$-component wave functions (Joos 1962, Weinberg 1964). For spin 1, for example, the boost on the states $\mathbf{e}(M)$ is a straightforward generalisation of (1-6.4), namely $\exp(\boldsymbol{\eta}\mathbf{S})$. As in (1-6.9), it may be rewritten as the square root of a polynomial

$$e^{\boldsymbol{\eta}\mathbf{S}} = (m^2 + 2\mathbf{pS}(\mathbf{pS} + E))^{\frac{1}{2}} = (m^2 + 2p\lambda(p\lambda + E))^{\frac{1}{2}}. \tag{8.12}$$

The last expression applies to helicity states. As in the case of spin-$\frac{1}{2}$ particles, one must distinguish between right-handed and left-handed spin functions. The boost for the latter ones is $\exp(-\boldsymbol{\eta}\mathbf{S})$. The formalism is not particularly elegant for the construction of covariant matrix elements.

Spin-2 states may be described by a symmetric tensor $\varepsilon^{\mu\nu} = \varepsilon^{\nu\mu}$ of which all possible 4-vectors and scalars are eliminated, $P_\mu \varepsilon^{\mu\nu} = g_{\mu\nu}\varepsilon^{\mu\nu} = 0$. It was noted by Auvil and Brehm (1966) that in the helicity representation, these states are composed of $\varepsilon^\mu$ and $\varepsilon^\nu$ simply by the Clebsch-Gordans appropriate for $1 + 1 \to 2$:

$$\begin{aligned}&\varepsilon^{\mu\nu}(\pm 2) = \varepsilon^\mu(\pm 1)\varepsilon^\nu(\pm 1), \qquad \varepsilon^{\mu\nu}(\pm 1) = 2^{-\frac{1}{2}}[\varepsilon^\mu(\pm 1)\varepsilon^\nu(0) + \varepsilon^\mu(0)\varepsilon^\nu(\pm 1)],\\ &\varepsilon^{\mu\nu}(0) = 6^{-\frac{1}{2}}[\varepsilon^\mu(1)\varepsilon^\nu(-1) + \varepsilon^\mu(-1)\varepsilon^\nu(1) + 2\varepsilon^\mu(0)\varepsilon^\nu(0)].\end{aligned} \tag{8.13}$$

The construction is obviously correct for particles at rest, and since the boosts commute with the helicity operator, it is generally correct. The corresponding antisymmetric tensor describes spin-1 particles and transforms as

(12). Similarly, particles of spin $\frac{3}{2}$ may be described by products of $u$ and $\varepsilon^\mu$ with $1 + \frac{1}{2} \to \frac{3}{2}$ Clebsch-Gordans:

$$u^\mu(\pm\tfrac{3}{2}) = u(\pm\tfrac{1}{2})\varepsilon^\mu(\pm 1),$$
$$u^\mu(\pm\tfrac{1}{2}) = 3^{-\frac{1}{2}}u(\mp\tfrac{1}{2})\varepsilon^\mu(\pm 1) + (\tfrac{2}{3})^{\frac{1}{2}}u(\pm\tfrac{1}{2})\varepsilon^\mu(0). \tag{8.14}$$

These "Rarita-Schwinger spinors" (1941) satisfy the Dirac equation, and their spin-$\frac{1}{2}$ components can also be covariantly eliminated:

$$(P \cdot \gamma - m)u^\mu = 0, \qquad P_\mu u^\mu = \gamma_\mu u^\mu = 0. \tag{8.15}$$

Finally, spin-1 states can also be constructed from the direct product of spinors. In the high-energy representation,

$$w_{rr}(\pm 1) = u_r(\pm\tfrac{1}{2})u_r(\pm\tfrac{1}{2}), \qquad w_{rr}(0) = 2^{-\frac{1}{2}}[u_r(\tfrac{1}{2})u_r(-\tfrac{1}{2}) + u_r(-\tfrac{1}{2})u_r(\tfrac{1}{2})], \tag{8.16}$$

and correspondingly for $w_{ll}$. These are the Auvil-Brehm analogues of the Joos-Weinberg states and are relatively convenient for particles such as deuterons which couple to two fermions. The dominant part of this coupling is $\bar{u}_p \bar{\bar{u}}_n w = \overset{+}{u}_{pl} \overset{+}{u}_{nl} w_{rr} + \overset{+}{u}_{pr} \overset{+}{u}_{nr} w_{ll}$, with $p =$ proton, $n =$ neutron. If one uses the conventional $\varepsilon^\mu$ for the deuteron spin, the coupling reads instead $\bar{u}_p(F\gamma_\mu + GP_{p\mu})v_n \varepsilon^\mu$, where $v_n = i\gamma^2 u_n^*$ is the charge conjugate spinor (1-6.26).

Comparison between (1) and Maxwell's equations in the Lorentz gauge (1-0.26) shows that the 4-potential $A^\mu$ describes massless spin-1 particles, namely "photons." The difficulty of having $m = 0$ in $\varepsilon^\mu(\mathbf{p}, 0)$ is settled by means of gauge invariance. Application of a gauge transformation (1-0.25) to (2) gives a new polarization vector

$$\varepsilon'^\mu(\lambda, K) = \varepsilon^\mu(\lambda, K) + cK^\mu, \qquad c = \text{arbitrary}, \tag{8.17}$$

which also satisfies $\varepsilon' \cdot K = 0$. Emission and absorption of photons of 4-momentum $K$ and helicity $\lambda$ will be described by matrix elements of the form $J^\mu \varepsilon_\mu^*$ and $J^\mu \varepsilon_\mu$ as we shall see, where $J^\mu$ has the same meaning as in (7.4). Invariance of these matrix elements under the gauge transformation (17) requires $K_\mu J^\mu = 0$, which is identical with our current conservation (7.8). In this sense we may say that current conservation is a consequence of gauge invariance, which in turn is necessary for the description of massless photons (similar statements hold for gravitons, which are massless spin-2 particles).

Because of $K_\mu J^\mu = 0$, $\varepsilon^\mu(\mathbf{p}, 0)$ can be used also in the limit $m \to 0$. To see this, we decompose

$$\varepsilon^\mu(\mathbf{p}, 0) = m^{-1}K^\mu - m^{-1}(E - k, 0, 0, k - E). \tag{8.18}$$

The diverging term $m^{-1}K^\mu$ does not contribute and the rest vanishes for $m \to 0$ because of $E - k = m^2/2k$.

Once we have convinced ourselves that longitudinal photons are not produced, we can restrict the helicity states $\varepsilon^\mu(\mathbf{k}, \lambda)$ by imposing the transversality condition

$$\boldsymbol{\varepsilon}(\mathbf{k}, \lambda) \cdot \mathbf{k} = 0 \to \nabla\mathbf{A} = 0. \tag{8.19}$$

This is called the Coulomb gauge. It is obtained from a field $A'^{\mu}$ having $\nabla\mathbf{A}' = 0$ by the transformation

$$\mathbf{A} = \mathbf{A}' + \nabla\Lambda, \qquad A^0 = A^{0\prime} - \partial_t\Lambda, \qquad \Lambda = (4\pi)^{-1}\int \frac{d^3x'}{|\mathbf{x}' - \mathbf{x}|}\nabla'\mathbf{A}'(\mathbf{x}', t), \tag{8.20}$$

which obviously yields $\nabla\mathbf{A} = \nabla\mathbf{A}' + \nabla^2\Lambda = 0$. Then Maxwell's equations read

$$\Box A^{\mu} - \partial^{\mu}\,\partial_t A^0 = ej^{\mu}, \qquad -\nabla^2 A^0 = e\rho. \tag{8.21}$$

In this gauge, the scalar field $A^0$ is that of an "instantaneous Coulomb interaction" (no retardation). This formulation is not covariant but is nevertheless useful, particularly for the connection between the classical and the quantum field, to be discussed in chapter 3. An alternative way is to introduce a fourth photon state (the "scalar photon"), which always cancels the longitudinal photon in the spin summation

$$\sum_{4\text{ states}} \varepsilon_{\mu}(S, M)\varepsilon_{\nu}^*(S, M) = -g_{\mu\nu}. \tag{8.22}$$

Polarized spin-1 particles are described by a $3 \times 3$ Hermitean and traceless density matrix $\rho_{MM'}$, which obviously requires 8 real parameter (see section 4-13). Five of these involve $M = 0$ or $M' = 0$ and are absent for photons. The remaining elements $\rho_{\lambda\lambda'}$ with $\lambda^2 = \lambda'^2 = 1$ form a $2 \times 2$ matrix which can be parametrized in analogy with (3.8):

$$\rho = \frac{1}{2}\begin{pmatrix} 1 + \xi_3 & \xi_1 - i\xi_2 \\ \xi_1 + i\xi_2 & 1 - \xi_3 \end{pmatrix} \tag{8.23}$$

The real $\xi_i$ are "Stokes' parameters." Unpolarized photons have $\xi_i = 0$, while completely polarized photons have $\xi_1^2 + \xi_2^2 + \xi_3^2 = 1$. In the latter case, one can eliminate $\rho$ in favour of the two helicity states:

$$\boldsymbol{\varepsilon}(\mathbf{k}) = a_+\boldsymbol{\varepsilon}(\mathbf{k}, 1) + a_-\boldsymbol{\varepsilon}(\mathbf{k}, -1) = 2^{-\frac{1}{2}}(0, a_- - a_+, -ia_+ - ia_-, 0). \tag{8.24}$$

The last expression is of course only valid when $\mathbf{k}$ points along the $z$-axis. The normalization $\boldsymbol{\varepsilon} \cdot \boldsymbol{\varepsilon}^* = 1$ requires $|a_+|^2 + |a_-|^2 = 1$. The matrix elements of processes involving a completely polarized photon of momentum $k$ in the initial state are of the form $T_{if}(a_+, a_-, \mathbf{k}, M_f) = \varepsilon_{\mu}(\mathbf{k})J_{if}^{\mu}(M_f)$, and $\overline{|T_{if}|^2}$ for such cases is simply

$$\begin{aligned} \overline{|T_{if}|^2} &= \sum_{M_f} \mathbf{J}_{if}(M_f) \cdot \boldsymbol{\varepsilon}(\mathbf{k})\boldsymbol{\varepsilon}^*(\mathbf{k})\mathbf{J}_{if}^*(M_f) \\ &= \sum_{\lambda\lambda'}\sum_{M_f} a_{\lambda}a_{\lambda'}^*\mathbf{J}_{if}(M_f)\boldsymbol{\varepsilon}(\mathbf{k}, \lambda)\boldsymbol{\varepsilon}^*(\mathbf{k}, \lambda')\mathbf{J}_{if}^*(M_f). \end{aligned} \tag{8.25}$$

Comparison with (3.6) and (23) shows that $\rho_{\lambda\lambda'} = a_{\lambda}a_{\lambda}'^*$ and

$$\begin{aligned} \xi_1 &= \rho_{+-} + \rho_{-+} = 2\,\mathrm{Re}\,a_+a_-^*, \qquad \xi_2 = i(\rho_{+-} - \rho_{-+}) = -2\,\mathrm{Im}\,a_+a_-^* \\ \xi_3 &= \rho_{++} - \rho_{--} = |a_+|^2 - |a_-|^2 = i(\boldsymbol{\varepsilon} \times \boldsymbol{\varepsilon}^*) \cdot \hat{\mathbf{k}}. \end{aligned} \tag{8.26}$$

Instead of $\rho_{\lambda\lambda'}$, one can also use a density matrix in the space of states of linear polarization. For unpolarized photons with **k** along the $z$-axis, $\rho$ is again $\frac{1}{2}$ of the unit matrix

$$\rho_{xy} = \rho_{yx} = 0, \qquad \rho_{xx} = \rho_{yy} = \tfrac{1}{2}, \qquad \overline{|T_{if}|^2} = \frac{1}{2}\sum_{M_f} (|J_{if}^x|^2 + |J_{if}^y|^2) \quad (8.27)$$

which may be easier to calculate than (3.7).

We can now extend the formulas of the last section to cases where $b$ or $d$ have spin 1. The spin-$\frac{1}{2}$ functions $u_b$ and $\bar{u}_d$ are replaced by the corresponding spin-1 functions $\varepsilon_b^\mu$ and $\varepsilon_d^{*\mu}$. If particle $b$ is spinless, $J_{bd}^\mu$ is of the form

$$J_{bd}^\mu = \varepsilon_{dv}^* 0^{\mu\nu}, \qquad (P_b - P_d)_\mu 0^{\mu\nu} = 0. \qquad (8.28)$$

For opposite parities of $b$ and $d$ ($0^+ \to 1^-$ and $0^- \to 1^+$),

$$0^{\mu\nu} = g^{\mu\nu} P_d(P_b - P_d) - P_d^\mu (P_b - P_d)^\nu. \qquad (8.29)$$

For equal parities, $0^{\mu\nu}$ must be a pseudotensor (see section 4-7 and table 4-9).

Elastic scattering on deuterons contains three vertex functions, the forms of which depend on the choice of Lorentz representation. Glaser and Jakšić (1957), Gourdin (1963, 1966) use (5):

$$J_{bb'}^\mu = -(P_b + P_b')^\mu (F_1 \varepsilon'^* \cdot \varepsilon - \tfrac{1}{2} m^{-2} F_2 \varepsilon'^* \cdot q \varepsilon \cdot q)$$
$$- G_1(\varepsilon'^{*\mu} \varepsilon \cdot q - \varepsilon^\mu \varepsilon'^* \cdot q). \quad (8.30)$$

For the interpretation of $F_1$, $F_2$ and $G_1$ in terms of the charge, magnetic and quadrupole form factors, $F_C$, $F_M$, $F_Q$, one uses the Breit frame, with **q** along the $z$-axis. Introducing $\tau \equiv q^2/4m^2 = -t/4m^2$ and using $E^2/m^2 = 1 + \tau$, we find

$$J_{bb'}^\mu = (P_b + P_b')^\mu [F_1 \mathbf{e}'^* \mathbf{e} + 2\tau e_z'^* e_z (F_1 + (1 + \tau) F_2)] + G_1 \frac{E}{m} q(\varepsilon'^{*\mu} e_z - \varepsilon^\mu e_z'^*) \quad (8.31)$$

The term $e_z'^* e_z$ is rewritten as $\frac{1}{3}\mathbf{e}'^*\mathbf{e} + \frac{1}{3}(2e_z'^* e_z - e_x'^* e_x - e_y'^* e_y)$, the second piece being the matrix element $\hat{Q}_{bb'}$ of the quadrupole operator. $F_C$ and $F_Q$ are defined as the coefficients of $\mathbf{e}'^*\mathbf{e}$ and $\hat{Q}_{bb'}/2\tau$ in $J_{bb'}^0/2E$. We find

$$F_Q = F_1 + (1 + \tau)F_2 + G_1, \qquad F_C = F_1 + \tfrac{2}{3}\tau F_Q, \qquad F_Q(0) \equiv m^2 Q. \quad (8.32)$$

$F_M$ is defined as the coefficient of $i\mathbf{q} \times \mathbf{S}_{bb'}/2m$ in $\mathbf{J}_{bb}/2E$. According to (4), the matrices **S** are antisymmetric. Remembering moreover $\mathbf{q} = (00, q)$, we find from (31) $F_M = G_1$. The deuteron magnetic moment is $\mu = eG_1(0)/2m$.

Equations of motion for spin-1 particles in an electromagnetic field of arbitrary strength have been given by Shay and Good (1968), among others. They are unnecessary for the construction of the Born approximation, as we have just seen.

## 2-9 Relativistic Treatment of Recoil in Coulomb Scattering

The results of the previous section enable us to modify the treatment of potential scattering in sections 1-8 to 1-10 such that it applies to the scattering of two particles on each other. We simply postulate that the Born approximations are identical in both formalisms. This will determine all "recoil corrections" if the target particle is nonrelativistic (see also Friar 1976).

In view of (5.7), we consider instead of (7.6) the function

$$f_{if} = (8\pi s^{\frac{1}{2}})^{-1} T_{if} = \alpha \tfrac{1}{2} t^{-1} s^{-\frac{1}{2}} J_{aa',\mu} J^{\mu}_{bb'}, \qquad \alpha = e^2/4\pi. \tag{9.1}$$

When $a$ and $b$ are spinless, each $J$ is of the form (7.9). Insertion of (7.9) into (1) gives, with $t \equiv -q^2$ and using (7.22)

$$f_{if}(q^2) = -Z_a Z_b \alpha \tfrac{1}{2} q^{-2} s^{-\frac{1}{2}} (4P_a P_b + t) F_a(q^2) F_b(q^2) \tag{9.2}$$

Our aim now is to redefine the parameters of the Born approximation (6.36), $-2\eta k q^{-2} F$ such that the expression becomes identical with (2). As we require the unitarity equation to be solved by the partial-wave expansion, clearly $k$ must be replaced by the cms momentum $p$. Comparison with (2) shows that for $t \ll 4P_a P_b$,

$$\eta = Z_a Z_b \alpha \varepsilon / p, \qquad \varepsilon = P_a P_b s^{-\frac{1}{2}}. \tag{9.3}$$

This formula replaces the definition (1-8.1) of the Coulomb parameter. It is interesting to note from (2.9) and (3.14) that

$$\varepsilon / p = v_L^{-1}, \qquad \varepsilon^2 = \mu^2 + p^2, \qquad \mu \equiv m_a m_b s^{-\frac{1}{2}}. \tag{9.4}$$

Thus $\mu$ is the relativistic generalization for the reduced mass, and $\varepsilon$ is the corresponding energy. This simplifies the necessary changes in chapter 1. For example, eqs. (1-8.20) and (1-9.17) are now relations between $\varepsilon$ and $\mu$:

$$\varepsilon_B - \mu = (2n^2)^{-1} Z^2 \alpha^2 \mu (1 + \delta). \tag{9.5}$$

With $\varepsilon - \mu = \frac{1}{2} s^{-\frac{1}{2}} (s - (m_a + m_b)^2)$, the true binding energy is

$$\begin{aligned} B_n \equiv m_a + m_b - s_B^{\frac{1}{2}} &= -2 s_B^{\frac{1}{2}} (\varepsilon_B - \mu)(s_B^{\frac{1}{2}} + m_a + m_b)^{-1} \\ &\approx -s_B^{\frac{1}{2}} \frac{\varepsilon_B - \mu}{m_a + m_b} \left( 1 - \frac{1}{2} \frac{B_n}{m_a + m_b} \right) \end{aligned} \tag{9.6}$$

Inserting $\varepsilon_B - \mu$ and observing that $s^{\frac{1}{2}} \mu (m_a + m_b)^{-1}$ is just the nonrelativistic reduced mass $\mu_{nr}$, we obtain the following relativistic binding energies, for small $Z\alpha/n$ and point nuclei:

$$B_n = \tfrac{1}{2} Z^2 \alpha^2 \mu_{nr} n^{-2} [1 + \tfrac{1}{4} Z^2 \alpha^2 n^{-2} \mu_{nr} (m_a + m_b)^{-1}](1 + \delta). \tag{9.7}$$

Similarly, the probability density of the bound state wave functions is obtained from (1-1.13) by replacing $m$ by $\mu$ and $E$ by $\varepsilon$:

$$\rho(\mathbf{r}) = \mu^{-1} (\varepsilon - V) |\psi(\mathbf{r})|^2. \tag{9.8}$$

The variable $\mathbf{r}$ is now interpreted as $\mathbf{r}_a - \mathbf{r}_b$ in the cms. Clearly (7) is only valid for high enough values of $L$ where the form factor has no influence. In general, we must include a form factor (6.36), which in view of (2) must be taken as

$$F(q^2) = [1 + t(4P_a \cdot P_b)^{-1}]F_a(t)F_b(t), \qquad q^2 = -t. \tag{9.9}$$

The "true potential" is now given by the Fourier transform (6.34):

$$V(r) = Z_a Z_b \frac{\alpha}{r}\frac{2}{\pi}\int q \frac{dq}{q^2} \sin qrF(q^2). \tag{9.10}$$

Next, we elaborate the amplitudes for spin-$\frac{1}{2}$ scattering on a spinless target. We take for $J^\mu_{bb'}$ once more the "spinless" current $Z_b(P_b + P'_b)^\mu F_b$ and for $J^\mu_{aa'}$ we take (7.11). Then (1) becomes, using $P'_b \cdot \gamma = P_b \cdot \gamma$ and (7.22)

$$f_{if} = \tfrac{1}{2}\alpha t^{-1} s^{-\frac{1}{2}} Z_b F_b \bar{u}'[2G_M P_b \cdot \gamma - \tfrac{1}{2}m_a^{-1}F_2(4P_a \cdot P_b + t)]u \tag{9.11}$$

From (A-3.19) and (A-3.27), we see that (11) is proportional to $d^{\frac{1}{2}}_{\lambda\lambda'}(\theta)$ and that we can put it into the parity-conserving version of (5.29):

$$f_{if} = d^{\frac{1}{2}}_{\lambda\lambda'}(\theta)(f_1 + 4\lambda\lambda' f_2). \tag{9.12}$$

For $\lambda' = -\lambda$ we obtain the "helicity-flip" amplitude by picking only the $\delta_{\lambda,-\lambda'}$-parts in (A-3.19) and (A-3.27)

$$f_1 - f_2 = \tfrac{1}{2}\alpha t^{-1} s^{-\frac{1}{2}} Z_b F_b[4G_M E_b m_a - F_2 m_a^{-1} E(4P_a \cdot P_b + t)] \tag{9.13}$$

Similarly, the "nonflip" amplitude follows from the $\delta_{\lambda\lambda'}$-parts:

$$f_1 + f_2 = \tfrac{1}{2}\alpha t^{-1} s^{-\frac{1}{2}} Z_b F_b[4G_M(E_b E + p^2) - F_2(4P_a P_b + t)] \tag{9.14}$$

$$= 2\alpha t^{-1} s^{-\frac{1}{2}} Z_b F_b(F_1 P_a \cdot P_b - \tfrac{1}{4}F_2 t) \equiv f^{\frac{1}{2}-0}_{\text{Born}}, \tag{9.15}$$

where we have inserted $E_b E - p^2 = P_a \cdot P_b$ and $G_M = F_1 + F_2$ from (7.13). Eq. (15) can be rewritten as

$$f^{\frac{1}{2}-0}_{\text{Born}} = -2\eta p q^{-2} F_b Z_a^{-1}[F_1 - F_2 t(4P_a \cdot P_b)^{-1}], \tag{9.16}$$

which agrees with expression (6.36) for the spinless case, except for the redefinition (3) of $\eta p$ and another definition of the form factor $F$. In view of this analogy, it also pays to rewrite (13) as

$$f_1 - f_2 = f^{\frac{1}{2}-0}_{\text{Born}} \cdot R,$$
$$R = [F_1 P_a \cdot P_b - \tfrac{1}{4}F_2 t]^{-1}[F_1 E_b m_a - F_2 m_a^{-1}(p^2 s^{\frac{1}{2}} + \tfrac{1}{4}tE_a)] \tag{9.17}$$

We then obtain the more traditional spin-nonflip and spin-flip amplitudes $f = f_1 + f_2 \cos\theta$ and $g = -f_2 \sin\theta$ as

$$f = f^{\frac{1}{2}-0}_{\text{Born}}\left(\cos^2\frac{\theta}{2} + R\sin^2\frac{\theta}{2}\right), \qquad g = -\tfrac{1}{2} f^{\frac{1}{2}-0}_{\text{Born}}(1 - R)\sin\theta. \tag{9.18}$$

These expressions resemble the point Coulomb amplitudes (1-10.22). In fact, for $F_2 = 0$, we find from (3) and (4)

$$R = E_b m_a s^{-\frac{1}{2}} \varepsilon^{-1} = E_b \mu (m_b \varepsilon)^{-1}, \tag{9.19}$$

which reduces to $\mu/\varepsilon$ in all cases where the recoil energy can be neglected relative to the target mass, $E_b \approx m_b$. In these cases, the results of sections 1-9 and 1-10 can be taken over with the new definitions (3), (4), and (5). For $F_2 \neq 0$, on the other hand, recoil corrections clearly appear in combinations other than $\mu$ and $\varepsilon$. These new corrections can only be included perturbatively. The relevant expression will be derived in the next section by a different method.

Since electrons are so much lighter than nuclei, recoil corrections become large only for the scattering at high energies on light nuclei (in particular protons). In such cases, recoil corrections outside the Born approximation are very small. For the relativistic treatment of bound states, on the other hand, it normally suffices to use the nonrelativistic reduced mass. This is no longer the case when electrons are replaced by muons, pions, kaons, or antiprotons. In particular, when $\mu^-$, $\pi^-$, or $K^-$ are stopped in matter, they form "mesic atoms." The most important primary atom formation process is the emission of a $K$-shell electron. The incident particle is then bound with an energy comparable to that of the $K$-electron; its Bohr orbit thus has principal quantum number $n \approx (\mu/m_e)^{\frac{1}{2}}$ according to (5). From these states (with $n > 10$), the mesic atom rapidly cascades down to states of $n = 1 - 3$, first by the emissions of Auger electrons, later by X-ray emission. The corresponding Bohr orbits become very small compared to electron orbits; electron screening becomes quite unimportant, and one can study simple mesic Bohr atoms throughout the periodic table. Whereas $\pi^-$ and $K^-$ are rapidly absorbed by the nucleus in states of low $L$ ($L = 0$ or 1 for light pionic atoms), muonic atoms reach the $1S$ ground state and live there for $\sim 10^{-6}$ seconds, which is long on the scale of atomic processes. (For reviews of various applications of muonic atoms, see Hughes and Wu, 1975). Energy shifts of hadronic atoms will be treated in section 6-11.

## 2-10 Breit Equations

The nonrelativistic Schrödinger equation for a two-particle system is of the form

$$[E - H_a - H_b - V_{ab}(\mathbf{r})]\psi(\mathbf{r}_a, \mathbf{r}_b) = 0, \qquad \mathbf{r} = \mathbf{r}_a - \mathbf{r}_b. \tag{10.1}$$

Following Breit (1929), we shall now determine the two-particle potential $V_{ab}$ such that it reproduces the covariant Born approximation (7.6). The resulting deviations from the Coulomb potential are reliable in lowest-order perturbation theory only, but for some purposes (for example the interaction between two electrons in an atom, see Mann and Johnson 1971) this is sufficient. A more powerful method will be derived in section 3-14.

To cover the most important applications, we assume spin-$\frac{1}{2}$ for particles $a$ and $b$ and write the Born approximation in the form

$$T_{if} = -4m_a m_b \psi_{P_a}'^{+} \psi_{P_b}'^{+} \tilde{V}(\mathbf{p}_a, \mathbf{p}_b, \mathbf{q}) \psi_{P_a} \psi_{P_b}, \qquad \mathbf{q} \equiv \mathbf{p}_a - \mathbf{p}_a'. \quad (10.2)$$

where the $\psi_P$ are Pauli spinors, and $\tilde{V}$ is the Fourier transform of the potential (the factor $-4m_a m_b$ follows from comparison with the leading term in (9) below).

For the elastic scattering of free particles in the cms, we have $\mathbf{q}(\mathbf{p}_a + \mathbf{p}_a') = p_a^2 - p_a'^2 = 0$. In the construction of $V$, however, one should not use this relation because $V$ must be able to describe transitions to virtual states, where the component $P_a^0$ of the 4-momentum $P_a^\mu$ is not given by $(m_a^2 + p_a^2)^{\frac{1}{2}}$. In order to ensure gauge invariance for this more general case, one must write $T_{if}$ in the form (7.16). $V$ depends (in the cms) on $\mathbf{p}_a$ and $\mathbf{p}_a'$, and its coordinate representation $V$ is defined as follows

$$\tilde{V}(\mathbf{p}_a, \mathbf{p}_a') = \int d^3r e^{-i\mathbf{p}_a'\mathbf{r}} V(\mathbf{r}, \mathbf{p}) e^{i\mathbf{p}_a\mathbf{r}}, \qquad \mathbf{p} \equiv -i\nabla. \quad (10.3)$$

For example, when particle $b$ is spinless, its 3-vector current matrix element is $(\mathbf{p}_b + \mathbf{p}_b')F_b(q^2)$. Since $F_b/q^2$ is the Fourier transform of the Coulomb potential $V_C$ (assuming that particle $a$ is pointlike), we have in the cms

$$\begin{aligned} -(\mathbf{p}_b + \mathbf{p}_b')F_b/q^2 &= (\mathbf{p}_a + \mathbf{p}_a') \int d^3r e^{-i\mathbf{p}_a'} V_C(r) e^{i\mathbf{p}_a\mathbf{r}} \\ &= \int d^3r e^{-i\mathbf{p}_a'\mathbf{r}} (\mathbf{p}V_C + V_C\mathbf{p}) e^{i\mathbf{p}_a\mathbf{r}}. \end{aligned} \quad (10.4)$$

The transverse current $J^\perp$ which enters (7.16) cannot be expressed in terms of $V_C$ because it involves a part $F_b/q^4$. However, it can be expressed in terms of a function $W_C$ which is the Fourier transform of $F_b/q^4$. (See Friar 1976 for a review.) Below, we shall use the simpler prescription (10), which is correct except for recoil corrections to the nuclear charge distribution.

We now return to (2) and restrict ourselves to first-order relativistic corrections, where $\psi_{P_a}$ is given by (1-7.3):

$$u_a = (2m_a)^{\frac{1}{2}} \begin{pmatrix} (1 - p_a^2/8m_a^2) \\ \boldsymbol{\sigma}_a \mathbf{p}_a/2m_a \end{pmatrix} \psi_{P_a}(\mathbf{r}_a, M_a). \quad (10.5)$$

To the same order, we find

$$\begin{aligned} \bar{u}_a'\gamma^0 u_a = u_a'^{+} u_a &= 2m_a \psi_{P_a}'^{+}[1 - (p_a^2 + p_a'^2)/8m_a^2 + \boldsymbol{\sigma}_a\mathbf{p}_a'\boldsymbol{\sigma}_a\mathbf{p}_a/4m_a^2]\psi_{P_a} \\ &= 2m_a \psi_{P_a}'^{+}[1 - q^2/8m_a^2 + i\boldsymbol{\sigma}_a(\mathbf{q} \times \mathbf{p}_a)/4m_a^2]\psi_{P_a}, \end{aligned} \quad (10.6)$$

$$\bar{u}_a'\gamma^\perp u_a = \psi_{P_a}'^{+}[\sigma_a^\perp(\boldsymbol{\sigma}_a\mathbf{p}_a) + (\boldsymbol{\sigma}_a\mathbf{p}_a')\sigma_a^\perp]\psi_{P_a} = \psi_{P_a}'^{+}[i\boldsymbol{\sigma}_a \times \mathbf{q} + 2p_a^\perp]\psi_{P_a}. \quad (10.7)$$

For the anomalous magnetic moment coupling, we need

$$\bar{u}'_a u_a = 2m_a \psi'^{+}_{P_a}\left[1 - \left(p_a^2 - \mathbf{p}_a\mathbf{q} + \tfrac{1}{4}q^2 + \frac{i}{2}\boldsymbol{\sigma}_a(\mathbf{q}\times\mathbf{p}_a)\right)/2m_a^2\right]\psi_{P_a}. \quad (10.8)$$

Admitting just one form factor $F(q^2)$ and one anomalous magnetic moment $\kappa_b$, we find

$$\begin{aligned}\frac{\tilde{V}}{Z_a Z_b e^2 F(q^2)} = \frac{1}{q^2} - \frac{1}{8m_a^2} - \frac{1}{8m_b^2} + \frac{i\boldsymbol{\sigma}_a}{2m_a q^2}\left(\frac{\mathbf{q}\times\mathbf{p}_a}{2m_a} - \frac{\mathbf{q}\times\mathbf{p}_b}{m_b}\right) \\ - \frac{i\boldsymbol{\sigma}_b}{2m_b q^2}\left[\frac{\mathbf{q}\times\mathbf{p}_b}{2m_b} - \frac{\mathbf{q}\times\mathbf{p}_a}{m_a} + \frac{\kappa_b}{Z_b}\mathbf{q}\times\left(\frac{\mathbf{p}_b}{m_b} - \frac{\mathbf{p}_a}{m_a}\right)\right] \quad (10.9) \\ - \frac{1}{4m_a m_b}\left[p_a^{\perp}p_b^{\perp}/q^2 + \sigma_a^{\perp}\sigma_b^{\perp}\left(1 + \frac{\kappa_b}{Z_b}\right)\right].\end{aligned}$$

$V_{ab}$ is approximated by the usual definition

$$V_{ab}(\mathbf{p}_a, \mathbf{p}_b, \mathbf{r}) = (2\pi)^{-3}\int d^3q e^{i\mathbf{qr}}\tilde{V}(\mathbf{p}_a, \mathbf{p}_b, \mathbf{q}), \qquad \mathbf{r} \equiv \mathbf{r}_a - \mathbf{r}_b. \quad (10.10)$$

The symbols $\mathbf{p}_a$ and $\mathbf{p}_b$ stand for $-i\nabla_a$ and $-i\nabla_b$. When they act on $\exp\{i\mathbf{q}(\mathbf{r}_a - \mathbf{r}_b)\}$ in (10), they produce factors $\mathbf{q}$ and $-\mathbf{q}$. One sees easily that these factors make no contribution to (9), such that in practice $\mathbf{p}_a$ and $\mathbf{p}_b$ operate only on $\psi(\mathbf{r}_a, \mathbf{r}_b)$. The necessary integrals for (10) are

$$(2\pi)^{-3}\int q^{-2}F(q^2)e^{i\mathbf{qr}}\,d^3q \equiv V_C \to (4\pi r)^{-1} \quad (10.11)$$

$$(2\pi)^{-3}\int q^{-2}F(q^2)\mathbf{q}e^{i\mathbf{qr}}\,d^3q = -i\nabla V_C \to i\mathbf{r}(4\pi r^3)^{-1} \quad (10.12)$$

$$\begin{aligned}(2\pi)^{-3}\int q^{-2}F(q^2)q_i q_j e^{i\mathbf{qr}}\,d^3q = -\partial_i\,\partial_j V_C = \tfrac{1}{3}\,\delta_{ij}\rho(r) \\ + (\delta_{ij} - 3\hat{r}_i\hat{r}_j)\frac{d}{r\,dr}V_C \to \tfrac{1}{3}\,\delta_{ij}\,\delta_3(\mathbf{r}) + (\delta_{ij} - 3\hat{r}_i\hat{r}_j)(4\pi r^3)^{-1}\end{aligned} \quad (10.13)$$

The expressions behind the arrows are valid for $F(q^2) = 1$. In (13), we have added and subtracted a term $\partial_i^2$ because $\partial_i^2 V = \frac{1}{3}\nabla^2 V = -\frac{1}{3}\rho(r)$ where $\rho(r)$ is the normalized charge density of particle $b$ at cms separation $r = |\mathbf{r}_a - \mathbf{r}_b|$. Finally, we need

$$\begin{aligned}(2\pi)^{-3}\int q^{-4}F(q^2)q_i q_j e^{i\mathbf{qr}}\,d^3q = (2\pi)^{-3}\tfrac{1}{2}i\,\partial_i\int d^3q F(q^2)e^{i\mathbf{qr}}\,\partial_{q_j}q^{-2} \\ \to \tfrac{1}{2}\,\partial_i r_j(4\pi r)^{-1} = \tfrac{1}{2}(\delta_{ij} - \hat{r}_i\hat{r}_j)(4\pi r)^{-1}. \quad (10.14)\end{aligned}$$

The last expression is obtained by partial integration. Thus we obtain from (9) with $e^2/4\pi = \alpha$ and $F(q^2) = 1$

$$\frac{V}{Z_a Z_b \alpha} = \frac{1}{r} - \frac{\pi}{2}(m_a^{-2} + m_b^{-2})\,\delta(\mathbf{r}) - \frac{1/r}{2m_a m_b}(\delta_{ij} - \hat{r}_i \hat{r}_j) p_{ai} p_{bj}$$
$$- \frac{\boldsymbol{\sigma}_a}{r^3}\left(\frac{\mathbf{r} \times \mathbf{p}_a}{4m_a^2} - \frac{\mathbf{r} \times \mathbf{p}_b}{2m_a m_b}\right) \tag{10.15}$$
$$+ \frac{\boldsymbol{\sigma}_b}{r^3}\left[\frac{\mathbf{r} \times \mathbf{p}_b}{4m_b^2}\left(1 + \frac{2\kappa_b}{Z_b}\right) - \frac{\mathbf{r} \times \mathbf{p}_a}{2m_a m_b}\left(1 + \frac{\kappa_b}{Z_b}\right)\right] + V_{SS}/Z_a Z_b \alpha,$$

$$V_{SS} = \frac{Z_a \alpha (Z_b + \kappa_b)}{4 m_a m_b r^3}\left[\boldsymbol{\sigma}_a \boldsymbol{\sigma}_b (1 - \tfrac{8}{3}\pi r^3\,\delta_3(\mathbf{r})) - 3(\boldsymbol{\sigma}_a \hat{r})(\boldsymbol{\sigma}_b \hat{r})\right]. \tag{10.16}$$

In the cms we have $\mathbf{p}_a = -\mathbf{p}_b = \mathbf{p}$, $\mathbf{r} \times \mathbf{p} = \mathbf{L}$, and the remaining spin-dependent potentials of (15) give the spin-orbit potentials

$$V_{LS_a} = \frac{-Z_a Z_b \alpha}{2 m_a m_b r^3}\mathbf{L}\boldsymbol{\sigma}_a\left(1 + \frac{m_b}{2m_a}\right),$$
$$V_{LS_b} = \frac{-Z_a Z_b \alpha}{2 m_a m_b r^3}\mathbf{L}\boldsymbol{\sigma}_b\left[1 + \frac{m_a}{2m_b} + \frac{\kappa_b}{Z_b}\left(1 + \frac{m_a}{m_b}\right)\right]. \tag{10.17}$$

When $a$ is an electron and $b$ a nucleus, $V_{LS_a}$ dominates over $V_{LS_b}$. In that case, the relevant mass combination in $V_{LS_a}$ can be rewritten as

$$(4m_a^2 m_b)^{-1}(2m_a + m_b) \approx (2m_a m_b)^{-2}(m_b + m_a)^2 = (2\mu_{nr})^{-2}, \tag{10.18}$$

where $\mu_{nr}$ is the normal reduced mass. This confirms our result of the previous section, that all recoil corrections to the fine structure are given by the reduced mass, provided $\kappa_a = 0$. The case $\kappa_a \neq 0$ follows by analogy with $V_{LS_b}$. One finds that $\kappa_a$ in (1-9.18) is multiplied by $1 - m_a/m_b$.

In the presence of a 4-potential $A^\mu$, $H_a$ is given by (1-4.2), and $\mathbf{p}_a$, $\mathbf{p}_b$ are replaced by $\boldsymbol{\pi}_a$, $\boldsymbol{\pi}_b$ everywhere in (15). However, the spinor (5) must also be changed. This is seen best by introducing the usual cms coordinate

$$\mathbf{R} = m^{-1}(m_a \mathbf{r}_a + m_b \mathbf{r}_b), \qquad m \equiv m_a + m_b, \tag{10.19}$$

and by expressing $\mathbf{p}_a$ and $\mathbf{p}_b$ in terms of $\mathbf{p} = -i\,\nabla$ and $\mathbf{P} = -i\,\nabla_R$

$$\mathbf{p}_a = \mathbf{p} + m^{-1} m_a \mathbf{P}, \qquad \mathbf{p}_b = -\mathbf{p} + m^{-1} m_b \mathbf{P}. \tag{10.20}$$

In the cms of $a$ and $b$, we have the product $u_a(\mathbf{p}, M_a) u_b(-\mathbf{p}, M_b)$, where to first order in $\mathbf{p}$

$$u_a(\mathbf{p}, M_a) = (2m_a)^{\frac{1}{2}}\begin{pmatrix} 1 \\ \boldsymbol{\sigma}_a \mathbf{p}_a / 2m_a \end{pmatrix}\psi_P(\mathbf{r}_a, M_a). \tag{10.21}$$

In this product, $u_a$ must now be Lorentz-transformed with momentum $\mathbf{P} m_a/m$, while $u_b$ must be Lorentz-transformed with momentum $\mathbf{P} m_b/m$. In section 1-6 we have elaborated the Lorentz transformations $A(\mathbf{p})$ and $A(-\mathbf{p})$ for right- and left-handed spinors respectively. To first order in $\mathbf{P}$, we have

$$A(\mathbf{P}m_a/m)\psi_r = (1 + \mathbf{P}\boldsymbol{\sigma}/2m)\psi_r. \tag{10.22}$$

When the Lorentz-transformations of $\psi_r$ and $\psi_l$ are performed, we find the transformed spinor in the low-energy representation from (1-5.8)

$$(u'_a)_{g,k} = \frac{1}{2}\left[\left(1 + \frac{\mathbf{P}\boldsymbol{\sigma}_a}{2m}\right)\left(1 + \frac{\mathbf{p}\boldsymbol{\sigma}_a}{2m_a}\right) \pm \left(1 - \frac{\mathbf{P}\boldsymbol{\sigma}_a}{2m}\right)\left(1 - \frac{\mathbf{p}\boldsymbol{\sigma}_a}{2m_a}\right)\right]\chi_{a0}, \tag{10.23}$$

$$(u'_a)_g = \left(1 + \frac{\mathbf{P}\boldsymbol{\sigma}_a\mathbf{p}\boldsymbol{\sigma}_a}{4mm_a}\right)\chi_{a0}, \qquad (u'_a)_k = \boldsymbol{\sigma}_a\left(\frac{\mathbf{P}}{2m} + \frac{\mathbf{p}}{2m_a}\right)\chi_{a0}. \tag{10.24}$$

The small components $(u'_a)_k$ are the same as in (5), since the bracket contains just $\mathbf{p}_a/2m_a$ according to (20). The spinor $\psi_{P_a}$ must now be introduced such that its normalization contains no operators. To first order in the corrections, we have

$$\chi_0 = (1 - \mathbf{P}\boldsymbol{\sigma}_a\mathbf{p}\boldsymbol{\sigma}_a/4mm_a - (\boldsymbol{\sigma}_a\mathbf{p}_a)^2/8m_a^2)\psi_{P_a}. \tag{10.25}$$

Now we may replace $\mathbf{p}_a$ by $\boldsymbol{\pi}_a$ and $\mathbf{p}_b$ by $\boldsymbol{\pi}_b$. For this purpose, the term $\mathbf{P}\boldsymbol{\sigma}_a\mathbf{p}\boldsymbol{\sigma}_a$ must be expressed in terms of $\mathbf{p}_a$ and $\mathbf{p}_b$:

$$\mathbf{P}\boldsymbol{\sigma}_a\mathbf{p}\boldsymbol{\sigma}_a = (\mathbf{p}_a + \mathbf{p}_b)\boldsymbol{\sigma}_a m^{-1}(\mathbf{p}_a m_b - \mathbf{p}_b m_a)\boldsymbol{\sigma}_a. \tag{10.26}$$

For particle $b$, $\mathbf{P}$ remains the same and $\mathbf{p}$ is reversed, such that the combination (26) enters with an extra minus sign, and with $\boldsymbol{\sigma}_a$ replaced by $\boldsymbol{\sigma}_b$. Substituting now $\mathbf{p}_a \to \boldsymbol{\pi}_a$, $\mathbf{p}_b \to \boldsymbol{\pi}_b$ and solving (1) by the methods of section 1-7, we find in addition to (1-4.5), (1-7.6), an interaction $V'_{ab}$ with the external electric field (Brodsky and Primack 1968, Brodsky 1971):

$$\begin{aligned} H_P - V &= \sum_{i=a}^{b}\left\{\boldsymbol{\pi}_i^2/2m_i - \mu_i\boldsymbol{\sigma}_i\mathbf{H}_i - \frac{e}{8}m_i^{-2}(Z_i + 2\kappa_i)[2(\mathbf{E}_i \times \boldsymbol{\pi}_i)\boldsymbol{\sigma}_i + \operatorname{div}\mathbf{E}_i]\right\} + V'_{ab}, \\ \mathbf{E}_i &\equiv \mathbf{E}(\mathbf{r}_i) \text{ etc.}, \qquad V'_{ab} = \tfrac{1}{4}em^{-1}(\boldsymbol{\sigma}_a/m_a - \boldsymbol{\sigma}_b/m_b)[Z_b\mathbf{E}_b \times \boldsymbol{\pi}_a - Z_a\mathbf{E}_a \times \boldsymbol{\pi}_b] \end{aligned} \tag{10.27}$$

If $\mathbf{E}$ is caused by the atomic nucleus, $V'_{ab}$ is effectively a 3-body force. It is due to the boost (22) and is therefore of kinematical origin (Krajcik and Foldy, 1970).

CHAPTER THREE

# Radiation and Quantum Electrodynamics

## 3-0 Second Quantization

So far in this book the 4-potential $A^\mu(t, \mathbf{x})$ was either classical or due to a nucleus. An essential extension of the formalism is necessary for processes in which photons are emitted or absorbed. For example, if an excited atom $|e_i\rangle$ has been formed by a rapid collision at a certain time, the probability amplitude of finding the atom in the state $|e_i\rangle$ decreases afterwards, and new states $|e_f\rangle|\gamma_\lambda(\mathbf{k})\rangle$ appear, in which the atom is in states of lower energy, the missing energy being carried away by a photon $|\gamma_\lambda(\mathbf{k})\rangle$ of momentum $\hbar\mathbf{k}$. The Hamilton operator (1-4.2) is capable of inducing such transitions, but only if $A^\mu$ is interpreted as an operator which can create and annihilate photons. As explained in section 2-8, gauge invariance permits us to impose the condition div $\mathbf{A} = 0$ on this operator. Since we wish to keep classical electric and magnetic fields as well, we put

$$A^\mu(x) = A^\mu_{\text{cl}}(x) + A^\mu_{\text{op}}(x), \qquad A^0_{\text{op}} = 0, \qquad \text{div}\,\mathbf{A}_{\text{op}} = 0. \tag{0.1}$$

This separation is not covariant; covariance will be established only at the level of matrix elements of scattering or decay processes. A classical field can be viewed as a limiting case of the covariant extension. In any case, the operator $\mathbf{A}_{\text{op}}$ corresponds to a classical observable and must therefore be Hermitean. In vacuum, it can be decomposed into plane waves as follows:

$$\mathbf{A}_{\text{op}}(j^\mu = 0) \equiv \mathbf{A}_0 = \int d^3_L k \sum_\lambda \left(a_\lambda(\mathbf{k})\boldsymbol{\varepsilon}(\mathbf{k}, \lambda)e^{-iK\cdot x} + a^+_\lambda(\mathbf{k})\boldsymbol{\varepsilon}^*(\mathbf{k}, \lambda)e^{iK\cdot x}\right),$$
$$\mathbf{k}\cdot\boldsymbol{\varepsilon}(\mathbf{k}, \lambda) = 0, \qquad K^0 \equiv \omega_k = |\mathbf{k}| = k, \tag{0.2}$$

where the expansion coefficients $a_\lambda(\mathbf{k})$ are now operators, and the second term in the sum follows from the first one by Hermiticity. We shall see later from energy conservation that the operator $a^+$ increases the photon number by one unit, while $a$ lowers it by one unit. The condition $K^0 = k$ follows from $\Box \mathbf{A}_0 = 0$, which is a consequence of (2-8.21).

Similarly, for the theory of electron-positron pair production and of $\beta$-decay, operators which can create or annihilate electrons and neutrinos (and their antiparticles) will be useful, although perhaps not absolutely necessary. Moreover, this "second quantization" is quite convenient for processes which involve several identical particles. (Even solid state physicists use it for this purpose). We shall therefore also interpret our good old Dirac wave function $\psi$ as an operator field $\psi_{\text{op}}$ in a Hilbert space of variable particle numbers. In fact, the term "quantum field theory" is frequently used as a synonym for "elementary particle theory" (Schweber 1961, Gasiorowicz 1966, and others).

A Hilbert space with variable particle number is called "Fock space." A state in this space may be denoted by $|n_1, n_2, n_3, n_4 \ldots\rangle$, where $n_i$ is the number of identical particles in the $i$th one-particle state. The vacuum and 1-particle states are in this notation

$$|0\rangle \equiv |0, 0, 0, 0 \ldots\rangle, \qquad |i\rangle = |0, 0 \ldots 1, 0 \ldots\rangle, \tag{0.3}$$

where the 1 stands in the $i$th position. All states can be constructed from the vacuum by repeated application of "particle creation" operators $a^+$:

$$a_i^+ |0\rangle = |i\rangle, \qquad a_i^+ |n_1 \ldots, n_i, \ldots\rangle = c_{n_i+1} |n_1 \ldots, n_i + 1, \ldots\rangle. \tag{0.4}$$

The $c_{n_i}$ may still depend on the other occupation numbers. The first equation in (4) says $c_1(0, 0 \ldots) = 1$ and normalizes the operator. From the state normalization $\langle i | i\rangle = 1$ (we consider only discrete states for the time being), it follows that the Hermitean conjugate operator $a_i$ is an "annihilation operator"

$$a_i |i\rangle = 0, \qquad a_i|n_1, \ldots, n_i, \ldots\rangle = c_{n_i}^* |n_1, \ldots, n_i - 1, \ldots\rangle, \qquad c_0 = 0. \tag{0.5}$$

In addition to $a_i^+$ and $a_i$, one needs number operators $N_i$, with eigenvalues $n_i$:

$$N_i|n_1 \ldots n_i \ldots\rangle = n_i|n_1 \ldots n_i \ldots\rangle, \qquad [N_i, N_j] = 0. \tag{0.6}$$

Our aim is to construct $N_i$ from $a_i^+$ and $a_i$. Since $a_j^+$ raises the numbers $n_j$ of particles in the $j$th one-particle state by one, we require:

$$N_i a_j^+ - a_j^+ N_i = \delta_{ij}. \tag{0.7}$$

Imagining $N_i$ as a polynomial in $a_i$ and $a_i^+$, we must know how to commute $a_j^+$ with $a_i^+$ and $a_i$, in order to exploit (7). The relation between $a_j^+ a_i^+$ and $a_i^+ a_j^+$ is determined by the particles' "statistics": For a system containing (among others) two identical particles, the Hamiltonian is invariant under the exchange of these particles (for the Pauli Hamiltonian for example, we have $H(\mathbf{r}_1, \mathbf{p}_1, \boldsymbol{\sigma}_1, \mathbf{r}_2, \mathbf{p}_2, \boldsymbol{\sigma}_2, \mathbf{r}_3 \ldots) = H(\mathbf{r}_2, \mathbf{p}_2, \boldsymbol{\sigma}_2, \mathbf{r}_1, \mathbf{p}_1, \boldsymbol{\sigma}_1, \mathbf{r}_3 \ldots)$). Consequently, if one permutes two identical particles in a wave function $\psi(\mathbf{r}_1, M_1, \mathbf{r}_2, M_2, \mathbf{r}_3 \ldots)$ in the coordinate representation, $\psi$ being an eigenfunction of $H$, one obtains a new wave function $\psi'$ which must describe the same state, $\psi' = \lambda\psi$. Permuting the particles once more, one has $\psi'' = \lambda^2\psi$. Since permuting the same particles twice gives the identity, this implies $\psi'' = \psi$, $\lambda^2 = 1$, i.e., $\lambda = 1$ ("bosons") or $\lambda = -1$ ("fermions"). For our creation operators in Fock space, we thus require

$$a_i^+ a_j^+ = \lambda a_j^+ a_i^+, \qquad \lambda = \pm 1. \tag{0.8}$$

For $i = j$ and $\lambda = -1$, (8) gives $(a_i^+)^2 = 0$, i.e., one cannot have two identical fermions in the same state (Pauli exclusion principle): $n_i = 0$ or 1. To define multifermion states uniquely, we require that all states are created from $|0\rangle$ by the necessary creation operators in the same order in which the single-particle states appear in Fock space. That means, $a_1^+$ operates first, if $n_1 = 1$. Otherwise $a_2^+$ operates first, if $n_2 = 1$, etc. This requires

$$c_{n_i+1} = (-1)^{n_{<i}} \qquad \text{for fermions,} \tag{0.9}$$

where $n_{<i}$ is the total number of creation operators which $a_i^+$ must pass before reaching its proper place.

The relation between $a_i a_j^+$ and $a_j^+ a_i$ is now determined by the desired spectrum of $N_i$ as well as by phase conventions. In order to fulfill (6), $N_i$ must contain equal numbers of creation and annihilation operators. If we know how to commute $a_i$ and $a_i^+$, the most general polynomial expansion of $N_i$ is

$$N_i = \sum_{\nu=0} \alpha_\nu (a_i^+)^\nu (a_i)^\nu \tag{0.10}$$

with real coefficients $\alpha_\nu$. From $N_i|0\rangle = 0$, we find $\alpha_0 = 0$. For fermions, the coefficients with $\nu > 1$ remain arbitrary, since $a_i^{+2} = 0$. We may therefore define $\alpha_{\nu>1} = 0$,

$$N_i = a_i^+ a_i \tag{0.11}$$

for fermions. Moreover, when $N_i$ operates on a state

$$|n_1 \ldots n_i \ldots n_j\rangle \sim (a_j^+)^{n_j} \ldots (a_i^+)^{n_i} \ldots (a_1^+)^{n_1}|0\rangle, \tag{0.12}$$

it must pass $n_{<i}$ creation operators before it reaches $a_i^+$. This accumulates the phase (9) from passing the factor $a_i^+$ of the product $a_i^+ a_i$. This phase must be cancelled by a corresponding phase from passing the factor $a_i$ of the same product. We thus require for fermions ($\lambda = -1$)

$$a_i a_j^+ - \lambda a_j^+ a_i = \delta_{ij}. \tag{0.13}$$

For bosons, we remember from the algebraic treatment of independent harmonic oscillators $i, j$ that if (13) is valid with $\lambda = +1$ and $N_i$ is defined by (11), then (7) is satisfied, such that $N_i$ has the eigenvalues 0, 1, 2 ... The normalization constant $c_{n_i}$ follows from (4), (5) and (6) as

$$|c_{n_i}|^2 = n_i, \qquad c_{n_i} = n_i^{\frac{1}{2}}. \tag{0.14}$$

In the programme of second quantization, we must solve the inverse problem, i.e., derive (13) from (7), (8), and (10). If $N_i$ is given by (11), we are clearly led to (13) with $\lambda = 1$, by the same argument which worked for the fermions. But how can we exclude terms with $v > 1$ in (10) for bosons? Suppose we choose a different one-particle basis in Fock space, with particle creation operators $b_q^+$. The $a_i^+$ may be expanded in terms of the $b_q^+$, and the orthonormalization of the 1-particle states requires unitary expansion coefficients

$$a_i^+ = \sum_q c_{qi}^* b_q^+, \qquad \sum_q c_{qi}^* c_{qj} = \delta_{ij}. \tag{0.15}$$

If we require the commutation relation of $a_i$ and $a_j^+$ to be independent of such unitary transformations, we are immediately restricted to the form $a_i a_j^+ - \lambda a_j^+ a_i = \alpha_1^{-1} \delta_{ij}$ and to (11), since the total particle number must be representation-independent, $N = \sum_i N_i = \sum_q N_q$.

In the case of particles with spin, the magnetic quantum number $M_i$ is part of the index $i$. For free particles, the single-particle basis for our Fock space can be taken as plane waves. The normalization (1-1.15) leads to

$$a_M(\mathbf{p})a_{M'}^+(\mathbf{p}') - \lambda a_{M'}^+(\mathbf{p})a_M(\mathbf{p}) = (2\pi)^3 2E_p\, \delta_3(\mathbf{p} - \mathbf{p}')\, \delta_{MM'}, \tag{0.16}$$

$$N = \int d_L^3 p \sum_M N_M(\mathbf{p}), \qquad N_M(\mathbf{p}) = a_M^+(\mathbf{p})a_M(\mathbf{p}). \tag{0.17}$$

We are now through with our programme. So far, physics has entered only via (8) and our identification of $a^+$ and $a$ of (2) with photon creation and annihilation operators. The method presented here is equivalent to the "canonical field quantization."

Since the total energy of a distribution of free photons is $\sum_\lambda \int d_L^3 k \omega_k n_\lambda(\mathbf{k})$, the free photon Hamiltonian must be

$$H_0^{(\gamma)} = \int d_L^3 k \omega_k \sum_\lambda N_\lambda(\mathbf{k}) = \int d^3k (16\pi^3)^{-1} \sum_\lambda a_\lambda^+(\mathbf{k}) a_\lambda(\mathbf{k}). \tag{0.18}$$

Here we have replaced the magnetic quantum number $M$ of (17) by the photon helicity $\lambda$. Classically, $H_0^{(\gamma)}$ should correspond to the free field Hamiltonian,

$$H(\mathbf{A}) = \frac{1}{2}\int d^3x(\mathbf{E}^2 + \mathbf{H}^2) = \frac{1}{2}\int d^3x[(\partial_t \mathbf{A})^2 + (\nabla \times \mathbf{A})^2], \tag{0.19}$$

where we have used (1-0.5) with $\phi = 0$. We also expect that replacing $\mathbf{A}$ by the free-field operator $\mathbf{A}_0$ in (19) brings us back to (18), apart from commutators. Insertion of (2) gives (h.c. = hermitean conjugate)

$$H(\mathbf{A}) = -\frac{1}{2}\int d^3x\, d^3_Lk\, d^3k'(16\pi^3\omega')^{-1}$$
$$\times \sum_{\lambda\lambda'} \{[\omega_k a_\lambda(\mathbf{k})\varepsilon_\lambda(\mathbf{k})e^{-iK\cdot x} - \text{h.c.}][\omega'_k a_{\lambda'}(\mathbf{k}')\varepsilon_{\lambda'}(\mathbf{k}')e^{-iK'\cdot x} - \text{h.c.}]$$
$$+ [\mathbf{k} \times \varepsilon_\lambda(\mathbf{k})a_\lambda(\mathbf{k})e^{-iK\cdot x} - \text{h.c.}][\mathbf{k}' \times \varepsilon_{\lambda'}(\mathbf{k}')a_{\lambda'}(\mathbf{k}')e^{-iK'\cdot x} - \text{h.c.}]\}. \tag{0.20}$$

Using the rule $(a \times b)(c \times d) = (ab)(cd) - (ad)(bc)$, we have $(\mathbf{k} \times \varepsilon)(\mathbf{k}' \times \varepsilon') = \mathbf{k}\mathbf{k}'\varepsilon\varepsilon' - (\mathbf{k}\varepsilon')(\mathbf{k}'\varepsilon)$ etc. The $\int d^3x$ gives $8\pi^3\, \delta_3(\mathbf{k} + \mathbf{k}')$ in the $aa'$ and $a^+a'^+$-terms and $8\pi^3\, \delta_3(\mathbf{k} - \mathbf{k}')$ in the $aa'^+$ and $a'^+a$-terms, such that we are left with $\mathbf{k}' = \pm\mathbf{k}$. Using $\mathbf{k} \cdot \varepsilon = 0$, we see that the final $\mathbf{k}$-integration involves $\omega_k$ and $k$ only in the combinations $\omega_k^2 \pm k^2$. Thus the terms with $\mathbf{k}' = -\mathbf{k}$ drop out, and the remaining terms combine into

$$2\omega_k^2\varepsilon^*_\lambda(\mathbf{k})\varepsilon_{\lambda'}(\mathbf{k})[a^+_\lambda(\mathbf{k})a_{\lambda'}(\mathbf{k}) + a_\lambda(\mathbf{k})a^+_{\lambda'}(\mathbf{k})] = 2\omega_k^2\, \delta_{\lambda\lambda'}(a^+a + aa^+). \tag{0.21}$$

If we now neglect the commutator, we can write $aa^+$ as $a^+a$, in which case we are in fact back at (18). One should also note that if photons were fermions, we would write $aa^+$ as $-a^+a$, and thus find $H_0^{(\gamma)} = 0$. Thus photons must be bosons.

Since the plane-wave decomposition (2) of the free-field operator $\mathbf{A}_0$ is linear in the creation and annihilation operators, its matrix elements between Fock states differing by more than one occupation number are all zero. For the remaining states, we find from (4), (5) and (13)

$$\langle n_1 \ldots n'_\lambda(\mathbf{k})|\mathbf{A}_0(x)|n_1 \ldots n_\lambda(\mathbf{k})\rangle = (n_\lambda(\mathbf{k}))^{\frac{1}{2}}\varepsilon_\lambda(\mathbf{k})e^{-iK\cdot x}\,\delta_{n_\lambda'(\mathbf{k}),\, n_\lambda(\mathbf{k})-1}$$
$$+ (n_\lambda(\mathbf{k}) + 1)^{\frac{1}{2}}\varepsilon^*_\lambda(\mathbf{k})e^{iK\cdot x}\,\delta_{n_\lambda'(\mathbf{k}),\, n_\lambda(\mathbf{k})+1}. \tag{0.22}$$

It may be confusing to note that $A^\mu_{\text{cl}}$ in (1) has the same dimension as the momentum, whereas the matrix elements of $\mathbf{A}_{\text{op}}$ are dimensionless. There is no contradiction here, because $\mathbf{A}_{\text{op}}$ has no diagonal matrix elements in Fock space, whereas $A^\mu_{\text{cl}}$ is like a number in this space.

## 3-1 Atomic Radiation and Thomson Scattering

We are now in a position to calculate processes with free photons in the initial or final states, using time-dependent perturbation theory. For nonrelativistic electrons, we can use the Pauli equation

$$i\,\partial_t\psi_P = (H_0 + H_{\text{per}}(t))\psi_P, \tag{1.1}$$

where $\psi_P$ has still the meaning of chapter 1, and $H_{\text{per}}$ is that part of $H_P$ (1-4.2) which involves $\mathbf{A}_{\text{op}}(t, \mathbf{x})$. In principle, $\mathbf{A}_{\text{op}}$ differs from $\mathbf{A}_0$ which satisfies Maxwell's equations only in vacuum. For our present examples, however, the difference is irrelevant. Thus we put

$$H_{\text{per}} = e(2m)^{-1}[-2i\mathbf{A}_0 \nabla + e\mathbf{A}_0 \mathbf{A}_0 + \boldsymbol{\sigma}(\nabla \times \mathbf{A}_0)] \tag{1.2}$$

Relativistic corrections are given by (1-7.6), (2-10.27), recoil corrections by (3-3.27) below.

To be specific, let us calculate the transition rate $\Gamma_{lm}$ between two atomic 1-electron states $l$ and $m$. With few exceptions, the term $e^2\mathbf{A}_0 \mathbf{A}_0/2m$ can be neglected in (2). The remaining terms change the photon number by exactly one unit. If the state $l$ is an excited state, we wish to know its de-excitation rate, and we expect from energy conservation that the de-excitation goes to final states with one additional photon of energy $\omega_k = E_l - E_m > 0$. Denoting the momentum and helicity of the additional photon by $\mathbf{k}$ and $\lambda$, we immediately pick the appropriate matrix element of (2) in Fock space, which we call $H^{\text{em}}$. According to (0.22), it is

$$\begin{aligned} H^{\text{em}}(t) &= (n_\lambda(\mathbf{k}) + 1)^{\frac{1}{2}} e^{i\omega_k t} \hat{H}^{\text{em}}, \\ \hat{H}^{\text{em}} &= e(2m)^{-1} e^{-i\mathbf{k}\mathbf{x}}[-2i\boldsymbol{\varepsilon}_\lambda^*(\mathbf{k}) \nabla - i\boldsymbol{\sigma}(\mathbf{k} \times \boldsymbol{\varepsilon}_\lambda^*(\mathbf{k}))]. \end{aligned} \tag{1.3}$$

Similarly, if $l$ is the ground state, we can effectively replace (2) by

$$\begin{aligned} H^{\text{abs}}(t) &= n_\lambda^{\frac{1}{2}}(\mathbf{k}) e^{-i\omega_k t} \hat{H}^{\text{abs}}, \\ \hat{H}^{\text{abs}} &= e(2m)^{-1} e^{i\mathbf{k}\mathbf{x}}[-2i\boldsymbol{\varepsilon}_\lambda(\mathbf{k}) \nabla - i\boldsymbol{\sigma}(\mathbf{k} \times \boldsymbol{\varepsilon}_\lambda(\mathbf{k}))]. \end{aligned} \tag{1.4}$$

In the operators $H^{\text{em}}$ and $H^{\text{abs}}$, the photon looks like an external plane wave. Let us now concentrate on de-excitation. The case $n_\lambda(\mathbf{k}) = 0$ is called "spontaneous emission", $n_\lambda(\mathbf{k}) > 0$ is called "induced emission." This latter type is particularly important in laser emission, where each excited atom contributes one photon to the avalanche. The time-dependent perturbation theory starts from the expansion of the electronic state $\psi_P$ in terms of the stationary unperturbed states

$$\begin{aligned} \psi(t, \mathbf{x}) &= \sum_n c_n(t) u_n(\mathbf{x}) e^{-iE_n{}^0 t}, \qquad c_n(0) = \delta_{nl}, \\ \partial_t c_m &= -i \sum_n \langle m | H_{\text{per}}(t, \mathbf{x}) | n \rangle e^{i(E_m{}^0 - E_n{}^0)t} c_n. \end{aligned} \tag{1.5}$$

Here $E_m^0$ and $E_n^0$ are the relevant eigenvalues of the unperturbed operator $H_0$. In the following, we merely write $E$ for $E^0$. We can now replace $H_{\text{per}}$ by (3). To start with, the photon momentum $\mathbf{k}$ is completely arbitrary, and our first aim is to show that the transition rate is zero unless energy is conserved. To lowest order in the perturbation, we put $c_n = \delta_{nl}$ at all times in (5):

$$\begin{aligned} c_m^{\text{em}(1)} &= -i \int_0^t dt' \langle m | H^{\text{em}}(t, \mathbf{x}) | l \rangle e^{i(E_m - E_l)t'} \\ &= \langle m | \hat{H}^{\text{em}}(\mathbf{x}) | l \rangle (E_m - E_l + \omega_k)^{-1}[1 - e^{i(E_m - E_l + \omega_k)t}] \end{aligned} \tag{1.6}$$

for $n_\lambda(\mathbf{k}) = 0$. The transition probability per unit time into the momentum interval $d^3k$ is, with $y = \frac{1}{2}(E_m - E_l + \omega_k)$,

$$d\Gamma_{lm} = t^{-1} |c_m^{\text{em}(1)}|^2 \, d_L^3 k = |\langle m | \hat{H}^{\text{em}} | l \rangle|^2 y^{-2} \sin^2(ty) \, d^3k (16\pi^3 \omega_k)^{-1}. \tag{1.7}$$

Since an accurate determination of the energy is only possible after a large time interval, we take $t \to \infty$ in (7):

$$\lim_{t\to\infty} (ty^2)^{-1} \sin^2(ty) = \pi \, \delta(y), \qquad \delta(x/a) = a \, \delta(x), \tag{1.8}$$

$$\begin{aligned} d\Gamma_{lm} &= 2\pi |\langle m | \hat{H}^{\text{em}} | l \rangle|^2 \, \delta(E_m - E_l + \omega_k) \, d^3k (16\pi^3 \omega)^{-1} \\ &= (8\pi^2)^{-1} |\langle m | \hat{H}^{\text{em}} | l \rangle|^2 k \, d\Omega_k, \qquad k = E_l - E_m. \end{aligned} \tag{1.9}$$

Equation (9) is known as Fermi's golden rule. Just to remind the reader, the dipole approximation puts $\exp(-i\mathbf{kx}) \approx 1$ in (3), and the electric approximation neglects the electron spin. Using moreover the fact that $H_0$ contains $\nabla$ only in the kinetic energy $-\nabla^2/2m$, we have

$$[\mathbf{r}, mH_0] = [\mathbf{r}, -\nabla^2/2] = \nabla, \tag{1.10}$$

$$\langle m | \hat{H}^{\text{em}}_{\text{el. dip.}} | l \rangle = -ie(E_l - E_m) \langle m | \mathbf{r} | l \rangle \boldsymbol{\varepsilon}_\lambda^*(\mathbf{k}) = -iek \langle m | \mathbf{r} | l \rangle \boldsymbol{\varepsilon}_\lambda^*(\mathbf{k}). \tag{1.11}$$

Summation over $\lambda$ and integration over $\Omega_k$ in (9) is done using

$$\sum_\lambda \int d\Omega_k \, \varepsilon_\lambda^{i*}(\mathbf{k}) \varepsilon_\lambda^j(\mathbf{k}) = \tfrac{2}{3} 4\pi \, \delta_{ij}, \tag{1.12}$$

$$\Gamma_{lm(\text{el. dip.})} = \tfrac{4}{3} \alpha k^3 |\langle m | \mathbf{r} | l \rangle|^2. \tag{1.13}$$

In this approximation, the integrand of the transition rate consists of a number of $\delta$-functions at energies $E_m - E_l$. In reality, however, the finite lifetime of each atomic level produces a natural line width. The width is obtained from second-order perturbation theory as follows

$$c_n(t) = \delta_{nl} + c_n^{(1)}(t) + c_n^{(2)}(t), \tag{1.14}$$

$$\begin{aligned} c_n^{(2)}(t) &= -i \sum_m \int_0^t dt' \langle n | H_{\text{per}}(t') | m \rangle e^{i(E_n{}^0 - E_m{}^0)t'} c_m^{(1)}(t') \\ &= -\sum_m \int_0^t dt' \int_0^{t'} dt'' \langle n | H_{\text{per}}(t') | m \rangle \\ &\quad \times \langle m | H_{\text{per}}(t'') | l \rangle \exp\{i(E_n^0 - E_m^0)t' + i(E_m^0 - E_l^0)t''\}. \end{aligned} \tag{1.15}$$

After a lengthy calculation (see e.g., Merzbacher 1970), one finds that the main effect is a weak exponential time-dependence of $c_l$,

$$c_l(t) = \exp(-i \, \Delta\tilde{E}_l t), \qquad \Delta\tilde{E}_l \equiv \Delta E_l - i\Gamma_l/2, \tag{1.16}$$

which can also be incorporated in the first-order calculation, simply by replacing $E_l^0$ in (5) by

$$\tilde{E}_l = E_l - i\Gamma_l/2 = E_l^0 + \Delta E_l - i\Gamma_l/2. \tag{1.17}$$

For the theory of the line width, second-order perturbation theory is actually unnecessary, because the imaginary part of (16) follows directly from probability conservation (2-3.2). With $\rho = |\psi_l|^2$ to first order in the perturbation, we have to second order

$$d|\psi_l|^2/dt = -\Gamma_l |\psi_l|^2, \qquad \Gamma_l \equiv \sum_m \Gamma_{lm}. \tag{1.18}$$

Evidently, a replacement analogous to (16) must also be made in the final state $m$. Instead of (6), the time integration now gives us for $t \to \infty$

$$c_m^{\mathrm{em}} = \langle m | \hat{H}^{\mathrm{em}} | l \rangle [\omega + E_m - E_l + i(\Gamma_l + \Gamma_m)/2]^{-1}. \tag{1.19}$$

Energy conservation is slightly violated at this point but is restored when the excitation of the state $|l\rangle$ (in a collision, say) and the decay of $|m\rangle$ are included in the calculation, as we shall see in section 4-8. The photon emission probability is now

$$\begin{aligned} dW_m(\gamma) &= |c_m^{\mathrm{em}}|^2 \, d_L^3 k \\ &= (2\pi)^{-1} \Gamma_{lm} \, d\omega [(\omega + E_m - E_l)^2 + \tfrac{1}{4}(\Gamma_l + \Gamma_m)^2]^{-1}. \end{aligned} \tag{1.20}$$

If $\Gamma_l$ and $\Gamma_m$ are sufficiently small, one can take $k = E_l - E_m \approx E_l^0 - E_m^0$ as in (9) in the evaluation of $\Gamma$. The resulting "Lorentz curve" is shown in fig. 3-1. For electric dipole transitions, one may use the approximations (11) and (13)

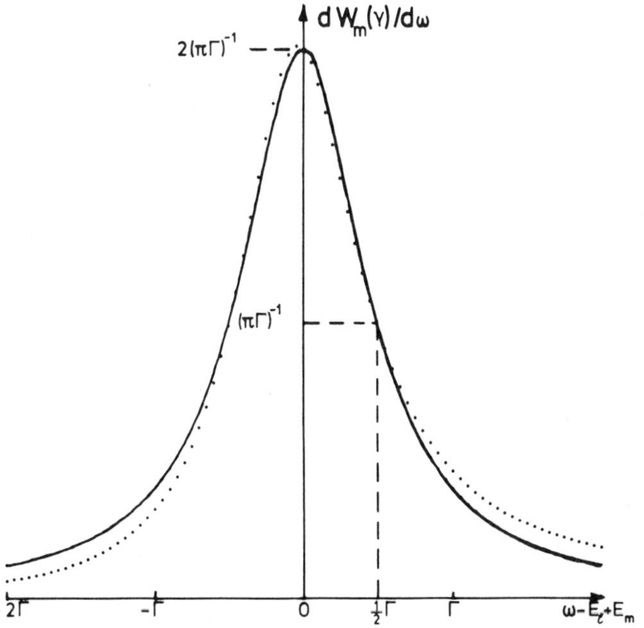


**Figure 3-1** The natural line width (20) for the case $\Gamma_m = 0$, $\Gamma_{lm} = \Gamma_l \equiv \Gamma$. Full curve: $\Gamma = \Gamma_R$ independent of $\omega$ (Lorentz curve). Dotted curve: $\Gamma = \omega^3 (E_l - E_m)^{-3} \Gamma_R$ ($p$-wave Breit-Wigner curve). Both curves assume $(E_l - E_m)/\Gamma_R = 10$. As this ratio is increased, the curves approach each other.

with $k$ equal to the integration variable of (20). This gives the "$p$-wave Breit-Wigner" resonance curve of fig. 3-1.

As a second application, we consider the scattering of photons $\gamma e \to \gamma' e'$ in the limit where the electron binding energy is negligible with respect to the photon energy $\omega_i$, but where the final electron is still nonrelativistic, $E_{e'} \approx m$. The second condition is fulfilled for $\omega_i \ll m$ ("Thomson scattering"). As the initial photon must be annihilated and at the same time the final photon must be created, the general treatment of $\gamma e \to \gamma' e'$ must include all terms of order $e^2 A^2$ (see for example Sakurai 1967, Landau and Lifshitz 1971, and section 3-7 below).

However, in the Thomson limit, (15) is negligible because it involves a product of gradients according to (2). $c_n^{(1)}$, on the other hand, is easily derived from the $\mathbf{A}_0 \cdot \mathbf{A}_0$-term in (2):

$$\begin{aligned}
&\langle \mathbf{k}_f \lambda_f, \mathbf{p}'M' | H_{\text{per}}(t, \mathbf{x}) | \mathbf{k}_i \lambda_i, \mathbf{0}, M \rangle \\
&\quad = 2m \int d^3x e^{-i\mathbf{p}'\mathbf{x}} \chi^+(M') \Phi_{\mathbf{k}_f\lambda_f} (2m)^{-1} e^2 \mathbf{A}_0(t, \mathbf{x}) \cdot \mathbf{A}_0(t, \mathbf{x}) \Phi_{\mathbf{k}_i\lambda_i} \chi(M) \\
&\quad = \int d^3x e^{-i\mathbf{p}'\mathbf{x}}\, \delta_{M'M}\, \boldsymbol{\varepsilon}^*(\mathbf{k}_f, \lambda_f) \cdot \boldsymbol{\varepsilon}(\mathbf{k}_i, \lambda_i) 2e^2 e^{i(\mathbf{k}_i - \mathbf{k}_f)\mathbf{x}} e^{i(\omega_f - \omega_i)t}.
\end{aligned} \tag{1.21}$$

Here the initial and final electron states are the plane waves $\chi(M) \exp(-imt + i\mathbf{0}\mathbf{x})$ and $\chi(M') \exp(-imt + i\mathbf{p}' \cdot \mathbf{x})$, respectively (a factor $2m$ appears because we shall use the relativistic phase space which requires the normalization (1-1.15) also in the nonrelativistic limit $E = m$), and the $\Phi$'s are the one-photon Fock states on which the $A$'s operate. To evaluate $\Phi_f^+ \mathbf{A}_0 \mathbf{A}_0 \Phi_i$, one must change the names of the integration and summation variables in the second $\mathbf{A}_0$, say to $\mathbf{k}'$ and $\lambda'$. There are then two identical contributions: one from $a_\lambda(\mathbf{k}) a_{\lambda'}^+(\mathbf{k}')$ at the point $\lambda = \lambda_i$, $\mathbf{k} = \mathbf{k}_i$, $\lambda' = \lambda_f$, $\mathbf{k}' = \mathbf{k}_f$, the other from $a_\lambda^+(\mathbf{k}) a_{\lambda'}(\mathbf{k}')$ at the point $\lambda = \lambda_f$, $\mathbf{k} = \mathbf{k}_f$, $\lambda' = \lambda_i$, $\mathbf{k}' = \mathbf{k}_i$. The integral over $d^3x$ produces a factor $(2\pi)^3 \delta_3(\mathbf{k}_i - \mathbf{k}_f - \mathbf{p}')$ which expresses momentum conservation. Contrary to (7), the initial condition in scattering is set at $t \to -\infty$ (see section 1-3), and the $S$-matrix element is

$$\begin{aligned}
S_{if}^{(1)} &= \lim_{t\to\infty} c_{if}^{(1)}(t) \\
&= -i(2\pi)^3 \delta_3(\mathbf{k}_i - \mathbf{k}_f - \mathbf{p}')\, \delta_{M'M}\, \boldsymbol{\varepsilon}^*(\mathbf{k}_f \lambda_f) \boldsymbol{\varepsilon}(\mathbf{k}_i \lambda_i) 2e^2 \int_{-\infty}^{\infty} dt' e^{i(\omega_f - \omega_i)t'} \\
&= -i(2\pi)^4 \delta_4(P_i - P_f)\, \delta_{M'M}\, \boldsymbol{\varepsilon}^*(\mathbf{k}_f, \lambda_f) \boldsymbol{\varepsilon}(\mathbf{k}_i \lambda_i) \cdot 2e^2,
\end{aligned} \tag{1.22}$$

which shows that $T_{if}$ of (2-1.4) is

$$T_{if} = -2e^2\, \delta_{M'M}\, \boldsymbol{\varepsilon}^*(\mathbf{k}_f \lambda_f) \boldsymbol{\varepsilon}(\mathbf{k}_i \lambda_i). \tag{1.23}$$

The differential cross section follows from (2-4.8), with $s = m^2$, $p = p' \to k$, and the definition (2-3.10) of $\overline{|T_{if}|^2}$:

$$d\sigma/d\Omega = (4m^2)^{-1}\overline{|T_{if}/4\pi|^2}$$
$$= r_e^2 \sum_{\lambda_i\lambda_i'} \sum_{\lambda_f} \rho_{\lambda_i\lambda_i'} \boldsymbol{\varepsilon}^*(\mathbf{k}_f \lambda_f) \cdot \boldsymbol{\varepsilon}(\mathbf{k}_i \lambda_i) \boldsymbol{\varepsilon}^*(\mathbf{k}_i' \lambda_i') \cdot \boldsymbol{\varepsilon}(\mathbf{k}_f, \lambda_f), \tag{1.24}$$

where $r_e$ is the "electron radius" (C-1.7), and $\rho$ is the density matrix for polarized photons. The electron density matrix is eliminated by the $\delta_{M'M}$ in (23). In the following, we first evaluate this expression for completely linearly polarized light, taking our $z$-axis along $\mathbf{k}_i$ and our $x$-axis along the polarization, such that $\boldsymbol{\varepsilon}(\mathbf{k}) = (0, 1, 0, 0)$ according to (2-8.24). The density matrix is then $\rho_{xx} = 1$, $\rho_{xy} = \rho_{yy} = 0$, and (24) becomes

$$d\sigma(\gamma_x e \to \gamma' e')/d\Omega = r_e^2 \sum_{\lambda_f} |\varepsilon_x(\mathbf{k}_f, \lambda_f)|^2. \tag{1.25}$$

$\varepsilon_x(\mathbf{k}_f, \lambda_f)$ is taken from (2-8.7), which gives

$$d\sigma(\gamma_x e \to \gamma' e')/d\Omega = r_e^2(\cos^2 \vartheta \cos^2 \phi + \sin^2 \phi). \tag{1.26}$$

For polarization along the $y$-axis, we must of course pick the $y$-component of (2-8.7):

$$d\sigma(\gamma_y e \to \gamma' e')/d\Omega = r_e^2(\cos^2 \vartheta \sin^2 \phi + \cos^2 \phi). \tag{1.27}$$

The differential cross section for unpolarized photons is the average of these two expressions and is independent of $\phi$ as expected:

$$d\sigma(\gamma e \to \gamma' e')/d\Omega = \tfrac{1}{2} r_e^2(\cos^2 \vartheta + 1), \qquad \sigma(\gamma e) = \tfrac{8}{3}\pi r_e^2. \tag{1.28}$$

For partly polarized light, (26) and (27) are added with different weights. Note that (21) is actually valid for photon scattering on particles of arbitrary masses, spins, and charges $Ze$, if the factor $e^2$ is replaced by $Z^2e^2$. Therefore if the electron is bound to a nucleus of mass $m_N$ and charge $Ze$, one must add a nuclear contribution $\frac{8}{3}\pi m_N^{-2} Z^4 \alpha^2$ to $\sigma$.

## 3-2 Particle Fields and Currents. Spin and Statistics

Second quantization can be applied to all kinds of particles: electrons, muons, pions, nucleons, and nuclei. Plane wave functions of free particles are interpreted in analogy with (0.22). The plane wave decomposition of the free pion field operator is

$$\psi_0(x) = \int d_L^3 p[a(\mathbf{p})e^{-iP \cdot x} + b^+(\mathbf{p})e^{iP \cdot x}], \tag{2.1}$$

$$\psi_0^+(x) = \int d_L^3 p[b(\mathbf{p})e^{-iP \cdot x} + a^+(\mathbf{p})e^{iP \cdot x}], \tag{2.2}$$

where $a$ and $a^+$ annihilate and create, say, free $\pi^+$ mesons. The terms proportional to $e^{iP\cdot x}$ are required by "causality," as we shall see below. The physics of such terms appears only after the inclusion of interactions. It will turn out that $b^+(\mathbf{p})$ either annihilates a $\pi^+$ of energy $-E_p$ or creates a $\pi^-$ of energy $+E_p$. Since free particles of negative total energies do not exist, the latter alternative is the correct one. It is then clear that $\psi_0^+$ describes the free $\pi^-$ field. For neutral mesons such as $\pi^0$ or $K^0$, one can have $a(\mathbf{p}) = b(\mathbf{p})$ (Hermitean field). This question is decided by the $CP$-properties of the interactions. It turns out that the $\pi^0$ field is Hermitean but that of $K^0$ is not.

The free field operators for electrons and positrons are expanded into plane waves as follows:

$$\begin{aligned}\Psi_0(x) &= \int d_L^3p \sum_M \{a_M(\mathbf{p})\psi(\mathbf{p}, M) + b_M^+(\mathbf{p})\psi_c(\mathbf{p}, M)\} \\ &= \int d_L^3p \sum_M \{a_M(\mathbf{p})u(\mathbf{p}, M)e^{-iP\cdot x} + b_M^+(\mathbf{p})v(\mathbf{p}, M)e^{iP\cdot x}\},\end{aligned} \tag{2.3}$$

$$\bar{\Psi}_0(x) = \int d_L^3p \sum_M \{b_M(\mathbf{p})\bar{v}_M(\mathbf{p})e^{-iP\cdot x} + a_M^+(\mathbf{p})\bar{u}(\mathbf{p}, M)e^{iP\cdot x}\}, \tag{2.4}$$

where $a$ and $a^+$ annihilate and create electrons and $b$ and $b^+$ annihilate and create positrons, respectively. The negative-energy spin function $v$ is defined in (1-6.26).

For electrons and also muons, the free field current operator is given by the expression corresponding to (1-5.24), with an extra minus sign to account for the negative charge of the electron:

$$j^\mu(x) = -\bar{\Psi}_0(x)\gamma^\mu\Psi_0(x). \tag{2.5}$$

(In principle, we should write $j_0^\mu$ to indicate that the current operator is constructed from free field operators.) $j$ is a Hermitean operator, with the following nonvanishing matrix elements between many-electron states:

$$\langle e_f | j^\mu | e_i \rangle = -\bar{\psi}_f\gamma^\mu\psi_i, \tag{2.6}$$

which is exactly what we found in (2-6.20). We now postulate that the current matrix element between the corresponding positron states must be opposite to that of (6):

$$\langle e_f^+ | j^\mu | e_i^+ \rangle = \bar{\psi}_{cf}\gamma^\mu\psi_{ci}. \tag{2.7}$$

This requires that the operators $a$, $b^+$, $b$ and $a^+$ satisfy anticommutation rules (eq. 0.16 with $\lambda = -1$ and $b_i$ treated like another operator $\neq a_i$). Thus electrons are fermions. This also explains why we got an inconsistent result in (1-6.25). The normalization $n^{\frac{1}{2}}$ is omitted in (6) because $n < 2$ for fermions. However, in electron scattering where (6) will be needed one must remember that $\langle e_f | j^\mu | e_i \rangle = 0$ for $n_f - 1 = 1$. Thus, certain matrix elements vanish in the presence of a second electron, even if the interaction between the electrons is omitted (Pauli exclusion principle).

The free pion current operator is given by the expression corresponding to (1-1.10) and (1-1.11). Using the bidirectional $\overleftrightarrow{\partial}^\mu$ of (2-6.22), we write

$$j^\mu = \psi_0^+(i\,\overleftrightarrow{\partial}^\mu - eA^\mu)\psi_0, \tag{2.8}$$

This operator, with $Z = +1$, gives the correct current matrix elements between $\pi^+$ states and also between $\pi^-$ states if and only if the operators $a, a^+$, $b, b^+$ satisfy commutation relations and not anticommutation relations. In this way one can conclude that all charged particles obey the spin-statistics theorem (see for example Streater and Wightman 1964):

$$a(\mathbf{p})a^+(\mathbf{p}') - \lambda a^+(\mathbf{p}')a(\mathbf{p}) = (2\pi)^3\,\delta_3(\mathbf{p}-\mathbf{p}')\cdot 2E_p, \qquad \lambda = (-1)^{2S}, \tag{2.9}$$

where $S$ is the spin of the particles described by the free field. For neutral particles, one can use their interactions with the charged ones to show that they also obey (9).

The matrix elements of (8) between single pion states are

$$\langle \pi_f^+ \,|\, j^\mu \,|\, \pi_i^+ \rangle = e^{iP_f\cdot x} i\,\overleftrightarrow{\partial}^\mu e^{-iP_i\cdot x} = e^{iP_f\cdot x}(P_i + P_f)^\mu e^{-iP_i\cdot x}, \tag{2.10}$$

in agreement with (2-6.26). However as explained in section 2-7, these matrix elements can be used in the Born approximation of electromagnetic scattering only for $P_i \approx P_f$. The current operator $j^\mu(\pi)$ is that of a hypothetical "point-like" pion which has no strong interactions. The quantitative use of the quantum field formalism is really restricted to electrons and muons.

The matrix elements of field operators between vacuum states ("vacuum expectation values") are obviously zero. The vacuum expectation values of the currents should also be zero, but with $b_M(\mathbf{p})b_M^+(\mathbf{p})\,|0\rangle = |0\rangle$ one finds a nonzero expectation value $\langle 0\,|\,j^\mu\,|\,0\rangle \neq 0$. However, this is easily cured by redefining $j^\mu$ as the commutator

$$j^\mu = -\tfrac{1}{2}[\overline{\Psi}_0, \gamma^\mu \Psi_0] = -\frac{1}{2}\sum_{\alpha\beta} \gamma^\mu_{\alpha\beta}(\overline{\Psi}_{0\alpha}\Psi_{0\beta} - \Psi_{0\beta}\overline{\Psi}_{0\alpha}), \tag{2.11}$$

which does not affect the one-particle matrix elements.

The vacuum expectation values of products of field operators are singular. From (1), (2), and (9) we find

$$\begin{aligned}\langle 0\,|\,\psi_0^+(x)\psi_0(y)\,|\,0\rangle &= \int d_L^3p'\, d_L^3p\, e^{-iP'\cdot x} e^{iP\cdot y} \langle 0\,|\,b(\mathbf{p}')b^+(\mathbf{p})\,|\,0\rangle \\ &= \int d_L^3p\, e^{-iP\cdot(x-y)} = -i\,\Delta_+(x-y),\end{aligned} \tag{2.12}$$

and from (3) and (4)

$$\langle 0\,|\,\Psi_0(x)\overline{\Psi}_0(y)\,|\,0\rangle = \int d_L^3p\, e^{-iP\cdot(x-y)} \sum_M u(\mathbf{p}, M)\bar{u}(\mathbf{p}, M) = iS_+(x-y). \tag{2.13}$$

These functions occurred already in (2-6.10) and (2-6.17). Similarly,

$$\langle 0|\psi_0(x)\psi_0^+(y)|0\rangle = \int d_L^3 p e^{iP(x-y)} = -i\,\Delta_-(x-y), \tag{2.14}$$

$$\langle 0|\bar{\Psi}_0(x)_\alpha \Psi_0(y)_\beta|0\rangle = \int d_L^3 p e^{-iP(x-y)}\left(\sum_M v\bar{v}\right)_{\alpha\beta} = -iS_-(y-x)_{\alpha\beta}. \tag{2.15}$$

We shall also need the commutator of a field with its conjugate at different space-time points. From (1), (2), and (9) we have

$$\begin{aligned}[\psi_0^+(x), \psi_0(y)] &= \int d_L^3 p d_L^3 p' \{[b(\mathbf{p}), b^+(\mathbf{p}')]e^{-iP\cdot x+iP'\cdot y} + [a^+(\mathbf{p}), a(\mathbf{p}')]e^{iP\cdot x-iP'\cdot y}\} \\ &= \int d_L^3 p[e^{-iP(x-y)} - e^{iP(x-y)}] \equiv -i\,\Delta(x-y).\end{aligned} \tag{2.16}$$

Comparison with (12) and (14) shows

$$\Delta = \Delta_+ - \Delta_-. \tag{2.17}$$

Note that (16) is a property of the operators, while (12) applies to vacuum expectation values. For fermion fields, it is the anticommutator which gives an operator-free function (a "*c*-number"):

$$\{\Psi_0(x), \Psi_0(y)\} \equiv iS(x-y) = iS_+(x-y) - iS_-(x-y). \tag{2.18}$$

In the case of the photon field, we have

$$a_\lambda(\mathbf{k})a_{\lambda'}^+(\mathbf{k}') - a_{\lambda'}^+(\mathbf{k}')a_\lambda(\mathbf{k}) = (2\pi)^3\,\delta_{\lambda\lambda'}\,\delta_3(\mathbf{k}-\mathbf{k}')\cdot 2\omega_k, \tag{2.19}$$

which gives

$$[A_0^\mu(x), A_0^\nu(y)] = -ig^{\mu\nu}D_+(x-y). \tag{2.20}$$

In (20) we have included a contribution from "longitudinal photons," applied the spin summation (2-8.9) and gauge invariance, or alternatively (2-8.22). Thus (20) forms the basis of the covariant extension of the photon field quantization.

The meaning of the operator $\psi_0^+(x) = \psi_0^+(x^0, \mathbf{x})$ is that it creates a free $\pi^+$ (or annihilates a free $\pi^-$) at time $x^0$ and position $\mathbf{x}$. Causality requires that no signal propagates faster than light, i.e., an extended physical system at another point $y$ cannot be influenced by the action of $\psi_0^+(x)$ if $x - y$ is spacelike, $(x-y)^2 < 0$. In perturbation theory, this is only possible if the commutator of $\psi_0^+$ and $\psi_0$ vanishes for spacelike separation

$$[\psi_0^+(x), \psi_0(y)] = 0 \qquad \text{for } (x-y)^2 < 0. \tag{2.21}$$

The condition is in fact satisfied by (16) (one can accomplish $x^0 = y^0$ by a Lorentz transformation and then check $\int d_L^3 p\,(\exp(i\mathbf{p}(\mathbf{x}-\mathbf{y})) - \exp(-i\mathbf{p}(\mathbf{x}-\mathbf{y}))) = 0$). On the other hand, if we had omitted the second half of the operator $\psi_0$ or taken different commutation relations for $b(p)$, we would not have obtained (21). We may therefore say that the second half of the field operator is required by causality.

## 3-3 Perturbation Theory with Field Operators

The quantum field theory of interacting particles is obtained by applying the old Klein-Gordon, Dirac (1-5.11), and Maxwell (1-0.26) equations to field operators instead of wave functions:

$$(i\,\partial^\mu - ZeA^\mu_{\text{op}})\gamma_\mu \Psi_{\text{op}} = m\Psi_{\text{op}}, \qquad \Box A^\mu_{\text{op}} = ej^\mu_{\text{op}},$$
$$j^\mu_{\text{op}} = -\tfrac{1}{2}[\bar{\Psi}_{\text{op}}, \gamma^\mu \Psi_{\text{op}}] + \cdots \tag{3.1}$$

The full current operator $j^\mu_{\text{op}}$ is obtained from the free field operator $j^\mu$ by replacing $\Psi_0 \to \Psi_{\text{op}}$. The dots in (1) indicate the contributions of the fields of other particles (see eq. 17 below). As far as electrons and photons are concerned, the equations can be obtained from a Lagrangian density $\mathscr{L}(A^\mu_{\text{op}}, \partial^\nu A^\mu_{\text{op}}, \Psi_{\text{op}}, \partial^\mu \Psi_{\text{op}}, \bar{\Psi}_{\text{op}}, \partial^\mu \bar{\Psi}_{\text{op}})$ by means of a variation principle. The derivative $\partial^\mu \Psi_i$ of a field $\Psi_i$ is varied independently of $\Psi_i$, and Euler's equations read

$$\delta\mathscr{L}/\delta\Psi_i = \partial^\mu(\delta\mathscr{L}/\delta(\partial^\mu \Psi_i)). \tag{3.2}$$

If we neglect the commutator nature of the current, we simply have

$$\mathscr{L} = -\tfrac{1}{4}F_{\mu\nu}F^{\mu\nu} + \bar{\Psi}_{\text{op}}[\gamma_\mu(\tfrac{1}{2}\overleftrightarrow{\partial}^\mu + eA^\mu_{\text{op}}) - m]\Psi_{\text{op}}. \tag{3.3}$$

Insertion of $\Psi_i = A^\mu_{\text{op}}$ gives Maxwell's equations, whereas $\Psi_i = \bar{\Psi}_{\text{op}}$ gives Dirac's equation.

Unfortunately, these equations cannot be solved except perturbatively. Despite some formal difficulties, this procedure gives excellent results in quantum electrodynamics.

As in section 2-6, the differential equations (1) are first rewritten as integral equations and then solved iteratively. There is however one difference: when the operators are fixed by free particle operators $\Psi_0$, $A^\mu_0$, $j_\mu$ at time $t = -\infty$, then the retarded propagators must be used:

$$\Psi_{\text{op}}(x) = \Psi_{\text{in}}(x) + Ze\int d^4y S_R(x-y)\gamma_\mu A^\mu_{\text{op}}(y)\Psi_{\text{op}}(y),$$
$$\Psi_{\text{in}} \equiv \Psi_0, \qquad S_R(x) = -(i\,\partial_\mu\gamma^\mu + m)\,\Delta_R(x), \qquad \Delta_R(x^0 < 0) = 0. \tag{3.4}$$

The use of $\Delta_R$ guarantees that only $y^0 < x^0$ contributes to the integral. Similarly, (2-7.1) is replaced by $(A_{\text{in}} \equiv A_0)$

$$A^\mu_{\text{op}}(x) = A^\mu_{\text{in}}(x) + e\int d^4y D_R(x-y) j^\mu_{\text{op}}(y), \qquad D_R = \Delta_R(m^2 = 0). \tag{3.5}$$

Equations (4) and (5) are sometimes called "Yang-Feldman equations." The Feynman propagators $\Delta_F$ will appear automatically in the construction of the $S$-matrix. Of course one can also express $\Psi_{op}$ and $A^\mu_{op}$ in terms of the free particle operators $\Psi_{out}(x)$, $A^\mu_{out}(x)$, to which $\Psi_{op}$ and $A^\mu_{op}$ reduce at time $t \to \infty$. In that case one needs the "advanced Green's functions" $\Delta_A$ and $S_A$:

$$\Psi_{op}(x) = \Psi_{out}(x) + Ze \int d^4y S_A(x-y)\gamma_\mu A^\mu_{op}(y)\Psi_{op}(y),$$
$$A^\mu_{op}(y) = A^\mu_{out}(y) + e \int d^4x D_A(y-x) j^\mu_{op}(x), \tag{3.6}$$

$$S_A = -(i\,\partial_\mu \gamma^\mu + m)\,\Delta_A, \qquad \Delta_A(x^0 > 0) = 0. \tag{3.7}$$

Obviously, the Fourier transform $\Phi_A(P)$ of $\Delta_A$ has both $P_0$-poles in the upper half plane:

$$\Phi_A = (m^2 - P^2 + i\varepsilon \text{ sign } P_0)^{-1},$$
$$S_A = -(2\pi)^{-4} \int d^4P(P\cdot\gamma + m)e^{-iP\cdot x}\Phi_A \tag{3.8}$$

Thus $\Psi_{op}$ is an "interpolating field" between $\Psi_{in}$ at $t = -\infty$ and $\Psi_{out}$ at $t = \infty$. But whereas $\Psi_{in}$ and $\Psi_{out}$ are linear in the free-particle creation and annihilation operators, $\Psi_{op}$ is not, and the formalism does contain matrix elements for the production of large numbers of particles.

The difference between $\Psi_{in}$ and $\Psi_{out}$ is that $\Psi_{in}$ creates and annihilates free particles in the initial state before the interaction is "adiabatically switched on," whereas $\Psi_{out}$ does the same operations in the final state after the interaction is "adiabatically switched off" (see Källén 1958). Otherwise both operators have the same commutation relations, so there must exist a unitary operator $S$ connecting them:[1]

$$\Psi_{out} = S^{-1}\Psi_{in}S, \qquad A_{out} = S^{-1}A_{in}S, \qquad S^+S = 1. \tag{3.9}$$

The adiabatic switching on and off of the interaction Hamiltonian can be formally achieved by taking the unit of electric charge $e$ as a function of time

$$e(t) = \lim\,(\alpha \to 0)\exp(-\alpha|t|)e(0), \tag{3.10}$$

We shall return to this problem in section 13.

The final state $|n_{out}\rangle$ into which an initial state $|n\rangle$ develops via the interaction is given by

$$|n_{out}\rangle = S^{-1}|n\rangle. \tag{3.11}$$

If $|n\rangle$ contains only one particle and no "external field," we have of course $S = 1$, similarly if we consider only particles that do not interact with each other. The probability amplitude to find a certain final state $|n'_{out}\rangle$ among the outcome of the interactions among particles in $|n\rangle$ is given by

$$\langle n'_{out}|n\rangle = \langle n'|(S^{-1})^+|n\rangle = \langle n'|S|n\rangle, \tag{3.12}$$

1. The Fock representation is unique up to unitary equivalence (Bogoliubov et al. 1975).

where the matrix elements of $S$ are taken between ingoing states alone. The operator $S$ can be found by applying (9) in the form $S\Psi_{\text{out}} = \Psi_{\text{in}} S$, $SA_{\text{out}} = A_{\text{in}} S$ to the $t \to \infty$ limits of (4) and (5):

$$\Psi_0(x)S - S\Psi_0(x) = ZeS \int d^4y S(x-y)\gamma_\mu A^\mu_{\text{op}}(y)\Psi_{\text{op}}(y), \qquad (3.13)$$

$$A^\mu_0(x)S - SA^\mu_0(x) = eS \int d^4y D(x-y) j^\mu_{\text{op}}(y), \qquad (3.14)$$

where the "retardation" index $R$ on $S(x-y)$ and $D(x-y)$ has become unnecessary, and the index "in" has been replaced again by the original index "0." Expanding now $S$ in powers of $e$,

$$S = 1 + eS^{(1)} + e^2 S^{(2)} + e^3 S^{(3)} + \ldots, \qquad (3.15)$$

we get $S^{(1)}$ by replacing $A^\mu_{\text{op}}$ $\Psi_{\text{op}}$ and $j^\mu_{\text{op}}$ on the right-hand sides of (13) and (14) by the incoming free fields and currents $A^\mu_0$, $\Psi_0$, and $j^\mu$. In view of (2.20), we see that (14) is satisfied by

$$S^{(1)} = i \int d^4y A^\mu_0(y) j_\mu(y). \qquad (3.16)$$

This expression satisfies also (13), when the component of $j_\mu$ which refers to electron and muon fields is given by (2.5) or better by (2.11). However, $j_\mu$ has also components in the fields of other particles and stable nuclei:

$$j_\mu = j_\mu(e) + j_\mu(\mu) + j_\mu(\text{nuclei}) + \ldots, \qquad (3.17)$$

where the operators $j_\mu$(nuclei) cannot be written down, due to the strong interactions of these particles. We only know that their matrix elements between single-particle states are of the general forms discussed in section 2-7. The matrix elements of products of such operators, on the other hand, are largely unknown.

Strictly speaking, the matrix elements of $S^{(1)}$ do not describe any physical process. However, they are approximations to processes which we have already studied:

i) The scattering of electrons and muons on nuclei is obtained if the nuclear current operator is removed from (17) and the free field operator $A^\mu_0$ in (16) is replaced by the corresponding classical potential $A^\mu_{\text{cl}}$ of (0.1). In that way one reproduces exactly the matrix elements (2-6.19) of the operator $S$.

ii) The emission and absorption of photons by bound electrons is obtained by replacing the free electron field operator $\Psi_0$ by the electron field operator in the presence of an external field $A^\mu_{\text{cl}}$ ("Furry picture"):

$$\Psi_F(x) = \int dE \sum_{j\pi\lambda} [a_{j\pi\lambda}(E)\psi_{j\pi\lambda}(E, \mathbf{r})e^{-iEt} + \overset{+}{b}_{j\pi\lambda}(E)\psi_{j\pi\lambda, c}(E, \mathbf{r})e^{iEt}], \qquad (3.18)$$

For spinless nuclei, one has $\mathbf{A}_{\text{cl}} = 0$, in which case the $\psi_{j\pi\lambda}$ are the Coulomb eigenstates (1-7.12) of energy $E$, angular momentum $j$, parity $\pi$, and helicity $\lambda$. The integration over $E$ includes now a summation over the bound states, and the bound state creation and annihilation operators produce matrix elements of the type (1.6).

To get the formal connection, we separate in (16) the space and time integrations:

$$eS^{(1)} = -i \int dt H_I(t), \qquad H_I(t) = \int d^3x H_I(x), \qquad H_I(x) = -eA_0^\mu(x) j_\mu(x), \tag{3.19}$$

$$H_I(x) = H(A_{\text{cl}}^\mu) + H_{\text{per}}, \qquad H(A_{\text{cl}}^\mu) = eA_{\text{cl}}^\mu j_\mu, \tag{3.20}$$

$$H_{\text{per}} = e\mathbf{A}_0(x)\mathbf{j}(x). \tag{3.21}$$

$H_I(x)$ is the interaction Hamiltonian density. The "external potential" part $H(A_{\text{cl}}^\mu)$ is now omitted because $\Psi_F$ satisfies the Dirac equation in the field $A_{\text{cl}}^\mu$ instead of the free particle Dirac equation (the Green's function $S_R$ in (2) must be modified accordingly). The remaining "perturbative part" $H_{\text{per}}$ is the relativistic, "full-operator" version of (1.2), with $\mathbf{j}(x)$ given by (2.5). It is more satisfactory than (1.2) in the sense that both electron and photon states are handled by creation and annihilation operators. However, it is still unsatisfactory in the sense that the nucleus is only represented by $A_{\text{cl}}^\mu$. For example, momentum is normally not conserved in this approximation. (If $A_{\text{cl}}^\mu$ is treated in the Born approximation, one can use the method of section 2-7 to establish momentum conservation, see section 3-5 below.)

We end this section with a few trivial but important comments on the derivation of electromagnetic properties of atoms and nuclei from the operator $H_I$ of (19). We first assume that the target is represented by one product of single-particle states of $N$ identical particles (eigenstates of an effective central Coulomb or shell-model potential):

$$|e\rangle = a_{nN}^+ \cdots a_{n_2}^+ a_{n_1}^+ |0\rangle, \tag{3.22}$$

where $|0\rangle$ is the vacuum state and $a_{nk}^+$ creates a particle in a state of quantum numbers $n_k$. $|e\rangle$ is totally antisymmetric. According to (2.6), the matrix element of $H_I(x)$ in Fock space is

$$\langle e_f | H_I(x) | e_i \rangle = eA_0^\mu(x_1)\bar{\psi}_f(x_1)\gamma_\mu \psi_i(x_1), \qquad f \neq i, \tag{3.23}$$

where $x_1$ is the coordinate of one of the $N$ particles, no matter which. Obviously, the antisymmetrization has no influence. Similarly, for $i = f$, we find

$$\langle e_i | H_I(x) | e_i \rangle = eA_0^\mu(x_1) \sum_{k=1}^{N} \bar{\psi}_{n_k}(x_1)\gamma_\mu \psi_{n_k}(x_1). \tag{3.24}$$

On the other hand, if one avoids second quantization by representing initial and final states by normalized determinants of wave functions, then (23) and (24) both amount to

$$\langle e_f | H_I(x) | e_i \rangle = (N!)^{-1} \operatorname{Det}_f (\bar{\psi}_{n_k}(x_k)) \left[ \sum_{k=1}^{N} eA_0^\mu(x_k, t)\gamma_\mu^{(k)} \right] \operatorname{Det}_i (\psi_{n_k}(x_k)). \tag{3.25}$$

This form exhibits the additivity of the contributions from the constituents, and is used also in the presence of configuration mixing (linear combinations of determinants). However, since such states represent only an ansatz for the true system of interacting particles, nothing has been proven. On the contrary, additivity is destroyed by relativistic corrections, as we have seen in (2-10.27). Only the nonrelativistic limit permits additivity. For a two-particle system, the wave function then separates as usual

$$\psi(\mathbf{r}_a, \mathbf{r}_b) = \psi(\mathbf{r})\psi(\mathbf{R}), \qquad \mathbf{r} = \mathbf{r}_a - \mathbf{r}_b, \tag{3.26}$$

and for wavelengths of $\mathbf{A}$ much longer then the atomic dimensions, that part of $H_{NR}$ which operates on $\mathbf{r}$ is, with $\mu^{-1} = m_a^{-1} + m_b^{-1}$ ($\mu$ = reduced mass)

$$H_{\text{per}}(\mathbf{r}, \nabla) = -ie\mathbf{A}_0(\mu^{-1} + (Z_b - 1)m_b^{-1})\,\nabla. \tag{3.27}$$

One must also pay attention to the nonrelativistic limit $\langle f|i\rangle = 2m_i\,\delta_{if}$ of our covariant normalization of states. If one employs bound states which are normalized to unity in (25), one must use the normalization $\bar{u}u = 1$ for free electrons and add a factor $2m_i$ afterwards, $m_i$ being the total mass of the bound system.

## 3-4 Second-Order Perturbation Theory and Electron-Electron Scattering

The expression for $S^{(2)}$ in the expansion (3.15) of the $S$-matrix is most conveniently (although perhaps not most convincingly) obtained by interpreting $S$ as the limit of a unitary time-shifting operator:

$$S = \lim_{t\to\infty} U(t), \qquad U(-\infty) = 1. \tag{4.1}$$

It is then clear from (3.19) that $U$ must fulfil the equation

$$U(t) = 1 - i\int_{-\infty}^{t} H_I(t')U(t'). \tag{4.2}$$

To second order in the interaction Hamiltonian $H_I$, we find

$$S^{(2)} = -\int_{-\infty}^{\infty} dt H_I(t)\int_{-\infty}^{t} dt' H_I(t'). \tag{4.3}$$

This is analogous to the second-order approximation to the coefficients $c_n(t)$ in the time-dependent perturbation theory (1.15). One can also derive (3) directly from (3.13) and (3.14), but that takes more time. Next, we insert $H_I$ from (3.19) and extend the integration formally over all space and time:

$$S^{(2)} = -\frac{1}{2}\int d^4x d^4y T A_0^{\mu}(x) j_\mu(x) A_0^{\nu}(y) j_\nu(y), \tag{4.4}$$

where the "time-ordering operator" $T$ exchanges the $x$-dependent operators with the $y$-dependent ones for $x^0 < y^0$. With the help of $T$, we can in fact write down $S$ in closed form

$$S = T \exp\left\{-i \int_{-\infty}^{\infty} H_I(t)\, dt\right\}. \tag{4.5}$$

We can now "derive" the results of section 2-7 by taking the matrix elements of (4) between two-particle states which contain no photon. As far as the photon Fock space is concerned, we have the vacuum expectation value of the time-ordered product, which is evaluated with the help of (2.20):

$$\begin{aligned} \langle 0 | T A_0^\mu(x) A_0^\nu(y) | 0 \rangle &= -ig^{\mu\nu}[\theta(x^0 - y^0) D_+(x-y) + \theta(y^0 - x^0) D_-(x-y)] \\ &= -ig^{\mu\nu} D_F(x-y). \end{aligned} \tag{4.6}$$

This derivation sheds new light on the Feynman propagator $\Delta_F$ of (2-6.15). The remaining factors of the matrix elements $S_{if}$ of (2-7.5) are now obvious, except perhaps for the missing factor $\frac{1}{2}$.

Let us show how this factor disappears, say, for elastic $a - b$ scattering. The currents which appear in (4) are both given by the sum (3.17), but obviously only the combinations $j_\mu(a, x) j_\nu(b, y)$ and $j_\mu(b, x) j_\nu(a, y)$ contribute to the matrix elements. Thus we have

$$\begin{aligned} S_{if} &- \langle f | i \rangle \\ &= \tfrac{1}{2} i \int d^4x d^4y D_F(x-y) \langle a'b' | j^\mu(a, x) j_\mu(b, y) + j^\mu(b, x) j_\mu(a, y) | ab \rangle \\ &= ie^2 \int d^4x d^4y D_F(x-y) \langle a' | j^\mu(a, x) | a \rangle \langle b' | j_\mu(b, y) | b \rangle. \end{aligned} \tag{4.7}$$

The situation is different for the scattering of identical particles. For electron-electron scattering, we have

$$S_{if} = \langle f | i \rangle + \tfrac{1}{2} ie^2 \int d^4x d^4y D_F(x-y) \langle e_c e_d | j^\mu(x) j_\mu(y) | e_a e_b \rangle, \tag{4.8}$$

with $j^\mu = -\frac{1}{2}[\bar{\Psi}_0, \gamma^\mu \Psi_0]$ = free electron current operator. In this case the electron in state $|a\rangle \equiv |\mathbf{p}_a, M_a\rangle$ can be annihilated either by $\Psi_0(y)$ or by $\Psi_0(x)$. The electron in state $|b\rangle = |\mathbf{p}_b, M_b\rangle$ is then annihilated by the remaining $\Psi_0$. As the $\Psi$'s anticommute, one gets a minus sign between these two contributions. Similarly, two possibilities occur in the construction of the final state from $\bar{\Psi}_0(x)$ and $\bar{\Psi}_0(y)$. Two of the 4 possibilities are again identical after a renaming of the integration variables, and the $T$-matrix element becomes

$$\begin{aligned} T_{if} &= e^2[t^{-1} \bar{u}_c \gamma^\mu u_a \bar{u}_d \gamma_\mu u_b - u^{-1} \bar{u}_d \gamma^\mu u_a \bar{u}_c \gamma_\mu u_b] = T_a + T_b, \\ t &= (P_a - P_c)^2, \qquad u = (P_a - P_d)^2. \end{aligned} \tag{4.9}$$

There are thus two Born terms in this case (fig. 3-4). It is easy to check that $T_{if}$ changes sign not only under the exchange of $a$ and $b$, but also under that

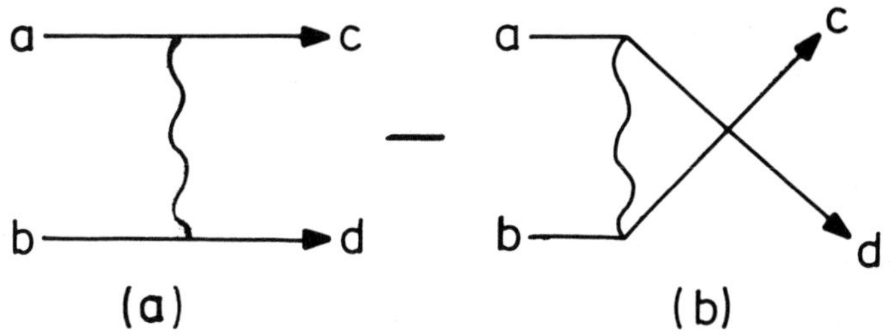


**Figure 3-4** The two graphs of electron-electron scattering.

of $c$ and $d$ as expected. For $t \ll u$, the first term dominates, and one has the normal Coulomb amplitude of distinguishable particles.

The cross section is again given by (2-4.9),

$$d\sigma/dt = (4\lambda)^{-1}\pi \sum_{M_a \cdots M_d} |T_a + T_b|^2, \tag{4.10}$$

but the integration over the phase space ends at $\theta = \frac{1}{2}\pi$ ($t = -2p^2$), as a state with $\theta > \frac{1}{2}\pi$ is identical to the state at $\pi - \theta$.

The $S$-operator (4) has nonzero matrix elements also for processes such as Compton scattering which involve two real photons, or the second-order Born approximation for the scattering by an external field. With the exception of the external field problem, which is not automatically invariant against translations, one can quite generally perform the integration over one of the two 4-coordinates $x^\mu$, $y^\nu$ and obtain an energy-momentum $\delta$-function in the matrix elements. As in the nonrelativistic case, one needs a formal operator $P^\mu_{\text{op}}$ which generates displacements in space and time of all field operators $O(y)$:

$$O(y) = e^{iP_{\text{op}}\cdot y}O(0)e^{-iP_{\text{op}}\cdot y}, \qquad e^{-iP_{\text{op}}\cdot y}|i\rangle = e^{-iP_i\cdot y}|i\rangle \tag{4.11}$$

where $P_i$ is the total 4-momentum of state $|i\rangle$. ($P^0_{\text{op}}$ is just the Hamilton operator.) Then, for any two operators $O_1(x)$ and $O_2(y)$, we have

$$\begin{aligned}\langle f|O_1(x)O_2(y)|i\rangle &= \langle f|O_1(x)e^{iP_{\text{op}}\cdot y}O_2(0)|i\rangle e^{-iP_i y} \\ &= \langle f|e^{iP_{\text{op}}\cdot y}e^{-iP_{\text{op}}\cdot y}O_1(x)e^{iP_{\text{op}}\cdot y}O_2(0)|i\rangle e^{-iP_i y} \\ &= e^{i(P_f-P_i)\cdot y}\langle f|O_1(x-y)O_2(0)|i\rangle\end{aligned} \tag{4.12}$$

Performing now the $y$-integration, we see that $S^{(2)}_{if}$ is of the required form (2-1.4). Introducing $x' = x - y$, we find

$$T^{(2)}_{if} = \tfrac{1}{2}i\int d^4x'\langle f|TA^\mu_0(x')j_\mu(x')A^\nu_0(0)j_\nu(0)|i\rangle \tag{4.13}$$

The generalization of this trick to triple operators $O_1(x)O_2(y)O_3(z)$ is obvious.

## 3-5 Matrix Elements for Bremsstrahlung and Photoabsorption. Feynman Rules

In section 3-1 we discussed the emission and absorption of photons by bound electrons. We now treat the corresponding processes for continuum states in the Furry picture (second quantization for electrons but not for nuclei), but this time the electron-nucleus interaction is treated in Born approximation. Therefore, instead of (3.18), we use

$$\Psi_A = \int d_L^3 p \sum_M [a_{AM}(\mathbf{p})\psi_A(\mathbf{p}, M) + b_{AM}^+(\mathbf{p})\psi_{Ac}(\mathbf{p}, M)],$$
$$\psi_A(\mathbf{p}, M) = \psi_0(\mathbf{p}, M) + \psi_{1A}(\mathbf{p}, M), \tag{5.1}$$

where $\psi_{1A}$ is given by (2-6.3)

$$\psi_{1A}(x) = -e\int d^4y S_F(x-y)\gamma^\mu\psi_0(y)A_\mu(y)$$
$$= e\int d^4y \int \frac{d^4P}{(2\pi)^4} e^{-iP(x-y)} \frac{P\cdot\gamma + m}{m^2 - P^2}\gamma_\mu\psi_0(y)t^{-1}e^{-i(P_b-P_{b'})y}J^\mu_{bb'}. \tag{5.2}$$

Here we have inserted $S_F$ from (2-6.16) and $A_\mu(y)$ from (2-7.7). If now the current operator is expressed in terms of $\Psi_A$ and $\bar{\Psi}_A$ with $\bar{\Psi}_A$ constructed correspondingly from $\bar{\Psi}_0$ and

$$\bar{\psi}_{1A}(x) = e\int d^4y \int \frac{d^4P}{(2\pi)^4} e^{iP(x-y)}\bar{\psi}_0(y)\gamma_\mu \frac{P\cdot\gamma + m}{m^2 - P^2} t^{-1}e^{-i(P_b-P_{b'})y}J^\mu_{bb'}, \tag{5.3}$$

then we can again start from the lowest order $S$-operator (3.16)

$$eS^{(1)} = i\int d^4x A_0^\mu(x)\bar{\Psi}_A(x)\gamma_\mu\Psi_A(x). \tag{5.4}$$

As in the case of bound electrons, this treatment neglects those amplitudes where the photon is emitted or absorbed by the nucleus. Such "recoil corrections" are added below.

The matrix elements of (4) between an incoming electron and an outgoing electron plus photon contain 4 pieces: one piece contains neither $\psi_{1A}$ nor $\bar{\psi}_{1A}$ and vanishes because of energy-momentum conservation ("a free electron does not radiate"). One piece is bilinear in $\psi_{1A}$ and $\bar{\psi}_{1A}$; it is of the order $e^4$ and must be dropped in the Born approximation. There remain two pieces of the form $\bar{\psi}_0\gamma_\mu\psi_{1A} + \bar{\psi}_{1A}\gamma_\mu\psi_0$, which give rise to the following expression:

$$\langle\mathbf{k}\lambda, \mathbf{p}'_a M'_a|eS^{(1)}|\mathbf{p}_a M_a\rangle = -ie^3\varepsilon_\lambda^{\nu *}(\mathbf{k})\int d^4x e^{iK\cdot x} j_{A\nu}(x), \tag{5.5}$$

$$j_{A\nu}(x) = \int d^4y (2\pi)^{-4}\, d^4P t^{-1}e^{-i(P_b-P_{b'})y}J^\mu_{bb'}\,\bar{u}'_a O_{\nu\mu} u_a, \tag{5.6}$$

$$O_{\nu\mu} = e^{iP_a'x}\gamma_\nu e^{-iP(x-y)} \frac{P \cdot \gamma + m}{m^2 - P^2} \gamma_\mu e^{-iP_a y} + e^{iP_a'y} e^{iP(x-y)} \gamma_\mu \frac{P \cdot \gamma + m}{m^2 - P^2} \gamma_\nu e^{-iP_a x}, \tag{5.7}$$

with $u_a = u(\mathbf{p}_a, M_a)$ etc.

The integration over $x$ and $y$ produces

$$(2\pi)^8\, \delta_4(K + P_a' - P)\, \delta_4(P_a - P + P_b - P_b')$$

in the first term of $O_{\nu\mu}$ and $(2\pi)^8\, \delta_4(K + P - P_a)\, \delta_4(P + P_b - P_b' - P_a')$ in the second term. Thus, we obtain the $S$-matrix element in the form (2-1.4):

$$\begin{aligned} S_{if} &= i(2\pi)^4\, \delta_4(P_a + P_b - P_a' - P_b' - K) T(ab \to \gamma a'b'), \\ T &= -e^3 \varepsilon_\lambda^{\nu *}(\mathbf{k}) \bar{u}_a' t^{-1} J_{bb'}^\mu \\ &\quad \times \left[ \gamma_\nu \frac{(P_a' + K) \cdot \gamma + m}{m^2 - (P_a' + K)^2} \gamma_\mu + \gamma_\mu \frac{(P_a - K) \cdot \gamma + m}{m^2 - (P_a - K)^2} \gamma_\nu \right] u_a . \end{aligned} \tag{5.8}$$

The $S$-matrix element of photoabsorption $\gamma ab \to a'b'$ (such processes can occur in a plasma) has the same structure. From $\mathbf{A}_{\text{op}}$ of (0.2) we see that we merely have to replace $K_\mu$ by $-K_\mu$ and $\boldsymbol{\varepsilon}_\lambda^*(\mathbf{k})$ by $\varepsilon_\lambda(\mathbf{k})$:

$$\begin{aligned} S_{if} &= i(2\pi)^4\, \delta_4(K + P_a + P_b - P_a' - P_b') T(\gamma ab \to a'b'), \\ T &= -e^3 \varepsilon_\lambda^\nu(\mathbf{k}) t^{-1} J_{bb'}^\mu \\ &\quad \times \bar{u}_a' \left[ \gamma_\nu \frac{(P_a' - K) \cdot \gamma + m}{m^2 - (P_a' - K)^2} \gamma_\mu + \gamma_\mu \frac{(P_a + K) \cdot \gamma + m}{m^2 - (P_a + K)^2} \gamma_\nu \right] u_a . \end{aligned} \tag{5.9}$$

Obviously, the same factors occur again and again in matrix elements. It is therefore useful to collect them in the form of rules for the construction of the $T$-matrix elements (Feynman rules), which in turn are expressed in terms of graphs (Feynman graphs). In lowest nonvanishing order perturbation theory (Born approximation), the rules go as follows:

1. Draw lines from left to right representing the in- and outgoing massive particles.
2. Mark as many points (= "vertices") on these lines as you need powers of $e$ to get a nonzero matrix element.
3. Draw from each point a wavy line, either to the left (for an incoming photon) or to the right (for an outgoing photon) or to another vertex (for a virtual photon).
4. Associate with each incoming (= left) fermion segment a factor $u_i = u(\mathbf{p}_i, M_i)$, with each outgoing (= right) fermion segment a factor $\bar{u}_f = \bar{u}(\mathbf{p}_f, M_f)$, with each incoming photon a factor $\varepsilon_\lambda(\mathbf{k})$, and with each outgoing photon a factor $\varepsilon_\lambda^*(\mathbf{k})$.
5. Associate with each electron vertex a factor $-e\gamma$, to be contracted with the $\varepsilon$ or $\varepsilon^*$ of the photon line ending at the vertex, or with the $-e\gamma$ from the other end of a virtual photon line. For a nucleon, $-\gamma^\mu$ is replaced by $0^\mu$ of (2-7.11) or (2-7.13), and for a spinless particle, it is replaced by $J_{bb'}^\mu$ of (2-7.9). But read also the derivation of (7.24) below!

6. Associate with each virtual photon line a factor $K^{-2}$ and with each inner fermion segment a factor $(P \cdot \gamma + m)(m^2 - P^2)^{-1}$, where the values of $K$ and $P$ follow from energy-momentum conservation at each vertex.
7. Multiply all factors together, starting with the incoming particles from the right (remember that matrix multiplication proceeds from right to left!).
8. Sum the contributions from the various graphs, obtained by permuting the photon lines at the vertices or by assigning a different number of vertices to one fermion line (still keeping the total number of vertices at its minimum).
9. In the case of identical particles, permute their quantum numbers according to their statistics either in the initial state or in the final state, but not in both.

An example for the last rule is given by the matrix element (4.9) for electron-electron scattering, the Feynman graphs of which are shown in fig. 3-4. For particle-nucleus scattering, there is only one graph of the type shown in fig. 2-7.2.

Bremsstrahlung and photoabsorption have 4 graphs each. The two graphs corresponding to (8) are shown in fig. 3-5.1, those corresponding to (9) are shown in fig. 3-5.2. For bremsstrahlung, the remaining two graphs are shown in fig. 3-5.3. They obviously arise when the nucleon current operator is added to the electron current operator in (4). The Feynman rules tell us their matrix elements as follows:

$$T_{\text{recoil}} = e^3 \varepsilon_\lambda^{\nu *}(\mathbf{k}) \bar{u}'_a \gamma^\mu u_a (P_a - P'_a)^{-2}$$
$$\times \bar{u}'_b \left[ O_\mu \frac{(P'_b + K) \cdot \gamma + m_b}{m_b^2 - (P'_b + K)^2} O_\nu + O_\nu \frac{(P_b - K) \cdot \gamma + m_b}{m_b^2 - (P_b - K)^2} O_\mu \right] u_b . \quad (5.10)$$

However, these formulas can hardly be exact, as the nucleon current operators are unknown. In (10) we have approximated the one-nucleon matrix elements of the operator product $J_\mu(x) J_\nu(y)$ by the product of the one-nucleon matrix elements of $J$. The expected deviations go under the name

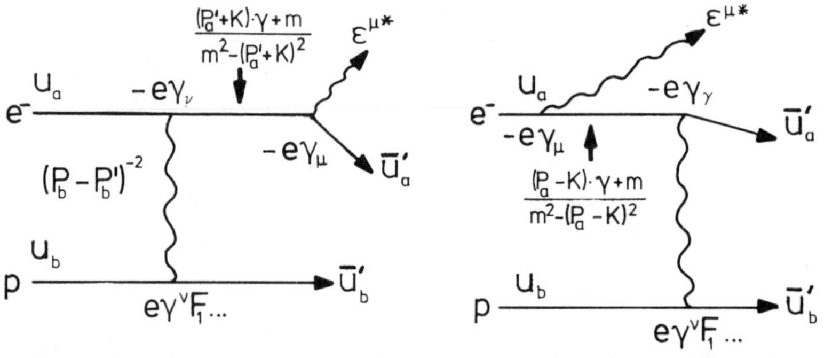


**Figure 3-5.1** The two dominant graphs for electron bremsstrahlung on protons.

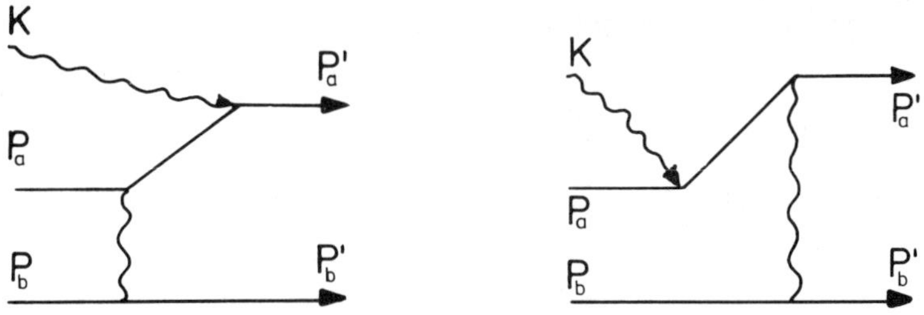


**Figure 3-5.2** The two dominant graphs for photoabsorption.

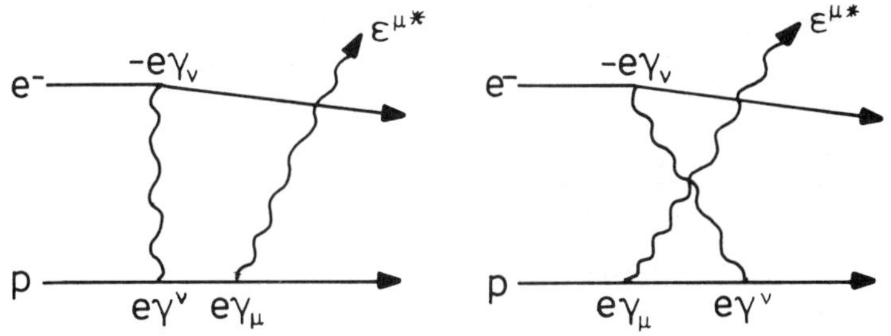


**Figure 3-5.3** The recoil graphs of bremsstrahlung.

"nuclear polarizability." Fortunately, the fact that nucleons are so heavy allows us to approximate, according to (2-7.13)

$$0_\mu \approx (2m_b)^{-1} Z_b (P_b + P_b')_\mu \approx Z_b \, \delta_{\mu 0} , \tag{5.11}$$

i.e., the nucleon spin can be neglected both in (10) and in (8). The largest error anyway is due to the use of Born distorted operators instead of proper Furry operators. We shall return to this point in section 8-3.

Suppose now we have calculated the covariant Born approximation for a certain process. In the next order perturbation approximation for the same process, two different types of diagrams occur. In the first type, one marks two more points on the through-going lines and connects them by a photon line (see fig. 3-5.4). In the second type, one inserts an "electron-positron-loop" as explained in section 3-10 below (for photon-photon scattering, an electron-positron loop which connects the two initial photons with the two final photons gives in fact the lowest nonvanishing contribution, i.e., the process has no Born approximation). In both cases, 4-momentum conservation at each vertex of the loop determines the loop momenta only if one loop momentum is given. This is treated according to the rule:

10. Integrate over that undetermined loop momentum and multiply by $-i(2\pi)^{-4}$. Keep the $-i\varepsilon$ of (2-6.7) in the denominators of the loop propagators.

The $i\varepsilon$ is of course common to all Feynman propagators, but in the "nonloop" propagators it is superfluous. Unfortunately, the integral is normally divergent. Convergent results for the last two graphs of fig. 3-5.4

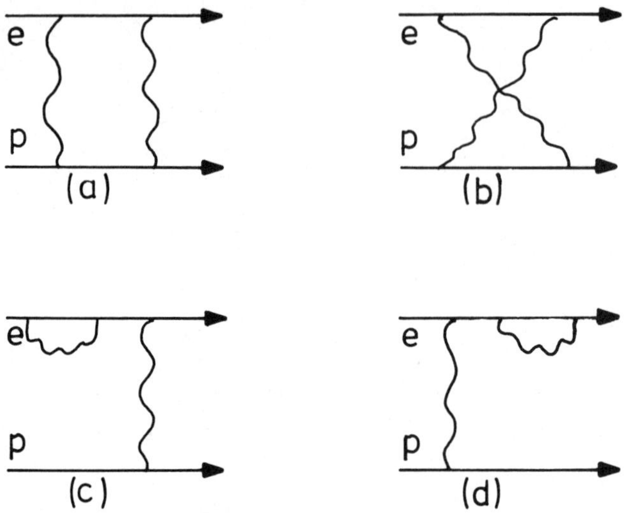


**Figure 3-5.4** Some 4th order graphs for $e^- p$ scattering.

are obtained by the addition of counterterms in the interaction Hamiltonian, resulting in a "renormalization" of charges and masses. Thus, although the tenth rule sounds simple, its practical application is not straightforward. We therefore prefer to develop higher-order perturbation theory from a different starting point, to be explained in section 3-11.[1]* The "self-energy" diagrams (c) and (d) of fig. 3-5.4 do not occur in that approach, and the problem of diverging integrals is clearer. However, the main merit of the new approach is its generalization to strong interactions. The connection between the two methods will be given in section 6-1.

## 3-6 Pair Creation. Substitution Rule

The negative-frequency part $b^+ v \exp(iP \cdot x)$ of the free electron field (2.3) gives an additional matrix element when inserted into (5.4). It is obtained from (5.9) by replacing $u_a$ by $v_a$ and $P_a$ by $-P_a$:

$$S^{(-)}(\gamma ab \to a'b') = i(2\pi)^4\, \delta_4(K + P_b - P_a - P'_a - P'_b)T^{(-)}(\gamma ab \to a'b'), \tag{6.1}$$

$$T^{(-)} = -e^3 t^{-1} J^{\mu}_{bb'} \varepsilon^{\nu} \times \bar{u}'_a \left[ \gamma_\nu \frac{(P'_a - K)\cdot\gamma + m}{m^2 - (P'_a - K)^2} \gamma_\mu + \gamma_\mu \frac{(-P_a + K)\gamma + m}{m^2 - (P_a - K)^2} \gamma_\nu \right] v_a . \tag{6.2}$$

1. An exception will be made in the summation of low-energy Coulomb scattering graphs by means of the Bethe-Salpeter equation, section 3-14.

Obviously it makes little sense to interpret this part as the photo-absorption on an electron of energy $-E_a < -m$. On the contrary, the $\delta_4$-function in (1) shows that the only possible physical interpretation is that of an *outgoing* particle of 4-momentum $+P_a$. However, this particle cannot be an electron, because the reaction $\gamma p \to eep$ would not conserve electric charge. We are forced to conclude that if (2) has a physical interpretation at all, it must refer to the production of a particle of 4-momentum $+P_a$ and charge $+e$, with the same mass and spin as an electron. Such particles, discovered by Anderson (1933), are called positrons, $e^+$ or $\bar{e}$. The reaction which is described by (2) is thus $\gamma p \to e^+ e^- p$.

We must also check the meaning of the negative energy spinor $v_a = v(\mathbf{p}_a, M_a) = iu(-P_a^\mu, -M_a)$. The reversal of the 4-momentum is now clear, but what about the reversal of $M_a$? When $-\mathbf{p}_a$ and $\mathbf{p}_a'$ are parallel, those parts of (2) which only contain $\gamma_0$ and/or $\gamma_3$ produce a factor $\delta_{M_a', -M_a}$, which must be interpreted as conservation of the angular momentum along $\mathbf{p}_a$, $0 = M_a + M_{\bar{e}}$. This shows that $M_{\bar{e}} = M_a$, i.e., the spinor $v(\mathbf{p}_a, M_a)$ does represent an outgoing positron of momentum $+\mathbf{p}_a$ and magnetic quantum number $+M_a$. We therefore arrive at the

**Feynman rule for positrons:** An outgoing positron of 4-momentum $P_\mu$ and magnetic quantum number $M$ contributes a factor $iu(-P^\mu, -M) = v(\mathbf{p}, M)$ to the matrix element. Correspondingly, an incoming positron of the same quantum numbers contributes a factor $-i\bar{u}\,(-P^\mu, -M)$. The momenta of inner lines follow again from energy-momentum conservation at each vertex. In the Feynman graphs, positrons are frequently represented by arrows pointing in the "wrong" time direction: out-going positrons point backwards in time (fig. 3-6.1). However, in later graphs (for example 3-6.3 and 3-6.4) we prefer to treat particles and antiparticles symmetrically.

In a propagator $(Q \cdot \gamma + m)(m^2 - Q^2)^{-1}$, we can interpret $Q_\mu$ either as the 4-momentum of a virtual electron or as the negative 4-momentum of a virtual positron. No arrows are therefore necessary on propagators (see fig. 3-6.2). In fact, returning to (2-6.8), we see that the main difference between the relativistic and nonrelativistic propagators lies in the negative energy pole, which gives rise to positrons. It is true that nonrelativistic

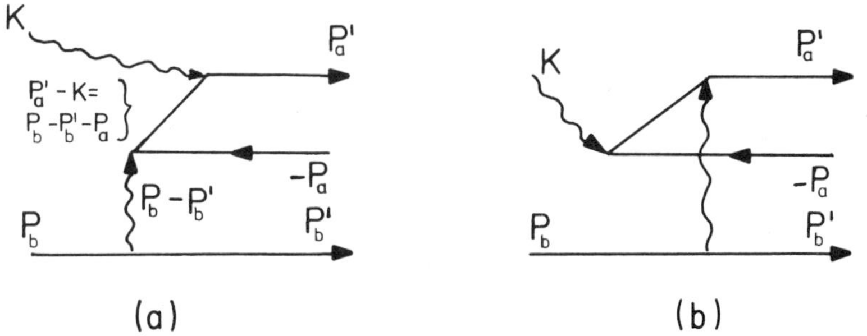


**Figure 3-6.1** The pair production graphs corresponding to the graphs of photoabsorption, fig. 3-5.2.

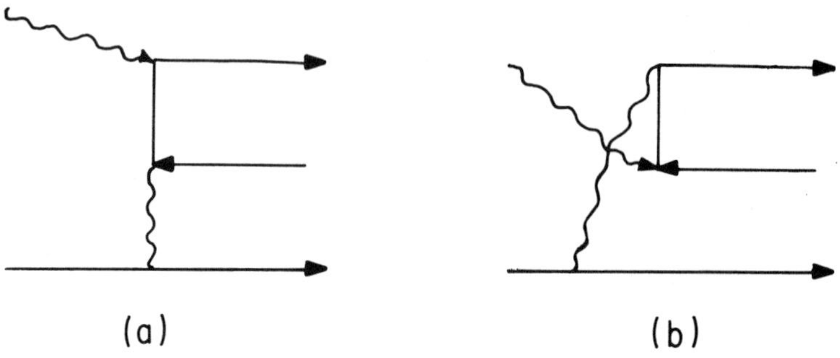

Figure 3-6.2 The graphs of fig. 3-6.1 in an equivalent drawing.

electrodynamics is charge symmetric, too: it allows the existence of positrons and predicts their scattering in terms of electron scattering. However, in nonrelativistic electrodynamics the positron world is separated from the electron world: there is no way to produce positrons, and a theory without positrons is self-consistent.

It is clear from section 3-2 that negative frequency parts occur for all particles (in fact, our starting point was the photon). Some particles carry "generalized charges," which are additively conserved like the electric charge. Protons and neutrons, for example, carry "baryon number" (see chapter 4). The electron itself carries a "lepton number," which appears to be additive (examples of nonadditive quantum numbers are parity and the square of angular momentum). It is clear that the conservation laws can only remain valid if the corresponding antiparticles have *all* additive quantum numbers reversed. Thus, the Feynman rule for positrons is immediately generalized to all kinds of antiparticles:

**Feynman rule for antiparticles.** ("*Substitution rule*"). An outgoing (incoming) antiparticle is formally treated as an incoming (outgoing) particle which has all additive quantum numbers reversed.

Note in this context that energies, momenta, and magnetic quantum numbers are additive, but helicities are not. In fact, the simultaneous reversal of **p** and $M$ does not affect the eigenvalues of $|p|^{-1}\mathbf{p}\boldsymbol{\sigma}$. Noting moreover that the polarization vector $\varepsilon^\mu$ of a photon satisfies $\varepsilon^*(\mathbf{k}, \lambda) = \varepsilon(-\mathbf{k}, \lambda)$, we see that the photon field (0.2) has the same structure as the electron field (2.3). Looking back at bremsstrahlung (photoemission) and photoabsorption of the previous section, we may say that photons are their own antiparticles. The photoabsorption matrix element (5.9) is obtained from (5.8) simply by replacing $K_\mu \to -K_\mu$ everywhere, without changing the helicity.

Although bremsstrahlung $ep \to \gamma ep$ and pair production $\gamma p \to e^+e^-p$ are formally described by the same matrix element (only the ranges of the variables differ), the cross sections for the two processes can nevertheless be quite different. The phase space for three particles of masses $m_1$, $m_2$, and $m_3$ in the final state vanishes below the threshold value

$$s_0(m_1, m_2, m_3) = (m_1 + m_2 + m_3)^2 = m_a^2 + m_b^2 + 2E_{\text{ath}}^{\text{lab}} m_b. \qquad (6.3)$$

Bremsstrahlung is possible at all electron energies $E_a^{\mathrm{iab}} > m_a$, whereas pair production becomes possible only for

$$k^{\mathrm{lab}} > k_{\mathrm{th}}^{\mathrm{lab}} = 2m(1 + 2m/m_b). \tag{6.4}$$

These processes will be discussed in chapter 8. We now discuss simpler pair creations reactions such as $\pi^+\pi^- \to e^+e^-$ or the decay of the excited $0^+$ state of $^4He$, $\alpha^* \to e^+e^-\alpha$ (fig. 3-6.3). The matrix element for such cases has the usual form

$$T_{if} = e^2 t^{-1} J^{\mu}_{\bar{a}a}(t) g_{\mu\nu} J^{\nu}_{bb'}(t), \tag{6.5}$$

only the definitions of $t$ and $J^{\mu}_{\bar{a}a}$ are modified according to our substitution rules for antiparticles:

$$t = (P_e + P_{\bar{e}})^2 = 2(m^2 + P_e P_{\bar{e}}), \qquad J^{\mu}_{\bar{a}a} = Z\bar{u}(\mathbf{p}_e, M_e)\gamma^{\mu} v(\mathbf{p}_{\bar{e}}, M_{\bar{e}}). \tag{6.6}$$

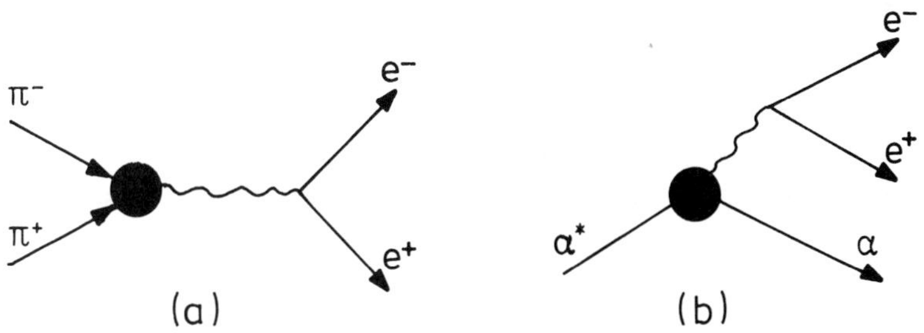


**Figure 3-6.3** Two pair production processes for which the full matrix element has the structure (6.5).

As $t$ plays now the role of the square of a cms energy, we shall replace it by the letter $s$. In the cms, we thus have (see fig. 2-4):

$$\begin{aligned} P_e &= (E, \mathbf{p}_e), \qquad P_{\bar{e}} = (E, -\mathbf{p}_e), \\ \mathbf{p}_e &= (p_e \sin\theta, 0, p_e \cos\theta), \qquad E = \tfrac{1}{2}s^{\frac{1}{2}}. \end{aligned} \tag{6.7}$$

The decay probability is proportional to

$$\begin{aligned} W &= \sum_{\lambda_e \lambda_{\bar{e}}} |J^{\mu}_{bb'} J_{\bar{a}a\mu}(\lambda_e, \lambda_{\bar{e}})|^2 \\ &= J^{\nu}_{bb'} J^{*\mu}_{bb'} \operatorname{trace} \gamma_{\mu}(P_e \cdot \gamma + m)\gamma_{\nu}(P_{\bar{e}}\gamma - m), \end{aligned} \tag{6.8}$$

according to (A-4.10) and (A-4.6). The trace is obtained from (2-7.19) by reversal of $m^2$:

$$W = 4J^{\nu}_{bb'} J^{*\mu}_{bb'}[P_{e\mu} P_{\bar{e}\nu} + P_{\bar{e}\mu} P_{e\nu} - \tfrac{1}{2}s g_{\mu\nu}]. \tag{6.9}$$

Whenever $\mu = 0$ or $\nu = 0$, the square bracket in (9) vanishes. This is, of course, a consequence of gauge invariance, $(P_e + P_{\bar{e}})_{\mu} J^{\mu}_{\bar{a}a} = 0$ in this system. Thus (9) is

$$W = 4J^{i}_{bb'} J^{*k}_{bb'}(\tfrac{1}{2}s\, \delta_{ik} - 2p_i p_k). \tag{6.10}$$

After averaging over the directions of the electron momentum, we can replace $p_i p_k \to p_i^2 \, \delta_{ik} \to \frac{1}{3} p_e^2 \, \delta_{ik}$ and obtain

$$W = 4\mathbf{J}_{bb'}\mathbf{J}^*_{bb'}(\tfrac{1}{2}s - \tfrac{2}{3}p_e^2) = \mathbf{J}_{bb'}\mathbf{J}^*_{bb'}\tfrac{4}{3}(s + 2m_e^2). \tag{6.11}$$

This result permits an interesting interpretation and generalization. Undoing the photon spin summation in (5) using (2-8.8) we arrive at

$$T_{if} = \sum_{M = \pm 1, 0} T(i \to \hat{\gamma}_M) s^{-1} T(\hat{\gamma}_M \to f), \tag{6.12}$$

$$T(i \to \hat{\gamma}_M) = eJ^\nu_{bb'} \varepsilon^*_\nu(M), \qquad T(\hat{\gamma}_M \to f) = e\varepsilon_\mu(M) J^\mu_{\bar{a}a}, \tag{6.13}$$

where $T(i \to \hat{\gamma}_M)$ and $T(\hat{\gamma}_M \to f)$ can be viewed as the matrix elements describing production and decay of a virtual photon of mass $s^{\frac{1}{2}} > 0$, respectively. Equation (11) tells us that after spin summation and angular integration of the final state, the corresponding probabilities factorize:

$$\sum_{\lambda_f} \int d\Omega_e \, | T_{if} |^2 = s^{-2} \sum_M | T(i \to \hat{\gamma}_M) |^2 \sum_{\lambda_f} \int d\Omega_e \, | T(\hat{\gamma}_M \to f) |^2, \tag{6.14}$$

where the second probability is in fact independent of the photon magnetic quantum number $M$. It is useful to define the "decay rate of a virtual photon" according to (2-3.9)

$$\begin{aligned} m_\gamma \Gamma_f &= \frac{1}{2} \sum_{\lambda_f} \int d \text{ Lips } (s; f) | T(\hat{\gamma}_M \to f) |^2 \\ &= (8\pi s^{\frac{1}{2}})^{-1} p' \sum_{\lambda_f} \int (4\pi)^{-1} \, d\Omega \, | T(\hat{\gamma}_M \to f) |^2. \end{aligned} \tag{6.15}$$

Insertion of (14) into the cross-section formula (2-4.8) gives for reactions $b\bar{b} \to f$ which proceed via a virtual photon:

$$\begin{aligned} \sigma(b\bar{b} \to \hat{\gamma} \to f) &= (2S_b + 1)^{-2} p'(16\pi p s)^{-1} \sum_M \sum_{\lambda_b \lambda_{\bar{b}}} | T(b\bar{b} \to \hat{\gamma}_M) |^2 \\ &\times s^{-2} \sum_{\lambda_f} \int \frac{d\Omega}{4\pi} | T(\hat{\gamma}_M \to f) |^2 = 12\pi p^{-2} s^{-2} (2S_b + 1)^{-2} m_\gamma \Gamma_{b\bar{b}} m_\gamma \Gamma_f . \end{aligned} \tag{6.16}$$

In (16) we have also introduced $m_\gamma \Gamma_i$ which is defined in analogy with (15), and made use of the fact that $\sum_{\lambda_b \lambda_{\bar{b}}} | T(b\bar{b} \to \gamma_M) |^2$ is independent of $M$, such that $\sum_M$ brings just a factor 3. With the first expression in (15) for $m_\gamma \Gamma_f$, (16) applies also to processes where the virtual photon decays into more than two particles (fig. 3-6.4). This structure will be discussed in section 4-8. For the $e\bar{e}$ final state, (14) together with (11) gives

$$m_\gamma \Gamma(\hat{\gamma} \to \bar{e}e) = \tfrac{2}{3} \alpha p_e s^{-\frac{1}{2}} (2m_e^2 + s). \tag{6.17}$$

For the $\pi^+\pi^-$ final state:

$$T(\hat{\gamma}_M \to \pi^+\pi^-) = e\varepsilon_\mu(M)(P_+ - P_-)^\mu F(s) = -2e\boldsymbol{\varepsilon}(M)\mathbf{p}_+ F(s), \tag{6.18}$$

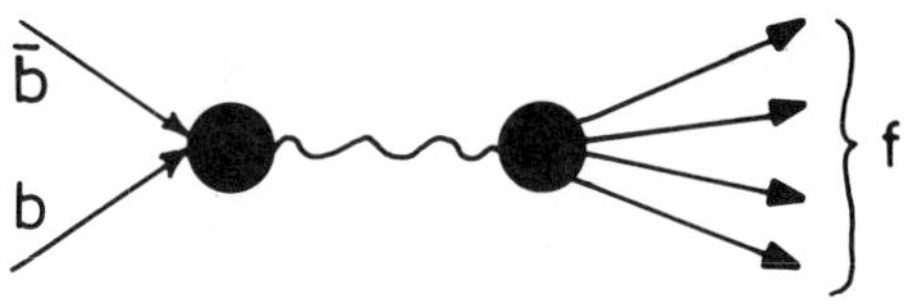


**Figure 3-6.4** The most general reaction to which (6.16) applies.

where $F$ is the pion form factor defined in (2-7.9). Choosing $M = 0$ for simplicity, we obtain from (14)

$$m_\gamma \Gamma(\hat{\gamma} \to \pi^+\pi^-) = (8\pi s^{\frac{1}{2}})^{-1} p_+ \frac{1}{2}\int_{-1}^{1} dx\, 4e^2 p_+^2 x^2 |F|^2 \\ = \tfrac{2}{3}\alpha p_+^3 s^{-\frac{1}{2}} |F(s)|^2. \tag{6.19}$$

These expressions can be used both for initial and final states in (16) but not for elastic scattering, i.e., they describe reactions such as $\pi^+\pi^- \leftrightarrow e^-e^+$, $ee^+ \to \mu^-\mu^+$ but not $e^-e^+ \leftrightarrow e^-e^+$. The reason is that the latter case has two graphs (fig. 3-6.5), one of which (graph $a$) is not of the type shown in fig. 3-6.4.

Next, we study the photon decay matrix elements for given values of the helicities $\lambda_e$ and $\lambda_{\bar{e}}$. We quantize the positron spin along $-\mathbf{p}_{\bar{e}}$, such that for the 2-component antiparticle spinor

$$i\chi(-P_{\bar{e}}, -M_{\bar{e}}) = im^{-\frac{1}{2}}(-E_{\bar{e}} - \mathbf{p}_{\bar{e}}\boldsymbol{\sigma})^{\frac{1}{2}}\chi_{0\bar{e}}(-M_{\bar{e}}) \\ = im^{-\frac{1}{2}}(-E_{\bar{e}} - 2M_{\bar{e}}p_{\bar{e}})^{\frac{1}{2}}\chi_{0\bar{e}}(-M_{\bar{e}}) \\ = -m^{-\frac{1}{2}}(E_{\bar{e}} - 2\lambda_{\bar{e}}p_{\bar{e}})^{\frac{1}{2}}\chi_{0\bar{e}}(\lambda_{\bar{e}}), \tag{6.20}$$

with $-M_{\bar{e}} = \lambda_{\bar{e}}$. We then get from (A-3.22)

$$J^\mu_{e\bar{e},r} = -(E_e + 2\lambda_e p_e)^{\frac{1}{2}}(E_{\bar{e}} - 2\lambda_{\bar{e}}p_{\bar{e}})^{\frac{1}{2}}\chi_{0e}^+(\lambda_e)\sigma^\mu\chi_{0\bar{e}}(\lambda_{\bar{e}}). \tag{6.21}$$

The advantage of using the states $\chi_{0\bar{e}}$ is that in the cms, $\mathbf{p}_{\bar{e}} = -\mathbf{p}_e$ implies that $\chi_{0e}$ and $\chi_{0\bar{e}}$ are quantized along the same direction, i.e., they are identical in that case, $\chi_{0e}(\lambda) = \chi_{0\bar{e}}(\lambda)$.

Once the positive-imaginary branch of the square root has been chosen for $(-E_{\bar{e}} + 2\lambda_{\bar{e}}p_{\bar{e}})^{\frac{1}{2}}$ in (20), the opposite branch must be chosen for $\chi_l$ in (A-3.23). This follows directly from (1-6.27) and the reversal of $P_\mu$ (compare

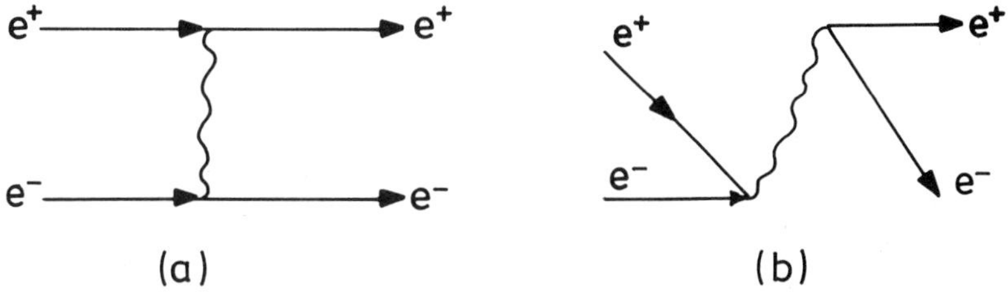


**Figure 3-6.5** The two graphs of $e^+e^-$ scattering ("Bhabha scattering"). They are related to graphs (a) and (b) of fig. 3-4 by the substitution rule.

(A-3.14) in the low-energy representation. Both square roots must be factored out). We thus have

$$im^{\frac{1}{2}}\chi_l(-P_{\bar{e}}, -M_{\bar{e}}) = (E_{\bar{e}} + 2\lambda_{\bar{e}}p_{\bar{e}})^{\frac{1}{2}}\chi_{0\bar{e}}(\lambda_{\bar{e}}),$$
$$J^{\mu}_{e\bar{e},\,l} = (E_e - 2\lambda_e p_e)^{\frac{1}{2}}(E_{\bar{e}} + 2\lambda_{\bar{e}}p_{\bar{e}})^{\frac{1}{2}}\chi_{0e}(\lambda_e)\sigma_\mu\chi_{0\bar{e}}(\lambda_{\bar{e}}), \tag{6.22}$$

and in the cms, dropping the indices $e$ and $\bar{e}$:

$$J_l^0 = F_+\,\delta_{\lambda\bar{\lambda}} = -J_r^0(-\lambda, -\bar{\lambda}), \qquad F_\pm \equiv (E - 2\lambda p)^{\frac{1}{2}}(\bar{E} \pm 2\lambda p)^{\frac{1}{2}}, \tag{6.23}$$

$$J_l^z = -2\lambda \cos\theta\, F_+\,\delta_{\lambda\bar{\lambda}} + \sin\theta\, F_-\,\delta_{\lambda,\,-\bar{\lambda}} = -J_r^z(-\lambda, -\bar{\lambda}, -\theta),$$
$$J_l^x = 2\lambda \sin\theta\, F_+\,\delta_{\lambda\bar{\lambda}} + \cos\theta\, F_-\,\delta_{\lambda,\,-\bar{\lambda}} = J_r^x(-\lambda, -\bar{\lambda}, -\theta), \tag{6.24}$$
$$J_l^y = 2i\lambda F_-\,\delta_{\lambda,\,-\bar{\lambda}} = -J_r^y(-\lambda, -\bar{\lambda}).$$

These formulas will be needed in chapter 5. For our present purpose, we insert $\bar{E} = E$, $J^\mu = J_r^\mu + J_l^\mu (= -\bar{u}\gamma^\mu v)$, arriving at

$$F_+ = m, \qquad F_- = E - 2\lambda p, \qquad J^0 = 0, \tag{6.25}$$

$$J^z = 4m\lambda \cos\theta\,\delta_{\lambda\bar{\lambda}} - 2E\sin\theta\,\delta_{\lambda,\,-\bar{\lambda}},$$
$$J^x = -4m\lambda \sin\theta\,\delta_{\lambda\bar{\lambda}} - 2E\cos\theta\,\delta_{\lambda,\,-\bar{\lambda}}, \qquad J^y = -4i\lambda E\,\delta_{\lambda,\,-\bar{\lambda}}. \tag{6.26}$$

From (26) we see that for $E \gg m$, electron and positron carry mainly opposite helicities.

As an application, we calculate the differential cross section for $e^-e^+ \to \mu^-\mu^+$:

$$d\sigma/d\Omega = p_\mu(4p_e s)^{-1}\alpha^2 s^{-2}\frac{1}{4}\sum_{\lambda_\mu\lambda_{\bar{\mu}}} W, \tag{6.27}$$

with $W$ defined by (8). The electron momentum which enters (10) refers now to the initial state and can be taken along the $z$-axis, such that (10) gives

$$\tfrac{1}{4}W = \tfrac{1}{2}s\mathbf{J}_{b\bar{b}}\mathbf{J}^*_{b\bar{b}} - 2p_e^2|J^z_{b\bar{b}}|^2 = 2m_e^2|J^z_{b\bar{b}}|^2 + \tfrac{1}{2}s(|J^x_{b\bar{b}}|^2 + |J^y_{b\bar{b}}|^2). \tag{6.28}$$

The vector $\mathbf{J}_{b\bar{b}}$ can now be taken from (26), with $m$, $E$ and $\theta$ referring to the muon mass, energy and momentum. After summation over the muon helicities, we have, with $4E^2 = s$:

$$\sum_{\lambda_f}|J^z_{b\bar{b}}|^2 = 8m_\mu^2\cos^2\theta + 2s\sin^2\theta, \qquad \sum_{\lambda_f}|J^x_{b\bar{b}}|^2 = 8m_\mu^2\sin^2\theta + s\cos^2\theta, \tag{6.29}$$

$$d\sigma/d\Omega = p_e^{-1}p_\mu\alpha^2 s^{-3}[m_e^2(4m_\mu^2\cos^2\theta + s\sin^2\theta) + s(m_\mu^2\sin^2\theta + \tfrac{1}{4}s\cos^2\theta) + \tfrac{1}{4}s^2], \tag{6.30}$$

and the angular integration gives

$$\sigma = \tfrac{4}{3}\pi\alpha^2 p_e^{-1}p_\mu s^{-1}(1 + 2s^{-1}m_e^2)(1 + 2s^{-1}m_\mu^2), \tag{6.31}$$

in agreement with (16).

## 3-7 Compton Scattering

Photon-electron scattering is called Compton scattering. Its matrix element $T_C$ is given in the Born approximation by graphs (a) and (b) of fig. 3-7. According to Feynman's rules

$$T_C = e^2 \bar{\mathbf{u}}\, Q u,$$
$$Q = \varepsilon'^* \cdot \gamma \frac{(P+K)\cdot\gamma + m}{m^2 - (P+K)^2}\varepsilon\cdot\gamma + \varepsilon\cdot\gamma\frac{(P-K')\cdot\gamma + m}{m^2-(P-K')^2}\varepsilon'^*\cdot\gamma. \tag{7.1}$$

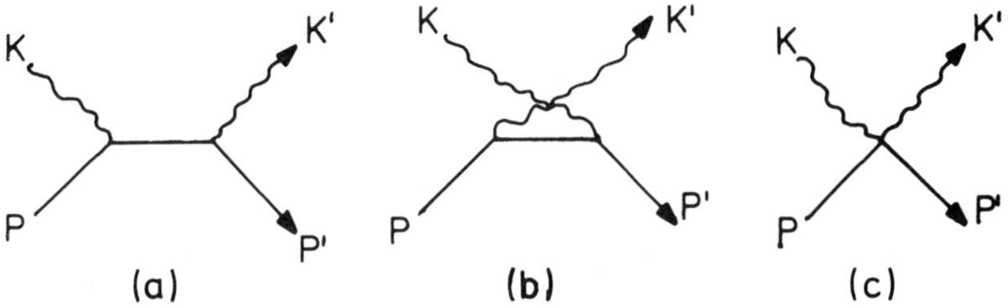


**Figure 3-7** The Feynman graphs of Compton scattering. Graph (c) has a factor $2e^2 g_{\mu\nu}$ at the vertex and is absent for photon scattering on electrons and muons.

Alternatively, one may replace $-eJt^{-1}$ by $\varepsilon'^*$ in (5.9). We calculate $T_C$ in the lab system $P^\mu = (m, \mathbf{0})$. With $P\cdot\varepsilon = 0$, we have $(P+K)\cdot\gamma\,\varepsilon\cdot\gamma = -\varepsilon\cdot\gamma(P+K)\cdot\gamma$ (compare A-4.13). In this form, $-P\cdot\gamma$ can be cancelled against $m$ because of Dirac's equation $P\cdot\gamma u = mu$. Applying the same procedure in the second term, we obtain in the lab system

$$\begin{aligned} Q &= \varepsilon'^*\cdot\gamma\,\varepsilon\cdot\gamma\, K\cdot\gamma/2mk + \varepsilon\cdot\gamma\,\varepsilon'^*\cdot\gamma\, K'\cdot\gamma/2mk' \\ &= \boldsymbol{\varepsilon}'^*\boldsymbol{\gamma}\boldsymbol{\varepsilon}\boldsymbol{\gamma}(\gamma_0 - \hat{\mathbf{k}}\boldsymbol{\gamma})/2m + \boldsymbol{\varepsilon}\boldsymbol{\gamma}\boldsymbol{\varepsilon}'^*\boldsymbol{\gamma}(\gamma_0 - \hat{\mathbf{k}}'\boldsymbol{\gamma})/2m. \end{aligned} \tag{7.2}$$

$Q$ is a $2\times 2$ matrix in the $(g, k)$-space of $u$, but because of $u_k = 0$ in the lab system, we only need its first column (compare A-3.7). From (1-5.11), we see that $\boldsymbol{\varepsilon}'^*\cdot\boldsymbol{\gamma}\boldsymbol{\varepsilon}\boldsymbol{\gamma} = -\boldsymbol{\varepsilon}'^*\boldsymbol{\sigma}\boldsymbol{\varepsilon}\boldsymbol{\sigma}$ is a multiple of the unit matrix in $(g, k)$-space. We thus obtain

$$Q_{gg} = -\boldsymbol{\varepsilon}'^*\boldsymbol{\varepsilon}/m, \tag{7.3}$$

$$\begin{aligned} 2mQ_{kg} &= -\boldsymbol{\varepsilon}'^*\boldsymbol{\sigma}\boldsymbol{\varepsilon}\boldsymbol{\sigma}\hat{\mathbf{k}}\boldsymbol{\sigma} - \boldsymbol{\varepsilon}\boldsymbol{\sigma}\boldsymbol{\varepsilon}'^*\boldsymbol{\sigma}\hat{\mathbf{k}}'\boldsymbol{\sigma} = i\boldsymbol{\varepsilon}'^*\boldsymbol{\sigma}\boldsymbol{\varepsilon}_x\boldsymbol{\sigma} + i\boldsymbol{\varepsilon}\boldsymbol{\sigma}\boldsymbol{\varepsilon}_x'^*\boldsymbol{\sigma} \\ &= i\boldsymbol{\varepsilon}\boldsymbol{\varepsilon}_x'^* + i\boldsymbol{\varepsilon}_x\boldsymbol{\varepsilon}'^* + \boldsymbol{\beta}\boldsymbol{\sigma}, \end{aligned} \tag{7.4}$$

$$\boldsymbol{\varepsilon}_x \equiv \hat{\mathbf{k}}\times\boldsymbol{\varepsilon}, \qquad \boldsymbol{\varepsilon}_x'^* \equiv \hat{\mathbf{k}}'\times\boldsymbol{\varepsilon}'^*, \qquad \boldsymbol{\beta} \equiv \boldsymbol{\varepsilon}_x\times\boldsymbol{\varepsilon}'^* - \boldsymbol{\varepsilon}\times\boldsymbol{\varepsilon}_x'^*. \tag{7.5}$$

using (1-4.3) and $\boldsymbol{\varepsilon}\mathbf{k} = \boldsymbol{\varepsilon}'\mathbf{k}' = 0$. Next, we put $T_C$ into the form $\chi_0'^+ R\chi_0$, where both spinors refer to electrons at rest. Using (A-3.7),

$$2mT_C e^{-2} = k_+\chi_0'^+\left[2mQ_{gg} - \boldsymbol{\sigma}\mathbf{p}'\frac{2m}{E'+m}Q_{kg}\right]\chi_0 \equiv \chi_0'^+(R_0 + \mathbf{R}\boldsymbol{\sigma})\chi_0, \tag{7.6}$$

$$R_0 = -(2m)^{-\frac{1}{2}}(E'+m)^{\frac{1}{2}}(2\boldsymbol{\varepsilon}\boldsymbol{\varepsilon}'^* + (E'+m)^{-1}\mathbf{p}'\boldsymbol{\beta}), \tag{7.7}$$

$$\mathbf{R} = -i(2m)^{-\frac{1}{2}}(E'+m)^{-\frac{1}{2}}[\mathbf{p}'(\boldsymbol{\varepsilon}_x\boldsymbol{\varepsilon}'^* + \boldsymbol{\varepsilon}\boldsymbol{\varepsilon}_x'^*) + \mathbf{p}'\times\boldsymbol{\beta}]. \tag{7.8}$$

For $E' \approx m \gg p'$, we recover the Thomson formula (1.23). The term $\mathbf{p}'\boldsymbol{\beta} = (\mathbf{k} - \mathbf{k}')\boldsymbol{\beta}$ in (7) can be simplified by using

$$\mathbf{k}(\boldsymbol{\varepsilon}_x \times \boldsymbol{\varepsilon}'^*) = -k\boldsymbol{\varepsilon}\boldsymbol{\varepsilon}'^*, \quad \mathbf{k}(\boldsymbol{\varepsilon} \times \boldsymbol{\varepsilon}_x'^*) = k\boldsymbol{\varepsilon}_x\boldsymbol{\varepsilon}_x'^* \quad \text{etc, and} \quad k - k' = E' - m:$$

$$\mathbf{p}'\boldsymbol{\beta} = -(E' - m)(\boldsymbol{\varepsilon}\boldsymbol{\varepsilon}'^* + \boldsymbol{\varepsilon}_x\boldsymbol{\varepsilon}_x'^*) \tag{7.9}$$

The matrix elements (7) and (8) allow the calculation of all possible polarization effects. Electron polarization effects can be calculated using (A-4.4) (see Tolhoek 1956, McMaster 1961, Olsen 1968). We assume unpolarized electrons, in which case we only need $|R_0|^2 + |\mathbf{R}|^2$:

$$\begin{aligned} \mathbf{p}' \cdot \boldsymbol{\beta} &= -(E' - m)(\boldsymbol{\varepsilon} \cdot \boldsymbol{\varepsilon}'^* + \boldsymbol{\varepsilon}_x \cdot \boldsymbol{\varepsilon}_x'^*) \\ 2m|R_0|^2 &= 8m|\boldsymbol{\varepsilon} \cdot \boldsymbol{\varepsilon}'^*|^2 - 4(E' - m)\,\mathrm{Re}\,(\boldsymbol{\varepsilon} \cdot \boldsymbol{\varepsilon}'^*\boldsymbol{\varepsilon}_x^* \cdot \boldsymbol{\varepsilon}_x') \\ &\quad + (E' + m)^{-1}|\mathbf{p}' \cdot \boldsymbol{\beta}|^2, \\ 2m\mathbf{R} \cdot \mathbf{R}^* &= (E' - m)|\boldsymbol{\varepsilon}_x \cdot \boldsymbol{\varepsilon}'^* + \boldsymbol{\varepsilon} \cdot \boldsymbol{\varepsilon}_x'^*|^2 + (E' - m)\boldsymbol{\beta}\boldsymbol{\beta}^* \\ &\quad - (E' + m)^{-1}|\mathbf{p}' \cdot \boldsymbol{\beta}|^2. \end{aligned} \tag{7.10}$$

The last two terms in (10) follow from the rule $(a \times b)(c \times d) = (ac)(bd) - (ad)(bc)$, and $p'^2$ has been rewritten as $(E' + m)(E' - m)$. We thus obtain

$$|R_0|^2 + \mathbf{R}\mathbf{R}^* = 4|\boldsymbol{\varepsilon} \cdot \boldsymbol{\varepsilon}'^*|^2 + \tfrac{1}{2}(E'/m - 1)F \tag{7.11}$$

$$F \equiv |\boldsymbol{\varepsilon}_x \cdot \boldsymbol{\varepsilon}'^* + \boldsymbol{\varepsilon} \cdot \boldsymbol{\varepsilon}_x'^*|^2 + \boldsymbol{\beta}\boldsymbol{\beta}^* - 4\,\mathrm{Re}\,(\boldsymbol{\varepsilon} \cdot \boldsymbol{\varepsilon}'^*\,\boldsymbol{\varepsilon}_x^* \cdot \boldsymbol{\varepsilon}_x') \tag{7.12}$$

At this point, it is convenient to identify $\boldsymbol{\varepsilon}$ and $\boldsymbol{\varepsilon}'$ with helicity states, for which (compare 2-8.6)

$$\boldsymbol{\varepsilon}_x \equiv \hat{\mathbf{k}} \times \boldsymbol{\varepsilon} = -i\lambda\boldsymbol{\varepsilon}, \qquad \boldsymbol{\varepsilon}_x' = -i\lambda'\boldsymbol{\varepsilon}'. \tag{7.13}$$

(Note, by the way, $\boldsymbol{\varepsilon} \cdot \boldsymbol{\varepsilon} = 0$). We then get

$$\begin{aligned} \boldsymbol{\varepsilon}_x \cdot \boldsymbol{\varepsilon}'^* + \boldsymbol{\varepsilon} \cdot \boldsymbol{\varepsilon}_x'^* &= -i(\lambda - \lambda')\boldsymbol{\varepsilon} \cdot \boldsymbol{\varepsilon}'^*, \\ \boldsymbol{\beta} &= -i(\lambda + \lambda')\boldsymbol{\varepsilon} \times \boldsymbol{\varepsilon}'^*, \qquad \boldsymbol{\varepsilon}_x^*\boldsymbol{\varepsilon}_x' = \lambda\lambda'\boldsymbol{\varepsilon}^*\boldsymbol{\varepsilon}'. \end{aligned}$$

$$\begin{aligned} F &= (\lambda - \lambda')^2|\boldsymbol{\varepsilon} \cdot \boldsymbol{\varepsilon}'^*|^2 + (\lambda + \lambda')^2|\boldsymbol{\varepsilon} \times \boldsymbol{\varepsilon}'^*|^2 - 4\lambda\lambda'|\boldsymbol{\varepsilon} \cdot \boldsymbol{\varepsilon}'^*|^2 \\ &= 2(1 - \lambda\lambda')|\boldsymbol{\varepsilon} \cdot \boldsymbol{\varepsilon}'^*|^2 + 2(1 + \lambda\lambda')(1 - |\boldsymbol{\varepsilon} \cdot \boldsymbol{\varepsilon}'|^2) - 4\lambda\lambda'|\boldsymbol{\varepsilon} \cdot \boldsymbol{\varepsilon}'^*|^2. \end{aligned} \tag{7.14}$$

Inserting now $4|\boldsymbol{\varepsilon} \cdot \boldsymbol{\varepsilon}'|^2 = 1 + \cos^2 \vartheta - 2\lambda\lambda' \cos \vartheta$ and $4|\boldsymbol{\varepsilon} \cdot \boldsymbol{\varepsilon}'^*|^2 = 1 + \cos^2 \vartheta + 2\lambda\lambda' \cos \vartheta$, we find

$$F = 2(1 - \cos \vartheta)(1 + \lambda\lambda' \cos \vartheta), \tag{7.15}$$

where $\vartheta$ is the angle between $\mathbf{k}$ and $\mathbf{k}'$ in the lab system. For pure photon states of arbitrary polarizations $\boldsymbol{\varepsilon}$ and $\boldsymbol{\varepsilon}'$, we thus have

$$|R_0|^2 + \mathbf{R}\mathbf{R}^* = 4|\boldsymbol{\varepsilon} \cdot \boldsymbol{\varepsilon}'^*|^2 + (E'/m - 1)(1 - \cos \vartheta)(1 + \xi_3\xi_3' \cos \vartheta), \tag{7.16}$$

where $\xi_3$ is defined in (2-8.26). The differential cross section (2-4.9) contains (for unpolarized electrons)

$$\overline{|T_C/4\pi|^2} = \alpha^2(|R_0|^2 + \mathbf{RR}^*) \tag{7.17}$$

There is nothing wrong in expressing a cms differential cross section in terms of lab quantities. However, in this case it is customary to quote the lab differential cross section $d\sigma/d\Omega_L$. Evaluating the two equations which involve $t$ in (2-4.12) in the lab system, we have

$$-\tfrac{1}{2}t = m(E' - m) = kk'(1 - \cos\vartheta) \tag{7.18}$$

and rewriting $E' - m$ as $k - k'$ we find

$$\cos\vartheta = 1 + m/k - m/k', \qquad d\cos\vartheta = m\, dk'/k'^2 = \tfrac{1}{2}\, dt/k'^2; \tag{7.19}$$

$$\begin{aligned} d\sigma_C(\boldsymbol{\varepsilon}, \boldsymbol{\varepsilon}')/d\Omega_L &= 2k'^2\, d\sigma/dt\, d\phi = (4k^2m^2)^{-1}k'^2\overline{|T_C/4\pi|^2} \\ &= \tfrac{1}{4}r_e^2\left(\frac{k'}{k}\right)^2\left\{\left(\frac{k'}{k} + \frac{k}{k'} - 2\right)(1 + \xi_3\xi_3'\cos\vartheta) + 4|\boldsymbol{\varepsilon}\cdot\boldsymbol{\varepsilon}'^*|^2\right\}. \end{aligned} \tag{7.20}$$

This is the "Klein-Nishina" formula for photons of well-defined polarizations both in the initial and final states. The term $\zeta_3\zeta_3'\cos\vartheta$ disappears when the initial photon has no circular polarization or the circular polarization of the final photon is not recorded. If no polarization component of the final photon is recorded, (20) must be summed over $\lambda_f = \pm 1$, in which case the first term inside the curly brackets is multiplied by 2, and the second term is treated as in (1.24) for Thomson scattering (Note incidentally that in Thomson scattering, lab system and cms are identical, i.e., $\vartheta = \theta$. The azimuthal angle $\phi$ is, of course, always the same, $\phi_L = \phi_{\text{cms}} = \phi$). When the incident photons are also unpolarized, one must average over the initial photon polarization states, which affects only the $|\boldsymbol{\varepsilon}\cdot\boldsymbol{\varepsilon}'^*|^2$-term in (20), compare (1.28):

$$\begin{aligned} d\sigma_C/d\Omega_L &= \tfrac{1}{2}r_e^2\left(\frac{k'}{k}\right)^2\left(\frac{k'}{k} + \frac{k}{k'} - 2 + \cos^2\vartheta + 1\right) \\ &= \tfrac{1}{2}r_e^2\left(\frac{k'}{k}\right)^2\left(\frac{k'}{k} + \frac{k}{k'} - \sin^2\vartheta\right) \end{aligned} \tag{7.21}$$

This expression can be integrated, using again (19) and (18)

$$\begin{aligned} \sigma_C &= \tfrac{1}{2}r_e^2\frac{2\pi m}{k^2}\int_{(1+2k/m)^{-1}k}^{k} dk'\left[\frac{k'}{k} + \frac{k}{k'} - 1 + \left(1 + \frac{m}{k} - \frac{m}{k'}\right)^2\right] \\ &= 2\pi r_e^2 m\left\{\frac{k^3 + 9k^2m + 8km^2 + 2m^3}{k^2(m + 2k)^2} + \frac{k^2 - 2km - 2m^2}{2k^3}\ln\left(1 + \frac{2k}{m}\right)\right\}. \end{aligned} \tag{7.22}$$

For $k \gg m$ this simplifies to

$$\sigma_C(k \gg m) = \pi r_e^2\frac{m}{k}\left\{\ln\frac{2k}{m} + \frac{1}{2}\right\}. \tag{7.23}$$

For $k \ll m$, we recover the Thomson limit (1.28) by expanding (22) to second order in $k/m$.

These formulas have been derived here for technical training and for applications. For physical understanding, on the other hand, it is better to consider the Compton scattering on spinless particles, $\gamma\pi \to \gamma'\pi'$ or $\gamma\alpha \to \gamma'\alpha'$. This case is in closer analogy with the nonrelativistic treatment of $\gamma e \to \gamma' e'$, because it contains a contribution from first-order perturbation theory, which is derived from the $A^2$-term in (2-6.18). The complete matrix element of second-order perturbation theory is thus (at least in the region where the form factors are negligible)

$$T_C = e^2 \varepsilon'^{\nu *} \varepsilon^{\mu} \left[ \frac{(2P' + K')_\nu (2P + K)_\mu}{m^2 - (P + K)^2} + \frac{(2P - K')_\nu (2P' - K)_\mu}{m^2 - (P' - K)^2} + 2g_{\mu\nu} \right] \qquad (7.24)$$

The first two terms follow from the Feynman rules and are again represented by graphs (a) and (b) of fig. 3-7. The last term follows directly from (2-6.18) and the discussion (essentially about the factor 2) which led to (1.22). It is represented by the "seagull graph" (c) of fig. 3-7. It can also be derived from the first two terms by the requirement of gauge invariance, i.e., if one writes $T_C = \varepsilon^\mu J_\mu$, one must have $K^\mu J_\mu = 0$ (the test is quite trivial, note that $K^\mu(2P + K)_\mu = 2K \cdot P = -m^2 + (P + K)^2$ etc.).

In the lab system, $\mathbf{p} = 0$, the first two terms in (24) are in fact zero, and the whole matrix element is simply $-2e^2\boldsymbol{\varepsilon}\boldsymbol{\varepsilon}'^*$ as in (1.23). The lab differential cross section is obtained from (20) simply by omitting the first term in the curly bracket, and for unpolarized photons, we arrive at

$$d\sigma_C/d\Omega_L = \tfrac{1}{2}(k'/k)^2(1 + \cos^2 \vartheta). \qquad (7.25)$$

The intelligent reader may now wonder why Compton scattering on electrons has no seagull graph, particularly as its nonrelativistic limit obviously contains such a graph. The answer is that this graph is formally eliminated by the use of 4-component spinors. If perturbation theory is developed from the 2-component spinors (1-5.4), then the seagulls become obvious. For fermions with inner structure, such as protons and neutrons, they may appear even in the 4-component formulation.

## 3-8 The Reactions $e^+e^- \leftrightarrow \gamma\gamma$

The matrix element $T_A$ for electron-position annihilation $e^+e^- \to \gamma\gamma$ is obtained from $T_C$ for Compton scattering by substituting (see fig. 3-8.1) $\bar{u} \to \bar{v}$,

$$P' \to -P_+, \quad M' \to -M_+, \quad K \to -K_1, \quad K' \to K_2, \quad \varepsilon \to \varepsilon_1^*, \quad \varepsilon'^* \to \varepsilon_2^*, \qquad (8.1)$$

according to the Feynman rules of section 3-6 for antiparticles:

$$T_A = e^2 \bar{v}(P_+, M_+) Q u,$$

$$Q = \varepsilon_2^* \cdot \gamma \frac{(P - K_1) \cdot \gamma + m}{m^2 - (P - K_1)^2} \varepsilon_1^* \cdot \gamma + \varepsilon_1^* \cdot \gamma \frac{(P - K_2) \cdot \gamma + m}{m^2 - (P - K_2)^2} \varepsilon_2^* \cdot \gamma \qquad (8.2)$$

$$= (2m)^{-1}[\boldsymbol{\varepsilon}_2^* \cdot \boldsymbol{\gamma}\ \boldsymbol{\varepsilon}_1^* \cdot \boldsymbol{\gamma}(\gamma_0 - \hat{\mathbf{k}}_1 \cdot \boldsymbol{\gamma}) + \boldsymbol{\varepsilon}_1^* \cdot \boldsymbol{\gamma}\ \boldsymbol{\varepsilon}_2^* \cdot \boldsymbol{\gamma}(\gamma_0 - \hat{\mathbf{k}}_2 \cdot \boldsymbol{\gamma})],$$

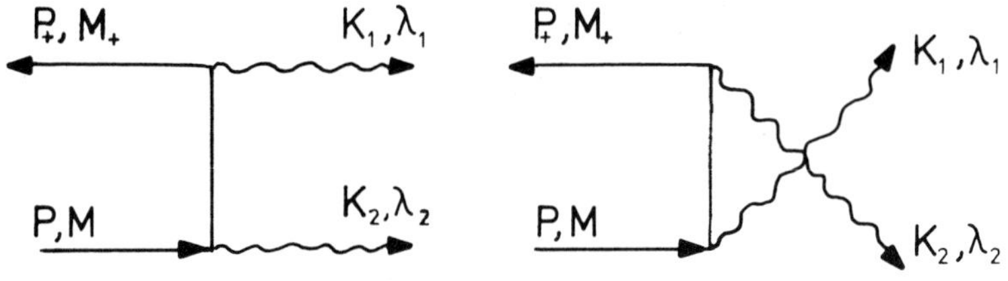


**Figure 3-8.1** The two Feynman graphs for $e^+e^- \to \gamma\gamma$.

where the second form refers again to the electron rest frame. The form (7.4) for $Q_{kg}$ may be used with $\hat{\mathbf{k}} \to \hat{\mathbf{k}}_1$, $\hat{\mathbf{k}}' \to \hat{\mathbf{k}}_2$ ; (7.6) is changed to

$$T_A = e^2(E_+ - m)^{\frac{1}{2}}\chi_0'^{+}(-M_+)[Q_{gg} - (E_+ - m)^{-1}\boldsymbol{\sigma}\mathbf{p}_+ Q_{kg}]\chi_0(M) \quad (8.3)$$

according to (A-3.15). The rest of the calculation proceeds as before and (7.16) is still valid, with $E' = -E_+$ and $k = -k_1$ and $\cos\vartheta$ defined by (7.19), i.e., $\cos\vartheta = 1 - m/k_1 - m/k_2$. We thus obtain

$$\overline{|T_A/4\pi|^2} = \tfrac{1}{2}\alpha^2(2 + k_1/k_2 + k_2/k_1 - 4|\varepsilon_1^*\varepsilon_2^*|^2), \quad (8.4)$$

where $k_1$ and $k_2$ are determined from $k_1 + k_2 = E_+ + m$ and $(P_+ - K_1)^2 = m^2 - 2E_+ k_1 + 2p_+ k_1 \cos\vartheta_A = (P - K_2)^2 = m^2 - 2mk_2$ ($\vartheta_A$ is the angle between $\mathbf{p}$ and $\mathbf{k}_1$). Eliminating $k_2$ by the first equation, we find

$$k_1 = m(m + E_+)(m + E_+ - p_+ \cos\vartheta_A)^{-1},$$
$$k_2 = m^{-1}k_1(E_+ - p_+ \cos\vartheta_A). \quad (8.5)$$

The factor $\frac{1}{2}$ in $|T_A|^2$ comes from the weight $\frac{1}{4}$ in front of the fermion spin summation.

The inverse reaction $\gamma\gamma \to e^+e^-$ (pair production) contains also $|T_A|^2$. The corresponding cross sections differ only in their phase space factors and spin summations. For linearly polarized photons, we define (Olsen, 1968)

$$\begin{aligned} U_C &= \frac{1}{4}\sum_{MM'} |T_C/e^2|^2 \\ &= 8\left(\frac{\varepsilon \cdot P'\, \varepsilon' \cdot P}{2P \cdot K'} - \frac{\varepsilon \cdot P\, \varepsilon' \cdot P'}{2P \cdot K} - \frac{1}{2}\varepsilon \cdot \varepsilon'\right)^2 \\ &\quad - 1 + \frac{1}{2}\left(\frac{P \cdot K}{P \cdot K'} + \frac{P \cdot K'}{P \cdot K}\right) \end{aligned} \quad (8.6)$$

and express the differential cross sections $d\sigma_C$, $d\sigma_A$ and $d\sigma_P$ for Compton scattering, pair annihilation and pair production in their respective cms, using (2-4.8):

$$d\sigma_C/d\Omega = U_C\alpha^2/2s = U_C\alpha^2\tfrac{1}{2}(E + p)^{-2} \quad (8.7)$$

$$d\sigma_A/d\Omega = -U_A\alpha^2 k/2ps, \qquad s = 4E^2, \qquad p = (E^2 - m^2)^{\frac{1}{2}}. \quad (8.8)$$

$$d\sigma_P/d\Omega = -U_A\alpha^2 p/2ks. \quad (8.9)$$

$U_A$ is obtained from $U_C$ by the replacement (1). The annihilation cross section when photon polarizations are not recorded becomes

$$d\sigma_A/d\Omega = \alpha^2(2ps^{\frac{1}{2}})^{-1}(1-\beta^2\cos^2\theta)^{-2}$$
$$\times\,[1-\beta^4+\beta^2\sin^2\theta(2-\beta^2\sin^2\theta)], \quad (8.10)$$

with $\beta = p/E$ and $\theta =$ cms angle. The total cross section is

$$\sigma_A = 2\pi\int_{-1}^{0} d\cos\theta\, d\sigma_A/d\Omega$$
$$= \pi\alpha^2(2p)^{-2}\left[(3-\beta^4)\ln\frac{1+\beta}{1-\beta} - 2\beta(2-\beta^2)\right]. \quad (8.11)$$

The integration ends at $\theta = \pi/2$ as the two photons are identical and go off in opposite directions in the cms. In the nonrelativistic limit, the square

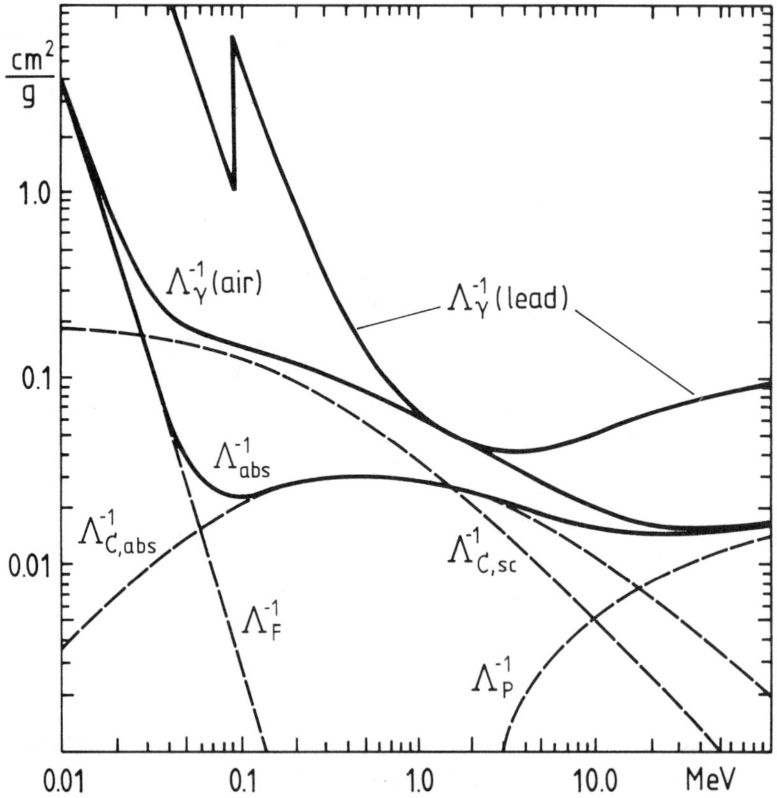


**Figure 3-8.2** Photon mass absorption coefficients $\Lambda_\gamma^{-1}$ (definition in 3.23) for lead (upper curves) and air (lower curves) as functions of photon energy. For air, $\Lambda_\gamma^{-1}$ is divided into $\Lambda_F^{-1}$ for photoabsorption, $\Lambda_P^{-1}$ for pair production and the "scattering" and "absorption" pieces $\Lambda_{C,\,sc}^{-1}$, $\Lambda_{C,\,abs}^{-1}$ of $\Lambda_C^{-1}$ (8.13) (broken lines). Also shown is $\Lambda_{abs}^{-1} = \Lambda_\gamma^{-1} - \Lambda_{C,\,sc}^{-1}$ (from Leipunskii et al., 1965).

bracket reduces to $2\beta = 2p/m$ and with $2mp = mp_L = m^2 v_L$, we obtain

$$\sigma_A(\beta \to 0) = \pi\alpha^2(2pm)^{-1} = \pi r_e^2 v_L^{-1}. \tag{8.12}$$

The absorption of photons by ordinary matter depends strongly on the photon energy (fig. 3-8.2). The photoelectric cross section $\sigma_F$ dominates at low energies, next comes the Compton cross section $\sigma_C$, and above 20 MeV the pair production $\sigma_P$ dominates. The Compton effect differs from the other absorption processes in the sense that the photon is re-emitted (at a different angle and frequency). It is sometimes useful to separate $\sigma_C$ into one part $\sigma_{C,\,\mathrm{sc}}$, which is proportional to the expectation value of the outgoing photon energy, and another part $\sigma_{C,\,\mathrm{abs}}$ which is proportional to the expectation value of the recoil energy:

$$\sigma_C = \sigma_{C,\,\mathrm{sc}} + \sigma_{C,\,\mathrm{abs}}, \qquad \sigma_{C,\,\mathrm{sc}} = \int \frac{k'}{k}\, d\sigma_C, \qquad \sigma_{C,\,\mathrm{abs}} = \int \left(1 - \frac{k'}{k}\right) d\sigma_C \tag{8.13}$$

The corresponding mass absorption coefficients are denoted by $\Lambda_{C,\,\mathrm{sc}}^{-1}$ and $\Lambda_{C,\,\mathrm{abs}}^{-1}$.

## 3-9 Selection Rules and Positronium Decay

Invariance under charge conjugation was shown in section 1-1 for the Klein-Gordon equation and in section 1-5 for the Dirac equation. As the 4-potential $A^\mu(x)$, the pion field $\psi(x)$ and the electron field $\Psi(x)$ are now interpreted as the operators (0.2), (2.1) and (2.3), we rewrite the charge conjugation (1-1.16), (1-6.24) as

$$A_c^\mu(x) = -A^\mu(x), \qquad \psi_c(x) = \eta_c \psi^+(x), \qquad \Psi_c(x) = i\eta_c \gamma^2 \Psi^+(x), \tag{9.1}$$

where the Hermitean conjugation in $\psi$ and $\Psi$ transforms creation operators into annihilation operators and vice versa. A state which contains nothing but $n$ photons is obtained by applying the operator $A^\mu$ $n$ times to the vacuum state. Therefore it is an eigenstate of the "charge conjugation" operator $C$, with eigenvalue $(-1)^n$:

$$A_c^\mu \equiv C A^\mu C^{-1}, \qquad C\,|\gamma_1 \cdots \gamma_n\rangle = (-1)^n\,|\gamma_1 \cdots \gamma_n\rangle. \tag{9.2}$$

We shall see that the strong interactions are also $C$-invariant, but not the weak ones. Therefore, an even number of photons cannot transform into an odd number of photons except possibly by a weak interaction. This statement remains obviously true for virtual photons, such that graphs of the type of fig. 3-9 vanish (Furry's theorem).

The Hermitean conjugation in (1) implies that a state containing one charged particle is transformed into the corresponding antiparticle state. Particle-antiparticle pair states with permutation symmetry $s = \pm 1$ are again eigenstates of $C$, with eigenvalues $\pm s$. Consider for example a $\pi^+\pi^-$ state of angular momentum $L$. Such a state is generated by $\psi^+(x)\psi(y)\,|0\rangle$. It

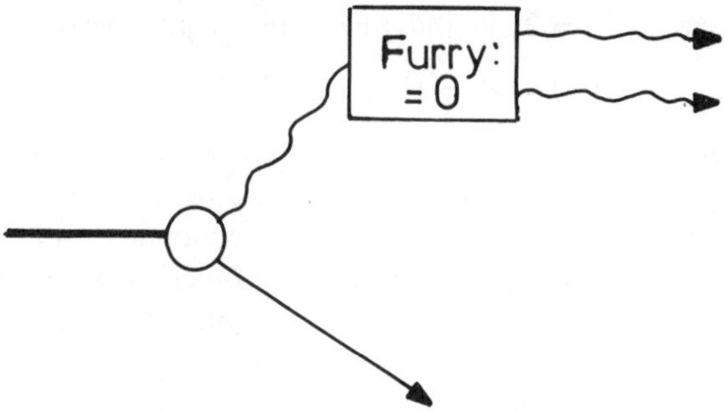


**Figure 3-9** A class of graphs which vanish by Furry's theorem.

is charge-conjugated into a state where the $\pi^+$ position is occupied by $\pi^-$ and vice versa (the phase $\eta_c$ of (1) drops out). That state differs from the original state only by a factor $(-1)^L$, such that we have

$$C\,|\pi^+, \pi^-, L, M\rangle = (-1)^L\,|\pi^+, \pi^-, L, M\rangle. \tag{9.3}$$

Electron-positron states are more complicated. The total angular momentum $\mathbf{J}$ can be split into an orbital part $\mathbf{L}$ and a total spin part, $\mathbf{S} = \mathbf{S}_1 + \mathbf{S}_2$ with eigenvalues $L(L+1)$ and $S(S+1)$ of $\mathbf{L}^2$ and $\mathbf{S}^2$, respectively. The total spin eigenfunctions are symmetric for $S = 1$ and antisymmetric for $S = 0$, such that the orbital spin permutation symmetry is $(-1)^{L+S+1}$. The charge conjugation eigenvalue has an extra minus sign, however, from the restoration of the original sequence of operators (compare 0.8):

$$C\,|e^+e^-, L, S, J, M\rangle = (-1)^{L+S}\,|e^+e^-, L, S, J, M\rangle. \tag{9.4}$$

The Coulomb bound states of $e^+e^-$ pairs are called "positronium." (2) and (4) imply that only states with $L + S$ even can annihilate into two photons. In particular, the triplet $S$-state $^3S_1$ can only annihilate into three photons (see table 3-9). This is also verified by direct calculation. From eq. (8.3) we see that for $\mathbf{p}_+ \to 0$

$$T_A \to -e^2(E_+ - m)^{-\frac{1}{2}}\chi_0'^{+}\mathbf{p}_+\boldsymbol{\sigma}Q_{kg}\chi_0 = -e^2\chi_0'^{+}(-M_+)\boldsymbol{\sigma}\hat{\mathbf{p}}Q_{kg}\chi_0(2m)^{\frac{1}{2}}, \tag{9.5}$$

and from (8.2) or (7.4) one can then derive the desired result, observing $\mathbf{k}_1 = -\mathbf{k}_2$. The full matrix element must be evaluated with the proper bound state wave function $\psi(r)$ instead of plane waves. The resulting decay rate is obtained from (2-3.21) and (8.12), with $\mu = \frac{1}{2}m$:

$$\Gamma_B = \pi\alpha^2(2pm)^{-1}\,|\psi(0)|^2 p/\mu = \tfrac{1}{8}m\alpha^5. \tag{9.6}$$

**Table 3-9** The decays of positronium ground slates (Stroscio 1975).

| Positronium state | $C$ | Decay channel | $\Gamma^{-1}$ [sec] |
|---|---|---|---|
| $^1S_0$ (paraposit.) | 1 | $\gamma\gamma$ | $1.2523 \times 10^{-10}$ |
| $^3S_1$ (orthoposit.) | $-1$ | $\gamma\gamma\gamma$ | $1.3808 \times 10^{-7}$ |

This result holds for uncorrelated spins. Out of the 4 possible spin combinations, only one has total spin 0. Thus the decay rate of parapositronium, which is the only $L = 0$ state that can decay into two photons, is $4\Gamma_B$. The decay rate of orthopositronium is given by $e^+e^- \to \gamma\gamma\gamma$, which is smaller by $\sim \alpha/\pi$ (see Akhiezer and Berestetskii (1965), Landau and Lifshitz (1971)). Theoretical lifetimes (including radiative corrections) are given in table 3-9.

It is also interesting to note that the relative intrinsic parity of a fermion and its antifermion is negative. In fact, application of the parity operator to the charge conjugate spin state $v$ of (1-6.26) gives

$$\begin{aligned} \gamma_0 v(\mathbf{p}, M) &= i\gamma_0 \gamma^2 u^*(\mathbf{p}, M) = -i\gamma^2[\gamma_0 u(\mathbf{p}, M)]^* \\ &= -i\gamma^2 u^*(-\mathbf{p}, M) = -v(-\mathbf{p}, M). \end{aligned} \tag{9.7}$$

Thus the parity of an electron-positron state of orbital angular momentum $L$ is $(-1)^{L+1}$. In particular, it is the triplet $S$-state and not the singlet $P$-state which couples to one virtual photon. This is also evident from the fact that the matrix elements (6.26) do not go to zero for $p \to 0$. A virtual photon at rest has of course negative parity, as the vector potential $\mathbf{A}(x)$ which creates photons is a proper vector and not a pseudovector. Thus the reaction $e^+e^- \to \pi^+\pi^-$ goes exclusively from an initial $s$-state to a final $p$-state!

Combination of (4) and the parity transformation gives

$$C\mathscr{P}\,|e\bar{e}, L, S, J, M\rangle = (-1)^{S+1}\,|e\bar{e}, L, S, J, M\rangle, \tag{9.8}$$

which shows that $S$ is also conserved in the presence of weak interactions.

## 3-10 Vacuum Polarisation and Charge Renormalization

So far we have only calculated the lowest nonvanishing order in the perturbation expansion (3.15) for the operator $S$. From the general expression (4.5) and the fact that $H_I = -eA_0^\mu j_\mu$ is linear in the photon creation and annihilation operators, it follows that the next nonvanishing order must contain two extra powers of $H_I$. For a two-particle reaction $ab \to cd$, this means that we must go from $S^{(2)}$ as given by (4.4) to $S^{(4)}$. This operator is so lengthy that we shall not write it down.

Some matrix elements of $S^{(4)}$ have already been indicated in fig. 3-5.4. Graph (a) is the iteration of the lowest-order graph and is included in the potential treatment. Graphs (c) and (d) give no observable effects. When the upper particle $e$ is replaced by a heavier charged particle $a$, then graph (b) is small in comparison with the "vacuum polarization" graph of fig. 3-10, at least for small values of $-t$. This follows from the fact that the former graphs contain two "massive propagators," whereas the vacuum polarization $S^{(4)}_{\text{vac}}$ contains two electron propagators instead.

A calculation of $S^{(4)}_{\text{vac}}$ from the Feynman rules is rather tedious (Bjorken and Drell 1965). We adopt the more elegant method of Källén (1958), which exploits the analyticity properties of $S^{(4)}_{\text{vac}}$. The first iteration $j_\mu^{(1)}$ of the full

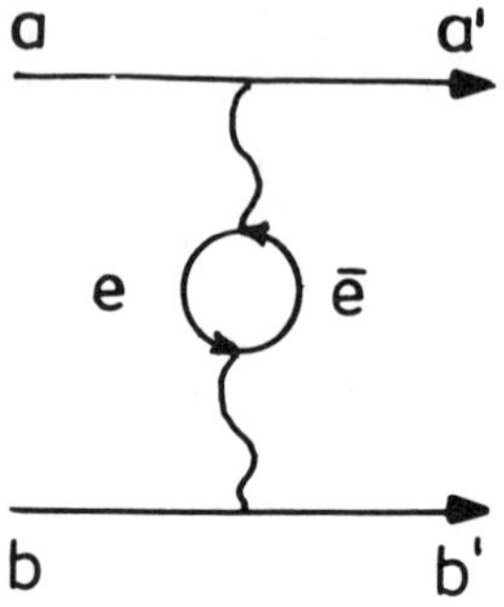


**Figure 3-10** The vacuum polarisation graph.

current operator $\frac{1}{2}Z[\bar{\Psi}, \gamma_\mu \Psi]$ is obtained by inserting (3.4) for $\Psi$, the corresponding equation for $\bar{\Psi}$ and by keeping first-order terms in $e$ only. With $Z^2 = 1$,

$$j_\mu^{(1)}(x) = e \int d^4x' \tfrac{1}{2}\{[\bar{\Psi}_0(x), \gamma_\mu S_R(x-x')\gamma_\nu \Psi_0(x')] + [\bar{\Psi}_0(x')\gamma_\nu \tilde{S}_R(x-x'), \gamma_\mu \Psi_0(x)]\}A^\nu(x'), \tag{10.1}$$

where $\tilde{S}_R(x - x')$ has the $i$ of (3.3) replaced by $-i$ according to the rules stated in connection with (A-4.8) ("pulling $\gamma_0$ through"). $S_{\text{vac}}^{(4)}$ is now obtained from $S_{if}^{(2)}$ of (2-7.3) by replacing $j_{\text{tar},\,\mu}(x)$ by $j_\mu^{(1)}(x)$ and inserting for the $A^\nu(x')$ of (1)

$$A^\nu(x') = e \int d^4z\, D_F(x'-z) j_{\text{tar}}^\nu(z). \tag{10.2}$$

Moreover, as the elastic scattering $ab \to a'b'$ contains no electrons or positrons among the in- and outgoing particles, the vacuum expectation value of the curly bracket in (1) is understood. We define the vacuum polarization tensor

$$\Pi_{\mu\nu}(x-x') = e^2 \tfrac{1}{2}\langle 0 | [\bar{\Psi}_0(x), \gamma_\mu S_R(x-x')\gamma_\nu \Psi_0(x')] + [\bar{\Psi}_0(x')\gamma_\nu \tilde{S}_R(x-x'), \gamma_\mu \Psi_0(x)] | 0 \rangle, \tag{10.3}$$

and obtain for $S_{\text{vac}}^{(4)}$

$$S_{\text{vac}}^{(4)} = ie^2 \int d^4y\, d^4z j_{aa'}^\mu(y) j_{bb'}^\nu(z) \int d^4x\, d^4x'\, D_F(y-x) D_F(x'-z) \Pi_{\mu\nu}(x-x'), \tag{10.4}$$

which is again symmetric in projectile and target. Introducing now the Fourier transform

$$\Pi_{\mu\nu}(x-x') = (2\pi)^{-4} \int d^4q e^{-iq(x-x')}\, \tilde{\Pi}_{\mu\nu}(q) \tag{10.5}$$

as well as the Fourier transforms (2-7.2) of $D_F(y-x)$ and $D_F(x'-z)$, we can perform the integrations over $y$, $z$, $x$, and $x'$. Each integration gives a $\delta_4$-function of 4-momenta. Three of these $\delta_4$-functions are cancelled by the Fourier integrals over $d^4q$, $d^4K$ and $d^4K'$, and one $\delta_4$-function is left, expressing energy-momentum conservation as in (2-7.5). The result is

$$T^{(4)}_{\text{vac}} = e^2 t^{-2} J^{\mu}_{aa'} J^{\nu}_{bb'} \tilde{\Pi}_{\mu\nu}(q), \qquad q^2 = t. \tag{10.6}$$

Current conservation implies $\partial^{\mu}\Pi_{\mu\nu} = \partial^{\nu}\Pi_{\mu\nu} = 0$. For $\tilde{\Pi}(q)$ this means $q^{\mu}\tilde{\Pi}_{\mu\nu} = q^{\nu}\tilde{\Pi}_{\mu\nu} = 0$ and shows that $\tilde{\Pi}_{\mu\nu}$ must be of the form

$$\tilde{\Pi}_{\mu\nu}(q) = (q_{\mu}q_{\nu} - g_{\mu\nu}t)\Pi(t). \tag{10.7}$$

For $t > 0$, $\Pi$ could also depend on the sign of $q_0$ (see the next section). The $q_{\mu}q_{\nu}$-term does not contribute to (6) because of $J^{\mu}q_{\mu} = 0$. We can therefore combine $T^{(4)}_{\text{vac}}$ with $T^{(2)}_{if}$ of (2-7.6) into

$$T'_{if} = T^{(2)}_{if} + T^{(4)}_{\text{vac}} = e^2 t^{-1} J^{\mu}_{aa'} J_{bb'\mu}[1 - \Pi(t)]. \tag{10.8}$$

The function $\Pi(t)$ will be evaluated in the next section. Here we only anticipate $\Pi(0) > 0$. This implies a strange "renormalization" of the elementary electric charge $e$. The measurable charge $e_R$ is defined by Coulomb's law, which follows from (8) in the limit $t \to 0$:

$$e_R^2 = e^2(1 - \Pi(0)) = 4\pi\alpha, \qquad \alpha = (136.036)^{-1}. \tag{10.9}$$

We must conclude that it is impossible to measure $e$ or $\Pi(0)$ independently by Coulomb's law. We therefore rewrite (8) as

$$T'_{if} = e_R^2 t^{-1} J^{\mu}_{aa'} J_{bb'\mu}[1 + \Pi(0) - \Pi(t)] + 0(e^6) \tag{10.10}$$

to first order in $\Pi$. The measurable quantities are now $e_R$ and $\Pi(0) - \Pi(t)$. The error of (10) is of the order of $e^6$, but unfortunately we do not know $e$. In any case, $e$ should be larger than $e_R$. A corresponding renormalization occurs in the 6th and higher orders of the perturbation expansion (see the textbooks mentioned in the preface), such that all observable corrections contain powers of $e_R^2$ instead of $e^2$. It is likely that the expansion is only semiconvergent, i.e., from a certain power of $e_R^2$ onwards, its terms increase again. In practice, however, the first few terms give excellent results, and the higher terms are too complicated for practical computations. In the following sections the index $R$ on $e_R$ is again omitted. Thus, from here on, $e$ denotes the observable elementary charge, and $T'_{if}$ is given by (10) in the form

$$T'_{if} = e^2 t^{-1} J_{aa'} \cdot J_{bb'}[1 + \Pi(0) - \Pi(t)] \tag{10.11}$$

It is instructive to consider what would have happened if gauge invariance $q^{\mu}\tilde{\Pi}_{\mu\nu}(q) = 0$ were broken. In that case, (7) would be of the general structure

$$\tilde{\Pi}_{\mu\nu}(q) = \Pi_1 q_{\mu}q_{\nu} + \Pi_2 g_{\mu\nu}, \tag{10.12}$$

the square bracket of (8) would contain $1 - t^{-1}\Pi_2$, and the $t$-dependence of the "modified propagator" in (8) would be

$$t^{-1}[1 + t^{-1}\Pi_2] = [t - \Pi_2]^{-1} + 0(e^4) \tag{10.13}$$

which shows that the pole in the photon propagator would be shifted from $t = 0$ to $t = \Pi_2(0)$, i.e., the photon would have a mass $\Pi_2(0)^{\frac{1}{2}}$. (The second form of (13) must be used, because its imaginary part near $t =$ (physical mass)$^2$ is given by unitarity, as we shall see in section 6-1). As the photon mass is known to be zero with very high accuracy, one can say that gauge invariance (and current conservation) are required by the masslessness of the photon. This should not be confused with the conservation of electric charge, which is an integral quantity similar to baryon number.

## 3-11 Unitarity and Analyticity. Uehling Potential

Because of the retarded Green's function $S_R$ of (3.3) in (10.3), $\Pi_{\mu\nu}(x - x')$ is zero for $x^0 - x^{0\prime} < 0$:

$$\Pi_{\mu\nu}(x^0 < 0, \mathbf{x}) = 0. \tag{11.1}$$

Inverting (10.5) and allowing complex $q_0$, we see that

$$\tilde{\Pi}_{\mu\nu}(q) = \int d^3x e^{-i\mathbf{q}\mathbf{x}} \int_0^\infty dx^0 e^{ix^0 \operatorname{Re} q_0} e^{-x^0 \operatorname{Im} q_0}\, \Pi_{\mu\nu}(x^0, \mathbf{x}) \tag{11.2}$$

is an analytic function of complex $q_0$ which falls off exponentially in the upper half of the $q_0$-plane ($\operatorname{Im} q_0 > 0$). We can thus apply the residue method (2-6.9) for an analytic function $f(z)$. Changing to a counterclockwise integration in the upper half plane, we have for $\operatorname{Im} z > 0$

$$f(z) = (2\pi i)^{-1} \oint dz' f(z')(z' - z)^{-1} = (2\pi i)^{-1} \int_{-\infty}^{\infty} dx f(x)(x - z)^{-1}, \tag{11.3}$$

since the integration over the upper semicircle with $|z'| = |q_0'| = \infty$ vanishes for an exponentially damped function. Next, we compute the imaginary part of $\Pi(t)$ for real $t$. We assume spinless particles $a$ and $b$ for convenience, in which case $J_{aa'}$ and $J_{bb'}$ of (10.11) are real:

$$\operatorname{Im} T'_{if} = -e^2 t^{-1} J_{aa'} J_{bb'} \operatorname{Im} \Pi(t). \tag{11.4}$$

Time-reversal invariance for the scattering of spinless particles requires simply $T_{if} = T_{fi}$, such that the unitarity condition (2-5.3) can be rewritten as

$$\operatorname{Im} T_{if} = \frac{1}{2} \sum_k \int d \text{ Lips } (s; k) T_{ik} T^*_{fk}, \tag{11.5}$$

where $k$ includes all open channels. Presently we are only interested in $T'_{if}$ which is that part of $T_{if}$ where the particles $a$ and $b$ are connected by a single photon line. We first assume $t > 0$, $q_0 > 0$. In that case we can go to the rest frame of the virtual photon ($\mathbf{q} = 0$) and consider instead of $ab \to a'b'$ the reaction $a\bar{a} \to b\bar{b}$ at $t = (P_a + P_{\bar{a}})^2$, which has the same matrix elements according to the substitution rule. Possible channels are then $k = e\bar{e}$, $\mu\bar{\mu}$, $\pi^+\pi^-$, ..., depending on the value of $t$ (fig. 3-11.1c). Thus

$$\text{Im } T'_{a\bar{a},\, b\bar{b}} = \frac{1}{2} \sum_{\lambda_e \lambda_{\bar{e}}} \int d \text{ Lips } (t, e\bar{e}) T_{a\bar{a},\, e\bar{e}} T^*_{e\bar{e},\, b\bar{b}} + \cdots \tag{11.6}$$

$$T_{a\bar{a},\, e\bar{e}} = -e^2 t^{-1} J^\mu_{a\bar{a}} \bar{u}(P_e) \gamma_\mu v(P_{\bar{e}}), \ldots \tag{11.7}$$

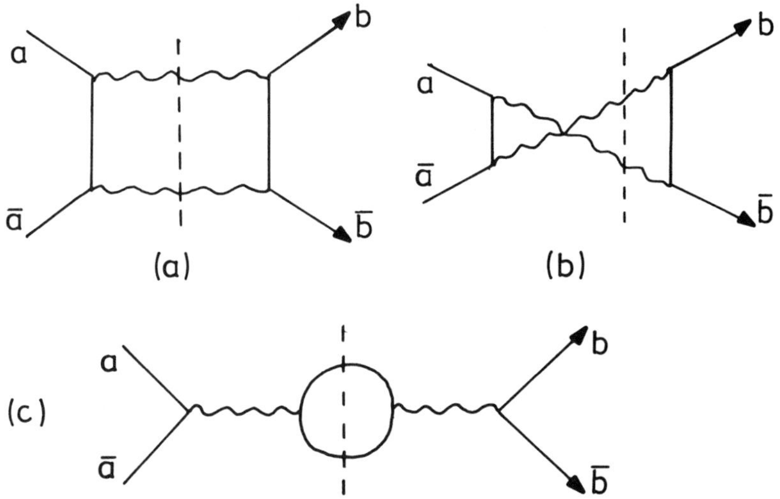


**Figure 3-11.1** Illustration of the unitarity equation (1) for $i = a\bar{a}$, $j = b\bar{b}$ by "cut" or "unitary" diagrams. Particles cut by the vertical dashed line are real. The dashed line implies integration over $d$ Lips $(s; k)$ and summation over the helicities of the cut particles.

$T_{a\bar{a},\, e\bar{e}}$ and $T_{e\bar{e},\, b\bar{b}}$ are used to second order in $e$, which gives (6) to fourth order. Performing now the spin summation as in (6.8), we find that the first term of (6) is in fact of the form (4), with

$$\text{Im } \Pi_{e\bar{e}}(t) = t^{-1} \tfrac{2}{3} \alpha p_e(t) t^{-\frac{1}{2}} (2m_e^2 + t) = t^{-1} m_\gamma \Gamma(\hat{\gamma} \to e\bar{e}) \tag{11.8}$$

in the notation of (6.17). Although we have only included the electron field in the current operator (10.1), it is obvious that the fields of heavier particles should also be included, and that (10.11) contains the full vacuum polarisation:

$$\Pi(t) = \sum_{k = e\bar{e},\, \mu\bar{\mu},\, \pi^+\pi^-} \Pi_k(t), \qquad \text{Im } \Pi_k(t) = t^{-1} m_\gamma \Gamma(\hat{\gamma} \to k). \tag{11.9}$$

Note that for $q_0 > 0$

$$\text{Im } \Pi(0 < t < 4m_e^2) = 0, \qquad \text{Im } \Pi(0 < t < 4m_\mu^2) = \text{Im } \Pi_{e\bar{e}}. \tag{11.10}$$

We can now find $\Pi(q_0 = t^{\frac{1}{2}})$ in the lower half-plane Im $q_0 < 0$, at least for Re $q_0 > 0$. An analytic function $f(z)$ can be decomposed into "real-analytic" and "imaginary-analytic" functions $h(z)$ and $g(z)$ as follows

$$f(z) = h(z) + g(z), \qquad h^*(z) = h(z^*), \qquad g^*(t) = -g(z^*). \quad (11.11)$$

For real $z$, $h$ is real and $g$ is imaginary. Moreover, if an analytic function is zero on a line, it is zero everywhere. On the real axis we may put $g = \text{Im}\,\Pi$. Rewriting (10) as $g(0 < q_0 < 2m_e) = 0$, we find $g = 0$. Consequently $\Pi(q_0)$ is real-analytic:

$$\Pi(q_0^*) = \Pi^*(q_0). \quad (11.12)$$

Finally, the substitution rule tells us that we may replace all ingoing particles by outgoing antiparticles and vice versa if we at the same time invert all 4-momenta. This leads us from $T'_{a\bar{a},\, b\bar{b}}(q_0)$ to $T'_{b\bar{b},\, a\bar{a}}(-q_0)$. Since $\Pi$ contains no reference to the external particles except through $q^\mu$, we have in the cms

$$\Pi(q_0) = \Pi(-q_0) \quad (11.13)$$

(also for complex $q_0$). With (10), (12) and (13), we can expand the integration contour into the lower half of the $q_0$-plane as indicated in fig. 3-11.2. The contour cannot cross the points $q_0 = \pm 2m_e$ because the function $p_e = (q_0^2/4 - m_e^2)^{\frac{1}{2}}$ of (8) has its branch points there. The analytic function $\Pi(q_0)$ is thus unique only in the cut $q_0$-plane. The cuts are made from $\pm 2m_e$ to $\pm\infty$. The integral along the lower semicircle at $|q_0| = \infty$ is again negligible. In the integrations above and below the cuts, Re $\Pi$ cancels out, and Im $\Pi$ is doubled. Identifying $f(z) = \Pi(q_0)$, we obtain from (3), (12) and (13)

$$\begin{aligned}\Pi(q_0 + i\varepsilon) &= \frac{1}{\pi}\left[\int_{-\infty}^{-2m_e} + \int_{2m_e}^{\infty}\right]\frac{dx}{x - q_0 - i\varepsilon}\,\text{Im}\,\Pi(x)\\ &= \frac{1}{\pi}\int_{2m_e}^{\infty} dx\left(\frac{1}{x - q_0 - i\varepsilon} + \frac{1}{x + q_0 - i\varepsilon}\right)\text{Im}\,\Pi(x).\end{aligned} \quad (11.14)$$

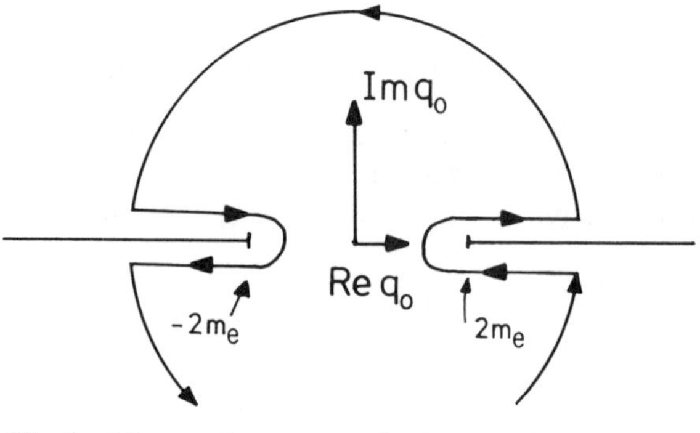


**Figure 3-11.2** The final integration contour in the complex $q_0$-plane for the Cauchy integral for $\Pi(q_0)$.

Returning now to the variable $t = q_0^2$, we can rewrite (14) as

$$\Pi(t + i\varepsilon) = \frac{1}{\pi}\int_{t_0}^{\infty} \frac{dt'}{t' - t - i\varepsilon}\,\mathrm{Im}\,\Pi(t'), \qquad t_0 = 4m_e^2. \tag{11.15}$$

Thus the two cuts in $q_0$ coalesce into one cut along $t > t_0$, and (15) can be used also for $t < 0$. The denominators of (15) can be rewritten as

$$(x - i\varepsilon)^{-1} = \frac{x}{x^2 + \varepsilon^2} + \frac{i\varepsilon}{x^2 + \varepsilon^2} = \frac{P}{x} + i\pi\,\delta(x), \tag{11.16}$$

where $P$ stands for the "principal value": in the integration over $x$, a vanishingly small interval $\delta$ is omitted on either side of the denominator. This is possible since

$$\int_{-\delta}^{\delta} x\,dx f(x)(x^2 + \varepsilon^2)^{-1} = \int_{-\delta/\varepsilon}^{\delta/\varepsilon} \frac{y\,dy}{y^2 + 1} f(\varepsilon y) \to f(0)\int_{-\delta/\varepsilon}^{\delta/\varepsilon} \frac{y\,dy}{y^2 + 1} = 0 \tag{11.17}$$

for any function $f(x)$ which is continuous at $x = 0$. Similarly, the $\delta$-function follows from

$$\int_{x_1}^{x_2} \frac{\varepsilon\,dx}{x^2 + \varepsilon^2} = \tan^{-1}\left(\frac{x}{\varepsilon}\right)\Bigg|_{x_1}^{x_2} = \begin{cases}\pi & \text{for } x_1 < 0 < x_2, \\ 0 & \text{otherwise.}\end{cases} \tag{11.18}$$

Thus the imaginary part of (15) is an identity for real $t$, and the real part is

$$\mathrm{Re}\,\Pi(t) = \frac{P}{\pi}\int_{t_0}^{\infty} dt'(t' - t)^{-1}\,\mathrm{Im}\,\Pi(t'), \tag{11.19}$$

with the understanding that the "physical" $\Pi(t)$ occurs at the upper edge of the cut. (19) is called a "dispersion relation" or a "spectral representation," $\mathrm{Im}\,\Pi(t)$ being the "spectral density." These names come from the Kramers-Kronig dispersion relations in optics.

Before evaluating (19) we make a few general comments:

Firstly, the unitarity equation (5) was defined originally for physical $i$ and $f$, i.e., $t > 4m_a^2$, $t > 4m_b^2$. Equation (10) on the other hand is independent of the external masses, which is seen from (10.3). Therefore, (17) can be used to define $\mathrm{Im}\,T_{a\bar{a},\,b\bar{b}}$ also below $4m_a^2$ or $4m_b^2$. This is called "extended unitarity" (see also section 6-1).

Secondly, (1) follows from causality (namely the use of retarded Green's functions) and the local nature of the interaction. The combination of these two properties is called "microcausality." It is believed that this property applies also to the contributions of $\pi^+\pi^-$ or $p\bar{p}$ pairs although these particles have strong form factors. For example, $\Pi_\pi$ can be calculated from $\Gamma(\hat{\gamma} \to \pi^+\pi^-)$ provided the pion form factor $F$ which enters (6.18) is known.

Thirdly, the prime in $T'_{a\bar{a},\,b\bar{b}}$ (which signals the one-photon exchange) can be dropped, if the sum in (6) includes also other open channels such as $\gamma\gamma$,

$e\bar{e}\gamma$, etc. (purely hadronic interactions are considered later). To fourth order in $e$, $e\bar{e}\gamma$ does not contribute, and we have

$$\operatorname{Im} T^{(4)}_{a\bar{a},\,b\bar{b}}(0 < t < 4m_e^2) = \frac{1}{2} \sum_{\lambda_1\lambda_2} \int d \operatorname{Lips}(t;\gamma\gamma) T^{(2)}_{a\bar{a},\,\gamma\gamma} T^{(2)*}_{\gamma\gamma,\,b\bar{b}} \quad (11.20)$$

where $T^{(2)}$ is the lowest-order Born approximation. Since each $T$ contains two diagrams (compare fig. 3-8), (20) contains 4 combinations of diagrams, which are pairwise identical, however. The resulting two inequivalent combinations are illustrated by the "cut" or "unitary" diagrams (a) and (b) of fig. 3-11.1. They represent parts of the "Feynman diagrams" (a) and (b) of fig. 3-5.4. We shall learn in chapter 6 that also such complicated diagrams can be calculated from analyticity and unitarity. Another relatively simple application is given in 3-12 below. In all these cases, the necessary analyticity can be derived from microcausality, at least for the pure QED problems. However, we shall simply assume in the following that the $T$-matrix elements are analytic functions and derive their singularities from the unitarity equation.

Fourthly, the method of calculating matrix elements from unitarity and analyticity applies to all orders of the perturbation theory and is hopefully independent of the expansion. Thus, if one knows that $e\bar{e}$ bound states exist (which is a nonperturbative result), such states must in principle be included in the unitarity equation (6). Examples of this type will be studied in chapter 6.

Finally, we must face the possibility that the integral (3) diverges. In fact, (8) can be rewritten as

$$\operatorname{Im} \Pi_{e\bar{e}}(t) = \frac{\alpha}{3}(1 - 4m_e^2/t)^{\frac{1}{2}}(1 + 2m_e^2/t), \quad (11.21)$$

which tends to $\alpha/3$ for $t \to \infty$ and shows that (19) does not exist. In such a case, one must first subtract from $f(z)$ the Cauchy integral for $f(z_1)$ at a fixed point $z_1$:

$$\begin{aligned} f(z) - f(z_1) &= (2\pi i)^{-1} \oint dz' f(z') \left( \frac{1}{z' - z} - \frac{1}{z' - z_1} \right) \\ &= \frac{z - z_1}{2\pi i} \oint \frac{dz' f(z')}{(z' - z)(z' - z_1)}. \end{aligned} \quad (11.22)$$

This expression has one extra power of $z'$ in the denominator. For the vacuum polarisation, we know already from (10.11) that $\Pi(0) - \Pi(t)$ is the physically relevant quantity. We therefore take $z_1 = 0$ in (22) and obtain the final result (for $t < 4m_e^2$):

$$\Pi_{e\bar{e}}(t) - \Pi_{e\bar{e}}(0) = \frac{\alpha}{3\pi} t \int_{4m_e^2}^{\infty} \frac{dt'}{t'} (1 - 4m_e^2/t')^{\frac{1}{2}} (1 + 2m_e^2/t')(t' - t)^{-1}. \quad (11.23)$$

Generally speaking, if $f(z)$ diverges as $|z|^{n-\alpha}$ for $|z| \to \infty$ $(0 < \alpha < 1, n = 1, 2, 3 \ldots)$, a dispersion relation with "$n$ subtractions" must be used. Each subtraction introduces an unknown constant. In our example this is $\Pi(0)$.

From (10.11) it is clear that the vacuum polarisation can be included in the construction of a potential, using the method of (2-6.34). The new term is called the Uehling potential, the total potential is called the "electric potential":

$$V_{el} = V + V_U, \; V_U = \frac{\alpha}{3\pi}\int_{4m^2}^{\infty}\frac{ds}{s}\left(1 + 2\frac{m^2}{s}\right)\left(1 - 4\frac{m^2}{s}\right)^{\frac{1}{2}} V(r, s) \quad (11.24)$$

where $V(r, s)$ is given by (2-9.10) but with $q^2$ in the denominator replaced by $s + q^2$. As $2m_e$ is much smaller than the inverse nuclear radius, there is a region in $r$ for which $V_U$ can be evaluated using $F(q^2) = 1$:

$$V(r, s) = Z_a Z_b \frac{\alpha}{r}\frac{2}{\pi}\int q\, dq (s + q^2)^{-1} \sin qr = Z_a Z_b \frac{\alpha}{r} \exp(-s^{\frac{1}{2}} r) \quad (11.25)$$

$$V_U = \tfrac{2}{3} Z_a Z_b \alpha^2 I(2m_e r)/\pi r, \qquad I(z) \equiv \int_1^{\infty} e^{-\xi z}(1 + \tfrac{1}{2}\xi^{-2})(1 - \xi^{-2})^{\frac{1}{2}} \frac{d\xi}{\xi} \quad (11.26)$$

$$I(z) = (1 + z^2/12)K_0(z) - \tfrac{5}{6}(1 + z^2/10) z K_1(z) + \tfrac{3}{4} z(1 + z^2/9)\int_z^{\infty} K_0(t)\, dt \quad (11.27)$$

$$= \begin{cases} -C - \frac{5}{6} + \ln(2/z) + \frac{3}{8}z(\pi - z) + \pi z^3/24 & \text{for } z \ll 1 \\ \frac{3}{4}(2\pi/z)^{\frac{1}{2}} e^{-z}/z & \text{for } z \gg 1 \end{cases} \quad (11.28)$$

with $C = 0.57772$ = Euler's constant. Note that $V$ and $V_U$ have the same sign.

## 3-12 The Anomalous Magnetic Moment of the Electron

When particle $a$ in the reaction $a\bar{a} \to b\bar{b}$ is an electron, $T_{a\bar{a},\, e\bar{e}}$ on the right-hand side of (11.6) contains two graphs, namely (a) and (b) of fig. 3-6.5. Graph (b) has been treated in the last two sections. We now turn our attention to graph (a). The resulting unitarity diagram is shown in fig. 3-12a. It is responsible for the electron form factors ("vertex corrections") and particularly for the electron's anomalous magnetic moment.

Since the spinors $u_a$ and $\bar{v}_a$ and $\gamma$-matrices occurring in $T_{a\bar{a},\, e\bar{e}}$ are complex, unitarity does not apply in the simple form (11.6) any longer. However, we can exploit the fact that the $e\bar{e}$ state is connected to the remaining external particles $b\bar{b}$ by just one photon, and that therefore $T^{(4)}_{a\bar{a},\, b\bar{b}}$ has the following structure according to (2-7.11):

$$T^{(4)}_{a\bar{a},\, b\bar{b}} = e^2 \bar{v}_a [G_M \gamma_\mu + (2m)^{-1} F_2 (P_{\bar{a}} - P_a)_\mu] u_a J^\mu_{b\bar{b}}/t. \quad (12.1)$$

The same structure must appear on the right-hand side of (11.6), and at the end we can identify the corresponding form factors with the imaginary parts

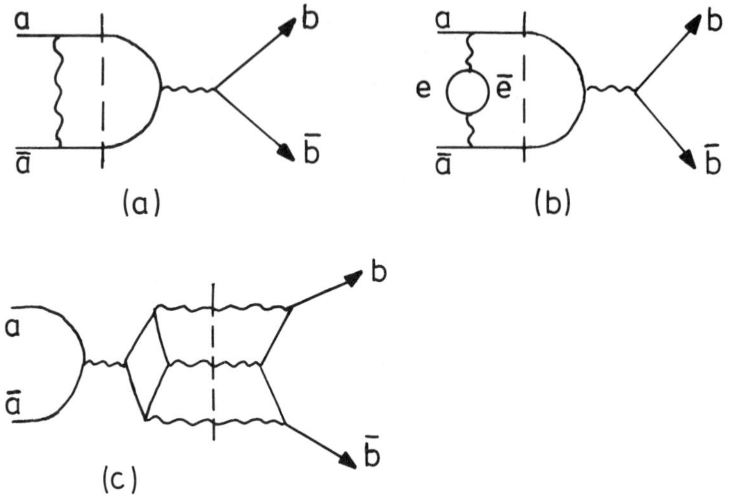


**Figure 3-12** Some "unitary diagrams" for the calculation of $\Lambda_\mu$. (a) the main contributor to the anomalous magnetic moment $\kappa$. (b) the main origin of $\kappa_e - \kappa_\mu$. (c) A "higher-order vacuum polarization" contribution, which is important in the electric potential of high-$Z$ nuclei at relatively large distances.

of $G_M$ and $F_2$, respectively. For the unitarity diagram (a) of fig. 3-12, we have

$$\sum_{\lambda_e\lambda_{\bar e}} T^{(a)}_{a\bar a,\, e\bar e} T_{e\bar e,\, b\bar b} = -e^4 \sum_{\lambda_e\lambda_{\bar e}} \bar v_a \gamma^\sigma v(\lambda_{\bar e})\bar u(\lambda_e)\gamma_\sigma u_a \bar v(\lambda_{\bar e})\gamma_\mu u(\lambda_e) J^\mu_{b\bar b}/tt' \tag{12.2}$$

with $t' = (P_a - P_e)^2$. The signs in (1) and (2) refer to the current $\bar v \gamma u$ which is the charge conjugate of (6.6) and therefore has $Z = +1$. From here on, we adopt the presentation of Landau and Lifschitz (Vol. IVb, 1975). After spin summation with the aid of (A-4.6), we can write

$$\frac{1}{2}\sum_{\lambda_e\lambda_{\bar e}} \int d\ \text{Lips}\ (t;\, e\bar e) T^{(2a)}_{a\bar a,\, e\bar e} T_{e\bar e,\, b\bar b} = e^2 \bar v_a \Lambda_\mu u_a J^\mu_{b\bar b}/t, \tag{12.3}$$

$$\Lambda_\mu = -\tfrac{1}{2}e^2 \int d\ \text{Lips}\ (t;\, e\bar e)\gamma^\sigma(P_{\bar e}\cdot\gamma - m)\gamma_\mu(P_e\cdot\gamma + m)\gamma_\sigma/t'. \tag{12.4}$$

In the part containing the $\gamma$-matrices, we pull both $\gamma^\sigma$ and $\gamma_\sigma$ towards $\gamma_\mu$ and use $\gamma^\sigma\gamma_\mu\gamma_\sigma = -2\gamma_\mu$:

$$\begin{aligned}
&[2P^\sigma_{\bar e} - (P_{\bar e}\cdot\gamma + m)\gamma^\sigma]\gamma_\mu[2P_{e\sigma} - \gamma_\sigma(P_e\cdot\gamma - m)] \\
&\quad = 2[2P_{\bar e}P_e\gamma_\mu - (P_{\bar e} - P_{\bar a})\gamma(P_e\gamma)\gamma_\mu - \gamma_\mu(P_{\bar e}\gamma) \\
&\qquad \times (P_e - P_a)\gamma - (P_{\bar e} - P_{\bar a})\gamma\gamma_\mu(P_e - P_a)\gamma] \\
&\quad = 4[(P_a P_{\bar a} + \tfrac{1}{2}t')\gamma_\mu - m(P_e - P_a)_\mu + (P_a - P_{\bar a})_\mu \\
&\qquad \times (P_e - P_a)\gamma + (P_e - P_a)_\mu(P_e - P_a)\gamma].
\end{aligned} \tag{12.5}$$

In the second expression we have also used $mu_a = P_a \gamma u_a$, $\bar{v}_a m = -\bar{v}_a P_{\bar{a}} \gamma$, and in the last expression we have used $P_{\bar{e}} - P_{\bar{a}} = -(P_e - P_a)$, $P_e P_{\bar{e}} = P_a P_{\bar{a}}$,

$$-(P_{\bar{e}} - P_{\bar{a}})\gamma P_e \gamma = (P_e - P_a)\gamma(P_e - P_a + P_a)\gamma = t' + (P_e - P_a)\gamma m,$$

etc.

The terms in (5) which multiply $\gamma_\mu$ depend only on $t$ and $t'$. For these, we can rewrite $d$ Lips according to (2-2.7) as

$$d \text{ Lips } (t; e\bar{e}) = (8\pi t^{\frac{1}{2}})^{-1} p \, d \cos \theta' = (16\pi t^{\frac{1}{2}} p)^{-1} \, dt', \qquad p \equiv (t/4 - m^2)^{\frac{1}{2}}. \tag{12.6}$$

The integral over $dt'/t'$ which multiplies the first term of (5) diverges at the lower limit $t' = 0$. The divergence arises from small-angle Coulomb scattering in $e\bar{e} \to e\bar{e}$ and is connected with the "infrared-divergence" of the next section. We treat it by attributing a small mass $\lambda$ to the photon, such that $t'$ in the denominator of the photon propagator is replaced by $t' - \lambda^2$. The divergent integral is then

$$I \equiv \int_{-4p^2}^{0} dt'(\lambda^2 - t')^{-1} = -\ln (\lambda^2/4p^2) = \ln (t/m^2 - 4) + 2 \ln (m/\lambda). \tag{12.7}$$

In the next term, we have $\frac{1}{2} \int dt' = 2p^2$. For the remaining terms, we use the cms with $\mathbf{p}_a = -\mathbf{p}_{\bar{a}} \equiv \mathbf{p}$, $(P_e - P_a)^\mu = (0, \mathbf{p}_e - \mathbf{p})$, $(P_a - P_{\bar{a}})^\mu = (0, 2\mathbf{p})$, and $t' = -2p^2(1 - \cos \theta)$. By taking the $z$-axis along $\mathbf{p}$ we get

$$\int \frac{d \cos \theta}{1 - \cos \theta} (\mathbf{p}_e - \mathbf{p}) = -2\mathbf{p}, \int \frac{d \cos \theta}{1 - \cos \theta} (\mathbf{p}_e - \mathbf{p})_i (\mathbf{p}_e - \mathbf{p})_j = p_i p_j + p^2 \delta_{ij}; \tag{12.8}$$

$$\begin{aligned} \Lambda = -\alpha(2pt^{\frac{1}{2}})^{-1} \Big\{ & (-P_a P_{\bar{a}} I + 2p^2)\gamma - \int \frac{d \cos \theta}{1 - \cos \theta} \\ & \times [-m(\mathbf{p}_e - \mathbf{p}) - 2\mathbf{p}((\mathbf{p}_e - \mathbf{p})\boldsymbol{\gamma}) - (\mathbf{p}_e - \mathbf{p})((\mathbf{p}_e - \mathbf{p})\boldsymbol{\gamma})] \Big\} \end{aligned} \tag{12.9}$$

$$\begin{aligned} &= \alpha(2pt^{\frac{1}{2}})^{-1}[(P_a P_{\bar{a}} I - 3p^2)\gamma + 2m\mathbf{p} + 4\mathbf{p}(\mathbf{p}\boldsymbol{\gamma}) - \mathbf{p}(\mathbf{p}\boldsymbol{\gamma})] \\ &= \alpha(t^2 - 4m^2 t)^{-\frac{1}{2}}[(P_a P_{\bar{a}} I - 3p^2)\gamma - m\mathbf{p}]. \end{aligned}$$

In the last expression, we have used $\bar{v}_a \mathbf{p}\boldsymbol{\gamma} u_a = -m\bar{v}_a u_a$ which follows from $P_a \gamma u_a = mu_a$ and $\bar{v}_a \gamma_0 u_a = 0$. We now see that (1) has the same form as (9) in the cms. Identification of the coefficients of $\gamma$ and $m\mathbf{p}$ gives, with $Z = +1$

$$\text{Im } F_2(t) = Z\alpha m^2 (t^2 - 4m^2 t)^{-\frac{1}{2}}, \tag{12.10}$$

$$\text{Im } G_M(t) = Z\alpha(t^2 - 4m^2 t)^{-\frac{1}{2}}[(\tfrac{1}{2}t - m^2)I - 3(\tfrac{1}{4}t - m^2)]. \tag{12.11}$$

The dispersion relation (11.19) for (10) reads

$$F_2(t) = Z\pi^{-1} \int_{4m^2}^{\infty} \alpha m^2 (t'^2 - 4m^2 t')^{-\frac{1}{2}} (t' - t - i\varepsilon)^{-1} \, dt'. \tag{12.12}$$

Here we are only interested in its value at $t = 0$, which is just the anomalous magnetic moment:

$$\begin{aligned}\kappa &= Z\pi^{-1}\alpha m^2 \int_{4m^2}^{\infty} (t'^2 - 4m^2 t')^{-\frac{1}{2}}\, dt'/t' \\ &= Z\pi^{-1}\alpha\tfrac{1}{2}\,(1 - 4m^2/t')^{\frac{1}{2}}\Big|_{4m^2}^{\infty} = Z\tfrac{1}{2}\pi^{-1}\alpha = 0.00116Z.\end{aligned} \tag{12.13}$$

This value must be included in the Pauli (1-4.5) or Dirac (1-5.14) equations for electrons and muons in an external field. Because of the high precision reached in the experimental determination of $\kappa$ (see table C-1), higher-order corrections have also been calculated (see the review of Lautrup et al., 1972). The 6th order contains 7 different graphs. One of these (Fig. 3-12b) has a vacuum polarization inserted in the photon line. Since this is always dominated by $e\bar{e}$-pairs in the vertex, it gives different contributions for electrons and muons.

For hadrons, strong interactions make themselves felt at short distances and there is in general no point in calculating the QED contributions to $\Lambda_\mu$ for these particles. An exception is again the limit $t \to 0$ which makes itself felt at large distances. In particular, when the target carries a large electric charge $Ze$, "higher-order vacuum polarization" graphs such as that of fig. 3-12c which are proportional to $(Z\alpha)^3$ may contribute. Moreover, the 3-photon state in the unitarity condition contributes already for $0 < t < 4m_e^2$. The corresponding potential has a representation similar to (11.25), and since the lower integration limit of $s$ is now zero, $V(r \to \infty)$ falls off like a power of $r$ rather than exponentially. The same remark applies to the potential of (11.20), which for neutral particles (atoms) dominates at large $r$ (van der Waals potential, see Feinberg and Sucher 1970).

## 3-13 Soft Photons and Electron Vertex Function

The matrix element (5.8) for bremsstrahlung in electron-nucleus scattering diverges in the limit of zero photon momentum, $K^\mu \to 0$ ("soft photon"). This is seen by rewriting the denominators of the electron propagators as $-2P'_a \cdot K$ and $2P_a \cdot K$, using $P'^2_a = P^2_a = m^2$. We also know from the discussion in section 2-10 (by going to the electron rest frame) that all spin dependence of the current matrix element $J^\mu$ disappears in the limit of zero 4-momentum transfer. Therefore, the soft photon limit $T^{\text{soft}}_{ab\gamma}$ of the bremsstrahlung matrix element can be expressed in terms of the elastic scattering amplitude $T_{ab}$:

$$T^{\text{soft}}_{ab\gamma} = Z_a e\varepsilon^*_\mu(k, \lambda)[P'^\mu_a(P'_a \cdot K)^{-1} - P^\mu_a(P_a \cdot K)^{-1}]T_{ab}. \tag{13.1}$$

Target recoil corrections are neglected. Actually, (1) is true for the soft photon emission in the elastic scattering of charged particles of arbitrary spins, and corresponding formulas apply also to particle decays (see section 5-6 for $\Lambda \to p\pi^-\gamma$ decays). The second term in (1) follows from the first one simply by gauge invariance $K_\mu J^\mu = 0$. In the following, we use the rest frame of the incident electron, $P_a^\mu = (m, \mathbf{0})$ and the $z$-axis along $\mathbf{p}_a'$. Then the second term in (1) vanishes and the first becomes by means of (2-8.6)

$$T_{ab\gamma}^{\text{soft}} = -Z_a e \lambda 2^{-\frac{1}{2}} \sin \theta_\gamma p_a' T_{ab}/m_a k, \tag{13.2}$$

which is the typical form of dipole radiation. The bad thing about (2) is not so much that it becomes infinite for $k \to 0$, but that also the cross section, integrated over a very small sphere of $k$-values, $|\mathbf{k}| < \Delta k$ is infinite ("infrared divergence"). With $Z_a^2 = 1$ and after summation over the photon helicity $\lambda$, we have

$$d\sigma_{ab\gamma}^{(\Delta k)} = e^2 p_a'^2 m_a^{-2}\, d\sigma_{ab} \int^{\Delta k} d_L^3 k \sin^2 \theta_\gamma / k^2. \tag{13.3}$$

Inserting $d_L^3 k = k^2\, dk\, d\Omega_\gamma (16\pi^3 \omega_k)^{-1}$ with $\omega_k = |\mathbf{k}|$, we see that (3) diverges logarithmically at the lower limit, $k = 0$. Formally, this can be treated by attributing a small mass $\lambda$ to a real photon and then taking $\lambda \to 0$ in the final formulas for measurable quantities. The integration over $\Omega_\gamma$ in (3) produces a factor $8\pi/3$, such that we have, with $e^2/4\pi = \alpha$,

$$\begin{aligned} d\sigma_{ab\gamma}^{(\Delta k)} &= \frac{2\alpha}{3\pi} p_a'^2 m_a^{-2}\, d\sigma_{ab} \int_0^{\Delta k} dk\, (k^2 + \lambda^2)^{-\frac{1}{2}} \\ &= \frac{2\alpha}{3\pi} p_a'^2 m_a^{-2}\, d\sigma_{ab} \ln \frac{2\Delta k}{\lambda}. \end{aligned} \tag{13.4}$$

With $\Delta k$ sufficiently small, the reaction will be mistaken as elastic scattering by experimental physicists. One should therefore consider (4) together with elastic scattering, which contains the divergent integral (12.7).

We can now obtain $G_M$ from (12.11) via a dispersion relation. Obviously, one subtraction is required. Since we know already $G_M(0) = Z(1 + \alpha/2\pi)$, we subtract at $t = 0$. According to (11.22), we have

$$\text{Re } G_M(t) = G_M(0) + \frac{t}{\pi} P \int dt' \text{ Im } G_M(t') t'^{-1} (t' - t)^{-1} \tag{13.5}$$

We are only interested in $F_1 = G_M - F_2$ near $t = 0$:

$$F_1(t) = Z + tF_1'(0), \tag{13.6}$$

$$\begin{aligned} F_1'(0) &= \frac{1}{\pi} \int_{4m^2}^{\infty} t^{-2}\, dt \text{ Im } F_1(t) \\ &= \frac{Z\alpha}{\pi} \int t^{-\frac{5}{2}}\, dt\, (t - 4m^2)^{-\frac{1}{2}} [(\tfrac{1}{2}t - m^2)I - \tfrac{3}{4}t + 2m^2]. \end{aligned} \tag{13.7}$$

After the anomalous magnetic moment $F_2(0)$, this gives the most important radiative correction for low-energy electrons. The infrared divergent part $F'_{1\lambda}$ is evaluated using

$$\int_{4m^2}^{\infty} t^{-\frac{5}{2}}\, dt\, (t - 4m^2)^{-\frac{1}{2}}$$

$$= -t^{-\frac{3}{2}}(t - 4m^2)^{\frac{1}{2}}\Big|_{4m^2}^{\infty} + \tfrac{1}{6}m^{-2} \int t^{-\frac{3}{2}}\, dt\, (t - 4m^2)^{-\frac{1}{2}} = \tfrac{1}{12}m^{-4}, \tag{13.8}$$

$$F'_{1\lambda}(0) = Z\frac{\alpha}{2\pi}\left(\tfrac{1}{2}m^{-2} - \tfrac{1}{6}m^2/m^4\right) \ln\frac{m}{\lambda} = Z\frac{\alpha}{3\pi} m^{-2} \ln\frac{m}{\lambda}.$$

The remaining terms give

$$\tilde{F}'_1(0) = F'_1(0) - F'_{1\lambda}(0) = Z\alpha(8\pi m^2)^{-1}. \tag{13.9}$$

In the construction of the differential elastic cross section $d\sigma_{ab}$ only terms linear in $\alpha$ must be kept in $F_1^2(t)$ at our stage of perturbation calculation. For small $t$, this gives

$$\begin{aligned} d\sigma_{ab} &= (1 + Z_a^{-1} t F'_1(0))^2\, d\sigma_{ab}^{\text{Born}} \\ &= \left(1 + \frac{2}{3\pi}\alpha t m^{-2} \ln\frac{m}{\lambda} + \frac{\alpha t}{4\pi} m^{-2}\right) d\sigma_{ab}^{\text{Born}}. \end{aligned} \tag{13.10}$$

Adding to this (4) and observing $t = -p_a'^2$ in that system, we find to our surprise that the $\ln \lambda$-terms cancel out. The remainder gives the experimental cross section

$$d\sigma_{ab}^{\text{ex}}(\Delta k) = d\sigma_{ab} + d\sigma_{ab\gamma}^{(\Delta k)} = \left[1 + \frac{2\alpha}{\pi} t m^{-2}\left(\frac{1}{8} - \frac{1}{3}\ln\frac{2\Delta k}{m}\right)\right]\sigma_{ab}^{\text{Born}} \tag{13.11}$$

where $\Delta k$ is the experimental energy resolution. In practice, the resolution will be given as a function of $\mathbf{k}_{\text{lab}}$, such that (11) cannot be used. However, it can be shown that the infrared divergence cancels out for arbitrary $t$, arbitrary differential cross sections $d\sigma_{ab}$ and arbitrary numbers of photons (see Grammer and Yennie 1973).

From (11) we see that $d\sigma_{ab}^{\text{ex}}$ decreases with decreasing $\Delta k$ and may even become negative. Obviously, the perturbation expansion diverges when the photons become sufficiently soft.

The origin of these difficulties can be traced back to the assumption that the asymptotic fields $\Psi_{\text{in}}$ and $\Psi_{\text{out}}$ of section 3 have no interactions whatsoever. This was achieved by the unphysical trick (3.10). In practice, however, a moving charge emits and absorbs photons of almost-zero frequencies even at large distances from a target. The classification into states of given photon numbers looses its meaning here, i.e., the appropriate

Hilbert space is not a Fock space. Instead, one may use "coherent states," which are eigenstates of the annihilation operator $a_i$ of (0.2)

$$|Z_i\rangle = \exp\left(-\tfrac{1}{2}|Z_i|^2\right) \sum_{n_i=0}^{\infty} (Z_i)^{n_i}(n_i!)^{-\frac{1}{2}} |n_i\rangle, \tag{13.12}$$

$$a_i |Z_i\rangle = Z_i |Z_i\rangle, \tag{13.13}$$

according to (0.14). Physical charged particles have a slightly smeared out mass in this formalism. See the review by Jauch and Rohrlich (1976).

We should also say a word about the required subtraction in the dispersion relation (5) or (7). It appears that this entails another redefinition of the electric charge, analogous to the one due to vacuum polarisation. However, it so happens that this effect is exactly cancelled by the "electron wave function renormalization" of graphs (c) and (d) of fig. 3-5.4. Mathematically, this is due to the identity

$$\partial S_F(p)/\partial p_\mu = S_F(p)\gamma_\mu S_F(p), \qquad S_F(P) = (P \cdot \gamma - m)^{-1}, \tag{13.14}$$

which is proved by differentiating $S_F S_F^{-1} = 1$ with respect to $P_\mu$:

$$(\partial S_F/\partial P_\mu)S_F^{-1} + S_F\, \partial\, (P \cdot \gamma - m)/\partial P_\mu = 0. \tag{13.15}$$

It is the basis of the so-called "Ward identity", to which we shall return in another context. Since graphs (c) and (d) of fig. 3-5.4 do not arise from unitarity diagrams, this problem simply disappears in the dispersion relation approach, at least to this order. The coupling constant of the subtracted dispersion relation is always the physical one.

Normally, (1) is valid only to zeroth order in $K^\mu$. However, one can always find a form of comparable simplicity which works to first order in $K^\mu$ (Low 1958). For this purpose one puts

$$T_{ab\gamma} = \varepsilon^{\mu *} J_\mu, \qquad J_\mu = J_\mu^{\text{ext}} + J_\mu^{(0)} + 0_\mu(k), \tag{13.16}$$

where $J_\mu^{\text{ext}}$ contains all terms of order $k^{-1}$ and is gauge invariant, $K^\mu J_\mu^{\text{ext}} = 0$, $J_\mu^{(0)}$ is of zeroth in $k$ and $0_\mu(k)$ contains all the rest, i.e., it is at least linear in $k$. Since $J_\mu^{\text{ext}}$ is separately gauge invariant, gauge invariance requires

$$k^\mu J_\mu^{(0)} = -k^\mu 0_\mu(k), \tag{13.17}$$

and since the right-hand side of (17) is of order $k^2$, $J_\mu^{(0)}$ is in fact zero:

$$T_{ab\gamma} = \varepsilon^{\mu *}(J_\mu^{\text{ext}} + 0_\mu(k)). \tag{13.18}$$

Normally, $J_\mu^{\text{ext}}$ gives the photon emission from all external legs of a process, according to the Feynman rules and including spin. This "Low-theorem" will be applied to photoproduction in section 7-4.

## 3-14 Bethe-Salpeter Equation

Let $T^{(2)}$ denote the Born approximation for the elastic scattering of two massive particles $a$ and $b$. The fourth-order square graph (fig. 3-5.4a) is of the form

$$\langle f | T^{sq} | i \rangle = -i(2\pi)^{-4} \int d^4 p'_a \sum_{M_{a'} M_{b'}} \langle f | T^{(2)} | a'b' \rangle \langle a'b' | T^{(2)} | i \rangle \Phi'_a \Phi'_b , \tag{14.1}$$

where the summation extends over possible magnetic quantum numbers of the virtual particles $a'$ and $b'$ and $\Phi'_a$, $\Phi'_b$ are the "spinless" propagators (2-6.7). At non-relativistic energies, $T^{sq}$ can be as large as $T^{(2)}$, even if the coupling constants are small. In that case, all "ladder graphs" (fig. 3-14.1) are in fact of comparable magnitude, and only their sum $T^{BS}$ can approximate the full scattering amplitude. An implicit expression for this sum is obviously (Bethe and Salpeter 1951)

$$\langle f | T^{BS} | i \rangle = \langle f | T^{(2)} | i \rangle - i(2\pi)^{-4} \int d^4 P'_a \sum_{M_{a'} M_{b'}} \times \langle f | T^{(2)} | a'b' \rangle \langle a'b' | T^{BS} | i \rangle \Phi'_a \Phi'_b \tag{14.2}$$

Other graphs such as the "crossed graph" $T^{cr}$ (fig. 3-5.4b) are less dangerous and can be included in $T^{(2)}$ if necessary (see below).

The full amplitude can have poles in $s$ even if neither $T^{(2)}$ nor any single ladder graph has such poles. Thus (2) provides a basis for the calculation of bound states. Near the pole, $T^{BS}$ tends to infinity, and since $T^{(2)}$ remains finite, the inhomogenous term in (2) is negligible there. The resulting homogenous equation contains the quantum numbers of the initial state only via $T^{BS} | i \rangle$, such that the dependence of $\langle f | T^{BS} | i \rangle$ on these quantum numbers remains arbitrary. We can therefore suppress these quantum numbers in the following.

For convenience, we assume that particle $a$ has spin $\frac{1}{2}$ and that a possible spin of particle $b$ can be treated nonrelativistically, i.e., by the states $\chi_0(M_b)$ which are independent of $P_b$. This is the case of one-electron atoms. We define

$$\begin{aligned} \langle a'b' | T^{BS} | i \rangle &\equiv \chi^+(M'_b) \bar{u}' T^{BS}(P'_a), \\ \langle f | T^{(2)} | a'b' \rangle &\equiv -\chi^+(M_{bf}) \bar{u}_f \tilde{V}(P'_a, P_f) u' \chi(M'_b) \cdot 2m_b . \end{aligned} \tag{14.3}$$

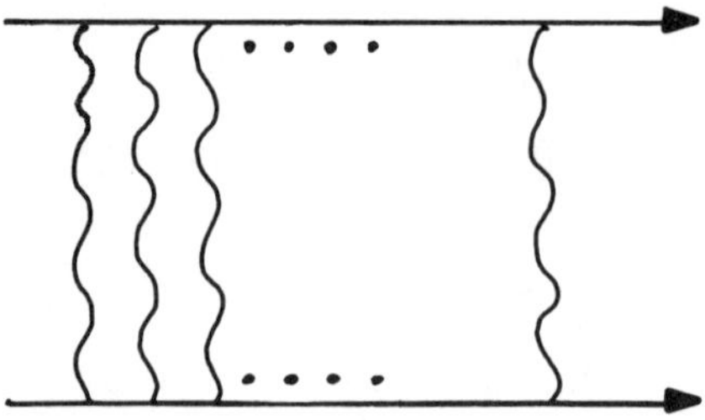

**Figure 3-14.1** A general ladder graph.

Since (2) applies for any combination of $M_{af}$ and $M_{bf}$, the factor $\chi^+(M_{bf})\bar{u}_f$ can be divided off from (2):

$$T^{BS}(P_f) = \frac{2im_b}{(2\pi)^4}\int d^4P'_a \tilde{V}(P'_a, P_f)(P'_a \cdot \gamma + m_a)T^{BS}(P'_a)\Phi'_a\Phi'_b. \quad (14.4)$$

The spin summations have been performed for $a$ by (A-4.6) and for $b$ by the general nonrelativistic completeness relation. To obtain a potential equation, we must somehow integrate over $P_a^{0'}$ in (4). To get an idea how this is best done, one inspects $T^{\text{sq}}$ and also $T^{\text{cr}}$ (Gross 1969). These graphs have poles in $P_a^{0'}$ from the propagators $\Phi'_a$ and $\Phi'_b$ and from the meson propagators of $T^{(2)}$, as well as other singularities due to form factors. In QED, gauge invariance ensures that the main part of $T^{(2)}$ has no pole in $P_a^{0'}$ (see 2-7.16). The most important poles are then due to $\Phi'_a$ and $\Phi'_b$. According to (2-6.8)

$$\Phi'_a\Phi'_b = (4E'_aE'_b)^{-1}\{(E'_a - i\varepsilon - P_a^{0'})^{-1}(E'_b - i\varepsilon - s^{\frac{1}{2}} + P_a^{0'})^{-1} + \cdots\} \quad (14.5)$$

The dots contain at least one antiparticle pole, which is located on the opposite side of the Re $P_a^{0'}$-axis (fig. 3-14.2). In the nonrelativistic domain and for $m_b > m_a$, $E'_b - m_b \approx p_b'^2/2m_b$ is smaller than $E'_a - m_a \approx p_a'^2/2m_a = p_b'^2/2m_a$ (we work in the cms where $\mathbf{p}'_a = -\mathbf{p}'_b \equiv \mathbf{p}'$). As a first approximation, one can therefore close the $P_a^{0'}$-integration contour in the upper half-plane and keep only the pole at $P_a^{0'} = s^{\frac{1}{2}} - E'_b + i\varepsilon$. According to (2-6.9), this gives a factor $+2\pi i$ times the function at $P_a'^0 = s^{\frac{1}{2}} - E'_b$. The result is

$$\begin{aligned} T^{BS}(\mathbf{p}_f) = \frac{-m_b}{(2\pi)^3}\int \frac{d^3p'}{E'_b}\tilde{V}(\mathbf{p}', \mathbf{p}_f)[(s^{\frac{1}{2}} - E'_b)\gamma_0 - \mathbf{p}'\boldsymbol{\gamma} + m_a] \\ \times T^{BS}(\mathbf{p}')[m_a^2 - (s^{\frac{1}{2}} - E'_b)^2 + p'^2 - i\varepsilon]^{-1}, \quad (14.6) \\ E'_b \equiv m_b + p'^2/2m_b. \end{aligned}$$

In other words, we have put the heavier particle on the mass shell. If the 3-dimensional integral in (6) is such that particle $b$ remains always nonrelativistic, this is better than putting the lighter particle on the mass shell, by a factor $E'_a/m_b$. The approximation is less convincing for $m_a = m_b$ (Caswell

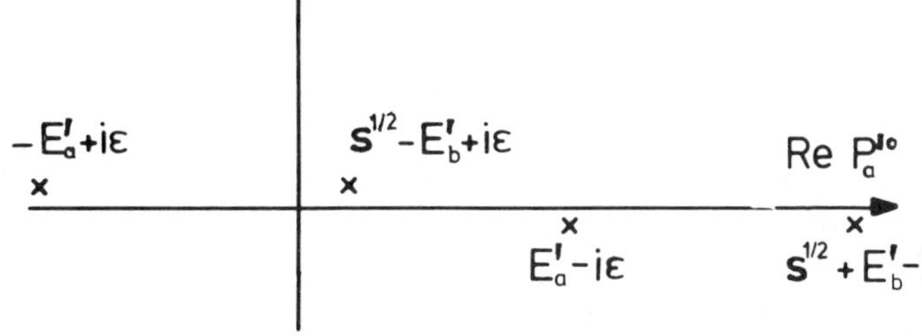


**Figure 3-14.2** The poles of the propagator product $\Phi'_a\Phi'_b = (E_a'^2 - P_a'^{02} - i\varepsilon)\times{}^{-1}(E_b'^2 - (s^{\frac{1}{2}} - P_a'^0)^2 - i\varepsilon)^{-1}$ in the complex $P_a'^0$-plane.

and Lepage, 1978). Since we have used a static spin summation for particle $b$, we can put $m_b/E_b' = 1$ in (6).

Equation (6) is a Lippmann-Schwinger equation and is transformed into a Schrödinger equation by putting

$$T^{BS}(\mathbf{p}) = [(s^{\frac{1}{2}} - E_b)\gamma_0 - \mathbf{p}\boldsymbol{\gamma} - m_a]\tilde{\psi}(\mathbf{p}). \tag{14.7}$$

All square brackets on the right-hand side of (6) combine into one minus sign with this substitution:

$$[\gamma_0(s^{\frac{1}{2}} - H_0(\mathbf{p}_f))]\tilde{\psi}(\mathbf{p}_f) = (2\pi)^{-3} \int d^3p' \tilde{V}(\mathbf{p}', \mathbf{p}_f)\tilde{\psi}(\mathbf{p}') \tag{14.8}$$

$$H_0(\mathbf{p}) = m_b + p^2/2m_b + \gamma_0 \mathbf{p}\boldsymbol{\gamma} + m_a\gamma_0 = H_a + H_b. \tag{14.9}$$

$\tilde{\psi}(\mathbf{p})$ can be identified with the Fourier transform of the wave function, and $H_0(\mathbf{p})$ with the sum of the free-particle Hamiltonians. Equation (8) has been discussed by Grotch and Yennie (1969) for the hydrogen atom. It forms the basis of one- and two-body equations and shows that in the two-body case, a simple relativistic equation obtains only for $m_a \ll m_b$.

Successive improvements of (8) can be found as follows: Firstly, one calculates the sum of the complete fourth-order graphs $T^{(4)} = T^{\text{sq}} + T^{\text{cr}}$. One then subtracts from $T^{(4)}$ the corresponding scattering amplitude obtained from (8) and adds the remaining part $T^{(4')}$ as a "two-photon-exchange potential" to $T^{(2)}$ in (3). It is essential here that the $b$-vertex is dominantly nonspinflip, such that the principal value integrals of $T^{\text{sq}}$ and $T^{\text{cr}}$ cancel each other in the static limit:

$$\begin{aligned} T^{(4)} &\approx 2m_b[m_b^2 - (P_b + q_1)^2 - i\varepsilon]^{-1} + 2m_b[m_b^2 - (P_b' - q_1)^2 - i\varepsilon]^{-1} \\ &\approx (-q_1^0 - i\varepsilon)^{-1} + (q_1^0 - i\varepsilon)^{-1} = 2\pi i\delta\,(q_1^0) \end{aligned} \tag{14.10}$$

For further details, see Brodsky (1971).

In QED, the vacuum polarisation is easily incorporated in $T^{(2)}$. Other corrections such as the ones due to soft photons (Lamb shift) are added later. Form factors and possible excited nuclear states $b'$ bring additional corrections which go under the name "nuclear polarisability."

Finally, we make the traditional transition from (8) to coordinate space. We put $\mathbf{p}' = \mathbf{p}_f + \mathbf{q}$ as in (2-10.2) and then drop the index $f$ on $\mathbf{p}_f$. The Fourier decomposition of $\tilde{\psi}(\mathbf{p})$ on the left-hand side of (8) is

$$\tilde{\psi}(\mathbf{p}) = (2\pi)^{-3} \int d^3r_1 \exp(-i\mathbf{p}\mathbf{r}_1)\psi(\mathbf{r}_1). \tag{14.11}$$

(8) is transformed into coordinate space by multiplying with $\exp(i\mathbf{p}\mathbf{r})$ and integrating over all $\mathbf{p}$. As $H_0(\mathbf{p})$ is a polynomial in $\mathbf{p}$, this gives

$$\begin{aligned} \int d^3p e^{i\mathbf{p}\mathbf{r}} H_0(\mathbf{p})\tilde{\psi}(\mathbf{p}) &= (2\pi)^{-3} \int d^3p H_0(-i\nabla)e^{i\mathbf{p}\mathbf{r}}e^{-i\mathbf{p}\mathbf{r}_1}\psi(\mathbf{r}_1)d^3r_1 \\ &= \int d^3r_1 H_0(-i\nabla)\,\delta_3(\mathbf{r} - \mathbf{r}_1)\psi(\mathbf{r}_1) = H_0(-i\nabla)\psi(\mathbf{r}) \end{aligned} \tag{14.12}$$

On the right-hand side of (8), we insert (2-10.3):

$$(2\pi)^{-6} \int d^3p e^{i\mathbf{p}\mathbf{r}}\, d^3r' e^{-i\mathbf{p}\mathbf{r}'}\, d^3p' V(\mathbf{r}', \nabla') e^{i\mathbf{p}'\mathbf{r}'}\, d^3r_1 e^{-i\mathbf{p}'\mathbf{r}_1} \psi(\mathbf{r}_1)$$
$$= \int d^3r'\, \delta\,(\mathbf{r} - \mathbf{r}') V(\mathbf{r}', \nabla')\, d^3r_1\, \delta\,(\mathbf{r}' - \mathbf{r}_1) \psi(\mathbf{r}_1) = V(\mathbf{r}, \nabla)\psi(\mathbf{r}) \qquad (14.13)$$

This is the usual result, thus confirming (2-10.3).

CHAPTER FOUR

# The Particle Zoo

## 4-1 Classification of Particles and Interactions

Every closed quantum mechanical system of total 4-momentum $P^\mu$ which exists only in a narrow interval of $P^2$ may be called a particle. The expectation value of $P^2$ is $m^2$, $m$ being the particle mass. For stable particles, the $P^2$-distribution is a $\delta$-function, whereas for unstable particles, it has a width of $2m\Gamma$, $\Gamma$ being the decay rate. This important point will be elaborated in section 4-8. By this definition, the ground state and excited states of an atom are different particles.

Atoms are composed of electrons ($e$) and nuclei ($\mathcal{N}$), the lightest nucleus being the proton ($p$). With the discovery of the neutron ($n$) by Chadwick (1932), it became clear that nuclei are composed of protons and neutrons. One may say that ordinary matter is composed of the "elementary particles" $e$, $p$, $n$. A common name for $p$ and $n$ is "nucleon" ($N$). More recently, the adjective "elementary" is withdrawn from the nucleons, in view of the possibility that they are composed of as yet unseen "quarks."

Since the discovery of the positron by Anderson (1933), it was suspected that protons and neutrons would also have antiparticles: the antiproton ($\bar{p}$) and antineutron ($\bar{n}$). After the construction of high-energy accelerators, the $\bar{p}$ was discovered in 1955 (Chamberlain et al.), and the $\bar{n}$ was discovered soon afterwards. The light antinuclei $\bar{d}$, $\bar{t}$, and $^3\overline{\text{He}}$ have also been seen (see Bozzoli et al. (1978) for references).

Although one should not say that photons ($\gamma$) are the carrier of the Coulomb force, their existence is certainly related to the Coulomb force. The theoretical description of spinless bosons is simpler in this respect. Yukawa (1935) postulated the existence of massive bosons as the carrier of nuclear forces. Soon afterwards, particles in the mass range of 100–150 MeV were found in Cosmic Rays, and they were tentatively identified with Yukawa's particles.

In 1927, the neutrino ($\nu$) was postulated by Pauli in order to save energy conservation in nuclear beta decay (it was positively identified much later, see section 5-1). From the neutron decay $n \to pe\nu$ it follows that $\nu$ must be a fermion. Later is was speculated that $\nu$ should also have a distinct antiparticle $\bar{\nu}$, and neutron decay was rewritten as $n \to pe\bar{\nu}$, in analogy with nuclear decays such as $\alpha^* \to \alpha e\bar{e}$ (see section 3-6).

Elementary particles were then classified into leptons ("light particles", $e$, $\bar{e}$, $\nu$, $\bar{\nu}$), mesons (particles of intermediate mass) and baryons (heavy particles, at that time only the nucleons $p$, $n$ and their expected antiparticles $\bar{p}$, $\bar{n}$). The mesons remained somewhat problematic because it appeared that most of them did not have the strong interaction with nuclei which Yukawa had postulated. This point was clarified by Lattes et al. (1947) in their discovery of the pi-mesons or pions ($\pi^\pm$, the $\pi^0$ was discovered later, see the review by Powell 1950). The other, not strongly interacting mesons were called $\mu$-mesons or muons ($\mu^\pm$). They are somewhat lighter than pions, and their existence in Cosmic Rays is almost entirely due to decays of the charged pions, $\pi^- \to \mu^-\bar{\nu}$, $\pi^+ \to \mu^+\nu$. It then became evident that the muon behaves like a heavy electron in its interactions; for this reason it was later moved from the meson class to the lepton class.

Before proceeding with the other particle classes, let us quickly complete the lepton class as we know it today. After the construction of the $\sim$ 30 GeV proton accelerators at Brookhaven and CERN, experiments with neutrinos from the decays in flight of high-energy pions became possible. It then turned out that in their collisions with nuclei, these neutrinos produced muons but not electrons (Danby et al. 1962). This showed that the neutrinos from $\pi \to \mu\bar{\nu}$ decays are different from those of $n \to pe\bar{\nu}$ decays. We must thus distinguish between electron-neutrinos ($\nu_e \equiv \nu$) and muon-neutrinos $\nu_\mu$. Pion decays are now written as $\pi^- \to \mu^-\bar{\nu}_\mu$, $\pi^+ \to \mu^+\nu_\mu$. At this point, it appears that the lepton class consists of two separate classes ($e$, $\nu$) and ($\mu$, $\nu_\mu$) plus the corresponding antiparticles. The problem is complicated by the neutrino helicity conservation (see section 5-5), but there exists at least one lepton number conservation law in the sense that the number of leptons minus that of antileptons is constant.

Analysis of the reaction $e^-e^+ \to \mu^\mp e^\pm$ plus neutral particles at total cms energies around 4 GeV (Perl 1975) indicated the existence of a lepton $\tau = \tau^-(\bar{\tau} = \tau^+)$ with a mass of about 1.8 GeV. The above reaction arises from $e^-e^+ \to \tau^-\tau^+$, followed immediately by the decays $\tau^- \to \mu^-\bar{\nu}_\mu\nu_\tau$ and $\tau^+ \to e^+\nu\bar{\nu}_\tau$ (in case of the sign combination $\mu^-e^+$), where the four neutrinos escape detection. Here we have assumed that $\tau$'s have their own type of neutrinos $\nu_\tau$, for which there is also some evidence.

It is now clear that the lepton class of today is characterized not by small masses, but by the absence of strong interactions: The charged leptons $e^{\mp}$, $\mu^{\mp}$, $\tau^{\mp}$ have electromagnetic and "weak" interactions, their neutrinos have only weak interactions, characterized by the "weak coupling constant"

$$G_\mu = 1.16631 \times 10^{-5}\ \text{GeV}^{-2}. \tag{1.1}$$

This number is obtained from $\mu$-decay $\mu \to e\bar{\nu}\nu_\mu$, eq. (5-5.22) below, but coupling constants of other weak decays such as $n \to pe\bar{\nu}$ and $\pi \to \mu\bar{\nu}_\mu$ are similar. The corresponding lifetimes depend strongly on the available phase space, however. For $n$ and $\pi^\pm$ they are 15.3 minutes and $2.5 \times 10^{-8}$ sec, respectively. The shortest lifetimes of weakly decaying particles measured so far are those of $K_S$ and $\Sigma^+$ ($\sim 0.8 \times 10^{-10}$ sec, see tables C-2 and C-4).

The electromagnetic interaction is characterized by the fine structure constant

$$\alpha = e^2/4\pi = 137.036^{-1} = 0.007297. \tag{1.2}$$

Particles have universal charges $Ze$, but this need not exclude the interactions of neutral particles ($Z = 0$) with photons. Neutral atoms, for example, may emit and absorb photons, and the $\pi^0$ decays into $\gamma\gamma$ with a lifetime of $\sim 10^{-16}$ sec. There are only two other particles in the most important tables (C-1, C-2, C-4) which decay dominantly electromagnetically, namely $\eta$ and $\Sigma^0$. Their lifetimes are $\sim 10^{-18}$ sec. Thus the lifetimes of electromagnetically decaying particles are considerably shorter than those of weakly decaying particles.

Next we come to the systematic discussion of the meson and baryon classes of elementary particles. Mesons are bosons, baryons are fermions. In addition to weak and electromagnetic interactions, these particles also have strong interactions. A common name for mesons and baryons is "hadrons," and the strong interaction is also called "hadronic."

Hadrons which decay dominantly by strong interactions have extremely short lifetimes. The $\eta'$ of table C-2 and all mesons of tables C-3 and C-6 belong to this category. The most long-lived vector meson of table C-3 is the $\phi$ meson, with a lifetime of $1.6 \times 10^{-22}$ sec. This is already unmeasurably short, only the width $\Gamma$ can be measured. Particles with such short lifetimes are more appropriately called "resonances." The decay $\rho \to \pi\pi$ of the $\rho$-resonance is characterized by a coupling constant $G_{\rho\pi\pi}$ which is quoted in a form analogous to (2),

$$G_{\rho\pi\pi}^2/4\pi = 2.85. \tag{1.3}$$

Nuclear forces are mainly mediated by the mesons $\pi$, $\rho$, and $\omega$. The mesons $\eta$, $\phi$, $A_2$, and $f$ make smaller contributions. However, also the $\pi N$ scattering amplitudes are essential in the theory of nuclear forces (section (7-5).

There are at least two groups of hadrons which have little to do with ordinary nuclear physics. These are the "strange" particles and the "new" particles. Here we mention only the strange particles; they are included in tables (C-2)–(C-6). In the meson class, these are the kaons ($K^+$, $K^0$), the

$K^*(K^{*+}, K^{*0})$ and the $K_N(K_N^+, K_N^0)$ and their antiparticles. Strange particles in the baryon "octet" are the Lambda hyperon ($\Lambda$), the Sigma hyperons ($\Sigma^+$, $\Sigma^0$, $\Sigma^-$) and the Xi-particles ($\Xi^0$, $\Xi^-$), and in the baryon "decuplet" (this multiplet structure refers to the group SU(3), see sects. 4-10 and 4-11) the $\Sigma^{*(+,0,-)}$, $\Xi^{*(0,-)}$ and $\Omega^-$. Other baryon resonances are collected in tables 7-2.2 (nonstrange) and 7-6.1 (strange).

The history of the discovery of the strange particles can be traced from the review by Dalitz (1957). These particles are produced by strong forces, but only in pairs. This takes a large amount of energy and thus explains their unimportance in nuclear physics. The strong or electromagnetic decay of a strange particle to nonstrange particles is forbidden by a new quantum number, "strangeness" or equivalently "hypercharge," which will be discussed in section 4-3. Here we only note that decays such as $K \to \pi\pi$ are weak, although they involve no leptons.

There are other approximate selection rules in the strong interaction, which result from isospin invariance (see the next section). In particular, the decays $\eta \to \pi\pi\pi$ of table C-2 are isospin-forbidden. For this reason, they are classified as electromagnetic decays. They are in fact comparable in strength to the decay $\eta \to \gamma\gamma$. Concluding then, from the types of particles involved it is not immediately obvious whether a certain decay is strong, electromagnetic, or weak.

As mentioned in section 1-2, weak interactions conserve $C\mathscr{P}\mathscr{T}$ and normally also $C\mathscr{P}$, but not $C$ and $\mathscr{P}$ separately. Christenson et al. (1964) discovered the decay $K_L \to \pi\pi$, where $K_L$ denotes the long-lived neutral kaon. The above-mentioned $K^0$ and its "antistate" $\bar{K}^0$ do not represent unstable particles; they cannot be assigned lifetimes in the sense of section 2-3. The states $K_L^0$ and $K_S^0$ that do have definite lifetimes (and masses) are linear combinations of $K^0$ and $\bar{K}^0$:

$$\begin{aligned} K_S &= 2^{-\frac{1}{2}}[(1+\varepsilon)K - (1-\varepsilon)\bar{K}], \\ K_L &= 2^{-\frac{1}{2}}[(1+\varepsilon)K + (1-\varepsilon)\bar{K}], \end{aligned} \tag{1.4}$$

$$|\varepsilon| = 2 \times 10^{-3}. \tag{1.5}$$

$C\mathscr{P}$-conservation would require $\varepsilon = 0$ and $\Gamma(K_L \to \pi\pi) = 0$, as we shall see. The main mixing of $K$ and $\bar{K}$ in (4) is of second order in the weak interaction. Wolfenstein (1964) suggested that $\varepsilon$ is a first-order effect of a new, "superweak" interaction with a coupling constant $\sim 10^{-8}\, G_\mu$ (see section 5-9). This would explain the absence of $C\mathscr{P}$-violation in other particle decays. However, it is quite possible that the decay Hamiltonian of the recently discovered "charmed" particles ($D$) violates $C\mathscr{P}$, with a relative magnitude $\sin\theta_2 \sin\theta_3 \sin\delta \approx 10^{-3}$ (section 5-11).

Weak and electromagnetic interactions have been united by Salam and Weinberg (section 5-10). "Heavy vector mesons" $W^\pm$ and $Z^0$ are predicted as the carriers of weak interactions, but so far these particles have not been discovered, due to their enormous masses.

Finally, all particles should have gravitation, characterized by

$$G_{\text{grav}} = 6.673 \times 10^{-8} \text{ cmg}^{-1} \text{ sec}^{-2} = 2.61 \times 10^{-66}, \qquad G_{\text{grav}}^2/4\pi \approx 10^{-132} \tag{1.6}$$

which is negligible in particle physics.

Nuclei are also hadrons. The adjective "composite" does not guarantee that they are composed of protons and neutrons only. We expect the presence of $\pi$, $\rho$, $\omega$, and $\Delta$ in nuclei and observe in fact these particles being produced at higher energies. At low energies, these additional hadrons give rise to "exchange-current effects" or "3-body forces." Only the deuteron can with some confidence be treated as a $pn$-bound state, because of its small binding energy. Frequently, it pays to treat nuclei on equal footing with the more "elementary" hadrons. For example, the reduced decay rates of $\pi^- \to \pi^0 e\bar{\nu}$ and $^{14}\text{O} \to {}^{14}\text{N}^* e\bar{\nu}$ decays are identical simply because the relevant hadron pairs $(\pi^-, \pi^0)$ and $(^{14}\text{O}, {}^{14}\text{N}^*)$ have identical spins and isospins. Also the dynamics which we shall develop in chapter 6 makes in principle no distinction between elementary and composite hadrons.

In addition to the lepton number conservation law, there exists an analogous baryon number conservation law: the number of baryons $(p, n, \Lambda \ldots)$ minus that of antibaryons $(\bar{p}, \bar{n}, \bar{\Lambda} \ldots)$ is constant. Antibaryons can only be produced pairwise with baryons. Reactions such as $pp \to ppp\bar{p}$ or $\pi^- p \to pn\bar{p}$ occur at energies above the corresponding threshold energies, but the cross sections remain always small. The best evidence for baryon number conservation comes from the stability of nuclei. The deuteron $d$ for example has baryon number $B = 2$ (this follows from breakup reactions such as $\gamma\, d \to np$). The decay $d \to \pi^+ \pi^0$ is only forbidden by baryon number conservation (decays such as $p \to \pi^+ \pi^0$ are forbidden by rotational invariance: matrix elements must be bilinear in spinors). Contrary to charge conservation which is connected with gauge invariance, the lepton and baryon number conservation laws are only empirical. In recent "grand unification schemes" of leptons and baryons, decays such as $p \to e^+ \pi^0$ in which a baryon decays into an antilepton are allowed in principle. Experimentally, the proton lifetime is $> 10^{30}$ years (Reines and Crouch, 1974).

## 4-2 Nucleons and Isospin, $\pi$- and $\eta$-Mesons

Proton ($p$) and neutron ($n$) have almost identical masses (see table C-4). This is the first indication of the charge independence of nuclear forces, which allows to treat $p$ and $n$ as different states of one particle, called the "nucleon" ($N$). Deviations from this symmetry are due to electromagnetic interactions, in particular to the Coulomb interaction between charged particles (see Wilkinson 1969, Blin-Stoyle 1973), and to a "quark mass splitting" (section 4-15).

Stable particles of baryon number 2 and 4 occur in only one charge state each, namely the deuteron $d$ and the $\alpha$-particle. These states are thus "singlets" in the new symmetry scheme, i.e., they transform into themselves under the transformations between $p$ and $n$. Consequently, the $p$-$d$ and $n$-$d$ scattering amplitudes must be identical. In fact, each amplitude has a pole in the $J = \frac{1}{2}$ state, $h$ (helion or $^3$He) and $t$ (triton or $^3$H) (see table 4-2.1).

The new symmetry group is now easily found. Neutrons and protons are represented by the unit vectors in a 2-dimensional space:

$$|p\rangle = \begin{pmatrix} 1 \\ 0 \end{pmatrix}, \qquad |n\rangle = \begin{pmatrix} 0 \\ 1 \end{pmatrix}. \tag{2.1}$$

**Table 4-2.1** Spins, isospins, masses and electric mean square radii of the lightest nuclei. Nuclear masses from Wapstra and Gove (1971). The nuclear mass is the atomic mass minus $Z$ electron masses. Radii from Höhler et al. (1976), De Jager et al. (1974).

| baryon number | symbols | spin | isospin | average mass [MeV] | $m_1 - m_2$ [MeV] | $\langle r_E^2\rangle^{\frac{1}{2}}$ [fm] |
|---|---|---|---|---|---|---|
| 1 | $\left.\begin{matrix} n \\ p \end{matrix}\right\} N$ | $\frac{1}{2}$ | $\frac{1}{2}$ | 938.927 | 1.2934 | — <br> 0.84 |
| 2 | $d = {}^2$H | 1 | 0 | 1875.628 | ($b$ = 2.2246) | 2.1 |
| 3 | $t = {}^3$H <br> $h = {}^3$He | $\frac{1}{2}$ | $\frac{1}{2}$ | 2808.679 | 0.530 | 1.70 <br> 1.84 |
| 4 | $\alpha = {}^4$He | 0 | 0 | 3727.408 | ($b_n$ = 20.578) | 1.67 |

We also introduce the charge operator $Q$ and the operator $\tau_1$ which exchanges $|p\rangle$ and $|n\rangle$:

$$Q = \begin{pmatrix} 1 & 0 \\ 0 & 0 \end{pmatrix} = \frac{1}{2}(1 + \tau_3), \qquad \tau_3 = \begin{pmatrix} 1 & 0 \\ 0 & -1 \end{pmatrix}, \qquad \tau_1 = \begin{pmatrix} 0 & 1 \\ 1 & 0 \end{pmatrix}. \tag{2.2}$$

As the scattering operator $S$ conserves charge, it must commute with $Q$. As its commutation with the unit matrix is trivial, it must also commute with $\tau_3$. Our new postulate is that it also commutes with $\tau_1$ which exchanges protons with neutrons. But if $S$ commutes both with $\tau_3$ and $\tau_1$, it must also commute with

$$(2i)^{-1}[\tau_3, \tau_1] = \tau_2 = \begin{pmatrix} 0 & -i \\ i & 0 \end{pmatrix}. \tag{2.3}$$

Thus $S$ commutes with 3 matrices $\tau_i$ which are formally identical with the Pauli spin matrices. We already know that these matrices generate the group $SU(2)$. The mathematical structure of the new symmetry is therefore again $SU(2)$; its transformation matrices are called "isospin rotations," although they have nothing to do with normal rotations. The standard generators of isospin rotations will be denoted by $\mathbf{I}$ ($[I_i, I_j] = i\varepsilon_{ijk} I_k$) such that $\mathbf{I} = \frac{1}{2}\boldsymbol{\tau}$ for nucleons. For a two-nucleon system, $\mathbf{I} = \mathbf{I}^{(1)} + \mathbf{I}^{(2)}$ is the total isospin operator. The charge states are expanded in terms of the eigenstates $|I, I_3\rangle$ of $\mathbf{I}^2$ and $I_3$ as follows

$$\begin{aligned} |pp\rangle &= |1, 1\rangle, & |pn\rangle &= 2^{-\frac{1}{2}}(|1, 0\rangle + |0, 0\rangle), \\ |nn\rangle &= |1, -1\rangle, & |np\rangle &= 2^{-\frac{1}{2}}(|1, 0\rangle - |0, 0\rangle). \end{aligned} \tag{2.4}$$

As usual, the states $|1, 0\rangle$ and $|1, -1\rangle$ can be obtained from $|1, 1\rangle$ by the step operator $I_- = I_1 - iI_2$. As these operators conserve the permutation symmetry of the state, all states $|1, I_3\rangle$ have the antisymmetry of the original $|pp\rangle$-state in orbital and spin space. The remaining state

$$|0, 0\rangle = 2^{-\frac{1}{2}}(|pn\rangle - |np\rangle) \tag{2.5}$$

must therefore be symmetric in orbital and spin space. Thus, if particles in the same isospin multiplet are considered identical, one can formulate a generalized symmetry principle: "States containing several identical fermions (bosons) are totally antisymmetric (symmetric) under the simultaneous exchange of orbital, spin and isospin quantum numbers."

Nuclear physicists frequently distinguish between "charge symmetry" (which is the equality of *pp* and *nn* scattering amplitudes) and "charge independence" (which says that the same amplitudes also describe *np* scattering in the antisymmetric states). From our derivation it is clear that such a distinction is basically meaningless. If one of these symmetries is broken, the other one is also broken. In practice, however, the breaking can be hard to calculate, such that one has to live with empirically different amounts of breaking.

The three pions $\pi^+$, $\pi^0$ and $\pi^-$ have almost identical masses, so we suspect that they belong to the representation "3" of $SU(2)$, i.e. they transform like an isovector. However, it must be remembered that $\pi^+$ and $\pi^-$ have automatically identical masses since they are antiparticles of each other. After all, we excluded the antiproton and antineutron when we stated that nucleons belong to the representation "2" of $SU(2)$. However, whereas nucleons and antinucleons differ by their baryon number, it turns out that $\pi^+$ and $\pi^-$ differ only by their charge. In particular, the $\pi^+ d$ and $\pi^- d$ scattering amplitudes are identical. This was in fact anticipated when pions were identified with the "carries of nuclear forces," as postulated by Yukawa. Thus a proton can emit a virtual $\pi^+$ or absorb a virtual $\pi^-$, thereby transforming itself into a neutron and vice versa. Clearly, this is only possible if $\pi^+$ and $\pi^-$ have $I_3 = 1$ and $-1$, respectively.

As the $S$-matrix commutes with all isospin rotations, we can apply Schur's lemma (B-2): $S$ is zero between states of different total isospin and a multiple of the unit matrix between states of equal isospin:

$$\langle I', I'_3|S|I, I_3\rangle = S(I)\, \delta_{II'}\, \delta_{I_3 I'_3}. \tag{2.6}$$

Let us now exploit (6) for the $2 \times 3 = 6$ charge combinations of $\pi N$ scattering. The eigenstates $|I, I_3\rangle$ of total isospin are given in table 2.2. Our sign

**Table 4-2.2** $N\pi$ states of total isospin $I$ and third component $I_3$ expressed in terms of charge states. Nucleon-first convention.

| $I$ | $I_3 = \frac{1}{2}$ | $I_3 = -\frac{1}{2}$ | $I_3 = \frac{3}{2}$ | $I_3 = -\frac{3}{2}$ |
|---|---|---|---|---|
| $\frac{3}{2}$ | $3^{-\frac{1}{2}} n\pi^+ + (\frac{2}{3})^{\frac{1}{2}} p\pi^0$ | $3^{-\frac{1}{2}} p\pi^- + (\frac{2}{3})^{\frac{1}{2}} n\pi^0$ | $p\pi^+$ | $n\pi^-$ |
| $\frac{1}{2}$ | $-(\frac{2}{3})^{\frac{1}{2}} n\pi^+ + 3^{-\frac{1}{2}} p\pi^0$ | $(\frac{2}{3})^{\frac{1}{2}} p\pi^- - 3^{-\frac{1}{2}} n\pi^0$ | — | — |

convention follows Condon and Shortley (1935) and the nucleon is written before the pion (the inverse order would give opposite signs for $I = \frac{1}{2}$). For $I_3 = \pm\frac{3}{2}$, (6) gives

$$S(\tfrac{3}{2}) = \langle p\pi^+ | S | p\pi^+ \rangle = \langle n\pi^- | S | n\pi^- \rangle. \tag{2.7}$$

The decomposition (0) of isospin eigenstates into charge combinations is an orthogonal transformation ($00_{\text{tr}} = 1$). This means that the coefficients in front of each charge combination also apply to the inverse transformation, for example

$$| p\pi^- \rangle = 3^{-\frac{1}{2}} | \tfrac{3}{2}, -\tfrac{1}{2} \rangle + (\tfrac{2}{3})^{\frac{1}{2}} | \tfrac{1}{2}, -\tfrac{1}{2} \rangle, \tag{2.8}$$

$$| n\pi^0 \rangle = (\tfrac{2}{3})^{\frac{1}{2}} | \tfrac{3}{2}, -\tfrac{1}{2} \rangle - 3^{-\frac{1}{2}} | \tfrac{1}{2}, -\tfrac{1}{2} \rangle. \tag{2.9}$$

Using now (6), we find

$$\langle p\pi^- | S | p\pi^- \rangle = \langle n\pi^+ | S | n\pi^+ \rangle = \tfrac{1}{3} S(\tfrac{3}{2}) + \tfrac{2}{3} S(\tfrac{1}{2}), \tag{2.10}$$

$$\langle n\pi^0 | S | p\pi^- \rangle = \langle p\pi^0 | S | n\pi^+ \rangle = \tfrac{1}{3} 2^{\frac{1}{2}} (S(\tfrac{3}{2}) - S(\tfrac{1}{2})). \tag{2.11}$$

Total and elastic cross sections for $\pi^{\pm} p$ scattering are shown in fig. 4-2. From $\sigma(\pi^+ p)$ one sees that the $I = \frac{3}{2}$ scattering amplitude $T^{\frac{3}{2}}$ has a maximum around $p_\pi = 300$ MeV (the $\Delta$ resonance). If the $I = \frac{1}{2}$ amplitude $T^{\frac{1}{2}}$ is much smaller in this region, we expect from (10) and (11)

$$\sigma_{\text{tot}}(\pi^- p) \approx 3\sigma_{\text{el}}(\pi^- p) \approx \tfrac{1}{3}\sigma(\pi^+ p), \tag{2.12}$$

which is approximately fulfilled.

Isospin states for two pions are given in table 2.3. Their charge permutation symmetry is indicated by the "Young tableau,"

$$\boxed{1|2} = \text{symmetric}, \qquad \begin{array}{|c|}\hline 1 \\ \hline 2 \\ \hline\end{array} = \text{antisymmetric}. \tag{2.13}$$

As pions are bosons, their complete state must always be totally symmetric.

3-pion states of definite permutation symmetries are given in table 2.4. Except for the first two $I = 1$ states, they coincide with the states $| I, I_3, I_{12} \rangle$

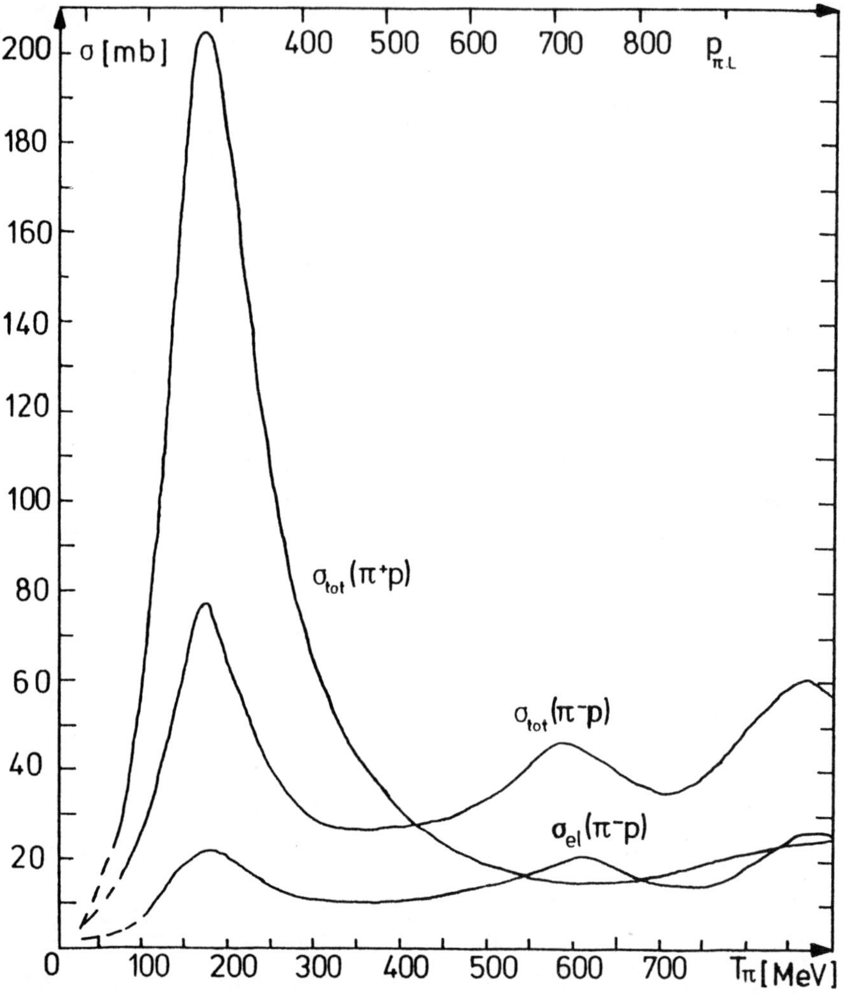


**Figure 4-2** Total $\pi^\pm p$ cross sections and elastic $\pi^- p$ cross section. $T_\pi = E_{\pi, l} - m_\pi$ is the pion lab kinetic energy. Below 500 MeV, the inelastic $\pi^+ p$ cross section is negligible, and the inelastic $\pi^- p$ cross section is due to charge exchange $\pi^- p \rightarrow \pi^0 n$. Up to 300 MeV, $\sigma_{tot}(\pi^\pm p)$ from Carter et al. (1971), at higher energies from (Höhler 1979), $\sigma_{el}(\pi^- p)$ from Berger (1972).

which couple first particles 1 and 2 to $I_{12}$ and then add particle 3. The value of $I_{12}$ follows from the sub-tableau of particles 1 and 2 by looking at table 2.3 and by observing that $I > 1$ requires $I_{12} > 0$.

For the first two $I = 1$ states, the situation is more complicated. The Young tableau $\boxed{1|2|3}$ is totally symmetric in all 3 indices, but the states $|1, I_3, 0\rangle$ and $|1, I_3, 2\rangle$ are not. However, the interested reader may check that the following linear combination is in fact totally symmetric:

$$|1, I_3, \boxed{1|2|3}\rangle = \tfrac{2}{3}|1, I_3, 2\rangle + \tfrac{1}{3}5^{\frac{1}{2}}|1, I_3, 0\rangle. \tag{2.14}$$

**Table 4-2.3** Isopin states for two pions.

| Symmetry | $I$ | $I_3 = 0$ | $I_3 = \pm 1$ | $I_3 = \pm 2$ |
|---|---|---|---|---|
| [1 2] | 2 | $6^{-\frac{1}{2}}(\pi^+\pi^- + \pi^-\pi^+ + 2\pi^0\pi^0)$ | $2^{-\frac{1}{2}}(\pi^\pm\pi^0 + \pi^0\pi^\pm)$ | $\pi^\pm\pi^\pm$ |
| | 0 | $3^{-\frac{1}{2}}(\pi^+\pi^- + \pi^-\pi^+ - \pi^0\pi^0)$ | | |
| [1 / 2] | 1 | $2^{-\frac{1}{2}}(\pi^+\pi^- - \pi^-\pi^+)$ | $\pm 2^{-\frac{1}{2}}(\pi^\pm\pi^0 - \pi^0\pi^\pm)$ | |

The other $I = 1$ state which is symmetric in 1 and 2 is now given by the orthogonal combination

$$\left| 1, I_3, \begin{matrix} \boxed{1}\boxed{2} \\ \boxed{3} \end{matrix} \right\rangle = -\tfrac{1}{3}5^{\frac{1}{2}} | 1, I_3, 2\rangle + \tfrac{2}{3} | 1, I_3, 0\rangle. \tag{2.15}$$

**Table 4-2.4** Isospin states of definite permutation symmetry for 3 pions. States of $I_3 < 0$ are obtained from those of $I_3 > 0$ by the substitution $\pi^+ \leftrightarrow \pi^-$ (for $I = 2$, an extra minus sign is needed).

| Symmetry | $I$ | | $\pi^+\pi^-\pi^0$ | $\pi^-\pi^+\pi^0$ | $\pi^+\pi^0\pi^-$ | $\pi^0\pi^+\pi^-$ | $\pi^-\pi^0\pi^+$ | $\pi^0\pi^-\pi^+$ | $\pi^0\pi^0\pi^0$ | | | | | |
|---|---|---|---|---|---|---|---|---|---|---|---|---|---|---|
| [1 2 3] | 3 | $10^{-\frac{1}{2}}$ | (1 | +1 | +1 | +1 | +1 | +1 | −2) | | | | | |
| | 1 | $15^{-\frac{1}{2}}$ | (1 | +1 | +1 | +1 | +1 | +1 | +3) | | | | | |
| [1 2 / 3] | 2 | $\frac{1}{2}$ | (0 | +0 | +1 | +1 | −1 | −1) | | | | | | |
| | 1 | $12^{-\frac{1}{2}}$ | (2 | +2 | −1 | −1 | −1 | −1) | | | | | | |
| [1 3 / 2] | 2 | $12^{-\frac{1}{2}}$ | (2 | −2 | +1 | −1 | −1 | +1) | | | | | | |
| | 1 | $\frac{1}{2}$ | (0 | +0 | +1 | −1 | +1 | −1) | | | | | | |
| [1 / 2 / 3] | 0 | $6^{-\frac{1}{2}}$ | (1 | −1 | −1 | +1 | +1 | −1) | | | | | | |
| | | | $\pi^+\pi^+\pi^-$ | $\pi^+\pi^-\pi^+$ | $\pi^-\pi^+\pi^+$ | $\pi^0\pi^0\pi^+$ | $\pi^0\pi^+\pi^0$ | $\pi^+\pi^0\pi^0$ | | | $\pi^+\pi^+\pi^0$ | $\pi^+\pi^0\pi^+$ | $\pi^0\pi^+\pi^+$ | $\pi^+\pi^+\pi^+$ |
| [1 2 3] | 3 | $15^{-\frac{1}{2}}$ | (1 | +1 | +1 | +2 | +2 | +2) | | $3^{-\frac{1}{2}}$ | (1 | +1 | +1) | 1 |
| | 1 | $15^{-\frac{1}{2}}$ | (2 | +2 | +2 | −1 | −1 | −1) | | | | | | |
| [1 2 / 3] | 2 | $12^{-\frac{1}{2}}$ | (2 | −1 | −1 | −2 | +1 | +1) | | $6^{-\frac{1}{2}}$ | (2 | −1 | −1) | |
| | 1 | $12^{-\frac{1}{2}}$ | (−2 | +1 | +1 | −2 | +1 | +1) | | | | | | |
| [1 3 / 2] | 2 | $\frac{1}{2}$ | (0 | +1 | −1 | +0 | −1 | +1) | | $2^{-\frac{1}{2}}$ | (0 | +1 | −1) | |
| | 1 | $\frac{1}{2}$ | (0 | −1 | +1 | +0 | −1 | +1) | | | | | | |

The 3-pion state with $I = 0$ finally is totally antisymmetric and can be written down directly as the normalized determinant of the three charge states.

The $\eta$-meson was discovered in the reaction $\pi^+ d \to pp\eta$ (Pevsner et al. 1961). It has no charged counterparts and therefore has the quantum numbers $I = I_3 = 0$. Apart from isospin, the $\eta$-meson is like a heavy $\pi^0$, as we shall see. The simplest production reaction is $\pi^- p \to \eta n$. Eta mesons play a minor role in pion-nuclear physics because their coupling constant to nucleons is relatively small (see section 7-6). Table C-2 contains one more particle, the $X$ or $\eta'$, which has exactly the quantum numbers of the $\eta$ except for its larger mass (957 MeV), which makes it still less important. Its main decay $X \to \eta\pi\pi$ proceeds via the strong interaction. The two pions in the final state must therefore carry $I = I_3 = 0$, which implies, according to table 2.3

$$\Gamma(X \to \eta\pi^+\pi^-)/\Gamma(X \to \eta\pi^0\pi^0) = 2. \tag{2.16}$$

The $X$-meson was discovered in the reaction $K^- p \to X\Lambda$ (Goldberg et al. 1964). The simpler reaction $\pi^- p \to Xn$ appears to be very weak, which indicates that $X$ couples more strongly to hyperons than to nucleons.

## 4-3 Strange Particles and Hypercharge

The four $K$-mesons or "kaons" $K^+$, $K^-$, $K_L^0$ and $K_S^0$ as well as the baryons $\Lambda$, $\Sigma$, $\Xi$ are called "strange particles." This name has historical reasons; from our present point of understanding, all hadrons are equally "strange." However, the physics of strange particles is rather disconnected from the theory of nuclear forces as well as from low-energy pion-nucleon interactions.

The strange baryons are the lambda hyperon $\Lambda$, the three sigma-hyperons $\Sigma^+$, $\Sigma^0$ and $\Sigma^-$, and the "cascade particles" $\Xi^0$ and $\Xi^-$. This name for $\Xi$ is motivated by the cascade of successive decays of these particles, $\Xi \to \Lambda\pi$, $\Lambda \to N\pi$. Sometimes the expression "Xi-hyperon" is also used, but we shall reserve the word "hyperon" for $\Lambda$ and $\Sigma$. Masses and decay modes of $N$, $\Lambda$, $\Sigma$, $\Xi$ are collected in table C-4, and the selection rules of strange particle decays are discussed at the end of sect. 4-11.

In strong interactions, strange particles always appear pairwise. In pion-proton collisions, for example, hyperons are produced only together with a $K^+$ or a $K^0$. In two-particle reactions, the following combinations are observed ("associated production"):

$$\pi^+ p \to K^+\Sigma^+, \qquad \pi^- p \to K^0\Lambda, \qquad \pi^- p \to K^0\Sigma^0, \qquad \pi^- p \to K^+\Sigma^-. \tag{3.1}$$

If the lab energy is high enough, production of additional pions is possible, e.g., $\pi^+ p \to K^+\pi^+\Lambda$. Reactions such as $\pi^- p \to \pi^0\Lambda$, which involve only one strange particle, have not been observed.

Another possible type of reactions is the kaon "pair production." Examples of this reaction are the following:

$$\pi^+ p \to K^+ \bar{K}^0 p, \qquad \pi^- p \to K^+ K^- n. \tag{3.2}$$

The states $K^0$ and $\bar{K}^0$ are related to the particles $K_L^0$ and $K_S^0$ by (1.4). Reactions of the type $\pi^+ p \to K^+ K^+ n$ have not been observed.

The first observations of strange particles came from cosmic ray experiments. As a rule, only the decays were observed. Already at that time, when nothing was known about associated production (1) or pair production (2) of strange particles, it was suggested that the production of strange particles was governed by an isospin selection rule. The successful assignment of $I$ and $I_3$ to these strange particles, together with the introduction of a new quantum number, called strangeness, is due to Gell-Mann (1956) and to Nishijima (1955).

First, the values of $I$ of the baryons are determined. If the interactions which are responsible for the existence of hadrons are isospin invariant, hadrons should always occur in multiplets of $2I + 1$ different charge states. We conclude that $\Lambda$, $\Sigma$ and $\Xi$ have $I = 0$, 1, and $\frac{1}{2}$, respectively. This assignment immediately explains the absence of reactions of the type pion + nucleon → hyperon + pions. The total isospin of the final state is integer, whereas that of the initial state is $\frac{1}{2}$ or $\frac{3}{2}$. On the other hand, in hyperon decays of the type $\Lambda \to N\pi$, isospin is clearly not conserved.

Next, the isospin properties of kaons are determined. Since there are four states, one might try $I = \frac{3}{2}$, but this assignment is ruled out by conservation of $I_3$ in (1) and (2) as follows. From the reactions

$$\pi^+ n \to K^+ \Lambda, \qquad \pi^- p \to K^0 \Lambda, \tag{3.3}$$

we learn that $K^+$ and $K^0$ have $I_3 = \frac{1}{2}$ and $-\frac{1}{2}$, respectively. The pair production (2) then requires $I_3 = \frac{1}{2}$ and $-\frac{1}{2}$ for $\bar{K}^0$ and $K^-$, respectively. Thus the kaons form two doublets of $I = \frac{1}{2}$. From (3), it also follows that $K^+$ and $K^0$ are in the same doublet. Since $\Lambda$ has zero isospin, the total isospin in (3) is $\frac{1}{2}$. From the last row in table 4-2 we find that the two reactions should have equal cross sections

$$\sigma(\pi^+ n \to K^+ \Lambda) = \sigma(\pi^- p \to K^0 \Lambda), \tag{3.4}$$

in agreement with experiment. Finally, the mesons $\bar{K}^0$ and $K^-$ must form the remaining doublet. The isospin assignments are summarized in table 4-3. Conservation of $I_3$ in (1) requires $I_3 = \pm 1$ and 0 for $\Sigma^\pm$ and $\Sigma^0$, respectively.

**Table 4-3** Assignment of $I$ and $I_3$ to hyperons and kaons.

| | $\Lambda$ | $\Sigma^+$ | $\Sigma^0$ | $\Sigma^-$ | $K^+$ | $K^0$ | $\bar{K}^0$ | $K^-$ |
|---|---|---|---|---|---|---|---|---|
| $I$ | 0 | | 1 | | $\frac{1}{2}$ | | $\frac{1}{2}$ | |
| $I_3$ | 0 | 1 | 0 | −1 | $\frac{1}{2}$ | $-\frac{1}{2}$ | $\frac{1}{2}$ | $-\frac{1}{2}$ |

High-energy beams of $K^-$ are the only practicable means of producing cascade particles. Possible two-particle reactions are

$$K^- p \to K^0 \Xi^0, \qquad K^- p \to K^+ \Xi^-, \tag{3.5}$$

from which one concludes that $\Xi^0$ and $\Xi^-$ have $I_3 = \frac{1}{2}$ and $-\frac{1}{2}$, respectively. Reactions such as $\pi^- p \to \pi^+ \Xi^-$ or $\pi^- p \to K^+ \Xi^-$ are all forbidden by conservation of $I_3$.

The selection rules in the production of strange particles are simplified by the introduction of a new quantum number called hypercharge (the older "strangeness" quantum number is mentioned below). For pions and hyperons, the electric charge $Q$ and the third component of isospin $I_3$ coincide. For other multiplets, the centre of the multiplet may have a net charge. In order to avoid half-integer values of the new quantum number, the charge of the centre of the multiplet is called $\frac{1}{2}Y$, where $Y$ is the hypercharge. We thus have

$$Q = I_3 + \tfrac{1}{2}Y. \tag{3.6}$$

Nucleons and kaons have $Y = 1$, while antikaons and cascade particles have $Y = -1$. The equivalent strangeness quantum number $S$ is defined by

$$Q = I_3 + \tfrac{1}{2}(S + B), \tag{3.7}$$

where $B$ is the baryon number defined at the end of section 4-1. For mesons, $S$ is the same as $Y$. The motivation for including the baryon number in the definition of $S$ was to obtain zero strangeness for the nucleons. From the point of view of isospin invariance, $Y$ is the more symmetric quantum number. A plot of $Y$ versus $I_3$ for baryons and mesons is shown in fig. 4-3.

One can easily convince oneself that the various forbidden reactions mentioned thus far do not conserve hypercharge.

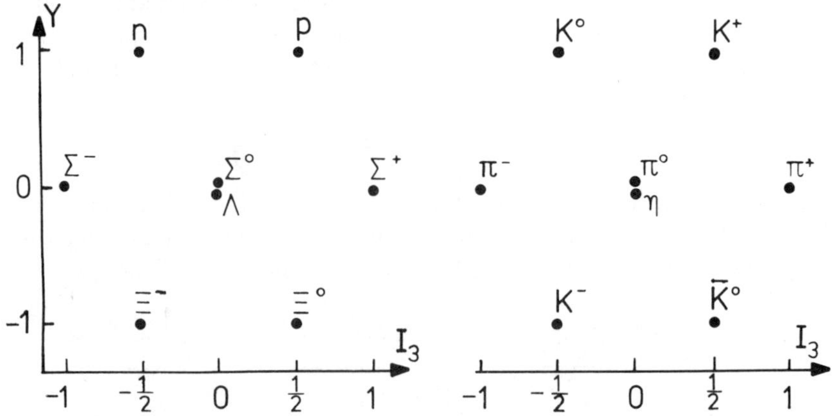


**Figure 4-3** Plot of hypercharge versus third component of isospin for baryons and mesons. The central point $Y = I_3 = 0$ is occupied by two particles, one of which has zero isospin.

In contrast to $I$, $Y$ is conserved in electromagnetic interactions. This follows from the absence of decays $K^0 \to \gamma\gamma$, $\Lambda \to \gamma n$ (as mentioned on p. 138, the corresponding decays $\pi^0 \to \gamma\gamma$, $\Sigma^0 \to \gamma\Lambda$ which do conserve $Y$ have extremely short lifetimes). From (6) and the fact that $Q$ is exactly conserved in all interactions, we see that $I_3$ is conserved in electromagnetic interactions as well. This statement was trivial as long as we considered only pions and nucleons, since for such "non-strange" particles, we have from (7) $I_3 = Q - \frac{1}{2}B$.

Hadrons with $Y = 0$ may be called "hyperneutral," those with $Y \neq 0$ "hypercharged." Thus $\pi$, $\eta$, $\Lambda$, $\Sigma$ are hyperneutral, whereas $K$, $N$, $\Xi$ are hypercharged.

Neutral kaons are really "strange particles" in the proper sense of the word. The "shortlived" $K_S$ decays mainly into two pions, while the "long-lived" $K_L$ decays into many channels (see table C-2). In the kaon rest frame, the time dependence of their state vectors is given by (1-1.15), with $\mathbf{p} = 0$ and $t \to$ proper time $\tau$:

$$K_S(\tau) = \exp\{-iM_S\tau\}K_S(0), \qquad K_L(\tau) = \exp\{-iM_L\tau\}K_L(0), \tag{3.8}$$

$$M_S = m_S - \tfrac{1}{2}i\Gamma_S, \qquad M_L = m_L - \tfrac{1}{2}i\Gamma_L, \qquad m_L - m_S \equiv \Delta m, \tag{3.9}$$

where the complex masses account for the fact that the probability of the state decreases as $\exp\{-\Gamma\tau\}$. To obtain the time dependence of $K$ and $\bar{K}$, we express these states in terms of $K_L$ and $K_S$ by inverting (1.4):

$$K = 2^{-\frac{1}{2}}(1 - \varepsilon)(K_S + K_L), \qquad \bar{K} = 2^{-\frac{1}{2}}(1 + \varepsilon)(K_L - K_S). \tag{3.10}$$

An arbitrary superposition $\psi$ of $K_S$ and $K_L$ has the following time dependence:

$$\begin{aligned}\psi(\tau) &= a_L e^{-iM_L\tau}K_L(0) + a_S e^{-iM_S\tau}K_S(0)\\ &= 2^{-\frac{1}{2}}[(1+\varepsilon)(a_L e^{-iM_L\tau} + a_S e^{-iM_S\tau})K\\ &\quad +(1-\varepsilon)(a_L e^{-iM_L\tau} - a_S e^{-iM_S\tau})\bar{K}],\end{aligned} \tag{3.11}$$

and the probability of finding a $K$ or $\bar{K}$ at time $\tau$ is, with $a_S \equiv a_L \rho e^{i\phi}$

$$\begin{aligned}W\begin{pmatrix}K\\ \bar{K}\end{pmatrix} &= \tfrac{1}{2}(1 \pm 2\ \mathrm{Re}\ \varepsilon)|a_L e^{-iM_L\tau} \pm a_S e^{-iM_S\tau}|^2\\ &= \tfrac{1}{2}(1 \pm 2\ \mathrm{Re}\ \varepsilon)|a_L|^2(e^{-\Gamma_L\tau} + \rho^2 e^{-\Gamma_S\tau}\\ &\quad \pm 2\rho e^{-(\Gamma_L+\Gamma_S)\tau/2}\cos(\Delta m\tau + \varphi)).\end{aligned} \tag{3.12}$$

When the initial state at $\tau = 0$ is pure $K^0$, we have $a_S = a_L$, $|a_L|^2 = \frac{1}{2} - \mathrm{Re}\ \varepsilon$ according to (10), and thus $\rho = 1$, $\varphi = 0$ in (12). Thus a state which is produced as a pure $K^0$ with $Y = +1$ in a strong interaction acquires a $\bar{K}^0$ component at later times, with a probability

$$W(K \to \bar{K}) = \tfrac{1}{4}(1 - 4\ \mathrm{Re}\ \varepsilon)[e^{-\Gamma_S\tau} + e^{-\Gamma_L\tau} - 2e^{-\frac{1}{2}(\Gamma_L+\Gamma_S)\tau}\cos\Delta m\tau]. \tag{3.13}$$

This probability can be measured, e.g., in a hydrogen bubble chamber by secondary interactions of the type $\bar{K}^0 p \to \pi^+ \Lambda$ which are excluded for $K^0$ by hypercharge conservation. In fact, as $\Delta m$ is as small as $3.5 \times 10^{-6}$ eV, it can only be measured by experiments of this type. More about this topic in section 5-8.

## 4-4 Two-Particle Spin States, Parity and Time-Reversal

For decays such as $\pi^0 \to \gamma\gamma$ or for nucleon-nucleon scattering, we shall need two-particle cms states where both particles carry spin. Such states can be written as $|p\lambda_1 \lambda_2 \hat{\mathbf{p}}\rangle$ where $\hat{\mathbf{p}}$ is the direction of $\mathbf{p} \equiv \mathbf{p}_1 = -\mathbf{p}_2$, characterized by the polar and azimuthal angles $\theta$ and $\phi$, and $\lambda_1, \lambda_2$ are the two particles' helicities. The total angular momentum operator is $\mathbf{J} = \mathbf{J}_1 + \mathbf{J}_2$, and

$$\mathbf{J}\mathbf{p}|p, \lambda_1 \lambda_2, \hat{\mathbf{p}}\rangle = \lambda p|p, \lambda_1\lambda_2, \hat{\mathbf{p}}\rangle, \qquad \lambda \equiv \lambda_1 - \lambda_2, \tag{4.1}$$

because of $\mathbf{J}\mathbf{p} = \mathbf{J}_1\mathbf{p}_1 - \mathbf{J}_2\mathbf{p}_2$. The 2-particle states (1) are constructed from the product of 1-particle states as follows:

$$|p, \lambda_1 \lambda_2, \hat{\mathbf{p}}\rangle = |\mathbf{p}, \lambda_1\rangle| -\mathbf{p}, \lambda_2\rangle(-1)^{S_2 - \lambda_2}. \tag{4.2}$$

The choice of phase in (2) is advocated by Jacob and Wick (1959), and corresponds to

$$|p, \lambda_1 \lambda_2, \hat{\mathbf{p}}\rangle = A_1(\mathbf{p})A_2(-\mathbf{p})|\lambda_1, \hat{\mathbf{p}}\rangle| -\lambda_2, \hat{\mathbf{p}}\rangle, \tag{4.3}$$

where $A_1$ and $A_2$ are the boost operators of particles 1 and 2, and $|\lambda_i, \hat{\mathbf{p}}\rangle$ are the spin states at rest, quantized along the common direction $\hat{\mathbf{p}}$. According to (B-3.11) we namely have

$$| -\lambda_2, \hat{\mathbf{p}}\rangle = (-1)^{S_2 - \lambda_2}|\lambda_2, -\hat{\mathbf{p}}\rangle, \tag{4.4}$$

where the transformation from $\hat{\mathbf{p}}$ to $-\hat{\mathbf{p}}$ has been achieved by a rotation through $\pi$ around the $y$-axis, for $\phi = 0$. Equation (3) has already been used in (3-6.21). The states (2) can be expanded in terms of states $|\lambda_1 \lambda_2 JM\rangle$ (we omit the $p$) which transform irreducibly under rotations:

$$\begin{gathered}|\lambda_1\lambda_2 \hat{\mathbf{p}}\rangle = \sum_{J, M} C_J D^J_{M\lambda}(\phi, \theta, -\phi)|\lambda_1 \lambda_2 JM\rangle, \\ C_J = \tfrac{1}{2}\pi^{-1}(2J + 1)^{\frac{1}{2}}.\end{gathered} \tag{4.5}$$

The normalization coefficient $C_J$ will be derived in section 7-3, where the connection with the $L$-basis is also developed. For a more detailed presentation, see Martin and Spearman (1970). Consider now the decay of a particle of spin $S$ and magnetic quantum number $M_S$ into the state (2). Rotational invariance requires

$$\langle\lambda_1 \lambda_2 JM|T|SM_S\rangle = \delta_{JS}\, \delta_{MM_S} T_S(\lambda_1, \lambda_2)(4\pi)^{\frac{1}{2}}, \tag{4.6}$$

where the factor $(4\pi)^{\frac{1}{2}}$ is included for later (in)convenience. We thus find from (5)

$$\langle p, \lambda_1 \lambda_2, \theta\phi \,|\, T \,|\, M\rangle = (2S+1)^{\frac{1}{2}} e^{i\phi(M-\lambda)}\, d^S_{M\lambda}(\theta) T_S(\lambda_1 \lambda_2). \tag{4.7}$$

Using the normalization (B-5.1) of the $d$-function, we now obtain from (2-3.9) and (2-2.7)

$$\begin{aligned} m\Gamma &= \frac{1}{2}\int d \text{ Lips } (s_d\,; f) \sum_{\lambda_1 \lambda_2} |T|^2 \\ &= p(8\pi s^{\frac{1}{2}})^{-1} \int d \cos\theta \sum_{\lambda_1\lambda_2} (d^S_{M\lambda})^2 \,|\, T_S\,|^2\, \tfrac{1}{2}(2S+1) \\ &= p(8\pi s^{\frac{1}{2}})^{-1} \sum_{\lambda_1\lambda_2} |\, T_S\,|^2, \end{aligned} \tag{4.8}$$

which shows explicitly that $\Gamma$ is independent of $M$.
Similarly, in a collision between two free particles, $\langle \lambda_c \lambda_d J_f M_f \,|\, T \,|\, \lambda_a \lambda_b JM\rangle$ vanishes except for $J = J_f$, $M = M_f$. Taking the $z$-axis along $\mathbf{p} = \mathbf{p}_a = -\mathbf{p}_b$ $(\theta_i = \phi_i = 0)$ and using $d_{M\lambda}(0) = \delta_{M\lambda}$, we find

$$\begin{gathered} T(\lambda_a \cdots \lambda_d, s, \Omega) = 8\pi s^{\frac{1}{2}} \sum_J (2J+1) e^{i\phi(\lambda-\lambda')}\, d^J_{\lambda\lambda'}(\theta) T_J(s, \lambda_a \cdots \lambda_d), \\ \lambda \equiv \lambda_a - \lambda_b,\ \lambda' \equiv \lambda_c - \lambda_d. \end{gathered} \tag{4.9}$$

This is a trivial generalization of (2-5.16). The partial wave unitarity (2-5.22) is now understood as a consequence of angular momentum conservation. If all open channels $k$ are two-particle channels with helicities $\lambda_{k1}$ and $\lambda_{k2}$, we find

$$\begin{aligned} &-i[T_{if,J}(\lambda_a\lambda_b, \lambda_c\lambda_d) - T^*_{fi,J}(\lambda_c\lambda_d, \lambda_a\lambda_b)] \\ &\qquad = 2\sum_k p_k \sum_{\lambda_{k1}\lambda_{k2}} T_{ik,J}(\lambda_a\lambda_b, \lambda_{k1}\lambda_{k2}) T^*_{fk,J}(\lambda_c\lambda_d, \lambda_{k1}\lambda_{k2}). \end{aligned} \tag{4.10}$$

This can be summarized by matrix notation, with $\tilde{p}$ = diagonal matrix of channel momenta:

$$T_J - T_J^+ = 2iT_J\tilde{p}T_J^+, \qquad \tilde{p} = \begin{pmatrix} p_1 & 0 & 0 \\ 0 & p_2 & 0 \\ 0 & 0 & \ldots \end{pmatrix} \tag{4.11}$$

The parity transformation has been defined in section 1-5. For the field operators of spin-0 and spin-$\frac{1}{2}$ particles, this implies

$$\psi_{\mathscr{P}}(t, \mathbf{x}) = \eta\psi(t, -\mathbf{x}), \qquad \Psi_{\mathscr{P}}(t, \mathbf{x}) = \eta'\gamma^0\Psi(t, -\mathbf{x}), \tag{4.12}$$

where the intrinsic parities $\eta$ and $\eta'$ can be different for different particles. In a two-particle reaction $ab \to cd$, parity conservation requires

$$\eta_a\eta_b\pi_{ab} = \eta_c\eta_d\pi_{cd}, \qquad \pi_{ab} = (-1)^{L_{ab}}, \tag{4.13}$$

where $L_{ab}$ and $L_{cd}$ are the orbital angular momenta of initial and final states. For leptons, we can always put $\eta' = 1$ because these particles are either created as particle-antiparticle pairs (example $\pi^+\pi^- \to e^+e^-$) in which case only the square of $\eta'$ enters, or they are created by weak interactions (example pion decay $\pi \to \mu\nu$) in which case parity is not conserved. For hadrons, three overall intrinsic parities remain free, for similar reasons: baryon number conservation, charge conservation and hypercharge conservation except in weak, parity-violating interactions. We remove the ambiguities by defining $\eta' \equiv 1$ for the baryon $\Lambda$, hypercharged baryon $n$ and charged, hypercharged baryon $p$:

$$\eta_\Lambda \equiv \eta_n \equiv \eta_p \equiv 1. \tag{4.14}$$

The parities of all other hadrons are now observable. Particles and their anti-particles have identical parities in the case of bosons and opposite parities in the case of fermions (see 3-9.7).

The parity of charged pions has been measured in the reaction $\pi^- d \to nn$, with the initial particles in a Coulomb bound state (pionic atom) with $L_{\pi d} = 0$ (Chinowsky and Steinberger 1954). In this case, $L_{cd} = L_{nn}$ must be 1 because the total angular momentum is 1 (remember that the deuteron has spin 1) and the triplet $S$ and $D$ states are excluded for two identical fermions. We thus have $\eta(\pi^-) = \eta(\pi^+) = -1$. As the pion has spin 0, it is said to be a "pseudoscalar" particle.

The parity of kaons has been determined from the production of hypernuclei from $K$-mesic atoms:

$$K^- \, {}^4\text{He} \to {}^4_\Lambda\text{He}\,\pi^-, \qquad K^- \, {}^4\text{He} \to {}^4_\Lambda\text{H}\,\pi^0. \tag{4.15}$$

The index $\Lambda$ indicates that one neutron is replaced by a $\Lambda$-hyperon. All particles participating in these two reactions have spin zero, and ${}^4\text{He}$, ${}^4_\Lambda\text{He}$ and ${}^4_\Lambda\text{He}$ have positive parity. Consequently, conservation of parity requires the same parity for $K$ and $\pi$.

From isospin invariance and the convention (3) we expect that $\pi^0$ and $K^0$ have the same $\eta$ as their charged counterparts. This is confirmed by the partial wave analyses of the charge exchange reactions $\pi^- p \to \pi^0 n$ and $K^- p \to K^0 n$, which require $L_{ab} = L_{cd} = L$.

The parity of the $\eta$-meson follows from the existence of the decay $\eta \to \pi^0\pi^0\pi^0$. The Bose-Einstein principle requires the $\pi^0\pi^0\pi^0$ states to be totally symmetric, which in turn implies that all relative angular momenta are even. Since the total intrinsic parity of the final state is negative, the $\eta$-meson has negative parity. Of course this argument requires parity conservation in $\eta$-decay. A similar argument cannot be applied to the decays $K_L \to \pi^0\pi^0\pi^0$ and $K_S \to \pi^0\pi^0$, since these decays are due to weak interactions as can be seen from the comparatively long lifetimes of $K_L$ and $K_S$. Collecting our results, we have

$$\eta_{\pi^\pm} = \eta_{\pi^0} = \eta_{K^\pm} = \eta_{K^0} = \eta_\eta = -1. \tag{4.16}$$

Next, we consider the parity ($\mathscr{P}$) and time-reversal ($\mathscr{T}$) transformations of single-particle plane wave states $|\mathbf{p}, M\rangle$ and $|\mathbf{p}, \lambda\rangle$, where the magnetic quantum number $M$ is the eigenvalue of the $z$-component of the spin operator $S$ and $\lambda$ is the eigenvalue of the pseudoscalar operator $\mathbf{pS}|\mathbf{p}|^{-1}$. From the classical correspondences between $\mathbf{p}$ and $m d\mathbf{r}/dt$ and $\mathbf{S}$ and $\mathbf{r} \times \mathbf{p}$, it is clear that $\mathbf{p}$ changes sign both under $\mathscr{P}$ and $\mathscr{T}$, $M$ changes sign under $\mathscr{T}$ only and $\lambda$ changes sign under $\mathscr{P}$ only. In addition, $\mathscr{T}$ implies a reversal of the time sequence in scattering experiments, i.e., it exchanges initial and final states. Since the $S$-matrix elements contain the final state in the "bra-part" $\langle\ |$, we indicate this exchange by transforming from "ket" $|\ \rangle$ to "bra" $\langle\ |$ notation in table 4-4. This transformation involves complex conjugation. Thus $\mathscr{T}$ is an "antilinear" operator, $\mathscr{T}c|\ \rangle = c^*\mathscr{T}|\ \rangle$, similar to charge conjugation.

**Table 4-4** The action of parity and time reversal on plane-wave states

| $\|\ \rangle$ | $\mathscr{P}\|\ \rangle$ | $\mathscr{T}\|\ \rangle$ |
|---|---|---|
| $\|\mathbf{p}, M\rangle$ | $\eta\|-\mathbf{p}, M\rangle$ | $\langle-\mathbf{p}, -M\|$ |
| $\|\mathbf{p}, \lambda\rangle$ | $\eta\|-\mathbf{p}, -\lambda\rangle$ | $\langle-\mathbf{p}, \lambda\|$ |

In principle, the particles could acquire phases under time reversal, e.g., $\mathscr{T}|\mathbf{p}, M\rangle = e^{i\varphi}\langle-\mathbf{p}, -M|$. However, since the relative phases between initial and final states cannot be measured, the factor $e^{i\varphi}$ may be absorbed in the definition of the outgoing particle state.

For a general reaction $ab \to cd$, the momentum reversal does not change the cms scattering angle $\theta$. One would think that the azimuthal angle $\phi$ (see fig. 2-5) is transformed into $\phi + \pi$ but this is not so. The point is that $\phi$ is defined in a right-handed coordinate system based on momenta, which becomes a left-handed system after time reversal. Thus $\phi$ is transformed into $2\pi - (\phi + \pi) = \pi - \phi$. Invariance of the $T$-matrix elements under $\mathscr{T}$ thus requires

$$T_{ab\to cd}(\lambda_a\lambda_b, \lambda_c\lambda_d, \theta, \phi) = T_{cd\to ab}(\lambda_c\lambda_d, \lambda_a\lambda_b, \theta, \pi - \phi). \tag{4.17}$$

The point is most clearly seen for $\phi = 0$ (fig. 4-4). Starting from the standard coordinate system in which the incident particle defines the $z$-axis and the outgoing particle has $p_y = 0$, $p_x > 0$, time reversal leads to a state in which the outgoing particle has $p_x < 0$, which corresponds to the standard system rotated by $\pi$ around the incident momentum. From (9), it follows that for $\phi = 0$

$$T_{ab\to cd}(\lambda_a\lambda_b, \lambda_c\lambda_d, \theta) = (-1)^{\lambda-\lambda'}T_{cd\to ab}(\lambda_c\lambda_d, \lambda_a\lambda_b, \theta). \tag{4.18}$$

Expanding now both sides of (18) as in (9) and eliminating $d_{\lambda'\lambda}$ on the right-hand side in favour of $d_{\lambda\lambda'}$ by means of (B-3.10), we find that partial-wave $T$-matrix elements are symmetric:

$$T_{if,J}(\lambda_a\lambda_b, \lambda_c\lambda_d) = T_{fi,J}(\lambda_c\lambda_d, \lambda_a\lambda_b). \tag{4.19}$$

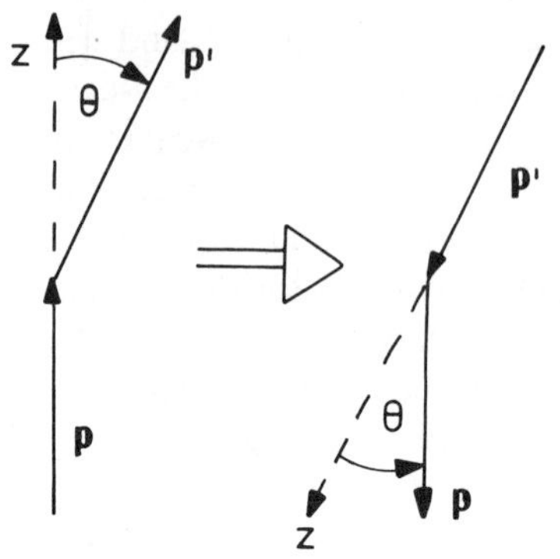


**Figure 4-4** Illustration of the scattering angle $\theta$ after time reversal.

For the matrix elements $T_S(\lambda_1 \lambda_2)$ of the decay $d \to 12$, time reversal invariance implies

$$T_S(d \to 12, \lambda_1 \lambda_2) = T_S(12 \to d, \lambda_1 \lambda_2), \tag{4.20}$$

where the second matrix element describes the inverse process, namely formation of the unstable particle $d$ in collisions of 1 and 2.

The consequences of parity conservation are most easily discussed for $\phi = 0$ by using the operator

$$\mathscr{Y} = e^{-i\pi J_y}\mathscr{P}, \tag{4.21}$$

which represents a reflection in the $x$-$z$-plane. It leaves all momenta in the $x$-$z$-plane unchanged and therefore commutes with all boosts $A(\mathbf{p})$ for which $p_y = 0$. We can thus apply $\mathscr{Y}$ to a 1-particle state as follows:

$$\begin{aligned} \mathscr{Y}|p, \lambda_1, \theta\rangle &= A(\mathbf{p})\mathscr{Y}|\lambda_1, \theta\rangle \\ &= A(\mathbf{p})e^{-i\pi J_y}\eta_1|\lambda_1, \theta\rangle = \eta_1(-1)^{S_1-\lambda_1}|p, -\lambda_1, \theta\rangle \end{aligned} \tag{4.22}$$

according to (B-3.11). For the 2-particle states (3), we find similarly

$$\mathscr{Y}|p, \lambda_1\lambda_2, \theta\rangle = \eta_1\eta_2(-1)^{S_1+S_2-\lambda_1+\lambda_2}|p, -\lambda_1, -\lambda_2, \theta\rangle. \tag{4.23}$$

For the decay matrix elements (5), this implies

$$T_S(-\lambda_1, -\lambda_2) = \eta\eta_1\eta_2(-1)^{S_1+S_2-S}T_S(\lambda_1, \lambda_2), \tag{4.24}$$

and for the partial-wave scattering amplitudes $T_J(\lambda_a \cdots \lambda_d)$ of (9):

$$\begin{gathered} T_J(-\lambda_a \cdots -\lambda_d) = \eta\eta'(-1)^{S_a+S_b-S_c-S_d}T_J(\lambda_a \cdots \lambda_d), \\ \eta = \eta_a\eta_b, \qquad \eta' = \eta_c\eta_d \end{gathered} \tag{4.25}$$

In (24) and (25), the $\lambda_i$-phases have been cancelled by the phases from the transformation (B-3.12).

A simple application of time-reversal invariance is the "principle of detailed balance"

$$(2S_a + 1)(2S_b + 1)\lambda(s, m_a^2, m_b^2)\, d\sigma\, (ab \to cd)/dt$$
$$= (2S_c + 1)(2S_d + 1)\lambda(s, m_c^2, m_d^2)\, d\sigma\, (cd \to ab)/dt, \qquad (4.26)$$

which follows from insertion of (18) into (2-4.9). Historically, this equation was applied to the pair of reactions $\pi^+ d \leftrightarrow pp$ to prove that the $\pi^+$ has spin 0.

## 4-5 Charge Conjugation and $G$-Parity

The operation of charge conjugation on vector, scalar and spinor fields has been discussed in 3-9. For an arbitrary one-particle state, $C$ is defined as the replacement of a particle by its antiparticle without change of momentum $\mathbf{p}$ or helicity $\lambda$:

$$C\,|a, \mathbf{p}, \lambda\rangle = \eta_c(a)\,|\bar{a}, \mathbf{p}, \lambda\rangle, \qquad (5.1)$$

where $\eta_c(a) = \pm 1$ is the particle's charge parity. For leptons we can put $\eta_c = 1$. Neutral hyperneutral mesons are eigenstates of $C$: according to (3-9.2), the eigenvalue is determined by the number of photons to which they may decay. In particular, the existence of the decays $\pi^0 \to \gamma\gamma$ and $\eta \to \gamma\gamma$ shows that these particles have

$$\eta_c(\pi^0) = \eta_c(\eta) = 1. \qquad (5.2)$$

For the charged members of isospin multiplets, the phases $\eta_c$ are fixed by the phase of the neutral member and the phase convention (B-3.7) for the isospin step operators $I_\pm$. Charge conjugation reverses the eigenvalues of $I_3$, and transforms a representation $D$ of $SU_2$ into its complex conjugate $D^*$ which implies for the generators $\mathbf{I}$ of $SU_2$ transformations (compare 1-4.12)

$$C\mathbf{I} = -\mathbf{I}^*C. \qquad (5.3)$$

On the other hand, it is advisable to use the representation $D$ also for antiparticles, because otherwise particles and antiparticles get different Clebsch-Gordan coefficients, which is confusing for pion and similar meson multiplets which combine particles and antiparticles. We therefore consider instead of $C$ the so-called "$G$-conjugation" operator

$$G = C \exp\,(i\pi I_2) = \exp\,(i\pi I_2)C, \qquad (5.4)$$

which commutes with all three generators $\mathbf{I}$ according to (1-4.12), (1-4.13) and (3). Thus, if the members of particle and antiparticle multiplets are denoted by $|m, I_3, \mathbf{p}, \lambda\rangle$ and $|\bar{m}, I_3, \mathbf{p}, \lambda\rangle$ respectively, we have

$$G\,|m, I_3, \mathbf{p}, \lambda\rangle = \eta_G\,|\bar{m}, I_3, \mathbf{p}, \lambda\rangle, \qquad (5.5)$$

where the "$G$-parity" $\eta_G$ is independent of $I_3$, and $|\bar{m}, I_3\rangle$ transforms under isospin in the same way as $|m, I_3\rangle$. The connection between $\eta_G$ and $\eta_c$ is

$$\eta_c(I, I_3) = \eta_G(-1)^{I+I_3} \tag{5.6}$$

according to (B-3.11). For $\pi^0$ and $\eta$ we have $I_3 = 0$ and $I = 1$ and 0, respectively, such that (2) gives

$$\eta_G(\pi) = -1, \qquad \eta_G(\eta) = 1, \tag{5.7}$$

and application of (6) to $I_3 = \pm 1$ gives $C|\pi^\pm\rangle = -|\pi^\mp\rangle$. For $\Lambda$ and nucleons $N$, we can define

$$G|\Lambda\rangle = |\bar{\Lambda}\rangle, \qquad G|N\rangle = |\bar{N}\rangle. \tag{5.8}$$

Remembering that the antineutron has $I_3 = +\frac{1}{2}$, (8) is explicitly

$$G|p\rangle = |\bar{n}\rangle, \qquad G|n\rangle = |\bar{p}\rangle. \tag{5.9}$$

Then we find from (6)

$$C|\Lambda\rangle = |\Lambda\rangle, \qquad C|n\rangle = |\bar{n}\rangle, \qquad C|p\rangle = -|\bar{p}\rangle. \tag{5.10}$$

Now $G$ and $C$ for the remaining baryons $\Sigma$, $\Xi$ and mesons $K$ are fixed by the requirement that the baryons can emit and absorb single (virtual) pions and kaons, thereby transforming themselves according to the isospin and hypercharge (and baryon number) selection rules: $\Sigma \to \pi\Lambda$ entails $\eta_G(\Sigma) = \eta_G(\pi)$, $N \to K\Lambda$ and $N \to K\Sigma$ both entail $\eta_G(K) = \eta_G(N)$, and $\Lambda \to K\Xi$ and $\Sigma \to K\Xi$ entail $\eta_G(\Xi) = \eta_G(N)$. The results are collected in table 4-5.1. Note that $G^2 = -1$ for $I = \frac{1}{2}$, because of $\exp(2i\pi J_y) = -1$ in such cases.

**Table 4-5.1** The action of $G$ and $C$ on single-particle states.

| | $\pi^+$ | $\pi^0$ | $\pi^-$ | $\eta$ | $K^+$ | $K^0$ | $\bar{K}^0$ | $K^-$ |
|---|---|---|---|---|---|---|---|---|
| $G$ | $-\pi^+$ | $-\pi^0$ | $-\pi^-$ | $\eta$ | $\bar{K}^0$ | $K^-$ | $-K^+$ | $-K^0$ |
| $C$ | $-\pi^-$ | $\pi^0$ | $-\pi^+$ | $\eta$ | $-K^-$ | $\bar{K}^0$ | $K^0$ | $-K^+$ |
| | $\Sigma^+$ | $\Sigma^0$ | $\Sigma^-$ | $\Lambda$ | $p$ | $n$ | $\bar{n}$ | $\bar{p}$ |
| $G$ | $-\bar{\Sigma}^+$ | $-\bar{\Sigma}^0$ | $-\bar{\Sigma}^-$ | $\bar{\Lambda}$ | $\bar{n}$ | $\bar{p}$ | $-p$ | $-n$ |
| $C$ | $-\bar{\Sigma}^-$ | $\bar{\Sigma}^0$ | $-\bar{\Sigma}^+$ | $\bar{\Lambda}$ | $-\bar{p}$ | $\bar{n}$ | $n$ | $-p$ |

The fact that hyperneutral mesons are eigenstates of $G$ leads to simple selection rules analogous to Furry's theorem for photons. From $G|\pi\rangle = -|\pi\rangle$ it follows that an $n$-pion state has $G$-parity $(-1)^n$ and cannot transform into $n \pm 1$ pion states by strong interactions. Similarly, $G|\eta\rangle = |\eta\rangle$ implies that the $\eta$-meson can couple strongly to 4 pions but not to 3. The decays $\eta \to \pi\pi\pi$ violate $G$-parity. They are of the order of second-order electromagnetic decays such as $\eta \to \gamma\gamma$ (decays $\eta \to \pi\pi$ are parity-forbidden).

$G$-parity is also useful for $K\bar{K}$ and $N\bar{N}$ states of given isospin. Analogous to (2.4) and (2.5), we have

$$\begin{aligned} |N\bar{N}, 1, 1\rangle &= |p\bar{n}\rangle, & |N\bar{N}, 0, 0\rangle &= 2^{-\frac{1}{2}}(|p\bar{p}\rangle - |n\bar{n}\rangle), \\ |N\bar{N}, 1, -1\rangle &= |n\bar{p}\rangle, & |N\bar{N}, 1, 0\rangle &= 2^{-\frac{1}{2}}(|p\bar{p}\rangle + |n\bar{n}\rangle); \end{aligned} \tag{5.11}$$

similarly for kaons. For the $I = 0$ state, we have $G = C$ according to the definition (4), and from section 3-9 we remember that particle-antiparticle states of orbital angular momentum $L$ and total spin $S$ are eigenstates of $C$ with eigenvalues $(-1)^{L+S}$. The $I_3 = 0, I = 1$ state must then be an eigenstate of $G$ with the opposite eigenvalue (compare eq. 6), and since the eigenvalues of $G$ are independent of $I_3$, we have in fact

$$\begin{aligned} G|N\bar{N}, I, I_3, L, S\rangle &= (-1)^I|\bar{N}N, I, I_3, L, S\rangle \\ &= (-1)^{L+S+I}|N\bar{N}, I, I_3, L, S\rangle. \end{aligned} \tag{5.12}$$

This formula also applies to $K\bar{K}$ states, with $S = 0$.

For $I > 0$, the operators $G$ and $C$ are not equivalent, because isospin conservation and therefore $G$ invariance is broken in electromagnetic interactions whereas $C$ is not. An even "better" operator is $C\mathscr{P}$, which is also conserved in the "normal weak" interactions, as we shall see. For example, from table 4-5.1 we see that the combinations

$$|K_1\rangle = 2^{-\frac{1}{2}}(|K^0\rangle - |\bar{K}^0\rangle), \qquad |K_2\rangle = 2^{-\frac{1}{2}}(|K^0\rangle + |\bar{K}^0\rangle \tag{5.13}$$

are eigenstates of $C$ with eigenvalues $-1$ and $+1$, respectively. As kaons at rest are also eigenstates of $\mathscr{P}$ with eigenvalue $-1$, it follows that

$$C\mathscr{P}|K_1\rangle = |K_1\rangle, \qquad C\mathscr{P}|K_2\rangle = -|K_2\rangle. \tag{5.14}$$

Looking now at the physical states $K_S$ and $K_L$ of (1.4), we find

$$|K_S\rangle = |K_1\rangle + \varepsilon|K_2\rangle, \qquad |K_L\rangle = |K_2\rangle + \varepsilon|K_1\rangle, \tag{5.15}$$

which shows that the small quantity $\varepsilon$ is a measure of $C\mathscr{P}$-violation.

$\pi^+\pi^-$ and $\pi^0\pi^0$ are also eigenstates of $C\mathscr{P}$ with eigenvalue $+1$. Thus in the limit of $C\mathscr{P}$ conservation, the decay $K_L \to \pi\pi$ is forbidden, which explains the long lifetime of $K_L$ relative to $K_S$.

In nucleon-antinucleon annihilation into mesons $\mathscr{P}$, $C$ and **I**-conservation lead to a number of selection rules. According to section 3-9, the eigenvalues of $\mathscr{P}$ are $(-1)^{L+1}$ for $N\bar{N}$ states. A state of two pseudoscalar mesons of angular momentum $L$ has parity $(-1)^L$. Thus reactions $N\bar{N} \to \pi\pi$, $N\bar{N} \to K\bar{K}$, $N\bar{N} \to \pi\eta$ are only possible if

$$(-1)^{L_{ab}+1} = (-1)^{L_{cd}} = (-1)^J \tag{5.16}$$

i.e., for $N\bar{N}$ states that have $J = L \pm 1$, which is only possible for $S = 1$. For $N\bar{N} \to K\bar{K}$, $G$-parity (12) is then automatically conserved, but in $N\bar{N} \to \pi\pi$ the final state has $G = +1$, in which case (12) requires that either $L$ or $I$ must be odd.

**Table 4-5.2** States of zero baryon number and hypercharge. For the $N\bar{N}$ states, the notation $^{2S+1}L_J$ is used. $C\mathscr{P}$ from (3-9.8). For further explanations, see text.

| | $I = 0$ | | | | $I = 1$ | | | | $I_3 = 0$ | |
|---|---|---|---|---|---|---|---|---|---|---|
| $N\bar{N}$ | $\mathscr{P}$ | $G$ | Meson | Pions | $\mathscr{P}$ | $G$ | Meson | Pions | $C\mathscr{P}$ | $K\bar{K}$ |
| $^1S_0$ | $-$ | $+$ | $\eta$, $X$ | $4P^3$ | $-$ | $-$ | $\pi$ | $3S^2$ | $-$ | — |
| $^3S_1$, $^3D_1$ | $-$ | $-$ | $\omega$, $\phi$ | $3P^2$ | $-$ | $+$ | $\rho$ | $2P$ | $+$ | $K^+K^-$, $K_1K_2$ |
| $^1P_1$ | $+$ | $-$ | — | $3SP$ | $+$ | $+$ | $(B)$ | $4SP^2$ | $-$ | — |
| $^3P_0$ | $+$ | $+$ | $(\varepsilon)$ | $2S$ | $+$ | $-$ | — | $5SP^3$ | $+$ | $K^+K^-$, $\begin{cases} K_1K_1 \\ K_2K_2 \end{cases}$ |
| $^3P_1$ | $+$ | $+$ | — | $4SP^2$ | $+$ | $-$ | — | $3P^2$ | $+$ | — |
| $^3P_2$, $^3F_2$ | $+$ | $+$ | $f$, $f'$ | $2D$ | $+$ | $-$ | $A_2$ | $3PD$ | $+$ | $K^+K^-\begin{cases} K_1K_1 \\ K_2K_2 \end{cases}$ |

In table 4-5.2 we have collected the $N\bar{N}$ states for $L < 2$. Also listed are $\eta$ and $\pi$ in the appropriate places, together with a few resonances, to be discussed later. The column labelled "pions" contains the lowest possible number of pions in $N\bar{N}$ annihilation, together with the lowest orbital angular momentum configuration. The notation $S$, $P$, $D$ in this column refers to the angular momenta of the subsystem of the first two, three, etc., pions. In one case we see that the allowed minimum of pions is five. The decay of a $0^+$ state into three pseudoscalar particles is forbidden by parity conservation. The coupling of three spinless particles to $J = 0$ is of the type $3p^2$, which has positive orbital parity. Due to the negative intrinsic parity, the system's total parity is negative. This argument forbids not only the decay into three pions, but also decay into $\bar{K}K\pi$, $\eta\pi\pi$ etc.

The states $p\bar{p}$ and $n\bar{n}$ have $I_3 = 0$ and are $CP$-eigenstates. If their decay into $K\bar{K}$ states are not parity-forbidden, they go either into $K^+K^-$ and $K_1K_2$ or into $K^+K^-$ and equal amounts of $K_1K_1$ and $K_2K_2$, because of $CP$ conservation. This is indicated in the third part of table 4-5.2.

## 4-6 Three Particles in the Final State

Many-body phase space differentials can be expressed in terms of products of two-body phase space differentials (Srivastava and Sudarshan 1958). We show how this works for 3 particles. Let $P_1$, $P_2$ and $P_3$ be the 4-momena of these particles. We introduce an auxiliary 4-momentum

$$P_d = P_1 + P_2, \qquad P_d^2 = s_d, \tag{6.1}$$

by multiplying the phase space differential by $1 = d^4P_d\,\delta_4(P_d - P_1 - P_2)$:

$$\begin{aligned} d\text{ Lips }(s; P_1, P_2, P_3) &= (2\pi)^4\,\delta_4(P_i - P_1 - P_2 - P_3)\prod_{k=1}^{3} d_L^3 p_k \\ &= \delta_4(P_i - P_d - P_3)\,d_L^3\mathbf{p}_3\,d^4P_d\,d\text{ Lips }(s_d; P_1, P_2). \end{aligned} \tag{6.2}$$

Now we make $s_d$ an explicit integration variable by writing

$$d^4P_d = d^4P_d\,\delta\,(P_d^2 - s_d)\,ds_d = (2\pi)^3\,d_L^3p_d\,ds_d. \tag{6.3}$$

Using this expression, we transform (3) into

$$d\text{ Lips }(s;\,P_1,\,P_2,\,P_3) = (2\pi)^{-1}\,d\text{ Lips }(s;\,P_d,\,P_3)\,d\text{ Lips }(s_d;\,P_1,\,P_2)\,ds_d, \tag{6.4}$$

which is the desired recurrence relation. It can be understood as follows: the first $d$ Lips on the right-hand side of (4) is that of particle 3 and a fictitious particle $d$ of effective mass $s_d^{\frac{1}{2}}$. The second $d$ Lips refers to the "decay" $d \to 1$, 2. We write "decay" because in a proper decay, $s_d$ would have a fixed value. In (4), however, it is an integration variable.

The obvious generalization of (4) to $n$ particles is

$$d\text{ Lips }(s;\,P_1\cdots P_m,\,P_{m+1}\cdots P_n) = (2\pi)^{-1}\,d\text{ Lips }(s;\,P_d,\,P_{m+1}\cdots P_n) \times d\text{ Lips }(s_d;\,P_1\cdots P_m)\,ds_d. \tag{6.5}$$

We return to (4) and insert the 2-particle phase spaces from (2-2.7):

$$d\text{ Lips }(s;\,P_1,\,P_2,\,P_3) = \frac{ds_d}{2\pi}\,\frac{p_3\,d\Omega_3}{16\pi^2 s^{\frac{1}{2}}}\,\frac{p\,d\Omega}{16\pi^2 s_d^{\frac{1}{2}}},\qquad p \equiv \tfrac{1}{2}s_d^{-\frac{1}{2}}\lambda^{\frac{1}{2}}(s_d,\,m_1^2,\,m_2^2). \tag{6.6}$$

This form is convenient for cases in which the final state is in fact reached in two steps via an intermediate $d$ (normally a resonance). In more general cases, a form which treats the 3 particles symmetrically is to be preferred. For this purpose, we use (2-2.16) for $d$ Lips $(s_d,\,P_1,\,P_2)$ in the over-all cms, with $\mathbf{p}_L = -\mathbf{p}_3$

$$(4\pi)^{-2}s_d^{-\frac{1}{2}}p\,d\Omega = (4\pi)^{-2}p_3^{-1}\,dE_1\,d\phi, \tag{6.7}$$

where $E_1$ now refers to the over-all cms. We also introduce the "squares of the effective masses,"

$$s_{jk} = (P_j + P_k)^2 = (P_i - P_l)^2 = s + m_l^2 - 2s^{\frac{1}{2}}E_l \qquad (jkl\text{ cyclic}), \tag{6.8}$$

and insert $ds_d = ds_{12} = 2s^{\frac{1}{2}}\,dE_3$ in (6):

$$\begin{aligned} d\text{ Lips} &= (4\pi)^{-4}\pi^{-1}\,dE_1\,dE_3\,d\Omega_3\,d\phi \\ &= (4\pi)^{-4}\pi^{-1}\,dE_1\,dE_2\,dE_3\,\delta\,(s^{\frac{1}{2}} - \textstyle\sum E_i)\,d\Omega_3\,d\phi. \end{aligned} \tag{6.9}$$

The last form exhibits the symmetry in $E_1$, $E_2$, $E_3$ and shows that the density of states is independent of these energies within the physical boundaries. A plot of reaction events in the $(E_1,\,E_2)$ plane or equivalently in the $(s_{23},\,s_{13})$ plane is called a Dalitz plot (Dalitz 1953, Fabry 1954). The cms condition $\mathbf{p}_1 + \mathbf{p}_2 + \mathbf{p}_3 = 0$ means that the 3 momenta are coplanar. The three angles $\Omega_3$ and $\phi$ define the orientation of the momentum plane in space. The boundary of the Dalitz plot corresponds to collinear momenta,

$$(E_1^2 - m_1^2)^{\frac{1}{2}} \pm (E_2^2 - m_2^2)^{\frac{1}{2}} \pm (E_3^2 - m_3^2)^{\frac{1}{2}} = 0. \tag{6.10}$$

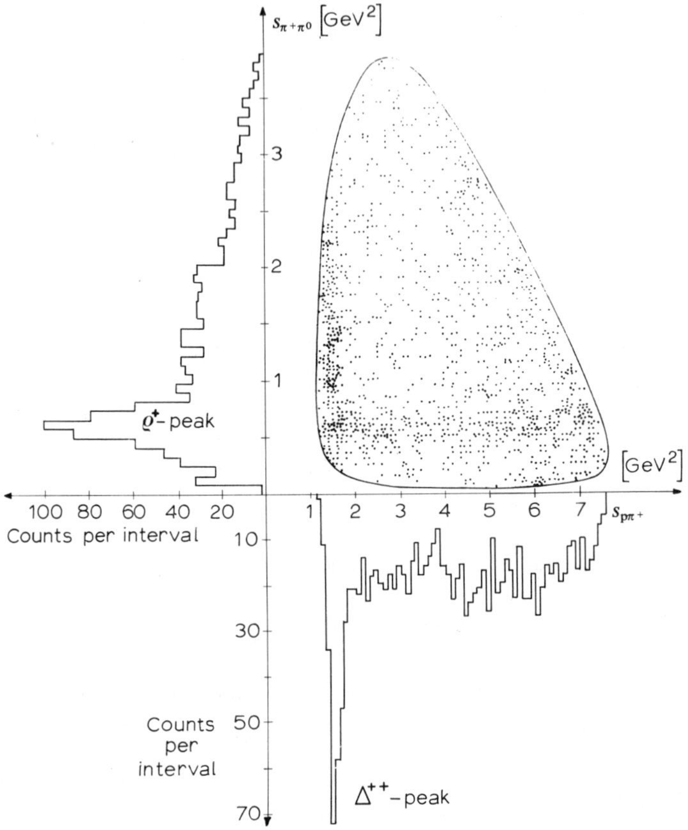


**Figure 4-6.1** Dalitz plot of the $\pi^0\pi^+p$ final state in 4 GeV $\pi^+p$ collisions, including the projections on the $s(\pi^+\pi^0)$ and $s(p\pi^+)$ axes (effective mass distributions). Figure by courtesy of N. Schmitz, see Aachen (1964).

Figure 4-6.1 shows a Dalitz plot of the $\pi^0\pi^+p$ final state in 4 GeV $\pi^+p$ collisions. Such plots are particularly useful for the analysis of 3-body decays of unpolarized particles, where they contain *all* accessible information. When the 3 particles in the final state have similar masses, a symmetric representation is achieved by plotting the events as functions of

$$X = 3^{\frac{1}{2}}(E_1 - E_2)Q^{-1}, \qquad Y = 3T_3Q^{-1} - 1, \qquad Q \equiv s^{\frac{1}{2}} - \sum_{i=1}^{3} m_i, \quad (6.11)$$

where $T_3 = E_3 - m_3$ is the kinetic energy of particle 3. If 2 of the 3 particles are identical as in $K^+ \to \pi^+\pi^+\pi^-$ decays, these are taken as particles 1 and 2, and one can define $X > 0$. For $m_1 = m_2 \approx m_3$ and mildly relativistic pions, the boundary curve (10) is close to a circle (fig. 4-6.2) in the variables (11).

Inside the boundary curve, the density of states is constant, and any variation of the density of events must be ascribed to a variation of $|T|^2$. Historically, the plot of $K^+ \to \pi^+\pi^+\pi^-$ decays played a role in the discovery

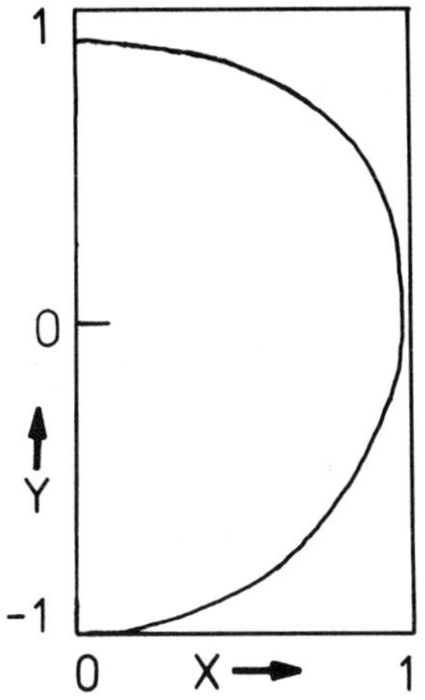


**Figure 4-6.2** Boundary (10) of the Dalitz plot in the variables $X$ and $Y$ of (11) for the decays $K^{\pm} \to \pi^{\pm}\pi^{\pm}\pi^{\mp}$, from Mast et al. (1969). The plot itself consists of an array of numbers corresponding to a total of 51,000 events.

of parity violation in weak interactions. Bose-Einstein symmetry requires the relative angular momentum of the two $\pi^+$ to be 0, 2, 4 etc, and the uniformity of events in $X$ shows that it is mainly 0. Similarly, the weak dependence of $|T|^2$ on $Y$ (in particular the fact that it does not go to zero for $Y \to -1$) shows that $\pi^-$ is dominantly in an $S$-state relative to the other two pions. (For a pion in angular momentum $L$, $|T|^2$ goes as $p^{2L}$-for $p \to 0$, see also table 4-9). The $\pi\pi\pi$-state is thus dominantly $S^2$ (in the notation of table 4-5.2), the $K^+$ must therefore have spin 0 and also negative parity (due to the total negative intrinsic parity of the three pions), if parity is conserved. But in that case $K \to \pi\pi$ obviously violates parity.

## 4-7 The Decays $\pi^0 \to \gamma\gamma$ and $\pi^0 \to \gamma e\bar{e}$

As applications of 2-particle spin states, 3-particle states and quantum electrodynamics, we calculate the decays of $\pi^0$, beginning with the dominant decay $\pi^0 \to \gamma\gamma$ (the decay $\eta \to \gamma\gamma$ is formally identical). The decay matrix element $T$ must be linear in the helicity states $\varepsilon_1^{\mu *}(\lambda_1)$ and $\varepsilon_2^{\mu *}(\lambda_2)$, with $\varepsilon_i^{\mu}(\lambda)$ given by (2-8.6) (the index $i$ refers to the momentum $K_i$ of the $i$th photon):

$$T(\lambda_1, \lambda_2) = -ie^2 0_{\mu\nu}\varepsilon_1^{\mu *}(\lambda_1)\varepsilon_2^{\nu *}(\lambda_2), \tag{7.1}$$

where $0_{\mu\nu}$ is a tensor containing the 4-vectors $K_1$ and $K_2$, and a factor $-ie^2$ has been taken out for later convenience. Gauge invariance requires

$$0_{\mu\nu}K_1^{\mu} = 0_{\mu\nu}K_2^{\nu} = 0. \tag{7.2}$$

If $\pi^0$ were a scalar particle, then $0_{\mu\nu}$ would be the proper tensor (2-8.29). For a pseudoscalar particle, we need a pseudotensor instead. Such an object can in fact be constructed from $K_1$ and $K_2$. To derive it, we first show that the determinant of 4 different 4-vectors $A$, $B$, $C$, $D$ is Lorentz invariant. It is conveniently expressed in terms of the Levi-Civita symbol $\varepsilon_{\alpha\beta\gamma\delta}$, which is antisymmetric under the exchange of any two adjacent indices:

$$\det(ABCD) = \varepsilon_{\alpha\beta\gamma\delta} A^\alpha B^\beta C^\gamma D^\delta, \qquad\qquad (7.3)$$
$$\varepsilon_{\alpha\beta\gamma\delta} = -\varepsilon_{\beta\alpha\gamma\delta}, \ldots, \qquad \varepsilon_{0ijk} \equiv \varepsilon_{ijk}.$$

Under a Lorentz transformation $A'^\mu = \Lambda^\mu_\alpha A^\alpha$ (1-0.14), we get

$$\det(A'B'C'D') = \varepsilon_{\mu\nu\rho\sigma} \Lambda^\mu_\alpha A^\alpha \Lambda^\nu_\beta B^\beta \Lambda^\rho_\gamma C^\gamma \Lambda^\sigma_\delta D^\delta$$
$$= \det(\Lambda) \det(ABCD), \qquad\qquad (7.4)$$

which shows that (3) is invariant under proper Lorentz transformation and changes sign under improper ones (a "Lorentz-invariant pseudoscalar").

It is now clear that the following combination of $K_1$ and $K_2$ is a pseudotensor:

$$0_{\mu\nu} = g_{\pi\gamma\gamma} \varepsilon_{\mu\nu\rho\sigma} K_1^\rho K_2^\sigma, \qquad\qquad (7.5)$$

where $g_{\pi\gamma\gamma}$ is a "decay constant." Due to the antisymmetry of the indices $\mu \cdots \sigma$, (2) is automatically fulfilled. We can also replace $K_2$ by $K_1 + K_2$ without changing the value of (5). This is convenient in the cms, where $K_1 + K_2$ has only a zero-component, namely $s^{\frac{1}{2}} = m_\pi$. We find in the cms

$$0_{\mu\nu} = g_{\pi\gamma\gamma} \varepsilon_{\mu\nu\rho 0} K_1^\rho m_\pi, \qquad T = ie^2 m_\pi g_{\pi\gamma\gamma} (\boldsymbol{\varepsilon}_1^* \times \boldsymbol{\varepsilon}_2^*) \cdot \mathbf{k}_1 \qquad\qquad (7.6)$$

As a consequence of the cross product in (6), only the components of $\boldsymbol{\varepsilon}_1^*$ and $\boldsymbol{\varepsilon}_2^*$ perpendicular to $\mathbf{k}_1 = -\mathbf{k}_2$ contribute, i.e., $T$ vanishes for $\lambda_1 = 0$ or $\lambda_2 = 0$, even if one or both photons are virtual. Adopting now the prescription (4.3), we simply have $\varepsilon_2^\nu(\lambda_2) = \varepsilon_1^\nu(-\lambda_2)$, since the boost $A(\mathbf{k})$ does not affect the states $\boldsymbol{\varepsilon}^\nu(\pm 1)$ at all. This gives us

$$T = ie^2 m_\pi g_{\pi\gamma\gamma} [\boldsymbol{\varepsilon}_1^*(\lambda_1) \times \boldsymbol{\varepsilon}_1^*(-\lambda_2)] \cdot \mathbf{k}_1 = e^2 g_{\pi\gamma\gamma} k \lambda_1 \, \delta_{\lambda_1, \lambda_2}. \qquad\qquad (7.7)$$

The last expression is most easily obtained by taking $\mathbf{k}_1$ along the $z$-axis, in which case $\boldsymbol{\varepsilon}^*(\lambda) = 2^{-\frac{1}{2}}(-\lambda, i, 0)$. No angular functions appear for the decays of spinless particles. In fact, (4.7) reduces for $S = 0$ to

$$T = T_0(\lambda_1, \lambda_2). \qquad\qquad (7.8)$$

The decay rate (4.8) becomes ($\alpha = e^2/4\pi$):

$$m\Gamma = 4\pi\alpha^2 g_{\pi\gamma\gamma}^2 k^3 m_\pi. \qquad\qquad (7.9)$$

This expression is valid for decay into two different particles. For identical particles, the phase space is only half as large (compare the remark following (3-8.11). For decays into two massless particles, we have moreover $k = \frac{1}{2}m_\pi$, such that the final result is

$$m\Gamma(\pi^0 \to \gamma\gamma) = \tfrac{1}{4}\pi\alpha^2 g_{\pi\gamma\gamma}^2 m_\pi^4. \qquad\qquad (7.10)$$

The dependence of (7) on $\lambda_1$ and $\lambda_2$ is evident: As $\pi^0$ is spinless, we must

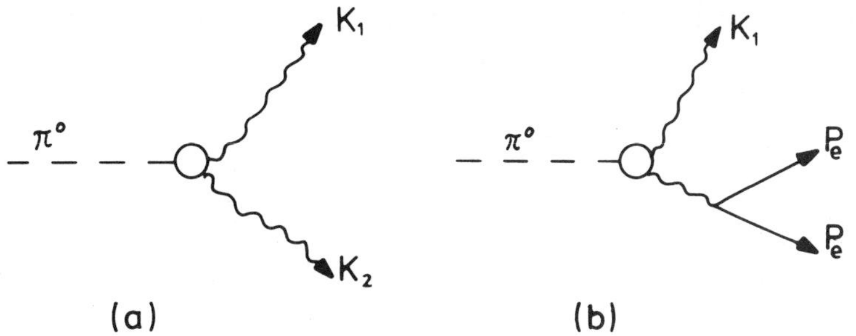


**Figure 4-7** Feynman diagrams for the decays (a) $\pi^0 \to \gamma\gamma$, (b) $\pi^0 \to \gamma e\bar{e}$.

have $\lambda = \lambda_1 - \lambda_2 = 0$, and the factor $\lambda_1$ is required by parity conservation (4.24), with $\eta = -1$.

Next, consider the decay $\pi^0 \to \gamma e\bar{e}$, which constitutes the remaining 1.2% of $\pi^0$ decays. It is due to the Feynman graph (b) of fig. 4-7, which can be viewed as the "inner conversion" of one (virtual) photon into an $e\bar{e}$ pair ("Dalitz pair"). Here we have for the first time 3 particles in the final state. The decay rate is obtained by inserting (6.4) into the general expression (2-3.9)

$$m_\pi \Gamma(\pi \to \gamma e\bar{e}) = \frac{1}{2} \int (2\pi)^{-1}\, d \text{ Lips } (m_\pi^2, P_d, K_1) \times d \text{ Lips } (s_d; P_e, P_{\bar{e}})\, ds_d \sum_{\lambda_1 \lambda_e \lambda_{\bar{e}}} |T_{if}|^2, \tag{7.11}$$

where $s_d = P_d^2 = (P_e + P_{\bar{e}})^2$ appears as integration variable. The matrix element is of the form

$$T_{if} = s_d^{-1} \sum_{\lambda_2} T(\pi^0 \to \gamma\hat{\gamma}) T(\hat{\gamma} \to e\bar{e}), \tag{7.12}$$

and inserting (3-6.15), we find

$$\begin{aligned} m_\pi \Gamma(\pi \to \gamma e\bar{e}) &= \frac{1}{2} \int (\pi s_d^2)^{-1}\, ds_d \int d \text{ Lips } (m_\pi^2; P_d, K_1) \\ &\quad \times \sum_{\lambda_1 \lambda_2} |T(\pi^0 \to \gamma\hat{\gamma})|^2 m_\gamma \Gamma(\hat{\gamma} \to e\bar{e}) \qquad (7.13) \\ &= \pi^{-1} \int s_d^{-2}\, ds_d m_\pi \Gamma(\pi \to \gamma\hat{\gamma}) m_\gamma \Gamma(\hat{\gamma} \to e\bar{e}). \end{aligned}$$

Using now $m_\gamma \Gamma(\hat{\gamma} \to e\bar{e}) = \frac{2}{3}\alpha p_e s_d^{-\frac{1}{2}}(2m_e^2 + s_d)$ from (3-6.17) and (9) for $m_\pi \Gamma(\pi \to \gamma\hat{\gamma})$, with $k = E_{\gamma 1} = \frac{1}{2}(m_\pi - s_d/m_\pi)$ (compare 2-2.18) we find the ratio

$$\begin{aligned} r &= \Gamma(\pi \to \gamma e\bar{e})/\Gamma(\pi \to \gamma\gamma) \\ &= \frac{2}{\pi} \int s_d^{-2}\, ds_d (1 - s_d/m_\pi^2)^3 \cdot \tfrac{2}{3}\alpha(\tfrac{1}{4} - m_e^2/s_d)^{\frac{1}{2}}(2m_e^2 + s_d) \qquad (7.14) \\ &= \frac{4\alpha}{3\pi}[\ln (m_\pi/m_e - \tfrac{7}{4})] = 0.0031 \cdot 3.83 = 0.0119, \end{aligned}$$

in good agreement with experiment. Note that in $\pi \to \gamma e\bar{e}$ decays, the "decay constant" $g_{\pi\gamma\gamma}$ can be a function of $s_d = (P_e + P_{\bar{e}})^2$. In analogy with the form factors of section 2-7, we put

$$g_{\pi\gamma\gamma}(s_d) = g_{\pi\gamma\gamma} F_{\pi\gamma\gamma}(s_d), \qquad F_{\pi\gamma\gamma}(0) = 1. \tag{7.15}$$

We shall see later that in the small interval $0 < s_d < m_\pi^2$, $F(s_d) \approx 1$ should be a good approximation. For $\eta$-decays, the approximation $F_{\eta\gamma\gamma}(s_d) \approx 1$ is probably poorer. Actually, the $\pi^0$ and $\eta$ decay rates have only been measured indirectly (see section 8-2).

## 4-8 Production and Decay of Shortlived Particles

The lifetime of unstable particles can be inferred from their decay distribution along their path $R$ as explained in (2-3.4). Figure 4-8.1 shows a schematic bubble chamber picture of the reaction $K^- p \to e\bar{e}\gamma\Lambda$, where the $\Lambda$ is identified by its decay into $p\pi^-$. (The photon $\gamma$ is normally not detected, but in this case its momentum can be inferred from energy-momentum conservation). We know that the reaction $K^- p \to \pi^0\Lambda$ is allowed by strong interactions, therefore we suspect that the particles $e$, $\bar{e}$, $\gamma$ in fig. 4-8.1 come from the

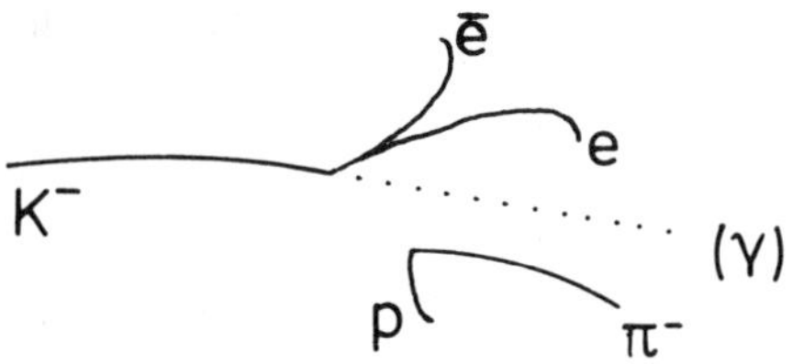


**Figure 4-8.1** Schematic bubble chamber picture of the reaction $K^- p \to \pi^0\Lambda \to e\bar{e}\gamma p\pi^-$. The photon direction is reconstructed.

decay $\pi^0 \to e\bar{e}\gamma$. But the $\pi^0$ lifetime is so short that the $\pi^0$ decays practically always at its point of production. How do we know that the $\pi^0$ exists at all, and how do we determine its mass? The answer is given by the effective mass distribution $d\sigma/ds_d$, with $s_d = (P_e + P_{\bar{e}} + K\gamma)^2$. (For the normal decay $\pi^0 \to \gamma\gamma$, one could in the reaction $K^- p \to \gamma\gamma\Lambda$ measure $s_d$ as $(P_K + P_p - P_\Lambda)^2$.) This distribution shows a sharp peak at $s_d = m_{\pi 0}^2$. Position, width and intensity of this peak follow from time-dependent perturbation theory as in 3-1. The final state $|m\rangle|\mathbf{k}\rangle$ of section 3-1 is replaced by a general 2-particle state $|\mathbf{p}_1\rangle|\mathbf{p}_2\rangle$, $\langle m|\hat{H}^{em}|l\rangle$ is replaced by $-T(d + 12)$ (the minus sign appears because we have put $S = 1 + iT$, whereas (3-4.5) has $S = 1 - iH$). Possible widths of the decay products 1 and 2 are neglected, such that (3-1.19) becomes

$$C_{12} = T(d \to 12)(E_d - E_1 - E_2 - i\Gamma_d/2)^{-1}. \tag{8.1}$$

Since the formalism of section 3-1 assumed the decaying system to be nonrelativistic, (1) refers to the cms of particles 1 and 2. Relativistically, we thus have

$$E_1 + E_2 = s_d^{\frac{1}{2}}, \qquad E_d = m_d = m_d^0 + \Delta m_d. \tag{8.2}$$

The *cd*-state is produced instantly at $t = 0$ on this time-scale, such that the full matrix element is

$$T(ab \to c12) = T(d \to 12)\phi_d(s_d)T(ab \to cd), \tag{8.3}$$

where $\phi_d$ is the denominator of (1). Actually, since our states are normalized by (1-1.15), we need an additional factor $2m_d$ in the denominator. Also, we add to (1) a second term in which $s_d^{\frac{1}{2}}$ is replaced by $-s_d^{\frac{1}{2}}$. Dropping the index *d*, we thus have

$$\begin{aligned} \phi(s) &= (2m)^{-1}[(m - s^{\frac{1}{2}} - i\Gamma/2)^{-1} + (m + s^{\frac{1}{2}} - i\Gamma/2)^{-1}] \\ &= (m^2 - s - im\Gamma)^{-1}. \end{aligned} \tag{8.4}$$

This is nothing but the relativistic propagator (2-6.7) for stable particles, but with $\varepsilon$ replaced by $m\Gamma$. For $\pi^0$, $\Gamma$ is 8 eV and the additional term in (4) is completely negligible. For the broad resonances of the next section, this is not quite true any longer. The second term includes the contribution of anti-particles, analogous to (2-6.8). However, the main point of (4) is that unstable particles are treated exactly as virtual particles, which are off their "mass shell," $s \neq m^2 - im\Gamma$ (Fig. 4-8.2). It is also clear that if particle *d* carries spin, (3) must include a summation over its magnetic quantum number:

$$T(ab \to c12) = \sum_M T(ab \to cd_M)T(d_M \to 12)(m_d^2 - s_d - im_d\Gamma_d)^{-1}. \tag{8.5}$$

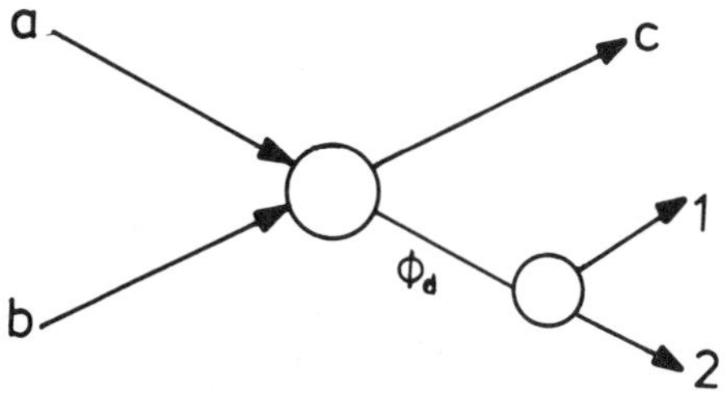


**Figure 4-8.2** Feynman graph for the reaction $ab \to c12$, with particles 1 and 2 originating from the decay of an unstable particle *d*.

We now study the differential cross section for the over-all process $ab \to c12$, starting from (2-3.16) and (5):

$$\begin{aligned} d\sigma(ab \to c12) &= \tfrac{1}{2}\lambda^{-\frac{1}{2}}(2S_a + 1)^{-1}(2S_b + 1)^{-1} \\ &\times \sum_{\lambda_a\lambda_b\lambda_c\lambda_1\lambda_2} \left| \sum_M T(ab \to cd_M)T(d_M \to 12)\right|^2 \\ &\times ((m_d^2 - s_d)^2 + m_d^2\Gamma_d^2)^{-1}\, d\,\text{Lips}\,(s; P_cP_1P_2). \end{aligned} \tag{8.6}$$

We insert again (6.4) for $d$ Lips and integrate over the decay angles of particles 1 and 2, which gives us

$$\frac{1}{2}\int d\text{ Lips }(s_d;\,12)\sum_{\lambda_1\lambda_2} T(d_M\to 12)T^*(d_{M'}\to 12)=\delta_{MM'}m_d\Gamma(d\to 12), \quad (8.7)$$

according to (4.7), the orthogonality of the $d$-functions, and (4.8). We thus obtain the important formula

$$d\sigma(ab\to c12)=\pi^{-1}\,ds_d\,d\sigma(ab\to cd)m_d\Gamma(d\to 12)((m_d^2-s_d)^2+m_d^2\Gamma_d^2)^{-1}. \quad (8.8)$$

It can be generalized to arbitrary numbers of particles in the initial and final states. In particular, if there is only one particle in the initial state, we obtain the formula for sequential decays,

$$T(a\to c12)=\sum_{\lambda_d} T(a\to cd)T(d\to 12)(m_d^2-s_d-im_d\Gamma_d)^{-1}, \quad (8.9)$$

$$d\Gamma(a\to c12)=\pi^{-1}\,ds_d\,d\Gamma(a\to cd)m_d\Gamma(d\to 12)((m_d^2-s_d)^2+m_d^2\Gamma_d^2)^{-1} \quad (8.10)$$

of which (7.13) is a special case, with $m_\gamma^2=0$, $m_\gamma\Gamma_\gamma\ll s_d$. If both particles $c$ and $d$ are unstable, we use

$$d\text{ Lips }(s;\,P_1P_2P_3P_4)=\tfrac{1}{4}\pi^{-2}\,ds_c\,ds_d\,d\text{ Lips }(s;\,P_cP_d)$$
$$\times\, d\text{ Lips }(s_d,\,P_1P_2)\,d\text{ Lips }(s_c,\,P_3P_4), \quad (8.11)$$

which is obtained by inserting the recurrence relation (6.4) into $d$ Lips ($s$; $P_dP_3P_4$) of (6.5). This gives us

$$d\sigma(ab\to 1234)=\pi^{-2}\,ds_c\,ds_d\,d\sigma(ab\to cd)$$
$$\times\, m_c\Gamma(c\to 12)((m_c^2-s_c)^2+m_c^2\Gamma_c^2)^{-1} \quad (8.12)$$
$$\times\, m_d\Gamma(d\to 12)((m_d^2-s_d)^2+m_d^2\Gamma_d^2)^{-1}.$$

For $m\Gamma\ll m^2$, (8) goes over into

$$d\sigma(ab\to c12)=d\sigma(ab\to cd)f_d(12), \qquad f_d(12)\equiv\Gamma(d\to 12)/\Gamma_d, \quad (8.13)$$

where $f_d$ is the ratio of the partial width to the total width, and $P_d^2=m_d^2$. This follows from (3-11.18) with $x=m_d^2-s_d$ and $\varepsilon=m_d\Gamma_d$.

It must be remembered that if particle $d$ carries spin, these formulas apply only after integration over the decay angles $\vartheta$, $\varphi$ of its decay products. If one is interested in the decay angular distribution, one must start from the more complicated expression (6). It is convenient to introduce a matrix notation

$$R(ab\to c\underline{d})_{MM'}=(2S_a+1)^{-1}(2S_b+1)^{-1}\sum_{\lambda_a\lambda_b\lambda_c} T(ab\to cd_M)T^*(ab\to cd_{M'}) \quad (8.14)$$

$$R(d\to 12)_{MM'}=\sum_{\lambda_1\lambda_2} T^*(d_M\to 12)T(d_{M'}\to 12) \quad (8.15)$$

$$d\sigma(ab\to c12)=\tfrac{1}{2}\lambda^{-\frac{1}{2}}\phi_d(s_d)\sum_{MM'} R(ab\to c\underline{d})_{MM'}R^*(d\to 12)_{MM'}\,d\text{ Lips}$$
$$=\tfrac{1}{2}\lambda^{-\frac{1}{2}}\phi_d(s_d)\text{ trace }[R(ab\to c\underline{d})R(d\to 12)]\,d\text{ Lips}. \quad (8.16)$$

The $R$-matrices are obviously Hermitean, and $R^+(d \to 12) = R(d \to 12)$ has been used in the last expression in (16). Normally, the differential cross section $d\sigma(ab \to cd)$ and the total width $\Gamma_d$ are already known when one starts looking at the decay angular distribution $W(\vartheta, \varphi)$. It is then convenient to normalize the distribution to 1 instead of $d\sigma(ab \to cd) f_d(12)$. For this purpose, the production and decay density matrices $\rho$ and $A$ are introduced

$$\rho_{MM'}(s, t) \equiv \sum_{\lambda_a \lambda_b \lambda_c} T(ab \to cd_M) T^*(ab \to cd_{M'}) \times \left[ \sum_{\lambda_a \cdots \lambda_d} |T(ab \to cd)|^2 \right]^{-1}, \tag{8.17}$$

$$A^S_{MM'}(\vartheta, \varphi) \equiv \frac{2S+1}{4\pi} \sum_{\lambda_1 \lambda_2} |T_S(\lambda_1 \lambda_2)|^2 e^{i\varphi(M'-M)} d^S_{M\lambda}(\vartheta) d^S_{M'\lambda}(\vartheta) \times \left[ \sum_{\lambda_1 \lambda_2} |T_S(\lambda_1 \lambda_2)|^2 \right]^{-1} \tag{8.18}$$

They are the normalized versions of (14) and (15):

$$\rho = \rho^+, \text{ trace } \rho = 1, \qquad A = A^+, \text{ trace } A = (2S+1)/4\pi. \tag{8.19}$$

The decay angular distribution is then (for collisions of unpolarized particles $a$, $b$)

$$W(s, t, \vartheta, \varphi) = \sum_{MM'} \rho_{MM'} A^S_{M'M} = \text{trace } (\rho A), \qquad \int d\cos\vartheta\, d\varphi W = 1. \tag{8.20}$$

The density matrix $\rho$ was already introduced in section 2-3. In particular, (2-3.6) is nothing but trace $(\rho R(d \to 12))$. Obviously, the same $\rho$ must be used in possible collisions of the unstable particle $d$ (for example if the $\Lambda$ in fig. 4-8.1 collides with a proton before decaying). The only difference to the $\rho$ of section 2-3 is that in "coincidence experiments" which we are presently discussing, $\rho$ is a function of $s$ and $t$ of the particular reaction $ab \to cd$, whereas the $\rho$ in section 2-3 referred to particles of unspecified origin. Naturally, that $\rho$ is obtained by averaging numerator and denominator in (17) over $s$ and $t$, possibly also over competing reactions which may produce the $\Lambda$.

The formalism of this section also applies to "resonance scattering," in which initial and final state are connected by a single unstable particle only, for example $\gamma\gamma$-scattering near $s = m^2_{\pi^0}$. In this case the phenomenon happens in the partial wave having $J = S$. Analogous to (5) we have

$$\begin{aligned} T(ab \to R \to cd) &= \sum_M T(ab \to R) T(R \to cd) \phi_R(s) \\ &= \sum_M D^J_{M\lambda}(\Omega_a) D^{J*}_{M\lambda'}(\varphi_c, \vartheta_c, -\varphi_c) \times (2J+1) T_S(ab \to R) T_S(R \to cd) \phi_R(s) \\ &= D^{J*}_{\lambda\lambda'}(\phi, \theta, -\phi)(2J+1) T_S(ab \to R) T_S(R \to cd) \times (m_R^2 - s - i m_R \Gamma_R)^{-1} \end{aligned} \tag{8.21}$$

In the first expression, $\Omega_a$ and $\varphi_c$, $\vartheta_c$ measure the azimuthal and polar angles of particles $a$ and $c$ with respect to some fixed direction, whereas in the second expression, $\phi$ and $\theta$ are the angles of particle $c$ relative to $a$, according to the group properties of the $D$-functions. The partial-wave amplitude in the expansion (4.9) becomes

$$T_J(s, \lambda_a \cdots \lambda_d) = (8\pi s^{\frac{1}{2}})^{-1} T_S(R \to ab) T_S(R \to cd)(m_R^2 - s - im_R \Gamma_R)^{-1}. \quad (8.22)$$

Here we have also used time-reversal invariance (4.20). As in (2-5.24) it is possible to write the total cross section as a sum of partial-wave cross sections:

$$\sigma = \sum_J \sigma_J,$$
$$\sigma_J = 4\pi(S_a + 1)^{-1}(S_b + 1)^{-1}(2J + 1) \sum_{\lambda_a \cdots \lambda_d} |T_J(\lambda_a \cdots \lambda_d)|^2 p'/p, \quad (8.23)$$

where the cms momentum of the final state is denoted by $p'$ as in section 2-4. Inserting now (22) and observing (4.8), we find

$$\sigma_J = 4\pi(2S_a + 1)^{-1}(2S_b + 1)^{-1}(2J + 1) \frac{m\Gamma(R \to ab) m\Gamma(R \to cd)}{(m^2 - s)^2 + m^2\Gamma^2} p^{-2}. \quad (8.24)$$

However, (22) and (24) do not hold in the case of strong interactions in the initial and final states. For example, although the $\Lambda$ peak in $\pi^- p$ elastic scattering has never been seen because it is too narrow, we still have $\sigma_{\frac{1}{2}}(\pi^- p) \neq 0$. Formulas for such cases will be given later.

It may be worth mentioning that neither (8) nor (24) have their maxima exactly at $s_d = m_d^2$, due to the $s_d$-dependence of $\Gamma$ (and in the case of (8), occasionally also of $\sigma(ab \to cd)$). Since $\Gamma$ increases with $p$ (see the next section), the maximum occurs always below $m_d^2$ (see also fig. 3-1).

Unitarity and time-reversal invariance determine the phase of $T_S(d \to 12)$: We define the unitary partial-wave $S$-matrix for 2-particle channels ($\pi N \to \pi N$, say):

$$S_J = 1 + 2i\tilde{p}^{\frac{1}{2}} T_J \tilde{p}^{\frac{1}{2}}, \qquad S_J S_J^+ = S_J^+ S_J = 1. \quad (8.25)$$

Obviously (4.11) implies unitarity of $S$ and vice versa. If we now add the one-particle channel $d$ (the $\Lambda$, say), then, to first order in $T(d \to 12)$ and $T(12 \to d)$, the full $S$-matrix is of the form

$$S_J^{\text{tot}} = \begin{pmatrix} S(d \to d) & S(d \to 12) \\ S(12 \to d) & S_J \end{pmatrix} \approx \begin{pmatrix} 1 & 2iT_J(d \to 12) p_{12}^{\frac{1}{2}} \\ 2iT_J(12 \to d) p_{12}^{\frac{1}{2}} & S_J \end{pmatrix} \quad (8.26)$$

where the "weak scattering" (22) can be neglected in $S_J$ because it is of second order. We find in the $d \to 12$-corner of the matrix equation $S_J^{\text{tot}} S_J^{\text{tot}+} = 1$:

$$-iT_J^*(12 \to d) p_{12}^{\frac{1}{2}} + iT_J(d \to 12) p_{12}^{\frac{1}{2}} S_J^+ = 0. \quad (8.27)$$

Using now time-reversal invariance (4.20), we obtain

$$T_J^*(d \to 12) p_{12}^{\frac{1}{2}} = T_J(d \to 12) p_{12}^{\frac{1}{2}} S_J^+. \quad (8.28)$$

Multiplying this equation by $T_J(d \to 12)$, we see that $T_J^2(d \to 12)S_J^+$ equals $|T_J(d \to 12)|^2$ which is a real quantity. In the one-channel case, $S_J = \exp(2i\,\delta_J)$ we thus have

$$T_J(d \to 12) = \pm |T_J(d \to 12)| \exp(i\,\delta_J) \tag{8.29}$$

i.e. $T_J(d + 12)$ has the phase which the elastic scattering amplitude has outside the "resonance." In $\Lambda \to N\pi$ decays, for example, one would decompose $T(\Lambda \to p\pi^-)$ and $T(\Lambda \to n\pi^0)$ into matrix elements for the $I = \frac{1}{2}$ and $I = \frac{3}{2} N\pi$ states, and each matrix element would have the phase of the corresponding elastic $\pi N$ amplitude near (but not strictly at) $s = m_\Lambda^2$. This will be done in section 4-11. For $\pi^0 \to \gamma\gamma$ decays, we have $S_J(\gamma\gamma \to \gamma\gamma) = 1$ outside the $\pi^0$ peak, and $g_{\pi\gamma\gamma}$ of the previous section is therefore real.

## 4-9 The Vector Mesons

The nine mesons $\pi^+, \pi^0, \pi^-, \eta, X, K^+, K^0, \bar{K}^0, K^-$ which we have discussed so far are all pseudoscalar particles, $0^-$ in the notation $J^{\mathscr{P}}$. The next important group of mesons are the vector mesons, which have $1^-$, i.e., $J = 1$ and negative parity. They can be subdivided into the "nonet" $\rho^+, \rho^0, \rho^-, \omega, \phi, K^{*+}, K^{*0}, \bar{K}^{*0}, K^{*-}$ (see table C-3) where $\phi$ is the heaviest particle with $m_\phi = 1020$ MeV, and the more recently discovered $\psi$-particles which are much heavier and can be neglected in the dynamics of "normal" strong interactions. The $\rho$-mesons have $Y = 0$, $I = 1$ as the pions, $\omega$ and $\phi$ have $Y = I = 0$ as $\eta$ and $X$, and the $K^*$ states have the same isospin and hypercharge as the kaons.

The decay $\rho \to \pi\pi$ is strong and has the enormous width of $\sim 150$ MeV, which requires corrections to the unstable-particle approximation of the previous section. For some time it was thought that the very existence of the $\rho$ state was due to the force between the two pions into which it decays, analogous to a resonance in potential scattering. The vector mesons and other strongly decaying states are often referred to as "resonances."

Historically, the $\rho^-$ and $\rho^0$ were detected as peaks in the $\pi\pi$-effective-mass projections of the Dalitz plot of the reaction $\pi N \to \pi\pi N$ (Erwin et al. 1961), similar to the situation in fig. 4-6 which shows the $\rho^+$-peak in $\pi^+ p \to \pi^0\pi^+ p$. The $\omega$ was detected as a rather sharp $\pi^+\pi^0\pi^-$-effective mass peak in the antiproton annihilation $\bar{p}p \to \pi^+\pi^+\pi^0\pi^-\pi^-$ (Maglic et al. 1961), and the $\phi$ was detected as a sharp peak in the $K^+K^-$ and $K_L K_S$ effective masses in $K^- p \to \Lambda K^+ K^-$ and $K^- p \to \Lambda K_L K_S$, respectively (Bertanza et al. 1962). The reaction $\pi^- p \to \omega n$ exists, too, and has been studied particularly at energies of a few GeV (Irving and Michael 1974). The reaction $\pi^- p \to \phi n$ on the other hand seems to be very weak, analogous to $\pi^- p \to Xn$.

The $K^*$ was discovered in its $K^{*-}$-state as a broad effective-mass peak in the reactions $K^- p \to K^0\pi^- p$ and $K^- p \to K^-\pi^0 p$ (Alston et al. 1961).

The isospin relations of $\rho \to \pi\pi$ decays follow from the last row of table 4-2.3. In particular, $\rho^0 \to \pi^0\pi^0$ is forbidden (this remains true also after inclusion of isospin violation, because two identical bosons cannot be in a $p$-state). The isospin relations of $K^* \to \pi K$ follow from the second row of table 4-2.3 with the substitution $p \to K^+$, $n \to K^0$. The isospin combinations into which $K^*$ decays are thus

$$|K^{*+}\rangle \to 3^{-\frac{1}{2}}|K^+\pi^0\rangle - (\tfrac{2}{3})^{\frac{1}{2}}|K^0\pi^+\rangle,$$
$$|K^{*0}\rangle \to (\tfrac{2}{3})^{\frac{1}{2}}|K^+\pi^-\rangle - 3^{-\frac{1}{2}}|K^0\pi^0\rangle, \tag{9.1}$$

and similarly for $\bar{K}^*$. Note that in all four states $K^*$ and $\bar{K}^*$,

$$\Gamma(K^* \to K\pi_{\text{charged}}) = 2\Gamma(K^* \to K\pi^0) \approx \tfrac{2}{3}\Gamma_{K^*}. \tag{9.2}$$

Let us try to understand the main features of vector meson decays, starting with the decays into two mesons, $\rho \to \pi\pi$, $\phi \to K\bar{K}$, $K^* \to K\pi$ (the decay $\omega \to \pi^+\pi^-$ is discussed later as it is isospin-forbidden). A common notation is $V \to 1, 2$ where particles 1 and 2 have 4-momenta $P_1$ and $P_2$, with $P_1 + P_2 = P_d$. The decay matrix element must be linear in the spin function $\varepsilon^\mu(M)$ of the vector meson. As it must also be Lorentz invariant, $\varepsilon^\mu$ can only occur in the combinations $\varepsilon \cdot P_1$ and $\varepsilon \cdot P_2$, but because of $(P_1 + P_2) \cdot \varepsilon = 0$ (2-8.2), the two combinations are in fact equivalent (except for "off-shell" effects, see below). We choose the form

$$T(V \to 1, 2) = G_{V12}\,\varepsilon_\mu(M)(P_1 - P_2)^\mu = -2G_{V12}\,\boldsymbol{\varepsilon}(M)\mathbf{p}_1. \tag{9.3}$$

The last expression is of course only valid in the cms. We define (3) for each charge combination separately. For $\rho^0 \to \pi^+\pi^-$, we take $\pi^+$ as particle 1, in analogy with (3-6.18). To the extent that isospin invariance is fulfilled, we expect for the "coupling constants" $G_{\rho\pi\pi}$

$$G_{\rho\pi\pi} \equiv G(\rho^0 \to \pi^+\pi^-) = G(\rho^+ \to \pi^+\pi^0) = -G(\rho^- \to \pi^-\pi^0). \tag{9.4}$$

For $K^* \to K\pi$, it is customary to take $G(K^* \to K^+\pi^0)$ as the standard coupling constant, analogous to $G(p \to p\pi^0)$ which is discussed in section 5-8. We thus expect

$$G_{K^*K\pi} \equiv G(K^{*+} \to K^+\pi^0) = -2^{-\frac{1}{2}}G(K^{*+} \to K^0\pi^+)$$
$$= -G(K^{*0} \to K^0\pi^0) = 2^{-\frac{1}{2}}G(K^{*0} \to K^+\pi^-). \tag{9.5}$$

Finally, for $\phi \to K\bar{K}$,

$$G_{\phi K\bar{K}} \equiv G(\phi \to K^+K^-) = -G(\phi \to K^0\bar{K}^0) = -G(\phi \to K_1K_2). \tag{9.6}$$

(the $K_1K_2$ state as opposed to $K_LK_S$ follows from the second row of table 4-5.2 and its explanation in section 4-5). For $M = 0$, we have $\boldsymbol{\varepsilon} \cdot \mathbf{p}_1 = p\cos\vartheta$ in (3), with $p$ = cms decay momentum. Comparing (3) with (4.7) and using (4.8), we find

$$T_1 = -(4\pi/3)^{\frac{1}{2}}2G_{V12}\,p, \qquad m\Gamma(V \to 1, 2) = \tfrac{2}{3}p^3 s^{-\frac{1}{2}} G_{V12}^2/4\pi. \tag{9.7}$$

The resulting coupling constants are collected in table C-3, for $s = m^2$. They are more uniform than the decay widths themselves. In particular, $\Gamma(\rho^0 \to \pi^+\pi^-)$ is more than 100 times larger than $\Gamma(\phi \to K_1K_2)$, whereas the corresponding coupling constants are comparable. This indicates that coupling constants are more fundamental than widths. In fact, later we shall derive a relation $G_{\rho\pi\pi} = G_{\rho K\bar{K}}$, although $\Gamma(\rho \to K\bar{K}) = 0$ because of lack of phase space. However, the question whether coupling constants or widths should be compared must be examined carefully. After all, Lorentz invariance alone allows $G_{V12}$ in (3) to be an arbitrary function of $(P_1 + P_2)^2 = s$. If it so happened that $G_{V12}$ were proportional to $(\frac{1}{4}s - m_1^2)^{-\frac{1}{2}}$ (which is $p^{-1}$ for $m_1 = m_2$), then $\Gamma/p$ would be more fundamental. The answer to this question is given in chapter 6. It turns out that these coupling "constants" are in fact functions of $s$, but analytic ones which have no unnecessary singularities. The main point for our present discussion follows equally well from potential theory. Consider the Klein-Gordon equation (1-3.10) for the radial wave function of the two-meson cms motion, and let $U$ include a strong potential $V_{\text{strong}}$ of short range. The logarithmic derivative of the wave function is independent of normalization and therefore apt for a parametrization of the $p$-dependence of the phase shift $\delta_L$ which is defined in (1-3.14). We choose the smallest $r = r_0$ at which $V$ is still negligible and define

$$f_L(\rho) = R_L^{-1}\, \partial_p \rho R_L = (e^{2i\delta_L} h_L + h_L^*)^{-1}(e^{2i\delta_L}\, \partial_\rho \rho h_L + \partial_\rho \rho h_L^*), \tag{9.8}$$

which is essentially the logarithmic derivative of $u = rR_L$. We then obtain the following form of exp $(2i\,\delta_L)$:

$$e^{2i\delta_L} = e^{2i\xi_L}(f_L - \Delta_L - i\rho_L v_L)^{-1}(f_L - \Delta_L + i\rho v_L), \tag{9.9}$$

$$e^{2i\xi_L} = -h_L^*/h_L, \qquad \Delta_L \equiv |h_L|^{-2}\,\text{Re}\,(h_L\, \partial_\rho \rho h_L^*), \tag{9.10}$$

$$v_L = \rho^{-2}|h_L|^{-2}\,\text{Re}\,(h_L\, \partial_\rho \rho h_L) = \rho^{-2}|h_L|^{-2}. \tag{9.11}$$

For local potentials, $f_L$ decreases with increasing $p$. When $f_L$ passes $\Delta_L$, a "resonance" occurs. We put

$$e^{2i\delta_L} = e^{2i\xi_L}(1 + 2ipT_{L,\,\text{res}}), \qquad T_{L,\,\text{res}} = r_0 v_L(f_L - \Delta_L - i\rho v_L)^{-1}. \tag{9.12}$$

We expand $f_L - \Delta_L$ in $s^{\frac{1}{2}}$ around the zero:

$$f_L - \Delta_L = (s^{\frac{1}{2}} - m)\, \partial_{s^{\frac{1}{2}}}(f_L - \Delta_L)_{s^{\frac{1}{2}}=m} = (m - s^{\frac{1}{2}})2r_0/\gamma \equiv (m - s^{\frac{1}{2}})\rho v_L 2/\Gamma, \tag{9.13}$$

where $\gamma$ is called the reduced width. The quantity $\Gamma$ which is defined in the last expression of (13),

$$\Gamma = pv_L(pr_0)\gamma, \tag{9.14}$$

is the width of the resonance since insertion into $T_{L,\,\text{res}}$ gives

$$T_{L,\,\text{res}} = \tfrac{1}{2}p^{-1}\Gamma(m - s^{\frac{1}{2}} - i\tfrac{1}{2}\Gamma)^{-1}, \tag{9.15}$$

which is the normal resonance formula for elastic scattering, without the antiparticle correction (8.4). From table A-1 we see that for $pr_0 \ll L$, $v_L \approx (pr_0)^{2L}$. In other words, for $pr_0 \ll L$, $\Gamma$ is proportional to $p^{2L+1}$ due to the singular nature of the centrifugal potential ("barrier penetration factor"). However, we can hardly take $r_0$ much smaller than 1 fm, according to our knowledge about the range of strong interactions. With $r_0 = 1$ fm we would already have $pr_0 = 1$ for $p = 0.1973$ GeV, which is smaller than the actual decay momenta of $\rho$ and $K^*$ (table C-3). For $L = 1$, (14) would require $pG \sim v_1^{\frac{1}{2}} = \rho(1+\rho^2)^{-\frac{1}{2}}$:

$$G_{V12}(s) = G_{V12}(m^2)F(s), \qquad F = (1 + p^2(m^2)r_0^2)^{\frac{1}{2}}(1 + p^2(s)r_0^2)^{-\frac{1}{2}}, \quad (9.16)$$

In practice, (16) may be used with $r_0^2 > 0$ as a free parameter. We see that for $r_0 p \gg L$, $\Gamma/p$ is in fact expected to be independent of $s$, rather than (7) with constant $G$. In an approximate "effective field theory" in which fields are defined also for unstable particles, the decay matrix elements would be the Born terms of a local interaction between the fields, where each 4-momentum in the matrix element corresponds to a derivative of the field operator as in (3-2.8). The rule there is to take as few derivatives as possible, and form factors of the type (16) are in fact difficult to include. In this sense, (3) may be called the "Born approximation" of the decay. The Born approximations for various meson decays are collected in table 4-9, and will be discussed in the appropriate places.

In principle, the $s$-dependence of $\Gamma$ can be measured in the partial-wave cross section (8.24), see the discussion for the $\rho$ below. For narrow resonances such as $\phi$, the interval in $s$ which can be studied (roughly $m^2 - 2m\Gamma < s < m^2 + 2m\Gamma$) is too small.

There is however also indirect evidence for the variation of $\Gamma$ with $p$, for example from the breaking of isospin invariance in the decays $\phi \to K^+K^-$ and $\phi \to K_1 K_2$. This invariance requires $m(K^+) = m(K^0)$ and therefore also equality of the decay momenta $p_+$ and $p_0$ in $\phi \to K^+K^-$ and $\phi \to K_1 K_2$. In reality, $K^0$ is 4 MeV heavier than $K^+$. This leads to $p_+^3/p_0^3 = 1.54$ and accounts roughly for the observed ratio $\Gamma(\phi \to K^+K^-)/\Gamma(\phi \to K_1 K_2)$ when inserted into (7). It shows that isospin invariance holds better on the coupling constant level than for the widths themselves. For a quantitative analysis of the $K^+K^-$ final state, the Coulomb potential between the two charged particles must be included in (1-3.10): The Hankel functions which enter (1-3.14) are then replaced by the corresponding Coulomb wave functions of appendix A-2, which gives the following modification of (14) (Pilkuhn 1974):

$$\Gamma = p[F_L^2(pr_0) + G_L^2(pr_0)]^{-1}\gamma. \quad (9.17)$$

The cleanest reactions for the study of $\rho^0$, $\omega$ and $\phi$ are $e\bar{e} \to \pi^+\pi^-$, $e\bar{e} \to \pi^+\pi^0\pi^-$ and $e\bar{e} \to K\bar{K}$, respectively. They contain the $J = 1$ partial wave only and no additional hadrons. The relevant Feynman graph for $e\bar{e} \to 12$ via a resonance is shown in fig. 4-9.1. It contains two one-particle states (the photon and the vector meson), one after the other. The matrix element for the virtual transition $V \to \hat{\gamma}$ is traditionally expressed in terms of the inverse of a constant $f_V$,

$$T(V_M \to \hat{\gamma}_{M'}) = -ef_V^{-1}m_V^2 \varepsilon_\mu(M)\varepsilon^{*\mu}(M') = ef_V^{-1}m_V^2\, \delta_{MM'}. \quad (9.18)$$

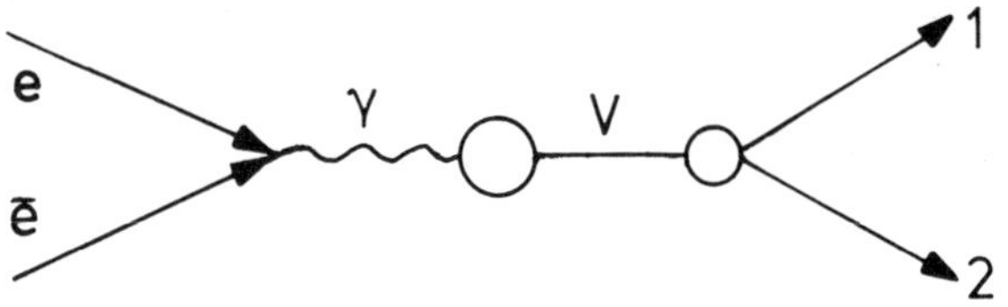


**Figure 4-9.1** The Feynman graph for $e\bar{e} \to 12$ via a vector meson.

The factor $m_V^2$ is added mainly in order to make $f$ dimensionless as we shall see, but for $s \sim m_V^2$, it nicely cancels the $s^{-1}$ from the photon propagator.

Apart from a radiative correction, the one-photon exchange approximation is in fact better than the unstable-particle approximation, at least for $e\bar{e} \to \pi^+\pi^-$, even when $s$ is much closer to $m_\rho^2$ than to zero. It is therefore advisable to start from the form (3-6.5) with $t \to s$, $J^\mu_{\bar{a}a} = -\bar{u}\gamma^\mu v$ and $J^\mu_{bb'} = F_1(s)(P_1 - P_2)^\mu$ according to (3-6.18):

$$T_{if} = -e^2 s^{-1} \bar{u}\gamma^\mu v F_1(s)(P_1 - P_2)_\mu = 2e^2 s^{-1} \bar{u}\boldsymbol{\gamma} v \mathbf{p}_1 F_1(s), \tag{9.19}$$

where $F_1$ is the electric form factor of particle 1 ($\pi^+$, $K^+$ or $K_1^0$). Inserting (3-6.26) for $-\bar{u}\boldsymbol{\gamma}v$ and expanding $T_{if}$ as in (4.9), we find, as expected, that only $J = 1$ contributes:

$$T_J(s, \lambda_e, \lambda_{\bar{e}}) = -\delta_{J,1} 2e^2 s^{-1} F_1(s) p (12\pi)^{-1} (m_e s^{-\frac{1}{2}} \delta_{\lambda_e \lambda_{\bar{e}}} + 2^{-\frac{1}{2}} \delta_{\lambda_e, -\lambda_{\bar{e}}}). \tag{9.20}$$

In the present reactions we have always $m_e^2 \ll s$, such that for unpolarized electrons and positrons according to (8.23)

$$\sigma(e\bar{e} \to 1, 2) = 3\pi \sum_{\lambda_e \lambda_{\bar{e}}} |T_1|^2 p_1 2 s^{-\frac{1}{2}} = \tfrac{8}{3}\pi\alpha^2 p_1^3 s^{-\frac{5}{2}} |F_1(s)|^2. \tag{9.21}$$

So far we have only exploited the one-photon intermediate state. In the vector meson approximation, on the other hand, we have the decay matrix element from (18) and (3-6.13)

$$T(V \to e\bar{e}) = -e^2 f_V^{-1} m_V^2 s^{-1} \varepsilon_\mu(M) \bar{u}(\lambda_e) \gamma^\mu v(\lambda_{\bar{e}}). \tag{9.22}$$

Combining this with (3) for $T(V \to 1, 2)$, using the decomposition (8.21) for a single vector meson intermediate state and comparing the result with (19), we find ("vector meson dominance model")

$$F_1(s) = G_{V12} f_V^{-1} m_V^2 (m_V^2 - s - i m_V \Gamma_V)^{-1}, \tag{9.23}$$

where this $\Gamma$ is the total width in contrast to $\Gamma(V \to 1, 2)$ of (7). We can also define the partial width $\Gamma(V \to e\bar{e})$. Comparison of (22) with the last row of table 4-9 gives, with $m_e^2 \ll s$:

$$m_V \Gamma(V \to e\bar{e}) = \tfrac{1}{3} 4\pi\alpha^2 f_V^{-2} m_V^6 s^{-2}. \tag{9.24}$$

Figure 4-9.2a shows $|F_{K^+}|^2$ as measured in the reaction $e\bar{e} \to K^+K^-$ (Gourdin 1974). The curve is given by (23), with the parameters $m_\phi$ and $\Gamma_\phi$ of table C-3. Figure 4-9.2b shows $|F_\pi|^2$ as measured in $e\bar{e} \to \pi^+\pi^-$. The curve is

**Table 4-9** "Born approximations" for the decay matrix elements (4.7). The resulting widths are given by (4.8).

| Decay | $\langle p, \lambda_1 \lambda_2, \theta\phi \mid T \mid M \rangle$ | $T_S(\lambda_1, \lambda_2)$ | $m\Gamma$ |
|---|---|---|---|
| $0^+ \to 0^- 0^-$ | $G_0$ | $G_0$ | $\frac{1}{2} p s^{-\frac{1}{2}} G_0^2/4\pi$ |
| $1^- \to 0^- 0^-$ | $G_1 \varepsilon_\mu(M)(P_1 - P_2)^\mu$ | $3^{-\frac{1}{2}} 2 G_1 p$ | $\frac{2}{3} p^3 s^{-\frac{1}{2}} G_1^2/4\pi$ |
| $2^+ \to 0^- 0^-$ | $G_2 \varepsilon_{\mu\nu}(M)(P_1 - P_2)^\mu (P_1 - P_2)^\nu$ | $(2/15)^{1/2} 4 G_2 p^2$ | $\frac{16}{15} p^5 s^{-\frac{1}{2}} G_2^2/4\pi$ |
| $3^+ \to 0^- 0^-$ | $G_3 \varepsilon_{\mu\nu\sigma}(M)(P_1 - P_2)^\mu (P_1 - P_2)^\nu (P_1 - P_2)^\sigma$ | $70^{-1/2} 16 G_3 p^3$ | $\frac{64}{35} p^7 s^{-\frac{1}{2}} G_3^2/4\pi$ |
| $0^- \to 1^- 1^-$ | $ig \varepsilon_{\alpha\beta\gamma\delta} P_1^\alpha \varepsilon_1^{*\beta} P_2^\gamma \varepsilon_2^{*\delta}$ | $\delta_{\lambda_1 \lambda_2} \lambda_1 g p s^{\frac{1}{2}}$ | $p^3 s^{\frac{1}{2}} g^2/4\pi$ |
| $1^\mp \to 1^- 0^\mp$ | $-ig \varepsilon_{\alpha\beta\gamma\delta} P_1^\alpha \varepsilon_1^{*\beta} P_d^\gamma \varepsilon_d^\delta(M)$ | $3^{-\frac{1}{2}} \lambda_1 g p s^{\frac{1}{2}}$ | $\frac{1}{3} p^3 s^{\frac{1}{2}} g^2/4\pi$ |
| $2^+ \to 1^- 0^-$ | $-ig_2 \varepsilon_{\alpha\beta\gamma\delta} P_1^\alpha \varepsilon_1^{*\beta} P_d^\gamma \varepsilon_d^{\delta\mu}(M) P_{1\mu} - e \varepsilon_d^{*\mu} \bar{u} \gamma_\mu v$ | $10^{-\frac{1}{2}} \lambda_1 g p^2 s^{\frac{1}{2}}$ | $\frac{1}{10} p^5 s^{\frac{1}{2}} g_2^2/4\pi$ |
| $1^- \to \frac{1}{2}\ \frac{1}{2}$ | | $-e 3^{-\frac{1}{2}} 2(m_1 \delta_{\lambda_1 \lambda_2} + 2^{-\frac{1}{2}} s^{\frac{1}{2}} \delta_{\lambda_1 - \lambda_2}$ | $\frac{2}{3} \alpha p s^{-\frac{1}{2}} (2m_1^2 + s)$ |

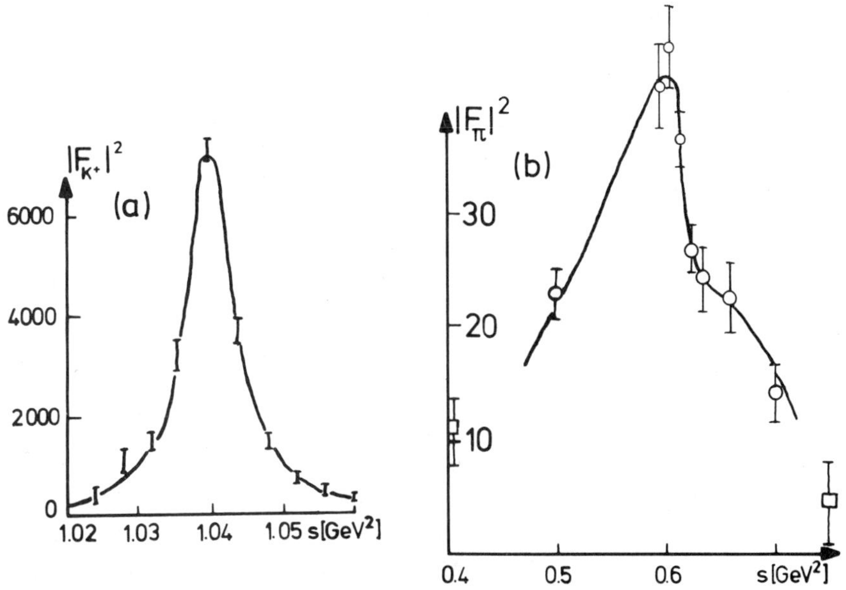


**Figure 4-9.2** The absolute squares of the form factors (a) (23) for $K^+K^-$ in the $\phi$ region, (b) (26) for $\pi^+\pi^-$ in the $\rho$-region. From Gourdin (1974).

again given by (23), but this time $\Gamma_\rho$ is not taken as a constant but varies with $s$ according to (7):

$$\Gamma_\rho(s) = \Gamma_\rho(m_\rho^2)m_\rho s^{-\frac{1}{2}}(s - 4m_\pi^2)^{\frac{3}{2}}(m_\rho^2 - 4m_\pi^2)^{-\frac{3}{2}}. \tag{9.25}$$

A small $\omega$-contribution from the isospin-violating decay $\omega \to \pi^+\pi^-$ is also added:

$$\begin{aligned} F_\pi(s) = {} & G_{\rho\pi\pi} f_\rho^{-1} m_\rho^2(m_\rho^2 - s - im_\rho\Gamma_\rho(s))^{-1} \\ & + G_{\omega\pi\pi} e^{i\,\delta\pi\pi} f_\omega^{-1} m_\omega^2(m_\omega^2 - s - im_\omega\Gamma_\omega)^{-1}, \end{aligned} \tag{9.26}$$

where $\delta_{\pi\pi} \approx 90^0$ is approximately the phase of the first term at $s = m_\omega^2$, according to (8.29) in the limit $m_\omega\Gamma_\omega \ll m_\rho\Gamma_\rho$, and $G_{\omega\pi\pi}$ is a real parameter, roughly consistent with $\Gamma(\omega \to \pi\pi)$ measured in other reactions (see table C-3). The main point for our present discussion is that the $s$-variation of (25) gives a better fit to the data than $\Gamma_\rho(s) = \text{const} \equiv \Gamma_\rho$. Other modifications will be added later, for example to ensure $F_\pi(0) = 1$.

Next, we discuss the main decay of the $\omega$ meson, $\omega \to \pi^+\pi^0\pi^-$. The negative $C$-parity of the $\omega$ requires the $\pi^+\pi^-$ pair to be in a $P$- or $F$-wave relative to each other, and conservation of parity and angular momentum requires the $\pi^0$ to be in the same partial wave relative to $\pi^+\pi^-$. Thus the decay matrix element must be of the form

$$T(M, P_+P_0P_-) = G_{\omega\pi 3}\,\varepsilon_{\alpha\beta\gamma\delta}P_+^\alpha P_0^\beta P_-^\gamma\,\varepsilon^\delta(M) = G_{\omega\pi 3}\,\boldsymbol{\varepsilon}(M)\mathbf{n}. \tag{9.27}$$

The last expression is valid in the cms only, with

$$\mathbf{n} = E_+(\mathbf{p}_0 \times \mathbf{p}_-) + E_0(\mathbf{p}_- \times \mathbf{p}_+) + E_-(\mathbf{p}_+ \times \mathbf{p}_0) \tag{9.28}$$

(compare 7.3). The vector **n** lies normal to the decay plane which is spanned by the pion momenta $\mathbf{p}_+$, $\mathbf{p}_0$ and $\mathbf{p}_-$ in their cms. It vanishes for collinear events (boundary of the Dalitz plot according to 6.10). The ansatz (27) with constant $G_{\omega\pi 3}$ gives good agreement with the Dalitz plot of $\omega \to \pi^+\pi^0\pi^-$ decays (Flatté et al. 1966). Nevertheless, it has been proposed by Gell-Mann, Sharp and Wagner (1962) that the decay is dominated by the $\rho\pi$ intermediate states, with $\rho^+\pi^-$, $\rho^0\pi^0$ and $\rho^-\pi^+$ added coherently. The "Born approximation" for $\omega \to \rho\pi$ has the form of the 6th row in table 4-9, with $g = g_{\omega\rho\pi}$ for each charge of the $\pi$ separately. A minus sign is included in the definition of $g$ such that $T(1^- + 1^-0^-)$ is the crossed version of $T(0^- \to 1^-1^-)$. Insertion into (8.9) gives

$$G_{\omega\pi 3} \approx -G_{\rho\pi\pi} G_{\omega\rho\pi}[\phi_\rho(s_{+0}) + \phi_\rho(s_{+-}) + \phi_\rho(s_{0-})],$$
$$\phi_\rho(s_{ij}) = (m_\rho^2 - (P_i + P_j)^2 - im_\rho\Gamma_\rho)^{-1}. \tag{9.29}$$

A rough evaluation gives $G^2_{\omega\rho\pi}/4\pi \approx 22\ \text{GeV}^{-2}$, for a partial width of 9 MeV. The model violates 3-particle unitarity and makes perhaps more sense for the decay $\phi \to \pi^+\pi^0\pi^-$, where the phase space is large enough to show the effects of the $\rho$-propagators.

Finally, we discuss the decays $\omega \to \gamma\pi^0$, $\phi \to \gamma\pi^0$ and $\phi \to \gamma\eta$, of which $\omega \to \gamma\pi^0$ actually accounts for 9% of $\Gamma_\omega$. The "Born approximation" is given by the 6th row of table 4-9, except for a factor $e$ to indicate the electromagnetic nature of the interaction, analogous to (18). With $p = \frac{1}{2}(s^{\frac{1}{2}} - m_\pi^2 s^{-\frac{1}{2}})$, we have

$$m\Gamma(\omega \to \gamma\pi^0) = \tfrac{1}{24}s^2(1 - m_\pi^2/s)4\pi\alpha g^2_{\omega\gamma\pi}/4\pi. \tag{9.30}$$

However, as $p$ is very large here, (30) with constant $g_{\omega\gamma\pi}$ can hardly reproduce a possible $s$-dependence.

A simple model for the coupling of photons to hadronic states says that the coupling is dominated by single vector-meson intermediate states ("vector meson dominance model"). Because of isospin conservation on the hadronic side, only the graph of fig. 4-9.3 contributes. With $s_a = 0$, we find from (23)

$$G_{\omega\rho\pi} = f_\rho g_{\omega\gamma\pi}. \tag{9.31}$$

Inserting the constants of table C-3, we get $G^2\rho\pi/4\pi \approx 24\ \text{GeV}^{-2}$, close to the above value.

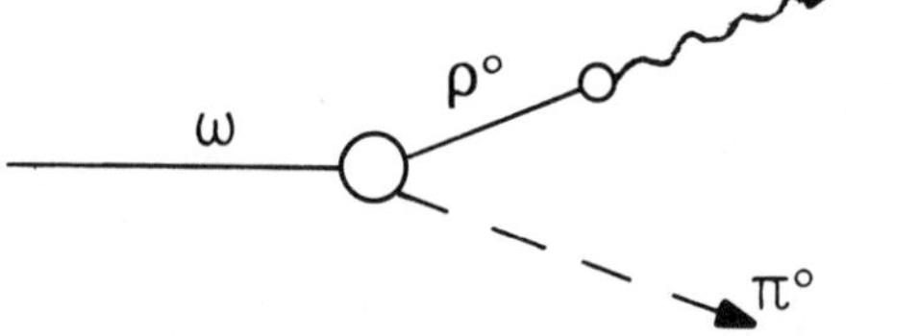


**Figure 4-9.3** The decay $\omega \to \pi^0\gamma$ in the vector meson dominance model.

## 4-10 $SU_3$-Symmetry

Soon after the discovery of kaons and hyperons, one started to look for approximate symmetry groups of hadrons which contained $SU_2$ as a subgroup. Sakata (1956) postulated invariance of the strong interactions under unitary transformations of the three states $p$, $n$ and $\Lambda$. The $\Sigma$ hyperons, for example, were regarded as $\bar{K}N$ or $\pi\Lambda$ bound states. Although the generalization from $SU_2$ to $SU_3$ turned out to be correct later, the Sakata model was unsuccessful because $p$, $n$, and $\Lambda$ do not form an $SU_3$-triplet. It was suggested by Gell-Mann (1961) and Ne'eman (1961) that $p$, $n$, and $\Lambda$ are part of a larger "supermultiplet," namely an $SU_3$-octet. The other 5 members are $\Sigma^+, \Sigma^0, \Sigma^-, \Xi^0$ and $\Xi^-$, i.e., all 8 baryons of fig. 4-3. Similarly the 8 pseudoscalar mesons of fig. 4-3 are grouped into one octet. This assignment is also called the "eightfold way." The $\eta$ meson was in fact predicted by the eightfold way.

Mathematically, the octet representation occurs in the product $3 \otimes \bar{3}$ of a triplet with its antitriplet (see B-6.13). A hypothetical triplet of "quarks" $q = (u, d, s)$ was postulated by Gell-Mann (1964), which allows one to interpret the mesons as $q\bar{q}$ bound states. Baryons cannot be $q\bar{q}$ states because of their baryon number and fermion number. If no other constituents of matter are allowed, one must interpret the baryons as $qqq$ bound states, in which case the $q$'s must be fermions of baryon number $\frac{1}{3}$. According to appendix B-6 we have

$$3 \otimes 3 \otimes 3 = (\bar{3} \oplus 6) \otimes 3 = 1 \oplus 8 \oplus 8 \oplus 10 \tag{10.1}$$

which contains the representation 8 even twice. Irrespective of the existence of such quarks, it is convenient to construct the octet representation from the representations 3 and $\bar{3}$. From the fact that $u$ and $d$ form an isospin doublet while $s$ is an isospin singlet, we find, using (2.4) and our sign conventions of sect 4-5:

$$\pi^+ = u\bar{d}, \qquad \pi^- = d\bar{u}, \qquad \pi^0 = 2^{-\frac{1}{2}}(u\bar{u} + d\bar{d}), \tag{10.2}$$

$$K^+ = u\bar{s}, \qquad K^0 = d\bar{s}, \qquad \bar{K}^0 = s\bar{d}, \qquad K^- = s\bar{u}. \tag{10.3}$$

The isosinglet state $\eta$ must be a combination of $s\bar{s}$ and $u\bar{u} - d\bar{d}$ according to (2.5). We determine its octet component $\eta_8$ by first constructing the $SU_3$-singlet state (which we call $X_1$ and which must constain equal amounts of $u\bar{u}$, $d\bar{d}$ and $s\bar{s}$) and by taking $\eta_8$ orthogonal to $X_1$:

$$X_1 = 3^{-\frac{1}{2}}(-u\bar{u} + d\bar{d} - s\bar{s}), \qquad \eta_8 = 6^{-\frac{1}{2}}(-u\bar{u} + d\bar{d} + 2s\bar{s}). \tag{10.4}$$

The indices "1" and "8" on $X$ and $\eta$ will be needed later, because the physical states $X$ and $\eta$ will slightly differ from $X_1$ and $\eta_8$. We use Condon-Shortley phase conventions both in $I$-space and $U$-space: Since $u\bar{u} - d\bar{d}$ is our $I = 0$ combination (compare 2.5), $X_1$ must contain the $U = 0$ combination $d\bar{d} - s\bar{s}$. Overall signs remain of course arbitrary. The hypercharge operator $Y$ commutes with all 3 isospin rotation generators $F_1$, $F_2$, $F_3$ of

(B-1.11). It must therefore be of the form $a \cdot 1 + b \cdot F_8$, where 1 is the unit operator and $F_8$ the generator corresponding to $\lambda^8$ of (B-6.4). From the fact that our supermultiplets have the expectation value $\langle Y \rangle = 0$, we find $a = 0$. According to (B-6.4), $u$ and $d$ have eigenvalues $3^{-\frac{1}{2}}$ and $-2 \cdot 3^{-\frac{1}{2}}$ of $F_8$, respectively, therefore $K^+$ of (3) has the eigenvalue $3^{\frac{1}{2}}$. On the other hand, kaons have hypercharge $+1$, which shows that

$$Y = 3^{-\frac{1}{2}} F_8, \qquad Q = \tfrac{1}{2}(F_3 + 3^{-\frac{1}{2}} F_8). \tag{10.5}$$

The second relation is (3.6) for the charge operator $Q$, expressed in terms of the generators of $SU_3$ (remember $F_3 = 2I_3$). It is easily checked from table B-6 that $Q$ commutes with $F_6$ and $F_7$, i.e., $Q$ is a multiplet of the unit operator in $U$-spin space. In fact, since it has the same spectrum of eigenvalues as $Y$, $Q$ plays the same role in $U$-space as $Y$ in $I$-space.

It is clear that (5) is independent of the quark-model. For the quarks themselves, if they exist, we have according to (B-6.4)

$$Y = \frac{1}{3}\begin{pmatrix} 1 & 0 & 0 \\ 0 & 1 & 0 \\ 0 & 0 & -2 \end{pmatrix}, \qquad Q = \frac{1}{3}\begin{pmatrix} 2 & 0 & 0 \\ 0 & -1 & 0 \\ 0 & 0 & -1 \end{pmatrix}. \tag{10.6}$$

Thus these fancy objects have noninteger charges. However, these conclusions need not apply to other models which contain, for example, two triplets with different charges or an additional $SU_3$-singlet which may carry the baryon number (Bacry et al. 1965).

To find the physical $U$-spin multiplets, one must rotate the multiplet in the $Y - I_3$-plot (fig. 4-3) such that the particles with $Y = \text{const}$ are replaced by those with $Q = \text{const}$ (fig. 4-10). Only the transformation between the two states in the centre ($Y = Q = 0$) is more complicated. To find the $U$-spin triplet "$t$" for the mesons, say, we use $U_- K^0 = 2^{\frac{1}{2}} t$, where $U_-$ is the $U_3$-lowering operator $U_1 - iU_2$ $(= \frac{1}{2}\lambda^6 - \frac{1}{2} i \lambda^7$ in the quark triplet). The factor $2^{\frac{1}{2}}$ follows from (B-3.7) for $j = 1$. Similarly we have $U_- K^+ = \pi^+$,

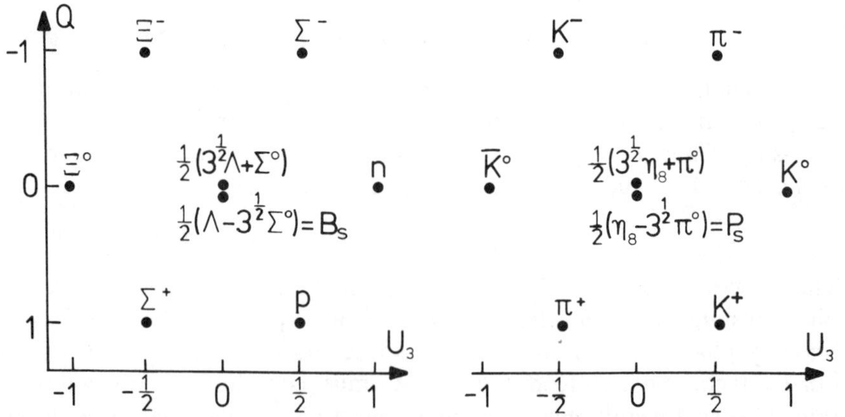


**Figure 4-10** The $U$-spin multiplets of the baryon and pseudoscalar meson octets.

and the isospin raising operator gives $I_+ K^0 = K^+, I_+ \pi^0 = 2^{\frac{1}{2}}\pi^+, I_+ \eta = 0$. We thus have $U_- I_+ K^0 = \pi^+$. On the other hand, one can convince oneself in appendix B-6 that $U_-$ and $I_+$ commute, such that we also have $\pi^+ = I_+ U_- K^0 = I_+ 2^{\frac{1}{2}} t = 2\langle \pi^0 | t \rangle \pi^+$, i.e., the coefficient of $\pi^0$ in $t$ must be $+\frac{1}{2}$. This leaves $|\langle \eta_8 | t \rangle|^2 = \frac{3}{4}$. The sign of $\langle \eta_8 | t \rangle$ is chosen positive, and the $U$-spin singlet state "$s$" is given by the orthogonal combination:

$$t = \tfrac{1}{2}(3^{\frac{1}{2}}\eta_8 + \pi^0), \qquad s = \tfrac{1}{2}(\eta_8 - 3^{\frac{1}{2}}\pi^0). \tag{10.7}$$

The members of $U$-spin multiplets are still eigenvalues of $Y$, with eigenvalues

$$Y = U_3 + \tfrac{1}{2}Q, \tag{10.8}$$

which is the "rotated version" of (3.6), $Q = I_3 + \frac{1}{2}Y$. It follows from (B-6.4) and is also evident from fig. 4-10.

Let us now discuss some consequences of $SU_3$-symmetry. As $SU_2$, the symmetry is an "internal" one, in the sense that its generators commute with all generators of the Lorentz group. Therefore, all particles within one $SU_3$-multiplet must have the same spin, parity and mass. Some of the mass equalities, for example $m_\pi = m_K$, are grossly violated, which shows that $SU_3$-symmetry can only be of limited value. Not only is it necessary to include "kinematical corrections" as in our treatment of $\phi$-decays in the previous section, but one must also allow for a "semistrong" symmetry-breaking interaction, analogous to the electromagnetic interaction for $SU_2$-symmetry breaking. The form of this interaction must be found empirically, and the scheme will be useful only if the $SU_3$-breaking interaction has simple $SU_3$-transformation properties. It was proposed by Gell-Mann (1961) and Okubo (1962) that the "semistrong" Hamiltonian should transform like a component of an octet, at least dominantly. Since it must commute with the isospin operators $\mathbf{I}$, it must transform as the 8- or $Y$-component of an octet operator. In lowest order perturbation theory for the semistrong interaction, the mass splitting is given by the expectation values of an operator $T_8^{(8)}$, which according to the Wigner-Eckart theorem (B-3.16) are proportional to the generalized Clebsch-Gordon-coefficients which describe the $SU(3)$-invariant coupling constants between the baryon octet and $\eta_8$, say. To find these coefficients, it is in principle necessary to reduce the direct product $8 \otimes 8$:

$$8 \otimes 8 = 1 \oplus 8 \oplus 8 \oplus 10 \oplus \overline{10} \oplus 27. \tag{10.9}$$

This way is described in most books on $SU_3$-symmetry (de Swart 1963, Carruthers 1966, Gourdin 1967, Lichtenberg 1970, Gasiorowicz 1966). It is complicated by the fact that $8 \otimes 8$ contains two octets according to (9). Consequently, the semistrong mass splitting and the strong ($= SU_3$-invariant) meson-baryon coupling constants (which will be discussed in chapter 6) contain two reduced matrix elements each for the baryon octet. This is in contrast to $SU_2$-symmetry where in the reduction of the direct product, every representation from the smallest spin, $|j_1 - j_2|$, to the largest one, $j_1 + j_2$, occurs exactly once. We shall therefore follow a different way, advocated by Lipkin (1965), which is entirely based on the $U$-spin subgroup of $SU_3$ (and on isospin, of course). For example, from (8) it follows that within a $U$-spin multiplet (where $Q$ is constant), the mass splitting operator and $\eta_8$ coupling constants are given by

$$\Delta m(U_3) = \Delta m_1 + U_3\, \Delta m_2, \qquad g_{BB\eta 8}(U_3) = g_{\eta 1} + U_3 g_{\eta 2}. \qquad (10.10)$$

Picking now the $U$-spin triplet $\Xi^0$, $\frac{1}{2}(3^{\frac{1}{2}}\Lambda + \Sigma^0)$, $n$, and using at the same time isospin invariance,

$$\begin{aligned} &m(\Xi^0) \approx m(\Xi^-) \approx m_\Xi, && m_n \approx m_p \approx m_N, \\ &m(\Sigma^+) \approx m(\Sigma^0) \approx m(\Sigma^-) \approx m_\Sigma, && m(\Lambda\Sigma) \approx 0, \end{aligned} \qquad (10.11)$$

we find from (10)

$$\Delta m_\Xi = \Delta m_1 - \Delta m_2, \qquad \Delta m_N = \Delta m_1 + \Delta m_2, \qquad \tfrac{3}{4}\,\Delta m_\Lambda + \tfrac{1}{4}\,\Delta m_\Sigma = \Delta m_1. \qquad (10.12)$$

For the experimental masses $m = m_0 + \Delta m$, we thus find the mass formula of Gell-Mann and Okubo ("GMO")

$$m_N + m_\Xi = \tfrac{1}{2}(3m_\Lambda + m_\Sigma). \qquad (10.13)$$

This is the end of this game, because in the isospin invariant approximation the baryon octet contains only 4 different masses, while the mass operator has 3 reduced matrix elements ($m_0$, $\Delta m_1$ and $\Delta m_2$).

Before comparing (13) with experiment, one must decide how to treat the (admittedly small) isospin mass splittings. For our present purpose, it is sufficient to define $m_N$, $m_\Sigma$ and $m_\Xi$ as the average masses of the isospin multiplets. According to table C-4, this gives

$$\begin{array}{rlrl} m_N = & 939.93 \text{ MeV} & m_\Lambda = & 1115.6 \text{ MeV} \\ m_\Xi = & 1318.1 \text{ MeV} & m_\Sigma = & 11193.1 \text{ MeV} \\ \hline & 2258.0 \text{ MeV} & = & 2269.9 \text{ MeV} - 19.9 \text{ MeV} \end{array} \qquad (10.14)$$

Thus the hypothesis that the mass-breaking operator transforms under $SU_3$ like the eighth component of an octet, together with lowest-order perturbation theory, works surprisingly well, at least for the baryon octet.

For the mesons, things are complicated both by much larger relative mass splittings and by the existence of two particles with $I = Y = 0$, namely $\eta$ and $X$ among the pseudoscalar mesons and $\phi$ and $\omega$ among the vector mesons. Before the discovery of the $X$ meson, $\eta_8$ was identified with the $\eta$ meson, in which case the Gell-Mann-Okubo mass formula analogous to (13)

$$3m_{\eta 8} = 4m_K - m_\pi \qquad (10.15)$$

gave $m_{\eta 8} = 619$ MeV, which is larger than the experimental $m_\eta = 549$ MeV. It was then argued that the formula should be used not for the masses but for their squares,

$$3m_{\eta 8}^2 = 4m_K^2 - m_\pi^2, \qquad (10.16)$$

because the operator $P^2$ with eigenvalues $m^2$ is simpler than $(P^2)^{\frac{1}{2}}$. In that case one gets $m_{\eta 8} = 569.5$ MeV (here we have used $m_K \equiv m(K^0)$, $m_\pi \equiv m(\pi^0)$). When the $X$-meson was discovered, it was obvious that there existed also an $SU_3$-singlet state $X_1$ (given by (4) in the quark model) and that the physical $\eta$ and $X$ could be mixtures of $\eta_8$ and $X_1$, because the semistrong interaction does not respect the $SU_3$-classification:

$$\eta = \eta_8 \cos\theta_p - X_1 \sin\theta_p, \qquad X = \eta_8 \sin\theta_p + X_1 \cos\theta_p. \quad (10.17)$$

The index $p$ on the mixing angle $\theta$ stands for "pseudoscalar." Similarly, for the vector mesons, we have

$$\phi = \phi_8 \cos\theta_v - \omega_1 \sin\theta_v, \qquad \omega = \phi_8 \sin\theta_v + \omega_1 \cos\theta_v \quad (10.18)$$

(The distribution of the indices 8 and 1 on the particle symbols is always done such that $|\theta| \leq 45^0$). The "rotation" (17) implies

$$m_{\eta 8} + m_{X1} = m_\eta + m_X, \qquad m_{\eta 8} = m_\eta \cos^2\theta_p + m_X \sin^2\theta_p, \quad (10.19)$$

while the $m$'s must be replaced by $m^2$ if the quadratic mass formula (16) is used. In this way, the GMO mass formula for mesons cannot be tested but, assuming that it is correct (either in the linear form or in the quadratic one), it can be used to determine the mixing angle (table 4-10). This angle can then be tested, for example, from the $SU_3$-predictions for meson decays. To anticipate the results, it appears that $\theta_p$ lies between $-10°$ and $-20°$ while $\theta_v$ lies around $35^0$. The difference in the values of $m_{\eta 8}$ in the two mass formulas indicates of course a breakdown of first-order perturbation theory, independent of the mixing problem.

**Table 4-10** Mass of the octet components and resulting mixing angles according to the versions (15) or (16) of the GMO mass formula.

| Mass formula | $m_{\eta 8}$ | $\lvert\theta_p\rvert$ | $m_{\phi 8}$ | $\lvert\theta_V\rvert$ |
|---|---|---|---|---|
| Linear | 619 MeV | $24^0$ | 939 MeV | $37^0$ |
| Quadratic | 569.5 MeV | $10.4^0$ | 935 MeV | $40^0$ |

We now relate the coupling constants for $\rho \to \pi\pi$, $K^* \to K\pi$ and $\phi \to K\bar{K}$ decays by means of $U$-spin invariance. The states $K^{*0}$, $V_t = \frac{1}{2}(3^{\frac{1}{2}}\phi_8 + \rho^0)$ and $\bar{K}^{*0}$ form a $U$-spin triplet. Among the decay products, $K^+, \pi^+$ as well as $\pi^-, K^-$ form $U$-spin doublets. Insertion of the appropriate Clebsch-Gordan coefficients gives

$$2^{-\frac{1}{2}}G(K^{*0} \to K^+\pi^-) = \tfrac{1}{2}3^{\frac{1}{2}}G(\phi_8 \to \pi^+\pi^-) + \tfrac{1}{2}G(\rho^0 \to \pi^+\pi^-) \quad (10.20)$$

$$= \tfrac{1}{2}3^{\frac{1}{2}}G(\phi_8 \to K^+K^-) + \tfrac{1}{2}G(\rho^0 \to K^+K^-). \quad (10.21)$$

In the second term, we use $G(\phi_8 \to \pi^+\pi^-) = 0$, due to conservation of $G$-parity. With (9.4) and (9.5) we obtain from (20)

$$G_{\rho\pi\pi}/G_{K*K\pi} = 2 \quad (10.22)$$

in rough agreement with the ratio $\pm(3.4)^{\frac{1}{2}}$ deduced from table C-3. To find $G_{\phi 8K\bar{K}}$, we must eliminate $G(\rho^0 \to K^+K^-)$ from (21). Isospin invariance tells us

$$G_{\rho K\bar{K}} \equiv G(\rho^0 \to K^+K^-) = G(\rho^0 \to K^0\bar{K}^0) \qquad (= G(\rho^0 \to K_1 K_2)). \quad (10.23)$$

$K^0$ and $\bar{K}^0$ are both members of the $U$-spin triplet (see fig. 4-10) and the $p$-wave $K^0\bar{K}^0$-state must have $U = 1$ (exactly as the $p$-wave $\pi^+\pi^-$-state must have $I = 1$). Therefore the $U = 0$ combination $\phi_8 - 3^{\frac{1}{2}}\rho$ cannot decay into $K^0\bar{K}^0$, and $\omega_1$ cannot decay into $K\bar{K}$ either:

$$-G_{\phi 8K\bar{K}} - 3^{\frac{1}{2}}G_{\rho K\bar{K}} = 0, \qquad G_{\omega 1K\bar{K}} = 0. \quad (10.24)$$

The minus sign in front of $G_{\phi 8K\bar{K}}$ comes from (9.6). Thus (21) gives

$$G_{K^*K\pi} = \tfrac{1}{2}3^{\frac{1}{2}}G_{\phi 8K\bar{K}} + \tfrac{1}{2}G_{\rho K\bar{K}} = 3^{-\frac{1}{2}}G_{\phi 8K\bar{K}} = 3^{-\frac{1}{2}}G_{\phi K\bar{K}}/\cos\theta_v \quad (10.25)$$

or $G^2_{K^*K\pi}/G^2_{\phi K\bar{K}} = \frac{1}{3}/\cos^2\theta_v = 0.57$ $(\theta_v = 40^0)$, again in agreement with the ratio 0.56 from table C-3. From the discussion following (9), it appears that the decays $8 \to 8 \times 8$ should have two reduced matrix elements. However, if all three octets are mesons, one reduced matrix element (essentially $G_{\phi 8\pi\pi}$) vanishes due to $G$-parity.

For completeness, we also calculate

$$G_{K^*K\eta 8} = G(K^{*+} \to K^+\eta_8) = G(K^{*0} \to K^0\eta_8). \quad (10.26)$$

This time we use the $U$-spin doublet $K^{*+}$ and $\rho^+$ in the initial state and the $U$-spin doublet $K^+$ and $\pi^+$ in the final state, together with $P_s = \frac{1}{2}(\eta_8 - 3^{\frac{1}{2}}\pi^0)$:

$$G(K^{*+} \to K^+\eta_8) - 3^{\frac{1}{2}}G(K^{*+} \to K^+\pi^0)$$
$$= G(\rho^+ \to \pi^+\eta_8) - 3^{\frac{1}{2}}G(\rho^+ \to \pi^+\pi^0) \quad (10.27)$$

with $G(\rho^+ \to \pi^+\eta) = 0$ by $G$-parity, $G(K^{*+} \to K^+\pi^0) = G_{K^*K\pi}$ and $G(\rho^+ \to \pi^+\pi^0) = G_{\rho\pi\pi} = 2G_{K^*K\pi}$, we get

$$G_{K^*K\eta 8} = -3^{\frac{1}{2}}G_{K^*K\pi}. \quad (10.28)$$

## 4-11 The Baryon Decuplet and Pionic Baryon Decays

The baryon decuplet consists of the $\Delta$ or $N^*_{33}$ which has $I = \frac{3}{2}$ (charge states $\Delta^{++}, \Delta^+, \Delta^0, \Delta^-$), the $\Sigma(1385)$ which we call $\Sigma^*$ and which has $I = 1$ (charge states $\Sigma^{*+}$, $\Sigma^{*0}$, $\Sigma^{*-}$, the $\Xi^*(1530)$ which we call $\Xi^*$ and which has $I = \frac{1}{2}$ (charge states $\Xi^{*0}$ and $\Xi^{*-}$), and the $\Omega^-$. The hypercharges of these particles are 1, 0, $-1$ and $-2$ respectively (see fig. 4-11). Their masses, widths and decay modes are collected in table C-5. The differences between the average masses of the multiplets from $\Delta$ to $\Sigma^*$ to $\Xi^*$ to $\Omega^-$ are 153 MeV, 150 MeV, and 138 MeV, respectively, the $\Omega^-$ being the heaviest particle. The heavier particles have the longer lifetimes, which is unusual but easy to explain: The

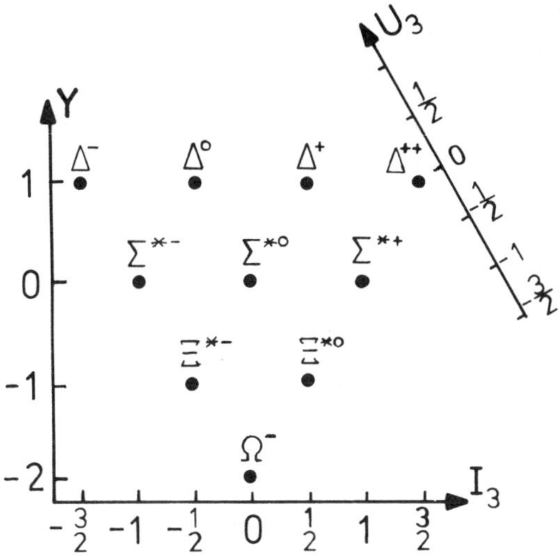


**Figure 4-11** The quantum numbers $Y$, $I_3$ and $U_3$ of the baryon decuplet; $\Delta \equiv N^*_{33}$, $\Sigma^* \equiv \Sigma(1385)$, $\Xi^* \equiv \Xi^*(1530)$ and $\Omega^-$.

$\Omega^-$ cannot decay strongly because the lightest channel of hypercharge $-2$ and baryon number 1 is $\Xi\bar{K}$ which is already "closed" (i.e., $m_\Xi + m_K > m_\Omega$). Because of its hypercharge, $\Omega^-$ is also very difficult to produce. One possible reaction is $K^-p \to K^+K^0\Omega^-$, which is also the reaction in which $\Omega^-$ was discovered in 1964. The other decuplet states do decay strongly, but the decay momentum in $\Xi^* \to \Xi\pi$ decays is relatively small, which implies a small decay rate, particularly because the decay is $p$-wave, $\Gamma \sim p^3$.

Historically, $\Delta^{++}$ and $\Delta^0$ were discovered as resonances in the total $\pi^+p$ and $\pi^-p$ cross sections (see fig. 4-2), long before the discovery of any other "resonance" (see Gell-Mann and Watson 1954). The maximum of $\sigma(\pi^+p)$ is 210 mb, which is close to $8\pi/p^2$. According to (8.24), the resonant cross section in the $J$th partial wave is

$$\sigma_{J,\,\mathrm{res}} = 2\pi(2J+1)m^2\Gamma^2[(m^2-s)^2+m^2\Gamma^2]^{-1}p^{-2} \tag{11.1}$$

for elastic scattering of spin-0 particles on spin-$\frac{1}{2}$ particles, and a maximum of $8\pi/p^2$ requires $J=\frac{3}{2}$. Thus the $\Delta$ is a spin-$\frac{3}{2}$ particle. Its isospin $\frac{3}{2}$ was already explained in (2.12). The older notation $N^*_{33}$ stands for $N^*_{2I,\,2J}$ (the name "isobar" is also used).

The three $\Sigma^*$ states have masses around 1384 MeV and cannot be seen as resonances in $K^-p$ scattering because $m_K + m_p = 1432$ MeV. They are produced in reactions such as $K^-p \to \pi^\pm\Sigma^{*\mp}$ and $K^-p \to \pi^0\Sigma^{*0}$ where the $\Sigma^*$ decays immediately, normally into $\pi\Lambda$. Similarly, $\Xi^*$ is produced in $K^-p \to K^0\Xi^{*0}$ and $K^-p \to K^+\Xi^{*-}$.

Today it is known that $\Delta$, $\Sigma^*$ and $\Xi^*$ are $\frac{3}{2}^+$-states, i.e., they have spin-$\frac{3}{2}$ and positive parity. As the pion in the decay carries negative intrinsic parity, the decay of these states occurs in the $p$-wave. In the notation of (2-5.28), the resonating partial wave in $\pi N$ scattering is $f_{1+}$. Because of its isospin $\frac{3}{2}$, the smallest $SU_3$-representation into which the $\Delta$ fits is 10 (the decuplet). When the eightfold way was proposed in 1961, the $\Omega^-$ was not yet discovered, but the $\Sigma^*$ and $\Xi^*$ fitted naturally into 5 of the remaining 6 places of the decuplet. The decuplet is mathematically simpler than the octet because no point in the $Y - I_3$ plot (fig. 4-11) is occupied more than once. Therefore, all states are also eigenstates of $U_3$, the $U$-spin multiplets being

$$\begin{aligned} &U = \tfrac{3}{2}\!: \quad \Delta^-, \Sigma^{*-}, \Xi^{*-}, \Omega; \\ &U = 1\!: \quad \Delta^0, \Sigma^{*0}, \Xi^{*0}; \qquad U = \tfrac{1}{2}\!: \quad \Delta^+, \Sigma^{*+}; \qquad U = 0\!: \quad \Delta^{++}. \end{aligned} \tag{11.2}$$

The GMO mass formula (10.10) for the $U = \frac{3}{2}$ multiplet requires equal spacing between the masses of $\Delta$, $\Sigma^*$, $\Xi^*$ and $\Omega^-$, which was not only tested by the known masses of $\Delta$, $\Sigma^*$ and $\Xi^*$, but also allowed a prediction of the $\Omega^-$ mass, accurate enough to predict that the $\Omega^-$ could only decay weakly.

The decay matrix elements for pionic decays of baryons, $B \to B_1 \pi$ are collected in table 4-11.1, for spins-$\frac{1}{2}$ and $\frac{3}{2}$ of the initial baryon, respectively. The construction principle is the same as in table 4-9, i.e., the matrix element must be linear in the spinor ($u(M)$ or $u^\mu(M)$) of the initial baryon and in the adjoint spinor $\bar{u}_1(\lambda_1)$ of the final baryon, and it must be Lorentz invariant. Parity conservation requires

$$A = 0, \qquad B^3 = 0, \tag{11.3}$$

but since the decays $\Lambda \to N\pi$, $\Sigma \to N\pi$, $\Xi \to \Lambda\pi$ and $\Omega \to \Xi\pi$, $\Omega \to \Lambda K^-$ are all weak, we do not impose parity conservation. From (1-6.17) we remember that the combinations $\bar{u}_1 u$ and $\bar{u}_1 \gamma_5 u$ are scalar and pseudoscalar, respectively. These quantities are evaluated in A-3 and the quantities $k_\pm$ of (A-3.8) are to be taken at $t = m_\pi^2 = m_2^2$:

$$k_\pm = [2m(E_1 \pm m_1)]^{\frac{1}{2}} = [(m \pm m_1)^2 - m_2^2]^{\frac{1}{2}}. \tag{11.4}$$

The angular dependence of the decay matrix element is given by (4.7):

$$\langle p, \lambda_1, \theta\phi \,|\, T \,|\, M \rangle = \begin{cases} 2^{\frac{1}{2}} e^{i\phi(M-\lambda_1)}\, d^{\frac{1}{2}}_{M\lambda_1}(\theta) T_{\frac{1}{2}}(\lambda_1) & \text{(octet)} \\ 2 e^{i\phi(M-\lambda_1)}\, d^{\frac{3}{2}}_{M\lambda_1}(\theta) T_{\frac{3}{2}}(\lambda_1) & \text{(decuplet)} \end{cases} \tag{11.5}$$

The matrix elements $T^s$, $T^p$ and $T^d$ which appear in $T_S$ and $m\Gamma$ in table 4-11.1 are

$$T^s = k_+ A, \qquad T^p = k_- B, \qquad T^{p3} = k_+ p A^3, \qquad T^d = k_- p B^3. \tag{11.6}$$

**Table 4-11.1** Matrix elements for the decays $B \to B_1 \pi$. The angular dependence is given by (5), and $T^s$, $T^p$, $T^{p3}$, $T^d$ are explained in (6). The Rarita-Schwinger spinor $u_\mu$ in the decay $\frac{3}{2} \to \frac{1}{2}0$ is defined in section 2-8.

| Decay | $\langle p, \lambda_1 \theta\phi \,\|\, T \,\|\, M \rangle$ | $T_S(\lambda_1)$ | $m\Gamma$ |
|---|---|---|---|
| $\frac{1}{2} \to \frac{1}{2}0$ | $\bar{u}_1(\lambda_1)(A - B\gamma_5)u(M)$ | $2^{-\frac{1}{2}}(T^s + 2\lambda_1 T^p)$ | $\frac{1}{2} p s^{-\frac{1}{2}}(\|T^s\|^2 + \|T^p\|^2)/4\pi$ |
| $\frac{3}{2} \to \frac{1}{2}0$ | $\bar{u}_1(\lambda_1)(A^3 - B^3\gamma_5)p_2^\mu u_\mu(M)$ | $6^{-\frac{1}{2}}(T^{p3} + 2\lambda_1 T^d)$ | $\frac{1}{6} p s^{-\frac{1}{2}}(\|T^{p3}\|^2 + \|T^d\|^2)/4\pi$ |

(The expressions for $T^s$ and $T^p$ follow from the formulas of A-3. For $T^{p3}$ and $T^d$, one must in addition use (2-8.14) and formulas such as (9.3). Note that $P_2^\mu u_\mu = \frac{1}{2}(P_2^\mu - P_1^\mu)u_\mu$ because of $P^\mu u_\mu = 0$, which explains the $\frac{1}{6}$ in the second row of table 4-11.1 as compared to the $\frac{2}{3}$ in the second row of table 4-9).

Contrary to the "decay coupling constants" $G$ and $g$ of table 4-9, the "invariant amplitudes" $A$ and $B$ may be complex. In the weak decays, they have the phases of elastic $B_1\,\pi$ scattering according to (8.29). In the strong decays $\Delta \to N\pi$, $\Sigma^* \to \Lambda\pi$, $\Sigma^* \to \Sigma\pi$ and $\Xi^* \to \Xi\pi$, there is only one decay amplitude (namely $A^3$), which should be approximately real (see sections 6-5 and 7-2).

The indices $s$, $p$ and $d$ in (6) stand for $s$-, $p$- and $d$-waves, of course, since they have parities $+, -, +$, not counting the intrinsic parity of the pion. It is somewhat less obvious that the amplitudes $T^s$, $T^p$ and $T^d$ have the threshold behaviour $p^L$. To see this, it is necessary to decompose $p$ according to (2-2.9) and (2-2.11) as

$$p = \tfrac{1}{2}s^{-\frac{1}{2}}k_+ k_-. \tag{11.7}$$

The threshold, $p \to 0$, corresponds to $m \to m_1 + m_2$, $k \to 0$. Therefore it is the power of $k_-$ that counts. $T^d$ for example can be written as

$$T^d = k_-^2 \tfrac{1}{2}s^{-\frac{1}{2}}k_+ B^3. \tag{11.8}$$

In the strong decuplet decays, $m$ must be replaced by $s^{\frac{1}{2}}$, and $A^3$ may depend on $s$. It was suggested by Brueckner (1952) that one should again use (9.16):

$$A^3_{\Delta N\pi} = G_{\Delta N\pi}(m^2)(1 + p^2(m^2)r_0^2)^{\frac{1}{2}}(1 + p^2(s)r_0^2)^{-\frac{1}{2}}. \tag{11.9}$$

Isospin invariance requires (first row of table 4-2.2):

$$\begin{aligned} G_{\Delta N\pi} &\equiv G(\Delta^{++} \to p\pi^+) = G(\Delta^- \to n\pi^-) \\ &= 3^{\frac{1}{2}}G(\Delta^+ \to n\pi^+) = 3^{\frac{1}{2}}G(\Delta^0 \to p\pi^-). \end{aligned} \tag{11.10}$$

Since the $\Lambda$ has $I = 0$, isospin invariance requires

$$G_{\Sigma^*\Lambda\pi} = G(\Sigma^{*+} \to \Lambda\pi^+) = G(\Sigma^{*0} \to \Lambda\pi^0) = G(\Sigma^{*-} \to \Lambda\pi^-). \tag{11.11}$$

The isospin relations for $\Xi^* \to \Xi\pi$ are quite analogous to those of $K^* \to K\pi$ (9.5), and the isospin relations for $\Sigma^* \to \Sigma\pi$ follow from the last row of table 4-2.3, with the first $\pi$ replaced by $\Sigma$:

$$\begin{aligned} G_{\Sigma^*\Sigma\pi} &\equiv G(\Sigma^{*+} \to \Sigma^+\pi^0) = -G(\Sigma^{*+} \to \Sigma^0\pi^+) \\ &= G(\Sigma^{*-} \to \Sigma^0\pi^-) = -G(\Sigma^{*-} \to \Sigma^-\pi^0) \\ &= G(\Sigma^{*0} \to \Sigma^+\pi^-) = -G(\Sigma^{*0} \to \Sigma^-\pi^+), \\ &G(\Sigma^{*0} \to \Sigma^0\pi^0) = 0. \end{aligned} \tag{11.12}$$

Table 4-11.2 collects coupling constants for the width formula

$$m\Gamma = \tfrac{1}{6}p^3 s^{-\frac{1}{2}}\frac{G^2}{4\pi}(1 + p^2(m^2)r_0^2)(1 + p^2(s)r_0^2)^{-1}. \tag{11.13}$$

**Table 4-11.2** Masses and widths of pionic decuplet decays from an isospin-constrained fit, and resulting coupling constants according to eq. (13) (from Nagels 1976).

| | $\Delta^{++} \to p\pi^+$, | $\Sigma^{*0} \to \Lambda\pi^0$, | $\Sigma^{*0} \to \Sigma^+\pi^-$, | $\Xi^{*0} \to \Xi^0\pi^0$ |
|---|---|---|---|---|
| $m$[MeV] | 1231.1 | 1385.2 | 1385.2 | 1531.7 |
| $\Gamma(m)$[MeV] | 111.5 | 32.7 | 2.23 | 3.18 |
| $G^2/4\pi$ | 14.6 | 6.74 | 2.03 | 1.98 |
| $SU_3$ | 13.8 | 6.9 | 2.3 | 2.3 |

Here a factor $k_+^2$ from (6) has been dropped for 3 reasons, none of which is compelling: (i) $G$ is now dimensionless, (ii) $k_+^2$ is fairly constant across the resonance even for the $\Delta$, but it slightly deteriorates the resonance shape (see section 7-2), (iii) the agreement withe $SU_3$-predictions below is improved.

Comparison between (13) and (9.14) shows that the reduced width is

$$m\gamma = \tfrac{1}{6}(r_0^{-2} + p^2(m^2))G^2/4\pi. \tag{11.14}$$

This quantity is not suited for $SU(3)$-comparisons; $r_0^2$ is not known for $\Xi^*$, and varies from 1 fm$^2$ for $\Delta$ to 4 fm$^2$ for $\Sigma^*$ (Holmgren et al. 1977). We shall return to this point in section 7-2.

The $SU_3$-predictions for the decays of table 4-11.2 follow from $U$-spin invariance in the decays $|U = \tfrac{3}{2}, U_3\rangle \to |U = 1, U_3 - \tfrac{1}{2}\rangle|\pi^-\rangle$, where the $\pi^-$ is the $U_3 = +\tfrac{1}{2}$-component of the $U$-spin doublet, and $U = 1$ states are $n$, $\tfrac{1}{2}(3^{\frac{1}{2}}\Lambda - \Sigma^0)$ and $\Xi^0$ (fig. 4-10). The appropriate $CG$-coefficients are again in the first row of table 4-2.2:

$$\begin{aligned} G(\Delta^- \to n\pi^-) &= 2^{\frac{1}{2}}G(\Sigma^{*-} \to \Lambda\pi^-) = -6^{\frac{1}{2}}G(\Sigma^{*-} \to \Sigma^0\pi^-) \\ &= 3^{\frac{1}{2}}G(\Xi^* \to \Xi^0\pi^-). \end{aligned} \tag{11.15}$$

Combining this with (10), (11), (12), $G(\Xi^* \to \Xi^0\pi^-) = 2^{\frac{1}{2}}G_{\Xi^*\Xi\pi}$, we get

$$G_{\Delta N\pi} = 2^{\frac{1}{2}}G_{\Sigma^*\Lambda\pi} = -6^{\frac{1}{2}}G_{\Sigma^*\Sigma\pi} = 6^{\frac{1}{2}}G_{\Xi^*\Xi\pi}. \tag{11.16}$$

Let us now discuss the weak pionic baryon decays $\Lambda \to N\pi$, $\Sigma \to N\pi$, $\Xi \to \Lambda\pi$, $\Omega^- \to \Xi\pi$. None of these conserves hypercharge or isospin, but (i) why are the decays $\Xi \to N\pi$, $\Omega^- \to \Lambda\pi$, $\Omega^- \to \Sigma\pi$, $\Omega^- \to n\pi^-$ absent and (ii) why decays $\Lambda$ twice as much into $p\pi^-$ as into $n\pi^0$, and why lives $\Xi^0$ about twice as long as $\Xi^-$? Obviously, (i) the hypercharge changes only by one unit ($|\Delta Y| = 1$ rule) and (ii) the isospin changes only by $\tfrac{1}{2}$($\Delta I = \tfrac{1}{2}$-rule). From the second row of table 4-2.2 and the phase theorem (8.29), we get for the invariant amplitudes $A$ and $B$ of $\Lambda$ decays

$$\begin{aligned} 2^{-\frac{1}{2}}A(\Lambda \to p\pi^-) &= -A(\Lambda \to n\pi^0) = |A_\Lambda| \exp(i\delta_{0+}^{(\frac{1}{2})}), \\ 2^{-\frac{1}{2}}B(\Lambda \to p\pi^-) &= -B(\Lambda \to n\pi^0) = |B_\Lambda| \exp(i\,\delta_{1-}^{(\frac{1}{2})}), \end{aligned} \tag{11.17}$$

**Table 4-11.3** The decay constants $A$ and $B$ of pionic decays in the baryon octet. (From Particle Data Group (1978)). Final state interaction phases neglected. $A$ and $B$ in $\Lambda \to p\pi^-$ are taken positive, the other signs are taken from (17), (18), (19), (21). $A$ and $B$ are obtained from the decay rates (table C-4) and asymmetry parameters (last row). The determination of $\alpha$ is explained in section 4-13. The bracketed digit indicates the error (see table C-1).

| | $\Lambda \to p\pi^-$ | $\Lambda \to n\pi^0$ | $\Sigma^+ \to n\pi^+$ | $\Sigma^+ \to p\pi^0$ | $\Sigma^- \to n\pi^-$ | $\Xi^0 \to \Lambda\pi^0$ | $\Xi^- \to \Lambda\pi^-$ |
|---|---|---|---|---|---|---|---|
| A × $10^7$ | 3.27(2) | −2.39(4) | −0.13(4) | −3.3(1) | −4.27(2) | 3.39(7) | 4.51(4) |
| B × $10^7$ | 22.5(5) | −16(1) | −42.2(4) | 27(1) | 1.4(2) | −13.(2) | −14.8(8) |
| $\alpha$ | 0.64(1) | 0.65(4) | 0.07(2) | −0.98(2) | 0.07(1) | −0.44(8) | −0.39(2) |

where the phase shifts are those of $I = \frac{1}{2}\pi N$ scattering in the $S_{\frac{1}{2}}$ and $P_{\frac{1}{2}}$ states at $s = m_\Lambda^2$. Deviations from $\Gamma(\Lambda \to p\pi^-)/\Gamma(\Lambda \to n\pi^0) = 2$ can arise from differences in $k_-$ and $p$ in the two decays (because of the $p - n$ and $\pi^0 - \pi^-$ mass differences) as well as from the Coulomb corrections in $\Lambda \to p\pi^-$ (which can be calculated from (9.17)), but the two sources tend to cancel each other, and $I = \frac{3}{2}$ amplitudes seem to occur at the 3%-level. Numerical values of $A$ and $B$ are collected in table 4-11.3, under the assumption that all phase shifts are negligible.

Now let us look at the decays $\Xi^0 \to \Lambda\pi^0$ and $\Xi^- \to \Lambda\pi^-$. The final state obviously has $I = 1$. The $\Delta I = \frac{1}{2}$-rule requires that the weak part of the Hamiltonian which induces these decays transforms under isospin rotations like the neutral component (because of charge conservation) of an isodoublet, i.e. $H_{\text{weak}}(\Delta Y = 1)$ is a tensor operator which transforms as a $K^0$ or $K^{*0}$. According to the Wigner-Eckart theorem (B-3.16), the decays into $\Lambda\pi^0$ and $\Lambda\pi^-$ are related by the Clebsch-Gordan coefficients for $\Xi^0 K^0 \to \Lambda\pi^0$ and $\Xi^- K^0 \to \Lambda\pi^-$ ("spurious-kaon rule"), which are $2^{-\frac{1}{2}}$ and 1, respectively (compare 2.4). Thus

$$2^{\frac{1}{2}} A(\Xi^0 \to \Lambda\pi^0) = A(\Xi^- \to \Lambda\pi^-) \equiv A_\Xi, \tag{11.18}$$

and correspondingly for $B$. For the $\Sigma$-decays, the $\Delta I = \frac{1}{2}$-rule requires

$$A(\Sigma^- \to n\pi^-) = A(\Sigma^+ \to n\pi^+) + 2^{\frac{1}{2}} A(\Sigma^+ \to p\pi^0). \tag{11.19}$$

This relation follows from isospin conservation in $K^0\Sigma \to N\pi$. There are two isospin amplitudes, $A(\frac{3}{2}) = A(\Sigma^- \to n\pi^-)$ and $A(\frac{1}{2})$. The isospin structure is the same as in elastic $n\pi$ scattering (replace $K^0$ by $n$, $\Sigma$ by $\pi$), and (19) is easily verified from (2.7), (2.10) and (2.11). We also find from (2.11)

$$A(\Sigma^0 \to p\pi^-) = A(\Sigma^+ \to p\pi^0), \tag{11.20}$$

but the decays $\Sigma^0 \to N\pi$ are extremely rare because they must compete with the electromagnetic decay $\Sigma^0 \to \Lambda\gamma$.

So much for the isospin selection rule. We now postulate that $H_{\text{weak}}$ has also the $SU_3$-properties of $K^0$ or $K^{*0}$, i.e., it belongs to an octet. One then obtains the "Lee-Sugawara"-relation

$$\tfrac{1}{2}[A(\Lambda \to p\pi^-) - 3^{\frac{1}{2}} A(\Sigma^+ \to p\pi^0)] = A(\Xi^0 \to p\pi^-). \tag{11.21}$$

According to (20), the left-hand side is the decay amplitude of the $U$-spin singlet $\frac{1}{2}(\Lambda - 3^{\frac{1}{2}}\Sigma^0)$. Contrary to eqs. (18)–(20), (21) is not valid for the corresponding amplitudes $B$ unless an additional assumption ($R$ invariance) is made (Gell-Mann 1964, Okubo 1964). See also Gourdin (1967).

## 4-12 The Tensor Mesons and the Quark Model

Next to the $0^-$ and $1^-$ meson nonets, there exists a nonet of $2^+$ mesons $A_2^+$, $A_2^0$, $A_2^-$, $f$, $f'$, $K_N^+$, $K_N^0$, $\bar{K}_N^0$, $K_N^-$ which were discovered in the reactions $\pi^\pm p \to A_2^\pm p$ $\pi^- p \to A_2^0 n$, $\pi^- p \to fn$, $K^- p \to f'\Lambda$, $K^- p \to f'\Sigma^0$ and $K^- p \to K_N^- p$ (see for example Gasiorowicz 1966). The masses and decay properties of these states are given in table C-6. Their spin states $\varepsilon^{\mu\nu}(M)$ (2-8.13) are tensors under Lorentz transformations and the states are therefore also called "tensor mesons". Matrix elements for the decays of table C-6 ($2^+ \to 0^-0^-$ and $2^+ \to 1^-0^-$) have already been included in table 4-9. The $SU_2$- and $SU_3$-classification of $A_2$, $f$, $f'$, $K_N$ is analogous to that of $\rho$, $\omega$, $\phi$, $K^*$, but the decays $A_2 \to \rho\pi$, $f \to \pi\bar{\pi}$, $f' \to K\bar{K}$ show that these states have $G$-parities opposite to $\rho$, $\omega$ and $\phi$, in agreement with the $J = 2$ $N\bar{N}$ states of table 4-5.2.

At this point, the reader may check whether he has understood the isospin formalism, by calculating the widths $\Gamma(f^0 \to \pi^+\pi^-)$ and $\Gamma(f^0 \to \pi^0\pi^0)$. From the second row of table 4-2.3, one finds

$$\Gamma(f \to \pi^+\pi^-)/\Gamma(f \to \pi^0\pi^0) = 2 \tag{12.1}$$

analogous to (2.16), because the three states $\pi^+\pi^-$, $\pi^0\pi^0$ and $\pi^-\pi^+$ all have equal weights. If the coupling constant $G_2$ for one of these states is called $G_{f\pi\pi}$, then the total $f \to \pi\pi$ width is

$$\Gamma(f \to \pi\pi) = \tfrac{8}{5}p^5 s^{-\frac{1}{2}} G_{f\pi\pi}^2/4\pi, \tag{12.2}$$

where a factor $\frac{16}{15}$ comes from the third row of table 4-9, a factor $\frac{1}{2}$ comes from the phase space for identical particles and a factor 3 comes from the summation over the three states. On the other hand, there are only 2 decay channels in $f \to \pi\pi$, namely the charged pion channel ($\pi^+\pi^-$) and the $\pi^0\pi^0$ channel. Question: How are the coupling constants for these two channels related by isospin invariance? Answer:

$$G(f \to \pi^+\pi^-) = -G(f \to \pi^0\pi^0) \equiv G_{f\pi\pi}. \tag{12.3}$$

Now try to derive (1) and (2) from (3) and the third row of table 4-9, which applies to the decay into two distinguishable particles only.

The states $f$ and $f'$ are linear combinations of $SU_3$-octet and $SU_3$-singlet states (compare 10.18):

$$f' = f_8 \cos\theta_t - f_1 \sin\theta_t, \qquad f = f_8 \sin\theta_t + f_1 \cos\theta_t. \tag{12.4}$$

The mixing angle is $32^0$, using the quadratic mass formula analogous to (10.16).

The decays $2^+ \to 1^-0^-$ are again $p$-wave. Therefore, their $SU_2$- and $SU_3$-properties follow from those of $1^- \to 0^-0^-$ by substituting

$$\begin{aligned} &K^* \to K_N, \qquad \phi_8 \to f_8, \qquad \rho \to A_2, \\ &\text{first } K \to K^*, \qquad \text{first } \pi \to \rho, \qquad \text{first } \eta_8 \to \phi_8. \end{aligned} \tag{12.5}$$

The resulting $SU_3$-values are collected in table 4-12, expressed in units of

$$G_F \equiv G(K_N \to K^*\pi) \equiv G(K_N^* \to K^{*+}\pi^0). \tag{12.6}$$

However, in this way we get only the coupling constants for $K_N \to K^*\pi$, $f_8 \to K^*\bar{K}$, $A_2 \to K^*\bar{K}$, $K_N \to K^*\eta_8$ and $A_2 \to \rho\pi$ decays. To obtain those of $K_N \to \rho K$, $f_8 \to \bar{K}^*K$, $A_2 \to \bar{K}^*K$ and $K_N \to \phi_8 K$, we must permute the two pseudoscalar mesons before substituting (5). Because of the antisymmetry of the decay matrix element $\varepsilon_\mu(M)(P_1 - P_2)^\mu$, the coupling constant changes

**Table 4-12** $SU_3$-predictions for the coupling constants of tensor meson decays (octet components only). The meson-baryon coupling constants on the right-hand side of the table are included for later convenience.

| Decay | $G_F$ | Decay | $G_D$ | " Decay " | $G_F = (1-\alpha)G,\ G_D = \alpha G$ |
|---|---|---|---|---|---|
| $K_N \to K^*\pi$ | $1$ | $K_N \to K\pi$ | $1$ | $N \to N\pi$ | $G_F + G_D = G$ |
| $K_N \to \rho K$ | $-1$ | | | $N \to \Sigma K$ | $-G_F + G_D = (2\alpha - 1)G$ |
| $f_8 \to K^*\bar{K}$ | $3^{1/2}$ | $f_8 \to K\bar{K}$ | $3^{-1/2}$ | $\Lambda \to N\bar{K}$ | $3^{1/2}G_F + 3^{-1/2}G_D = 3^{-1/2}(3 - 2\alpha)G$ |
| $f_8 \to \bar{K}^*K$ | $-3^{1/2}$ | | | $\Lambda \to \Xi K$ | $-3^{1/2}G_F + 3^{-1/2}G_D = 3^{-1/2}(4\alpha - 3)G$ |
| $A_2 \to K^*\bar{K}$ | $-1$ | $A_2 \to K\bar{K}$ | $1$ | $\Sigma \to N\bar{K}$ | $-G_F + G_D = (2\alpha - 1)G$ |
| $A_2 \to \bar{K}^*K$ | $1$ | | | $\Sigma \to \Xi K$ | $G_F + G_D = G$ |
| $K_N \to K^*\eta$ | $-3^{1/2}$ | $K_N \to K\eta_8$ | $3^{-1/2}$ | $N \to N\eta_8$ | $-3^{1/2}G_F + 3^{-1/2}G_D = 3^{-1/2}(4\alpha - 3)G$ |
| $K_N \to \phi_8 K$ | $3^{1/2}$ | | | $N \to \Lambda K$ | $3^{1/2}G_F + 3^{-1/2}G_D = 3^{-1/2}(3 - 2\alpha)G$ |
| $A_2 \to \rho\pi$ | $2$ | | | $\Sigma \to \Sigma\pi$ | $2G_F = 2(1-\alpha)G$ |
| | | $f_8 \to \pi\pi$ | $2 \cdot 3^{-1/2}$ | $\Lambda \to \Sigma\pi$ | $2 \cdot 3^{-1/2}G_D = 2 \cdot 3^{-1/2}\alpha G$ |
| | | $A_2 \to \pi\eta_8$ | $-2 \cdot 3^{-1/2}$ | $\Sigma \to \Sigma\eta_8$ | $-2 \cdot 3^{-1/2}G_D$ |
| | | | | $\Sigma \to \Lambda\pi$ | |
| | | $f_8 \to \eta_8\eta_8$ | $2 \cdot 3^{-1/2}$ | $\Lambda \to \Lambda\eta_8$ | $2 \cdot 3^{-1/2}G_D$ |

sign under permutation. Thus, if we define in analogy with (9.5), (9.6), (10.23), (10.26):

$$\begin{aligned} G_{KN\rho K} &\equiv G(K_N^+ \to \rho^0 K^+) = -2^{-\frac{1}{2}} G(K_N^+ \to \rho^+ K^0) \\ &= -G(K_N^0 \to \rho^0 K^+) = 2^{-\frac{1}{2}} G(K_N^0 \to \rho^- K^+), \\ G_{f8\bar{K}^*K} &\equiv G(f_8 \to K^{*-} K^+) = -G(f_8 - \bar{K}^{*0} K^0), \qquad (12.7) \\ G_{A\bar{K}^*K} &\equiv G(A_2^0 \to K^{*-} K^+) = G(A_2^0 \to \bar{K}^{*0} K^0) \\ &= 2^{-\frac{1}{2}} G(A_2^+ \to \bar{K}^{*0} K^+) = 2^{-\frac{1}{2}} G(A_2^- \to K^{*-} K^0), \\ G_{KN\phi 8K} &\equiv G(K_N^+ \to \phi_8 K^+) = G(K_N^0 \to \phi_8 K^0), \qquad (12.8) \end{aligned}$$

these coupling constants differ from those of $K_N \to K^*\pi$, $f_8 \to K^*\bar{K}$, $A_2 \to K^*\bar{K}$ and $K_N \to K^*\eta$ just by a minus sign.

The $SU_3$-properties of $2^+ \to 0^-0^-$ decays are different because the final state is now symmetric under exchange of the two mesons. We consider again the $U$-spin multiplets of section 4-10 and replace $K^* \to K_N$, $\phi_8 \to f_8$, $\rho \to A_2$. Calling the $K_N \to K\pi$ decay constant $G_D$ for later convenience, we get from (10.20) and (10.21)

$$G_D \equiv 2^{-\frac{1}{2}} G(K_N^0 \to K^+\pi^-) = \tfrac{1}{2}3^{\frac{1}{2}} G(f_8 \to \pi^+\pi^-) \qquad (12.9)$$

$$= \tfrac{1}{2}3^{\frac{1}{2}} G(f_8 \to K^+K^-) + \tfrac{1}{2} G(A_2^0 \to K^+K^-). \qquad (12.10)$$

Here we have already used $G(A_2 \to \pi\pi) = 0$, due to $G$-parity. From (9), we obtain a relation for $G_{f\pi\pi}$ (defined in (3)) which has no analogue in $1^- \to 0^-0^-$ decays. This has to do with the fact that (9) contains only $8_F$, while $2^+ \to 0^-0^-$ decays contain only $8_D$.

Also the determination of $G(A_2^0 \to K^+K^-)$ goes somewhat differently. In analogy with (9.6) and (10.23) we define

$$\begin{aligned} G_{f8K\bar{K}} &\equiv G(f_8 \to K^+K^-) = -G(f_8 \to K^0\bar{K}^0), \qquad (12.11) \\ G_{AK\bar{K}} &\equiv G(A_2^0 \to K^+K^-) = G(A_2^0 \to K^0\bar{K}^0) \\ &= 2^{-\frac{1}{2}} G(A_2^+ \to K^+\bar{K}^0) = 2^{-\frac{1}{2}} G(A_2^- \to K^0K^-). \end{aligned}$$

This time it is the $U$-spin triplet state $\frac{1}{2}(3^{\frac{1}{2}} f_8 + A_2^0)$ which cannot decay into $K^0\bar{K}^0$, such that (10.24) is replaced by

$$-3^{\frac{1}{2}} G_{f8K\bar{K}} + G_{AK\bar{K}} = 0, \qquad G_{f1K\bar{K}} = -G_{f1\pi\pi} \equiv G_1 \neq 0. \qquad (12.12)$$

Thus the $SU_3$-singlet state $f_1$ of (4) gets its own coupling constant. From (10) and (13), we obtain not only $G_{AK\bar{K}}$ and $G_{f8K\bar{K}}$ as listed in table 4-12, but also the decay constants of the physical states (4):

$$\begin{aligned} G_{f'K\bar{K}} &= 3^{-\frac{1}{2}} \cos\theta_t G_D - \sin\theta_t G_1, \\ G_{fK\bar{K}} &= 3^{-\frac{1}{2}} \sin\theta_t G_D + \cos\theta_t G_1 \end{aligned} \qquad (12.13)$$

$$\begin{aligned} G_{f'\pi\pi} &= 2 \cdot 3^{-\frac{1}{2}} \cos\theta_t G_D + \sin\theta_t G_1 \\ G_{f\pi\pi} &= 2 \cdot 3^{-\frac{1}{2}} \sin\theta_t G_D - \cos\theta_t G_1 \end{aligned} \qquad (12.14)$$

Equation (10.27) which previously allowed us the determination of $G_{K^*K\eta 8}$ now reads

$$G(K_N^+ \to K^+\eta_8) - 3^{\frac{1}{2}}G(K_N^+ \to K^+\pi^0) = G(A_2 \to \pi\eta), \qquad (12.15)$$

where we have again used $G_{A\pi\pi} = 0$. This time, we need one more equation to determine both $G_{KNK\eta 8}$ and $G_{A\pi\eta}$: We take all 3 particles in $U$-spin triplets and the properties $C^{111}_{101} = C^{111}_{1\text{-}10}$, $C^{111}_{000} = 0$ of the Clebsch-Gordans for $1 \to 1 + 1$ give us

$$3^{\frac{1}{2}}G(K_N^0 \to K^0\eta_8) + G(K_N^0 \to K^0\pi^0) = 3^{\frac{1}{2}}G(f_8 \to K^0\bar{K}^0) + G(A_2^0 \to K^0\bar{K}^0), \qquad (12.16)$$

$$G(3^{\frac{1}{2}}f_8 + A_2^0) \to (3^{\frac{1}{2}}\eta_8 + \pi^0)(3^{\frac{1}{2}}\eta_8 + \pi^0) = 0. \qquad (12.17)$$

(16) says $3^{\frac{1}{2}}G_{KNK\eta 8} = G_D$, which inserted into (15) gives $G_{A\pi\eta} = -2 \cdot 3^{-\frac{1}{2}}G_D$. Finally, since both $A_2 \to \eta_8\eta_8$ and $A_2 \to \pi^0\pi^0$ vanish, (17) gives $G_{f8\eta 8\eta 8} - G_{f8\pi\pi} + 2G_{A\pi\eta} = 0$.

$SU_3$-symmetry seems to work well by and large, at least on the level of coupling constants, but some questions remain: (i) why do vector and tensor mesons have so large mixing angles, and (ii) why do $\phi$ and $f'$ decay dominantly into $K\bar{K}$? These questions are related to the small decay widths of the new bosons, which we shall discuss later. They are answered by the "naive", nonrelativistic quark model, which says that not only are mesons $q\bar{q}$ bound states and baryons $qqq$ bound states, but also are they described by wave functions of nonrelativistic motion. As for loosely bound systems, current operators are taken as the sum of the current operators of the constituent quarks. Reviews of the model have been given by Feld (1969), Kokkedee (1969), Mopurgo (1970), and Lipkin (1973). As quarks have not yet been seen, it is unlikely that they exist as free particles. A quantitative treatment in terms of nonrelativistic bound states may then be hard to justify (except possibly for the heavier "charmonium" states, see section 4-16).

The historical starting point was the observation, expressed for the pions by Fermi and Yang (1949), that all $0^-$ and $1^-$ mesons can be interpreted as $S$-wave bound states of octet baryon-antibaryon pairs $(B_8 \bar{B}_8)_s$. The $2^+$ mesons can be interpreted as the corresponding $^3P_2$ bound states. For the hyperneutral mesons, the identification is made in table 4-5.2. The "crossed version" of this statement merely says that baryons can emit and absorb mesons "singly," such that these mesons can serve as the carrier of "nuclear forces." It restricts the meson parity to $\mathscr{P} = (-1)^{J+1}$ if the neutral hyperneutral member of the multiplet has $C\mathscr{P} = -1$ ($S = 0$ according to 3-9.8), and excludes also $J^{\mathscr{P}C} = 0^{--}$ (see 3-9.4 for $L = 0$). The isospin of mesons could be as large as 2 in this model. The quark model replaces $B\bar{B}$ by $q\bar{q}$, and since $q$ has $I = \frac{1}{2}$ or zero, mesons are restricted by $I < \frac{3}{2}$ in the quark model. Moreover, as baryons are taken as $qqq$-states, they must have $I < 2$ and $Y < 2$. Other states are not allowed by the naive quark model and are called "exotic" ones (for example possible $KN$ resonances).

The next step was the introduction of approximate $SU_6$-symmetry. In nuclear physics, the energy levels of light nuclei can be classified into "Wigner supermultiplets", which are representations of $SU_4$, which arises as a combination of the isospin $SU_2$ with an assumed spin independence of nuclear forces. The basic quartet in that case is $(p_+, p_-, n_+, n_-)$, where the sign indicates the eigenvalue of $\sigma_3$. The $SU_6$-model merely replaces the states $p$, $n$ by the three quarks $u$, $d$, $s$. The $SU_6$-reduction of $q\bar{q}$-states is

$$6 \otimes \bar{6} = 1 \oplus 35. \tag{12.18}$$

Here spin is counted explicitly, i.e., the spin-triplet $q\bar{q}$ states are counted 3 times. For the $S$-wave $q\bar{q}$ states, the "1" in (18) must be the $X_1$ of (10.17). The "35" contains one octet-singlet (8 states), one octet-triplet ($8 \times 3 = 24$ states) and one singlet-triplet (3 states) in the classification corresponding to the subgroup $SU_3 \otimes SU_2$ of $SU_6$. For the $(q\bar{q})_s$-states, this makes just the $0^+$ octet, the $1^-$ octet (24 states) and the $\omega_1$ (3 states). Thus $\omega_1$ and $\phi_1$ are "$SU_6$-degenerate", but $X_1$ and $\eta_8$ are not. The $\omega - \phi$ mixing angle is now derived from the postulate that $SU_6$-symmetry is first broken in its $SU_3$-subgroup, i.e., the nonrelativistic quark model applies also to broken $SU_3$ (Zweig 1964, Iizuka 1966). The breaking is simply introduced by making the $s$-quark heavier than the $u$- and $d$-quarks. The heavier of the physical vector mesons must then be a pure $s\bar{s}$ state:

$$\phi = s\bar{s}, \qquad \omega = 2^{-\frac{1}{2}}(-u\bar{u} + d\bar{d}), \qquad \sin\theta_v = 3^{-\frac{1}{2}}. \tag{12.19}$$

The second equation follows from $\langle\omega\,|\,\phi\rangle = 0$, and the last one from (10.4) for the pair of states $(\omega_1, \phi_8)$ and (10.18). It fixes the mixing angle both in magnitude ($35.3^0$) and sign. $\theta_v = \arcsin 3^{-\frac{1}{2}}$ is called the "ideal mixing angle." The same mixing occurs for the $^3P_2$ bound states ($f - f'$-mixing):

$$f' = s\bar{s}, \qquad f = 2^{-\frac{1}{2}}(-u\bar{u} + d\bar{d}). \tag{12.20}$$

Moreover, Zweig and Iizuka postulated that in strong 2-particle decays, the quarks of the initial state must also appear as constituents of each of the two final particles (fig. 4-12). This rule forbids $\phi \to \rho\pi$, $f' \to \pi\pi$, but not $f \to K\bar{K}$, $A_2 \to K\bar{K}$.

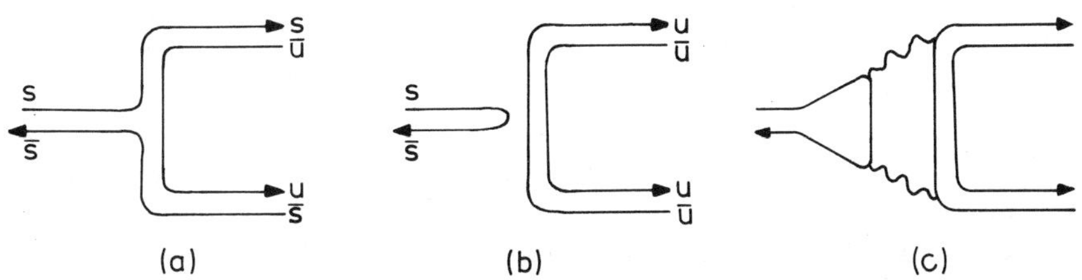


**Figure 4-12** Application of the Zweig-Iizuka rule to meson decays: the decay (a) ($\phi \to K\bar{K}$, $f' \to K\bar{K}$) is allowed, while (b) ($f' \to \pi\pi$) is forbidden. In QCD, (b) is replaced by (c) which contains 2 (3 in the case of negative $C$-parity) "hard" gluons.

For the $qqq$ states we have

$$6 \otimes 6 \otimes 6 = 20 \oplus 56 \oplus 2 \cdot 70. \tag{12.21}$$

The $SU_3 \otimes SU_2$-content of 56 is $8 \otimes 2$ (spin-$\frac{1}{2}$) and $10 \otimes 4$ (spin-$\frac{3}{2}$). Thus the baryon octet and decuplet are accommodated in one single representation of $SU_6$, and the decuplet decay constants can be related to the meson-baryon coupling constants which will be discussed later. Also for such cases, the Zweig-Iizuka rule appears to be fulfilled.

Even if the quark-quark interaction should contain an appreciable spin-orbit coupling, the $SU_6$-classification may still be good for the $S$-states, where this coupling vanishes. However, since the 3-quark bound states which we have discussed so far are totally symmetric in spin-isospin space (which is most easily seen for the $\Delta^{++}$ which consists of 3 $u$-quarks with parallel spins), the generalized Pauli principle would require totally antisymmetric orbital wave functions if quarks were identical particles. This is a very unlikely configuration for a ground state. Therefore, Greenberg (1964) postulated that quarks have an additional "hidden" quantum number, which is nowadays called "color" (see the reviews by Greenberg and Nelson 1977, Greenberg 1978). Thus quarks are supposed to exist in three different colors, and all physical bound states are singlets in color space. The forces between quarks are mediated by vector bosons called "gluons," which form a color octet. (color transformations form once more the group $SU(3)$, which is supposed to be an exact symmetry group). These speculations are incorporated in nonabelian gauge field theory called "quantum chromodynamics" (QCD), alluding to the highly respectable QED. The interested reader is referred to recent reviews (Marciano and Pagels 1978, Appelquist et al. 1978). An example of nonabelian gauge transformations will be discussed in section 5-10. In QCD, the interaction between quarks becomes quite weak at short distances ("asymptotic freedom"), and the Zweig-Iizuka rule follows from the suppression of $q\bar{q}$ annihilation into gluons (fig. 4-12.c). Note in this context that states of negative charge conjugation cannot annihilate into one gluon because of color conservation.

## 4-13 Decay Angular Distributions

In the past, angular distributions of decays $d \to 12$ in reactions such as $ab \to cd$ have been used to determine the spin of particle $d$ (in the case of $S = \frac{1}{2}$ baryons also the $s/p$ ratio, and for $S > \frac{1}{2}$ baryons the parity). Today, they are used to measure the density matrices which characterize the production process.

The matrix $A$ which determines the decay angular distribution (8.20) has been given in (8.18). For decays into two spinless particles, we have $\lambda_1 = \lambda_2 = 0$, and the expression simplifies to

$$\begin{aligned} A^S_{MM'}(\vartheta, \varphi) &= (4\pi)^{-1}(2S+1)e^{i\varphi(M'-M)}\, d^S_{M0}(\vartheta)\, d^S_{M'0}(\vartheta) \\ &= Y^{S*}_M(\vartheta, \varphi) Y^S_{M'}(\vartheta, \varphi). \end{aligned} \tag{13.1}$$

For parity-conserving decays into particles of arbitrary spin, we find from (4.24) and the symmetry $d_{M\lambda} = (-1)^{M-\lambda}\, d\lambda M$

$$A^S_{-M,\,-M'} = (-1)^{M-M'} A^{S*}_{MM'}. \tag{13.2}$$

Combining this with $A = A^+$, we get for $S = 1$:

$$A^1 = \begin{pmatrix} A_{11} & A_{10} & A_{1-1} \\ A^*_{10} & A_{00} & -A_{10} \\ A^*_{1-1} & -A^*_{10} & A_{11} \end{pmatrix} \qquad \begin{aligned} A_{11} &= \tfrac{3}{8}\pi^{-1} \sin^2 \vartheta, \\ A_{00} &= \tfrac{3}{4}\pi^{-1} \cos^2 \vartheta, \\ A_{10} &= -\tfrac{3}{4}\pi^{-1} 2^{-\frac{1}{2}} \sin \vartheta \cos \vartheta e^{-i\varphi}, \\ A_{1-1} &= -\tfrac{3}{8}\pi^{-1} \sin^2 \vartheta e^{-2i\varphi}. \end{aligned} \tag{13.3}$$

The explicit expressions follow from table B-4. The decay distribution trace $(\rho A)$ is, observing $\rho = \rho^+$:

$$\begin{aligned} W(\vartheta, \varphi) &= A_{00}\rho_{00} + A_{11}(\rho_{11} + \rho_{-1-1}) + 2 \operatorname{Re} (A_{1-1}\rho^*_{1-1}) \\ &\quad + 2 \operatorname{Re} (A_{10}\rho^*_{10}) - 2 \operatorname{Re} (A^*_{10}\rho_{0-1}) \\ &= \tfrac{1}{4}\pi^{-1}(1 + a_1), \\ a_1 &= \tfrac{1}{2}(3\rho_{00} - 1)(3 \cos^2 \vartheta - 1) - 3 \sin^2 \vartheta \operatorname{Re} (e^{-2i\varphi}\rho^*_{1-1}) \\ &\quad + 3 \cdot 2^{\frac{1}{2}} \sin \vartheta \cos \vartheta \operatorname{Re} (\rho_{0-1} e^{i\varphi} - \rho^*_{10} e^{-i\varphi}) \end{aligned} \tag{13.4}$$

In the last expression, we have eliminated $\rho_{11} + \rho_{-1-1}$ by the trace condition, $\rho_{11} + \rho_{-1-1} + \rho_{00} = 1$. (4) can be further simplified if the decaying particle $d$ is produced in a (parity-conserving) collision of unpolarized particles $ab \to cd$. (4.22) remains true for all spin states that are quantized along directions in the production plane, and we get

$$T(M_a \cdots M_d, \theta) = \eta\eta'(-1)^{S_a - M_a \cdots S_d - M_d} T(-M_a, \ldots -M_d, \theta). \tag{13.5}$$

Insertion of this symmetry into (8.17) gives

$$\rho_{-M,\,-M'} = (-1)^{M-M'} \rho_{M,\,M'}. \tag{13.6}$$

(The simplest way to see this is to write the sign factor of $T^*(M_a \cdots M_d, \theta)$ as the inverse of that of (5), with $M_d$ replaced by $M'_d$). This gives us

$$\rho^*_{1-1} = \rho^*_{-11} = \rho_{1-1}, \qquad \rho_{0-1} = -\rho_{01} = -\rho^*_{10}, \tag{13.7}$$

i.e., $\rho_{1-1}$ is real and (4) becomes

$$\begin{aligned} a_1 &= \tfrac{1}{2}(3\rho_{00} - 1)(3 \cos^2 \vartheta - 1) - 3\rho_{1-1} \sin^2 \vartheta \cos 2\varphi \\ &\quad - 3 \cdot 2^{\frac{1}{2}} \operatorname{Re} \rho_{10} \sin 2\vartheta \cos \varphi. \end{aligned} \tag{13.8}$$

These formulas apply only in the absence of other partial waves, of course. In the reactions $\pi^- p \to \rho^0 n$ and $K^- p \to K^{*0} n$, for example, the two mesons are partly in the $S$-wave, in which case two more terms, $2 \cdot 3^{\frac{1}{2}} \operatorname{Re} \rho_{0S} \cos \vartheta - 2 \cdot 6^{\frac{1}{2}} \operatorname{Re} \rho_{1S} \sin \vartheta \cos \varphi$ must be added to (8), where $\rho_{MS}$ refers to the combination $T(ab \to c\rho_M)T^*(ab \to c(\pi\pi)_S)$ (Hyams et al., 1968).

Tensor meson decay distributions are already lengthy (Pilkuhn 1972,

$2^+ \to 0^-0^-$ on p. 10, and $2^+ \to 1^-0^-$ on p. 14). The matrix element for $\omega \to \pi^+\pi^0\pi^-$ decays (9.27) is obtained from that of $\omega \to \pi^+\pi^-$ (9.3) essentially by the replacement $\mathbf{p}_1 \to \mathbf{n}$. Therefore, the angular distributions of $\omega \to \pi\pi\pi$ or $\phi \to \rho\pi \to \pi\pi\pi$ decays are again given by (4), if $\vartheta$ and $\varphi$ are interpreted as the angles of the normal to the decay plane.

Next we discuss the decays $1^- \to 1^-0^-$, such as $\omega \to \gamma\pi^0$ or $\phi \to \rho\pi$. According to table 4-9, the matrix element $T_1(\lambda_1)$ is proportional to $\lambda_1$, i.e., $|T_1(1)|^2 = |T_1(-1)|^2$, $T_1(0) = 0$, and (8.18) gives

$$A^1_{MM'} = \tfrac{3}{8}\pi^{-1} e^{i\varphi(M'-M)}(d^1_{M1}\, d^1_{M'1} + d^1_{M-1}\, d^1_{M'-1}). \tag{13.9}$$

The matrix form of $A$ is as in (3), of course, only the expressions for the elements of the first row are different:

$$A_{11} = \tfrac{3}{16}\pi^{-1}(1 + \cos^2 \vartheta), \qquad A_{00} = \tfrac{3}{8}\pi^{-1} \sin^2 \vartheta, \tag{13.10}$$

$$A_{10} = \tfrac{3}{8}\pi^{-1} 2^{-\frac{1}{2}} \sin \vartheta \cos \vartheta e^{-i\varphi}, \qquad A_{1-1} = \tfrac{3}{16}\pi^{-1} \sin^2 \vartheta e^{-2i\varphi}. \tag{13.11}$$

$A_{10}$ and $A_{1-1}$ are just $-\frac{1}{2}$ of the corresponding expressions in (3), and if one writes $\cos^2 = 1 - \sin^2$ in $A_{11}$ and $\sin^2 = 1 - \cos^2$ in $A_{00}$, one sees that the $\vartheta$-dependent parts are also here $-\frac{1}{2}$ of the corresponding expressions in (3). One thus obtains an angular distribution

$$W(\vartheta, \varphi) = \tfrac{1}{4}\pi^{-1}(1 - \tfrac{1}{2}a_1), \tag{13.12}$$

i.e., the anisotropic part is $-\frac{1}{2}$ of that of $1^- \to 0^-0^-$ decays.

For $1^- \to \frac{1}{2}\frac{1}{2}$ decays, $A$ follows from the last row of table 4-9 as

$$\begin{aligned} A^1_{MM'} &= \tfrac{3}{4}\pi^{-1}(2m_1^2 + s)^{-1} e^{i\varphi(M'-M)} \\ &\quad \times [2m_1^2\, d^1_{M0}\, d^1_{M'0} + \tfrac{1}{2}s(d^1_{M-1}\, d^1_{M'-1} + d^1_{M1}\, d^1_{M'1})]. \end{aligned} \tag{13.13}$$

The resulting decay distribution is

$$W(\vartheta, \varphi) = \tfrac{1}{4}\pi^{-1}[1 - a_1(s - 4m_1^2)(s + 2m_1^2)^{-1}] \tag{13.14}$$

It is isotropic at the threshold $s = 4m_1^2$, as expected.

Next, we consider pionic baryon decays. (8.18) reads here

$$\begin{aligned} A^S_{MM'} &= \tfrac{1}{4}\pi^{-1}(2S + 1) e^{i\varphi(M'-M)} \sum_{\lambda_1} |T_S(\lambda_1)|^2\, d^S_{M\lambda_1}\, d^S_{M'\lambda_1} \\ &\quad \times [|T_S(\tfrac{1}{2})|^2 + |T_S(-\tfrac{1}{2})|^2]^{-1} \end{aligned} \tag{13.15}$$

We introduce the asymmetry parameter

$$\alpha = (|T_S(\tfrac{1}{2})|^2 - |T_S(-\tfrac{1}{2})|^2)(|T_S(\tfrac{1}{2})|^2 + |T_S(-\tfrac{1}{2})|^2)^{-1}, \tag{13.16}$$

in terms of which (15) becomes

$$\begin{aligned} A^S_{MM'} &= \tfrac{1}{8}\pi^{-1}(2S + 1) e^{i\varphi(M'-M)} \\ &\quad \times [d^S_{M\frac{1}{2}}\, d^S_{M'\frac{1}{2}} + d^S_{M-\frac{1}{2}}\, d^S_{M'-\frac{1}{2}} + \alpha(d^S_{M\frac{1}{2}}\, d^S_{M'\frac{1}{2}} - d^S_{M-\frac{1}{2}}\, d^S_{M'-\frac{1}{2}})] \end{aligned} \tag{13.17}$$

From table 4-11.1 we find

$$\begin{aligned} \alpha(S = \tfrac{1}{2}) &= 2 \operatorname{Re} (T^{s*} T^p)(|T^s|^2 + |T^p|^2)^{-1}, \\ \alpha(S = \tfrac{3}{2}) &= 2 \operatorname{Re} (T^{d*} T^{p3})(|T^d|^2 + |T^{p3}|^2)^{-1}. \end{aligned} \tag{13.18}$$

Experimental values of $\alpha$ for hyperon decays have been included in table 4-11.3. Table B-4 gives

$$A^{\frac{1}{2}}_{MM} = \tfrac{1}{4}\pi^{-1}(1 + 2M\alpha \cos \vartheta),$$
$$A^{\frac{1}{2}}_{M-M} = \tfrac{1}{4}\alpha \sin \vartheta\, e^{-2iM\varphi}, \tag{13.19}$$

$$W(\vartheta, \varphi) = \tfrac{1}{4}\pi^{-1}[1 + \alpha \cos \vartheta(\rho_{\frac{1}{2}\frac{1}{2}} - \rho_{-\frac{1}{2}-\frac{1}{2}})]$$
$$+ 2\alpha \sin \vartheta \operatorname{Re}(e^{i\varphi}\rho_{\frac{1}{2}-\frac{1}{2}}) \tag{13.20}$$
$$= \tfrac{1}{4}\pi^{-1}(1 + \alpha \mathbf{P} \cdot \hat{\mathbf{p}}_1). \tag{13.21}$$

In the last expression, we have used (2-3.8) for $\rho$, and $\hat{\mathbf{p}}_1$ is the unit vector along the momentum of the final baryon. In parity-conserving reactions $ab \to cd$, we have $P_z = P_x = 0$, $\rho_{\frac{1}{2}-\frac{1}{2}} = -\frac{1}{2}iP_y$ according to (6), and (20) becomes

$$W(\vartheta, \varphi) = \tfrac{1}{4}\pi^{-1}(1 + \alpha \sin \vartheta \sin \varphi P_y). \tag{13.22}$$

For nucleon-pion scattering, $P_y$ was already given in (1-7.30). For $S = \frac{3}{2}$, we find from table B-4:

$$A_{\pm\frac{3}{2},\pm\frac{3}{2}} = \tfrac{3}{8}\pi^{-1} \sin^2 \vartheta(1 \pm \alpha \cos \vartheta),$$
$$A_{\pm\frac{1}{2},\pm\frac{1}{2}} = \tfrac{1}{8}\pi^{-1}[1 + 3\cos^2 \vartheta \pm \alpha \cos \vartheta(9 \cos^2 \vartheta - 5)],$$
$$A_{\pm\frac{3}{2},\pm\frac{1}{2}} = \tfrac{1}{8}\pi^{-1}e^{\mp i\varphi}3^{\frac{1}{2}} \sin \vartheta[\mp 2 \cos \vartheta + \alpha(1 - 3\cos^2 \vartheta)],$$
$$A_{\pm\frac{3}{2},\mp\frac{1}{2}} = -\tfrac{1}{8}\pi^{-1}e^{\mp 2i\varphi}3^{\frac{1}{2}} \sin^2 \vartheta(1 \pm 3\alpha \cos \vartheta), \tag{13.23}$$
$$A_{\frac{3}{2}-\frac{3}{2}} = A^*_{-\frac{3}{2}\frac{3}{2}} = -\tfrac{3}{16}\pi^{-1}\alpha e^{-3i\varphi} \sin^3 \vartheta,$$
$$A_{\frac{1}{2}-\frac{1}{2}} = A^*_{-\frac{1}{2}\frac{1}{2}} = \tfrac{1}{8}\pi^{-1}\alpha e^{-i\varphi} \sin \vartheta(9\cos^2 \vartheta - 1).$$

Here we merely give the distribution for the strong decays of the $\frac{3}{2}^+$ states, where $\alpha = 0$, and we also assume (6):

$$W = \tfrac{3}{4}\pi^{-1}[\tfrac{1}{3} - \cos^2 \vartheta(\rho_{\frac{3}{2}\frac{3}{2}} - \tfrac{1}{3}\rho_{\frac{1}{2}\frac{1}{2}})$$
$$- 2 \cdot 3^{-\frac{1}{2}}(\sin^2 \vartheta \cos 2\varphi \operatorname{Re} \rho_{\frac{3}{2}-\frac{1}{2}} + \sin 2\vartheta \cos \varphi \operatorname{Re} \rho_{\frac{3}{2}\frac{1}{2}})]. \tag{13.24}$$

One can also derive "joint decay distributions" for the simultaneous decays of particles $c$ and $d$ (see Pilkuhn 1967), or decay distributions for the case that one incident particle is polarized.

## 4-14 $SU_2$- and $SU_3$-Properties of the Electromagnetic Current

According to (3-3.17), the electromagnetic current operator is composed additively of a leptonic part and a hadronic part (to lowest order in the electromagnetic interaction):

$$j^\mu(x) = j^\mu(e, x) + j^\mu(\mu, x) + j^\mu(\text{hadrons}, x). \tag{14.1}$$

From the definition of $j^0$, from (3.6) and (10.5) we have

$$\int d^3xj^0(\text{hadrons}, t, \mathbf{x}) = Q(\text{hadrons}) = I_3 + \tfrac{1}{2}Y = \tfrac{1}{2}(F_3 + 3^{-\frac{1}{2}}F_8). \quad (14.2)$$

We now assume that $j^\mu$ (hadrons) itself transforms as in (2), i.e., under $SU_2$ it transforms like a combination of neutral isovector and isoscalar operators, and under $SU_3$ it transforms like the $U$-spin scalar component of an octet operator. Using the Wigner-Eckart theorem of B-3, this means that the $SU_3$-properties of matrix elements for the emission and absorption of single photons by hadrons are identical to those of the $U$-spin scalar combination $\frac{1}{2}(\phi_8 - 3^{\frac{1}{2}}\rho^0)$.

At the $SU_2$-level, there are not many tests of this assumption. In the decays $\Delta^+ \to p\gamma$ and $\Delta^0 \to n\gamma$, the isoscalar photon contribution vanishes, and the Clebsch-Gordans of the isovector part are identical with those of $\pi^0$ in the first row of table 4-2.2, i.e.,

$$\Gamma(\Delta^+ \to p\gamma) = \Gamma(\Delta^0 \to n\gamma). \quad (14.3)$$

This relation can be tested in pion photoproduction $\gamma N \to \Delta \to \pi N$, but the test gets complicated by non-resonant amplitudes (section 7-4). More generally, pion photoproduction is described by 3 isospin amplitudes; $T^S$ refers to isoscalar absorption of the photon and populates only the $I = \frac{1}{2}N\pi$ state, $T^{v1}$ and $T^{v3}$ refer to isovector photon absorption and populate the $I = \frac{1}{2}$ and $\frac{3}{2}N\pi$ states, respectively. Their contributions to charged pion photoproduction follow from the second row of table 4-2.2, and from (2.11):

$$T\left(\gamma\begin{pmatrix}p\\n\end{pmatrix} \to \pi^\pm\begin{pmatrix}n\\p\end{pmatrix}\right) = \tfrac{1}{3}2^{\frac{1}{2}}(T^{v3} - T^{v1} \mp 3^{\frac{1}{2}}T^0). \quad (14.4)$$

The isovector $\pi^0$ photoproduction amplitude has the same total isospin decomposition as $\pi^0 N \to \pi^0 N$ which is found from (2.9), say. Adding the isoscalar amplitude, we find

$$T\left(\gamma\begin{pmatrix}p\\n\end{pmatrix} \to \pi^0\begin{pmatrix}p\\n\end{pmatrix}\right) = \tfrac{2}{3}T^{v3} + \tfrac{1}{3}T^{v1} \pm 3^{\frac{1}{2}}T^s. \quad (14.5)$$

When the amplitudes for all $4N\pi$ charge combinations are known, the assumed absence of an isotensor component in the photon couplings (which would affect the $I = \frac{3}{2}$-$N\pi$ states) can be tested.

Consider next the matrix elements of $j^\mu(x)$ between single spin-$\frac{1}{2}$ baryon states, which are given by (2-7.4), (2-7.11) or (2-7.13). For protons and neutrons, the isoscalar ($s$) and isovector ($v$) contributions to the form factors are

$$F_i\begin{pmatrix}p\\n\end{pmatrix} = F_i^s \pm F_i^v, \qquad F_1^s(0) = F_1^v(0) = \tfrac{1}{2},$$
$$F_2^s(0) = -0.061, \qquad F_2^v(0) = 1.853 \equiv \kappa^v. \quad (14.6)$$

Here a possible isotensor current component cannot couple because initial and final nucleons have only isospin $\frac{1}{2}$. The 6 form factors $F_i(\Sigma^Z)$ for $\Sigma$-hyperons of charge $Z = \pm 1, 0$, on the other hand, are expressible in terms of their isoscalar and isovector components as follows (compare the last line of table (4-2.3)

$$F_i(\Sigma^Z) = F_i^s(\Sigma) + ZF_i^v(\Sigma), \qquad F_1^s(\Sigma, 0) = 0,\ F_1^v(\Sigma, 0) = 1. \tag{14.7}$$

Unfortunately, the absence of an isotensor current cannot be tested here because nothing is known about $F_i(\Sigma^0)$. The decays $\Sigma^* \to \Sigma\gamma$ are also $SU_2$-related but have hardly been seen.

The $SU_3$-structure of the current operator can be tested in several places. Consider first the magnetic moments $\mu$ of the baryons, which are defined in (1-4.5). Their $SU(3)$-comparison is complicated by the factor $(2m)^{-1}$, which varies by 40% between nucleons and $\Xi$. It arises naturally in the "Dirac part" $(2m)^{-1}Ze$ of (1-4.5), but in the "anomalous part" $(2m)^{-1}\kappa e$ it has no dynamical justification. Therefore, this factor should be kept fixed at $m = m_p$ in $SU_3$-comparisons. The quantity with simple $SU_3$-relations is thus

$$\tilde{\mu} = \mu - Ze/2m_i, \qquad \tilde{\mu}[e/2m_p] = \mu[e/2m_p] - Zm_p/m_i. \tag{14.8}$$

The second relation is given because magnetic moments are frequently quoted in units of $e/2m_p$ ("nuclear magnetons").

Experimental values and $SU_3$-predictions for $\tilde{\mu}$ are collected in table 4-14. If $j^\mu$ is a $U$-spin scalar, baryons within the same $U$-spin multiplets (see fig. 4-10) must have identical form factors:

$$F(\Sigma^+) = F(p), \qquad F(\Xi^-) = F(\Sigma^-), \tag{14.9}$$

$$F(\Xi^0) = F(n) = \tfrac{1}{4}[3F(\Lambda) + 2 \cdot 3^{\frac{1}{2}}F(\Lambda\Sigma) + F(\Sigma^0)], \tag{14.10}$$

$$0 = \tfrac{1}{4}[3^{\frac{1}{2}}F(\Lambda) - 2F(\Lambda\Sigma) - 3^{\frac{1}{2}}F(\Sigma^0)]. \tag{14.11}$$

The last equation comes from the requirement that matrix elements of $j^\mu$ between the $U$-spin singlet and the neutral component of the $U$-spin triplet vanish. Another relation follows from the fact that $\Sigma^0 \to \Sigma^0\phi_8$ and $\Sigma^0 \to \Lambda\rho^0$ have identical coupling constants because their final states differ only in the order of the isovector and isoscalar particles (compare also table 4-12). Since $\Sigma^0 \to \Sigma^0\rho^0$ and $\Sigma^0 \to \Lambda\phi_8$ are both zero by isospin conservation, $\Sigma^0 \to \Sigma^0(\phi_8 - 3^{\frac{1}{2}}\rho^0)$ and $\Sigma^0 \to \Lambda(\rho^0 - 3^{-\frac{1}{2}}\phi_8)$ have also identical coupling constants, i.e.

$$F(\Lambda\Sigma) = -3^{\frac{1}{2}}F(\Sigma^0). \tag{14.12}$$

Inserting this into (11), we see that $F(\Lambda) = -F(\Sigma^0)$, (10) gives now $F(n) = -2F(\Sigma^0)$, and $F(\Sigma^-)$ follows from (7) and the first equation of (9) as

$$F(\Sigma^-) = 2F(\Sigma^0) - F(\Sigma^+) = -F(n) - F(p) = -2F^s. \tag{14.13}$$

Replacing now $F \to \tilde{\mu}$, we have derived all $SU_3$-predictions of table 4-14. Their experimental tests are again of low accuracy.

**Table 4-14** Experimental values and $SU_3$-predictions for $\tilde{\mu} = \mu - Ze/2m_i$. $\mu_{\Lambda\Sigma}$ is the constant for $\Sigma^0 \to \Lambda\gamma$ decay, with $m = \frac{1}{2}(m_\Sigma + m_\Lambda)$ according to (2-7.14). $\mu$ from table C-4.

| | $\tilde{\mu} = \mu - Ze/2m$ | $[e/2m_p]$ |
|---|---|---|
| Baryon | exp | $SU(3)$ |
| $p$ | 1.792846 | $2\tilde{\mu}^s = \tilde{\mu}_p + \tilde{\mu}_n = -0.1203$ |
| $n$ | $-1.91315$ | |
| $\Lambda$ | $-0.614 \pm 0.005$ | $\frac{1}{2}\mu_n = -0.957$ |
| $\Sigma^+$ | $1.83 \pm 0.40$ | $\tilde{\mu}_p = 1.79$ |
| $\Sigma^0$ | — | $-\frac{1}{2}\mu_n$ |
| $\Sigma^-$ | $0.34 \pm 0.37$ | $-2\tilde{\mu}^s$ |
| $(\Lambda\Sigma^0)$ | $\pm 1.90 \pm 0.26$ | $\frac{1}{2}3^{\frac{1}{2}}\mu_n = -1.66$ |
| $\Xi^0$ | — | $\mu_n$ |
| $\Xi^-$ | $-0.94 \pm 0.75$ | $-2\tilde{\mu}^s$ |

The $SU_3$-prediction for $\mu_{\Lambda\Sigma}$ can be converted into a prediction for the $\Sigma^0$ decay rate, using (2-7.14). The term proportional to $F_3$ vanishes for real photons, and the remaining term is conveniently rewritten by (2-6.29), where the term $(P_i + P_f)^\mu$ can be omitted in the cms ($= \Sigma$ rest frame):

$$\begin{aligned}\langle p\lambda_1\lambda_\gamma\theta\phi|T|M\rangle &= e\kappa_{\Lambda\Sigma}\bar{u}_1(\lambda_1)\varepsilon_\mu^*(\lambda_\gamma)\gamma^\mu u_\Sigma(M)\\ &= e\kappa_{\Lambda\Sigma}k_-\chi_0^{\frac{1}{2}+}(\lambda_1)(\hat{\mathbf{p}}_\Lambda\cdot\boldsymbol{\varepsilon}^* - i\boldsymbol{\sigma}(\hat{\mathbf{p}}_\Lambda\times\boldsymbol{\varepsilon}^*))\chi_0(M)\\ &= -e\kappa_{\Lambda\Sigma}k_-\lambda_\gamma\chi_0^{\frac{1}{2}+}(\lambda_1)\boldsymbol{\sigma}\cdot\boldsymbol{\varepsilon}^*\chi_0(M).\end{aligned} \tag{14.14}$$

(A-3.11) has been used to derive the second row, ($k_- = m_\Sigma - m_\Lambda$ from A-3.8), and $\hat{\mathbf{p}}_\Lambda \cdot \boldsymbol{\varepsilon}^* = 0$, $\hat{\mathbf{p}}_\Lambda \times \boldsymbol{\varepsilon}^* = -\hat{\mathbf{k}} \times \boldsymbol{\varepsilon}^* = -i\lambda_\gamma\boldsymbol{\varepsilon}^*$ (3-7.13) has been used in the last expression in (14). We recall from (4.7) that the angular dependence of (14) is $\exp\{i\phi(M-\lambda)\}\, d^{\frac{1}{2}}_{M\lambda}(\theta)$, with $\lambda = \lambda_1 - \lambda_\gamma$. Since $\lambda$ is restricted to $\pm\frac{1}{2}$, the matrix element must vanish unless $\lambda_1$ and $\lambda_\gamma$ have the same sign. It is therefore sufficient to evaluate (14) for $\phi = \theta = 0$ and $\lambda_\gamma = 2\lambda_1$. However, if the photon is taken as the second particle in the 2-particle helicity state (4.2), one must replace $\lambda_\gamma$ by $-\lambda_\gamma$ as in (7.7) if one quantizes all spins along the same axis. We thus find for $T_{\frac{1}{2}}(\lambda_1, \lambda_\gamma)$ and $m\Gamma$ of (4.7) and (4.8)

$$\begin{aligned}T_{\frac{1}{2}} &= -e\kappa_{\Lambda\Sigma}(m_\Sigma - m_\Lambda)\,\delta_{\lambda_\gamma,\,2\lambda_1},\\ \Gamma(\Sigma^0\to\Lambda\gamma) &= \alpha\kappa_{\Lambda\Sigma}^2 m_\Sigma^{-2}p(m_\Sigma - m_\Lambda)^2\end{aligned} \tag{14.15}$$

We now turn to the matrix elements of current operators for mesons, starting with the decays $\pi^0 \to \gamma\gamma$ and $\eta \to \gamma\gamma$. From the assumption that photons are $U$-spin scalars, it follows that these decays proceed via the $U = 0$ components of $\pi^0$ and $\eta$ only. The $U = 1$ combination $3^{\frac{1}{2}}\eta_8 + \pi^0$ cannot decay into $\gamma\gamma$, and consequently

$$3^{\frac{1}{2}}g_{\eta 8\gamma\gamma} + g_{\pi\gamma\gamma} = 0. \tag{14.16}$$

Using now (7.10) and the physical $\eta$ state (10.17), we get

$$\Gamma(\eta \to \gamma\gamma)/\Gamma(\pi^0 \to \gamma\gamma) = m_\pi^{-3} m_\eta^3 (3^{-\frac{1}{2}} \cos\theta_p + g_{X1\gamma\gamma}/g_{\pi\gamma\gamma} \sin\theta_p)^2. \quad (14.17)$$

Unfortunately, $\Gamma(X)$ is not known. With $g_{X1\gamma\gamma} = 0$ and $\theta_p = 10.4^0$, (17) gives 21.7, which is only half the experimental ratio of table C-2. To get agreement with experiment, we obviously need $|g_{X1\gamma\gamma}| > |g_{\pi\gamma\gamma}|$ and constructive interference in (17). $\Gamma(X \to \gamma\gamma) = 10$ keV (corresponding to $\Gamma(X) = \frac{1}{2}$ MeV) would do the job.

$CP$-invariance guarantees that particles and antiparticles have opposite form factors (see section 4-5). Consequently, $\pi^0$, $\eta$ and $X$ have $F = 0$, i.e., they do not couple to a single photon at all (this was our motivation for extracting a factor $Z_b$ in (2-7.9)). The only non-trivial $SU_3$-relations are those corresponding to the first equations in (9) and (10),

$$F(\pi^+) = F(K^+), \qquad F(\bar{K}^0) = F(K^0). \quad (14.18)$$

Since $CP$ requires $F(\bar{K}^0) = -F(K^0)$, $SU_3$ predicts in fact that also $K^0$ and $\bar{K}^0$ have zero form factors. However, the simple vector meson dominance model (9.23) does not fulfill (18).

For the $V \to \gamma$ coupling constants $f_V$ of (9.18), the argument analogous to (16) gives

$$3^{\frac{1}{2}} f_{\phi 8}^{-1} + f_\rho^{-1} = 0. \quad (14.19)$$

Since $\omega_1$ does not couple to $\gamma$ either, we obtain

$$3^{\frac{1}{2}} f_\rho = -f_{\phi 8} = f_\phi \cos\theta_v = f_\omega \sin\theta_v. \quad (14.20)$$

These relations are roughly fulfilled with the coupling constants of table C-3. For a quantitative comparison, one should include finite widths corrections to (9.24) (Renard 1970):

$$\Gamma(V \to e\bar{e}) = \tfrac{1}{3} 4\pi\alpha f^{-2} m_V a_V, \qquad a_\rho \approx 1.07, \qquad a_\omega \approx 1, \qquad a_\phi \approx 0.85, \quad (14.21)$$

which result from a coupled-channel $N/D$ calculation (see also section 6-8). With (21), we get

$$f_\rho^2/4\pi = 2.42, \qquad f_\omega^2/4\pi = 18.4, \qquad f_\phi^2/4\pi = 12.2 \quad (14.22)$$

With these numbers, (20) is even better satisfied.

## 4-15 Electromagnetic Mass Differences and State Mixing

In section 4-10 we investigated the $SU_3$-properties of the semistrong ($= SU_3$-breaking) mass splitting operator $\Delta m$ between particles in one $SU_3$-multiplet. We now allow also an $SU_2$-breaking part in the mass operator, with small matrix elements $\delta m$. At least a part of these matrix elements must be due to second order electromagnetic interactions (emission and reabsorption of a virtual photon), which involve the electromagnetic current twice. We therefore assume that the operator is a $U$-spin scalar, but not necessarily an octet operator. In this case we can use the analogues of eqs. (14.9)–(14.11), but not those of (14.7) or (14.12):

$$\delta m(\Sigma^+) = \delta m(p), \qquad \delta m(\Xi^-) = \delta m(\Sigma^-), \tag{15.1}$$

$$\delta m(\Xi^0) = \delta m(n) = \tfrac{1}{4}[3\ \delta m(\Lambda) + 2 \cdot 3^{\frac{1}{2}} m_{\Lambda\Sigma} + \delta m(\Sigma^0)], \tag{15.2}$$

$$\delta m(\Lambda) - 2 \cdot 3^{\frac{1}{2}} m_{\Lambda\Sigma} - \delta m(\Sigma^0) = 0. \tag{15.3}$$

Here $m_{\Lambda\Sigma}$ is a matrix element between the $I = 0$ and $I = 1$ states. Clearly, if isospin invariance is broken, the physical $\Lambda$ and $\Sigma^0$ need not be isospin eigenstates. We parametrize in analogy with (10.17)

$$\Sigma^0 = \Sigma_3 \cos\theta_{\Lambda\Sigma} - \Lambda_1 \sin\theta_{\Lambda\Sigma},$$

$$\Lambda = \Sigma_3 \sin\theta_{\Lambda\Sigma} + \Lambda_1 \cos\theta_{\Lambda\Sigma}, \tag{15.4}$$

where the indices 3 and 1 denote the isospin eigenstates. Actually, the mixing angle is small enough that we can put $\cos\theta = 1$, $\sin\theta = \theta$. Similarly, we put for $\pi^0$ and $\eta$:

$$\pi^0 = \pi_3 - \eta_1 \theta_{\pi\eta}, \qquad \eta = \pi_3 \theta_{\pi\eta} + \eta_1. \tag{15.5}$$

By adding the two equations in (1) and subtracting the first equation of (2), we get the mass formula of Coleman and Glashow (1961)

$$m_{\Xi^-} - m_{\Xi^0} + m_n - m_p = m_{\Sigma^-} - m_{\Sigma^+}, \tag{15.6}$$

which is well satisfied experimentally (7.7 MeV = 8.0 MeV). Inserting (3) into (2), we get $\delta m(n) = \delta m(\Sigma^0) + 3^{\frac{1}{2}} m_{\Lambda\Sigma}$. Subtracting from this the first equation of (1), we get

$$m_n - m_p + m_{\Sigma^+} - m_{\Sigma^0} = 3^{\frac{1}{2}} m_{\Lambda\Sigma}. \tag{15.7}$$

The mixing angle $\theta_{\Lambda\Sigma}$ is now derived from the requirement that $\Lambda$ and $\Sigma^0$ be eigenstates of the mass matrix:

$$m = \begin{pmatrix} m_1 & m_{\Lambda\Sigma} \\ m_{\Lambda\Sigma} & m_3 \end{pmatrix}, \qquad \Lambda = \begin{pmatrix} 1 \\ \theta_{\Lambda\Sigma} \end{pmatrix}, \qquad \Sigma^0 = \begin{pmatrix} -\theta_{\Lambda\Sigma} \\ 1 \end{pmatrix}. \tag{15.8}$$

To lowest order in $m_{\Lambda\Sigma}(m_\Lambda - m_\Sigma)^{-1}$, we have $m_1 = m_\Lambda$, $m_3 = m_\Sigma$,

$$\theta_{\Lambda\Sigma} = m_{\Lambda\Sigma}(m_1 - m_3)^{-1} = 3^{-\frac{1}{2}}(m_{\Sigma^0} - m_{\Sigma^+} + m_p - m_n)(m_{\Sigma^0} - m_\Lambda)^{-1}. \tag{15.9}$$

The corresponding formulas for mesons are obtained from (6) and (9) by substituting $\Xi^- \to K^-$, $\Xi^0 \to \bar{K}^0$, $n \to K^0$, $p \to K^+$, $\Sigma \to \pi$, $\Lambda \to \eta$. This transforms (6) into the identity $0 = 0$, and (9) into

$$\theta_{\pi\eta} = 3^{-\frac{1}{2}}(m_{\pi^0}^2 - m_{\pi^+}^2 + m_{K^+}^2 - m_{K^0}^2)(m_{\pi^0}^2 - m_\eta^2)^{-1}. \tag{15.10}$$

Here we have again preferred a quadratic mass formula as in (10.16). For the baryon decuplet, a $U$-spin scalar mass splitting operator requires

$$m_{\Delta^-} - m_{\Delta^0} = m_{\Sigma^{*-}} - m_{\Sigma^{*0}} = m_{\Xi^{*-}} - m_{\Xi^{*0}}, \tag{15.11}$$

$$m_{\Delta^0} - m_{\Delta^+} = m_{\Sigma^{*0}} - m_{\Sigma^{*+}}. \tag{15.12}$$

In principle, electromagnetic mass differences can be calculated by the method indicated in (3-10.13). We show how this is done for pseudoscalar particles (pions). The Feynman diagram for the exchange of a virtual pion contains a factor $(m_0^2 - t)^{-1}$ from the pion propagator, where $m_0$ is the mass of a fictitious pion which cannot interact with photons. Including now the photon interaction in the form of fig. 4-15(a), we get an extra term, which contains $-(m_0^2 - t)^{-2}\Pi(t)$ instead of $(m_0^2 - t)^{-1}$, where $i^{-1}\Pi$ accounts for the emission and absorption of the photon by a pion of 4-momentum $P^\mu$, with $P^2 = t$. To order $e^2$, the sum of the two terms is rewritten as

$$(m_0^2 - t)^{-1}(1 - \Pi(t)(m_0^2 - t)^{-1}) = (m_0^2 + \Pi(t) - t)^{-1} \qquad (15.13)$$

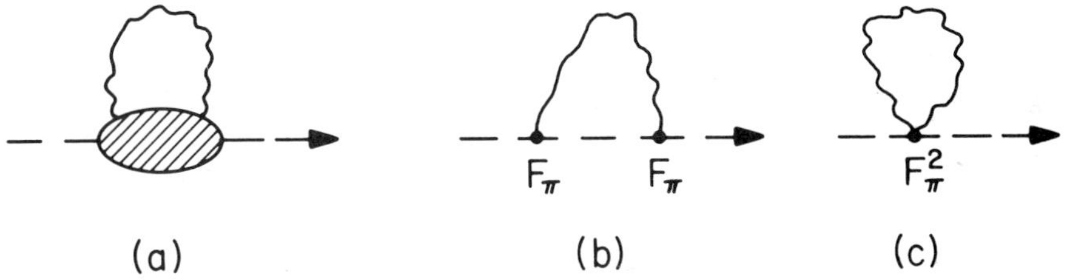


**Figure 4-15** (a) The electromagnetic self-energy graph, (b) the particle remains in its ground state, (c) the contact term for pions.

The square of the physical pion mass follows from (13) in the limit $t \to m_0^2 + \Pi(m_0^2)$,

$$m_\pi^2 = m_0^2 + \Pi(m_0^2). \qquad (15.14)$$

In the case at hand, we approximate $\Pi$ by graphs (b) and (c) of fig. 4-15. Graph (b) follows from (2-7.9) and the Feynman rules. For $\pi^\pm$,

$$i^{-1}\Pi_b = e^2 \int d^4q(2\pi)^{-4}[m_0^2 - (P-q)^2 - i\varepsilon]^{-1}[q^2 + i\varepsilon]^{-1}(2P - q)^2 F_\pi^2(q^2). \qquad (15.15)$$

$F_\pi$ is the pion form factor and is essential in this calculation because otherwise the integral would diverge. We approximate $F_\pi$ by the $\rho$-pole of (9.26), normalized to $F_\pi(0) = 1$:

$$F_\pi(q^2) \approx m_\rho^2(m_\rho^2 - q^2)^{-1}. \qquad (15.16)$$

The full matrix element $i^{-1}\Pi$ can be obtained from the forward Compton scattering amplitude $T_C$ of a virtual photon of 4-momentum $q^\mu$ as follows:

$$i^{-1}\Pi = \frac{1}{2}\sum\begin{pmatrix}\text{4 photon}\\ \text{states}\end{pmatrix}\int d^4q(2\pi)^{-4}T_C(k^\mu = k'^\mu = q^\mu) \times [m_0^2 - (P-q)^2 - i\varepsilon]^{-1}(-q^2 - i\varepsilon)^{-1}. \qquad (15.17)$$

The factor $\frac{1}{2}$ appears because the emitted photon is identical with the absorbed one. The form (17) applies also to particles with spin; the summation over the photon states replaces the factors $\varepsilon^{\nu*}\varepsilon^{\mu}$ in (3-7.1) and (3-7.24) by $-g^{\mu\nu}$ according to (2-8.22). Except for the form factors, (15) is obtained in this way from the first two terms of (3-7.24). Using now $g_{\mu\nu}g^{\mu\nu} = 4$, we get the full matrix element

$$\begin{aligned} i^{-1}\Pi = e^2 \int d^4q(2\pi)^{-4}(q^2 + i\varepsilon)^{-1} \\ \times [(2P - q)^2[m_0^2 - (P - q)^2 - i\varepsilon]^{-1} - 4]F_\pi^2(q^2). \end{aligned} \tag{15.18}$$

Graph (c) of fig. 4-15 may contain more complicated expressions; (18) is merely the gauge invariant extension of (15) (Riazuddin 1959). In the limit $F_\pi = 1$, the $-4$ follows from the $-\psi_0^+ eA^\mu\psi_0$-term in the current operator (3-2.8).

The integral (18) is conveniently evaluated by the parametrization method indicated in section 6-6 below. For details, see for example Appendix A5 of Jauch and Rohrlich (1976). To first order in $m_0^2/m_\rho^2$, we find

$$\begin{aligned} \Pi = \frac{\alpha}{4\pi}\{3m_\rho^2 + m_0^2 + \tfrac{1}{2}t - 2t^2m_\rho^2 + (2t - m_0^2)\ln(m_\rho^2 m_0^{-2}) \\ - 2t^{-1}(t^2 - m_0^4)\ln(1 - tm_0^{-2})\} \end{aligned} \tag{15.19}$$

$$\Pi(t = m_0^2) = \frac{\alpha}{4\pi}\{3m_\rho^2 + m_0^2[\ln(m_\rho^2 m_0^{-2}) + \tfrac{3}{2}]\}. \tag{15.20}$$

For $\pi^0$, we have $J^\mu(\pi^0) = 0$ and therefore also $\Pi = 0$. Assuming now that the $\pi^+ - \pi^0$ mass difference is entirely given by our calculation, we have $m(\pi^0) = m_0$, and $\Pi(\pi^+) = 0.0011\ \text{GeV}^2$, in good agreement with the experimental value $0.00126\ \text{GeV}^2$.

The sign of the mass difference can be understood classically: the electrostatic self-energy makes the charged members of a multiplet heavier than the neutral one. This remains true also in the calculations of $K^0 - K^+$ and $n - p$ mass differences (see Zee 1972 for a review). Unfortunately, it is contradicted by experiment. The origin of this discrepancy is presently unknown, it must be due to short-range effects. When the calculated mass splittings are subtracted, it turns out that the remaining discrepancies $\Delta m_q$ (which are rather large in most cases) are entirely due to an octet operator. For these, we can therefore again use the equivalents of (14.7) and (14.12), particularly

$$\Delta m_q(\Sigma^Z) = Z\,\Delta m_q(\Sigma), \tag{15.21}$$

which is the equal-spacing rule for $m(\Sigma^+), m(\Sigma^0), m(\Sigma^-)$, after subtraction of the "self-energies." This also explains the exceptional success of our calculation of $m_{\pi^+}^2 - m_{\pi^0}^2$: The CPT-theorem requires $m_{\pi^+} = m_{\pi^-}$ and therefore $\Delta m_q(\pi^Z) = 0$ according to (21). In the quark model, $\Delta m_q$ is incorporated simply by taking $m_d > m_u$. See also Nasrallah and Schilcher (1978).

## 4-16 The New Particles and Charm

In 1974, an extremely narrow $e\bar{e}$ resonance at 3098 MeV was found in the reaction $p^9Be \to e\bar{e} \ldots$ by Aubert et al., and in colliding beams of electrons and positrons by Augustin et al. The resonance was called $J$ by the first group and $\psi$ by the other one. Most of the subsequent particle discoveries came from $e\bar{e}$ colliding beam experiments, see Perez-y-Jorba and Renard (1977), Wilik and Wolf (1979).

The $J/\psi$-discovery was followed by that of $\psi'$ at 3685 MeV within a month. From the large $e\bar{e} \to \psi$ and $e\bar{e} \to \psi'$ cross sections it was expected (and soon confirmed) that these particles carry the quantum numbers of the photon, $J = 1$, $P = C = 1$. The relevant Feynman graph is shown in fig. 4-9.1. The resonant cross section for a final state $f$ is given by (8.24) with $J = 1$ and $S_a = S_b = \frac{1}{2}$:

$$\sigma(e\bar{e} \to f) = 3\pi m\Gamma_{e\bar{e}} m\Gamma_f[(m^2 - s)^2 + m^2\Gamma^2]^{-1} p_e^{-2}, \qquad p_e^{-2} \approx 4/s. \tag{16.1}$$

The results for $f = e\bar{e}$, $\mu\bar{\mu}$ and $f =$ hadrons are shown in fig. 4-16.1. Because of the limited energy resolution, the actual widths of fig. 4-16.1 are considerably larger than $\Gamma$. Therefore only the integrated cross section was compared with (1). Using (3-11.18), we have

$$\int \sigma(e\bar{e} \to f)\, ds = 12\pi^2\Gamma_{e\bar{e}}\Gamma_f/m\Gamma. \tag{16.2}$$

Assuming now that the observed cross sections for $f = e\bar{e}$, $\mu\bar{\mu}$ and hadrons add up to the total $e\bar{e} \to \psi$ cross section ($\Gamma_{e\bar{e}} + \Gamma_{\mu\bar{\mu}} + \Gamma_{\text{had}} = \Gamma$), one obtains from (2)

$$\sum_f \int \sigma(e\bar{e} \to f)\, ds = 12\pi^2\Gamma_{e\bar{e}}/m, \tag{16.3}$$

from which $\Gamma_{e\bar{e}}$ is obtained. The remaining widths follow now from the measured branching ratios.

The decay properties of $J/\psi$, $\psi'$ and further new particles are collected in table C-7. A $1^{--}$ state which decays into $\pi^+\pi^-\pi^0$ must have $I = 0$ (see the discussion of $\omega$-decay in 4-9 and table 4-2.4). The $\psi$ thus has $G$-parity $-1$ according to (5.4). This is confirmed by the dominance of odd pion numbers in the other decay channels. However, decays to $G = +1$ final states do occur. In fact, the fraction of $G$-violating $\psi$-decays is estimated to 17% from a comparison with $e\bar{e} \to$ hadrons just outside the resonance.

The decay $\psi' \to \psi\eta$ and the ratio $\Gamma(\psi' \to \psi\pi^+\pi^-)/\Gamma(\psi' \to \psi\pi^0\pi^0) \approx 2$ show that $\psi'$ also has $I = 0$. Today it is also known that $\psi$ has no charged counterparts. Using isospin invariance, one can correct for a number of unseen hadronic decays and thus improve the determination of $\Gamma$.

In about $\frac{1}{4}$ of all $\psi'$ decays, $\gamma$-rays of energies 261 MeV, 170 MeV or 128 MeV are emitted. From this one concludes that new hadrons of masses 3413, 3510 and 3554 MeV exist. These states are denoted by $\chi$ and $P_c$ in table

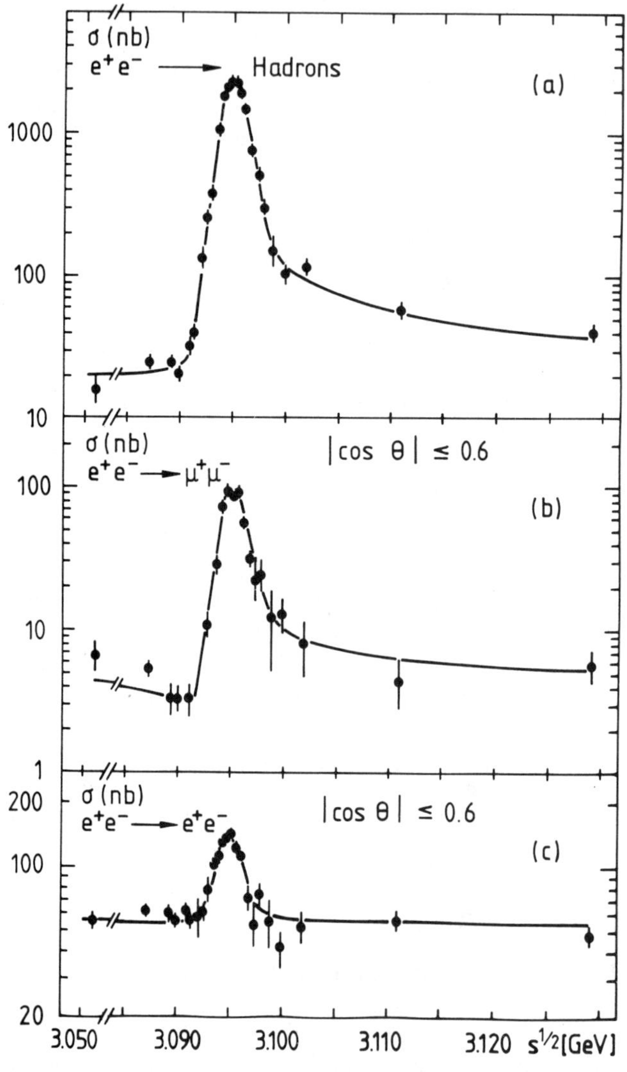


**Figure 4-16.1** The cross sections $\sigma(e\bar{e} \to f)$ across the $\psi$ resonance (from Augustin et al., 1974).

C-7. They are also seen as sharp peaks in a number of hadronic decay channels. The $P_c$ is exceptional in the sense that it decays dominantly into $\psi\gamma$. Thus the total (= direct + indirect) $\psi$ production in $\psi'$ decays is about 60%. The decays of the new particles are indicated schematically in fig. 16.2; their photonic decays are reminiscent of transitions between the various states of a composite system. In analogy with positronium ($e\bar{e}$), the name "charmonium" ($c\bar{c}$) is used.

The existence of a new additive quantum number was postulated as early as 1970 (Glashow et al.) to explain the absence of neutral currents in $K$-decay (section 5-11.). In the quark model, a fourth quark in addition to

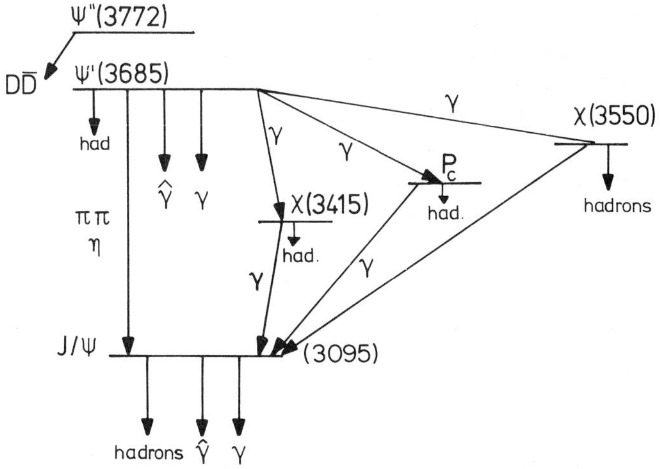


**Figure 4-16.2** Decay scheme of charmonium states.

the $u$, $d$, $s$ quarks is introduced, namely the "charmed quark" $c$. It has $I = Y = 0$ and carries the "charm" quantum number $C = +1$ (not to be confused with charge conjugation!). Since the explanation of Glashow et al. requires equal electric charges for $u$ and $c$ quark, the extension of the charge operator formula (3.6) to charmed states is

$$Q = I_3 + \tfrac{1}{2}(Y + C). \tag{16.4}$$

Moreover, $C$ must be conserved in strong and electromagnetic decays. The $\psi$ and $\chi$ states must therefore have $C = 0$. Thus the strong decays of $\psi$ and $\chi$ are not forbidden by any quantum number, and the only remaining suppression mechanism is the Zweig-Iizuka rule of section 4-12, which states that the quark and antiquark which form a meson must go into two of its decay products. From this we conclude that all $\psi$ and $\psi'$ states are pure $c\bar{c}$ states, with no components of $u\bar{u}$, $d\bar{d}$ or $s\bar{s}$. In a charmonium potential model, $\psi$, $\psi'$ and $\psi''$ correspond presumably to the $1S$, $2S$ and ($2D$, $J = 1$) spin-triplet $c\bar{c}$ bound states.

The quark model predicts the existence of pseudoscalar bound states

$$D^0 = c\bar{u},\ D^+ = c\bar{d},\ \bar{D}^0 = u\bar{c},\ D^- = d\bar{c}, \tag{16.5}$$

and of corresponding vector resonances $D^*$. In fact, since the $c$-quark is obviously heavier than the $u$ and $d$ quarks, one expects $m_D < m_\psi$. These particles were found in 1976 (Goldhaber et al.) in reactions of the type $e\bar{e} \to D\bar{D}$, $e\bar{e} \to D\bar{D}^*$, $D^*\bar{D}$, $e\bar{e} \to D^*\bar{D}^*$. Their masses and decay properties are included in table C-7. In analogy with $\phi \to K\bar{K}$ decays, charmonium states should decay dominantly into $D\bar{D}$, $D\bar{D}^*$ etc. whenever there is phase space available. This is in fact the case.

At a first sight it seems surprising that the Zweig-Iizuka rule works so accurately for $\psi$ and $\psi'$ decays when it fails by $0.16 \times 4$ MeV $\approx 0.6$ MeV in $\phi$ decays (in the $\phi \to \pi\pi\pi$ mode). However, a certain violation of this rule in

$\phi$-decays must be due to unitarity: if there is a strong matrix element for $K\bar{K} \to \pi\pi\pi$ in the $J = 1$ partial wave, then $\phi \to \pi\pi\pi$ occurs automatically via the chain of reactions $\phi \to K\bar{K}$, $K\bar{K} \to \pi\pi\pi$. For $\psi$ and $\psi'$, on the other hand, all decays are forbidden by the rule, and unitarity cannot produce a large width.

From the existence of the states (5) and $SU(2)$-symmetry it follows that also the states

$$F^+ = c\bar{s}, \qquad F^- = \bar{c}s \tag{16.6}$$

must exist. These $F$-mesons were discovered by Brandelik et al. (1977, 1979). Note that $\bar{D}^0$, $D^-$ and $F^-$ form an $SU(3)$-triplet, whereas $D^0$, $D^+$, $F^+$ form an antitriplet $(\bar{3})$:

$$(3) = (\bar{D}^0, D^-, F^-), \qquad (\bar{3}) = (D^0, D^+, F^+). \tag{16.7}$$

In addition to the charmed mesons, charmed baryons are also expected, but their experimental evidence is presently meagre.

The first evidence for charmed baryons came from the neutrino reaction $\nu p \to \mu^- \Lambda \pi^+ \pi^+ \pi^+ \pi^-$, the hadrons arising from the decay of a state of mass $2425 \pm 11$ MeV called $\Sigma_c^{++}$, the corresponding quark structure being *cuu*. The $\Sigma_c^{++}$ decays into $\pi^+ \Lambda_c$, where $\Lambda_c$ has $2257 \pm 10$ MeV and an isosinglet charmed baryon structure, $2^{-\frac{1}{2}} c(ud - du)$. The above reaction thus arises from the sequence $\nu p \to \mu^- \Sigma_c^{++}$, $\Sigma_c^{++} \to \pi^+ \Lambda_c$, $\Lambda_c \to \Lambda \pi^+ \pi^+ \pi^-$. The clearest evidence for $\Lambda_c$ comes presently from its $\Lambda \pi^+$ decay mode (see Baltay et al., 1979, Phys. Rev. Letters *42*, 1721, and references therein).

The first indication of a further additive quantum number came from the observation of two narrow $e\bar{e}$ peaks near 10 GeV in high-energy proton-nucleus collisions (Herb et al., 1977), which were later resolved into three peaks in $e\bar{e}$ collision experiments. These peaks are denoted by $\Upsilon$ in table C-7. The new quantum number is called "beauty" (b). In the quark model, the $\Upsilon$ states are described as $b\bar{b}$ bound states. Finally, theoretical speculations (see section 5-11) need a sixth quark ($t$), for which there is presently no experimental evidence.

CHAPTER FIVE

# Weak Interactions

## 5-1 Historical Introduction. Pion Decays

In the early studies of natural radioactive materials, radiation was classified by its electric charge into $\alpha$-rays ($Z = 2$, $^4$He-nuclei), $\beta$-rays ($Z = -1$, electrons) and $\gamma$-rays ($Z = 0$, photons). Two years after the discovery of the positron in cosmic rays, Curie and Joliot (1934) demonstrated the emission of positrons ("$\beta^+$-radiation") from artificial radioactive material. A process which frequently competes with the $\beta^+$-decay of nuclei is the capture of an atomic electron by the nucleus, which was discovered by Alvarez (1938).

The prototype of $\beta$-decay is the decay of a free neutron $n \to pe^-\bar{\nu}$. $\bar{\nu}$ is the antiparticle of the neutrino $\nu$ (see below). The interactions of these particles are so weak that their direct observation in a given decay event is not possible. However, absorption and scattering of neutrinos can be studied when they are available in large quantities. The first experiment of this type was the detection of $\bar{\nu}$-induced $\beta$-transitions, using a nuclear reactor as an intense source of decaying neutrons (Reines and Cowan 1959).

The existence of the neutrino was postulated as early as 1927 by Pauli (unpublished) on the basis of the law of energy conservation. It is known (for example from $\gamma$-ray transitions) that low-lying nuclear excited states have well-defined energies (the decay widths are negligible, $10^{-18}$ eV for the neutron, see table C-4). Electrons from $\beta$-decay, on the other hand, show a continuous spectrum (see fig. 5-2.1a below). The missing energy is carried away by the neutrino. The electron spectrum extends all the way up to the nuclear excitation energy (in $n \to pe\bar{\nu}$ decays, $E_{max} = m_n - m_p = 1.293$ MeV apart from a small proton recoil energy), which shows that sometimes neutrinos of energy $\ll m_e$ are emitted, i.e., the neutrino has a negligible mass. (The most precise measurement of $m_\nu$ comes from a careful investigation of the upper end of the electron spectrum in triton decay ${}^3\mathrm{H} \to {}^3\mathrm{He}e\bar{\nu}$ (Bergkvist (1972), and gives $m_\nu < 60$ eV. Today, one can easily check that inclusion of the massless neutrino leads to conservation of both energy and momentum in decays such as $K_L \to \pi^+ e\bar{\nu}$, $K_L \to \pi^- e^+ \nu$. The same is true for the muon's neutrino $\nu_\mu$ in the decays $\pi^+ \to \mu^+ \nu_\mu$, $\pi^- \to \mu\bar{\nu}_\mu$, the kinematics of which was already discussed in (1-0.24) and fig. 2-3 for $m_2 = 0$).

After the discovery of the neutron in 1932, Fermi (1934) proposed that $\beta$-decay is due to a small piece in the Hamilton operator (the "weak Hamiltonian" $H_\beta$) which transforms protons into neutrons and vice versa. At the same time, it must create $e^+\nu$ and $e\bar{\nu}$ pairs, respectively, and it must be Lorentz-invariant and Hermitean. The simplest operator with the desired properties is (compare 3-3.19)

$$H_S(t) = \int d^3x H_S(x), \qquad H_S(x) = G_S[\bar{\Psi}_p(x)\Psi_n(x)\bar{\Psi}_e(x)\Psi_\nu(x) + \text{h.c.}] \quad (1.1)$$

where h.c. stands for the Hermitean conjugate part $\bar{\Psi}_n \Psi_p \bar{\Psi}_\nu \Psi_e$, and the $\Psi$'s contain creation and annihilation operators as in section 3-2. One can also insert Dirac matrices between $\bar{\Psi}_p$ and $\Psi_n$ and the corresponding Dirac matrices between the lepton spinors, to conserve Lorentz invariance. In fact, Fermi himself chose a 4-vector interaction

$$H_V = G_V[\bar{\Psi}_p \gamma_\mu \Psi_n \bar{\Psi}_e \gamma^\mu \Psi_\nu + \text{h.c.}] \quad (1.2)$$

because of its closer analogy with the electromagnetic interaction. Other possible Hamiltonians are obtained by replacing the $\gamma^\mu$'s by $\gamma_5$ (pseudoscalar interaction "$P$"), $\gamma^\mu\gamma_5$ (axial vector interaction "$A$"), or $\sigma^{\mu\nu}$ (tensor interaction "$T$"). Until 1956, the experimental and theoretical effort was concentrated on the determination of the 5 coupling constants $G_S$, $G_V$, $G_P$, $G_A$, $G_T$.

Let us consider the meaning of (2) for a moment. The full interaction Hamiltonian is supposed to be composed additively of strong, electromagnetic and weak interactions

$$H_I = H_{\text{strong}} - eA_0^\mu j_\mu(\text{el. mag.}) + H_\beta \quad (1.3)$$

Because the weak interactions are so weak, $\beta$-decay matrix elements can be calculated by perturbation theory to first order in $H_\beta$, whereas the strong interaction and frequently also the Coulomb part of the electromagnetic interaction must be included non-perturbatively. For the Coulomb part, this is achieved with sufficient accuracy by using (3-3.18) for the electron field operator. The strong interactions introduce very small complications for neutron decay, but for the $\beta$-decays of all other nuclei, they can be included only approximately by the use of nucleon shell-model wave functions (in the CVC theory, one can do better, as we shall see). It is customary here to work with normalized determinants of wave functions for the initial $|i\rangle$ and final

$|f\rangle$ nuclear states, and to use an isospin formalism, where the transformation from $|n\rangle$ to $|p\rangle$ is achieved by the isospin raising operator $\tau_+ = \frac{1}{2}(\tau_1 + i\tau_2)$, compare section 4-2. Then the matrix elements of (2) are given by a form analogous to (3-3.25)

$$\langle fe\bar{\nu}|H_V(t)|i\rangle = G_V\langle f|\sum_{k=1}^{N}\tau_+^{(k)}\gamma_\mu^{(k)}\bar{\psi}_e(\mathbf{x}_k)\gamma^\mu\psi_{\nu,c}(\mathbf{x}_k)|i\rangle e^{it(E_e+E_{\bar{\nu}}+E_f-E_i)} \quad (1.4)$$

but here our notation $\langle f|\cdots|i\rangle$ includes an integration over all nuclear coordinates, also $\mathbf{x}_k$ (compare 3-3.19), and the time dependence has been written separately. The nucleons inside a nucleus are essentially nonrelativistic, which means that the 5 Lorentz covariant operators $1, \gamma^\mu, \gamma_5, \gamma_5\gamma^\mu, \sigma^{\mu\nu}$ reduce either to $2m\chi_{0p}'^{+}\chi_{0n}$ ("Fermi transitions", 1 and $\gamma_0$) or to $2m\chi_{0p}'^{+}\sigma^k\chi_{0n}$ ("Gamow-Teller-transitions," $\gamma_5\gamma^k$ and $i\sigma^{ij}$) or to zero (see A-3.9 to A-3.13). Thus the pseudoscalar interaction is not on equal footing with the other ones, because it vanishes in the limit $p/m \to 0$..

The Fermi transitions do not change the nuclear spin, while the Gamow-Teller transitions can change it by one unit. All other transitions are called "forbidden" and occur only because the lepton pair can carry away orbital angular momentum, analogous to the higher multipoles in the emission of $\gamma$-rays. A study of the "allowed" transition showed immediately that Fermi- and Gamow-Teller matrix elements have comparable magnitudes, so that at least $G_V$ and $G_A$ or $G_S$ and $G_T$ must be nonzero. Thus the real problem was to distinguish between $S$ and $V$ in the Fermi matrix elements, and between $A$ and $T$ in the Gamow-Teller matrix elements. This was done by analyzing the angular distribution between $e$ and $\bar{\nu}$, particularly in the decay $^6\text{He} \to {^6\text{Li}}\, e\bar{\nu}$ (since $\mathbf{p}_\nu$ must be inferred from the nuclear recoil which is very small, such angular measurements are possible only in noble gases, where the recoil is not absorbed by chemical binding). The outcome of these experiments in 1956 was that $H_\beta$ was mainly $S$ and $T$. This conclusion was wrong, unfortunately. Already at that time, it was known that muon decay $\mu \to e\bar{\nu}\nu_\mu$ was mainly $V$ and $A$, and that pion decay $\pi^- \to \mu\bar{\nu}_\mu$ could only be $P$ or $A$, because of the pseudoscalarity of the pion. Although these decays are not directly related to $\beta$-decay, the similarity of the decay constants suggested that they should be due to a common "universal 4-fermion interaction" between the pairs of particles $(p, n)$, $(\nu, e)$ and $(\nu_\mu, \mu)$ (Pontecorvo 1947, and others). In particular, $\pi \to \mu\bar{\nu}_\mu$ was thought to proceed via a virtual $\bar{p}n$ state as in the Fermi-Yang model (1949).

The breakthrough to the correct theory of $\beta$-decay was initiated by a remark by Lee and Yang (1956) that perhaps these decays did not conserve parity. Their remark was motivated by the then quite apparent parity violation in pionic decays of kaons (see section 4-6). The reason that nobody had cared to test parity conservation in $\beta$-decay was probably because parity was taken to represent space inversion $S$ which is a Lorentz transformation. As mentioned already in section 1-3, one cannot tell on formal grounds whether $S$ should be represented by $P$ or by $CP$.

Within one year, the measurement of the electron angular asymmetry in the decays of polarized nuclei (Wu et al. 1957), the measurement of the longitudinal electron polarization (helicity) in $\beta$-decay (Frauenfelder et al. 1957) and the electron angular asymmetry in the decay $\mu \to e\bar{v}v_\mu$ (Garwin et al. 1957) demonstrated the violation of $P$ and $C$ invariance in all kinds of weak interactions.

Let us now calculate the decays $\pi^- \to \mu\bar{v}_\mu$ and $\pi^- \to e\bar{v}$ under the assumption that the Hamiltonian density $H_\beta(x)$ factorizes into one part $h(x)$ which acts only between hadron states and another part $l^+(x)$ which acts only between lepton states and which is identical for electrons and muons (lepton universality). The Hermitean conjugation in $l^+$ is a matter of convention, $\pi^+$ decays will be included by $h^+l$ instead. To start with, we take $h$ and $l^+$ as 4-vectors $h^\mu$ and $l^{+\mu}$. Lorentz invariance requires

$$\langle 0 | h^\mu(x) | \pi \rangle = e^{-iP_\pi \cdot x} \langle 0 | h^\mu(0) | \pi \rangle = e^{-iP_\pi \cdot x} f_\pi P_\pi^\mu . \tag{1.5}$$

since the pion's 4-momentum is the only available 4-vector on the hadronic side ($f_\pi$ is a constant). The simplest choice for $l^{+\mu}$ is a linear combination of $j_l^\mu$ and $j_r^\mu$ of (1-5.20). The plane-wave matrix elements of these operators have already been evaluated in section 3-6. In the pion rest frame $\mathbf{p}_\pi = 0$, $P_\pi^0 = m_\pi$ we only need (3-6.23) and we can put $\bar{E} = p = \frac{1}{2}(m_\pi - m_1^2/m_\pi)$ $(m_1 = m_\mu$ or $m_e)$, since the neutrino is massless. This gives us

$$F_+ = (E - 2\lambda p)^{\frac{1}{2}} p^{\frac{1}{2}} (1 + 2\lambda)^{\frac{1}{2}} = (E - p)^{\frac{1}{2}} (2p)^{\frac{1}{2}} \, \delta_{\lambda, \frac{1}{2}}, \tag{1.6}$$

i.e., $j_l^0$ contributes only to muon helicity $+\frac{1}{2}$ and $j_r^0$ only to $-\frac{1}{2}$. Experimentally, the muon helicity is $+\frac{1}{2}$ in $\pi^-$-decay, which shows that $j_r^0$ is absent. We thus have, with $E - p = m_1^2/m_\pi$

$$\begin{aligned} 2^{\frac{1}{2}} G_V^{-1} T(\pi^- \to \mu\bar{v}_\mu) &= \langle 0 | h_\mu(0) | \pi \rangle \langle \mu\bar{v} | 2j_e^\mu(0) | 0 \rangle \\ &= 2 f_\pi m_\pi \, \delta_{\lambda\bar{\lambda}} \, \delta_{\lambda, \frac{1}{2}} (2pm_1^2/m_\pi)^{\frac{1}{2}} \\ &= 2 f_\pi \, \delta_{\lambda\bar{\lambda}} \, \delta_{\lambda, \frac{1}{2}} (m_\pi^2 - m_1^2)^{\frac{1}{2}} m_1 \end{aligned} \tag{1.7}$$

$$\Gamma(\pi^- \to \mu\bar{v}_\mu) = p(8\pi m_\pi^2)^{-1} \sum_{\lambda\bar{\lambda}} |T|^2 = (m_\pi^2 - m_1^2)^2 m_\pi^{-3} m_1^2 G_V^2 f_\pi^2 / 2\pi \tag{1.8}$$

The factor $2^{\frac{1}{2}} G_V^{-1}$ in front of $T$ is introduced in view of (11) below, and the factor 2 in $\langle \mu\bar{v} | 2_{jl}^\mu | 0 \rangle$ comes from the fact that most people still prefer 4-component spinors. The expression for $\Gamma$ is taken from (4-4.8). The total lepton current operator is

$$\begin{aligned} l^{\mu+}(x) &= 2j_l^\mu(ve, x) + 2j_e^\mu(v_\mu\mu, x) \\ &= 2\Psi_{el}^+(x)\sigma_l^\mu \Psi_{vl}(x) + 2\Psi_{\mu l}^+(x)\sigma_l^\mu \Psi_{v\mu l}(x) \\ &= \bar{\Psi}_e(x)\gamma^\mu(1 - \gamma_5)\psi_v(x) + \bar{\Psi}_\mu(x)\gamma^\mu(1 - \gamma_5)\Psi_{v_\mu}(x) \end{aligned} \tag{1.9}$$

remembering from (1-6.18) that $\frac{1}{2}(1 - \gamma_5)$ is the projection operator for left-handed spinors. The most striking feature of (8) is the factor $m_1^2$, which is 50,000 times larger for the $\mu\bar{v}_\mu$ state than for the $e\bar{v}$ state. Repeating now the calculation for $\pi \to e\bar{v}$ decays, using the universal lepton current (9), we find a branching ratio of $1.28 \times 10^{-4}$ for $\pi \to e\bar{\nu}$ decays, in beautiful agreement with the value of table C-2.

For completeness, we now consider the case that $h$ and $l^+$ are scalars under proper Lorentz transformations. Then a matrix element equivalent to (7) which also gives the correct muon helicity is

$$\begin{aligned} T_P &= 2g_P\langle \mu\bar{\nu}_\mu | \Psi^+_{\mu r}(0)\Psi_{\nu_\mu l}(0)|0\rangle \\ &= 2g_P\, \delta_{\lambda\bar{\lambda}}(E + 2\lambda p)^{\frac{1}{2}} p^{\frac{1}{2}}(1 + 2\lambda)^{\frac{1}{2}} \\ &= 2g_P(E + p)^{\frac{1}{2}}(2p)^{\frac{1}{2}}\, \delta_{\lambda\bar{\lambda}}\, \delta_{\lambda,\frac{1}{2}} \end{aligned} \tag{1.10}$$

The factor $E - p = m_1^2/m_\pi$ of (7) is replaced by $E + p = m_\pi$, and (10) is larger for the $e\bar{\nu}$ final state than for the $\mu\bar{\nu}_\mu$ final state. Thus already a minute admixture of a Lorentz scalar interaction would change the $\pi \to e\bar{\nu}$ branching ratio. This encourages us to draw two conclusions from one number: (i) lepton universality is true, (ii) the interaction is of the type $h_\mu l^{+\mu} + h^+_\mu l^\mu$. The same analysis applies of course to the decays $K^- \to l\bar{\nu}$, $K^+ \to \bar{l}\nu$, with $f_\pi$ replaced by another decay constant $f_K$.

Returning now to nuclear $\beta$-decay and postulating that the interaction Hamiltonian should contain the hadronic operators in one single product $h \cdot l^+$ plus its Hermitean conjugate, we see that Fermi and Gamow-Teller operators can be united in one operator $h$ only if they are $V$ and $A$, respectively. $S$ and $T$ are excluded because of their different Lorentz transformation properties. Note also that a unification of $V$ and $A$ implies that parity does not represent a Lorentz transformation. In principle, one could thus have predicted parity violation (but of course not the elegant form (9) of purely lefthanded currents) from the mere existence of Fermi and Gamow-Teller decays!

The Hamiltonian density for $\beta$-decay is now written as

$$H_\beta = -2^{-\frac{1}{2}}G_V(h^\mu l^+_\mu + h^{+\mu} l_\mu), \qquad h^\mu = v^\mu + a^\mu \tag{1.11}$$

where $v^\mu$ and $a^\mu$ have the following matrix elements in Fock space between single-nucleon states, with $f_A \equiv G_A/G_V$

$$\Phi^+_p v^\mu(x)\Phi_n = \bar{\psi}_p(x)\gamma^\mu\psi_n(x), \qquad \Phi^+_p a^\mu(x)\Phi_n = f_A\bar{\psi}_p(x)\gamma_5\gamma^\mu\psi_n(x) \tag{1.12}$$

The factor $2^{-\frac{1}{2}}$ ensures that (2) and (11) give identical decay probabilities after summation over the lepton helicities (remember that the lepton operator in (2) is $j^\mu_e + j^\mu_r$ and that the matrix elements (A-3.22, A-3.24) of $j^\mu_l$ and $j^\mu_r$ are related by helicity reversal, whereas (9) contains a factor 2 instead). A minus sign is also included (11) because it will disappear in the $T$-matrix elements, according to (2-1.4) and (3-4.5). The numerical values of the decay constants are (Nagels et al. 1979)

$$\begin{aligned} G_V = 1.1147 \times 10^{-5}\ \text{GeV}^{-2}, \qquad f_A = 1.26, \\ f_\pi = 134\ \text{MeV} = 0.960\ m_{\pi^+}. \end{aligned} \tag{1.13}$$

The closeness of $f_A$ to 1 indicates that $h^\mu$ contains the nucleon fields (or better the quark fields) in the combination $\bar{\Psi}_p \gamma^\mu (1 - \gamma_5) \Psi_n$ and that the deviations in the matrix elements (12) are due to the strong interactions (meson fields). The coupling of only left-handed spinors went under the name "chiral invariance" (Sudarshan and Marshak 1957) and finally led to the rediscovery of 2-component spinors by Feynman and Gell-Mann (1958). Under the impact of this theoretical development, those experiments which had previously contradicted the universal $V - A$ theory were repeated and were now shown to agree with this theory. For example, the $\pi \to e\bar{\nu}$ branching ratio for which an upper limit of $2 \times 10^{-5}$ was established in 1957, jumped to $1.03 \times 10^{-4}$ (Anderson 1959)!

Let us now take a closer look at the neutrino and anti-neutrino helicities in $\beta^\pm$-decays. In $\beta$-decay, we need the following matrix element of the current operator

$$\begin{aligned} \langle e^- \bar{\nu} | l^{\mu +} | 0 \rangle &= 2 \langle e^- \bar{\nu} | \Psi_{el}^+(x) \sigma_l^\mu \Psi_{\nu l}(x) | 0 \rangle \\ &= 2 \psi_{el}^+ (\bar{p}_e , M_e) \sigma_l^\mu \psi_{\nu l, c} (\mathbf{p}_{\bar{\nu}}, M_{\bar{\nu}}) \end{aligned} \tag{1.14}$$

where $\psi_{\nu l, c} = v_{\nu l}(\mathbf{p}_{\bar{\nu}}, M_{\bar{\nu}}) e^{iP_{\bar{\nu}} x}$ is the left-handed component of the charge conjugate spinor in (3-2.3). $v_{\nu l}$ is given in (3-6.22), with $p_{\bar{e}}$ and $E_{\bar{e}}$ both replaced by $E_{\bar{\nu}}$. As in (6) above, it is nonzero only for $\lambda_{\bar{\nu}} = \frac{1}{2}$:

$$\langle e^- \bar{\nu} | l^{\mu +} | 0 \rangle = 2 \psi_{el}^+ (p_e , M_e) \sigma_l^\mu \chi_{0 \bar{\nu}}(\tfrac{1}{2}) \, \delta_{\lambda_{\bar{\nu}}, \frac{1}{2}} (2E_{\bar{\nu}})^{\frac{1}{2}} e^{iP_{\bar{\nu}} \cdot x} \tag{1.15}$$

Thus the antineutrino helicity in $\beta^-$-decay is always positive. Similarly, in $\beta^+$-decay, we need the following matrix element

$$\begin{aligned} \langle e^+ \nu | l^\mu | 0 \rangle &= 2 \langle e^+ \nu | \Psi_{\nu l}^+(x) \sigma_l^\mu \Psi_{el}(x) | 0 \rangle \\ &= 2 \, \delta_{\lambda_\nu, -\frac{1}{2}} \chi_{0\nu}^+(-\tfrac{1}{2}) \sigma_l^\mu \psi_{el, c}(\mathbf{p}_e, M_e, x) (2E_\nu)^{\frac{1}{2}} e^{iP_\nu \cdot x} \end{aligned} \tag{1.16}$$

with $\psi_{\nu l}^+ = e^{iP_\nu \cdot x} m_\nu^{\frac{1}{2}} \chi_l^+$ and $\chi_l$ given by (1-6.14). Thus the neutrino helicity in $\beta^+$-decay is always negative. In the following sections on $\beta$-decay, we omit the Kronecker-$\delta$ and possible summations over the neutrino and antineutrino helicity states, always understanding that the matrix element refers to the "allowed helicity."

The decay of a free muon $\mu^+ \to e^+ \nu \nu$ cannot be due to $H_\beta$ of (11). There must be another piece $H_{\mu e}$ in the weak Hamiltonian which couples all four lepton field operators. It is customary to take this as one piece of the symmetric expression

$$H_{\text{lept}}(x) = 2^{-\frac{1}{2}} G_\mu l^\mu(x) l_\mu^+(x) = H_{ee} + H_{\mu\mu} + H_{\mu e} \tag{1.17}$$

The pieces $H_{ee}$ and $H_{\mu\mu}$ induce elastic scattering of antineutinos on electrons and muons and will be discussed in section 5-10. $G_\mu$ is a common coupling constant which is determined from the muon decay rate. Its value (4-1.1) is so close to $G_V$ that one is tempted to write the "universal 4-fermion-interaction" in the form of a charged current interacting with itself:

$$H_{\text{ch}} \approx 2^{-\frac{1}{2}} G_\mu (h^\mu - l^\mu)(h_\mu - l_\mu)^+, \; \approx H_{nl} + H_\beta + H_{\text{lept}}. \tag{1.18}$$

The new nonleptonic piece $H_{nl}$ contributes to decays such as $K \to \pi\pi$, but quantitative calculations are impeded by the strong interactions between the hadrons of $h^\mu$ and $h_\mu^+$, such that the hypothesis of a (4-vector) × (4-vector)-interaction cannot be tested in these cases. In beta decays, on the other hand, the strong interactions affect only one of the two currents of the product $h^\mu l_\mu^+$. Repeating the arguments of section 2-7, it is clear that the most general form of the matrix elements $v^\mu(x)$ between *free* nucleons is

$$\Phi_p v^\mu(x)\Phi_n = \exp\{-i(P_n - P_p)x\}V_{np}^\mu \tag{1.19}$$

$$V_{np}^\mu = \bar{u}_p[f_1\gamma^\mu - \tfrac{1}{2}m_N^{-1}f_2\sigma^{\mu\nu}q_\nu + f_3 q^\mu]u_n, \qquad q^\mu \equiv P_n^\mu - P_p^\mu. \tag{1.20}$$

where $f_1$, $f_2$ and $f_3$ are functions of $t = q^2$. The term $f_3 q^\mu$ appears because gauge invariance is no longer required. A similar expression for the free-nucleon matrix elements of $a^\mu(x)$ is given in (4.5) below. The nonrelativistic limit of (20) is given by (12), if one in addition postulates $f_1(0) = 1$ (which is equivalent to $G_V = G_\mu$ incorporated in (18)). It was shown by Gerstein and Zeldovitch (1955) and later by Feynman and Gell-Mann (1958) that the only natural basis for such a postulate is a conserved vector current, $\partial_\mu v^\mu(x) = 0$ as in the electromagnetic case. However, the argument cannot be applied to that part of the operator which changes the hypercharge ("hypercharged current") and which is responsible for decays such as $K \to \mu\nu$ or $K \to \pi\mu\nu$. It was also qualitatively evident that this hypercharged current couples only with $\sim \frac{1}{4}$ of the strength of the "hyperneutral" current, not only in its vector piece, but also in its axial piece. This was incorporated by Cabibbo (1963) by writing instead of (18)

$$H_{\text{ch}} = 2^{-\frac{1}{2}}G_\mu(h_C^\mu - l^\mu)(h_{C\mu} - l_\mu)^+, \qquad h_{C\mu} \equiv \cos\theta_C h_\mu^{(0)} + \sin\theta_C h_\mu^{(1)}, \tag{1.21}$$

where $h_\mu^{(0)}$ and $h_\mu^{(1)}$ denote the hyperneutral and hypercharged pieces of the current operator, and $\theta_C$ is the "Cabibbo angle." Comparison between (21) and (11) gives (including a 1% correction from (6.14) below):

$$0.99G_V = G_\mu \cos\theta_C = 0.974G_\mu. \tag{1.22}$$

The numerical value of $\cos\theta_C$ follows from (13) and (4-1.1).

Within the framework of $SU(3)$-symmetry, Cabibbo was able to restore universality of the full current operator. Another important feature of weak interactions which is included in (21) is the absence of currents $h_\mu^{(2)}$, $h_\mu^{(3)}$ ... which change the hypercharge by more than one unit. Its experimental basis is the absence of decays such as $\Xi^- \to ne^-\bar{\nu}$, $\Omega^- \to ne^-\bar{\nu}$.

The hypothesis of a partially conserved axial current (PCAC) gave a relation between the constants $f_A$ and $f_\pi$ of axial current matrix elements, and the "current algebra" finally gave a relation between the axial coupling constant $G_A = G_V f_A$ and the vector coupling constant $G_V$. It also explained the fact that the Cabibbo angles of $v^\mu(x)$ and $a^\mu(x)$ are practically identical,

$$\theta_A = \theta_V \equiv \theta_C = 13.3^\circ, \qquad \sin\theta_C = 0.230 \pm 0.003. \tag{1.23}$$

This number results from a best fit to all baryon decays (table 5-8 below), but we get an estimate from the $K \to \mu\nu$ decay, which in view of (5) and (21) has the matrix element

$$\langle 0 | h^\mu(x) | K \rangle = \exp\{-iP_K \cdot x\} P_K^\mu f_\pi \tan\theta_C . \tag{1.24}$$

We can now use (8) in the form

$$\tan\theta_C = (1.339 m_\pi / m_K)^{\frac{1}{2}} (1 - m_\mu^2/m_\pi^2)(1 - m_\mu^2/m_K^2)^{-1} = 0.275, \tag{1.25}$$

which gives $\theta_C = 15.4^0$, in rough agreement with (23).

Many of the pioneering contributions to the theory of weak interactions are collected in a reprint volume by Kabir (1963). We have followed here the "historical summary" of Marshak et al. (1968). Topics such as lepton number versus muon number, neutral and charmed currents will be discussed later.

There exist several books on the theory of weak interactions with emphasis on nuclei and particularly on nuclear beta decay (Schopper 1966, Konopinski 1966, Wu and Moszkowski 1966). Morita (1973) discusses also muon capture in nuclei, and Blin-Stoyle (1973) includes other "fundamental interactions" in nuclei. In addition, there exist many excellent reviews, for example on electron capture (Bambynek et al. 1977) and muon capture (Mukhopadhyay 1977). Books with emphasis on particle physics are those of Marshak et al. (1968) and Commins (1973), Bailin (1977).

## 5-2 Allowed Beta Decay and Electron Capture

The Hamilton operator of beta decay is the volume integral of the Hamiltonian density (1.11). We first formulate the kinematics such that it includes recoil effects. A fully covariant formulation is very complicated because of the electromagnetic interaction in the final state. Fortunately, that interaction is large only when recoil effects are small, such that the electron field operator is well approximated by the Coulomb distorted operator $\Psi_F(\bar{v})$ of (3-3.18). (Some possible exceptions are mentioned by Stech and Schülke 1964). We can then restore momentum conservation by writing the true coordinate and initial nuclear state as

$$\mathbf{r}_{\text{true}} = \mathbf{r} + \mathbf{R}, \qquad |i\rangle_{\text{true}} = \exp(-iE_i t + i\mathbf{p}_i \mathbf{R}) |i\rangle, \tag{2.1}$$

where $\mathbf{R}$ is the nuclear cms coordinate. With a corresponding notation for the final nuclear state, we get from (3-3.19)

$$\begin{aligned} S_{ife\bar{v}} &= -i \int dt\, d^3R \langle fe\bar{v} | H_\beta(t, \mathbf{r} + \mathbf{R}) | i \rangle_{\text{true}} \\ &= i(2\pi)^4\, \delta_4 (P_i - P_f - P_e - P_{\bar{v}}) T_{ife\bar{v}}, \\ T_{ife\bar{v}} &= 2^{-\frac{1}{2}} G_V \langle fe\bar{v} | \int d^3r h^\mu(0, \mathbf{r}) l_\mu^+(0, \mathbf{r}) | i \rangle, \end{aligned} \tag{2.2}$$

where we have inserted (1.11) for $H_\beta$; $\mathbf{r}$ is now the distance from the nuclear cms, and the states $|i\rangle$ and $|f\rangle$ describe only the inner degrees of freedom of the nucleus. The integration over the nucleon coordinates $\mathbf{r}_k$ in a configuration space expression such as (1.4) must now include a function $\delta_3(\sum_k \mathbf{r}_k)$.

We can now use our covariant formalism of chapter 2 for the calculation of decay rates etc. The differential decay rate in the rest frame of the initial nucleus is given by (2-1.13):

$$d\Gamma_f(e\bar{\nu}) = (2m_i)^{-1} \int d \text{ Lips } (m_i^2; fe\bar{\nu}) \sum_{M_f \lambda_e} |T|^2, \tag{2.3}$$

where $d$ Lips is the 3-particle phase space differential (2-1.8),

$$\begin{aligned} d \text{ Lips } (m_i^2; fe\bar{\nu}) &= (2\pi)^{-5}\, \delta\, (m_i - E_f - E_e - E_{\bar{\nu}})\, d^3p_e\, d^3p_{\bar{\nu}}\, (8E_e E_{\bar{\nu}} E_f)^{-1} \\ &= (2\pi)^{-5} p_e^2\, dp_e\, d\Omega_e\, E_{\bar{\nu}}^2\, d\Omega_{\bar{\nu}} (8E_e E_{\bar{\nu}} E_f)^{-1}. \end{aligned} \tag{2.4}$$

In the last expression, we have used $d^3p_{\bar{\nu}} = p_{\bar{\nu}}^2\, dp_{\bar{\nu}}\, d\Omega_{\bar{\nu}} = E_{\bar{\nu}}^2\, dE_{\bar{\nu}}\, d\Omega_{\bar{\nu}}$ and integrated $dE_{\bar{\nu}}$ over the $\delta$-function. We now approximate $E_f \approx m_f$ (nuclear recoil will be included at the end of section 5-3):

$$\begin{aligned} d\Gamma_f(e\bar{\nu}) &= \tfrac{1}{2}\pi^{-3} p_e^2\, dp_e\, E_{\bar{\nu}}^2 \int (4\pi)^{-2}\, d\Omega_e\, d\Omega_{\bar{\nu}} \sum_{M_f \lambda_e} |T|^2\, (16 E_e E_{\bar{\nu}} m_i m_f)^{-1} \\ E_{\bar{\nu}} &= m_i - m_f - E_e \equiv E_{\max} - E_e\,. \end{aligned} \tag{2.5}$$

Note that nuclear data tables quote the masses of neutral atoms, and that $E$ in beta decay frequently denotes the kinetic energy $E_e - m_e$. The resulting $E_{\max}$ coincides with our definition (5) in $\beta^-$ decay but is smaller by $2m_e$ in $\beta^+$ decay. The lepton part of the operator in (2) has the following matrix elements according to (1.15):

$$\langle \mathbf{p}_e \lambda_e \mathbf{p}_{\bar{\nu}} \lambda_{\bar{\nu}} | l_\mu^+(0, \mathbf{r}) | 0 \rangle = 2\psi_{el}^+(\mathbf{p}_e, \lambda_e, \mathbf{r}) \sigma_{\mu l} \chi_{0\bar{\nu}}(\tfrac{1}{2}) (2E_{\bar{\nu}})^{\frac{1}{2}} e^{-i\mathbf{p}_{\bar{\nu}}\mathbf{r}}. \tag{2.6}$$

Due to the kinematical limitations on $\mathbf{p}_e$ and $\mathbf{p}_{\bar{\nu}}$ and the shallowness of the Coulomb potential $V$ (for a uniformly charged nucleus of radius $R$ and charge $Z$, the minimum of $V$ is $-\frac{3}{2}Z\alpha R$), the expression (6) varies very little across the nuclear volume and is well approximated by its value at $r = 0$. It is then convenient to introduce the Fourier transform of $h^\mu$:

$$\langle f | h^\mu(0, \mathbf{r}) | i \rangle = (2\pi)^{-3} \int d^3q e^{i\mathbf{q}\mathbf{r}} H_{if}^\mu(\mathbf{q}). \tag{2.7}$$

The integration over $d^3r$ in (2) produces approximately a factor $(2\pi)^3\, \delta_3(\mathbf{q})$, and integration over $d^3q$ gives

$$T_{ife\bar{\nu}} \approx 2G_V H_{if}^\mu(0) \psi_{el}(\mathbf{p}_e, \lambda_e, \mathbf{0}) \sigma_{\mu l} \chi_{0\bar{\nu}}(\tfrac{1}{2}) E_{\bar{\nu}}^{\frac{1}{2}}. \tag{2.8}$$

This is the "allowed approximation." $H_{if}^\mu(0)$ is zero unless the initial and final nuclear states have identical parities and differ in their spins at most by one unit. It is maximal if the two states belong to the same isospin multiplet ("superallowed transitions").

Next, we study $\psi_e(\mathbf{p}_e, \lambda_e, 0)$ by inserting the partial-wave decomposition (1-7.31)

$$\psi(\mathbf{p}, \lambda, \mathbf{r}) = \sum_{\kappa} \alpha_{\kappa\lambda}\psi_{\kappa\lambda} = \sum_{\kappa, \lambda} C^{L\,\frac{1}{2}\,j}_{0\,\lambda\,\lambda}\, i^L[4\pi(2L+1)]^{\frac{1}{2}} \begin{pmatrix} g_\kappa \chi_L^{j\lambda} \\ i f_\kappa \chi_{\bar{L}}^{j\lambda} \end{pmatrix} \tag{2.9}$$

The coupled differential equations (1-7.18) and (1-7.19) must now be solved for an extended Coulomb potential $V(r)$:

$$\begin{aligned} \partial_r g &= -r^{-1}(1+\kappa)g + (E + m - V)f, \\ \partial_r f &= r^{-1}(\kappa - 1)f + (m + V - E)g. \end{aligned} \tag{2.10}$$

Since $V(0)$ is finite, the $r \to 0$-behavior of $g$ and $f$ is given by the first term in each equation:

$$g_\kappa(r \to 0) = (E+m)^{\frac{1}{2}}\hat{g}_\kappa r^{-\kappa-1}, \qquad f_\kappa(r \to 0) = (E-m)^{\frac{1}{2}}\hat{f}_\kappa r^{\kappa-1}. \tag{2.11}$$

The energy square roots have been separated off in view of (1-9.10).

The integrability condition requires $\hat{g}_\kappa = 0$ for $\kappa > 0$ and $\hat{f}_\kappa = 0$ for $\kappa < 0$; thus for $r \to 0$ only $g_{-1}$ and $f_1$ survive, and (9) reduces to

$$\begin{aligned} \psi(\mathbf{p}, \lambda, 0) &= (4\pi)^{\frac{1}{2}}[C^{0\,\frac{1}{2}\,\frac{1}{2}}_{0\,\lambda\,\lambda}\psi_{-1,\lambda}(p, 0) + i3^{\frac{1}{2}}C^{1\,\frac{1}{2}\,\frac{1}{2}}_{0\,\lambda\,\lambda}\psi_{1,\lambda}(p, 0)] \\ &= (4\pi)^{\frac{1}{2}} \begin{pmatrix} \hat{g}_{-1}\sqrt{E+m} \\ 2\lambda \hat{f}_1 \sqrt{E-m} \end{pmatrix} \chi_0^{j\lambda}. \end{aligned} \tag{2.12}$$

Here we have inserted 1 and $-2\lambda 3^{-\frac{1}{2}}$ for the two Clebsch-Gordans. Moreover, from (1-7.11) we have $(4\pi)^{\frac{1}{2}}\chi_0^{j\lambda} = \chi(\lambda)$, such that the only difference between (12) and a free-particle spinor (A-3.2) resides in the weights $g_{-1}$ and $f_1$ of the large and small components of the spinor. Thus the leptonic part of (8) is

$$L^{\text{Coul}}_{\mu e \bar{\nu}} \equiv 2\psi^+_{el}(0)\sigma^\mu \chi_{0\,\bar{\nu}}(\tfrac{1}{2})(2E_{\bar{\nu}})^{\frac{1}{2}} = \tfrac{1}{2}(\hat{g}_{-1} + \hat{f}_1)L_{\mu e \bar{\nu}} + \tfrac{1}{2}(\hat{g}_{-1} - \hat{f}_1)L'_{\mu e \bar{\nu}}, \tag{2.13}$$

where $L_{\mu e \bar{\nu}}$ is given by twice the lefthanded current matrix element (3-6.22) with $p_{\bar{\nu}} = E_{\bar{\nu}}$:

$$L_{\mu e \bar{\nu}} = 2(E - 2\lambda p)^{\frac{1}{2}}(2E_{\bar{\nu}})^{\frac{1}{2}}\chi^+_{0e}(\lambda)\sigma^\mu \chi_{0\,\bar{\nu}}(\tfrac{1}{2}), \tag{2.14}$$

and $L'_{\mu e \bar{\nu}}$ is the corresponding quantity with a righthanded 2-component electron spinor, which differs from (14) only by the sign in front of $2\lambda p$ in the first square root:

$$L^{\text{Coul}}_{\mu e \bar{\nu}} = \{(\hat{g}_{-1} + \hat{f}_1)(E - 2\lambda p)^{\frac{1}{2}} + (\hat{g}_{-1} - \hat{f}_1)(E + 2\lambda p)^{\frac{1}{2}}\}(2E_{\bar{\nu}})^{\frac{1}{2}}\chi^+_{0e}(\lambda)\sigma^\mu\chi_{0\,\bar{\nu}}(\tfrac{1}{2}). \tag{2.15}$$

Consider next the nuclear matrix element $H^\mu_{if}(0)$ of (8) for the case of spinless states $|i\rangle$ and $|f\rangle$. The only 4-vectors which we can form out of the nuclear degrees of freedom are $(P_i + P_f)^\mu$ and $(P_i - P_f)^\mu$, but since we neglect recoil effects we have

$$P^\mu_f \approx P^\mu_i = (m_i, \mathbf{0}), \qquad H^\mu_{if}(0) = M_F\, 2m_i\, \delta_{\mu 0}, \tag{2.16}$$

$$T_{ife\bar{\nu}} \approx 2^{-\frac{1}{2}} G_V M_F\, 2m_i \{\;\}(2E_{\bar{\nu}})^{\frac{1}{2}} R_0, \tag{2.17}$$

$$R_0 = \chi^+_{e0}(\lambda)\chi_{\bar{\nu}0}(\tfrac{1}{2}) = d_{\lambda\frac{1}{2}}(\vartheta)\exp\{i\varphi(\tfrac{1}{2} - \lambda)\}, \qquad \vartheta = \pi - \vartheta_{\bar{\nu}}, \tag{2.18}$$

where $\vartheta$ and $\varphi$ are the angles of $-\mathbf{p}_{\bar{\nu}}$ according to (3-6.20), in a coordinate system where $\mathbf{p}_e = \mathbf{p}$ points along the $z$-axis (compare (1-4.17) for $m \to \lambda$, $\lambda \to \frac{1}{2}$). The proportionality constant $M_F$ in (16) is called the "Fermi matrix element." In the decay rate, we need $|R_0|^2$:

$$d_{\lambda\lambda}^2 = \cos^2 \tfrac{1}{2}\vartheta = \tfrac{1}{2}(1 + \cos\vartheta), \qquad d_{-\lambda\lambda}^2 = \sin^2 \tfrac{1}{2}\vartheta = \tfrac{1}{2}(1 - \cos\vartheta),$$
$$|R_0|^2 = \tfrac{1}{2} - \lambda_e \cos\vartheta_{\bar{\nu}}. \tag{2.19}$$

Integration over the normally not observed neutrino angles eliminates the $\cos\vartheta$-contribution in (19), and one obtains

$$\int (4\pi)^{-1}\, d\Omega_{\bar{\nu}} \sum_{\lambda\,\bar{\nu}} |T|^2(16EE_{\bar{\nu}}m_i^2)^{-1} = \tfrac{1}{4}G_v^2|M_F|^2|\{\}|^2(2E)^{-1}$$
$$= \tfrac{1}{4}G_v^2|M_F|^2[|\hat{g}_{-1}|^2 + |\hat{f}_1|^2 - 4\lambda v \operatorname{Re} \hat{g}_{-1}^* \hat{f}_1 + (|\hat{g}_{-1}|^2 - |\hat{f}_1|^2)m/E], \tag{2.20}$$

where $v = p/E$ is the electron velocity. After summation over the electron helicity $\lambda$, the interference between $\hat{g}_{-1}$ and $\hat{f}_1$ drops out, and the differential decay rate (5) becomes

$$d\Gamma_f(e\bar{\nu}) = \tfrac{1}{2}\pi^{-3}p^2\, dpE_{\bar{\nu}}^2\tfrac{1}{2}[|\hat{g}_{-1}|^2 + |\hat{f}_1|^2]G_V^2|M_F|^2(1 + \varepsilon m/E),$$
$$\varepsilon \equiv [|\hat{g}_{-1}|^2 - |\hat{f}_1|^2][|\hat{g}_{-1}|^2 + |\hat{f}_1|^2]^{-1}. \tag{2.21}$$

We shall see below that the main part of $\hat{g}_{-1}$ and $\hat{f}_1$ is strongly energy-dependent and universal for all allowed decays and, to a less extent, even for forbidden decays. It has therefore become customary to write $d\Gamma_f$ for *arbitrary* $\beta$-decays as (with $p\, dp = E\, dE$)

$$d\Gamma_f(e\bar{\nu}) = \tfrac{1}{2}\pi^{-3}pE\, dE(E_{\max} - E)^2F(Z, E)C(E), \tag{2.22}$$
$$F \equiv \tfrac{1}{2}[|\hat{g}_{-1}|^2 + |\hat{f}_1|^2], \tag{2.23}$$

where $F(Z, E)$ is the "Fermi-function," and the "spectrum shape factor" $C$ contains all the rest. For allowed decays $C$ is fairly energy-independent (see fig. 5-2.c), but for forbidden decays it is not (fig. 5-2.d). A plot of $(d\Gamma/pEFdE)^{\frac{1}{2}}$ versus $E$ is called a "Kurie plot" or "Fermi plot." For allowed decays, it gives nearly a straight line (fig. 5-2.b), which is convenient for an experimental determination of the endpoint energy $E_{\max}$.

A good estimate of $F$ is obtained from (1-9.10) which applies for a point Coulomb potential, provided the factor $\rho^{\gamma_0 - 1}$ with $\gamma_0 = (1 - Z^2\alpha^2)^{\frac{1}{2}}$ according to (1-9.5) is taken as $(pR)^{\gamma_0 - 1}$ where $R$ is some "nuclear radius." The hypergeometric function is 1 at the origin, and the residual phase is unessential. We can thus put

$$F = F_0 L_0, \; F_0 = 4|\Gamma(\gamma_0 + i\eta)|^2\Gamma^{-2}(2\gamma_0 + 1)e^{-\pi\eta}(2pR)^{2\gamma_0 - 2}, \tag{2.24}$$

where again $L_0$ takes care of the rest. The main effect comes from the normalization factor $N$ of (1-9.13). For light nuclei ($\gamma_0 \approx 1$) and nonrelativistic electrons ($E \sim m$), $F_0$ reduces to Gamow's penetration factor $C_0$ of (A-2.5):

$$F_0(Z^2\alpha^2 \ll 1, E \sim m) = 2\pi\eta(e^{2\pi\eta} - 1)^{-1}, \tag{2.25}$$

which is an important factor also in $C_L$ for $L > 0$.

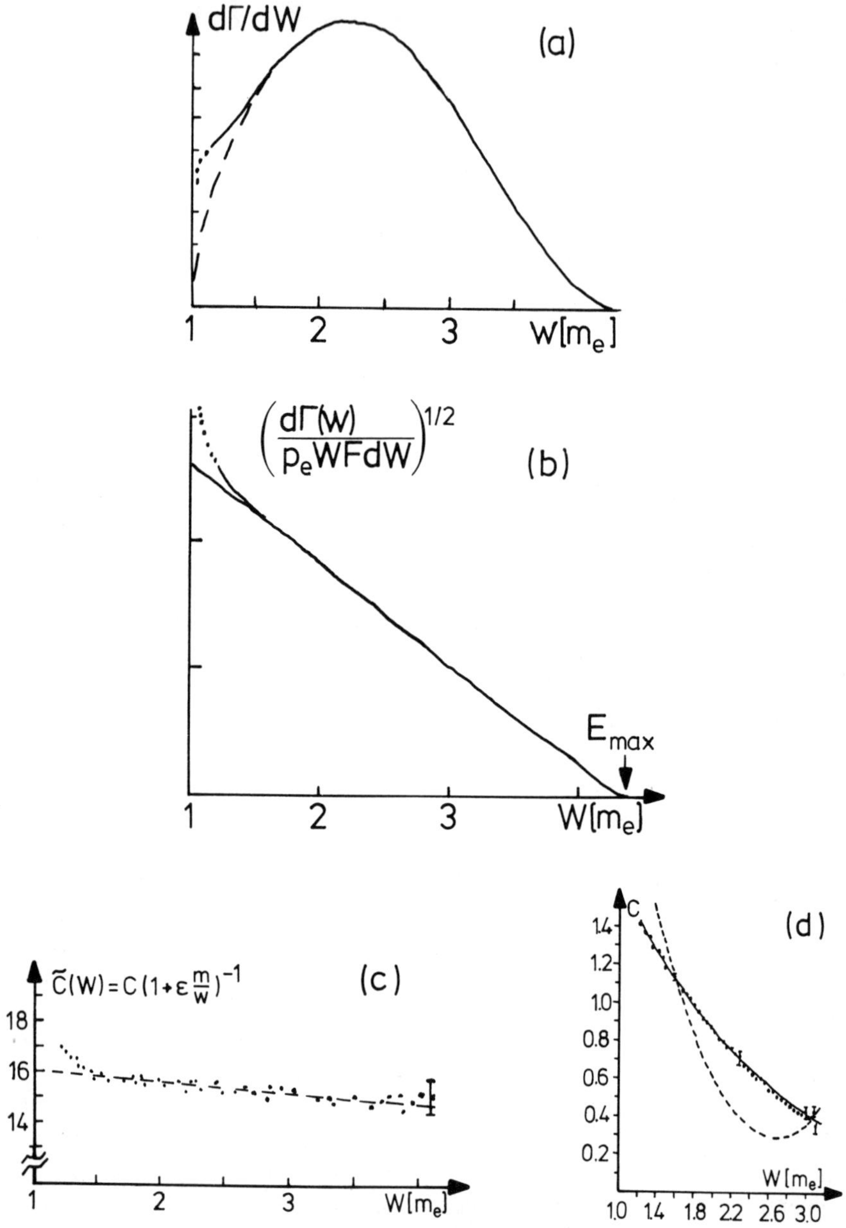


**Figure 5-2,** a-d The decay $^{32}P \to {}^{32}Se\bar{\nu}$. (a) The electron energy spectrum in arbitrary units. The dashed curve indicates a corresponding positron emission spectrum. (b) The Kurie plot in arbitrary units. Also shown is a straight line corresponding to $C = \text{const}$. (c) The spectrum shape factor $\tilde{C}(W)$. The factor $(1 + \varepsilon m/w)$ of (36) has been divided off such that one expects $\tilde{C} = \text{const}$. The dashed curve is $C_0(1 - 0.027\ W)\ W = E_e$. From Flothmann et al. (1969) (after detector corrections due to Bremsstrahlung, $\tilde{C} = C_0'(1 - 0.019\ W)$, Wiesner 1973). (d) shape factor of the non-unique forbidden $1^- \to 0^+$ decay of $^{210}$Bi (Figure from Behrens and Szybisz, 1974).

The total decay rate of allowed decays is well approximated by the so-called "Fermi integral"

$$\Gamma \approx \tfrac{1}{2}\pi^{-3} m_e^5 fC, \qquad f = \int_{m_e}^{E_{\max}} dE F(Z, E) pE(E_{\max} - E)^2 m_e^{-5}. \quad (2.26)$$

The powers of $m_e$ are introduced to make $f$ dimensionless. $\beta$-decay is an old field of physics of practical importance, and is a bit overloaded with concepts. Instead of $\Gamma$ or $C$, one normally quotes "$ft_{\frac{1}{2}}$-values" or "reduced halflives" $ft_{\frac{1}{2}}$, where $f$ is given by (26) also for forbidden decays, and $t_{\frac{1}{2}}$ is the "half-life"

$$t_{\frac{1}{2}} = \Gamma^{-1} \ln 2. \quad (2.27)$$

For allowed decays

$$ft_{\frac{1}{2}} = 2\pi^3 \ln 2\, m_e^{-5} C^{-1} \equiv KC^{-1}. \quad (2.28)$$

So far we have concentrated our attention on the allowed decays between spinless nuclear states. The most general allowed nuclear matrix element can be written in the form

$$H^{\mu}_{if}(0) = \int d^3r \langle f | h^{\mu}(0, \mathbf{r}) | i \rangle \equiv 2m_i(M_F, \mathbf{M}_{GT}), \quad (2.29)$$

where the "Gamow-Teller" part $\mathbf{M}_{GT}$ is a vector which is constructed out of spin states only. For example, if the initial nucleus has spin 1, (29) must be proportional to its spin state $\boldsymbol{\varepsilon}_i(M_i)$. If the final nucleus is spinless, we have

$$M_F = 0, \qquad \mathbf{M}_{GT} = M_{GT}\, \boldsymbol{\varepsilon}_i(M_i). \quad (2.30)$$

If initial and final nuclei have spins $S_i = S_f > 0$, we have

$$M_F = M_V \langle M_f | M_i \rangle, \qquad \mathbf{M}_{GT} = f_A M_A [S(S+1)]^{-\frac{1}{2}} \langle M_f | \mathbf{S} | M_i \rangle, \quad (2.31)$$

where $|M\rangle$ only refers to the spin part of the nuclear state. In $\mathbf{M}_{GT}$, a square root has been included because when all spin summations and angular integrations are done, the spin operator $\mathbf{S}$ enters the decay rate as $\mathbf{S}^2 = S(S+1)$. The factor $f_A$ has been included for the convenience of models which calculate $H^{\mu}_{if}$ from the single-nucleon matrix elements (1.12). Most important is the case $S = \frac{1}{2}$, which includes the free neutron decay itself:

$$M_F = M_V \chi^+_{f0}(M_f) \chi_{i0}(M_i), \qquad \mathbf{M}_{GT} = 3^{-\frac{1}{2}} f_A M_A \chi^+_{f0}(M_f) \boldsymbol{\sigma} \chi_{i0}(M_i). \quad (2.32)$$

The full decay matrix element is a generalization of (17):

$$T_{ife\bar{\nu}} \approx 2^{-\frac{1}{2}} G_V 2m_i \{\ \} (2E_{\bar{\nu}})^{\frac{1}{2}} (M_F R_0 + \mathbf{M}_{GT} \mathbf{R}), \quad (2.33)$$

$$\mathbf{R} = \chi^+_{e0}(\lambda) \boldsymbol{\sigma} \chi_{\bar{\nu}0}(\tfrac{1}{2}), \qquad R_z = 2\lambda R_0, \qquad R_x = 2i\lambda R_y = R_0(-\lambda), \quad (2.34)$$

where the curly brackets are taken from (15), and the coordinate system for $\mathbf{R}$ is chosen as for $R_0$ in (18). The only nontrivial point in (33) is the plus sign in front of the Gamow-Teller part. There is no metric minus sign in this case because (15) represents a lefthanded current with lower index $\mu$, i.e., $\sigma^{\mu}$ stands for $\sigma_{l\mu}$ (compare 1-5.2).

The differential decay rate of unpolarized spin-$\frac{1}{2}$ nuclei involves $|M_V R_0|^2 + \frac{1}{3}|f_A M_A|^2 |\mathbf{R}|^2$ according to (33) and (A-4.4). Using (34) and (18), we find

$$|\mathbf{R}|^2 = \tfrac{3}{2} + \lambda_e \cos \vartheta_{\bar{\nu}}, \tag{2.35}$$

and the rate is again given by (20), (21) or (22), but with $|M_F|^2$ replaced by $|M_V|^2 + |f_A M_A|^2$. In particular, we have

$$C = G_V^2(|M_V|^2 + |f_A M_A|^2)(1 + \varepsilon m/E). \tag{2.36}$$

The longitudinal electron polarization (i.e., its helicity expectation value) follows from (20). In the approximation $\hat{g}_{-1} = \hat{f}_1$, we have

$$P_{e,l} = (|T(\tfrac{1}{2})|^2 - |T(-\tfrac{1}{2})|^2)\left(\sum_{\lambda_e} |T(\lambda_e)|^2\right)^{-1} = -v = -p/E. \tag{2.37}$$

The antineutrino helicity is always $\frac{1}{2}$. Leptonic matrix elements of $\beta^+$-decay $A_z \to A_{z-1}\bar{e}\nu$ are easily obtained from (3-6.22) by replacing $E_e \to E_\nu$, $p_e \to p_\nu = E_\nu$, and by using the Coulomb amplitudes $g_{-1}^+$ and $f_1^+$ of positively charged particles in (14):

$$L_{\mu\bar{e}\nu}^{\mathrm{Coul}} = \{(\hat{g}_{-1}^+ + \hat{f}_1^+)(E + 2\lambda p)^{\frac{1}{2}} + (\hat{g}_{-1}^+ - \hat{f}_1^+)(E - 2\lambda p)^{\frac{1}{2}}\}$$
$$(2E_\nu)^{\frac{1}{2}} \chi_{0\nu}^{\dagger}(-\tfrac{1}{2})\sigma^{\mu}\chi_{0\bar{e}}(\lambda), \tag{2.38}$$

i.e., the neutrino helicity is always $-\frac{1}{2}$ and the positron helicity expectation value for $\hat{g}_{-1}^+ = \hat{f}_1^+$ is $+v$. The Fermi function $F^+$ of positrons differs vastly from $F$ if $|\eta| > 1$. This follows already from (25) where $\eta$ is now positive, $F_0(\eta > 1) \approx 2\pi\eta e^{-2\pi\eta} \ll 1$. Electron capture $e^- A_Z \to A_{Z-1}\nu$ can frequently compete with $\beta^+$ decay because it has $2m_e$ more energy available and no Coulomb repulsion. In this case the matrix element follows already from (A-3.24) without application of the substitution rule. Here we need only the limit of nonrelativistic electrons $E = m$, $p = 0$. In that case the curly bracket in (38) reduces to $2g_{-1} m_e^{\frac{1}{2}}$, and $g_{-1}$ is essentially $\psi(0)$ as in (2-3.20), except for the different normalization of bound states. The Lorentz-invariant phase space is now $(4\pi)^{-2}\, d\Omega_\nu E_\nu m_i^{-1}$ according to (2-2.7), and the capture rate for 2(!) electrons in the $s$-state follows from (2-3.19) as

$$\Gamma(2e^- A_Z \to \nu A_{Z-1}) = (4m_i m_e)^{-1} |\psi(0)|^2 \sum_{\lambda_e} |T|^2 \frac{E_\nu}{4\pi m_i}$$
$$= \pi^{-1} C |\psi(0)|^2 E_\nu^2, \tag{2.39}$$

with $\psi$ normalized as in (2-3.20).

Note however that a relativistic evaluation of $\psi(0)$ may be necessary even when the electron energy is nonrelativistic.

## 5-3 Superallowed Decays, Neutron Decay

When the nucleus is described by $N$ nucleons bound in an effective central potential, insertion of the single-nucleon matrix element (1.12) into (2.29) gives

$$H^{\mu}_{if}(0) = \langle f \,|\, \sum_{k=1}^{N} (1 + f_A \gamma_5^{(k)}) \gamma^{(k)\mu} \tau_+^{(k)} \,|\, i \rangle 2m_i, \tag{3.1}$$

where $\tau_+^{(k)}$ is the isospin raising operator of the $k$th nucleon as in (1.4). The factor $2m_i$ appears in (1) because from here on, the states $|\, i\rangle$ are normalized to 1 instead of $2m_i$. Compare the remark at the end of section 3-3. For allowed transitions, only the static limit of (1) is needed, which has already been given in the discussion following (1.4). With the notation introduced in (2.29), it is

$$M_F = \langle f \,|\, \sum_{k=1}^{N} \tau_+^{(k)} \,|\, i \rangle, \qquad \mathbf{M}_{GT} = f_A \langle f \,|\, \sum_{k=1}^{N} \boldsymbol{\sigma}^{(k)} \tau_+^{(k)} \,|\, i \rangle. \tag{3.2}$$

A cms-constrained integration over all nuclear coordinates is understood. The operator entering $M_F$ is just the total isospin raising operator $I_+ = (I_1 + iI_2)$, and to the extent that nuclear states are pure isospin states, we have

$$M_F \approx \langle I' I'_3 S_f M_f \,|\, I_+ \,|\, I I_3 S_i M_i \rangle = \delta_{S_i S_f}\, \delta_{M_i M_f}\, \delta_{II'},\, \delta_{I_{3'}, I_3 - 1} C_+, \tag{3.3}$$

where the matrix element $C_+$ is the analogue of (B-3.7)

$$C_+ = [I(I+1) - I'_3(I'_3 - 1)]^{\frac{1}{2}}. \tag{3.4}$$

Thus $M_F$ is $\approx C_+$ if initial and final states belong to the same isospin multiplet (superallowed transitions), and zero otherwise.

The operator entering $\mathbf{M}_{GT}$ is also independent of the nucleon coordinates. For nuclei in which the open-shell nucleons are in $s$-states, the wave function factorizes approximately into a space-dependent part $\psi_r$ and a spin-isospin dependent part $\psi_{SI}$, and we get again $\mathbf{M}_{GT} = 0$ if the space-dependent parts are orthogonal, and

$$\mathbf{M}_{GT} = f_A \psi^+_{S'I'} \sum_{k=1}^{N} \vec{\sigma}^{(k)} \tau_+^{(k)} \psi_{SI} \tag{3.5}$$

if they are equal. These formulas can be applied to the decay of the free neutron itself, where comparison with (2.32) gives

$$M_V = 1, \qquad M_A = 3^{\frac{1}{2}}. \tag{3.6}$$

In this case it is of course unnecessary to use nuclear models, because no integration is left in (2.2) after integration over the cms variable $\mathbf{R}$ (= neutron position). We thus have

$$T_{npe\bar{\nu}} = 2^{-\frac{1}{2}} G_V \langle p \,|\, h^{\mu}(0, \mathbf{0}) \,|\, n \rangle L^{\text{Coul}}_{\mu e \bar{\nu}}, \tag{3.7}$$

where the leptonic part is given by (2.13) or (2.15). However, (7) cannot be quite exact either, since $\bar{\psi}_e(r=0)$ is infinite when the final proton is treated as a point particle. In practice, the difficulty is cured in the approximation (2.24) by inserting for $R$ the charge radius of the proton, and the finite proton size disappears completely in the approximation (2.25). Since $\eta = -\alpha E/p$ is $\ll 1$ for nearly all $E$, we can also approximate

$$F_0 \approx 1 - \pi\eta = 1 + \pi\alpha E/p, \tag{3.8}$$

which allows us to evaluate the Fermi integral (2.26) analytically:

$$\begin{aligned} f \approx \tfrac{1}{4}m_e^{-1}\left[E_{\max}\ln\frac{E_{\max}+p_{\max}}{m_e} - p_{\max} - \frac{1}{3}\frac{p_{\max}^3}{m_e^2}\left(1-\frac{2}{5}\frac{p_{\max}^2}{m_e^2}\right)\right] \\ + \pi\alpha\left[\tfrac{1}{30}E_{\max}^5 m_e^{-5} - \tfrac{1}{3}E_{\max}^2 m_e^{-2} + \tfrac{1}{2}E_{\max}/m_e - \tfrac{1}{5}\right]. \end{aligned} \tag{3.9}$$

Thus the neutron decay rate is given by (2.26) with $C$ defined by (2.36), and $\varepsilon \approx 0$:

$$C = G_V^2(1 + 3|f_A|^2). \tag{3.10}$$

Since $G_V$ is obtained from pure Fermi transitions using (3), the neutron decay rate is used to determine $|f_A|$ as given in (1.13). The phase of $f_A$ can be obtained from the electron angular distribution in the decay of polarized neutrons (the weak decay constant $G_V$ is defined to be real, which is possible to lowest order in the weak interaction). With neutron polarization $\mathbf{P} \neq 0$, the absolute square of the decay matrix element (2.33), summed over $M_f$, is given by (A-4.4):

$$\begin{aligned} \overline{|T|^2} &= 4G_V^2 E_{\bar\nu}(E - 2\lambda p)[|M_V|^2|R_0|^2 + \tfrac{1}{3}|f_A M_A|^2|\mathbf{R}|^2 + \mathbf{Z}\cdot\mathbf{P}], \\ \mathbf{Z} &= i|f_A M_A|^2\mathbf{R}^* \times \mathbf{R} - 2\,\mathrm{Re}\,(3^{-\frac{1}{2}}M_V f_A M_A R_0^*\mathbf{R}). \end{aligned} \tag{3.11}$$

Helicity $+\frac{1}{2}$ of the antineutrino is understood, and a common Coulomb factor is neglected. In the following, we evaluate only those parts of $\mathbf{Z}$ which remain after averaging over the azimuthal angle $\varphi$ of the neutrino. According to (2.34) $R_0R_x^*$, $R_0R_y^*$, $R_zR_x^*$, $R_zR_y^*$ are all proportional to $R_0(\lambda)R_0^*(-\lambda)$, which has a $\varphi$-dependence $\exp(-2i\varphi\lambda)$ according to (2.18). Therefore we have $Z_x = Z_y = 0$ after $\varphi$-integration. For $Z_z$ we find after insertion of (6)

$$\begin{aligned} Z_z &= i|f_A|^2(R_x^*R_y - R_y^*R_x) - 4\lambda|R_0(\lambda)|^2\,\mathrm{Re}\,f_A \\ &= 4\lambda|f_A R_0(-\lambda)|^2 - 4\lambda\,\mathrm{Re}\,f_A|R_0(\lambda)|^2 \\ &= |f_A|^2(2\lambda + \cos\vartheta_{\bar\nu}) - \mathrm{Re}\,f_A(2\lambda - \cos\vartheta_{\bar\nu}). \end{aligned} \tag{3.12}$$

After averaging over $\vartheta_{\bar\nu}$, (12) reduces to $2\lambda|f_A|^2 - 2\lambda\,\mathrm{Re}\,f_A$, $|R_0|^2$ and $|\mathbf{R}|^2$ are $\frac{1}{2}$ and $\frac{3}{2}$ according to (2.19) and (2.35), respectively, and (11) becomes

$$\overline{|T|^2} = 4G_v^2 E_{\bar\nu}(E - 2\lambda p)[\tfrac{1}{2} + \tfrac{3}{2}|f_A|^2 + 2\lambda P_z(|f_A|^2 - \mathrm{Re}\,f_A)]. \tag{3.13}$$

After summation over $\lambda$, this expression simplifies further, with $4\lambda^2 = 1$

$$\sum_{\lambda} \overline{|T|^2} = 4G_V^2 E_\nu E[1 + 3|f_A|^2 - 2P_z v(|f_A|^2 - \mathrm{Re}\, f_A)]. \qquad (3.14)$$

Since we have chosen the $z$-axis along $\mathbf{p}$ in section 5-2, $P_z$ stands for $\mathbf{P}\hat{\mathbf{p}}$. Experimentally one finds $\mathrm{Re}\, f_A = f_A$ or with $f_A = |f_A| \exp(i\phi)$: $\phi = (1.1 \pm 1.3)^0$ (see Kropf 1974). Thus $f_A$ is real, and $\beta$-decay is time-reversal invariant according to our general discussion at the end of section 4-8. The phase $\delta_J$ of (4-8.29) is of course the Coulomb phase which we have neglected from eq. (11) onwards.

Another decay to which (5) applies is triton decay $t \to he\bar{\nu}$ ($h \equiv$ helion $\equiv$ $^3$He). The space-dependent part $\psi_r$ of these nuclear states is totally symmetric here (all 3 nucleons are in relative $S$- or $D$-states), such that $\psi_{SI}$ must be totally antisymmetric in order to satisfy the generalized Pauli principle of section 4-2. There are exactly 4 such states, two of which contain two protons (the helion states), the other two containing two neutrons (the triton states). The two states of each kind are distinguished by the values $\pm\frac{1}{2}$ of $M$. Denoting the single-particle states by $pM$ and $nM$ respectively, we have

$$\begin{aligned} |hM\rangle &= A|pM, p, -M, nM\rangle \\ &= 6^{-\frac{1}{2}}\{|pM, p, -M, nM\rangle - |p, -M, pM, nM\rangle \\ &\qquad - |pM, nM, p, -M\rangle + - \cdots\} \end{aligned} \qquad (3.15)$$

where $A$ stands for antisymmetrization $(3!)^{-\frac{1}{2}} \det\{\ \}$, and the signs are arranged as in the $\pi\pi\pi$ state of $I = 0$ in table 4-2.4 which is also antisymmetric. Note that the antisymmetrization determines spin and isospin of this 3-nucleon system, $S = I = \frac{1}{2}$. The initial triton state can be obtained from (15) by the isospin-lowering operator $I_-$:

$$|tM\rangle = I_-|hM\rangle = \sum_{i=1}^{3} \tau_-^{(i)}|hM\rangle = A\left\{\sum_{i=1}^{3} \tau_-^{(i)}|pM, p, -M, nM\rangle\right\}. \qquad (3.16)$$

The operation of $\tau_+^{(k)}$ of (5) on (16) is simplified by the commutation relation

$$\begin{aligned} &[\tau_+^{(k)}, \tau_-^{(i)}] = \delta_{ik}\tau_3^{(k)}: \\ &\tau_+^{(k)}|tM\rangle = \left(\tau_3^{(k)} + \sum_i \tau_-^{(i)}\tau_+^{(k)}\right)|hM\rangle = \tau_3^{(k)}|hM\rangle. \end{aligned} \qquad (3.17)$$

$\tau_+^{(k)}|hM\rangle = 0$ because $\tau_+$ transforms $|nM\rangle$ into $|pM\rangle$ which is already occupied. Taking next $M_i = M_f = \frac{1}{2}$ in (5), we get

$$M_{GTx} = M_{GTy} = 0, \qquad M_{GTz} = f_A\langle h\tfrac{1}{2}|\sum_k \sigma_z^{(k)}\tau_3^{(k)}|h\tfrac{1}{2}\rangle = -f_A, \quad (3.18)$$

since $p\frac{1}{2}$, $p-\frac{1}{2}$, $n\frac{1}{2}$ are all eigenstates of $\sigma_z\tau_3$ with eigenvalues $1, -1, -1$ respectively. Comparison with (2.32) (in which $t$ and $h$ are treated as "elementary particles" of spin $\frac{1}{2}$) shows that

$$M_A = -3^{\frac{1}{2}}, \qquad (3.19)$$

which differs from the free neutron case (6) just by a sign. In (14), this means replacing $-\mathrm{Re}\, f_A$ by $+\mathrm{Re}\, f_A$ and gives a large correlation between the triton polarization and electron emission angle.

When spin-orbit coupling is included, we can no longer put $\psi = \psi_r \psi_{SI}$ and must return to the more general expression (2) for $\mathbf{M}_{GT}$. In $N = 4n + 1$ nuclei ($n = 1, 2, \ldots 9$) it is essentially one nucleon with $j = S$, $L = j \pm \frac{1}{2}$ which makes the $\beta$-transition. The matrix elements can then be computed from the states $\chi_L^{jM}$ of (1-7.11). Details are given by Konopinski (1966).

Finally, we indicate how to include nuclear recoil corrections. With a recoil kinetic energy $\frac{1}{2}m_f^{-1}p_f^2$ and $p_f^2 = (\mathbf{p}_e + \mathbf{p}_\nu)^2$, the maximal electron energy and the neutrino energy become

$$E_{\max} = \Delta m - \tfrac{1}{2}m_f^{-1}p_{\max}^2, \qquad \Delta m \equiv m_i - m_f, \tag{3.20}$$

$$\begin{aligned} E_\nu &= \Delta m - \tfrac{1}{2}m_f^{-1}(p_e^2 + 2\mathbf{p}_e\mathbf{p}_\nu + E_\nu^2) - E \\ &= E_{\max} - E + m_f^{-1}(E\,\Delta m - E^2 - \mathbf{p}_e\mathbf{p}_\nu). \end{aligned} \tag{3.21}$$

These two formulas replace the last line of (2.5). The $\mathbf{p}_e\mathbf{p}_\nu$-term vanishes after integration over $\Omega_\nu$ and spin summations (compare 2.19 and 12), and the Fermi integral is given by (2.26), with $(E_{\max} - E)^2$ replaced by $E_\nu^2$ (recoil corrections to $F(Z, E)$ are neglected). The integral is still elementary. In applications it is mainly needed for highly relativistic electrons, i.e., $p_{\max} \approx E_{\max} - \frac{1}{2}E_{\max}^{-1}m_e^2$. For this case we obtain

$$m_e^5 f \approx \tfrac{1}{30}(\Delta m)^5(1 - \tfrac{3}{2}m_f^{-1}\,\Delta m - 5(\Delta m)^{-2}m_e^2 - \cdots). \tag{3.22}$$

## 5-4 General Discussion of Nuclear Beta Decay

The nuclear (or more general "hadronic") weak current matrix element $H_{if}^\mu(\mathbf{q})$ in (2.7) can be reduced to a few scalar form factors $f(q^2)$. We begin with the covariant matrix elements of the operator $h^\mu(0) = h^\mu(x_\nu = 0)$ of (1.11). The results are similar to those found in section 2-7 for the matrix elements $J_{if}^\mu(\mathbf{q})$ of the electromagnetic current operator. Gauge invariance $q_\mu J^\mu = 0$ is no longer required, and parity conservation is also abandoned. When particles $i$ and $f$ are spinless, we have

$$\begin{aligned} \langle f\,|v^\mu(0)|\,i\rangle &= (P_i + P_f)^\mu f_+(t) + q^\mu f_-(t), \\ q^\mu \equiv (P_i - P_f)^\mu, &\qquad t = q_0^2 - \mathbf{q}^2, \end{aligned} \tag{4.1}$$

$$\langle f\,|a^\mu(0)|\,i\rangle = 0. \tag{4.2}$$

(1) is the covariant generalization of (2.16). Neglecting the Coulomb distortion in the lepton current, we have

$$L_{\mu e\bar\nu} = \bar u_e\,\gamma_\mu\,(1 - \gamma_5)v_{\bar\nu}, \qquad q^\mu L_{\mu e\bar\nu} = m_e \bar u_e(1 - \gamma_5)v_{\bar\nu}, \tag{4.3}$$

i.e., the second term in (1) is a correction of the order $m_e(m_i + m_f)^{-1}$. It is important only in muon capture or in the decay $K \to \pi\mu\nu$, where $m_e$ is replaced by $m_\mu$. For spin-$\frac{1}{2}$ hadrons, one splits $H^\mu_{if}$ into a vector part $V^\mu_{if}$ and an axial part $A^\mu_{if}$. For nucleons,

$$\langle p|v^\mu(0)|n\rangle \equiv V^\mu_{np}, \qquad \langle p|a^\mu(0)|n\rangle \equiv A^\mu_{np}, \tag{4.4}$$

$$A^\mu_{np} = \bar{u}_p\gamma_5[g_1\gamma^\mu - g_2\sigma^{\mu\nu}q_\nu + g_3q^\mu]u_n, \qquad g_1(0) = f_A. \tag{4.5}$$

$V^\mu$ has been given in (1.20), and $A^\mu$ has an extra factor $\gamma_5$.

The extension of this formalism to transitions between higher spin states has been reviewed by Holstein (1974). Here we merely wish to derive the number of form factors for transitions between arbitrary spin states. For this purpose, it is useful to consider an auxiliary " $W$-boson," such that $\beta$-decay becomes a two-step process $i \to fW$, $W \to e\bar{\nu}$. This particle differs from a kind of virtual heavy photon not only by its charge, but also by the fact that it contains a spin-zero component. Neglecting the Coulomb scattering of its decay products, we can use our formula (4-8.9) for sequential decays in the limit $m_d = m_w \to \infty$. Including a factor $-g_{\mu\nu}$ from the $W$ spin summation (2-8.9), we have

$$T_{ife\bar{\nu}} = (2^{-\frac{1}{2}}G_V m_W^2)H^\mu_{if}(-g_{\mu\nu})m_W^{-2}(-L^\nu_{e\bar{\nu}}). \tag{4.6}$$

We can now split $H^\mu_{if}$ into one part which is proportional to $q^\mu$ and which couples to scalar $W$ only, and a rest which is gauge invariant and which couples to vector $W$ only. Without loss of generality we assume $S_i \geq S_f$ for the nuclear spins. Then the number of independent form factors for the decay into $f$ plus a scalar $W$ equals the number of helicity amplitudes $T_{S_i}(\lambda_f)$, i.e., $2S_f + 1$. Similarly, the number of independent form factors for the decay into $f$ plus a vector $W$ equals that of helicity amplitudes $T_{S_i}(\lambda_f, \lambda_W)$, which is $3(2S_f + 1)$ for $S_i > S_f$ and $3(2S_f + 1) - 2$ for $S_i = S_f$. In the latter case namely, the requirement

$$|\lambda| = |\lambda_f - \lambda_W| \leq S_i \tag{4.7}$$

for the $d$-functions $d^{S_i}_{M\lambda}(\vartheta)$ excludes the combinations $\lambda = \pm(S_i + 1)$.

We now return to the general analysis of nuclear $\beta$-decay. As the electron's Coulomb interaction is more important than possible recoil effects, we start again from (2.2) and neglect in the following possible relations between $h^0$ and $\mathbf{h}$. Inserting the matrix elements (2.6) and (2.7) into (2.2), we have

$$T_{fe\bar{\nu}} = 2G_V E_{\bar{\nu}}^{\frac{1}{2}} \int d^3r(2\pi)^{-3}\, d^3q e^{i\mathbf{q}\mathbf{r}} H^\mu_{if}(\mathbf{q})\psi^+_{\mathrm{el}}(\mathbf{r})\sigma^\mu\chi_{\bar{\nu}0}(\tfrac{1}{2})e^{-i\mathbf{p}_{\bar{\nu}}\mathbf{r}}. \tag{4.8}$$

The first step is a multipole decomposition of $H^\mu_{if}\sigma^\mu = H^0_{if} + \mathbf{H}_{if}\boldsymbol{\sigma}$ (Weidenmüller 1961). $H^0_{if}$ is constructed out of the spin states of the initial and final nuclei plus the necessary powers of $\mathbf{q}$ to make $H^0_{if}$ a scalar. One expands these powers of $\mathbf{q}$ into spherical harmonics $Y^m_l(\hat{\mathbf{q}})$ and uses the Wigner-Eckart theorem (B-3.16), getting

$$H^0_{if} = \sum_{l,m} Y^m_l(\hat{\mathbf{q}})C^{S_i\, l\, S_f}_{M_i\, m\, M_f}F_{ll0}(q^2)(qR)^l. \tag{4.9}$$

Here we have extracted a factor $q^l$ such that $F(0) \neq 0$ and introduced the same radius $R$ as in (2.24). Similarly, $\mathbf{H}_{if}\boldsymbol{\sigma}$ can be decomposed into irreducible tensor operators with $S = 1$:

$$T^m_{klS}(\hat{\mathbf{q}}) = \sum_v C^{S\ \ l\ K}_{m-v\ v\ m}\, i^l Y^v_l(\hat{\mathbf{q}})(\sigma^{m-v})^S, \tag{4.10}$$

where $\sigma^n (n = -1, 0, 1)$ are the spherical components of the Pauli matrices. (10) has been written in such a form that it can also be used in the expansion (9) with $S = 0$:

$$H^\mu_{if}\sigma^\mu = \sum_{S=0}^{1} \sum_{Klm} (-i)^l C^{S_i\ K\ S_f}_{M_i\ m\ M_f}\, T^m_{KlS}(\hat{\mathbf{q}})(qR)^l F_{KlS}(q^2). \tag{4.11}$$

Due to the factor $(Rq)^l$ and corresponding factors from the lepton wave functions below, only the smallest possible $l$-value in (11) is normally of importance. It occurs for $S = 1$ and is $l_{\min} = K - 1$ according to (10). The minimum of $K$, in turn, is $K_{\min} = |S_i - S_f| \equiv \Delta S$ according to (11). However, the operator (10) has parity $(-1)^l$; in the static limit it contributes only if the product of the parities of the initial and final nuclear states is $\pi_i \pi_f = (-1)^l$. Consequently, for $\pi_i \pi_f = (-1)^{\Delta S}$, $l = \Delta S - 1$ contributes only via the "small nuclear components," and $l = \Delta S = K$ or $K - 1$ can compete. For $\Delta S > 0$ such transitions are called "$\Delta S^{\text{th}}$ nonunique forbidden transitions," and $\pi_i \pi_f = -(-1)^{\Delta S}$-transitions are called "$(\Delta S - 1)^{\text{th}}$ unique-forbidden transitions." The complete classification is given in table 5-4.

**Table 5-4** Classification of nuclear beta decays. Log $ft_{\frac{1}{2}}$-values of forbidden decays from Raman and Gove (1973).

| Class | $l$ | $\pi_i \pi_f$ | $K$ | log $ft_{\frac{1}{2}}$ (sec) |
|---|---|---|---|---|
| superallowed | 0 | 1 | 0, 1 | 2.9–3.6 |
| allowed | 0 | 1 | 0, 1 | 4.4–10.6 |
| 1st forbidden, nonunique | 0, 1 | −1 | 0, 1, 2 | > 5.1 |
| 1st forbidden, unique | 1 | −1 | 2 | 8.5–12.7 |
| 2nd forbidden, nonunique | 1, 2 | 1 | 2, 3 | ≥ 11.9* |
| 2nd forbidden, unique | 2 | 1 | 3 | ≥ 12.8 |
| 3rd forbidden | 2, 3 | −1 | 3, 4 | ≥ 17.6 |

* With the exception of $^{59}\mathrm{Fe}(\frac{3}{2}^-) \to {}^{59}\mathrm{Co}(\frac{7}{2}^-)e^-\bar{\nu}$ which has log $ft_{\frac{1}{2}} = 11.0$

All other nuclear physics is contained in the form factors $F$. For unique forbidden transitions, only $F_{K,K-1,1}$ contributes, with $K = \Delta S$. For non-unique forbidden transitions of $\Delta S > 0$, $F_{\Delta S, \Delta S - 1, 1}$, $F_{\Delta S, \Delta S, 0}$, $F_{\Delta S, \Delta S, 1}$ and $F_{\Delta S + 1, \Delta S, 1}$ can contribute, which explains the name "nonunique." For $\Delta S = 0$ non-unique forbidden transitions, the following six form factors can contribute: $F_{101}$, $F_{000}$, $F_{110}$, $F_{011}$, $F_{111}$, $F_{211}$. The integration in (8) over the directions of $\mathbf{q}$ is now performed using (1-3.11) and (B-5.4)

$$e^{i\mathbf{qr}} = 4\pi \sum_{lm} i^l j_l(qr) Y_l^m(\hat{\mathbf{q}}) Y_l^{-m}(\hat{\mathbf{r}}), \tag{4.12}$$

$$\int d\Omega_q e^{i\mathbf{qr}} T_{KlS}^m(\hat{\mathbf{q}}) = 4\pi i^l j_l(qr) T_{KlS}^{-m}(\hat{\mathbf{r}}), \tag{4.13}$$

$$\begin{aligned} T_{ife\bar{\nu}} = G_V E_{\bar{\nu}}^{\frac{1}{2}} \pi^{-2} \int d^3r q^2\, dq C_{M_i\, m\, M_f}^{S_i\ l\ S_f} j_K(qr) \\ \times (qR)^l F_{KlS}(q^2) \psi_{el}^{+} T_{KlS}^{-m}(\hat{\mathbf{r}}) \chi_{\bar{\nu}0}(\tfrac{1}{2}) e^{-i\mathbf{p}_{\bar{\nu}}\mathbf{r}}. \end{aligned} \tag{4.14}$$

Next, we perform the integration over the angles $\hat{\mathbf{r}}$ of $\mathbf{r}$. For this purpose, $e^{-i\mathbf{p}_\nu\mathbf{r}}$ and $\psi_{el}(\mathbf{r})$ must be partial-wave expanded to orders such that (14) does not vanish after the angular integration. For example, for $l = 1$, the pairs (1, 0) and (0, 1) of $L_e$ and $L_\nu$ will give comparable contributions. For the antineutrino plane wave, one can always use the threshold behaviour $p_{\bar{\nu}}^{L_\nu}$ of the Bessel function. Insertion of (A-1.1) into the expansion analogous to (12) gives

$$e^{-i\mathbf{p}_{\bar{\nu}}\mathbf{r}} = 4\pi \sum_{L_\nu m_\nu} (-i)^{L_\nu} [(2L_\nu + 1)!!]^{-1} (p_{\bar{\nu}} r)^{L_\nu} Y_{L_\nu}^{-m_\nu}(\hat{\mathbf{p}}_{\bar{\nu}}) Y_{L_\nu}^{m_\nu}(\hat{\mathbf{r}}) \tag{4.15}$$

The expansion of $\psi_e(\mathbf{r})$, with $z$-axis along $\mathbf{p}_e$, has been given in (2.9). We need its left-handed component $\psi_{el}(\mathbf{r})$ as defined in (1-5.11)

$$\begin{aligned} \psi_{el} &= 2^{-\frac{1}{2}}(\psi_g - \psi_k) \\ &= 2^{\frac{1}{2}} \pi^{\frac{1}{2}} \sum_{\kappa,\lambda} C_{0\ \lambda\ \lambda}^{L\ \frac{1}{2}\ j_e} i^L (2L+1)^{\frac{1}{2}} (g_\kappa \chi_L^{j_e\lambda} - i f_\kappa \chi_{\bar{L}}^{j_e\lambda}). \end{aligned} \tag{4.16}$$

It is now clear in principle how the $\hat{\mathbf{r}}$-integration is done: $T_{KlS}^m(\hat{\mathbf{r}})$ contributes functions $Y_l^\nu(\hat{r})$ according to (10), the neutrino plane wave contributes $Y_{L_\nu}^{m_\nu}(\hat{\mathbf{r}})$, and $\psi_{el}$ contributes either $Y_L^{m_1}(\hat{\mathbf{r}})$ or $Y_{\bar{L}}^{m_1}(\hat{\mathbf{r}})$ according to (1-7.11), for fixed $j_e$. It cannot contribute both because the integral $\int Y_l^\nu Y_L^{m_1} Y_{L_\nu}^{m_\nu} d\Omega \hat{r}$ vanishes for parity reasons if $L + L_\nu = l + 1$. We refer to Schülke (1964) and Schopper (1966) for the explicit formulas.

Next comes the integration over $r$ in (14). In the past, this has been done by using a power series expansion in $r$ for the functions $g_\kappa$ and $f_\kappa$ in (16). In a way, this is analogous to the expansion (15), which was justified by the smallness of $p_{\bar{\nu}}$. However, it has been shown by Behrens and Bühring (1971) that this is a poor approximation, at least for many nonunique forbidden transitions. Moreover, it can be avoided with little extra work. The starting point for the calculation of $g_\kappa$ and $f_\kappa$ is the pair of differential equations (2.10), and the $r \to 0$ behavior (2.11). The integrability depends on the sign of $K$. For $\kappa = -L - 1$, $j_e = L + \frac{1}{2}$ (remember 1-7.12) we have $\hat{f}_\kappa = 0$,

$$\begin{aligned} g_{L+} &= r^L (E+m)^{\frac{1}{2}} \left[ \hat{g}_{L+} + \int_0^r r'^{-L} dr' (E + m - V) f_{L+}(r'), \right. \\ f_{L+} &= r^{-L} \int_0^r r'^L dr' (V + m - E) g_{L+}(r'), \end{aligned} \tag{4.17}$$

while for $\kappa = L, j_e = L - \frac{1}{2}$, we have $\hat{g}_\kappa = 0$,

$$f_{L-} = r^{L-1}(E-m)^{\frac{1}{2}}\left[\hat{f}_{L-} + \int_0^r r'^{1-L}\,dr'(V+m-E)g_{L-}(r')\right],$$
$$g_{L-} = r^{-L-1}\int_0^r r'^{L+1}\,dr'(E+m-V)f_{L-}(r'). \tag{4.18}$$

It is easily verified by differentiation that these equations solve (2.10). Since $V$, $E+m$ and $E-m$ are all small in $\beta$-decay, the integral equations (17), (18) can be solved by iteration, and the lowest nonvanishing iteration is in fact sufficient. The solutions are most conveniently written for fixed $j_e$,

$$g_{L+} = C_{L+}(E+m)^{\frac{1}{2}}(p_e r)^L, \qquad f_{(L+1)-} = C_{(L+1)-}(E-m)^{\frac{1}{2}}(p_e r)^L, \tag{4.19}$$

$$g_{L-} = C_{L-}(E+m)^{\frac{1}{2}}(p_e r)^L(2L+1)^{-1}[1 - I_L(E+m)^{-1}],$$
$$f_{(L-1)+} = -C_{(L-1)+}(E-m)^{\frac{1}{2}}(p_e r)^L(2L-1)^{-1}[1 - I_{L-1}(E-m)^{-1}], \tag{4.20}$$

$$I_L \equiv r^{-1}(2L+1)\int_0^r dr'(r'/r)^{2L}V(v'). \tag{4.21}$$

Apart from a constant, $I_L$ is the $I(k, 1, 1, 1, r)$ of Behrens and Bühring. It is insensitive to the detailed shape of $V$ and is easily computed for a uniformly charged nucleus,

$$V(r > R) = Z_e Z\alpha/r, \qquad V(r < R) = Z_e Z(\tfrac{3}{2} - \tfrac{1}{2}r^2R^{-2})/R. \tag{4.22}$$

The $r$-integration

$$J(q) = \int_0^\infty r^2\,dr j_K(qr)(p_{\bar{\nu}}r)^{L_\nu}\cdot(g_\kappa \text{ or } f_\kappa) \tag{4.23}$$

can now be performed and only the integration over $q$ remains. Due to the oscillations in $j_K(qr)$, $J(q)$ is a rapidly decreasing function of $q$, and we can use the expansion of the form factor $F(q^2)$ around $q = 0$:

$$F_{KlS}(q^2) = F_{KlS}(0) - \tfrac{1}{2}(2l+3)^{-1}(qR)^2F^{(1)}_{KlS}. \tag{4.24}$$

Normally the first term suffices. One numerical problem remains, namely the computation of the amplitudes $C_{L\pm}$ in (19). For this purpose, $g_\kappa$ and $f_\kappa$ must be expressed in terms of the regular $(g_{\kappa p}, f_{\kappa p})$ and irregular $(\tilde{g}_{\kappa p}, \tilde{f}_{\kappa p})$ point Coulomb wave functions in the region $r_>$ where $V = Z_e Z\alpha/r$:

$$Ag_\kappa(r_>) = Bg_{\kappa p} + C\tilde{g}_{\kappa p}, \qquad Af_\kappa(r_>) = Bf_{\kappa p} + C\tilde{f}_{\kappa p}. \tag{4.25}$$

The coefficients $B$ and $C$ are related by the normalization of the wave function for $r \to \infty$. The asymptotic form of the regular point Coulomb wave function is given by (1-9.12) and (1-9.14) as

$$\begin{pmatrix} g \\ f \end{pmatrix} = (E \pm m)^{\frac{1}{2}}\rho^{-1}a_{\kappa\lambda}\,{\sin \atop \cos}(\rho - \eta\ln 2\rho - \tfrac{1}{2}L\pi + \sigma_\kappa). \tag{4.26}$$

Bhalla and Rose (1961) define the irregular Coulomb wave function as that in which $\gamma$ of (1-9.5) is replaced by $-\gamma$. It is clearly a solution of the differential equation (1-9.4) which contains only $\gamma^2$. From (1-9.14) we have

$$\begin{aligned}&\sigma_\kappa(-\gamma) - \sigma_\kappa(\gamma)\\ &\quad = \pi\gamma + \xi(-\gamma) - \xi(\gamma) + \arg\Gamma(-\gamma + i\eta) - \arg\Gamma(\gamma + i\eta) \equiv \Delta_\kappa .\end{aligned} \tag{4.27}$$

Thus $\tilde{g}_{\kappa p}$ behaves asymptotically as $\rho^{-1}\sin(\rho - \alpha + \Delta_\kappa)$, and

$$\begin{aligned}\sin(\rho - \alpha + \Delta_\kappa) &= \sin(\rho - \alpha)\cos\Delta_\kappa + \cos(\rho - \alpha)\sin\Delta_\kappa ,\\ \alpha &\equiv \eta \ln 2\rho - \tfrac{1}{2}L\pi + \sigma_\kappa ,\end{aligned} \tag{4.28}$$

and averaging of the intensity over one oscillation gives the condition

$$1 = B^2 + C^2 + 2BC\cos\Delta_\kappa . \tag{4.29}$$

On the other hand, by taking the quotient of the two equations (25), we find

$$\frac{g}{f} = \frac{g_p + \tilde{g}_p C/B}{f_p + \tilde{f}_p C/B} \quad \text{or} \quad \frac{C}{B} \equiv H = \frac{fg_p - gf_p}{g\tilde{f}_p - f\tilde{g}_p}. \tag{4.30}$$

Combination with (29) allows one to express both $C$ and $B$ in terms of $H$ which is independent of $A$. $H$ must be computed at a radius $r$ which is large enough that it includes all nuclear charge, but small enough that (19) and (20) are still valid. Further corrections arise from the screening due to atomic electrons (Bühring 1965). In this case an asymptotic form of the type (26) is valid outside the atomic electron cloud, but the factor $Z$ in the $\eta$ in front of the $\ln\rho$ is reduced to 1 for $\beta^-$-decay and $-1$ for $\beta^+$-decay. For allowed decays, the Thomas-Fermi form of a completely screened Coulomb potential can be solved exactly for the Klein-Gordon equation (Durand 1964).

In all these calculations, care is necessary in the evaluation of the phase $\xi$ of (1-9.9). Bhalla and Rose define $\xi$ and $\tilde{\xi}$ to lie in the first quadrant for $\kappa > 0$, and in the second quadrant for $\kappa < 0$, which implies

$$e^{i(\xi - \tilde{\xi})} = -Z_a(-\eta + i\gamma)(\gamma^2 + \eta^2)^{-\frac{1}{2}}. \tag{4.31}$$

A sign error for $-Z_a = -1$ (positrons) is eliminated in later tables by Behrens and Jänecke (1969).

## 5-5 Lepton Number Conservation. Muons and Their Neutrinos. $\mu$-Decay

In $\beta^-$-decay, the electron is produced together with an antineutrino, and in $\beta^+$-decay, the positron is produced together with a neutrino. If we define a number $L$

$$L(e^-) = L(\nu) = 1, \qquad L(e^+) = L(\bar{\nu}) = -1, \tag{5.1}$$

then the total lepton number $L$ is conserved. This conservation law can be tested in secondary interactions of the neutrino. Thus, the antineutrino from neutron decay $n \to pe^- \bar{\nu}$ can make "inverse $\beta$-decay" in hydrogen, $\bar{\nu}p \to ne^+$ (Reines and Cowan 1959) but not reactions such as $\bar{\nu}^{37}\text{Cl} \to {}^{37}\text{A}e^-$. Conversely, solar neutrinos, particularly from the decay ${}^8\text{B} \to {}^8\text{Be}^* e^+ \nu$ should induce the reaction $\nu^{37}\text{Cl} \to {}^{37}\text{A}e^-$ (Davis 1968). Another example is double $\beta$-decay (change of nuclear charge by two units in cases where the change of charge by one unit is energetically impossible). Here the $e^-e^-\bar{\nu}\bar{\nu}$ final state is allowed by lepton number conservation, whereas the neutrinoless $e^-e^-$ state is forbidden. Since this latter state has the larger phase space, the existence or nonexistence of neutrinoless double $\beta$-decay can be deducted from the lifetime of the nuclear state. A measured half-life of $10^{21.4}$ years for ${}^{130}\text{Te} \to {}^{130}\text{Xe}$ favours the ${}^{130}Te \to {}^{130}\text{Xe}_1 e^- e^- \bar{\nu}\bar{\nu}$ interpretation (see review by Bryman and Picciotto 1978).

However, although all these experiments are consistent with lepton number conservation, they do not really test it. The point is that the antineutrino is only emitted and absorbed with helicity $\frac{1}{2}$, while the neutrino is only emitted and absorbed with helicity $-\frac{1}{2}$. In other words, a $\bar{\nu}$ can never produce $e^-$ because it has the wrong helicity. Only if a right-handed current $2\,\delta_e j_r^\mu(\nu e, x)$ is added in (1.9), do the above reactions become possible for $\nu = \bar{\nu}$ ($\Psi_\nu$ is a Hermitean operator in this case), with a probability $\delta_e^2$.

Similar considerations apply to the neutrinos which are produced together with muons. Interactions of such neutrinos are studied systematically at the high-energy accelerators where the decays in flight $\pi^+ \to \mu^+ \nu_\mu$ and $K^+ \to \mu^+ \nu_\mu$ allow the construction of "neutrino beams." These neutrinos produce $\mu^-$ but no $\mu^+$ in secondary collisions. Conversely, the antineutrinos from $\pi^- \to \mu^- \bar{\nu}$, $K^- \to \mu^- \bar{\nu}$ produce $\mu^+$ but not $\mu^-$. The present experimental upper limit for neutrino interactions in an iron target is $\mu^+/\mu^- < 1.6 \times 10^{-4}$ (Holder et al., 1978). Thus $\delta_\mu^2 < 2 \times 10^{-4}$, if $\Psi_{\nu\mu}$ is Hermitean.

The first real test of the conservation of a possible lepton number arises when muons, electrons and their neutrinos are considered simultaneously. From the absence of the decay $\mu \to e\gamma$, Konopinski and Mahmoud (1953) concluded that muons and electrons of equal charge have opposite lepton numbers, which in view of (1) requires $L(\mu^-) = -L(\mu^+) = -1$. Denoting particles of negative lepton number by a bar, we have $\mu^+ = \mu$, $\mu^- = \bar{\mu}$ according to Konopinski and Mahmoud, and pion decays must be written as $\pi^+ \to \mu\bar{\nu}$, $\pi^- \to \bar{\mu}\nu$ if lepton number is conserved. From (1.6), we remember $\lambda_\nu = +\frac{1}{2}$ in $\pi^-$-decay, i.e., this neutrino has the same helicity as the antineutrino which can be absorbed in the $\beta^+$-interaction, $\bar{\nu}p \to e^+n$. Thus the sequence ($\pi^- \to \bar{\mu}\nu$, $\nu p \to e^+ n$) is helicity-allowed, but violates lepton number conservation. The first accelerator neutrino experiment (Danby et al. 1962) clearly showed that this sequence does not occur in nature, thus confirming lepton number conservation.

In the above reasoning, we have assumed that leptons have at most two inner quantum numbers which are conserved, namely electric charge $Q$ and lepton number $L$. However, the experimental absence of neutrinoless $\tau$-decays (in particular $\tau \to e\rho^0$ and $\tau \to \mu\rho^0$) tells us that leptons have at least one more inner quantum number. The scheme of Konopinski and Mahmoud is therefore not compelling. In fact, it has become customary to define

$$L(e^-) \equiv L(\mu^-) \equiv L(\tau^-) \equiv 1, \tag{5.2}$$

and to forbid the unobserved decays by additional "muonic" and "tauonic" charges. Thus $\pi^-$-decay is written as $\pi^- \to \mu\bar{\nu}_\mu$, and the reaction $\bar{\nu}_\mu n \to e^- p$ is forbidden by muonic charge conservation. This has the formal advantage that only left-handed currents are required in weak interactions, at least for electrons and muons (in $\tau$ decays, the relative amounts of $j_r$ and $j_l$ have not yet been determined). In the Konopinski-Mahmoud scheme, on the other hand, $\Psi_\mu^{KM}$ denotes a field which annihilates a $\mu^+$, and the total lepton current operator (1.9) must be rewritten as

$$\begin{aligned} l^{\mu +} &= 2j_l^\mu(\nu e) + 2j_r^\mu(\mu\nu), \\ j_r^\mu(\mu\nu) &= \Psi_{\nu r}^{KM+}\sigma^\mu\Psi_{\mu r}^{KM} = \tfrac{1}{2}\bar{\Psi}_\nu^{KM}\gamma^\mu(1+\gamma_5)\Psi_\mu^{KM}. \end{aligned} \tag{5.3}$$

The change in notation from particles to antiparticles is accompanied by a change from "right-handed" to "left-handed."

Historically, (2) was already introduced in 1957 (Landau, Lee and Yang, Salam) when parity violation was discovered. It was then thought that the very existence of the neutrino should be the origin of parity violation, in the sense that the neutrino's right-handed spinor $\psi_r$ should not exist in nature. This is only possible for a massless particle (see 1-5.13). The resulting equation $\pi_\nu \sigma_l^\nu \psi_l = 0$ is called the "Weyl equation," and the nonexistence postulate for $\psi_r$ goes under the name "2-component neutrino theory." Under the impact of this speculation the $\mu^+$ was immediately stamped "antilepton" when it turned out that it had helicity $-\frac{1}{2}$ in $\pi^+$-decay. However, this argument cannot compelling, since it is contradicted by the nonleptonic weak interactions.

The difficulty of unambiguously assigning additive quantum numbers to leptons is connected with the fact that many possible currents do not interact. For example, if there were doubly charged leptonic and hadronic currents, the Konopinski–Mahmoud scheme could be tested by the decay $K^+ \to \pi^-\mu^+e^+$ (Diamant–Berger 1976) or by the following muon capture in nuclei: $\mu^- pp \to e^+ nn$ (Badertscher et al. 1978).

As an exercise, let us now calculate muon decay in the formalism of Konopinski and Mahmoud, where it is denoted by $\mu^+ \to e^+\nu\nu$. The relevant muon current matrix element is obtained from (3)

$$\langle \nu | l^{\mu+}(x) | \mu^+ \rangle = 2(2E_\nu)^{\frac{1}{2}}\, \delta_{\lambda_\nu, \frac{1}{2}} \chi_{\nu 0}'^{+}(\tfrac{1}{2})\sigma^\mu \psi_r(\mathbf{p}_\mu, M_\mu, x) e^{iP_\nu \cdot x}. \tag{5.4}$$

The decay is induced by $H_{\mu e}$ of (1.17):

$$H_{\mu e}(x) = 2^{-\frac{1}{2}} G_\mu 4[j_l^\mu(\nu e) j_{\mu r}^+(\mu\nu) + j_r^\mu(\mu\nu) j_{\mu l}^+(\nu e)]. \tag{5.5}$$

The $T$-matrix element for this case is

$$T_{\mu\bar{e}\nu\nu} = -\langle \bar{e}\nu_1\nu_2 | H_{\mu e}(0) | \mu^+ \rangle = -2^{-\frac{1}{2}} G_\mu L^\mu_{\mu+\nu} L_{\mu\bar{e}\nu}, \tag{5.6}$$

$$L^\mu_{\mu+\nu} = \langle \nu_1 | l^{\mu+}(0) | \mu^+ \rangle = 2(2E_1)^{\frac{1}{2}}\, \delta_{\lambda_1, \frac{1}{2}} \chi^+_{10}(\tfrac{1}{2})\sigma^\mu \chi_0(\lambda)(E + 2\lambda p)^{\frac{1}{2}}, \tag{5.7}$$

$$L_{\mu\bar{e}\nu} = \langle \bar{e}\nu_2 | l_\mu(0) | 0 \rangle = 2(2E_2)^{\frac{1}{2}}\, \delta_{\lambda_2, -\frac{1}{2}} \chi^+_{20}(-\tfrac{1}{2})\sigma^\mu \chi_{0\bar{e}}(\lambda_{\bar{e}})(E_{\bar{e}} + 2\lambda_{\bar{e}} p_{\bar{e}})^{\frac{1}{2}}, \tag{5.8}$$

(7) is obtained from (4) by insertion of (1-6.14), and (8) is the plane-wave version of (2.38), $g^+_{-1} = f^+_1 = 1$. No index $\mu$ for "muon" is used in (7), and (8) has an index $\bar{e}$ for the positron.

From (7) and (8), we see that (6) vanishes unless $\lambda_1 = -\lambda_2$. From now on we understand that the state $|\nu_1 \nu_2\rangle$ is arranged such that the neutrino with positive helicity comes first. We thus have

$$T_{\mu\bar{e}\nu\nu} = -4 \cdot 2^{\frac{1}{2}} G_\mu E_1^{\frac{1}{2}} E_2^{\frac{1}{2}} (E_{\bar{e}} + 2\lambda_{\bar{e}} p_{\bar{e}})^{\frac{1}{2}} (E + 2\lambda p)^{\frac{1}{2}} T', \tag{5.9}$$

$$T' = \chi^+_{10}(\tfrac{1}{2})\sigma^\mu \chi_0(\lambda)\chi^+_{20}(-\tfrac{1}{2})\sigma^\mu \chi_{\bar{e}0}(\lambda_{\bar{e}}). \tag{5.10}$$

The decay matrix-element is Lorentz-invariant and can be evaluated in any frame of reference. Since one is mainly interested in the electron's energy and angular distribution, it turns out to be convenient to use the cms of the two neutrinos, fig. 5-5, with $\mathbf{p}_\mu = \mathbf{p}_{\bar{e}}$ pointing along the $z$-axis. The angles of $\nu_1$ are denoted by $\vartheta_1$, $\varphi_1$. In the second part of (10), we can express $\chi^+_{20}(-\frac{1}{2})$ in terms of $\chi^+_{10}(\frac{1}{2})$ and $\chi_{\bar{e}0}(\lambda_{\bar{e}})$ in terms of $\chi(-\lambda_{\bar{e}})$ (remembering that the positron spin states $\chi_{\bar{e}0}$ are quantized along $-\mathbf{p}_{\bar{e}}$) by means of (4-4.4):

$$\chi^+_{20}(-\tfrac{1}{2}) = \chi^+_{10}(\tfrac{1}{2}), \qquad \chi_{\bar{e}0}(\lambda_{\bar{e}}) = (-1)^{\frac{1}{2}+\lambda_{\bar{e}}}\chi_0(-\lambda_{\bar{e}}) \tag{5.11}$$

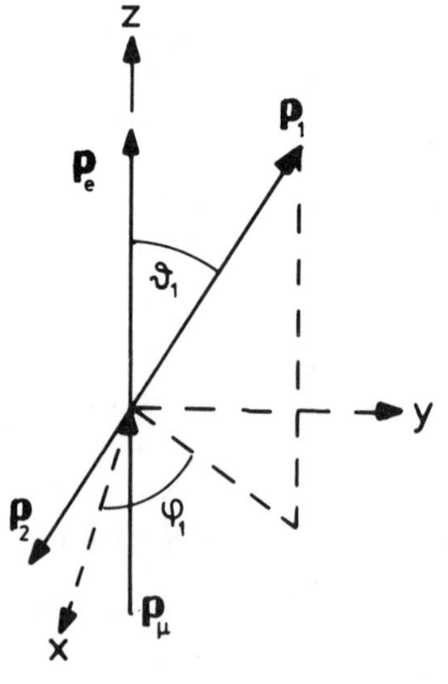


**Figure 5-5** The rest system of the two neutrinos in $\mu$-decay. $\mathbf{p}_\mu = \mathbf{p}_{\bar{e}}$ point along the $z$-axis.

Using now (1-4.21), we obtain

$$\begin{aligned} T' &= [d_{\lambda\frac{1}{2}}\, d_{-\lambda_{\bar{e}}\frac{1}{2}}\, e^{-i(1-\lambda+\lambda_{\bar{e}})\varphi_1}(1-4\lambda\lambda_{\bar{e}}) + d_{-\lambda\frac{1}{2}}\, d_{\lambda_{\bar{e}}\frac{1}{2}}\, e^{-i(1+\lambda-\lambda_{\bar{e}})\varphi_1}(1+4\lambda\lambda_{\bar{e}})] \\ &\quad \times (-1)^{\frac{1}{2}+\lambda_{\bar{e}}} \qquad (5.12) \\ &= 2(-1)^{\frac{1}{2}-\lambda} e^{-i(1-2\lambda)\varphi_1}\, d^2_{\lambda\frac{1}{2}}\, \delta_{\lambda,-\lambda_{\bar{e}}} + 2(-1)^{\frac{1}{2}+\lambda} e^{-i\varphi_1}\, d_{-\lambda\frac{1}{2}}\, d_{\lambda\frac{1}{2}}\, \delta_{\lambda\lambda_{\bar{e}}}. \end{aligned}$$

Due to $\delta_{\lambda,-\lambda_{\bar{e}}}\,\delta_{\lambda\lambda_{\bar{e}}} = 0$, $|T'(\lambda, \lambda_{\bar{e}})|^2$ contains no interference between the two terms in (12). Moreover, from the explicit form (table B-5) of the $d$-functions,

$$4\, d^2_{-\lambda\frac{1}{2}}\, d^2_{\lambda\frac{1}{2}} = \sin^2\vartheta_1, \qquad 4\, d^4_{\lambda\frac{1}{2}} = (1 + 2\lambda\cos\vartheta_1)^2, \tag{5.13}$$

$$|T'(\lambda, \lambda_{\bar{e}})|^2 = (1 + 2\lambda\cos\vartheta_1)^2\,\delta_{\lambda,-\lambda_{\bar{e}}} + \sin^2\vartheta_1\,\delta_{\lambda,\lambda_{\bar{e}}}. \tag{5.14}$$

In the differential decay rate, $|T|^2$ must be summed over $\lambda_{\bar{e}}$. With (14) and (9), we find

$$\begin{aligned} \sum_{\lambda_{\bar{e}}} |T_{\mu\bar{e}\nu\nu}|^2 &= 4^3 G_\mu^2 E_1 E_2 \{E_{\bar{e}}E - p^2\cos^2\vartheta_1 + 2\lambda p E_{\bar{e}} - 2\lambda p E\cos^2\vartheta_1 \\ &\quad + \cos\vartheta_1[2\lambda(E_{\bar{e}}E - p^2) - p(E - E_{\bar{e}})]\}. \end{aligned} \tag{5.15}$$

The neutrino emission angles $\vartheta_1$, $\varphi_1$ in this frame of reference are unobservable. The observable quantity is

$$\int d\Omega_1 \sum_{\lambda_{\bar{e}}} |T_{\mu\bar{e}\nu\nu}|^2 = 4^3\pi G_\mu^2 s_d\{E_{\bar{e}}E - \tfrac{1}{3}p^2 + 2\lambda p(E_{\bar{e}} - \tfrac{1}{3}E)\}. \tag{5.16}$$

Here we have also inserted $4E_1E_2 = (P_1 + P_2)^2 = s_d$, since the two neutrinos are massless. According to (2-3.6) and (2-3.8), the differential decay rate of muons with polarization $\mathbf{P}$ contains the combination $\frac{1}{2}\sum_{\lambda\lambda'}(\delta_{\lambda\lambda'} + \mathbf{P}\boldsymbol{\sigma}_{\lambda\lambda'})T(\lambda)T^*(\lambda')$. In principle, this expression contains both $|T(\lambda)|^2$ and $T(\lambda)T^*(-\lambda)$. However, one finds from (12) that $T'(\lambda)T'^*(-\lambda)$ is proportional to $\exp(2i\lambda\varphi_1)$, which vanishes after $\varphi_1$-integration. Thus only the single sum $\frac{1}{2}\sum_\lambda(1 + 2\lambda P_z)|T(\lambda)|^2$ remains. Also the 3-body phase space differential (4-6.6) simplifies, with $m_1 = m_2 = 0$:

$$d\,\text{Lips}\,(m_\mu^2;\, P_1, P_2, P_3) = ds_d\, p_3\, d\Omega_3\, d\Omega_1\, m_\mu^{-1}\pi^{-5}4^{-5}, \tag{5.17}$$

$$\begin{aligned} d\Gamma &= (2m_\mu)^{-1}\sum_\lambda(\tfrac{1}{2} + \lambda P_z)\sum_{\lambda_{\bar{e}}}|T_{\mu\bar{e}\nu\nu}|^2\, d\,\text{Lips} \\ &= 2^{-5}m_\mu^{-2}\, ds_d\, p_3\, d\Omega_3\,\pi^{-4}G_\mu^2 s_d\{E_{\bar{e}}E - \tfrac{1}{3}p^2 + P_z p(E_{\bar{e}} - \tfrac{1}{3}E)\}. \end{aligned} \tag{5.18}$$

In these expressions, $p_3$ and $\Omega_3$ refer to the electron momentum and solid angle in the muon rest frame, and $\mathbf{P}$ is the muon polarization vector in the same frame, since $\chi_0(\lambda)$ of (7) refers to a particle at rest. It is therefore meaningful to express also $E$, $E_{\bar{e}}$, $p$ and $s_d$ in terms of the electron energy $E_3$ in the muon rest frame. We have

$$\begin{aligned} (P_e + (P_1 + P_2))^2 &= m_e^2 + s_d + 2E_{\bar{e}}s_d^{\frac{1}{2}} = m_\mu^2, \\ (P_\mu - (P_1 + P_2))^2 &= m_\mu^2 + s_d - 2Es_d^{\frac{1}{2}} = m_e^2, \\ p = p_3 m_\mu s_d^{-\frac{1}{2}}, &\qquad s_d = m_\mu^2 + m_e^2 - 2m_\mu E_3. \end{aligned} \tag{5.19}$$

The last two equations follow from (2-4.14) and (4-6.8), respectively. One can also use $E_{\bar{e}}E - p^2 = P_\mu P_e = m_\mu E_3$ in (18). One thus finds

$$\begin{aligned} d\Gamma_\mu &= 2^{-4} m_\mu^{-1}\, dE_3 p_3\, d\Omega_3 \pi^{-4} G_\mu^2 \\ &\quad \times \{s_d m_\mu E_3 + \tfrac{2}{3} s_d p^2 + P_z p_3 m_\mu \tfrac{1}{2}[m_\mu^2 - s_d - m_e^2 - \tfrac{1}{3}(m_\mu^2 + s_d - m_e^2)]\} \\ &= \tfrac{1}{48}\pi^{-4} G_\mu^2 m_\mu^2\, dE_3 p_3\, d\Omega_3 \\ &\qquad \{3E_3(1+\varepsilon^2) - 4E_3^2 m_\mu^{-1} - 2\varepsilon^2 m_\mu + \mathbf{P}\mathbf{p}_3(4E_3 m_\mu^{-1} - 1 - 3\varepsilon^2)\} \end{aligned} \tag{5.20}$$

with $\varepsilon \equiv m_e m_\mu^{-1}$. Integration over $\Omega_3$ and $E_3$ gives the muon decay rate

$$\Gamma(\mu^+ \to e^+ \nu_l \nu_r) = \tfrac{1}{192}\pi^{-3} G_\mu^2 m_\mu^5 (1 - 8\varepsilon^2 + 0(\varepsilon^3)), \tag{5.21}$$

The maximum of $E_3$ is very close to $\frac{1}{2}m_\mu$. In this case the curly bracket of (20) is approximately $\frac{1}{2}m_\mu(1 + \mathbf{P}\hat{\mathbf{p}}_3)$. The positron is preferentially emitted parallel to the $\mu^+$ polarization. In $\mu^-$-decay, $\mathbf{p}_3$ of (20) is replaced by $-\mathbf{p}_3$.

The emission and possible reabsorption of virtual photons (radiative corrections) introduces additional terms of the order $\alpha/\pi$ in (20). To order $\varepsilon^0$, these lead to a *total* muon width

$$\Gamma_\mu \equiv \Gamma(\mu \to \bar{e}\nu\nu) + \Gamma(\mu \to \bar{e}\nu\nu\gamma) = \tfrac{1}{192}\pi^{-3} G_\mu^2 m_\mu^5 (1 - 8\varepsilon^2)\left[1 - \frac{\alpha}{2\pi}\left(\pi^2 - \tfrac{25}{4}\right)\right]. \tag{5.22}$$

Insertion of the experimental muon lifetime of table C-1 into this formula gives the value (4-1.1) for the fundamental constant $G_\mu$.

## 5-6 Conserved Vector Current (CVC)

The original argument for the conserved vector current hypothesis (Gerstein and Zeldovitch 1955, Feynman and Gell-Mann 1958) was the near equality of the muon decay constant $G_\mu$ and the vectorial neutron decay constant $G_V$ in the Hamiltonian (1.18). The situation may be compared with the matrix elements of the electromagnetic current operator. Although these elements are complicated and contain unknown functions of $t = q^2$, they reduce (except for kinematical factors) to $Ze$ at $t = 0$, where $e$ is the elementary charge. This is a consequence of (i) charge conservation and (ii) the fact that at least one baryon and one lepton have $Z = 0$. Consider for example neutron decay $n \to pe^-\bar{\nu}$. Charge conservation alone requires

$$e_n - e_p = e(e^-) + e(\bar{\nu}). \tag{6.1}$$

Adding the requirement $e_n = 0$, $e(\nu) = 0$, we get $-e_p = e(e^-)$, from which the universality of electric charges of baryons and leptons follows. For the weak interactions, the situation is basically the same, but a complication arises from the Cabibbo angle $\theta_C$, which will be treated in section 5-11. In the limit

$\theta_C \to 0$, we may say that the hyperneutral current has a weak charge $Z_W = 1$, and the hypercharged current has zero weak charge. The practical importance of CVC, on the other hand, resides in its ability to relate all kinds of matrix elements of the hyperneutral vector current operator $v_\mu^{(0)}(x)$. From section 1-1 we remember that current conservation implies charge conservation. We therefore postulate

$$\partial^\mu v_\mu^{(0)}(x) = 0. \tag{6.2}$$

We can now define a conserved weak charge operator $Q_+$:

$$Q_+ = \int d^3x v_0^{(0)}(t, \mathbf{x}), \qquad \partial_t Q_+ = -\int \mathbf{v}^{(0)}\, d\mathbf{f} = 0 \tag{6.3}$$

The last integral is a surface integral over an infinitely remote sphere and vanishes for localized systems. Since $Q_+$ is time-independent, it commutes with $S$ exactly as the electric charge operator $Q$. Moreover, $Q_+$ raises the charge by one unit and conserves the hypercharge. There is only one such operator in the hadronic Hilbert space, namely the isospin raising operator $I_+$. Therefore, $Q_+$ must be identical with the $(1 + i2)$-component of the isospin current. We noticed already in (3.3) that the Fermi operator for a system of loosely bound nucleons is $I_+$. The "CVC-theory" tells us that this result applies to all hadrons. For example, $n \to p$, ${}^3\text{H} \to {}^3\text{He}$, $\Xi^- \to \Xi^0$ and $K^0 \to K^+$ decays all have $C_+ = 1$ according to (3.4), while the decays within isotriplets such as $\pi^- \to \pi^0$, $\Sigma^- \to \Sigma^0$ or ${}^{14}\text{O} \to {}^{14}\text{N}^*$ all have $C_+ = 2^{\frac{1}{2}}$. Moreover, since $\pi$ and ${}^{14}\text{O}$ are both spinless, their decay matrix elements are both of the form (4.1). From (2), we find the following general property of matrix elements of the hyperneutral vector current:

$$i\langle f \mid \partial^\mu v_\mu^{(0)}(x) \mid i\rangle = i\, \partial^\mu e^{-iq\cdot x}\langle f \mid v_\mu^{(0)}(0) \mid i\rangle = q^\mu V_{if,\mu} = 0. \tag{6.4}$$

Applying this to (4.1) and neglecting the mass differences between $\pi^-$ and $\pi^0$ and between ${}^{14}\text{O}$ and ${}^{14}\text{N}$, we get for both cases

$$\langle f \mid v_\mu^{(0)}(0) \mid i\rangle = 2^{\frac{1}{2}}(P_i + P_f)^\mu F_{bi}(t), \qquad F_{bi}(0) = 1, \tag{6.5}$$

where $F_{bi}$ is the electric form factor (2-7.9) of $\pi^-$ or ${}^{14}\text{O}$. Consequently, both decays have identical $ft_{\frac{1}{2}}$-values with $M_A = 0$, $M_V = 2^{\frac{1}{2}}$ in (2.36), in good agreement with experiment (including the recoil correction (3.22) for $\pi^- \to \pi^0 e\bar{\nu}$).

On the other hand, if the mass differences within isospin multiplets are not neglected, application of (4) leads to wrong results. The reason is that isospin invariance is only an approximate symmetry, and therefore (2), (3) and (4) cannot be exact. A simple improvement is obtained by observing that $v_\mu$ carries unit charge and is affected by gauge transformations (1-1.9). The simplest gauge-invariant extension of (2) is

$$(i\,\partial^\mu - eA^\mu)v_\mu^{(0)}(x) = 0. \tag{6.6}$$

Assuming now an electromagnetic origin of the mass splitting, one is led to the prescription that the splitting must be neglected in the application of (4).

(6) can also be used for the calculation of certain forbidden nuclear decays (see Blin-Stoyle 1973).

The most general form of $V^{\mu}_{if}$ between spin-$\frac{1}{2}$ states has been given in (1.20). When $i$ and $f$ belong to the same isospin multiplet, CVC requires $f_3 = 0$,

$$V^{\mu}_{if} = C_{+}\bar{u}_f(2F^{v}_{1}\gamma^{\mu} - m^{-1}F^{v}_{2}\sigma^{\mu\nu}q_{\nu})u_i, \tag{6.7}$$

where $F^{v}_{i}(t)$ are the isovector parts of the electromagnetic form factors defined in (4-14.6). The factor 2 appears only for isodoublets, to compensate for $F^{v}_{1}(0) = \frac{1}{2}$ (see also 4-14.7). The "weak magnetism" term $\kappa^{v}\sigma^{\mu\nu}q_{\nu}$ is only a small correction in allowed decays and is best measured in transitions between different isospin multiplets where the analogue of (2-7.14) is the appropriate gauge-invariant form. The form factors are unimportant, and the interference between the weak magnetism and the axial decay matrix element can be measured, for example in $\Sigma^{-} \to \Lambda e\bar{\nu}$ decays where $\kappa^{v}_{\Lambda\Sigma} = \kappa_{\Lambda\Sigma}$ is known from the $\Sigma^0$ lifetime (4-14.15). The classical experiment for the identification of weak magnetism is the pair of decays $^{12}\text{B} \to {}^{12}\text{C}\, e\bar{\nu}$, $^{12}\text{N} \to {}^{12}\text{C}\, e^{+}\nu$ (Lee, Mo, Wu, 1963), due to its combination of precision and large energy release. Together with the $1^{+}$ excited state of $^{12}$C, $^{12}$B and $^{12}$N form an isotriplet with $J^{\pi} = 1^{+}$, while the ground state of $^{12}$C has $I = 0$, $J^{\pi} = 0^{+}$. (see fig. 5-6.1). From (2.20) we recall that the allowed part of these decays is pure Gamow-Teller. The magnitude of the magnetic moment is known from the $^{12}\text{C}^{*} \to {}^{12}\text{C}\gamma$ decay rate. The necessary formulas are contained in table 4-9. Note in particular $T_S(\lambda_{e\bar{\nu}}) \sim p = |\mathbf{p}_e + \mathbf{p}_{\bar{\nu}}|$. Moreover, the interference between the Gamow-Teller matrix element and the weak magnetism changes sign between $^{12}$B and $^{12}$N. The nuclear model dependence in the evaluation of $\mathbf{M}_{GT}$ using (3.2) is avoided if $\mathbf{M}_{GT}$ is determined from the decay rate using (2.30).

Working with operators which cannot be written down explicitly is necessarily somewhat abstract. However, one can derive relations such as (6) directly from matrix elements, using gauge invariance (4) and the dominance

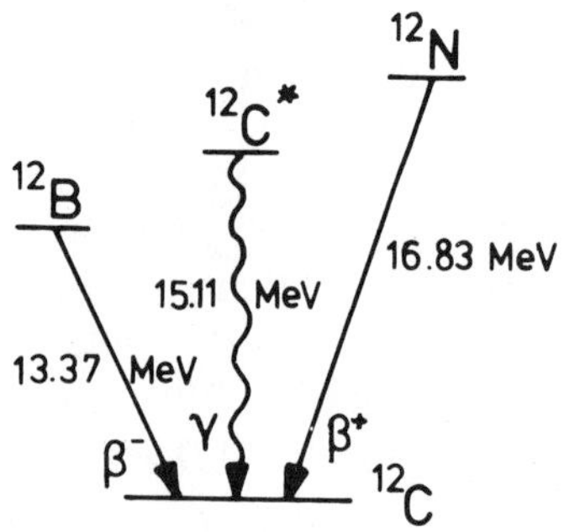


**Figure 5-6.1** The decays of the $1^{+}$ isotriplet $^{12}$B, $^{12}$C* and $^{12}$N to the $^{12}$C ground state which is $0^{+}$. The numbers give the total release of nuclear energy $m_i - m_f$ as defined in (2.5).

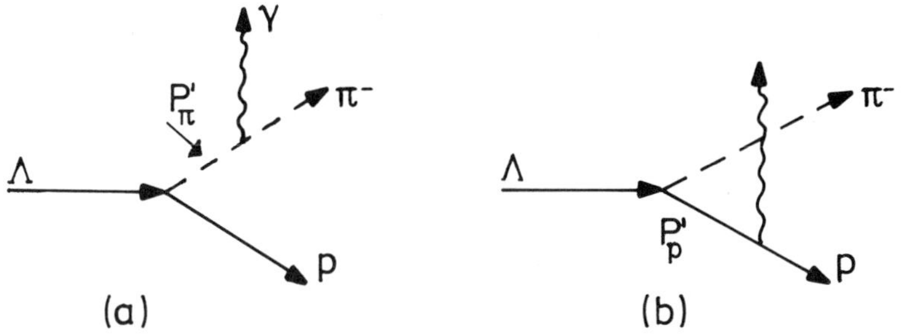


**Figure 5-6.2** The dominant graphs for $\Lambda \to p\pi^-\gamma$ decays.

of poles in certain kinematical limits. Consider first the electromagnetic interaction. For definiteness, we take the decay $\Lambda \to p\pi^-\gamma$, the matrix element of which has the form $\varepsilon_\mu^*(\lambda)J^\mu$. We need only the limiting case in which the photon 4-momentum $q_\mu$ goes to zero ("soft photon"). In this case $J^\mu$ is given by the contributions $J^\mu$(a) and $J^\mu$(b) of fig. 5-6.2, because these are the only ones which contain vanishing denominators in the limit $q_\mu \to 0$ $(P_\pi'^2 - m_\pi^2 = 2P_\pi \cdot q,\ P_p'^2 - m_p^2 = 2P_p \cdot q)$:

$$\lim_{q\to 0} (J^\mu(a)) = e_{\pi^-}(P_\pi' + P_\pi)^\mu (m_\pi^2 - P_\pi'^2)^{-1}\bar{u}_p(A - B\gamma_5)u_\Lambda,$$
$$\lim_{q\to 0} (J^\mu(b)) = e_p\bar{u}_p\gamma^\mu(P_p' \cdot \gamma + m)(m^2 - P_p'^2)^{-1}(A - B\gamma_5)u_\Lambda. \tag{6.8}$$

The weak part of these matrix elements is taken from table 4-11.1, but its details are unimportant here. The electric charges of $\pi^-$ and $p$ are denoted by $e_{\pi^-}$ and $e_p$, and we want to derive $e_{\pi^-} + e_p = 0$ from the requirement $q_\mu J^\mu = 0$. Writing $q = P_\pi' - P_\pi$ in (a) and $q = P_p' - P_p$ in (b) of (8), together with $\bar{u}_p(P_p' - P_p)\cdot\gamma(P_p'\cdot\gamma + m) = (P_p'\cdot\gamma - m)(P_p'\cdot\gamma + m) = P_p'^2 - m^2$, we see that the denominators in $q_\mu J^\mu(a)$ and $q_\mu J^\mu(b)$ disappear and that it is in fact necessary to have $e_{\pi^-} + e_p = 0$, provided of course $e_\Lambda = 0$.

Consider now the vector part of the matrix element for the reaction $\bar{\nu}n \to \mu^+\pi^-n$, and imagine that the initial neutron is somewhat off-shell such that it can almost decay into $n\pi^0$ and $p\pi^-$. In this case the graphs of fig. 5-6.3 make the dominant contributions, again due to their almost vanishing denominators. The graphs are analogous to those of fig. 5-6.2. The photon is

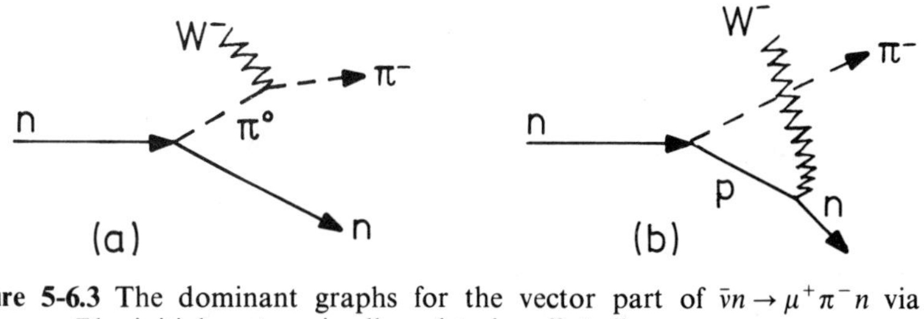


**Figure 5-6.3** The dominant graphs for the vector part of $\bar{\nu}n \to \mu^+\pi^-n$ via $W^-$ exchange. The initial neutron is allowed to be off-shell.

replaced by the charged $W$, and the strong $n \to \pi^0 n$ and $n \to \pi^- p$ amplitudes are parity- and isospin-conserving, $A = 0$ (4-11.3) and

$$2^{-\frac{1}{2}} B(n \to p\pi^-) = -B(n \to n\pi^0) \equiv -F_n(p_n^2)G, \qquad F_n(m_n^2) = 1, \quad (6.9)$$

as in (4-11.17). The constant $G$ is the famous pion-nucleon coupling constant and is defined as the value of $B$ when all three particles are on their mass shells, in particular $P_n^2 = m_n^2$. As real neutrons are stable against decays into real protons plus pions, the "on-mass-shell point" lies outside the physical region and is in fact unessential for our present problem. All we need is the ratio of charged to neutral pion emission from the neutron. Finally, the electric charges of $\pi^-$ and $p$ are replaced by the *weak* beta decay constants $G_\pi$ and $G_V$ of $\pi^+ \to \pi^0 \bar{e}\nu$ and $n \to pe\bar{\nu}$ decays. Thus the graphs of fig. 5-6.3 give

$$\begin{aligned} \lim_{q\to 0} (G_V V^\mu(a)) &= -G_\pi(P_{\pi^0} + P_\pi)^\mu (m_\pi^2 - P_{\pi^0}^2)^{-1} \bar{u}_n' F_n G\gamma_5 u_n \\ \lim_{q\to 0} (G_V V^\mu(b)) &= G_V \bar{u}_n' \gamma^\mu (P_p \cdot \gamma + m)(m^2 - P_p^2)^{-1} 2^{\frac{1}{2}} F_n G\gamma_5 u_n \end{aligned} \quad (6.10)$$

and their sum is only gauge invariant for $G_\pi = 2^{\frac{1}{2}} G_V$, which is exactly what we found in (5).

In nuclear physics, deviations from CVC arise mainly through isospin impurities of initial and final states. Consider first a $\beta^-$ decay between two $0^+$ states belonging to different isospin multiplets ("isospin-hindered Fermi transitions"). The initial state has $I_3 = -I$, and its isobaric analogue state (i.e., the member of the same isomultiplet) lies normally higher in energy (as in fig. 5-6.1) and is therefore inaccessible. The final state differs from that analogue state only by its isospin $I' < I$, and as $I'$ is only an approximate quantum number, the physical final state is in fact a mixture of these and other possible $0^+$ states. The situation differs from that described in section 4-15 only by the fact that frequently more than two states can mix. Let $\alpha_I$ be the admixture of $I$ in the final state which has approximately isospin $I' = I - 1$. We then have instead of (3.3)

$$M_F = \delta_{I_3', I_3 - 1} C_+ \alpha_I, \qquad C_+ = (2I)^{\frac{1}{2}}. \quad (6.11)$$

In the expression for $C_+$, we have inserted (3.4) with $I_3' = I_3 + 1 = -I + 1$. In $0^+ \to 0^+$ $\beta^+$ decay, the mixing occurs in the initial state for the following reason: The initial state has still $I_3 < 0$ in the cases studied, and since the nuclear charge is lowered in $\beta^+$-decay, we have $I_3' = I_3 - 1$. For low-lying nuclear states, $I$ has always the smallest value consistent with $I_3$, i.e., $I' = |I_3'| = |I_3 - 1| = I + 1$. Thus, now it is the final state which has an isobaric analogue state lying above the initial state. The log $ft_{\frac{1}{2}}$-values of isospin-hindered $0^+ \to 0^+$ transitions range from 6.5 to 10.2 (Raman and Gove 1973). See also Blin-Stoyle (1973).

For superallowed $\beta^\mp$ transitions, the isospin impurity corrections to (3.3) can be written (using $M_V$ of 2.31) as

$$|M_V(\beta^\mp)|^2 = [I(I+1) - I_3'(I_3' \mp 1)](1 - \delta_C), \quad (6.12)$$

where $\delta_C$ is obviously quadratic in the mixing coefficients $\alpha_{iI'}$ and $\alpha_{fI''}$ of initial and final states. Since there can also be several states of identical spin, parity and isospin, it is necessary to introduce an additional index for such states. The explicit formula can be found in Blin-Stoyle (1973).

Because of the relativistic electron, radiative corrections $\delta_R$ must also be included in the analysis of $\beta$-decay. For the examples of table 5-6, we have in fact $\delta_C \ll \delta_R$. The final formula for the "half-life" $t_{\frac{1}{2}}$ is

$$\tilde{f}t_{\frac{1}{2}}(1+\delta_R) = KG_V^{-2}(|M_V|^2 + |f_A M_A|^2)^{-1}, \qquad \tilde{f} \approx f(1+\varepsilon m/E), \quad (6.13)$$

using (2.28) and (2.36) (the precise calculation of $\tilde{f}$ uses the method of section 5-4 instead of the more standard definition (2.26)). Even so, there remains a model-dependent radiative correction $\Delta_\beta$ in $G_V$. A similar correction $\Delta_\mu$ occurs in $\mu$-decay, and only the difference $\Delta_\beta - \Delta_\mu$ is relevant for a comparison between $G_V^2$ and $G_\mu^2$. We therefore define

$$G_\beta \equiv G_V(1 - \tfrac{1}{2}(\Delta_\beta - \Delta_\mu)) \quad (6.14)$$

as the basic coupling constant of $\beta$-decay. In the Salam-Weinberg model (section 5-11),

$$\Delta_\beta - \Delta_\mu \approx 2\alpha\pi^{-1} \ln (m_Z/m_p) \approx 0.021, \qquad G_\beta = 0.989\, G_V. \quad (6.15)$$

**Table 5-6** *ft*-values of some important superallowed decays. The first two cases are $0^+ \to 0^+$ transitions; their numbers are taken from Raman et al. (1975).

| Decay | $t_{\frac{1}{2}}$[sec] | $\tilde{f}_{\frac{1}{2}}$ | $\delta_R$[%] | $\lvert M_V\rvert^2$ | $\lvert M_A\rvert^2$ |
|---|---|---|---|---|---|
| $^{14}O \to {}^{14}N$ | 70.57 | 3050 | 1.57 | 1.997 | 0 |
| $^{26m}Al \to {}^{26}Mg$ | 6.347 | 3042 | 1.62 | 1.996 | 0 |
| $n \to p$ | 637 | 1100 | 1.5 | 1 | 3 |
| $^3H \to {}^3He$ | $3.870 \times 10^8$ | 1150 | 1.5 | 1 | 2.9 |

## 5-7 Second Class Currents, PCAC

In the previous sections we occasionally implied that the matrix element $\langle n|h_\mu^+|p\rangle$ of the Hermitean conjugated hadronic current has the same form as $\langle p|h_\mu|n\rangle$. We now wish to check this point, starting from the definition of Hermiticity and the explicit expressions (1.20) and (4.5). Restricting ourselves to the axial part, we have

$$\begin{aligned} A_{pn\mu} &= \langle n(P_f)|a_\mu^+(0)|p(P_i)\rangle = \langle p(P_i)|a_\mu(0)|n(P_f)\rangle^* \\ &= \bar{u}(P_f)[g_1^*\widetilde{\gamma_5\gamma_\mu} - g_2^*\widetilde{\gamma_5\sigma_{\mu\nu}}(P_f - P_i)^\nu + g_3^*\tilde{\gamma}_5(P_f - P_i)_\mu]u(P_i) \quad (7.1) \\ &= \bar{u}(P_f)\gamma_5[g_1^*\gamma_\mu + g_2^*\sigma_{\mu\nu}(P_i - P_f)^\nu + g_3^*(P_i - P_f)_\mu]u(P_i) \end{aligned}$$

Here $\tilde{Q}$ denotes $\gamma_0 Q^+ \gamma_0$ and the last line is obtained using (A-4.9), in particular $\tilde{\gamma}_5 = -\gamma_5$ and $\widetilde{\gamma_5\gamma_\mu} = \gamma_5\gamma_\mu$. Since our $g_i$ are real by time-reversal invariance (compare 4-8.29), we see that $A_{pn}^\mu$ and $A_{np}^\mu$ of (4.5) are in fact identical (if one neglects the small difference between $u_p(P_i)$ and $u_n(P_i)$ for given $P_i$) in the first (= axial) and third (= "induced pseudoscalar") coupling. The $g_2$-term, on the other hand, has opposite signs in $\beta^+$ and $\beta^-$ decays. Repeating the argument for the vector current (1.20), we find equal signs for $f_1$ and $f_2$ but opposite signs for $f_3$. Fortunately, $f_3$ is zero by CVC, so we are only left with the opposite signs of $g_2$. These terms contribute in "forbidden order" only and are difficult to measure, but the sign change is so ugly that we postulate $g_2 = 0$. More generally speaking, we can decompose the hyperneutral part $h_\mu^{(0)}$ of the full hadronic current operator $h_\mu$ of (6.1) into an isovector and an isotensor piece. The isospins involved in $\beta$-decay are normally so small that the isotensor piece does not contribute anyway, and in that sense we can say that $h_\mu^{(0)}$ has $I = 1$. Next, we can also split $h_\mu^{(0)}$ into pieces of opposite charge symmetry (Lee and Yang 1960),

$$h_\mu^{(0)} = h_\mu^{(f)} + h_\mu^{(s)}, \qquad h_\mu^{(f)} = -e^{-i\pi I_2} h_\mu^{(f)+} e^{i\pi I_2}, \qquad h_\mu^{(s)} = e^{-i\pi I_2} h_\mu^{(s)+} e^{i\pi I_2}, \quad (7.2)$$

which are called first- and second-class currents, respectively. We now postulate the absence of second-class currents, $h_\mu^{(s)} = 0$. Taking (2) between proton and neutron states and using (B-3.11), we get the desired relation

$$\langle p(P_f) | h_\mu^{(0)} | n(P_i) \rangle = \langle n(P_f) | h_\mu^{(0)+} | p(P_i) \rangle. \quad (7.3)$$

In addition, we also get relations between current matrix elements between different isospin states, for example for the decays $\Sigma^\pm \to \Lambda e \bar{\nu}$:

$$\langle \Lambda | h_\mu^{(f)} + h_\mu^{(s)} | \Sigma^- \rangle = \langle \Lambda | -h_\mu^{(f)+} + h_\mu^{(s)+} | \Sigma^+ \rangle. \quad (7.4)$$

Let us now treat these decays in the allowed approximation, i.e., we neglect $f_2, f_3, g_2$ and $g_3$ in (4.4) and (4.5). $f_1$ is zero by CVC and only $g_1$ remains, which is $g_1^{(f)} \pm g_1^{(s)}$ in $\Sigma^\mp$-decays. In the following, we also neglect a possible form factor and put $g_1(t) \approx g_1(0) = f_{A\Sigma\Lambda}$. Using (2.26) with $C = 3G_V^2 f_A^2$ and (3.22), we have

$$\begin{aligned} \Gamma(\Sigma^\pm \to \Lambda e \nu) &\approx \tfrac{1}{2}\pi^{-3}\tfrac{1}{30}(\Delta m_\pm)^5(1 - \tfrac{3}{2}m_\Lambda^{-1}\,\Delta m_\pm)3G_V^2(f_A^\pm)^2, \\ \Delta m_\pm &\equiv m(\Sigma^\pm) - m(\Lambda), \qquad f_A^{(\pm)} = f_{A\Sigma\Lambda}^{(f)} \pm f_{A\Sigma\Lambda}^{(s)}. \end{aligned} \quad (7.5)$$

The resulting values of $f_A$ are computed in table 5-7. They are in fact consistent with $f_A^{(s)} = 0$, i.e., second-class currents appear to be absent. The best value of $f_{A\Sigma\Lambda}$ in the absence of second-class currents is also given in table 5-7. Similar analyses can of course be made for nuclear beta decays between different isospin multiplets as in fig. 5-6.3, but here nuclear models without

**Table 5-7** The hyperneutral $\Sigma^\pm$-decays under the assumption of pure axial coupling, eq. (7.5).

| | $\Gamma$ [$10^{-10}$ eV] | $\Delta$m [MeV] | $f_{A\Sigma\Lambda}^2$ | Assuming $f_{A\Sigma\Lambda}^{(s)} = 0$ |
|---|---|---|---|---|
| $\Sigma^+$ | $1.66 \pm 0.40$ | 73.77 | $0.42 \pm 0.10$ | $\lvert f_{A\Sigma\Lambda} \rvert = 0.64 \pm 0.03$ |
| $\Sigma^-$ | $2.66 \pm 0.27$ | 81.75 | $0.409 \pm 0.042$ | $= (0.624 \pm 0.03)(\tfrac{2}{3})^{\frac{1}{2}} f_A$ |

exchange currents automatically give $f_A^{(+)} = f_A^{(-)}$, except for isospin-breaking effects.

The hyperneutral vector current $v_\mu^{(0)}$ is of course first class by CVC. From (2), it also follows that $v_\mu^{(0)}$ and $a_\mu^{(0)}$ carry opposite $G$-parities:

$$Gv_\mu^{(0)}G^{-1} = v_\mu^{(0)}, \qquad Ga_\mu^{(0)}G^{-1} = -a_\mu^{(0)}, \qquad G \equiv Ce^{i\pi I_2}. \tag{7.6}$$

For $v_\mu^{(0)}$, this is verified from the $I = 1$ column of table 4-5.2, where the only $N\bar{N}$ state of spin 1 and negative parity has $G = +1$ like the $\rho$-meson. For $a_\mu^{(0)}$, an extra minus sign appears because opposite parity implies opposite charge parity by CP-conservation. A possible second-class axial current would again have $G = +1$, like the $B$-meson mentioned in table 4-5.2 ($B$ is an $\omega\pi$ resonance with $m = 1228$ MeV, $\Gamma = 125$ MeV).

The axial current matrix elements get contributions from pion exchange according to the graph of fig. 5-7, which are derived as follows: we assume the existence of an interacting pion field operator $\psi^+(x)$ which satisfies an equation of motion analogous to $A_\mu$ of (3-3.1):

$$(\Box + m_\pi^2)\psi^+(x) = j_\pi^+(x). \tag{7.7}$$

The "pion source operator" $j_\pi$ need not be specified. To lowest order in this operator, the $S$-matrix element is given by the analogue of (2-7.5), with $(P_a' - P_a)^\mu$ replaced by $q^\mu = (P_e + P_{\bar{\nu}})^\mu$ in the $\delta_4$-function and a scalar matrix element $J_{\pi e \bar{\nu}} = f_\pi q^\mu L_\mu^*(t)$ at the upper vertex, according to (1.5). Neglecting now a possible $t$-dependence of $f_\pi$, we get the following one-pion exchange (OPE) contribution to the matrix element of the axial current operator:

$$A_{if}^\mu(\pi) = f_\pi q^\mu (m_\pi^2 - t)^{-1} J_{\pi if}, \qquad J_{\pi if} \equiv \langle f \,|\, j_\pi^+(0) \,|\, i \rangle. \tag{7.8}$$

When $i$ and $f$ belong to the baryon octet, $J_{\pi if}$ can obviously be written in the form

$$J_{\pi bd} = \bar{u}_d \gamma_5 u_b G_{bd\pi} F_{bd\pi}(t), \qquad F_{bd\pi}(m_\pi^2) \equiv 1. \tag{7.9}$$

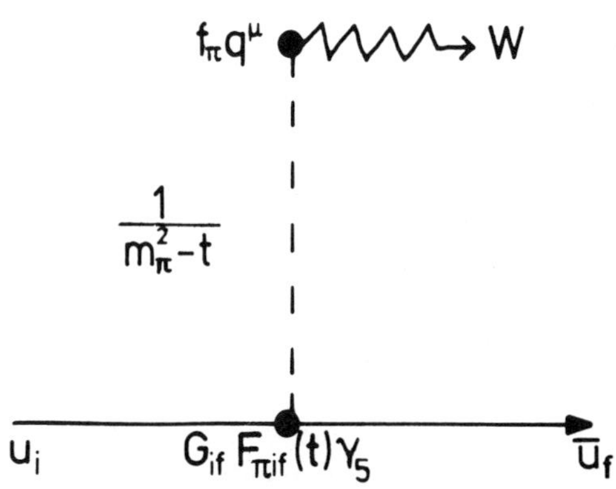


**Figure 5-7** The $\pi$-exchange contribution to the axial current matrix element between spin-$\frac{1}{2}$ hadrons of equal hypercharge.

The $\gamma_5$ appears because of the negative parity of the pion (since we work to lowest order in the weak interaction and the upper vertex is already weak, the lower one must conserve parity, contrary to the situation of table 4-11.1). For $b = n$, $d = p$, we have $G_{np} = 2^{\frac{1}{2}}G$ according to (6.9). Neglecting other possible contributions to the induced pseudoscalar coupling $g_3$ of (4.5), we find for nucleons

$$g_3 = 2^{\frac{1}{2}}GF_{NN\pi}(t)(m_\pi^2 - t)^{-1}f_\pi. \tag{7.10}$$

The induced pseudoscalar is again difficult to measure in $\beta$-decay because of the smallness of $q^\mu$. However, it has been measured in muon capture $\mu^- p \to n\nu$ and appears to agree with (10) (Vitale et al. 1975).

Now we come to our main point, the hypothesis of a partially conserved axial current (PCAC, Gell-Mann and Levy 1960, Nambu 1960). We cannot have $\partial^\mu a_\mu^{(0)}(x) = 0$, because for pion decay we find from (1.5)

$$\langle 0 | i\, \partial^\mu a_\mu^{(0)}(x) | P_\pi\rangle = f_\pi P_\pi^2 e^{-iP_\pi \cdot x} = f_\pi m_\pi^2 \langle 0 | \psi^+(x) | P_\pi\rangle. \tag{7.11}$$

However, we may postulate that all matrix elements of $\partial^\mu a_\mu$ (the divergence of the axial current) are proportional to $m_\pi^2$, in which case the current is conserved in the hypothetical limit of zero pion mass.

In general, the matrix elements $A_{if}^\mu$ of $a^{\mu(0)}(0)$ may be decomposed

$$A_{if}^\mu = A_{if}^{\mu\prime} + A_{if}^\mu(\pi), \tag{7.12}$$

where $A_{if}^{\mu\prime}$ contains no OPE. For the spin-$\frac{1}{2}$ baryons in the absence of second-class currents, with $i = b, f = d$,

$$A_{bd}^{\mu\prime} = \bar{u}_d \gamma_5 g_{1bd}(t)\gamma^\mu u_b. \tag{7.13}$$

We can then rewrite the postulate as

$$\begin{aligned} 0(m_\pi^2) = \langle f | i\, \partial^\mu a_\mu^{(0)}(x) | i\rangle &= q_\mu e^{i(P_f - P_i)x} A_{if}^\mu \\ &= e^{i(P_f - P_i)\cdot x}\left(q_\mu A_{if}^{\mu\prime} + f_\pi \frac{t}{m_\pi^2 - t} J_{\pi if}\right), \end{aligned} \tag{7.14}$$

or, going to the limit $m_\pi^2 \to 0$,

$$q_\mu A_{if}^{\mu\prime} = f_\pi J_{\pi if}. \tag{7.15}$$

For the baryon octet, we use $q = P_b - P_d$ and the Dirac equations $P_b \cdot \gamma u_b = m_b u_b$, $\bar{u}_d P_d \cdot \gamma = \bar{u}_d m_d$ for $q_\mu A_{bd}^{\mu\prime}$ and (9) for $J_{\pi bd}$, getting

$$g_{1bd}(t)(m_b + m_d) = f_\pi G_{bd\pi} F_{bd\pi}(t). \tag{7.16}$$

Concluding then, the final form for $A_{bd}^\mu$ of (4.5) in the absence of second-class currents is

$$A_{bd}^\mu = f_\pi G_{bd\pi} F_{bd\pi}(t) \bar{u}_d \gamma_5 [\gamma^\mu (m_b + m_d)^{-1} + q^\mu (m_\pi^2 - t)^{-1}] u_b. \tag{7.17}$$

In neutron beta decay, we have in addition $t \approx 0$, $g_1(0) = f_A$, $m_b = m_d = m_N$, in which case (16) gives the famous relation of Goldberger and Treiman (1958):

$$2m_N f_A = 2^{\frac{1}{2}} f_\pi G_\pi F_{NN\pi}(0). \tag{7.18}$$

The strong NN$\pi$-form factor $F_{NN\pi}(t)$ can be calculated only if a model for the source function $j_\pi^+(x)$ of (7) is given. In the corresponding formalism for nuclei, some properties of this function are known (M. Ericson et al., 1973). For example, $j_\pi^+(x)$ is zero outside the nucleus. In principle, $f_A$ can then be calculated for nuclei. Electromagnetic effects can also be incorporated by replacing $i\,\partial^\mu \to i\,\partial^\mu - eA^\mu$ in (14), in analogy with (6.6).

## 5-8 The Hypercharged Current, Cabibbo Theory

The subjects of this section have been reviewed extensively. In addition to the book of Marshak et al. (1969), the reviews of Bailin (1971) and Chounet et al. (1972) should be mentioned.

In (1.21) we have split the hadronic current operator into hypercharge conserving and changing pieces $h_\mu^{(0)}$ and $h_\mu^{(1)}$, with a relative strength $\tan\theta_C$ which was extracted from $\Gamma(K \to \mu\bar{\nu})/\Gamma(\pi \to \mu\bar{\nu})$ in (1.25). This is meaningful only if we have models both for $h_\mu^{(0)}$ and $h_\mu^{(1)}$. Since the strong interactions are approximately $SU_3$-invariant and $h_\mu^{(0)}$ and $h_\mu^{(0)+}$ transform like the charged members of a mesonic isotriplet, it follows that $h_\mu^{(1)}$ and $h_\mu^{(1)+}$ must represent the remaining two charged components of an octet. In particular, $h_\mu^{(1)}$ follows from $h_\mu^{(0)}$ by a rotation by $\pi$ about the 2-axis in $U$-space

$$h_\mu^{(1)} = e^{-i\pi U_2} h_\mu^{(0)} e^{i\pi U_2}, \qquad h_\mu^{(0)} = -e^{-i\pi U_2} h_\mu^{(1)} e^{i\pi U_2}, \qquad U_2 = \tfrac{1}{2}F_7 \quad (8.1)$$

(compare B-1.9, B-6.3 and B-3.11). Cabibbo (1963) postulated universality for the hadronic current entering (1.21) in the form

$$h_C = e^{-2i\theta_C U_2} h^{(0)} e^{2i\theta_C U_2} = \cos\theta_C h^{(0)} + \sin\theta_C h^{(1)}, \tag{8.2}$$

i.e., the actual strengths of $h^{(0)}$ and $h^{(1)}$ are due to a rotation by $2\theta_C$ around the $U_2$-axis. Since $\cos\theta_C$ is close to 1, this universality aspect is practically less important. What counts is (1), which allows calculation of hypercharge changing decays such as $\Lambda \to pe\bar{\nu}$ from hypercharge conserving ones such as $n \to pe\bar{\nu}$ and $\Sigma \to \Lambda e\bar{\nu}$, at least in the limit of $SU(3)$-symmetry. Loosely speaking, since $v_\mu^{(0)}$ and $a_\mu^{(0)}$ have the isospin properties of $\rho^-$ and $\pi^-$, $v_\mu^{(1)}$ and $a_\mu^{(1)}$ must have the isospin properties of $K^{*-}$ and $K^-$, and $v_\mu^{(1)+}$ and $a_\mu^{(1)+}$ must have those of $K^{*+}$ and $K^+$. This leads immediately to the following selection rules in hypercharge-changing decays

$$\Delta Y = \pm 1, \qquad \Delta Q = \Delta Y, \qquad \Delta I = \tfrac{1}{2}. \tag{8.3}$$

Here $\Delta X \equiv X_i - X_f$ of the hadrons, and $\Delta I$ denotes the isospin property of the reduced matrix element. Decays such as $\Omega \to \Lambda e\bar{\nu}$ ($\Delta Y = -2$, $\Delta I = 0$), $\Omega \to ne\bar{\nu}(\Delta Y = -3)$, $\Sigma^+ \to ne^+\nu$ and $K^0 \to \pi^+e^-\bar{\nu}$, $\bar{K}^0 \to \pi^-e^+\nu(\Delta Y = -\Delta Q,\ \Delta I_3 = \mp\frac{3}{2})$ are all forbidden by (3) and are in fact absent experimentally.

Before giving the matrix elements of $h_\mu^{(1)}$, we wish to define the $\pi$-baryon and $K$-baryon coupling constants for each charge combination separately and give their $SU_2$-relations. The coupling constant approximation to the pseudoscalar vertex function is $G_{if}\bar{u}_f\gamma_5 u_i$ (compare fig. 5-7). The pion-nucleon coupling constant $G$ is defined for the vertex $p \to p\pi^0$. Since the initial nucleon has isospin $\frac{1}{2}$, the second row of table 4-2.2 tells us

$$G \equiv G(p, p\pi^0) = -2^{-\frac{1}{2}}G(p, n\pi^+) = -G(n, n\pi^0) = 2^{-\frac{1}{2}}G(n, p\pi^-). \quad (8.4)$$

Inverse processes have equal coupling constant, e.g., $G(n\pi^+ \to p) = G(p \to n\pi^+) \equiv G(p, n\pi^+)$. This conforms with replacing an incoming $\pi^+$ by an outgoing "anti-$\pi^+$" remembering the negative charge parity of $\pi^+$ (table 4-5.1):

$$G(n\pi^+ \to p) = G(n \to p\overline{\pi^+}) = -G(n \to p\pi^-). \quad (8.5)$$

Next, we define the $\Sigma\Sigma\pi$ and $\Lambda\Sigma\pi$ coupling constants in analogy with $G_{\rho\pi\pi}$ of (4-9.4) and $G_{f\pi\pi}$ of (4-12.3), in order to be able to adopt their SU(3)-relations from those of meson decay constants (table 4-12):

$$\begin{aligned} G_{\Sigma\Sigma\pi} &\equiv G(\Sigma^0, \Sigma^+\pi^-) = G(\Sigma^+, \Sigma^+\pi^0) = -G(\Sigma^-, \Sigma^-\pi^0) = G(\Sigma^-, \Sigma^0\pi^-) \\ &= -G(\Sigma^0, \Sigma^-\pi^+) = -G(\Sigma^+, \Sigma^0\pi^+), \qquad G(\Sigma^0, \Sigma^0\pi^0) = 0. \end{aligned} \quad (8.6)$$

$$\begin{aligned} G_{\Lambda\Sigma\pi} &\equiv G(\Lambda, \Sigma^+\pi^-) = G(\Lambda, \Sigma^-\pi^+) = -G(\Lambda, \Sigma^0\pi^0) \\ &= -G(\Sigma^+, \Lambda\pi^+) = -G(\Sigma^0, \Lambda\pi^0) = -G(\Sigma^-, \Lambda\pi^-) \equiv -G_{\Sigma\Lambda\pi}. \end{aligned} \quad (8.7)$$

The minus signs of the last row are ugly but difficult to avoid: $\alpha = G_D/G = \frac{1}{2}3^{\frac{1}{2}}G_{\Lambda\Sigma\pi}/G$ is positive. Similarly, we define in analogy with (4-9.6), (4-10.23), $G_{f8\bar{K}^*K}$ (4-12.7) and $G_{A\bar{K}^*K}$ (4-12.8)

$$G_{\Lambda N\bar{K}} \equiv G(\Lambda, pK^-) = -G(\Lambda, n\bar{K}^0) = -G(n, \Lambda K^0) = -G(p, \Lambda K^+), \quad (8.8)$$

$$\begin{aligned} G_{\Sigma N\bar{K}} &\equiv G(\Sigma^0, pK^-) = G(\Sigma^0, n\bar{K}^0) = 2^{-\frac{1}{2}}G(\Sigma^+, p\bar{K}^0) \\ &= 2^{-\frac{1}{2}}G(\Sigma^-, nK^-) = -G(p, \Sigma^0K^+) = G(n, \Sigma^0K^0) \\ &= 2^{-\frac{1}{2}}G(p, \Sigma^+K^0) = -2^{-\frac{1}{2}}G(n, \Sigma^-K^+), \end{aligned} \quad (8.9)$$

$$G_{\Lambda\Xi K} \equiv G(\Lambda, \Xi^-K^+) = -G(\Lambda, \Xi^0K^0) = -G(\Xi^-, \Lambda K^-) = G(\Xi^0, \Lambda\bar{K}^0), \quad (8.10)$$

$$\begin{aligned} G_{\Sigma\Xi K} &\equiv G(\Sigma^0, \Xi^-K^+) = G(\Sigma^0, \Xi^0K^0) = 2^{-\frac{1}{2}}G(\Sigma^+, \Xi^0K^+) \\ &= 2^{-\frac{1}{2}}G(\Sigma^-, \Xi^-K^0) = -G(\Xi^-, \Sigma^0K^-) \\ &= G(\Xi^0, \Sigma^0\bar{K}^0) = -2^{-\frac{1}{2}}G(\Xi^0, \Sigma^+K^-) = 2^{-\frac{1}{2}}G(\Xi^-, \Sigma^-\bar{K}^0). \end{aligned} \quad (8.11)$$

The crossing $K^+ \leftrightarrow K^-$ includes again an extra minus sign from charge conjugation.

Having thus defined the necessary meson-baryon coupling constants on the right-hand side of table 4-12, we obtain the matrix elements of $h_\mu^{(1)}$ of table 5-8, by expressing the appropriate meson coupling constant in terms of $G$ and $\alpha$ and substituting in $V_\mu^{(1)}$ $G \to 2^{-\frac{1}{2}}$, $\alpha \to 0$ (according to CVC) and in $A_\mu^{(1)}$ $G \to 2^{-\frac{1}{2}}f_A$. The factor $2^{-\frac{1}{2}}$ arises because the charged pion has a coupling constant $2^{\frac{1}{2}}G$ in $n \rightarrow p$, whereas the matrix element of $h_\mu^{(0)}$ by definition has couplings $\gamma_\mu$ and $f_A \gamma_5 \gamma_\mu$ in this case.

Hyperon decays with $\Delta Y = -1$ are well described by the matrix elements of table 5-8. Instead of taking $\theta_C$ from $K \rightarrow \mu\bar{\nu}$ decays (1.25) and $\alpha = 0.624$ from $\Sigma \rightarrow \Lambda e\nu$ decays (table 5-7), one normally makes a best fit to all reactions of table 5-8. Including form factors etc, the result is (Nagels et al., 1979)

$$\sin \theta_C = 0.229 \pm 0.004, \qquad \alpha = 0.65 \pm 0.01. \tag{8.12}$$

If one distinguishes between two angles $\theta_A$ and $\theta_V$, then $\sin \theta_A$ from baryon decays increases by $\sim 2\%$. In the calculation of decay rates (3.21) must be used, and (3.22) suffices for electronic decays but not for muonic decays such as $\Lambda \rightarrow p\mu\bar{\nu}$. In view of the enormous mass splitting, the accuracy of $SU(3)$-models is of course questionable. For the matrix elements of $v_\mu$ at zero momentum transfer, it can be shown that first-order $SU(3)$-breaking effects vanish (Ademollo and Gatto 1964).

**Table 5-8** The operators $0^\mu_{if}$ for the matrix elements $\bar{u}_f\, 0^\mu_{if}\, u_i$ of the Cabibbo current (2) between states of the baryon octet. Form factors, weak magnetism and induced pseudoscalar couplings are omitted. The numerical coefficients are obtained from eqs. (4)–(11), the $SU(3)$-relations of table 4-12 and from the substitution mentioned in the text.

| Decay | $0^\mu_{if}$ |
|---|---|
| $n \rightarrow p$ | $\cos \theta_C(1 + f_A\gamma_5)\gamma^\mu$ |
| $\Sigma^\pm \rightarrow \Lambda$ | $-(\frac{2}{3})^{\frac{1}{2}} \cos \theta_C \alpha f_A \gamma_5 \gamma^\mu$ |
| $\Sigma^- \rightarrow \Sigma^0$ | $2^{\frac{1}{2}} \cos \theta_C[1 + (1-\alpha)\gamma_5 f_A]\gamma^\mu$ |
| $\Lambda \rightarrow p$ | $6^{-\frac{1}{2}} \sin \theta_C[3 + (3-2\alpha)f_A\gamma_5]\gamma^\mu$ |
| $\Sigma^- \rightarrow n$ | $\sin \theta_C[-1 + (2\alpha - 1)f_A\gamma_5]\gamma^\mu$ |
| $\Xi^- \rightarrow \Lambda$ | $6^{-\frac{1}{2}} \sin \theta_C[3 + (3-4\alpha)f_A\gamma_5]\gamma^\mu$ |
| $\Xi^- \rightarrow \Sigma^0$ | $-2^{-\frac{1}{2}} \sin \theta_C(1 + f_A\gamma_5)\gamma^\mu$ |
| $\Xi^0 \rightarrow \Sigma^+$ | $-\sin \theta_C(1 + f_A\gamma_5)\gamma^\mu$ |

The matrix elements of the decays $K^- \rightarrow \pi^0 e\bar{\nu}$, $K^- \rightarrow \pi^0\mu^-\bar{\nu}_\mu$ and $K_L^0 \rightarrow \pi^+ e\bar{\nu}$, $K_L^0 \rightarrow \pi^+\mu^-\bar{\nu}_\mu$ have the same form as (4.1) for $\pi^-$ $\beta$-decay. No kinematical approximations are generally possible here; the calculation contains a number of effects which are negligible in $\beta$-decay. The matrix element of the divergence of the current is

$$\langle\pi| i\, \partial_\mu v^\mu(0)|K\rangle = (m_K^2 - m_\pi^2)f_+(t) + tf_-(t) \equiv (m_K^2 - m_\pi^2)f_0(t). \tag{8.13}$$

We go to the lepton cms and take $\mathbf{p}_K = \mathbf{p}_\pi$ along the $z$-axis and $-\mathbf{p}_{\bar{\nu}} = \mathbf{p}_e = \mathbf{p}_1$ as in fig. 5-5. Then (4.1) becomes

$$\langle\pi|v^\mu(0)|K\rangle = \delta_{\mu 0}t^{-\frac{1}{2}}(m_K^2 - m_\pi^2)f_0 + \delta_{\mu 3}2p_K f_+ . \tag{8.14}$$

The first term is easily obtained from (14), with $E_K - E_\pi = t^{\frac{1}{2}}$ in this system. The full matrix element is given by (2.2) and (2.14) (the Coulomb effects of $K_L^0$-decays would be included by using (2.13) but are neglected here):

$$\begin{aligned} T_{ife\bar{\nu}} &= 2^{-\frac{1}{2}} G_V \langle \pi | v_\mu(0) | K \rangle L^\mu_{e\bar{\nu}} \\ &= 2^{\frac{1}{2}} G_V (E - 2\lambda p)^{\frac{1}{2}} (2E_{\bar{\nu}})^{\frac{1}{2}} \chi_0^+(\lambda, \vartheta_1 \varphi_1) \\ &\quad \times [f_0 t^{-\frac{1}{2}}(m_K^2 - m_\pi^2) - 2\sigma_z p_K f_+] \chi_0(\tfrac{1}{2}, \vartheta_1, \varphi_1), \end{aligned} \tag{8.15}$$

where $E$, $p$, $\lambda$ denote energy, momentum and helicity of the $e^-$ or $\mu^-$. A distinction between $\chi_{0e}$ and $\chi_{0\bar{\nu}}$ is unnecessary according to the comment following (3-6.21). Insertion of $\chi_0$ and $\chi_0^+$ from (1-4.18) gives

$$\begin{aligned} T_{ife\bar{\nu}} &= 2^{\frac{1}{2}} G_V (E - 2\lambda p)^{\frac{1}{2}} (2E_{\bar{\nu}})^{\frac{1}{2}} \\ &\quad \times [f_0 t^{-\frac{1}{2}}(m_K^2 - m_\pi^2)\, \delta_{\lambda\frac{1}{2}} - 2p_K f_+ (\cos \vartheta_1\, \delta_{\lambda\frac{1}{2}} - \sin \vartheta_1 e^{i\varphi_1} \delta_{\lambda, -\frac{1}{2}})] \\ &= 2^{\frac{1}{2}} G_V (1 - m^2/t)^{\frac{1}{2}} \\ &\quad \times [\delta_{\lambda\frac{1}{2}} m (f_0 t^{-\frac{1}{2}}(m_K^2 - m_\pi^2) - 2p_K f_+ \cos \vartheta_1) + 2p_K f_+ \sin \vartheta_1 e^{i\varphi_1} t^{\frac{1}{2}}]. \end{aligned} \tag{8.16}$$

In the last expression we have used $E - p = t^{-\frac{1}{2}} m^2$ and $E + p = t^{\frac{1}{2}}$. This part of the kinematics is the same as in $\pi$ decays (see the discussion following 1.6), but with $m_\pi$ replaced by $t^{\frac{1}{2}}$. The decay rate is given by (2-3.9) and (4-6.6),

$$\begin{aligned} d\Gamma &= d \text{ Lips } (m_K^2; P_\pi P_e P_\nu) \sum_\lambda | T_{ife\bar{\nu}} |^2 \\ &= \frac{dt}{4\pi} \frac{p_3\, d\Omega_3}{16\pi^2 m_K^2} \frac{p\, d\Omega_1}{16\pi^2 t^{\frac{1}{2}}} 2G_V^2 \left(1 - \frac{m^2}{t}\right) p_K^2 \\ &\quad \times \left[ m^2 \left( f_0 \frac{m_K^2 - m_\pi^2}{p_K t^{\frac{1}{2}}} - 2f_+ \cos \vartheta_1 \right)^2 + 4f_+^2 t \sin^2 \vartheta_1 \right] \end{aligned} \tag{8.17}$$

$$\begin{aligned} &= dt\, d \cos \vartheta_1 2^{-5} \pi^{-3} p_3^3 G_V^2 \left(1 - \frac{m^2}{t}\right) \\ &\quad \times \left[ m^2 \left( f_0 \frac{m_K^2 - m_\pi^2}{2p_K t^{\frac{1}{2}}} - f_+ \cos \vartheta_1 \right)^2 + 4f_+^2 t \sin^2 \vartheta_1 \right] \\ &= dt 2^{-4} \pi^{-3} G_V^2 p_3^3 (1 - m^2/t)^2 \\ &\quad \times [m^2 f_0^2 (m_K^2 - m_\pi^2)^2 / \lambda(m_K^2, m_\pi^2, t) + \tfrac{1}{3} f_+^2 (m^2 + 2t)]. \end{aligned} \tag{8.18}$$

Here we have used $p_K = p_3 m_K t^{-\frac{1}{2}}$ with $p_3$ = pion lab momentum (compare 5.19) and $p = \frac{1}{2} t^{\frac{1}{2}} - \frac{1}{2} m^2 t^{-\frac{1}{2}}$, and in (18) we have also integrated over $\cos \vartheta_1$. In the case of electrons, one can of course put $m = 0$, in which case $d\Gamma$ simplifies considerably. The integration over $t$ can also be performed if the $t$-dependence of $f$ is approximated by a linear form (Donaldson 1974):

$$f_{\substack{+\\0}}(t) = f_{\substack{+\\0}}(0)(1 + \lambda_{\substack{+\\0}} t/m_\pi^2) \tag{8.19}$$

$$\Gamma_{e3} \equiv \Gamma(K^- \to \pi^0 e \bar{\nu}) = G_V^2 \tfrac{1}{3} 2^{-8} \pi^{-3} m_K^5 \, | f_+(0) |^2 (0.573 + 0.138 \lambda_+ m_K^2 / m_\pi^2) \tag{8.20}$$

$$\Gamma_{\mu 3}/\Gamma_{e3} = (1 + 3.700\lambda_+ + 5.478\lambda_+^2)^{-1}$$
$$(0.6457 + 2.228\lambda_+ + 4.321\lambda_+^2 + 1.573\lambda_0 + 3.405\lambda_0^2 - 0.914\lambda_0\lambda_+) \quad (8.21)$$

From the Cabibbo theory and the Ademollo-Gatto theorem,

$$G_V f_+(0) = G_V \tan\theta_C \cdot \tfrac{1}{2} f_+(\pi^-) = 2^{-\frac{1}{2}} G_\mu \sin\theta_C \quad (8.22)$$

using (4-10.22) and CVC (6.5). The corresponding formulas for $\bar{K}^0 \to \pi^+ e\bar{\nu}$ and $\bar{\bar{K}}^0 \to \pi^+\mu^-\bar{\nu}_\mu$ decays are (Fearing et al. 1970, Donaldson et al. 1974)

$$\Gamma_{e3}^0 \equiv \Gamma(\bar{K}^0 \to \pi^+ e^- \bar{\nu}) = \tfrac{1}{3} 2^{-8} \pi^{-3} m_{K^0}^5 G_\mu^2 \sin^2\theta_C (0.5634 + 0.153\lambda_+). \quad (8.23)$$

$$\Gamma_{\mu 3}^0/\Gamma_{e3}^0 = (1 + 3.457\lambda_+ + 4.779\lambda_+^2)^{-1}$$
$$(0.6452 + 2.081\lambda_+ + 3.885\lambda_+^2 + 1.465\lambda_0 + 3.074\lambda_0^2 - 1.027\lambda_+\lambda_0). \quad (8.24)$$

(23) is roughly twice as large as (20), the factor 2 coming from (4-9.5). Experimental values of the expansion coefficients are $\lambda_+ = 0.028$ for $K^\pm$ and 0.030 for $K_L^0$, $\lambda_0 \approx -0.009$ for $K^\pm$ and 0.021 for $K_L^0$.
In practice, $\bar{K}^0$ does not decay exponentially as explained in section 4-3. Since $T(K^0 \to \pi^+ e\bar{\nu}) = 0$ by our selection rules, the time distribution $N^-(\tau)$ of $K \to \pi^+ e\bar{\nu}$ decays is proportional to $W(\bar{K})$ of (4-3.12). If the initial state is pure $K_L^0$ ($\rho = 0$), we get

$$\Gamma(K_L^0 \to \pi^\mp e^\pm \nu) = \tfrac{1}{2}(1 \pm 2 \operatorname{Re}\varepsilon)\Gamma_{e3}^0, \quad (8.25)$$

and since $\operatorname{Re}\varepsilon > 0$, the long-lived neutral kaon decays more into $e^+$ than into $e^-$. The difference is due to CP-breaking and allows an absolute definition of leptons as opposed to antileptons. (Contrary to a far-spread belief, one can find out whether other galaxies consist of matter or antimatter, simply by listening to the radio: the more intelligent inhabitants of galaxies state from time to time (on a frequency which our own astronomers have not yet discovered) whether the charged leptons preferentially produced in $K_L$-decays have the same charge as their atomic leptons, or the opposite one).

## 5-9 Purely Hadronic Weak Interactions. CPT-Theorem and $K \to \pi\pi$ Decays

The hyperons of table C-4 as well as $\Omega^-$ and $K_S$ decay almost exclusively without the emission of lepton pairs, i.e., purely hadronically, while the other kaons decay at least partly that way. The piece $-2^{-\frac{1}{2}} G_\mu h_C^\mu h_{C\mu}^+$ of (8.2) is probably responsible for these decays, but because of the presence of strong interactions, theoretical predictions are essentially restricted to selection rules. Since the above-mentioned decays all change the hypercharge, the relevant piece of (8.2) is

$$H_{nl}^{(1)} = 2^{-\frac{1}{2}} G_\mu \cos\theta_C \sin\theta_C (h^{(0)} h^{(1)+} + h^{(1)} h^{(0)+}). \quad (9.1)$$

Since $h^{(0)}$ transforms as an isovector and $h^{(1)}$ as an isospinor, we expect $\Delta I = \frac{1}{2}$ or $\frac{3}{2}$ in these decays. The experimental situation for hyperon decays has been described in section 4-11: Only $\Delta I = \frac{1}{2}$ occurs, and the operator belongs to an $SU(3)$-octet. This fact goes under the name "octet dominance," in analogy with the octet dominance in the mass splitting operator (section 4-10). It is basically not yet understood. In kaon decays, there is one obvious exception to this rule, namely the decay $K^+ \to \pi^+\pi^0$. Angular momentum conservation requires the two pions to be in an $S$-state, in which case their isospin is either 0 or 2. But $I = 0$ is excluded by $I_3 = +1$, therefore the decay has $\Delta I = \frac{3}{2}$ or even $\frac{5}{2}$. However, the decay is weak in comparison with $K_S^0 \to \pi\pi$, $\Gamma(K^+ \to \pi^+\pi^0)/\Gamma(K_S^0) = 0.0015$.

The accuracy of the $\Delta I = \frac{1}{2}$ rule can also be tested in $K_S \to \pi\pi$ decays alone. Allowing for the difference in the decay momenta $p_{+-}$ and $p_{00}$ in the $\pi^+\pi^-$ and $\pi^0\pi^0$ channels, we get from the first row of table 4-9 and the pion isospin decomposition

$$\begin{gathered}\Gamma(K_S \to \pi^0\pi^0)/\Gamma(K_S \to \pi^+\pi^-) = (\tfrac{1}{2} - 3 \operatorname{Re} \omega)p_{00}/p_{+-}, \\ \omega \equiv 2^{-\frac{1}{2}}\langle \pi\pi, I = 2|T|K_S\rangle\langle \pi\pi, I = 0|T|K_S\rangle^{-1}.\end{gathered} \tag{9.2}$$

Next, we discuss the $K \to 3\pi$ decays of table 5-9. All information is contained in the respective Dalitz plots, as discussed in section 4-6. At least two of the three pions have identical masses; these are taken as particles 1 and 2. The decay matrix element $T(E_1, E_2 E_3)$ must be symmetric in 1 and 2,

$$T(E_1, E_2, E_3) = T(E_2, E_1, E_3). \tag{9.3}$$

This is trivial for $1 = 2$ as in the states $\pi^+\pi^+\pi^-$, $\pi^0\pi^0\pi^+$ and $\pi^0\pi^0\pi^0$. For $K_L^0 \to \pi^+\pi^-\pi^0$, it is true only to the extent that $K_L^0$ is approximated by the state $K_2^0$ of (4-5.13) which has $C\mathscr{P} = -1$. The $\pi^0$ has $C\mathscr{P} = -(-1)^L$ where $L$ is the angular momentum between $\pi^0$ and the $\pi^+\pi^-$-pair. The $\pi^+\pi^-$ pair has $C\mathscr{P} = 1$ in all possible states, therefore $L$ must be even. On the other hand, since the system's total angular momentum is zero, the spin of the $\pi^+\pi^-$ pair $L_{+-}$ equals $L$ and is therefore also even.

The Dalitz plot of $K$ decays is so small that only a linear expansion of $T$ in $E_i$ can be determined. Because of (3) and $E_1 + E_2 + E_3 = m$, only $E_3$ enters the expansion:

$$\begin{gathered}T = a + b\hat{s}, \\ \hat{s} = \tfrac{1}{3}m_{\pi^+}^{-2}(2s_{12} - s_{13} - s_{23}) = \tfrac{2}{3}(m^2 + m_3^2 - 3mE_3 - m_1^2)m_{\pi^+}^{-2}.\end{gathered} \tag{9.4}$$

The variable $\hat{s}$ is equivalent to $Y$ of (4-6.11) and is preferred because it is easier to express in terms of Lorentz-invariants. The amplitude $a$ is totally symmetric in all 3 particles and must have $I = 1$ according to table 4-2.4 (The part with $I = 3$ is neglected because it requires at least $\Delta I = \frac{5}{2}$). Since $K_L^0$ has $I = \frac{1}{2}$, both $\Delta I = \frac{1}{2}$ and $\Delta I = \frac{3}{2}$ amplitudes can contribute to $I = 1$. The relative weights of these amplitudes depend only on the total pion charge, both for $a_{\frac{1}{2}}$, $a_{\frac{3}{2}}$ and $b_{\frac{1}{2}}$, $b_{\frac{3}{2}}$. The amplitudes $b$ and $c$ lead to those

**Table 5-9** The $\pi\pi\pi$-decays of $K^+$ and $K_L^0$. The "weight" in the first row multiplies the phase space differential (4-6.9) such that identical states are counted only once. The coefficients of the matrix elements for pion states of $I = 1$ and 2 are explained in the text, the factor $g$ is defined in (4), its experimental values are taken from the Particle Data group (1976). The last row gives the relative Dalitz plot areas, including $g(\pi^+\pi^+\pi^-) = -0.2$ and a Coulomb correction according to Mast et al. (1969).

| | $\pi^+\pi^+\pi^-$ | $\pi^0\pi^0\pi^+$ | $\pi^+\pi^-\pi^0$ | $\pi^0\pi^0\pi^0$ |
|---|---|---|---|---|
| Weight | $\frac{1}{2}$ | $\frac{1}{2}$ | 1 | $\frac{1}{6}$ |
| $T(I = 1)$ | $2a_{\frac{1}{2}} - 2a_{\frac{3}{2}} + b_{\frac{1}{2}}\hat{s} - b_{\frac{3}{2}}\hat{s}$ | $a_{\frac{1}{2}} + 2a_{\frac{3}{2}} + b_{\frac{1}{2}}\hat{s} + 2b_{\frac{3}{2}}\hat{s}$ | $a_{\frac{1}{2}} + 2a_{\frac{3}{2}} + b_{\frac{1}{2}}\hat{s} + 2b_{\frac{3}{2}}\hat{s}$ | $3a_{\frac{1}{2}} + 6a_{\frac{3}{2}}$ |
| $T(I = 2)$ | $c_{\frac{3}{2}}\hat{s}$ | $c_{\frac{3}{2}}\hat{s}$ | 0 | 0 |
| $g$ | $-0.21$ | 0.55 | 0.65 | 0 |
| Area of D.P. | $\equiv 1.000$ | 1.155 | 1.268 | 1.451 |

remaining $I = 1$ and $I = 2$ final states of table 4-2.4 which are symmetric in 1 and 2. Assuming $b$ and $c$ to be small, we have approximately

$$|T|^2 = |a|^2(1 + 2\hat{s}\ \mathrm{Re}\ ab^*/|a|^2) = |T(\hat{s} = 0)|^2(1 + g\hat{s}). \tag{9.5}$$

The experimental values of $g$ are included in table 10. For $g = 0$ and equal Dalitz plot areas, one would get

$$\Gamma(K^+ \to \pi^+\pi^+\pi^-) : \Gamma(K^+ \to \pi^0\pi^0\pi^+) : \Gamma(K_L^0 \to \pi^+\pi^-\pi^0) : \Gamma(K_L^0 \to \pi^0\pi^0\pi^0) = 4 : 1 : 2 : 3. \tag{9.6}$$

Two more purely hadronic weak interactions are observed in nuclei, namely "stimulated $\Lambda$ decay" in hypernuclei, $\Lambda p \to np$ and $\Lambda n \to nn$ (see for example Davis and Sacton, in Burhop 1967) and parity-violating transitions in nuclei (see for example Blin-Stoyle, 1973). The former are induced by (1) and presumably observe the $\Delta I = \frac{1}{2}$ rule, except for Coulomb and other corrections. The latter have $\Delta Y = 0$ and are induced partly by (8.2) and partly by the neutral current $j^\mu$ of (11.12) below:

$$H_{nl}^{(0)} = 2^{-\frac{1}{2}}G_\mu(\cos^2\theta_C h^{(0)}h^{(0)+} + \sin^2\theta_C h^{(1)}h^{(1)+} + h^{(0n)}h^{(0n)}). \tag{9.7}$$

They are normally described in terms of an internucleon potential which is parity-violating. Although (7) contains a $\Delta I = 2$-piece at least from $h^{(1)}h^{(1)+}$, one expects octet dominance also in the $\Delta Y = 0$ hadronic weak interaction, i.e., the $\Delta I = 2$-piece should be negligible in comparison with the $\Delta I = 0$ and $\Delta I = 1$ pieces which are components of the same $SU(3)$-octet as the $\Delta Y = 1$, $\Delta I = \frac{1}{2}$ piece.

Finally, we come to a fascinating topic, namely the decays $K_L^0 \to \pi\pi$ and the more general question of CPT-invariance. Consider first stable particles: how sure can we be that the antiproton has the same mass as the proton, and that is cannot decay? When we introduced the free field operators in section 3-2, we pointed out that their negative-energy parts are required by causality, (3-2.21) in the case of spinless particles. Consequently, if we rebaptize the ominous operator $b_M^+(\mathbf{p})$ (for particles of any spin) in front of a negative-energy state as $a(-P^\mu, -M)$, each field transforms into itself under the substitution $P^\mu \to -P^\mu$, $M \to -M$. This is the basis of the CPT-theorem. In its final version, the theorem is independent of perturbation theory (see the books by Streater and Wightman 1964, Jost 1965). The interpretation of the above substitution as the CPT transformation has been derived in section 3-6, the essential point being the physical absence of negative energy states. We saw that for every particle which carries an absolutely conserved charge of some kind, there exists a particle of opposite charge and exactly the same mass. To the extent that unstable particles decay exponentially (which is always the case for long lifetimes and absence of degeneracy), they can be treated as particles of complex mass $m - i\Gamma/2$, in which case particles and antiparticles must have identical lifetimes (Lee, Oehme and Yang 1957). When the substitution rule is applied to all particles in a reaction $i \to f$ and the spin dependence is taken out, one gets

$$T_{\bar{f}\bar{\imath}}(s, t) = T_{if}(s, t). \tag{9.8}$$

Because of $\bar{s} = (-P_a - P_b)^2 = s$, $\bar{t} = (-P_c + P_a)^2 = t$ etc., no extrapolation is required in these variables, and (8) is a direct consequence of the CPT theorem. In many cases, this is sufficient to determine partial decay rates of antiparticles. For notational convenience, we assume two-particle final states. From (8), we have $T_J(\bar{d} \to \overline{12}) = T_J(12 \to d)$, where $J$ is the spin of $d$. Inserting this into (4-8.27), we find

$$T_J^*(\bar{d} \to \bar{c})p_c^{\frac{1}{2}} = \sum_{c'} T_J(d \to c')p_c^{\frac{1}{2}}S_J^*(c \to c'), \tag{9.9}$$

$$\begin{aligned}\sum_c |T_J^*(\bar{d} \to \bar{c})|^2 p_c &= \sum_{cc'c''} T_J(d \to c')p_c'^{\frac{1}{2}}S_J^*(c \to c')S_J(c \to c'')T_J(d \to c'')p_c''^{\frac{1}{2}} \\ &= \sum_{c'} |T_J(d \to c')|^2 p_c' = \sum_c |T_J(d \to c)|^2 p_c,\end{aligned} \tag{9.10}$$

using the unitarity of the partial-wave $S$-matrix, $S_J^+ S_J = 1$. The summation over the channels $c, c', c''$ in (10) includes all open channels, but in practice $S_J$ consists frequently of a string of smaller matrices which are separately unitary. Looking for example at the long list of $K^\pm$-decays, we can safely put $S(\mu\nu \to \pi\pi) = S(\mu\nu \to \pi\pi\pi) = S(\mu\nu \to \pi e\nu) = 0$, because these matrix elements involve the weak interaction once more. The only reasonably large $S$-matrix element is presumably $S(\pi^\pm\pi^\pm\pi^\mp \to \pi^\pm\pi^0\pi^0)$ in this case. Clearly, the sum over $c$ can be restricted to channels which couple sufficiently strong. In particular, this implies particle-antiparticle equality of partial decay rates (and of energy spectra as well!) of practically all hadron decays to final states which contain a neutrino, apart from possible Coulomb corrections. Unfortunately, these decays are also CP-invariant, such that our ingenious considerations are superfluous. The only interesting case is presently $K^0$ decays, and here the formalism must be modified because the weak interaction mixes $K^0$ with $\bar{K}^0$.

The physical states $K_S$ and $K_L$ which have definite lifetimes are linear combinations of $K$ and $\bar{K}$. To first order in the deviations from the states $K_1$ and $K_2$ of (4-5.13) we have

$$2^{\frac{1}{2}}K_S = (1 + \varepsilon)K - (1 - \varepsilon)\bar{K}, \qquad 2^{\frac{1}{2}}K_L = (1 + \varepsilon')K + (1 - \varepsilon')\bar{K}. \tag{9.11}$$

In chapter 4, we assumed $\varepsilon' = \varepsilon$, which is a consequence of CPT, as we shall see. The time development of the state is given by (4-3.11), with $(1 \pm \varepsilon)a_L$ replaced by $(1 \pm \varepsilon')a_L$. Next, we invert (11):

$$2^{\frac{1}{2}}K = (1 - \varepsilon')K_S + (1 - \varepsilon)K_L, \quad 2^{\frac{1}{2}}\bar{K} = -(1 + \varepsilon')K_S + (1 + \varepsilon)K_L. \tag{9.12}$$

If the initial state was pure $K$, we have $a_L = 1 - \varepsilon$ and $a_S = 1 - \varepsilon'$, and the probability amplitude for finding $K$ at time $\tau$ is

$$(1 + \varepsilon')(1 - \varepsilon) \exp(-iM_L\tau) + (1 + \varepsilon)(1 - \varepsilon') \exp(-iM_L\tau). \tag{9.13}$$

If the initial state was pure $\bar{K}$, we have $a_L = 1 + \varepsilon$, $a_S = -(1 + \varepsilon')$ and the probability amplitude for finding $\bar{K}$ at time $\tau$ is

$$(1 - \varepsilon')(1 + \varepsilon) \exp(-iM_L\tau) + (1 - \varepsilon)(1 + \varepsilon') \exp(-iM_S\tau). \tag{9.14}$$

CPT requires equality of these two amplitudes at any time, which is only possible for $\varepsilon' = \varepsilon$. We put $\varepsilon' = \varepsilon$ for the time being and return to the question of CPT-violation below.

We now proceed to the analysis of $K_L \to \pi\pi$ decays. The decay amplitudes to $\pi\pi$ states of isospin $I = 0$ or 2 are denoted by $T_I$:

$$\langle \pi\pi, I | T | K \rangle \equiv T_I = G_I e^{i\delta_I}, \tag{9.15}$$

where $\delta_I$ is the $\pi\pi$-scattering phase shift in the channel of isospin $I$. $G_I$ is the "coupling constant" $G_0$ of table 4-9, according to (4-8.29) it would be real if time-reversal invariance were fulfilled. From (8) and the complex conjugate of (9) in the one-channel case, $S = \exp(2i\,\delta_I)$,

$$\langle \pi\pi, I | T | \bar{K} \rangle \equiv \bar{T}_I = -\langle K | T | \pi\pi, I \rangle = -G_I^* e^{i\delta_I}. \tag{9.16}$$

The minus sign comes from our choices of phases under CP, $CP|\bar{K}^0\rangle = -K^0$, table 4-5.1, which is not standard nowadays.

Since a common phase of all decay matrix elements cannot be measured, we can define $G_0$ to be real. In that case, we get from (4-1.4) and (16), (2)

$$\langle \pi\pi, 0 | T | K_L \rangle \langle \pi\pi, 0 | T | K_S \rangle^{-1} = \varepsilon, \qquad \omega \approx 2^{-\frac{1}{2}} G_0^{-1} e^{i(\delta_2 - \delta_0)} \operatorname{Re} G_2 . \tag{9.17}$$

$$\begin{aligned} \langle \pi\pi, 2 | T | K_L \rangle \langle \pi\pi, 0 | T | K_S \rangle^{-1} &\equiv 2^{\frac{1}{2}} \varepsilon_2 \\ = [G_2 - G_2^* + \varepsilon(G_2 + G_2^*)]\tfrac{1}{2} G_0^{-1} e^{i(\delta_2 - \delta_0)} &\approx i(\operatorname{Im} G_2) G_0^{-1} e^{i(\delta_2 - \delta_0)}. \end{aligned} \tag{9.18}$$

Here we have neglected $\varepsilon$ compared to unity. Experimentally, one measures the ratios

$$\begin{aligned} \eta_{+-} &\equiv \langle \pi^+\pi^- | T | K_L \rangle \langle \pi^+\pi^- | T | K_S \rangle^{-1} \approx (\varepsilon + \varepsilon_2)(1 + 2^{-\frac{1}{2}}\omega)^{-1}, \\ \eta_{00} &\equiv \langle \pi^0\pi^0 | T | K_L \rangle \langle \pi^0\pi^0 | T | K_S \rangle^{-1} \approx (\varepsilon - 2\varepsilon_2)(1 - 2^{\frac{1}{2}}\omega)^{-1}, \end{aligned} \tag{9.19}$$

where the final expressions follow from (17)–(18) and the isospin decomposition of $\pi\pi$-states (first row of table 4-2.3).

Actually one can neglect $\omega$ in (19), because the isospin relation

$$T_2 = (\tfrac{2}{3})^{\frac{1}{2}} \langle \pi^+\pi^0 | T | K^+ \rangle \tag{9.20}$$

(which also follows from table 4-2.3) tells us

$$|\omega|^2 + |\varepsilon_2|^2 = \tfrac{1}{3}\Gamma(K^+ \to \pi^+\pi^0)/\Gamma(K_S^0), \tag{9.21}$$

and shows that both $\omega$ and $\varepsilon_2$ are small.

The experimental situation in $K^0$ decays has been reviewed by Kleinknecht (1976). We have

$$\eta_{+-} \approx 2.27 e^{i45^0} \times 10^{-3}, \qquad \eta_{00} \approx 2.3 e^{i48^0} \times 10^{-3}, \tag{9.22}$$

i.e., $\varepsilon_2 = 0$ within the experimental accuracy. According to (18), this means that $G_2$ is real relative to $G_0$, and all CP-violation is due to the state-mixing parameter $\varepsilon$. This is exactly what is expected on the basis of the "superweak model" of Wolfenstein (1964), where the CP-violating part of the decay is due to a very small Hamiltonian $H_{SW}$ with direct $K^0 \to \bar{K}^0$ ($\Delta Y = 2$) transitions. If our current-current interaction (1) is correct, the CP-conserving part of the $K_L - K_S$ mass splitting must be a second order effect of the $\Delta Y = 1$-Hamiltonian (in particular, $\Gamma_S \sim G_\mu^2 \sin^2 \theta_C$). For an order-of-magnitude estimate, one may take

$$\langle \bar{K} | H_{SW} | K \rangle \sim g_{SW} G_\mu, \qquad \langle \bar{K} | H_{nl}^{(1)} H_{nl}^{(1)} | K \rangle \sim G_\mu^2 m_p^2, \tag{9.23}$$

where the proton mass $m_p$ has been inserted for dimensional reasons only. Then, with $\langle \bar{K} | H_{SW} | K \rangle / \langle \bar{K} | H_{nl}^{(1)} H_{nl}^{(1)} | K \rangle \approx \varepsilon$, we find

$$g_{SW} \approx \varepsilon G_\mu m_p^2 \approx 10^{-8}, \tag{9.24}$$

i.e., the superweak interaction is $10^{-8}$ of the normal weak interaction. If this is correct, it will be extremely difficult to detect in the decays of other "old" particles.

The same remark applies to a possible CPT-violating interaction. The success of the superweak model indicates that such an interaction would be still weaker than the superweak one. From the $K_L - K_S$ mass difference alone, one may conclude

$$(m_K - m_{\bar{K}}) m_K^{-1} < (m_L - m_S) m_K^{-1} \approx 0.7 \times 10^{-14}. \tag{9.25}$$

Also the superposition principle is accurately tested in $K_L^0$, $K_S^0$ decays.

## 5-10 Neutrino Experiments, Neutral Currents, Salam–Weinberg Model

The cross section for neutrino interactions with protons and heavier nuclei is extremely small at low energies but rises linearly with the neutrino lab energy. Accelerator experiments are done with neutrinos from $\pi$ and $K$ decays, which are almost entirely "muon neutrinos", $\nu_\mu$ and $\bar{\nu}_\mu$. The resulting cross sections with a muon in the final state (plus of course nucleons and possibly pions) are quoted per nucleon in medium-size nuclei (Eichten 1973)

$$\begin{aligned} \sigma(\nu_\mu N \to \mu^- \ldots) &= 0.74 \times 10^{-38} E_\nu^{\text{lab}} \text{ cm}^2/\text{GeV}, \\ \sigma(\bar{\nu}_\mu N \to \mu^+ \ldots) &= 0.28 \times 10^{-38} E_\nu^{\text{lab}} \text{ cm}^2/\text{GeV}. \end{aligned} \tag{10.1}$$

The cross sections for "quasi-elastic neutrino scattering" $\nu_\mu n \to \mu^- p$ and $\bar{\nu}_\mu p \to \mu^+ n$ flatten out around 1 GeV; the rise of (1) at higher energies is entirely due to meson production.

The matrix elements of neutrino collisions with hadrons are of the form

$$T = 2^{-\frac{1}{2}} G_V L_\mu H^\mu, \qquad L_\mu = \bar{u} \gamma_\mu (1 \mp \gamma_5) u_\nu. \tag{10.2}$$

$L_\mu$ can be used both for neutrinos and antineutrinos, with the $-$ sign for $\nu$ and the $+$ sign for $\bar{\nu}$. Explicit forms of $L_\mu$ are given in (A-3.21)–(A-3.24).

Here we are only interested in the differential cross section for 2-particle reactions $\nu b \to cd$ ($c = e$ or $\mu$) after spin summation, which is given by

$$d\sigma(\nu b \to cd) = \frac{\pi}{\lambda}(2S_b + 1)^{-1} \sum_{\lambda_\nu \lambda_c M_b M_d} |T/4\pi|^2 \, dt. \tag{10.3}$$

This expression differs from (2-4.9) by a factor 2. We assume that the incident $\nu$ is in the "correct" helicity state $\lambda_\nu$. The summation over $\lambda_\nu$ in (3) is harmless since the matrix element for the "wrong" helicity vanishes anyway. This method allows one to use again the covariant spin summation (A-4.10):

$$d\sigma/dt = \frac{\pi}{\lambda} G_V^2 S_{\mu\nu}(2S_b + 1)^{-1} \sum_{M_b M_d} J_{bd}^{\mu} J_{bd}^{\mu *}(4\pi)^{-2}, \tag{10.4}$$

$$\begin{aligned} S_{\mu\nu} &\equiv \tfrac{1}{2} \text{ trace } \gamma_\mu(1 \pm \gamma_5)P_a \cdot \gamma(1 \mp \gamma_5)\gamma_\nu(P_c \cdot \gamma + m_c) \\ &= \text{trace } \gamma_\mu(1 \pm \gamma_5)P_a \cdot \gamma\, \gamma_\mu(P_c \cdot \gamma + m_c) \\ &= 4(P_{a\mu} P_{c\nu} + P_{a\nu} P_{c\mu} - g_{\mu\nu} P_a P_c \pm i\varepsilon_{\mu\beta\nu\delta} P_a^\beta P_c^\delta). \end{aligned} \tag{10.5}$$

In the second line of (5) we have used the projector property $(1 \pm \gamma_5)^2 = 2(1 \pm \gamma_5)$, and in the last line we have used the traces of A-4. Note the similarity between (4) and the corresponding equations (2-7.19), (2-7.20) for electron scattering. As in the case of inelastic electron scattering, the spin-summed hadron current bilinear can only depend on the available tensors which can be constructed from the hadron 4-momenta, namely $P_b^\mu q^\nu$, $P_b^\nu q^\mu$, $q^\mu q^\nu$, $P_b^\mu P_b^\nu$, $g^{\mu\nu}$ and $i\varepsilon^{\mu\nu\rho\sigma} P_{b\rho} q_\sigma$. We explained already in (4.2) that the terms proportioned to $q^\mu$ or $q^\nu$ give matrix elements proportional to the charged lepton mass. They can be neglected at high energies, also in the case of outgoing muons, the errors being of the order of $m_\mu/E_\nu$ in this case. Thus only 3 of the possible 6 tensors need to be considered at high energies

$$(2S_b + 1) \sum_{M_a M_b} J_{bd}^{\mu} J_{bd}^{\nu *} = -W_1 g^{\mu\nu} + m_b^{-2} W_2 P_b^\mu P_b^\nu + \tfrac{1}{2} m_b^{-2} i\varepsilon^{\mu\nu\rho\sigma} P_{b\rho} q_\sigma W_3. \tag{10.6}$$

The weak structure functions $W_1$ and $W_2$ are quite analogous to those of electron scattering (2-7.30), only the parity-violating function $W_3$ is new. Insertion of (6) and (5) into (4) gives

$$\begin{aligned} \frac{d\sigma^{(\pm)}}{dt} = G_V^2(4\lambda\pi m_b^2)^{-1}[&-m_b^2 t W_1 + W_2(\tfrac{1}{2} t m_b^2 + 2(P_a P_b)(P_c P_b)) \\ &\mp W_3(P_a P_b P_c P_d - P_b P_c P_a P_d)]. \end{aligned} \tag{10.7}$$

To obtain the last term, we have used

$$\varepsilon_{\mu\nu\beta\delta}\varepsilon^{\mu\nu\alpha\gamma} = 2\,\delta_\beta^\gamma\,\delta_\delta^\alpha - 2\,\delta_\beta^\alpha\,\delta_\delta^\gamma, \tag{10.8}$$

where $\delta_\beta^\alpha = \delta_{\alpha\beta}$ is the Kronecker-delta. One may verify (8) by inserting specific values of the indices $\alpha \ldots \delta$. For the reactions $\bar{\nu}_\mu p \to \mu^+ n$ and $\nu_\mu n \to \mu^- p$, we get from (4.4) and (6)

$$d\sigma^{(\pm)}/dt = \frac{G_V^2}{2\pi} \cos^2 \tfrac{1}{2}\theta_{\text{lab}} E_\mu^{\text{lab}}/E_\nu^{\text{lab}} \Big[ G_E^{V\,2}(1+\tau)^{-1} + \Big( g_1^2 + \frac{\tau}{1+\tau} G_M^{V\,2} \Big) \times (1 + 2(1+\tau)\tan^2 \tfrac{1}{2}\theta_{\text{lab}}) \pm 2(E_\nu^{\text{lab}} + E_\mu^{\text{lab}}) m^{-1} G_M^V g_1 \tan^2 \tfrac{1}{2}\theta_{\text{lab}} \Big] \tag{10.9}$$

where the definitions are as in (2-7.27)–(2-7.29) for the isovector parts (6.7). The measured form factors have been reviewed by Gourdin (1974).

All weak processes which we have discussed so far involved neutrinos only in conjunction with $e^\pm$ or $\mu^\pm$, i.e., in the current-current interaction (8.2), the currents are "charged" (they are charge-changing currents, to be precise). In the case of hypercharged currents, direct evidence against the existence of neutral weak currents comes from the absence of decays such as $K^+ \to \pi^+ e^+ e^-$ or $K^+ \to \pi^+ \nu\bar{\nu}$. However, in neutrino experiments, a large number of neutrino collisions with nuclei (including the production of pions) were found which did not contain any charged lepton in the final state (Hasert et al. 1973). On an Fe-target, one finds (see Musset and Vialle 1978)

$$\begin{aligned} \sigma(\nu_\mu \text{Fe} \to \nu_\mu \ldots)/\sigma(\nu_\mu \text{ Fe} \to \mu^- \ldots) &\approx 0.26, \\ \sigma(\bar{\nu}_\mu \text{ Fe} \to \bar{\nu}_\mu \ldots)/\sigma(\bar{\nu}_\mu \text{ Fe} \to \mu^+ \ldots) &\approx 0.40 \end{aligned} \tag{10.10}$$

where the notation $\nu_\mu \ldots$ and $\bar{\nu}_\mu \ldots$ really means that the final states contain no charged leptons. Among the many possible final states, the reaction $\nu N \to \nu\Delta$ has been identified, which shows that the new hadronic current contains an isovector piece. More detailed information on the structure of the new interaction is presently still lacking.

Elastic neutrino-electron scattering $\nu_\mu e^- \to \nu_\mu e^-$ has also been seen (Gargamelle collaboration 1976). Interactions of this type were in fact anticipated on rather theoretical grounds:

In a complete theory of weak interactions, the local 4-fermion interaction is not acceptable because it cannot be renormalized. For example, the elastic scattering amplitude violates the unitarity requirement (4-4.10) at high energies. With coupling constant $G_\mu$ and a factor $S^{\mu\nu}(bd)$ analogous to $S_{\mu\nu}$ of (5) from the target spin summation

$$d\sigma(e\nu)/dt = (16\pi\lambda)^{-1} G_\mu^2 S_{\mu\nu} S^{\mu\nu}(bd) = 4(\pi\lambda)^{-1} G_\mu^2 (P_a P_b)(P_c P_d), \tag{10.11}$$

where we have again used (8). We shall discuss the complete treatment of spin in the partial-wave expansion in connection with nucleon-nucleon scattering. Here we only need to know that if the scattering occurs only in the $S$-wave, then we have (for $p \approx p'$)

$$d\sigma(\nu b \to cd)/d\Omega \le \tfrac{1}{4} p^{-2}. \tag{10.12}$$

(The limitation of partial-wave cross sections by unitarity in the spinless elastic case follows from 2-5.13 and 2-5.14). Inserting now $d\Omega = \pi\, dt/pp'$, we get $d\sigma/dt \le \frac{1}{4}\pi p^{-4}$. On the other hand, with $\lambda = (s - m_b^2)^2 = 4(P_a P_b)^2$ and $P_a P_b \approx P_c P_d$ at high energies (or for elastic scattering at all energies), (11) reduces to $\pi^{-1} G_\mu^2 = \text{const}$. There is thus a cms momentum $p_{\max}$ at which (11) becomes inconsistent with (12):

$$p_{\max} = (\pi/2G_\mu)^{\frac{1}{4}} \approx 300 \text{ GeV}. \tag{10.13}$$

To avoid the conflict with unitarity, the interaction must have a finite range. The simplest model which provides such a range is that of $W$-boson exchange, fig. 5-10a. The limit $m_W \to \infty$ of this model was mentioned already in (4.6). For finite $m_W$, we have for purely leptonic processes

$$T_{\nu e}^{(W)} = \tfrac{1}{8} g^2 L_{\nu e}^{\mu} L_{e\nu}^{\nu} (-g_{\mu\nu} + q_\mu q_\nu m_w^{-2})(m_W^2 - u)^{-1}, \qquad \tfrac{1}{8} g^2 = 2^{-\frac{1}{2}} G_\mu m_W^2. \tag{10.14}$$

When $u$ becomes comparable to $m_W^2$, (14) contributes to several partial waves and thus need not violate unitarity, if $m_W$ is not too large.

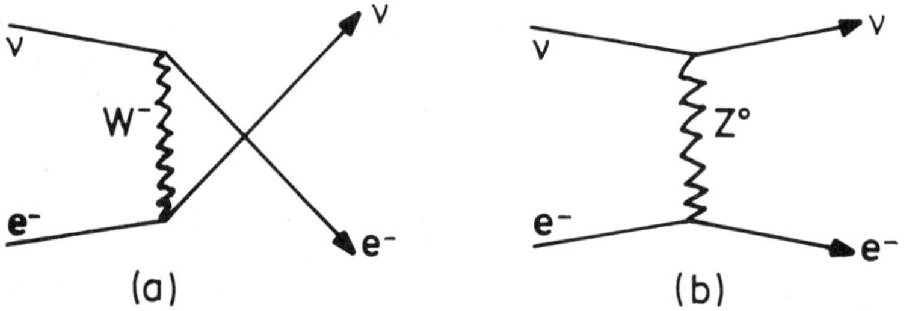


**Fig. 5-10.** (a) Neutrino-electron scattering in the $W$-boson hypothesis, (b) the additional neutral current contribution in the Salam–Weinberg model.

Unfortunately, this extension of the formalism introduces other reactions such as $\nu\bar{\nu} \to W^+ W^-$, which now show a bad high-energy behaviour. This again can be compensated either by the introduction of a new heavy lepton $l^+$ in addition to $e^-$ (Glashow et al.). But since neither $\mu^+$ nor $\tau^+$ couple to $W^+ \nu_e$, they cannot accomplish this compensation.

A different compensation mechanism was proposed by Glashow (1961), Salam and Ward (1964) and by Weinberg (1967). It requires an additional neutral boson $Z$ and combines weak and electromagnetic interactions in a way which produces neutral weak currents that are not purely lefthanded. The $W$ and $Z$ masses are predicted in terms of an angle (see (46) below) and have not yet been discovered. However, the neutral currents have been verified in some detail (in particular parity nonconservation in inelastic $ed$-scattering has been discovered by Prescott et al. (1978)), lending substantial support to the correctness of this "Salam–Weinberg theory." Early theoretical reviews have been given by Abers and Lee (1973), Beg and Sirlin (1974), Taylor (1976).

We begin with a world of $\nu_e$, $e$, $\bar{\nu}_e$, $\bar{e}$ and bosons. Other lepton pairs are added at the end of this section, hadrons are included in the next section. The left-handed two-component spinor fields of electron and neutrino are denoted by $e_l$ and $\nu_l$. They are grouped into a doublet $l_l$, and we require invariance of their interactions under unitary transformations,

$$l_l' = e^{-i\boldsymbol{\omega}\boldsymbol{\tau}/2} l_l, \qquad l_l \equiv \begin{pmatrix} \nu_l \\ e_l \end{pmatrix}, \tag{10.15}$$

quite analogous to the isospin rotations of protons and neutrons. Moreover, we allow $\boldsymbol{\omega}$ to vary in space and time, $\boldsymbol{\omega} = \boldsymbol{\omega}(x^\mu)$, such that (15) actually is a generalized gauge transformation analogous to (1-1.9). A triplet of 4-vector fields $\mathbf{W}^\mu$ is introduced by the "minimal coupling"

$$D^\mu \equiv \partial^\mu - ig\mathbf{W}^\mu\boldsymbol{\tau}/2 \tag{10.16}$$

in order to have a renormalizable interaction. $iD^\mu$ is the generalization of $-i\pi^\mu$ (1-1.4) to the group $SU(2)$. How do the $\mathbf{W}^\mu$ transform under the gauge group? In analogy with (1-1.7), we must have

$$D'^\mu l_l' = \exp(-i\boldsymbol{\omega}(x)\boldsymbol{\tau}/2)D^\mu l_l. \tag{10.17}$$

Insertion of (16) gives

$$(\partial^\mu - ig\mathbf{W}'^\mu\boldsymbol{\tau}/2)l_l' = e^{-i\boldsymbol{\omega}\boldsymbol{\tau}/2}(\partial^\mu - ig\mathbf{W}^\mu\boldsymbol{\tau}/2)e^{i\boldsymbol{\omega}\boldsymbol{\tau}/2}l_l' \tag{10.18}$$

for arbitrary $l_l'$. The $\partial^\mu l_l'$-terms cancel, and from the remaining terms $l_l'$ may be divided off. The transformation is already determined by an infinitesimal $\boldsymbol{\omega}$:

$$\begin{aligned} -ig\mathbf{W}'^\mu\boldsymbol{\tau}/2 &= (1 - i\boldsymbol{\omega}\boldsymbol{\tau}/2)(\partial^\mu - ig\mathbf{W}^\mu\boldsymbol{\tau}/2)(1 + i\boldsymbol{\omega}\boldsymbol{\tau}/2) \\ &= \partial^\mu i\boldsymbol{\omega}\boldsymbol{\tau}/4 - ig\mathbf{W}^\mu\boldsymbol{\tau}/2 + ig\varepsilon_{ijk} W_i^\mu \omega_j \tau_k/2. \end{aligned} \tag{10.19}$$

Since the $\boldsymbol{\tau}$ are linearly independent $2 \times 2$ matrices, (19) requires

$$\mathbf{W}'^\mu = W^\mu - g^{-1}\,\partial^\mu\boldsymbol{\omega} - \mathbf{W}^\mu \times \boldsymbol{\omega}. \tag{10.20}$$

The first two terms generalize (1-0.25). The last term is new and appears because the gauge transformations are no longer Abelian. The equations of motion are best derived from a Lagrangian density $\mathscr{L}(\psi_i(x), \partial^\mu\psi_i(x))$, the $\psi_i$ being the various fields $e_l, \nu_l, e_l^+, \nu_l^+, W_i^\mu$. The Lagrangian of QED has been given in (3-3.3). If we are not interested in the equation of motion for $\bar{\psi}_e$, we may replace $\frac{1}{2}\,\partial_\mu$ by $\partial_\mu$:

$$\mathscr{L}_{\text{QED}} = -\tfrac{1}{4}F_{\mu\nu}F^{\mu\nu} + \bar{\Psi}_e[\gamma_\mu(i\,\partial^\mu + eA^\mu) - m_e]\Psi_e, \tag{10.21}$$

with $F^{\mu\nu} = \partial^\mu A^\nu - \partial^\nu A^\mu$. Using now the same formalism for our doublet of lefthanded lepton spinors $l_l$ and triplet of 4-vector fields $\mathbf{W}^\mu$, we find that the more complicated transformation property (20) requires the following definition

$$\mathbf{F}^{\mu\nu} = \partial^\mu\mathbf{W}^\nu - \partial^\nu\mathbf{W}^\mu - g\mathbf{W}^\mu \times \mathbf{W}^\nu \tag{10.22}$$

to produce the gauge-invariant Lagrangian

$$\mathscr{L}_l = -\tfrac{1}{4}\mathbf{F}_{\mu\nu}\mathbf{F}^{\mu\nu} + il_l^+\,\sigma_l^\mu D_\mu l_l. \tag{10.23}$$

So far, the formalism was proposed by Yang and Mills (1954) as a dynamical basis of isospin invariance. They had the proton-neutron doublet instead of $l$ and thought of $\mathbf{W}^\mu$ as a kind of $\rho$-meson field. However, the $\rho$-meson had to be massless because a mass term $-m^2\mathbf{W}_\mu\mathbf{W}^\mu$ in the Lagrangian would destroy gauge invariance. For the same reason, there is no such term in the part $\mathscr{L}_l$ of the Salam-Weinberg Lagrangian. In addition, *both* lepton masses are exactly zero in (23). It was shown by t'Hooft (1971) that a "spontaneous breakdown" of gauge invariance can lead to massive mesons (and electrons) without destroying renormalizability. Such a breakdown is not a small violation of the invariance as in the case of isospin. $\mathscr{L}$ remains strictly invariant, but at least one of the fields $\psi_i$ has a nonzero vacuum expectation value $\langle\psi_i\rangle_0$, which drastically changes the physical content. Normally, massless "Goldstone" bosons appear in such cases (the pion is a Goldstone boson in the limit $\partial^\mu a_\mu = 0$, compare (7.14)). In the presence of gauge fields $\mathbf{W}^\mu$, the situation is more complicated (Higgs 1964). Salam and Weinberg introduced an additional scalar isodoublet field $\phi$, with the following self-interaction:

$$V_\phi = \mu^2\phi^+\phi + \lambda(\phi^+\phi)^2, \qquad \phi \equiv \begin{pmatrix}\phi^{(+)}\\ \phi^{(0)}\end{pmatrix}, \qquad \lambda > 0. \tag{10.24}$$

For $\mu^2 > 0$, one would interpret $\mu$ as the mass of the corresponding scalar bosons. In a potential theory with $V$ as a potential, the state of lowest energy would have $\langle\phi\rangle = 0$. For $\mu^2 < 0$, the minimum occurs at

$$|\phi|_{\min} = 2^{-\frac{1}{2}}v, \qquad v = (-\mu^2/\lambda)^{\frac{1}{2}}. \tag{10.25}$$

This is characteristic of a spontaneous breakdown of symmetry. It occurs for example in a ferromagnet which has $\langle\mathbf{H}\rangle \neq 0$. There, the appearance of a macroscopic magnetic field $\mathbf{H}$ breaks the rotational invariance of the system, among other things. The corresponding problem in quantum field theory is admittedly less clear. An additional complication arises from the electromagnetic field $A^\mu$, which according to (21) couples to right- and left-handed electrons with equal strengths but not at all to neutrinos. If we leave the electromagnetic interaction as an afterthought, it will destroy our left-handed $SU(2)$-symmetry of $\mathscr{L}$ and thus destroy renormalizability. We therefore admit another Hermitean 4-vector field $B^\mu$, which is an $SU(2)$-scalar. The $\phi$-dependent Lagrangian is thus

$$\mathscr{L}_\phi = \left(D_\mu^+\phi^+ + \frac{i}{2}g'B_\mu\phi^+\right)\left(D^\mu\phi - \frac{i}{2}g'B^\mu\phi\right) - V_\phi, \tag{10.26}$$

$$D_\mu^+ = \partial_\mu + ig\mathbf{W}^+\boldsymbol{\tau}/2.$$

where $g'$ is a second independent coupling constant. In the complete version, we also need right-handed electrons $e_r$, but not right-handed neutrinos. Thus $e_r$ is also an $SU(2)$-scalar and can couple to $B^\mu$. The complete Lagrangian is

$$\mathscr{L} = \mathscr{L}_l + \mathscr{L}_\phi + \mathscr{L}_B + ie_r^+\sigma_\mu(\partial^\mu + ig'B^\mu)e_r - \tfrac{1}{2}g'l_l^+\sigma_l Bl_l + \mathscr{L}_{\text{int}}, \tag{10.27}$$

$$\mathscr{L}_B = -\tfrac{1}{4}B_{\mu\nu}B^{\mu\nu}, \qquad B^{\mu\nu} = \partial^\mu B^\mu - \partial^\nu B^\mu. \tag{10.28}$$

The last term $\mathscr{L}_{\text{int}}$ is a coupling between left- and right-handed leptons, which must be free of derivatives and $SU(2)$-invariant:

$$\mathscr{L}_{\text{int}} = -G(e_r^+ \phi^+ l_l + l_l^+ \phi e_r). \tag{10.29}$$

$\mathscr{L}$ is now invariant under the group of $SU(2) \times U(1)$ gauge transformations. To find its physical content, one expresses the Higgs field $\phi$ in terms of a Hermitean scalar field $\eta$ which has zero vacuum expectation value. In view of (25) we put

$$\phi = \exp\,(i\xi\tau/2v)2^{-\frac{1}{2}}\begin{pmatrix} 0 \\ v+\eta \end{pmatrix}, \qquad \eta = \eta^+, \qquad \langle\eta\rangle_0 = 0. \tag{10.30}$$

We then make a gauge transformation

$$l_l' = \exp\,(-i\xi\tau/2v)l_l, \qquad \phi' = 2^{-\frac{1}{2}}\begin{pmatrix} 0 \\ v+\eta \end{pmatrix}. \tag{10.31}$$

In the following, we drop the prime: $l_l$, $e$ and $\mathbf{W}^\mu$ refer to the transformed fields. $\mathscr{L}_{\text{int}}$ of (29) becomes

$$\mathscr{L}_{\text{int}} = -2^{-\frac{1}{2}}Gv(e_r^+ e_r + e_l^+ e_l) + \cdots \tag{10.32}$$

Comparison with (1-6.17) (apart from a factor $m_v^{-1}$ which is ugly here but does not change Euler's equations) shows that (32) is equivalent to the mass term of (22), with an electron mass

$$m_e = Gv2^{-\frac{1}{2}}. \tag{10.33}$$

$\mathscr{L}_\phi$ of (26) is transformed into

$$\mathscr{L}_\phi = \tfrac{1}{2}(\partial^\mu\eta)^2 + \tfrac{1}{8}(v+\eta)^2\chi_-(g'B + g\boldsymbol{\tau}\mathbf{W})^2\chi_- - \mu^2(v+\eta)^2/16 - \lambda\left(\frac{v+\eta}{4}\right)^4,$$

$$\chi_- \equiv \begin{pmatrix} 0 \\ 1 \end{pmatrix}. \tag{10.34}$$

Thus the remaining scalar field $\eta$ has mass $-2\mu^2$. The quadratic term in the vector meson field is

$$\tfrac{1}{8}v^2[(g'B - gW_3)^2 + g^2(W_1^2 + W_2^2)]. \tag{10.35}$$

Its physical interpretation is that $W_1$ and $W_2$ describe particles of mass $gv/2$, whereas $W_3$ and $B$ cannot describe particles because of nondiagonal terms in the "mass operator" (35). On the other hand, $g'B - gW_3$ can be written as $(g^2 + g'^2)^{\frac{1}{2}}$ times the lower component of the orthogonal transformation

$$\begin{pmatrix} A \\ Z \end{pmatrix} = \begin{pmatrix} \cos\theta & \sin\theta \\ \sin\theta & -\cos\theta \end{pmatrix}\begin{pmatrix} B \\ W_3 \end{pmatrix}, \qquad \sin\theta = g'(g^2 + g'^2)^{-\frac{1}{2}},$$
$$\cos\theta = g(g^2 + g'^2)^{-\frac{1}{2}}. \tag{10.36}$$

The upper component $A$ of this transformation does not occur in (35), therefore it must describe massless particles (photons). Fields which create and annihilate particles of definite masses (and charges) are therefore $A^\mu$, $Z^\mu$ and $W^\mu_\pm = 2^{-\frac{1}{2}}(W^\mu_1 \mp iW^\mu_2)$:

$$m_W = \tfrac{1}{2}gv, \qquad m_Z = \tfrac{1}{2}v(g^2 + g'^2)^{\frac{1}{2}}, \qquad m_A = 0. \tag{10.37}$$

The remaining parts of $\mathscr{L}$ which involve leptons are

$$\begin{aligned}\mathscr{L}_{\text{lept}} &= 2^{-\frac{1}{2}}g(v_l^+\sigma_l^\mu e_l W_\mu^+ + e_l^+\sigma_l^\mu v_l W_\mu^-)\\ &\quad + gg'(g^2 + g'^2)^{-\frac{1}{2}}A_\mu(e_r^+\sigma^\mu e_r + e_l^+\sigma_l^\mu e_l) + \tfrac{1}{2}(g^2 + g'^2)^{-\frac{1}{2}}\\ &\quad \times Z_\mu[g'^2(2e_r^+\sigma^\mu e_r + e_l^+\sigma_l^\mu e_l + v_l^+\sigma_l^\mu v_l) - g^2(e_l^+\sigma_l^\mu e_l - v_l^+\sigma_l^\mu v_l)].\end{aligned} \tag{10.38}$$

The terms involving $W_\mu^\pm$ contain the weak charged currents; comparison with (14) shows

$$2^{-\frac{1}{2}}G_\mu \equiv g^2/8m_w^2 = (2v^2)^{-1}. \tag{10.39}$$

The terms involving $A_\mu$ contain the electromagnetic current; comparison with (22) shows

$$e = gg'(g^2 + g'^2)^{-\frac{1}{2}} = g \sin \theta. \tag{10.40}$$

The terms involving $Z_\mu$ contain the weak neutral currents. They produce matrix elements of a structure analogous to (14)

$$T_{ab}^{(Z)} = -g_Z^2 L_a^\mu L_b^\nu g_{\mu\nu}(m_Z^2 - t)^{-1}, \qquad L_a^\mu \equiv \langle a' \,|\, j^\mu(0) \,|\, a\rangle. \tag{10.41}$$

(fig. 5-9b). The neutrino current is purely left-handed:

$$j_\nu^\mu = v_l^+\sigma_l^\mu v_l = \tfrac{1}{2}\overline{\Psi}'_\nu\gamma^\mu(1 - \gamma_5)\Psi_\nu, \qquad g_Z = \tfrac{1}{2}(g^2 + g'^2)^{\frac{1}{2}}, \tag{10.42}$$

and the weak electron current is

$$\begin{aligned} j_e^\mu &= (g^2 + g'^2)^{-1}[g'^2(2e_r^+\sigma^\mu e_r + e_l^+\sigma_l^\mu e_l) - g^2 e_l^+\sigma_l^\mu e_l]\\ &= 2\sin^2\theta\bar{e}\gamma^\mu e - e_l^+\sigma_l^\mu e_l,\end{aligned} \tag{10.43}$$

using (36) and (1-5.19). If the electron current is written in the form $\overline{\Psi}_e\gamma^\mu(C_{Ve} - C_{Ae}\gamma_5)\Psi_e$ this corresponds to

$$C_{Ve} = 2\sin^2\theta - \tfrac{1}{2}, \qquad C_{Ae} = -\tfrac{1}{2}. \tag{10.44}$$

At low energies, we may neglect $t$ in (41) and $u$ in (14). Using $m_Z^{-2} = g^2(g^2 + g'^2)^{-1}m_W^{-2}$ from (37), we can write $T_{ab}^{(Z)}$ in a form analogous to (14)

$$T_{ab}^{(Z)} \approx -\tfrac{1}{4}g^2 L_a^\mu L_b^\nu g_{\mu\nu} m_W^{-2}. \tag{10.45}$$

For elastic $\nu e$ scattering, this must be added to (14). For the experimentally more important case of $\nu_\mu e$-scattering, (45) gives the full matrix element. One assumes that the electromagnetic and weak interactions of muons and their neutrinos are obtained from the above formulas simply by replacing $\nu \to \nu_\mu$, $e \to \mu$.

Experimentally, $v$ is determined by $G_\mu$, $G$ by $m_e$ and $g \sin \theta$ by $e$. The model contains only the "Weinberg angle" $\theta$ as a free parameter, which is determined by the strength of the neutral current. Presently, $\sin^2 \theta \approx 0.2$

$$m_W = \frac{e}{2 \sin \theta}(G_\mu 2^{\frac{1}{2}})^{-\frac{1}{2}} \approx 83 \text{ GeV}, \qquad m_Z = m_W/\cos \theta \approx 93 \text{ GeV}. \quad (10.46)$$

## 5-11 Neutral Hadron Currents. Decays of Charmed Particles

In the Salam-Weinberg model, the weak hadron current is treated in far-reaching analogy with the lepton current. This is possible in the quark model, where the Cabibbo current $h_C^\mu$ of (1.21) is expressed in terms of the quark fields $u(x)$, $d(x)$ and $s(x)$ as follows:

$$h_C^\mu(x) = 2u_l^+ \sigma_l^\mu d_l', \qquad d' \equiv d \cos \theta_C + s \sin \theta_C. \quad (11.1)$$

In other words, the quark current is purely left-handed, just as the lepton current (see also section 6-9). The main differences arise from the additional strong interactions and fractional quark charges, as we shall see.

We introduce a hadronic weak left-handed doublet $h_l$, which suffers the same gauge transformation as $l_l$ (10.15):

$$h_l' = e^{-i\omega(x)\tau/2} h_l, \qquad h_l = \begin{pmatrix} u_l \\ d_l' \end{pmatrix}. \quad (11.2)$$

All right-handed fields $u_r$, $d_r$, $s_r$ are singlets under this transformation. The interaction with $W^\pm$ is obtained from the corresponding terms in (10.38) simply be substituting $v_l^+ \to u_l^+$, $e_l \to d_l'$. The contribution of $d_l'$ to the $Z$-coupling is proportional to

$$d_l'^+ \sigma_l^\mu d_l' = \cos^2 \theta_C \, d_l^+ \sigma_l^\mu \, d_l + \sin^2 \theta_C s_l^+ \sigma_l^\mu s_l + \cos \theta_C \sin \theta_C (d_l^+ \sigma_l^\mu \, d_l + \text{h.c.}). \quad (11.3)$$

The last term is a hypercharged neutral current and must be wrong because no such current is observed experimentally. As there is no way of compensating this current with the 3 quark fields ($u$, $d$, $s$), the existence of a fourth "charmed" quark field $c(x)$ was postulated by Glashow, Iliopoulos, and Maiani (1970). They added to (2) another left-handed doublet,

$$h_{lC} = \begin{pmatrix} c_l \\ s_l' \end{pmatrix}, \qquad s' \equiv -d \sin \theta_C + s \cos \theta_C, \quad (11.4)$$

where $s'$ is orthogonal to $d'$ of (1). This entails a piece $s_l'^+ \sigma_l^\mu s_l'$ in addition to (3):

$$d_l'^+ \sigma_l^\mu d_l' + s_l'^+ \sigma_l^\mu s_l' = d_e^+ \sigma_l^\mu \, d_l + s_l^+ \sigma_l^\mu s_l. \quad (11.5)$$

The unwanted terms have disappeared. This "GIM-mechanism" is in fact independent of the Salam-Weinberg model; it works also in the simpler $W$-boson model (10.1) (however, there it is cutoff-dependent). In the above derivation, the cancellation of the hypercharged neutral current may look somewhat artificial. It is quite general, however. The "physical" fields $u$, $d$, $s$ and $c$ are selected by the strong interactions. By the universality postulate, the weak interaction may choose any norm-conserving combinations

$$\begin{pmatrix} u_1 \\ u_2 \end{pmatrix} = U \begin{pmatrix} u_l \\ c_l \end{pmatrix}, \quad \begin{pmatrix} d_1 \\ d_2 \end{pmatrix} = D \begin{pmatrix} d_l \\ s_l \end{pmatrix}, \qquad U^+U = D^+D = 1 \tag{11.6}$$

in its left-handed doublets

$$h_1 = \begin{pmatrix} u_1 \\ d_1 \end{pmatrix}, \qquad h_2 = \begin{pmatrix} u_2 \\ d_2 \end{pmatrix}. \tag{11.7}$$

These doublets are the analogues of the lepton doublets $(\nu_e, e^-)$ and $(\nu_\mu, \mu^-)$. The neutral weak current involves the combination

$$d_1^+ \sigma_l^\mu \, d_1 + d_2^+ \sigma_l^\mu \, d_2 = d_l^+ \sigma_l^\mu \, d_l + s_l^+ \sigma_l^\mu s_l, \tag{11.8}$$

which is always diagonal, independent of the matrix $D$. To complete this argument, we must show that the Cabibbo current (1) is also of the most general form. The coupling of the hadron doublets to $W^+$ is

$$\mathscr{L}_{hW^+} = 2^{-\frac{1}{2}} g(u_1^+ \sigma_l^\mu \, d_1 + u_2^+ \sigma_l^\mu \, d_2) W_\mu^+ = 2^{-\frac{1}{2}} g(u_l^+, c_l^+) U^+ \sigma_l^\mu D \begin{pmatrix} d_l \\ s_l \end{pmatrix} W_\mu^+. \tag{11.9}$$

The terms $u_1^+ \sigma_l^\mu \, d_1$ and $u_2^+ \sigma_l^\mu \, d_2$ correspond to $\nu_l^+ \sigma_l^\mu e_l$ and $\nu_\mu^+ \sigma_l^\mu \mu_l$, respectively. The matrices $U$ and $D$ do not operate in spin space and therefore commute with $\sigma_l^\mu$. The product $U^+D$ is again a unitary matrix, which can be made real by an appropriate choice of the phase between the fields:

$$U^+D = \begin{pmatrix} \cos \theta_C & \sin \theta_C \\ -\sin \theta_C & \cos \theta_C \end{pmatrix}, \qquad \mathscr{L}_{hW^+} = 2^{-\frac{1}{2}} g(u_l^+ \sigma_l^\mu \, d_l' + c_l^+ \sigma_l^\mu s_l') W_\mu^+. \tag{11.10}$$

In the last form, we have used the definitions (1) and (4) of $d'$ and $s'$. Thus, we have not only found the Cabibbo current, but we have also uniquely predicted the weak charm-changing current. In particular, its hypercharged and hyperneutral components are proportional to $\sin \theta_C$ and $\cos \theta_C$, respectively: $D$-decays which do not have a kaon in the final state are suppressed by a factor $\tan^2 \theta_C \approx 0.076$. The $F$ mesons on the other hand are $\bar{c}s$ and $\bar{s}c$ (4-16.6) and decay dominantly to $\bar{s}s$, i.e., they frequently contain an $\eta$-meson (4-10.4) among their decay products.

Next, we must find the amounts of $u_l^+ \sigma_l^\mu u_l$, $c_l^+ \sigma_l^\mu c_l$ and right-handed currents in the neutral weak current. Here it suffices to discuss the two "flavours" $u$ and $d$ and add in the final result a pieces in which $u$ and $d$ are replaced by the other two "flavours" $c$ and $s$. The amounts of right-handed currents are fixed by the electric charges. The generalization of the right-handed piece in (10.28) is thus

$$\mathscr{L}_r = i(u_r^+, d_r^+) \sigma_\mu (\partial^\mu - ig'Q) B^\mu \begin{pmatrix} u_r \\ d_r \end{pmatrix}, \qquad Q = \begin{pmatrix} Z_u & 0 \\ 0 & Z_d \end{pmatrix}. \tag{11.11}$$

We have $Z_u = \frac{2}{3}$ and $Z_d = -\frac{1}{3}$ in the quark model, whereas the leptons have $Z_u = 0$, $Z_d = -1$. The same generalization must be used for the left-handed currents which are proportional to $g'$ in (10.28), because they are needed to make the electromagnetic current parity-invariant. The remaining left-handed current comes from $\mathscr{L}_l$ in (10.24) and is always proportional to $g\tau/2$ according to (10.16). Therefore, the only possible generalization of (10.42), (10.43) to arbitrary electric charges is

$$j^\mu = -2\sin^2\theta j^\mu_{\text{em}} + (u_l^+, d_l^+)\sigma_l^\mu \tau_3 \begin{pmatrix} u_l \\ d_l \end{pmatrix}, \tag{11.12}$$

where $j^\mu_{\text{em}}$ is the (parity-conserving) electromagnetic current. For elastic scattering on spinless nuclei, only the vector part of $j^\mu$ contributes to $T^{(Z)}_{ab}$ of (10.45), with $-L^\mu_b$ replaced by the corresponding hadron current matrix element $H^\mu_b(\mathbf{q})$. Its static limit is given by (2.16), with $M_F = M_V = Z - N - 4Z\sin^2\theta$, $N$ being the number of neutrons in the nucleus $b$.

By analogy with the third lepton doublet $(\nu_\tau, \tau)$, we may speculate there exists also a third quark doublet, called $(t, b)$ ("truth" and "beauty"). The $b$ quark is tentatively identified with the constituent of the $\Upsilon$ resonance, because the production rate in $e\bar{e}$-collisions indicates that $Z_b^2$ is $\sim \frac{1}{9}$ rather than $\frac{4}{9}$. There is presently no indication of $t$-quark bound states. If our speculations are correct, small modifications of the Cabibbo theory will become necessary: the matrices $U$ and $D$ of (6) will be $3 \times 3$ matrices, and the interaction Lagrangian (9) must be replaced by

$$\mathscr{L}_{hW^+} = 2^{-\frac{1}{2}} g (u_l^+, c_l^+, t_l^+) \sigma_l^\mu A \begin{pmatrix} d \\ s \\ b \end{pmatrix}_l W_\mu^+, \tag{11.13}$$

where $A$ is a $3 \times 3$ unitary matrix. It has $3 \times 3 = 9$ real parameters (see section B-1), but 5 of these can be included in the relative phases between the 6 fields $d, s, b, u, c, t$. Consequently, $A$ can be expressed in terms of 4 angles (Kobayashi and Maskawa 1973):

$$A = \begin{pmatrix} c_1 & s_1 c_3 & s_1 s_3 \\ -s_1 c_2 & c_1 c_2 c_3 - s_2 s_3 e^{i\delta} & c_1 c_2 s_3 + s_2 c_3 e^{i\delta} \\ s_1 s_2 & -c_1 s_2 c_3 - c_2 s_3 e^{i\delta} & -c_1 s_2 s_3 + c_2 c_3 e^{i\delta} \end{pmatrix}, \quad \begin{aligned} c_j &\equiv \cos\theta_j, \\ s_j &\equiv \sin\theta_j. \end{aligned} \tag{11.14}$$

The upper left submatrix of $A$ is identical with $U^+D$ of (10) only for $\theta_2 = \theta_3 = 0$, in which case the $(t, b)$ quarks decouple from the rest. This is unlikely because it would imply an absolutely stable new meson, against which there is some experimental evidence. In the general case $\theta_j \neq 0$, the hyperneutral current remains unchanged (with $\theta_C = \theta_1$), but the hypercharged current is proportional to $\sin\theta_C \cos\theta_3$ instead of $\sin\theta_C$ as in (1.21). Inserting the numbers of section 5-1, one finds that $\cos\theta_3$ must be quite close to 1.

An elegant feature of (14) is its CP-violation, which enters naturally via the phase $\delta$. The decay $K^0 \to \pi\pi$ contains current products leading to $c$ and $t$ quarks in the intermediate states, with angular coefficients $A_{21}A_{22}$ and $A_{31}A_{32}$, respectively. The CP-violating amplitude $\langle \pi\pi | T | K_L \rangle$ is proportional to the imaginary parts of these products, which are $\pm c_2 s_1 s_2 s_3 \sin \delta$ according to (14). In the limit of equal $c$ and $t$ masses, the two contributions just cancel. This is so because CP-violation appears as an interference between amplitudes of different phase, and in the limit of degenerate intermediate states, one can always find one linear combination of these states which decouples. One thus finds

$$\langle \pi\pi | T | K_L \rangle \sim c_2 s_1 s_2 s_3 \sin \delta (m_t^2 - m_c^2). \tag{11.15}$$

Larger CP-breaking effects are expected in the weak decays of charmed and beautiful particles (Ellis et al. 1976. See also Harari 1978).

Quark masses are generated by an interaction analogous to (10.29). However, whereas the neutrino was massless, we now require an additional term which can produce an up-quark mass. This is achieved by

$$\mathscr{L}_{\text{int}}, h = -(G_d h_l^+ \phi \, d_r + G_u h_l^+ \phi_c u_r + \cdots + \text{h.c.}); \quad \phi_c \equiv c\phi_{\text{tr}}^+ \tag{11.16}$$

The new term contains the charge-conjugate field $\phi_c$ (compare 1-4.13) because of charge conservation.

## 5-12 Inclusive Lepton-Hadron Collisions

High-energy electron-nucleon and neutrino-nucleon collisions are used to study the innermost structure of nucleons. Normally only the lab scattering angle $\theta_L$ and energy loss $q_L^0$ of the lepton are measured. This is sufficient to determine the invariant mass of the outgoing hadrons, $s_d = (q + p_b)^2$:

$$s_d - t - m_b^2 = 2qP_b = 2m_b(E_a - E_a')_L = 2m_b q_L^0 \equiv 2m_b \nu \tag{12.1}$$

In the following, we neglect the lepton mass:

$$t = -2E_{aL}E_{aL}'(1 - \cos \theta_L), \qquad d \cos \theta_L \, dq_L^0 = dt \, ds_d (4m_b E_{aL} E_{aL}')^{-1}. \tag{12.2}$$

We also introduce the dimensionless variable

$$x = -t/2m_b \nu, \qquad 0 < x \leq 1. \tag{12.3}$$

Near $x = 1$, $s_d \approx m_b^2$ allows only elastic scattering, which is of the general form (2-7.31) for electrons and (10.7) for neutrinos. For $x < 1$, $d^2\sigma/dt \, ds_d$ integrates over a continuum of hadron states of given $s_d$. According to (4-6.4), one must then multiply (2-7.31) and (10.7) by $(2\pi)^{-1}$ and integrate over $d$ Lips $(s_d; P_1, P_2 \cdots)$. This shows, by the way, that the 1-particle phase space must be

$$d \text{ Lips } (s; e) = 2\pi \, \delta(s - m_e^2), \tag{12.4}$$

which follows also directly from (2-1.8) and (1-2.19).

The parton model claims that in the limit $-t \to \infty$ and fixed $x$, a nucleon appears as a collection of parallel-moving, pointlike free particles called "partons," in any system in which $\mathbf{p}_b \to \infty$ (for example the Breit system, $q_B^0 = 0$).

Partons are quarks, antiquarks, and gluons, characterized by functions $f_i(x')$ which give the probability of having a parton of type $i$ and momentum $x'\mathbf{p}_b$. The current interaction ejects this parton with 4-momentum $x'P_b + q$ ($P^0 = p$ for $m/p \to 0$). Since free partons are not detected, final state interactions must somehow produce quark-antiquark pairs to transform the partons into real meson, hopefully without major changes of their 4-momenta. In electron collisions, the parton polarization average is then

$$(2S_i + 1)^{-1} \sum_{\lambda\lambda'} \int dx' J_i^\mu J_i^\nu \, \delta[(x'P_b + q)^2 - m_i^2](2m_b)^{-1} f_i(x'), \qquad (12.5)$$

where the $\delta$-function comes from the 1-quark phase space $2\pi\,\delta(s_i - m_i^2)$. Rewriting $m_i^2 \approx (x'P_b)^2$, the argument is simplified:

$$\delta(2x'P_b q + t) = \delta(x' 2m_b \nu - 2m_b \nu x) = (2m_b \nu)^{-1}\,\delta(x' - x) \qquad (12.6)$$

for $\nu \to \infty$. For nucleons, this must be compared with (2-7.30). Terms proportional to $q^\mu$ and $q^\nu$ do not contribute:

$$(x 4m_b^2 \nu)^{-1}(2S_i + 1)^{-1} \sum_{\lambda\lambda' i} J_i^\mu J_i^\nu f_i(x) = m_b^{-2} W_2 P_b^\mu P_b^\nu - W_1 g^{\mu\nu}. \qquad (12.7)$$

The extra factor $x^{-1}$ accounts for $ds_i = x\, ds_d$, at fixed $t$. We can now make a few general statements about $W_2(x, \nu)$ and $W_1(x, \nu)$: gluons do not contribute to (7), because they carry no charge. Quarks and antiquarks have spin-$\frac{1}{2}$, such that the left-hand side of (7) can be evaluated according to (2-7.19):

$$(4m_b^2 x\nu)^{-1} \sum_i Z_i^2 S^{\mu\nu}(xP_b) f_i = (4m_b^2 x\nu)^{-1}(-2xm_b \nu g^{\mu\nu} + 4x^2 P_b^\mu P_b^\nu) \sum_i Z_i^2 f_i \qquad (12.8)$$

Comparison with the right-hand side of (7) gives

$$2m_b W_1 = \sum_i Z_i^2 f_i(x), \qquad \nu W_2 = x \sum_i^2 Z_i^2 f_i(x) = 2m_b x W_1 \equiv F_2(x). \qquad (12.9)$$

The prediction that $W_1$ and $\nu W_2$ are independent of $\nu$ (Bjorken's "scaling hypothesis," 1969) is experimentally violated by $\sim 20\%$, while the "Callan-Gross" relation (1969) $F_2 = 2m_b x W_1$ agrees with experiment for $x \geq 0.3$ (see Greenberg 1978). Using these relations and (2), one finds

$$d^2\sigma/dx\,d\nu = 4m_b(\alpha E/t)^2 F_2(x)(2 - 2y + y^2), \qquad y \equiv \nu/E \qquad (12.10)$$

Below the charm threshold, the proton contains $u$ and $d$ quarks as well as pairs of $u\bar{u}$, $d\bar{d}$ and $s\bar{s}$, with $Z_i = \frac{2}{3}$ and $-\frac{1}{3}$:

$$x^{-1} F_2^{ep} = \tfrac{4}{9}(f_u + f_{\bar{u}}) + \tfrac{1}{9}(f_d + f_{\bar{d}}) + \tfrac{1}{9}(f_s + f_{\bar{s}}) \qquad (12.11)$$

$$\int dx(f_u - f_{\bar{u}}) = 2, \qquad \int dx(f_d - f_{\bar{d}}) = 1, \qquad \int dx(f_s - f_{\bar{s}}) = 0. \qquad (12.12)$$

By isospin invariance, the $d$-quark distribution $f_d^n$ in the neutron equals the $u$-quark distribution $f_u$ in the proton, and correspondingly for $f_u^n$ and $f_s^n$:

$$x^{-1}F_2^{en} = \tfrac{1}{9}(f_u + f_{\bar{u}}) + \tfrac{4}{9}(f_d + f_{\bar{d}}) + \tfrac{1}{9}(f_s + f). \tag{12.13}$$

For $\mu^+$ and $\mu^-$ production in neutrino collisions with protons, one obtains from (10.7)

$$d^2\sigma^{(\pm)}/dx\,dv = G_V^2 m_N^2 E^2 \pi^{-1}[(1 - y + \tfrac{1}{2}y^2)F_2(x) \pm xy(1 - \tfrac{1}{2}y)F_3(x)] \tag{12.14}$$

where below the charm threshold

$$\begin{aligned}
F_2^{p\mu^-} &= 2x(f_d \cos^2\theta_C + f_s \sin^2\theta_C + f_{\bar{u}}),\\
F_3^{p\mu^-} &= 2(f_{\bar{u}} - f_d \cos^2\theta_C - f_s \sin^2\theta_C)\\
F_2^{p\mu^+} &= 2x(f_{\bar{d}} \cos^2\theta_C + f_{\bar{s}} \sin^2\theta_C + f_u),\\
F_3^{p\mu^+} &= -2(f_u - f_{\bar{d}} \cos^2\theta_C - f_{\bar{s}} \sin^2\theta_C)
\end{aligned} \tag{12.15}$$

Neglecting the small contributions $f_s \sin^2\theta_C$ and using $\cos^2\theta_C \approx 1$,

$$d^2\sigma^{p\mu^-}/dx\,dv = G_V^2 m_N E^2 \pi^{-1}[f_d + f_{\bar{u}}(1-y)^2]. \tag{12.16}$$

For $\bar{v}p \to \mu^+ \ldots$, $f_d$ is replaced by $f_{\bar{d}}$ and $f_{\bar{u}}$ by $f_u$. Note the typical $y$-dependences of $vq$, $\bar{v}\bar{q}$, $v\bar{q}$ and $\bar{v}q$ scattering:

$$d\sigma/dv\,dx \sim 1 \quad \text{for} \quad vq, \bar{v}\bar{q} \qquad \text{and} \qquad (1-y)^2 \quad \text{for} \quad \bar{v}q, v\bar{q}. \tag{12.17}$$

For collisions on neutrons, $u$ and $d$ as well as $\bar{u}$ and $\bar{d}$ are again interchanged. Denoting $F^{eN} = \tfrac{1}{2}F^{ep} + \tfrac{1}{2}F^{en}$, $F^- = \tfrac{1}{2}F^{p\mu^-} + \tfrac{1}{2}F^{n\mu^-}$, one finds

$$\tfrac{1}{9}x(f_s + f_{\bar{s}}) = F_2^{eN} - \tfrac{5}{18}F_2^- \tag{12.18}$$

Experimentally, this combination is very small, which shows that nucleons contain practically no $s\bar{s}$ pairs.

Finally, the parton momenta should add up to $\mathbf{p}_b$, $\int x\,dx \sum_i f_i = 1$. Experimentally, the contribution of quarks and antiquarks to this integral is

$$\int dx(9F_2^{eN} - \tfrac{3}{2}F_2^-) = 0.55. \tag{12.19}$$

The remaining momentum fraction 0.45 must be carried by gluons.

We have mentioned the quark-parton model because it is simple and qualitatively correct. Presently, one tries to understand the deviations from this model in terms of QCD, but the connection with the more familiar region of small momentum transfer is still obscure.

CHAPTER SIX

# Analyticity and Strong Interactions

## 6-1 Crossing Relations, Cuts and Poles

It is presently impossible to calculate hadron scattering from a few fundamental constants. Perturbation expansions are inappropriate because the relevant coupling constants are too large. One assumes that the scattering amplitude $T_{ij}(s, t)$ is obtained from an analytic function $T_{ij}(z, t)$ in the limit $z = s + i\varepsilon$. The basis of this assumption is microcausality, or the existence of equations such as (3-3.4) which contain retarded Greens functions. In some favorable cases such as elastic $\pi N$ scattering in a $t$-strip including $t = 0$, one can derive the analyticity properties of $T$ from the postulates of local field theory (see for example Källén 1964). The derivation is more complicated than that of section 3-11 because it must avoid the perturbation expansion.

An analytic function is determined by its singularities. Their positions in scattering amplitudes are given by the rules of Landau (1959) and Cutkosky (1960), which were found in perturbation theory (see Eden et al. (1966) for an excellent summary). We shall derive the singularities from the unitarity equation (2-5.3) and from the "crossing relations" or "crossing symmetry" (see also Barut 1967). The connection between the two methods is given at the end of this section.

In section 3-6 we formulated the substitution rule, which says that an outgoing antiparticle is formally treated as an incoming particle of opposite additive quantum numbers. In QED, the rule is used for the evaluation of matrix elements of "antiparticle-process" such as $e\bar{e} \to \gamma'\gamma$ (section 3-8) from that of the original "particle-process" $\gamma e \to \gamma' e'$. For strong interactions, on the other hand, the rule is used to impose restrictions on the unknown matrix elements. Normally one considers a two-particle reaction $ab \to cd$, in which all four particles $a$, $b$, $c$, and $d$ are at least stable against decays of the type $b \to \bar{a}cd$, although they need not be absolutely stable. The useful antiparticle substitutions in this case are those which substitute two particles simultaneously, i.e., which lead to the reactions $a\bar{c} \to \bar{b}d$, $\bar{c}b \to \bar{a}d$, plus 3 time-reversed and 6 CPT-conjugate reactions. They are called "crossing," the resulting relations being the crossing relations. When all 4 particles are spinless, the reaction $ab \to cd$ is described by one scalar function $T(s, t, u)$, with $u = \sum m^2 - s - t$ according to (2-4.4). The amplitudes for the 3 above channels are defined as follows

$$\begin{aligned}
&ab \to cd \text{ ("}s\text{-channel"): } T(s, t, u),\\
&\qquad s = (P_a + P_b)^2, \qquad t = (P_a - P_c)^2, \qquad u = (P_b - P_c)^2.\\
&a\bar{c} \to \bar{b}d \text{ ("}t\text{-channel"): } T_t(s_t, t_t, u),\\
&\qquad s_t = (P_a + P_{\bar{c}})^2, \qquad t_t = (P_a - P_{\bar{b}})^2.\\
&\bar{c}b \to \bar{a}d \text{ ("}u\text{-channel"): } T_u(s_u, t, u_u),\\
&\qquad s_u = (P_{\bar{c}} + P_b)^2, \qquad u_u = (P_b - P_{\bar{a}})^2.
\end{aligned} \tag{1.1}$$

The crossing relations read

$$T(s, t, u) = T_t(s_t = t, t_t = s, u) = T_u(s_u = u, t, u_u = s). \tag{1.2}$$

They are only meaningful for analytic functions because the physical regions of the variables of the 3 reactions are disjoint. This point is illustrated in fig. 6-1.1 for the case of an elastic $s$-channel, $m_a = m_c$, $m_b = m_d$. The limiting curves correspond to $\sin\theta = 0$, $\sin\theta_t = 0$ and $\sin\theta_u = 0$ where $\theta$, $\theta_t$ and $\theta_u$ are the cms scattering angles in the three channels. The curves are the zeros of a common function, called "Kibble's boundary function" $\phi$ (Kibble 1960). We find it by first discussing the 4-vector

$$K_\mu = \varepsilon_{\mu\nu\rho\sigma} P_a^\nu P_b^\rho P_c^\sigma \tag{1.3}$$

Because of the antisymmetry of $\varepsilon_{\mu\nu\rho\sigma}$ and $P_d = P_a + P_b - P_c$, $K_\mu$ changes at most its sign if $P_a$, $P_b$ or $P_c$ in (3) are replaced by $P_d$. We evaluate $K_\mu$ in the $s$-channel cms by replacing $P_a$ by $P_a + P_b$:

$$K_\mu = \varepsilon_{\mu 0\rho\sigma} s^{\frac{1}{2}} P_b^\rho P_c^\sigma, \quad K_0 = 0, \quad \mathbf{K} = -s^{\frac{1}{2}} \mathbf{p}_b \times \mathbf{p}_c, \quad |\mathbf{K}| = s^{\frac{1}{2}} pp' \sin\theta \tag{1.4}$$

$$4K_\mu K^\mu = 4sp^2p'^2 \sin^2\theta \equiv \phi \tag{1.5}$$

Obviously, $\phi = 0$ gives the desired boundaries for all 3 channels. Another useful form is

$$\phi = s(t - t_{\max})(u - u_{\max}), \qquad u - u_{\max} = t_{\min} - t. \tag{1.6}$$

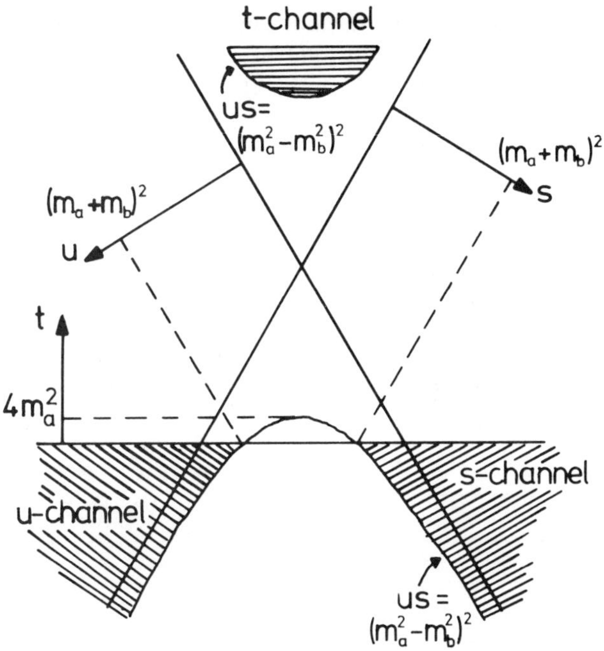


**Figure 6-1.1** The physical regions of the $s$-, $t$- and $u$-channels for $m_a = m_c$, $m_b = m_d$. The boundaries of the regions are given by the zero lines of $\phi$ (1.7).

For $m_a = m_c$, $m_b = m_d$, we have $t_{\max} = 0$, $t_{\min} = -4p^2 = -s^{-1}\lambda(s, m_a^2, m_b^2)$ according to (2-4.10), and thus

$$u_{\max} = 2m_a^2 + 2m_b^2 - s - t_{\min} = (m_a^2 - m_b^2)^2/s, \qquad \phi = t[su - (m_a^2 - m_b^2)^2] \tag{1.7}$$

After removal of possible spin functions and assuming time-reversal invariance (4-4.18), the unitarity equation (2-5.3) reads

$$\operatorname{Im} T_{ij}(s, t) = \frac{1}{2}\sum_k \int d \text{ Lips } (s; k) T_{ik}(s, t') T_{jk}^*(s, t'') \tag{1.8}$$

as in (3-11.5). The index $k$ runs over all open channels. We postulate that apart from limitations due to crossing, (8) can be used also for $s$ below the thresholds $(m_a + m_b)^2$ and $(m_c + m_d)^2$ of $i$ and $j$. In cases such as elastic $\pi N$ or $NN$ scattering, $k = i = j$ is in fact the lowest channel, the right-hand side of (8) is zero below $s = (m_a + m_b)^2$, and $T_{ii}$ is thus real there. For $\bar{N}N$-scattering, on the other hand, multipion states extend below $s = 4m_N^2$, and the right-hand side becomes zero only at $4m_\pi^2$ where the $\pi\pi$-states end.

At fixed $t \le 0$, the unitarity condition (8) introduces a cut in $s$ in the amplitude $T(s, t)$, which starts at the threshold $s_0$ of the lowest channel of $k$. The singularity at $s = s_0$ is a square root branch point since $d$ Lips $\sim \lambda^{\frac{1}{2}}(s,$

$m_1^2$, $m_2^2$) near $s_0 = (m_1 + m_2)^2$. In addition, the crossing relation (2) introduces a "crossed" or "left-hand" cut from the unitarity condition in the $u$-channel. This cut starts at $u_0 > 0$, i.e., at

$$m_a^2 + m_b^2 + m_c^2 + m_d^2 - t - u_0 \equiv \bar{s}_0 \tag{1.9}$$

For $\pi N$ and $NN$ scattering, we have

$$\pi N\colon s_0 = (m_N + m_\pi)^2 = u_0, \qquad \bar{s}_0 = (m_N - m_\pi)^2 - t, \tag{1.10}$$

$$NN\colon s_0 = 4m_N^2, \qquad u_0 = 4m_\pi^2, \qquad \bar{s}_0 = 4(m_N^2 - m_\pi^2) - t. \tag{1.11}$$

Thus, even for $t = 0$, there is only a small piece of the real $s$-axis where $T_{ij}$ is real (fig. 6-1.2). For $t < \sum m^2 - u_0 - s_0$, we have in fact $\bar{s}_0 > s_0$, and the two cuts overlap.

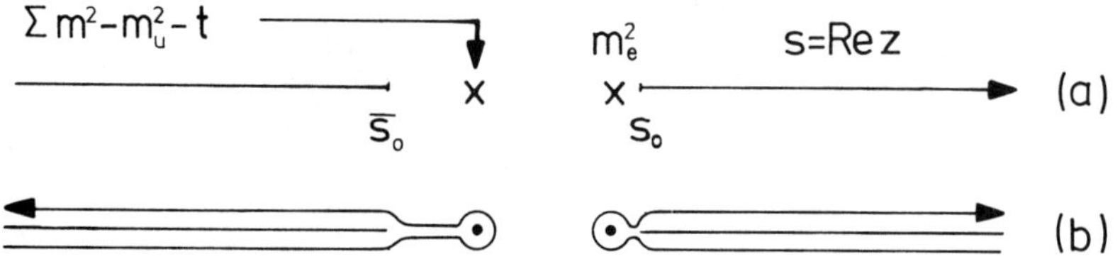


**Figure 6-1.2** (a) The cuts and poles of $T(z, t)$ in the complex $z$-plane at fixed $t$. (b) The path of the Cauchy integral.

Occasionally, single-particle states $|e\rangle$ can contribute to the states $k$ in (8), for example the nucleon in $\pi N$ scattering, the deuteron in $NN$ scattering and the pion in $\bar{N}N$-scattering, provided the spin and isospin quantum numbers allow this. Neglecting the electromagnetic and weak interactions which induce decays of $\pi^0$, $\pi^\pm$ and $n$, we may say that these single-particle states are eigenstates of the full Hamilton operator. They may be included in the complete set of states between which the scattering operator acts, with matrix elements

$$S_{ee'} = \langle e' | e \rangle, \qquad T_{ee'} = 0 \tag{1.12}$$

according to (2-1.4), with $P_i = P_e$, $P_f = P'_e$.

In the gap between $\bar{s}_0$ and $s_0$, the contribution of $e$ to the unitarity integral (8) is, for a spinless particle $e$

$$\text{Im } T_{ij}(\bar{s}_0 < s < s_0) = \tfrac{1}{2} \int d \text{ Lips } (s; e) T_{ie} T_{je}^*. \tag{1.13}$$

For $j = e$, we find from (12) and (13)

$$\text{Im } T_{ie}(\bar{s}_0 < s < s_0) = 0 \tag{1.14}$$

i.e., $T_{ie}$ is real there. Its value at $s = m_e^2$ is called a coupling constant and denoted by $G_{ie}$ in the following (when $i$ contains more than 2 particles, it is still a function of subenergies). From the definitions (2-1.8) and (1-2.19) of $d$ Lips and $d_L^3\, p_e$, we have

$$d \text{ Lips } (s; e) = (2\pi)^4\, \delta_4(P_i - P_e)(2\pi)^{-3}\, d^4P_e\, \delta(P_i^2 - m_e^2)\theta(P_0) \\ = 2\pi\, \delta(s - m_e^2) \tag{1.15}$$

The $\theta$-function can be omitted as long as $P_0 > 0$. Insertion of (15) into (13) gives

$$\text{Im } T_{ij} = G_{ie}G_{je}\pi\, \delta(s - m_e^2). \tag{1.16}$$

The full amplitude $T_{ij}$ gets a pole at $s = m_e^2$ from (16).

To see this, we start again from the Cauchy integral formula (3-11.3) and expand the integration contour until it winds along the singularities as indicated in fig. 6-1.2b. For the right-hand contour, we end up with a dispersion integral as in (3-11.15), but with the lower limit of integration below $m_e^2$. The contribution from the integration up to $s_0$ is

$$T_{ij}^{(e)} = \pi^{-1} \int G_{ie}G_{je}\pi\, \delta(s' - m_e^2)(s' - s - i\varepsilon)^{-1}\, ds' = G_{ie}G_{je}(m_e^2 - s - i\varepsilon)^{-1}, \tag{1.17}$$

where the $i\varepsilon$ is unnecessary in the physical region, $s > m_e^2$. The complete contribution of the right-hand contour is thus

$$T_{ij}(z, \text{right}) = G_{ie}G_{je}(m_e^2 - z)^{-1} + \pi^{-1} \int_{s_0}^{\infty} ds'(s' - t)^{-1} \text{ Im } T_{ij}(s'). \tag{1.18}$$

We now specify $i = ab$, $j = cd$, $T_{ij} = T$. If the $u$-channel reaction $\bar{c}b \to \bar{a}d$ contains a similar pole at $s_u = m_u^2$, with coupling constants $G_{\bar{c}bu}$ and $G_{\bar{a}du}$, then $T$ must also have a pole $G_{\bar{c}bu}G_{\bar{a}du}(m_u^2 - u)^{-1}$ according to (2). It may also have a pole in $t$, but since we are presently interested in the analyticity properties at fixed $t$, this pole is not shown explicitly. Finally, the unitarity cut of $T_u$ in the variable $s_u = u$ contributes the left-hand cut of $T$ in $s$, according to the relation $s = \sum m^2 - t - u$. Because of the opposite sense of the integration path as shown in fig. 6-1.2b, this contribution comes with a minus sign:

$$\begin{aligned} -(2\pi i)^{-1} \int_{-\infty}^{\bar{s}_0} \frac{ds'}{s' - z} [T(s' + i\varepsilon) - T(s' - i\varepsilon)] \\ = (2\pi i)^{-1} \int_{u_0}^{\infty} \frac{du'}{\sum m^2 - t - u' - z} [T_u(u' - i\varepsilon) - T_u(u' + i\varepsilon)] \\ = \frac{1}{\pi} \int_{u_0}^{\infty} \frac{du'}{z + t + u' - \sum m^2} \text{ Im } T_u(u') \\ = \frac{1}{\pi} \int_{u_0}^{\infty} \frac{ds'}{z + t + s' - \sum m^2} \text{ Im } T_u(s'). \end{aligned} \tag{1.19}$$

In the second expression, we have used (2) for complex $s$, observing that $\sum m^2$ and $t$ are real:

$$T(s + i\varepsilon, t) = T_u(u - i\varepsilon, t). \tag{1.20}$$

Using now $(s' - s - i\varepsilon)^{-1} = P(s' - s)^{-1} + i\pi\,\delta(s' - s)$ from (3-11.16) and assuming for simplicity exactly one pole in each channel, we get the complete dispersion relation as

$$\begin{aligned}\operatorname{Re} T(s, t) &= \frac{G_{abe} G_{cde}}{m_e^2 - s} + \frac{G_{\bar{c}bu} G_{\bar{a}du}}{m_u^2 - u} \\ &\quad + \frac{P}{\pi}\left[\int_{s_0}^{\infty} \frac{\operatorname{Im} T(s', t)}{s' - s} + \int_{u_0}^{\infty} \frac{\operatorname{Im} T_u(s', t)}{s + t + s' - \sum m^2}\right] ds'\end{aligned} \tag{1.21}$$

Note that for $t < 0$, both integrals still contain unphysical regions at small $s'$, even for elastic $\pi N$ scattering where $s_0 = u_0 = (m_a + m_b)^2$.

Elastic scattering has $t = -2p^2(1 - \cos\theta)$ and requires $\cos\theta \to -\infty$ for $p^2 \to 0$ and negative $t$. This complication is avoided only for $t = 0$. Poles in the scattering amplitude as a function of energy occur also in potential theory, where they correspond to bound states. For example, in section 1-8 we derived atomic binding energies from the positions of the poles of the Coulomb scattering amplitude. A similar but approximate interpretation of the deuteron pole in pn scattering will be given in sections 6-4 and 7-8. On the other hand, although the nucleon appears as a pole in $\pi N$ scattering, nobody would claim that $N$ is a $\pi N$ bound state.

Poles also appear in a perturbation treatment of strong interactions. This was in fact our argument in the derivation of the "induced pseudoscalar" (5-7.8) in $\beta$-decay. However, the residue of the pole is given by the product of the coupling constants of the relevant interaction Hamiltonians only in lowest nonvanishing order perturbation theory. Higher orders introduce "renormalized coupling constants" into the residues, as discussed in section 3-10. Thus, in the terminology of perturbation theory, the coupling constants of (16) etc. are "renormalized" ones. The question of convergence of the dispersion integrals is discussed for each case separately, frequently on the basis of experimental information.

When the external particles carry spin, (8) must be replaced by the original unitarity equation (2-5.3). (12) remains valid, but the left-hand side of (13) is replaced by $(2i)^{-1}(T_{ij} - T_{ji}^*)$, and (14) becomes

$$T_{ie}(s < s_0) = T_{ei}^*(s < s_0). \tag{1.22}$$

The product in (13) can be rewritten as $T_{ej} T_{ei}^*$, and instead of (17) we obtain

$$T_{ij}^{(e)} = \sum_M T_{ej}(M) T_{ei}^*(M)(m_e^2 - s)^{-1}. \tag{1.23}$$

The matrix element $T_{ej}$ must be linear in the spin function of $e$ and must be Lorentz invariant. If the two particles in the state $j$ are both spinless, the resulting matrix elements for spin 0, 1, 2, 3 can be taken from the first four rows of table 4-9, which was originally derived for resonance decays. The sum over $M$ in (23) can be performed covariantly by formulas such as (2-8.9). With $T_{ej} = G_{ecd}\,\varepsilon_\mu(M)(P_c - P_d)^\mu$ for spin 1 and $u = \sum m^2 - s - t$, we have

$$\begin{aligned} T_{ij}^{(e)} &= \sum_M G_{ecd}(P_c - P_d)^\mu \varepsilon_\mu(M)\varepsilon_\nu^*(M)(P_a - P_b)^\nu G_{eab}(m_e^2 - s)^{-1} \\ &= G_{ecd}G_{eab}(P_c - P_d)^\mu[-g_{\mu\nu} + (P_c + P_d)_\mu(P_e + P_b)_\nu m_e^{-2}] \\ &\quad \times (P_a - P_b)^\nu(m_e^2 - s)^{-1} \\ &= G_{ecd}G_{eab}[t - u + (m_c^2 - m_d^2)(m_a^2 - m_b^2)m_e^{-2}](m_e^2 - s)^{-1}. \end{aligned} \tag{1.24}$$

In the last square bracket we can replace $s$ by $m_e^2 + (s - m_e^2)$ and forget the $s - m_e^2$ since it vanishes at the pole. Although such terms may be relevant in dynamical models, they must clearly be omitted in an analysis which is based on analytic properties only. In analogy with perturbation theory, (24) is called a "Born term" and is illustrated by a Born graph as in fig. 6-1.3. The single particle pole is obtained from the Born term by replacing $s$ by $m_e^2$ in the numerator (see also 4.17 below). This replacement also sets possible form factors equal to unity. The corresponding problem for spinor particles will be treated in the next section. In the cms, the contribution of the single-particle state $e$ of arbitrary spin $S$ can be calculated from (4-8.21), with $R = e$ and $\Gamma_R = 0$.

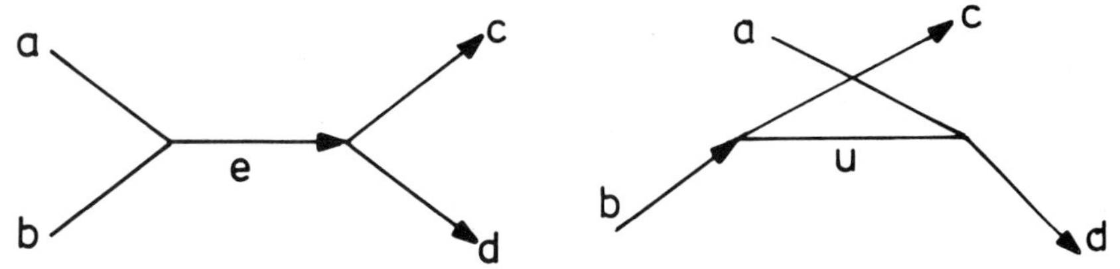


**Figure 6-1.3** Born graphs for $s$- and $u$-channel poles of $T(ab \to cd)$.

In addition to (21), we can also write down a dispersion relation for $T_u$:

$$\begin{aligned} \text{Re}\; T_u &= \frac{G_{\bar{c}bu}G_{\bar{a}du}}{m_u^2 - s} + \frac{G_{abe}G_{cde}}{m_e^2 - u} \\ &\quad + \frac{P}{\pi}\left[\int_{u_0}^{\infty} \frac{\text{Im}\; T_u(s')}{s' - s} + \int_{s_0}^{\infty} \frac{\text{Im}\; T(s')}{s + t + s' - \sum m^2}\right] ds'. \end{aligned} \tag{1.25}$$

It is then easy to see that the crossing symmetric and antisymmetric combinations

$$T^{(\pm)} = \tfrac{1}{2}T(s, t) \pm \tfrac{1}{2}T_u(s, t), \qquad T^{(\pm)}(u, t) = \pm T^{(\pm)}(s, t) \tag{1.26}$$

satisfy separate dispersion relations

$$\text{Re } T^{(\pm)}(s) = \tfrac{1}{2}G_{abe}G_{cde}\left(\frac{1}{m_e^2 - s} \pm \frac{1}{m_e^2 - u}\right)$$

$$+ \tfrac{1}{2}G_{\bar{c}bu}G_{\bar{a}du}\left(\frac{1}{m_u^2 - u} \pm \frac{1}{m_u^2 - s}\right) + I^{(\pm)}, \tag{1.27}$$

$$I^{(\pm)} = \frac{P}{\pi}\int_{s_{\min}}^{\infty} \text{Im } T^{(\pm)}(s')\left(\frac{1}{s' - s} \pm \frac{1}{s + t + s' - \sum m^2}\right) ds',$$

where $s_{\min}$ now is the smaller of the two limits $s_0$ and $u_0$.

The asymptotic properties of $I^{(\pm)}$ are particularly evident if one uses the crossing-antisymmetric variable

$$\nu = \tfrac{1}{4}m_b^{-1}(s - u) = \tfrac{1}{4}m_b^{-1}(2s - \textstyle\sum m^2 + t)$$

$$I^{(\pm)}(\nu, t) = \frac{P}{\pi}\int_{\nu_{\min}}^{\infty} \text{Im } T^{(\pm)}(\nu', t)\left(\frac{1}{\nu' - \nu} \pm \frac{1}{\nu' + \nu}\right) d\nu' \tag{1.28}$$

Obviously, $I^{(-)}$ converges one power of $\nu'$ faster than $I^{(+)}$.

Finally, we show the connection between our postulated "extended unitarity" and the Feynman rules for higher order perturbation theory. As an example, we consider a square diagram where all particles (virtual or real) have different masses (fig. 6-1.4). According to the tenth Feynman rule of section 3-5, we have for spinless particles $a$, $b$, $c$, $d$, $a'$, $b'$, $e$, $e'$

$$T^{\text{sq}}(s, t) = -ig_{aa'e}\,g_{a'e'c}\,g_{beb'}\,g_{b'e'd}(2\pi)^{-4}\int d^4P'_a\,\Phi'_a\,\Phi'_b\,\Phi_e\,\Phi'_e \tag{1.29}$$

We assume that the masses $m'_a$, $m'_b$, $m_e$ and $m'_e$ in the propagators $\Phi'_a = (m'^2_a - P'^2_a - i\varepsilon)^{-1}$, etc., are the physical masses, i.e., the "mass renormalization" of perturbation theory has been carried out. (To arrive at (29), one must assume a local field for every stable particle, irrespective of whether it is "elementary" or a "bound state"). We wish to show that for $s > (m'_a + m'_b)^2$ and independently of $m_a$, $m_b$, $m_c$, $m_d$

$$T^{\text{sq}}(s + i\varepsilon') - T^{\text{sq}}(s - i\varepsilon') = i\int d\text{ Lips }(s; a', b')T(ab \to a'b')T^*(cd \to a'b') \tag{1.30}$$

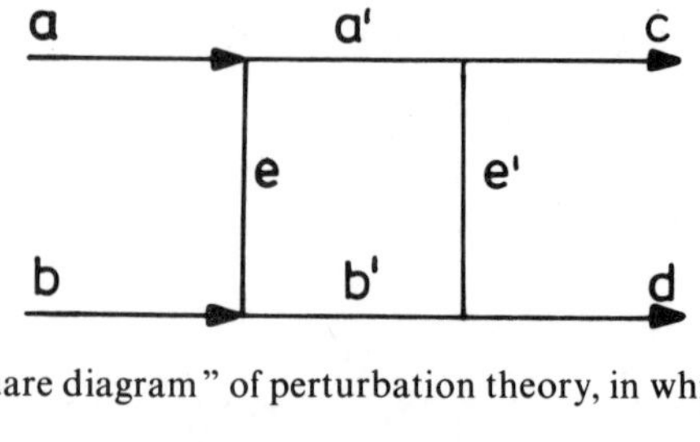


**Figure 6-1.4** A "square diagram" of perturbation theory, in which all masses may be different.

which is just twice the contribution of the state $k = a'b'$ to the unitarity equation (8).

As in section 3-14, we work in the cms $P_a + P'_a = (s^{1/2}, \mathbf{0})$, put $d^4P'_a = dP'^0_a\, d^3\mathbf{p}'$ and perform the integration over $P'^0_a$. The integrand consists of a number of poles in this variable, 4 of which are shown in fig. 3-14.2 (with $E'_a \equiv (m'^2_a + p'^2)^{1/2}$, $E'_b \equiv (m'^2_b + p'^2)^{1/2}$). In principle, the integral over $P'^0_a$ extends from $-\infty$ to $\infty$ along the real axis, but since the integrand is analytic in $P'^0_a$, the path of integration may be deformed, provided it avoids the poles. Figure 3-14.2 refers to real $s$, but we now see that we may move $s$ a little bit out into the complex plane without changing $T^{\text{sq}}$. In other words, the difference (30) for fixed $p'$ is normally zero. An exception occurs when two poles happen to lie exactly opposite each other on either side of the real axis, because in that case the path of integration is pinched between the poles. For the poles shown in fig. 3-14.2, this happens for $s^{1/2} - E'_b = E'_a$, $s^{1/2} = E'_a + E'_b$, which means that particles $a'$ and $b'$ are both on their "mass shells."

To evaluate (30) for this particular case, we first take all $i\varepsilon \to 0$ in the propagators and deform the path of integration as follows (Landau and Lifshitz 1975):

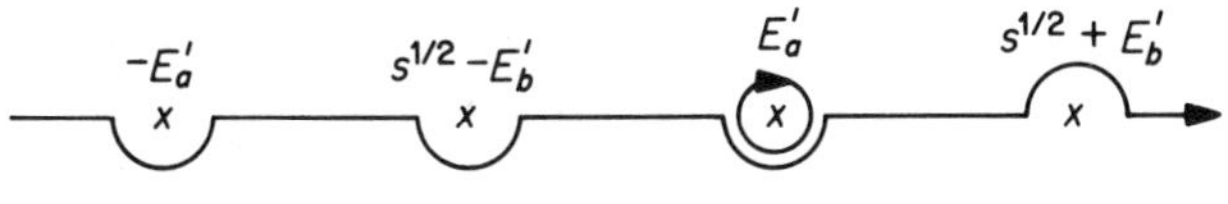


The path now includes a small clockwise circle around $E'_a$, whose contribution is obtained from the integrand by the substitution

$$(E'^2_a - P'^{02}_a)^{-1} \to 2\pi i\, \delta(E'^2_a - P'^{02}_a) = 2\pi i\, \delta(m'^2_a - P'^2_a)$$

(compare 2-6.9), and the understanding that the $P'^0_a$-integration is restricted to positive values. For the second propagator $\Phi'_b(s \pm i\varepsilon')$, we may now use (3-11.16):

$$\begin{aligned} &\Phi'_b(s + i\varepsilon') - \Phi'_b(s - i\varepsilon') \\ &\quad = [E'^2_b - (s^{1/2} - P'^0_a + i\varepsilon')^2]^{-1} - [E'^2_b - (s^{1/2} - P'^0_a - i\varepsilon')^2]^{-1} \\ &\quad = (E'^2_b - (s^{1/2} - P'^0_a)^2 - i\varepsilon'')^{-1} - (E'^2_b - (s^{1/2} - P'^0_a)^2 + i\varepsilon'')^{-1} \\ &\quad = 2\pi i\, \delta[E'^2_b - (s^{1/2} - P'^0_a)^2] = 2\pi i\, \delta(m'^2_b - P'^2_b), \end{aligned} \tag{1.31}$$

understanding that $s^{1/2} - P'^0_a$ is also positive. Collecting our results, we find that taking the discontinuity (30) amounts to the introduction of additional factors $-4\pi^2\, \delta(m'^2_a - P'^2_a)\, \delta(m'^2_b - P'^2_b)$ and understanding that both $P'^0_a$ and $s^{1/2} - P'^0_a$ must be positive. On the other hand, this is just what is needed to convert $-(2\pi)^{-4} \int d^4P'_a$ into $\int d$ Lips $(s;\, P'_a\, P'_b)$.

To complete our proof of (30), we must check that no other pair of poles can pinch the original path of integration. The poles cannot belong to two adjacent propagators because then all three particles at one vertex would be on their mass shells, which would mean that one of these particles is unstable against decay into the other two. The poles of particles $e$ and $e'$ can pinch the path only for $t > (m_e + m_e')^2$. However, this point is outside the physical region for $s > (m_a' + m_b')^2$. A derivation for arbitrary processes is given by Eden et al. (1966).

## 6-2 Application to Spin-$\frac{1}{2}$–Spin-0 Scattering

We know from (2-5.29) that the general amplitude $F(s, t, \lambda, \lambda')$ for reactions with one spinless particle and one spin-$\frac{1}{2}$ particle in initial and final states can be expressed in terms of 4 amplitudes $f_1 \cdots f_4$ which are independent of the helicities $\lambda$ and $\lambda'$. When parity is conserved, two of these 4 amplitudes are zero. We treat only the case when the product of intrinsic parities is identical in initial and final states. In that case we have $f_3 = f_4 = 0$. The remaining two amplitudes $f_1$ and $f_2$ can also be expressed in terms of $f$ and $g$ which were introduced already in (1-7.24). The crossing relations of these amplitudes are complicated by the $(E \pm m)^{\frac{1}{2}}$ which come from the relativistic spinors. To exhibit these, it is best to start from the form (A-3.1), $T = \bar{u}'Qu$ with 4-component spinors $u$ and $\bar{u}'$, for which the substitution rule has been discussed in section 3-6. The operator $Q$ must be scalar under parity transformation, and must contain exactly two arbitrary functions $A(s, t)$ and $B(s, t)$ according to our above considerations. The most convenient form is

$$T(\lambda, \lambda') = \bar{u}'[A(s, t) + \tfrac{1}{2}(P_a + P_c) \cdot \gamma B(s, t)]u, \tag{2.1}$$

where $a$ and $c$ denote the spinless particles. (This notation is appropriate for meson-nucleon collisions, where the projectile $a$ is spinless). $A$ and $B$ are called "invariant amplitudes."* It is now sufficient to apply the crossing relation to the operator $Q = A + \frac{1}{2}(P_a + P_c)\gamma B$. In particular, crossing from the $s$-channel to the $u$-channel gives

$$Q(s, t) = Q_u(u, t), \qquad A(s, t) = A_u(u, t), \qquad B(s, t) = -B_u(u, t). \tag{2.2}$$

The minus sign in front of $B$ comes from the fact that $P_a + P_c$ corresponds to $-P_{\bar{a}} - P_{\bar{c}}$ in the $u$-channel. To find the connection between (1) and (2-5.29), $T = 8\pi s^{\frac{1}{2}} \chi_0'^{+} (f_1 + 4\lambda\lambda' f_2)\chi_0$, we must express $T$ in terms of $\chi_0$ and $\chi_0'^{+}$ according to ($A$-3.4) and (A-3.17). We have

$$\left.\begin{matrix} Q_{gg} \\ Q_{kk} \end{matrix}\right\} = A \pm \tfrac{1}{2}(E_a + E_c)B, \qquad Q_{gk} = -Q_{kg} = -\tfrac{1}{2}(\mathbf{p}_a + \mathbf{p}_c)\boldsymbol{\sigma} B = (\lambda p + \lambda' p')B \tag{2.3}$$

in the cms helicity basis. Inserting these expressions into (A-3.17), we find, with $4\lambda^2 = 4\lambda'^2 = 1$

$$R = (E + m)^{\frac{1}{2}}(E' + m')^{\frac{1}{2}}\left\{A + \frac{1}{2}\left(E_a + E_c + \frac{p^2}{E + m} + \frac{p'^2}{E' + m'}\right)B\right.$$
$$\left. + \frac{4\lambda\lambda' pp'}{(E + m)(E' + m')}[-A + \tfrac{1}{2}(E_a + E_b + E' + m' + E + m)B]\right\}, \tag{2.4}$$

* They have no singularities due to spin ("kinematical singularities").

or with $(E+m)^{-1}p^2 = E - m$, etc., $E_a + E = E_c + E' = s^{\frac{1}{2}}$:

$$8\pi s^{\frac{1}{2}} f_1 = (E+m)^{\frac{1}{2}}(E'+m')^{\frac{1}{2}}[A + B(s^{\frac{1}{2}} - \tfrac{1}{2}m - \tfrac{1}{2}m')],$$
$$8\pi s^{\frac{1}{2}} f_2 = (E-m)^{\frac{1}{2}}(E'-m')^{\frac{1}{2}}[-A + B(s^{\frac{1}{2}} + \tfrac{1}{2}m + \tfrac{1}{2}m')]. \tag{2.5}$$

The matrix element for the "decay" of a single baryon $e$ into another baryon $d$ and a pseudoscalar meson $c$ is given by the first row of table 4-11.1, with $A = 0$ and $-B = G_{ecd}$ (compare also 5-7.9):

$$T_{e,\,cd} = G_{ecd}\bar{u}_d\gamma_5 u(M), \qquad T^*_{e,\,ab} = G_{eab}(\bar{u}_b\gamma_5 u(M))^* = -G_{eab}\bar{u}(M)\gamma_5 u_b. \tag{2.6}$$

The minus sign of the last expression comes from (A-4.9). The baryon exchange Born term is then given by (1.23)

$$T^{(e)}_{ab\to cd} = -\sum_M G_{ecd}G_{eab}\bar{u}_d\gamma_5 u(M)\bar{u}(M)\gamma_5 u_b(m_e^2 - s)^{-1}$$
$$= G_{ecd}G_{eab}\bar{u}_d(P_e\cdot\gamma - m_e)u_b(m_e^2 - s)^{-1}. \tag{2.7}$$

where we have performed the spin summation by means of (A-4.6) and used $\gamma_5^2 = 1$. The form $P_e \cdot \gamma$ may be replaced by $\frac{1}{2}(P_a + P_b + P_c + P_d)\cdot\gamma = \frac{1}{2}(m_b + m_d) + \frac{1}{2}(P_a + P_c)\gamma$ in order to find the contributions of (7) to $A$ and $B$:

$$A^{(e)} = G_{ecd}G_{eab}(\tfrac{1}{2}m_b + \tfrac{1}{2}m_d - m_e)(m_e^2 - s)^{-1},$$
$$B^{(e)} = G_{ecd}G_{eab}(m_e^2 - s)^{-1}. \tag{2.8}$$

In particular, for the nucleon pole in $\pi N$ scattering, we have $m_b = m_d = m_e$ and therefore $A^{(e)} = 0$.

The form (6) is unique only in the sense that it is the simplest matrix element. Apart from a possible form factor as in (5-6.9), the next simple matrix element is the socalled "pseudovector" coupling

$$T_{e,\,cd} = (4\pi)^{\frac{1}{2}} m_\pi^{-1} f_{ecd}\bar{u}_d\gamma_5\gamma_\mu P_c^\mu u(M) \tag{2.9}$$

where the factor $m_\pi^{-1}$ is inserted for dimensional reasons, and the $(4\pi)^{\frac{1}{2}}$ is introduced to avoid the $4\pi$-denominators in measurable quantities such as decay width (see tables 4-9 and 4-11.1). As long as $e$ and $d$ are both real, we can of course use the Dirac equations $P_e \cdot \gamma u = m_e u$, $\bar{u}_d P_d \cdot \gamma = m_d \bar{u}_d$, and with $P_c = P_e - P_d$, we find that (6) and (9) are identical for

$$G_{ecd} = (4\pi)^{\frac{1}{2}} f_{ecd}(m_e + m_d)m_\pi^{-1}. \tag{2.10}$$

On the other other hand, when $e$ is virtual, (6) and (9) give different Born terms. Starting from (9), we get the pseudovector Born term analogous to (7)

$$T^{(e,\,pv)} = \bar{u}_d\gamma_5(P_e - P_d)\cdot\gamma(P_e\cdot\gamma + m)(m_e^2 - s)^{-1}\ldots, \tag{2.11}$$

where the rest is presently unimportant. With $-\bar{u}_d\gamma_5 P_d\cdot\gamma = \bar{u}_d P_d\cdot\gamma\gamma_5 = m_d\bar{u}_d\gamma_5$ we can rewrite (11) as

$$T^{(e,\,pv)} = \bar{u}_d\gamma_5(P_e\cdot\gamma + m_d)(P_e\cdot\gamma + m_e)(m_e^2 - s)^{-1}\ldots$$
$$= \bar{u}_d\gamma_5[(P_e\cdot\gamma + m_e)(m_e^2 - s)^{-1}(m_e + m_d) - 1]\ldots \tag{2.12}$$

To obtain the last expression, we have put $m_d = (m_e + m_d) - m_e$ and used $(P_e \cdot \gamma - m_e)(P_e \cdot \gamma + m_e) = s - m_e^2$. We see that $T^{(e)}$ of (7) and $T^{(e, pv)}$ of (12) give identical pole contributions. The difference between the various possible Born terms resides in the non-pole contributions and is irrelevant for the dispersion relation analysis.

The amplitudes $A$ and $B$ satisfy separate dispersion relations of the type (1.21), except for the additional factor $\frac{1}{2}m_b + \frac{1}{2}m_d - m_e$ in front of $A^{(e)}$ and an extra minus sign from (2) whenever $B$ is crossed. For charged mesons on nucleons or isospin-$\frac{1}{2}$ nuclei, the $s$-channel reaction can be chosen such that it contains only $s$-channel poles (the neutron in $\pi^- p$ scattering, the $\Lambda$ and $\Sigma^0$ in $K^- p$-scattering). Then (1.21) becomes, for fixed $t$

$$\begin{aligned}
\operatorname{Re} A(s) &= \sum_e G_{ecd} G_{eab} (\tfrac{1}{2}m_b + \tfrac{1}{2}m_d - m_e)(m_e^2 - s)^{-1} \\
&\quad + \frac{P}{\pi} \left[ \int_{s_0}^{\infty} \frac{\operatorname{Im} A(s')}{s' - s} + \int_{u_0}^{\infty} \frac{\operatorname{Im} A_u(s')}{s + t + s' - \sum m^2} \right] ds', \\
\operatorname{Re} B(s) &= \sum_e G_{ecd} G_{eab} (m_e^2 - s)^{-1} \\
&\quad + \frac{P}{\pi} \left[ \int_{s_0}^{\infty} \frac{\operatorname{Im} B(s')}{s' - s} - \int_{u_0}^{\infty} \frac{\operatorname{Im} B_u(s')}{s + t + s' - \sum m^2} \right] ds'.
\end{aligned} \tag{2.13}$$

For the combinations $A^{(\pm)}$ and $B^{(\pm)}$ corresponding to (1.26), we have

$$\begin{aligned}
\operatorname{Re} A^{(\pm)} &= \frac{1}{2} \sum_e G_{ecd} G_{eab} \left( \frac{1}{m_e^2 - s} \pm \frac{1}{m_e^2 - u} \right) (\tfrac{1}{2}m_b + \tfrac{1}{2}m_d - m_e) \\
&\quad + \frac{P}{\pi} \int_{\nu_{\min}}^{\infty} \operatorname{Im} A^{(\pm)}(\nu') \left( \frac{d\nu'}{\nu' - \nu} \pm \frac{d\nu'}{\nu' + \nu} \right), \\
\operatorname{Re} B^{(\pm)} &= \frac{1}{2} \sum_e G_{ecd} G_{eab} \left( \frac{1}{m_e^2 - s} \mp \frac{1}{m_e^2 - u} \right) \\
&\quad + \frac{P}{\pi} \int_{\nu_{\min}}^{\infty} \operatorname{Im} B^{(\pm)}(\nu') \left( \frac{d\nu'}{\nu' - \nu} \mp \frac{d\nu'}{\nu' + \nu} \right).
\end{aligned} \tag{2.14}$$

For elastic $\pi^\pm p$ and $K^\pm p$ scattering, we use (5-8.5), (5-8.8) and (5-8.9) for the coupling constants, finding the following pole contributions:

$$\begin{aligned}
\pi^\mp p\colon A_{\text{pole}}^{(\pm)} &= 0, \qquad B_{\text{pole}}^{(+)} = \frac{G^2(s - u)}{(m_N^2 - s)(m_N^2 - u)}, \\
B_{\text{pole}}^{(-)} &= \frac{G^2(t - 2m_\pi^2)}{(m_N^2 - s)(m_N^2 - u)}.
\end{aligned} \tag{2.15}$$

$$K^{\mp}p\colon\ B_{\text{pole}}^{(+)} = \tfrac{1}{2}\sum_e G_{eN\bar K}^2 \frac{s-u}{(m_e^2 - s)(m_e^2 - u)},$$
$$B_{\text{pole}}^{(-)} = \sum_e G_{eN\bar K}^2 \frac{\tfrac{1}{2}t - m_K^2 - m_p^2 + m_e^2}{(m_e^2 - s)(m_e^2 - u)}, \tag{2.16}$$
$$A_{\text{pole}}^{(\pm)} = B_{\text{pole}}^{(\mp)}(m_N - m_e).$$

If violation of isospin invariance is admitted, formulas such as (16) must be used also for $\pi N$ scattering.

The simplest application of (13) is that to forward elastic $\pi^{\pm}p$ and $K^{\pm}p$ scattering, where according to (1)

$$T(\lambda\lambda') = 2m_b\ \delta_{\lambda\lambda'}(A + \omega B) = 8\pi s^{\frac{1}{2}}\ \delta_{\lambda\lambda'} f(\omega),$$
$$\omega \equiv E_a^{\text{lab}} = \tfrac{1}{2}m_b^{-1}(s - m_a^2 - m_b^2) \tag{2.17}$$

Here we have $t = 0$, and $\omega$ coincides with the variable $\nu$ of (1.28). The optical theorem (2-5.5) gives us $\operatorname{Im} T_{\lambda\lambda} = \lambda^{\frac{1}{2}}(s, m_a^2, m_b^2)\sigma_{\text{tot}}$, or

$$\operatorname{Im} f(\omega) = (4\pi)^{-1}p\sigma_{\text{tot}}, \tag{2.18}$$

which has the form used in potential theory. The resulting dispersion relations for the amplitudes $f_{\pm}$ of elastic $\pi^{\pm}p$ and $K^{\pm}p$ scattering are

$$\operatorname{Re} f_{\pm}(\omega) = \sum_e \frac{R_e}{\omega_e \pm \omega} + \frac{P}{4\pi^2}\int_{m_a}^{\infty} p(\omega')\,d\omega'\left[\frac{\sigma_+(\omega')}{\omega' \mp \omega} + \frac{\sigma_-(\omega')}{\omega' \pm \omega}\right]$$
$$+ \frac{1}{\pi}\int_{\omega_{\min}}^{m_a} \frac{d\omega'}{\omega' \mp \omega} \operatorname{Im} f_-(\omega') \tag{2.19}$$
$$R_e = G_{eab}^2[(m_e - m_b)^2 - m_a^2](16\pi m_b^2)^{-1},$$
$$\omega_e = \tfrac{1}{2}m_b^{-1}(m_e^2 - m_b^2 - m_a^2).$$

The last integral in (19) gives the contribution of the cut below threshold, $\omega_{\min} = \frac{1}{2}m_b^{-1}(s_0 - m_a^2 - m_b^2)$ with $s_0 = (m_n + m_{\pi 0})^2$ in $\pi^- p$ scattering and $(m_\Lambda + m_\pi)^2$ in $K^- p$ scattering.

The dispersion relation for $F^{(-)} \equiv \frac{1}{2}(f_- - f_+)$ becomes

$$\operatorname{Re} F^{(-)}(\omega) = \omega \sum_e \frac{R_e}{\omega_e^2 - \omega^2} + \frac{\omega}{4\pi^2} P \int p(\omega')\,d\omega' \frac{\sigma_- - \sigma_+}{\omega'^2 - \omega^2}$$
$$+ \frac{\omega}{\pi}\int_{\omega_{\min}}^{m_a} \frac{d\omega'}{\omega'^2 - \omega^2} \operatorname{Im} f_{(\omega')}^{(-)} \tag{2.20}$$

We also consider the $t$-channel amplitude $T(\bar b d \to a\bar c)$, which according to (1) and the antiparticle substitution can be written as

$$T(M, \bar M) = -i\bar u'(-M_{\bar b}, -P_{\bar b})[A + \tfrac{1}{2}(P_a - P_{\bar c}) \cdot \gamma B]u(M_d, P_d). \tag{2.21}$$

This time we specialize directly to elastic scattering in the $t$-channel cms, where $E_{\bar{c}} = E_a$, $\mathbf{p}_{\bar{c}} = -\mathbf{p}_a$, such that (3) transforms into

$$Q_{gg} = Q_{kk} = A, \qquad Q_{gk} = -Q_{kg} = -\mathbf{p}_a \boldsymbol{\sigma} B. \tag{2.22}$$

Inserting this into (A-3.29), we find, with $p_{az} = p_\pi \cos\theta_t$, $p_{ax} = p_\pi \sin\theta_t$, $p_{ay} = 0$,

$$\begin{aligned} T(\lambda\bar{\lambda}, \theta_t) &= \chi_0^+(\lambda) p_b \left[ A(1 + 4\lambda\bar{\lambda}) - B\mathbf{p}_a \boldsymbol{\sigma} \left( \frac{2\lambda p_b}{E_b + m_b} - \frac{2\bar{\lambda} p_b}{E_b - m_b} \right) \right] \chi_0(\lambda) \\ &= 2\, \delta_{\lambda\bar{\lambda}} (p_b A + m_b p_\pi B \cos\theta_t) - 4\lambda\, \delta_{\lambda, -\bar{\lambda}} p_\pi E_b B \sin\theta_t . \end{aligned} \tag{2.23}$$

Its partial-wave decomposition is given by (4-4.9) with $\lambda_c = \lambda_{\bar{c}} = 0$:

$$T(\tfrac{1}{2}, \tfrac{1}{2}, \theta_t, \phi_t) = 8\pi t^{\frac{1}{2}} \sum_J (2J+1) P_J(\cos\theta_t) T_+^J(t), \tag{2.24}$$

$$T(\tfrac{1}{2}, -\tfrac{1}{2}, \theta_t, \phi_t) = -8\pi t^{\frac{1}{2}} e^{i\phi_t} \sum_J [J(J+1)]^{-\frac{1}{2}} (2J+1) \sin\theta_t P_J' (\cos\theta_t) T_-^J(t). \tag{2.25}$$

Here we have used $d_{00}^J = P_J$ in (24) and (B-5.7) in (25). Later, we shall also need the inversions, which by means of (B-5.1), and (B-5.12) become

$$8\pi t^{\frac{1}{2}} T_+^J = \frac{1}{2} \int_{-1}^{1} dx P_J(x) T(\tfrac{1}{2}, \tfrac{1}{2}, \theta_t) = \int_{-1}^{1} dx [P_J p_b A + x P_J m_b p_\pi B], \tag{2.26}$$

$$\begin{aligned} 8\pi t^{\frac{1}{2}} T_-^J &= -\frac{J^{\frac{1}{2}}(J+1)^{\frac{1}{2}}}{4J+2} \int_{-1}^{1} dx\, (P_{J-1} - P_{J+1}) T(\tfrac{1}{2}, -\tfrac{1}{2}, \theta_t)/\sin\theta_t \\ &= (2J+1)^{-1} J^{\frac{1}{2}} (J+1)^{\frac{1}{2}} \int_{-1}^{1} dx (P_{J-1} - P_{J+1}) p_\pi E_b B. \end{aligned} \tag{2.27}$$

## 6-3 Poles and Anomalous Thresholds in cos $\theta$. Convergence of Partial-Wave Expansion

In analogy with (1.21), we may write a dispersion relation for the $s$-channel amplitude $T(s, t)$ as a function of $t$ or $u$ at fixed $s$:

$$\operatorname{Re} T(s,t) = \frac{-R_t}{m_e^2 - t} + \frac{-R_u}{m_u^2 - u} + \frac{P}{\pi} \left[ \int_{t_0}^{\infty} \frac{\operatorname{Im} T_t(s,t')}{t'-t} + \int_{u_0}^{\infty} \frac{\operatorname{Im} T_u(s,t')}{t'-u} \right] dt', \tag{3.1}$$

$$-R_t \equiv G_{ea\bar{c}} G_{e\bar{b}d}, \qquad -R_u \equiv G_{\bar{c}bu} G_{\bar{a}du}. \tag{3.2}$$

Possible $s$-channel poles do not appear now. The quantities $R_t$ and $R_u$ are the "residues" of the $t$- and $u$-channel poles, and spin is again neglected. We wish to discuss partial-wave amplitudes and therefore express $t$ and $u$ in terms of $x = \cos\theta$. For $t$ this has been done in (2-4.7); the corresponding formula for $u$ is

$$u = m_c^2 + m_b^2 - 2E_b E_c - 2pp' \cos\theta. \tag{3.3}$$

We abbreviate

$$m_e^2 - t \equiv 2pp'(z_e - x), \qquad z_e \equiv [E_a E_c - \tfrac{1}{2}(m_a^2 + m_c^2 - m_e^2)]/pp', \tag{3.4}$$

and similarly for the other 3 denominators in (1). The denominators are now expanded in terms of Legendre polynomials as follows:

$$(z - x)^{-1} = \sum_{L=0}^{\infty} (2L + 1)P_L(x)Q_L(z), \tag{3.5}$$

where the $Q_L$ are the "Legendre functions of the second kind." The orthogonality relations (2-5.10) provide us with the integral representation

$$Q_L(z) = \frac{1}{2}\int_{-1}^{1} dx P_L(x)(z - x)^{-1}, \qquad Q_0 = \tfrac{1}{2} \ln \frac{z + 1}{z - 1}, \qquad Q_1 = zQ_0 - 1. \tag{3.6}$$

For $L > 1$, these functions are obtained from the recurrence relation (B-5.9),

$$Q_L(z) = z(2 - L^{-1})Q_{L-1}(z) - (1 - L^{-1})Q_{L-2}(z). \tag{3.7}$$

For complex $x$, the natural domain of convergence for Legendre polynomial expansions is an ellipse of foci $\pm 1$. As (5) is obviously singular at $x = z$, the partial-wave expansion of (1) converges inside that ellipse (= Lehmann ellipse) which passes through $z_n$, where $z_n$ is the $z$ of smallest magnitude occurring in (1) (see fig. 6-3.1). Obviously $z$ is real and must not lie between $-1$ and $+1$. This condition is satisfied for physical $s$ if the external particles $a$, $b$, $c$, $d$ are stable against strong decays. Since the poles in $t$ and $u$ occur below the corresponding cuts, $z_n$ is determined by the poles, provided there are any. In nucleon-nucleon scattering, there are several $t$-channel poles, corresponding to $e = \pi, \eta, \rho, \omega \ldots$. For elastic scattering, (4) simplifies to

$$z_e(ab \to a'b') = 1 + \tfrac{1}{2}m_e^2/p^2, \tag{3.8}$$

and $z_n$ is given by the pion pole, since the pion is the lightest meson. In potential theory, $z_n$ determines the long-range part of the potential. In fact, the potential is defined as the Fourier transform of the Born approximation,

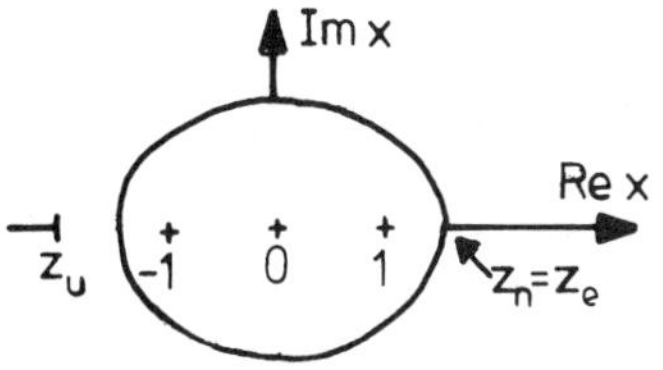


**Figure 6-3.1** The ellipse of convergence of the partial-wave expansion in the complex $x$-plane ($x = \cos\theta$), for $z_n > 0$.

which in the spinless case is just $-R_t(m_e^2 - t)^{-1}$. According to (2-6.34) and (3-11.25),

$$U(r) = 2EV = \frac{-2}{\pi r}\int q\,dq\,\sin qr(-R_t/8\pi s^{\frac{1}{2}})(m_e^2 + q^2)^{-1} = r^{-1}e^{-m_e r}R_t/8\pi s^{\frac{1}{2}}, \tag{3.9}$$

which is the famous "Yukawa potential." (Since we are considering spinless particles, the coupling constants in $R_t$ have the dimensions of a mass. The potential between two baryons will be discussed in section 7-5.) For $\pi^+ p$ scattering, the neutron pole in the $u$-channel competes with the $\rho^0$-pole in the $t$-channel. In analogy with (4), we put

$$m_u^2 - u = -2pp'(z_u - x), \qquad z_u = -[E_b E_c - \tfrac{1}{2}(m_b^2 + m_c^2 - m_u^2)]/pp' \tag{3.10}$$

according to (3). With $u = b =$ nucleon, $c =$ pion, $z_u = -(EE_n - \frac{1}{2}m^2)/p^2$ has in fact $|z_u| < z_e$ at low energies. In other words, the high partial waves in low-energy $\pi^+ p$ scattering are given by the neutron pole.

In nuclear reactions such as elastic $pd$-scattering ($a = c = p$, $b = d$, $u = n$ in fig. 6-1.3), $z_u$ can be quite close to $-1$, causing a sharply backward peaked scattering amplitude already at low energies. The nuclear physics terminology for such cases is "pickup" or "stripping." Here we put

$$m_b^2 \equiv (m_c + m_u)^2 - k_+^2, \qquad k_+^2 \equiv 2m_b b, \tag{3.11}$$

$$z_u = -(E_b E_c - (m_c + m_u)m_c + m_b b)/pp'. \tag{3.12}$$

In the nonrelativistic limit, $b$ in (11) is the binding energy of the target nucleus, and if it is small enough, $z_u$ comes close to $-1$.

In general, the partial-wave expansions (2-5.7) and (5) provide us with the following representation of partial-wave amplitudes:

$$T_L(s) = (16\pi s^{\frac{1}{2}} pp')^{-1}[-R_t Q_L(z_e) + \pi^{-1}P\int_{t_0}^{\infty} dt' Q_L(z)\,\text{Im}\,T_t(s, t'(z))] + \ldots \tag{3.13}$$

where the dots indicate the $u$-channel contributions. This expression is called the "Froissart-Gribov projection," apart from a complication due to the $u$-channel cut, which will be included in (10.16) below.

When poles and cuts are inserted into the right-hand side of the unitarity equation, they generate new singularities. The simplest of these is the normal threshold branch point at $s_0$. The next simple is the "triangle singularity" of fig. 6-3.2 which arises from the combination of a constant with a $t$-channel

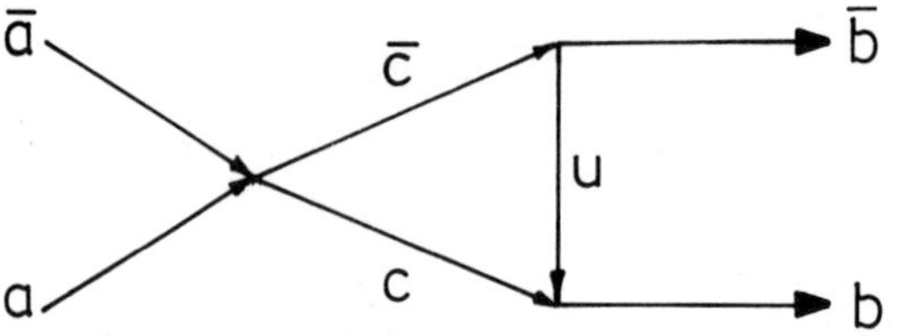


**Figure 6-3.2** The graph which produces the triangle singularity in elastic scattering $ab \rightarrow a'b'$.

pole. In nuclear physics, this singularity produces an "anomalous threshold" which gives the first indication of a long range in processes like pion-nucleus scattering or electromagnetic form factors of nuclei. We first consider elastic scattering $ab \to a'b'$, where $b$ is a nucleus.* The derivation starts conveniently from the partial-wave unitarity equation for the $t$-channel reaction $a\bar{a} \to b'\bar{b}$ below $4m_b^2$. In this region we have

$$p_b = p_{\bar{b}} = i\kappa_b, \qquad \kappa_b = (m_b^2 - t/4)^{\frac{1}{2}}. \tag{3.14}$$

We also assume that only one channel $k = c\bar{c}$ contributes to the unitarity equation. In the beginning we study a situation with real $p_c = (t/4 - m_c^2)^{\frac{1}{2}}$ and $p_a = (t/4 - m_a^2)^{\frac{1}{2}}$. Keeping

$$s = (P_a - P_{\bar{b}})^2 = -p_a^2 - p_b^2 + 2p_a p_b \cos\theta_{a\bar{b}} \tag{3.15}$$

real, we see that $\cos\theta_{a\bar{b}}$ is imaginary. If one inserts the ansatz (2-5.7) into the original unitarity equation (2-5.6), one must observe

$$P_L^*(\cos\theta_{a\bar{b}}) = (-1)^L P_L(\cos\theta_{a\bar{b}}). \tag{3.16}$$

Instead of (2-5.11), one thus finds

$$T_{a\bar{a},\, b\bar{b},\, L} - (-1)^L T^*_{b\bar{b},\, a\bar{a},\, L} = 2\pi i(-1)^L p_c T_{a\bar{a},\, c\bar{c},\, L} T^*_{b\bar{b},\, c\bar{c},\, L}. \tag{3.17}$$

In $T^*_{b\bar{b},\, c\bar{c},\, L}$, only the $u$-channel pole of elastic $cb \to c'b'$ scattering is kept, which now appears as a pole in the momentum transfer of $c\bar{c} \to b\bar{b}$:

$$T^{(u)}_{b\bar{b},\, c\bar{c},\, L} = (16\pi t^{\frac{1}{2}} p_c i\kappa_b)^{-1} G^2_{buc} Q_L(z), \tag{3.18}$$

$$\begin{aligned} z &= (p_c p_b)^{-1}[E_c E_b - \tfrac{1}{2}(m_c^2 + m_b^2 - m_u^2)] \\ &= (4ip_c\kappa_b)^{-1}[t - 2(m_c^2 + m_b^2 - m_u^2)] \equiv -i\zeta. \end{aligned} \tag{3.19}$$

From (6), we have

$$Q_0(-i\zeta) = \tfrac{1}{2}\ln\frac{1 + i/\zeta}{1 - i/\zeta} = i\tan^{-1}(\zeta^{-1}), \qquad Q_1 = \zeta\tan^{-1}(\zeta^{-1}) - 1, \ldots \tag{3.20}$$

We see that (18) is real for $t > 4m_b^2$ but imaginary for $4m_c^2 < t < 4m_b^2$ and odd $L$. For $m_b^2 < m_u^2 + m_c^2$, the unitarity equation (17) ends at $p_c = 0$, $t = 4m_c^2$ where $\zeta \to \infty$, $\tan^{-1}(\zeta^{-1}) \to 0$. The opposite case

$$m_b^2 > m_u^2 + m_c^2 \tag{3.21}$$

causes an anomalous threshold. In this case, $z$ passes through zero before $t$ reaches $4m_c^2$. When $t$ approaches $2(m_b^2 + m_c^2 - m_u^2)$ from above, $\tan^{-1}(\zeta^{-1})$ reaches $\pi/2$. The analytic continuation of $Q_0$ below this point is

$$Q_0(t < 2(m_b^2 + m_c^2 - m_u^2)) = i[\pi - \tan^{-1}(-\zeta^{-1})], \tag{3.22}$$

* Arbitrary masses are treated in section 7-8.

because it has the same value ($= i\pi/2$) as (20) at $\zeta^{-1} = \infty$. At $t = 4m_c^2$ $Q_0$ becomes $i\pi$, and below $4m_c^2$, it becomes complex:

$$T^{(u)}_{b\bar{b},\, c\bar{c},\, L}(t < 4m_c^2) = -(16\pi t^{\frac{1}{2}}\kappa_c\kappa_b)^{-1}G^2_{buc}Q_L, \qquad Q_0 = i\pi + \tfrac{1}{2}\ln\frac{z+1}{z-1}. \tag{3.23}$$

The $i\pi$ is the usual discontinuity of $\frac{1}{2}$ ln across its left-hand cut. The variable $z$ is negative in this range and reaches $-1$ at

$$m_u^2 t_{\text{tr}} = -(m_b^2 + m_c^2 - m_u^2)^2 + 4m_b^2 m_c^2 = -\lambda(m_u^2, m_b^2, m_c^2) \tag{3.24}$$

according to (2-2.9). Thus $T^{(u)}_{b\bar{b},\, c\bar{c},\, L}$ has a cut down to

$$t_{tr} = 4m_c^2 - m_u^{-2}(m_b^2 - m_c^2 - m_u^2)^2, \tag{3.25}$$

which is the "anomalous threshold." This cut must be included in the Cauchy integral occurring in the dispersion relation (1)

$$\frac{P}{\pi}\int_{t_0}^{\infty}\frac{\operatorname{Im} T_t(s, t')\,dt'}{t'-t} = \frac{P}{\pi}\left[\int_{t_{\text{tr}}}^{4m_c^2}\frac{\operatorname{disc} T_{a\bar{a},\, b\bar{b}}(s, t')}{2i(t'-t)} + \int_{4m_c^2}^{\infty}\frac{\operatorname{Im} T_{a\bar{a}b\bar{b}}(s, t')}{t'-t}\right]dt'. \tag{3.26}$$

But what is "disc $T_{a\bar{a},\, b\bar{b}}$"? Let $f(\tau)$ denote the analytic extension of the function defined in (17). Normally, $f$ is imaginary-analytic (compare 3-11.11); at $\tau = t + i\varepsilon$ it is imaginary. ($= 2i$ Im $T$). At the anomalous cut, $f$ is complex, i.e., it contains a real-analytic part $h$, $f = h + g$. The discontinuity across the cut is entirely in $g$, however. Using Im $Q_L = \pi P_L$, we have in our example

$$\operatorname{disc} T_{a\bar{a},\, b\bar{b},\, L}/2i = -T_{a\bar{a},\, c\bar{c},\, L}(16\pi t^{\frac{1}{2}}\kappa_b)^{-1}G^2_{buc}\pi P_L(z). \tag{3.27}$$

The triangle singularity was first discovered in perturbation theory (Karplus et al. 1958). Mandelstam (1960) discussed the development of the anomalous threshold as $m_b^2$ increases. Defining

$$x = m_b^2 - m_c^2 - m_u^2 - i\varepsilon, \qquad t_{\text{tr}} = 4m_c^2 - m_u^{-2}(x + i\varepsilon)^2, \tag{3.28}$$

one finds that the singularity lies on the second Riemann sheet for $x < 0$, winds around the branchpoint at $t = 4m_c^2$ for $x \sim 0$ and continues downwards on the physical sheet for $x > 0$ (see fig. 6-3.3). Its subsequent cut must

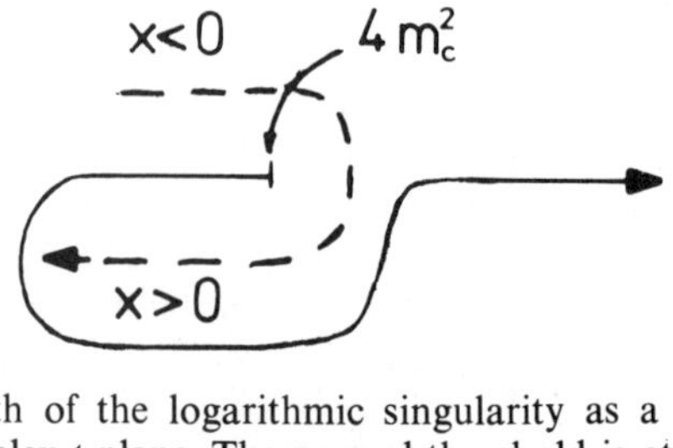


**Figure 6-3.3** The path of the logarithmic singularity as a function of $x = m_b^2 - m_c^2 - m_u^2$, in the complex $t$-plane. The normal threshold is at $4m_c^2$.

be circumvented in the Cauchy integral, which gives rise to the extra piece in (26). As $m_b$ approaches $m_u + m_c$, it corresponds to a nonrelativistic bound state with binding energy $b$ as in (11). It is then convenient to decompose $\lambda$ of (24) as in (2-2.11)

$$\begin{aligned} t_{\text{tr}} &= -m_u^{-2}[m_b^2 - (m_u + m_c)^2][m_b^2 - (m_u - m_c)^2] \\ &= k_+^2 m_u^{-2}[m_b^2 - (m_u - m_c)^2] \\ &\approx 8bm_b m_c m_u^{-1}. \end{aligned} \tag{3.29}$$

Mathematically, it may be worth mentioning that $\frac{1}{4}(-\lambda)^{\frac{1}{2}}$ is the area of a triangle with sides $m_b$, $m_c$, $m_u$, which explains the name "triangle function" of $\lambda$. In nuclear physics, the amplitudes $T(a\bar{a} \to c\bar{c})$ and $T(b\bar{b} \to c\bar{c})$ are needed at $t$-values which are so small that there is practically no connection to the values of these amplitudes in their physical regions. On the contrary, $T(b\bar{b} \to c\bar{c})$ is closely related to the elastic $cd$-scattering amplitude at backward angles, which is near the $u$-channel pole. This point will become still clearer at the end of section 6-6 and in section 7-8.

We have already encountered a triangle graph, namely in the calculation of the electron anomalous magnetic moment (fig. 3-12.a). This graph has $m_b = m_c = m_e$, $m_u = 0$ and is required only for $t > 4m_b^2$. We now consider the triangle graph with $c =$ pion, $b =$ baryon, which occurs (among others) in the calculation of baryon form factors. From (2.8) we have

$$A^{(u)} = (m_b - m_u)B^{(u)}, \qquad B^{(u)} = G_{bu}^2(m_u^2 - u)^{-1}. \tag{3.30}$$

Inserting these expressions into (2.26) and (2.27), we find for $J = 0$ and 1:

$$\begin{aligned} T_+^{0(u)} &= (8\pi t^{\frac{1}{2}} i\kappa_b)^{-1} G_{bu\pi}^2[p_\pi^{-1} p_b Q_0(m_b - m_u) + m_b Q_1], \\ T_+^{1(u)} &= (8\pi t^{\frac{1}{2}} i\kappa_b)^{-1} G_{bu\pi}^2[p_\pi^{-1} p_b(m_b - m_u) + m_b z]Q_1, \\ T_-^{1(u)} &= (8\pi t^{\frac{1}{2}} i\kappa_b)^{-1} G_{bu\pi}^2 \tfrac{4}{3} 2^{\frac{1}{2}} E_b(Q_0 - Q_2). \end{aligned} \tag{3.31}$$

Here we have used $xP_0 = P_1$ and $\frac{1}{2}\int x\, dx P_J(z - x) = zQ_J$ for $J > 0$ (to derive the latter integral, put $x = (x - z) + z$ and observe $\int dx P_J = 0$ for $J > 0$).

## 6-4 Poles and Residues in Potential Theory

Because of the great achievements of potential theory, we must establish the connection between $s$-channel poles and their coupling constants on the one side and the bound state wave function on the other. We start with an investigation of the partial-wave equation (1-3.9) for the scattering of two spinless particles, including the recoil corrections (2-9.4):

$$[\partial_r^2 + p^2 - U - r^{-2}L(L+1)]\phi_L = 0, \qquad p^2 = \varepsilon^2 - \mu^2, \qquad \varepsilon = P_a P_b s^{-\frac{1}{2}}, \tag{4.1}$$

$$\phi_L \equiv prR_L(pr)f_L(p). \tag{4.2}$$

A dimensionless function $f_L(p)$ (the "Jost function") has been included in (2) to give $\phi$ nice analytic properties as a function of $p$. The regular solution of (1) is, with $\rho = pr$

$$\phi_L = \rho j_L(\rho) - \int_0^r r'\,dr'\rho[j_L(\rho)n_L(\rho') - n_L(\rho)j_L(\rho')]U(r')\phi_L(r'). \tag{4.3}$$

This is verified by differentiation,

$$\begin{aligned}\partial_r\phi_L &= \left(1 - \int_0^r r'\,dr' n_L(\rho')U(r')\phi_L(r')\right)\partial_r(\rho j_L)\\ &\quad + \int_0^r r'\,dr' j_L(\rho')U(r')\phi_L(r')\,\partial_r(\rho n_L)\end{aligned} \tag{4.4}$$

$$\partial_r^2\phi_L = \left(1 - \int\cdots\right)\partial_r^2(\rho j_L) + U\phi_L + \int\cdots\partial_r^2(\rho n_L).$$

In the second equation we have used (A-1.7). The advantage of (3) is that it can be solved iteratively, which leads to analyticity properties in $p$. The proof is based on the power series expansions (A-1.1) and (A-1.2) of $j_L$ and $n_L$, neglecting a possible $p$-dependence of $U$ (see Taylor 1972). Since (1) is invariant under $p \to -p$, $\phi_L(r, -p)$ must be proportional to $\phi_L(r, p)$. From (3) and (A-1.6), we find $\phi_L(r, -p) = -(-1)^L\phi_L(r, p)$. We can also express (3) in terms of $h_L(\rho)$ and $h_L(-\rho)$, using $j_L = \frac{1}{2}h_L(\rho) + \frac{1}{2}(-1)^L h_L(-\rho)$, $n_L = (2i)^{-1}(h_L - \cdots)$:

$$\begin{aligned}\phi_L &= \tfrac{1}{2}\rho h_L(\rho)\left[1 - i\int_0^r r'\,dr'(-1)^L h_L(-\rho')U(r')\phi_L(r')\right]\\ &\quad + \tfrac{1}{2}\rho(-1)^L h_L(-\rho)\left[1 + i\int_0^r r'\,dr' h_L(\rho')U(r')\phi_L(r')\right].\end{aligned} \tag{4.5}$$

For $r \to \infty$, we find the asymptotic form of (5)

$$\phi_L(r \to \infty) = \tfrac{1}{2}\rho h_L(\rho)f_L(-p) + \tfrac{1}{2}\rho(-1)^L h_L(-\rho)f_L(p), \tag{4.6}$$

$$f_L(p) = 1 + i\int_0^\infty r\,dr h_L(pr)U(r)\phi_L(r). \tag{4.7}$$

Comparison with (1-8.16) shows that

$$S_L = f_L(-p)/f_L(p). \tag{4.8}$$

For real $p$ and $U$, we have $f_L(-p) = f_L^*(p)$, such that $S_L$ is just $\exp(2i\,\delta_L)$. For imaginary $p$, $S_L$ gets a pole when $f_L(p)$ passes through zero. Because of the analyticity in $p^2$, we can expand $\phi(r, p^2)$ around the zero of $f_L$ as follows:

$$\phi_L(r \to \infty, p^2 \to p_0^2) = Ne^{-\kappa r} + \beta(p^2 - p_0^2)e^{\kappa r}, \qquad p_0 = i\kappa. \tag{4.9}$$

The coefficients $N$ and $\beta$ are related by the normalization of the bound state wave function

$$\int \phi_L^2(r)\, dr \bigg|_{p=p_0} = 1. \tag{4.10}$$

We differentiate (1) with respect to $p^2$, multiply the resulting equation by $\phi$, and subtract this from ((1) multiplied by $\partial_p^2 \phi$). We thus find (assuming again $\partial_p^2 U = 0$)

$$(\partial_p^2\phi)(\partial_r^2\phi) - \phi\, \partial_r^2\, \partial_p^2\phi = \phi^2. \tag{4.11}$$

Taking this at $p^2 = p_0^2$ and integrating over $r$, we get $[(\partial_p^2\phi)(\partial_r\phi) - \phi\, \partial_r\, \partial_p^2\phi] = 1$ at $r = \infty$, $p^2 = p_0^2$. With $\partial_p^2\phi = \beta e^{\kappa r}$ and $\partial_r\phi = -N\kappa e^{-\kappa r}$, this gives $\beta = -1/N\kappa$. Next, we bring (9) into the asymptotic form of (6). With $\rho h_L(\rho) \to (-i)^{L+1} e^{-\kappa r}$ and $\rho h_L(-\rho) \to -(-i)^{L+1} e^{\kappa r}$, we have

$$f_L(-p) = 2Ni^{L+1}, \qquad f_L(p) = -2\beta(p^2 - p_0^2)i^{L+1}(-1)^L, \tag{4.12}$$

$$S_L = f_L(-p)/f_L(p) = -(-1)^L \frac{N/\beta}{p^2 - p_0^2} = (-1)^L \frac{\kappa}{p^2 + \kappa^2} N^2, \tag{4.13}$$

$$T_L = -\tfrac{1}{2} i p^{-1}(S_L - 1) \approx \tfrac{1}{2}(-1)^L N^2 (p_0^2 - p^2)^{-1} \tag{4.14}$$

near $p = i\kappa$. This must be compared with the dominance of one partial-wave amplitude in the full scattering amplitude $T(s, \cos\theta)$ of (2-5.7)

$$T/P_L = 8\pi m_e (2L + 1) T_L. \tag{4.15}$$

For $L = 0$, we have $P_0 = 1$, $T = G^2(m_e^2 - s)^{-1}$. Using $p_0^2 - p^2 = (m_e^2 - s)\mu/m_e$ in the nonrelativistic domain, we find

$$G^2 = 4\pi N^2 m_e^2/\mu. \tag{4.16}$$

For $L = 1$, we have $P_1 = \cos\theta$ and according to (1.24)

$$(m_e^2 - s)T/P_1 = 4G^2 p^2 = -12\pi N^2 m_e^2/\mu. \tag{4.17}$$

(The form of $T/P_1$ is most easily obtained from the second row of (1.24) in the cms by observing that the only nonzero components of the square bracket are $\delta_{ij}$). Note that the negative sign on the right-hand side of (17) corresponds to the fact that $p^2$ is negative at the pole.

In nuclear physics, one can get a qualitative understanding of the magnitude of $N$ by assuming a square-well potential of radius $R$ (Ericson and Locher 1970). For $L = 0$ we have exactly $\phi(r > R) = Ne^{-\kappa r}$ (see fig. 6-4.1). For $\kappa R > 1$, the value of $\phi$ at $R$ does not depend critically on $\kappa$ or $R$, which means that $N$ varies with $\kappa$ and $R$ roughly as $e^{\kappa R}$.

In section 6-3 we saw that the $t$-channel pole corresponds to a Yukawa potential. We now show for the reaction $ab \to a'cu$ that the exchange of a loosely bound particle $e$ (particle $b$ being the $e + u$ bound state, see fig. 6-4.2) corresponds to the "impulse approximation" of potential theory. We take $b$,

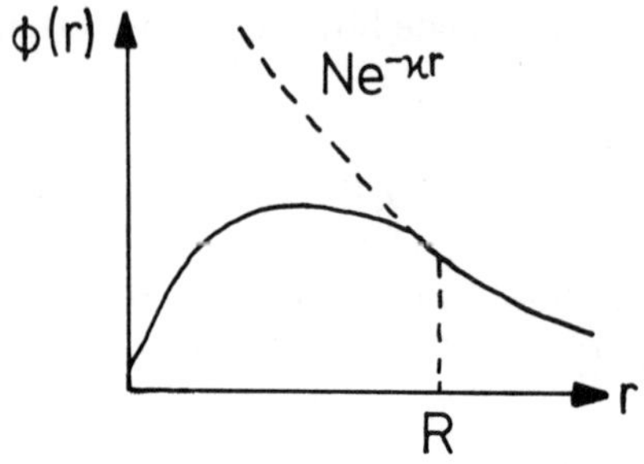


**Figure 6-4.1** The bound state wave function $\phi(r)$ for a large system, $\kappa R > 1$ (after Ericson and Locher 1970).

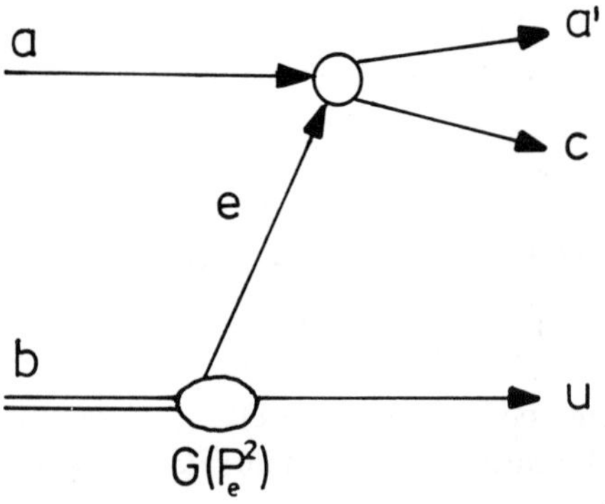


**Figure 6-4.2** The impulse approximation for collisions with a loosely bound particle $b$.

$e$ and $u$ as spinless particles, but allow a vertex function $G(t)$ with $t = P_e^2$ at the lower vertex, and an arbitrary scattering amplitude $T(ae \to a'c)$ for scattering on the virtual particle $e$ at the upper vertex:

$$T = G(t)(m_e^2 - t)^{-1}T(ae \to a'c). \tag{4.18}$$

In the lab system, we have $\mathbf{p}_u = -\mathbf{p}_e \equiv \mathbf{p}_k$, $E_u = (m_u^2 + p_k^2)^{\frac{1}{2}}$ is given by an expression analogous to (2-2.8), and

$$P_e^2 = m_b^2 + m_u^2 - 2m_b E_u \approx (m_b - m_u)^2 - m_b p_k^2/m_u, \tag{4.19}$$

$$m_e^2 - P_e^2 \approx 2m_e b + m_b p_k^2/m_u = (\kappa^2 + p_k^2)m_b/m_u, \qquad \kappa^2 \equiv 2\mu b, \tag{4.20}$$

where we have inserted $m_b \approx m_e + m_u - b$ and $\mu =$ reduced mass. We can now rewrite (18) as

$$T(ab \to a'cu) = T(ae \to a'c)\tilde{\psi}(\mathbf{p}_k)m_u/m_b, \qquad \tilde{\psi}(\mathbf{p}) = G(p^2)(\kappa^2 + p^2)^{-1}, \tag{4.21}$$

where $\tilde{\psi}(\mathbf{p})$ may be interpreted as the Fourier transform of the bound state wave function $\psi(\mathbf{r})$, $\mathbf{r} = \mathbf{r}_e - \mathbf{r}_u$. For this purpose, it suffices to show that $\psi$

goes asymptotically as $r^{-1}N \exp(-\kappa r)$. For a bound state of angular momentum $L$, we have

$$\psi_L(\mathbf{r}) = r^{-1}\phi_L(r)Y^L_M(\hat{\mathbf{r}}) = (2\pi)^{-3}\int d^3p e^{-i\mathbf{p}\mathbf{r}}\tilde{\psi}(\mathbf{p}\}, \tag{4.22}$$

where $\phi_L$ is normalized as in (10). For the asymptotic form we can take $p^2 = -\kappa^2$ in $G(p^2)$. From (3.9) we remember that the resulting integral (22) gives $(4\pi r)^{-1}G(-\kappa^2) \exp(-\kappa r)$, which agrees with (9) if we identify

$$N = G(-\kappa^2)(4\pi)^{-\frac{1}{2}} \tag{4.23}$$

(a factor $(4\pi)^{-\frac{1}{2}}$ is taken by $Y^0_0$ in 22).

It must be emphasized that (18) can be insufficient to describe data even if the function $G(-\kappa^2)$ is kept arbitrary. More complicated graphs (sometimes classified into initial or final state interactions) will also contribute. A particular case of final state interaction will be discussed in section 7-8.

## 6-5 Threshold Behavior of Partial-Wave Amplitudes. *K*-Matrix and Resonance Scattering

The thresholds of a reaction $ab \to cd$ are the energies at which $p = 0$ or $p' = 0$. At these points, $z_e$ of (3.4), $z_u$ of (3.10) and $z$ in the Froissart-Gribov projection (3.13) become infinite. The explicit expression for $z$ in (13) is

$$t' - t = 2pp'(z - x), \qquad z \equiv [E_a E_c - \tfrac{1}{2}(m_a^2 + m_c^2 - t')]/pp'. \tag{5.1}$$

In this case, the $Q_L$ given by (3.6) can be expanded as follows:

$$Q_0 = z^{-1} + \tfrac{1}{3}z^{-3} + \tfrac{1}{5}z^{-5} \ldots, \qquad Q_1 = \tfrac{1}{3}z^{-2} + \tfrac{1}{5}z^{-4} + \tfrac{1}{7}z^{-6} \ldots \tag{5.2}$$

The expansions for $L > 1$ are found from (3.7). The leading terms are proportional to $z^{-1-L}$ and $z^{-3-L}$ respectively. We therefore get

$$\text{Re } T_L(s) \sim (pp')^L \quad \text{for} \quad p \to 0 \quad \text{and/or} \quad p' \to 0. \tag{5.3}$$

The corrections to (3) are of order $p^2$ or $p'^2$. For elastic scattering, the correction from $Q_L$ would be of order $z^{-2} = p^4$, but the $s$-dependence of $T(s, t)$ brings $p^2$-terms. Such terms are particularly large in the case of a nearby resonance or bound state or zero of the scattering amplitude.

The behavior of Im $T_L(s)$ near $p' = 0$ is $p'^L$ according to the unitarity equation (2-5.12) and (3). For elastic scattering in the lowest channel near its threshold, Im $T_L$ can be computed from Re $T_L$ as in (2-5.13):

$$T_L = p^{2L}a_L(1 - ia_L p^{2L+1})^{-1} = p^{2L}a_L(1 + ia_L p^{2L+1})(1 + a_L^2 p^{4L+2})^{-1}. \tag{5.4}$$

In the last expression, one may of course omit the denominator, except perhaps for $L = 0$ in situations where $a_L$ is very large (nucleon-nucleon scattering). $a_L$ is called the "scattering length" and for $L = 0$ it has in fact the dimension of a length. The next power of $p^2$ in a threshold expansion is frequently given in the form (compare (2-5.13))

$$pT_L = (\cot \delta_L - i)^{-1}, \qquad p^{2L+1} \cot \delta_L = a_L^{-1} + \tfrac{1}{2} r_L p^2, \tag{5.5}$$

where $r_L$ is called the "effective range." This form is particularly useful near a pole or resonance of $T_L$. Taking $L = 0$ for example, we can write $T_0$ in the form

$$T_0 = (a^{-1} + \tfrac{1}{2}rp^2 - ip)^{-1} = \frac{2/r}{(i\kappa - p)(i\kappa' - p)}, \qquad \begin{pmatrix}\kappa \\ \kappa'\end{pmatrix} \equiv \frac{1}{r} \mp \left(r^{-2} + \frac{2}{ar}\right)^{\frac{1}{2}} \tag{5.6}$$

For $0 < -2r/a < 1$, the pole at $p = i\kappa$ lies close to the physical region, with residue $-2ir^{-1}(\kappa' - \kappa)^{-1} = -i(1 - r\kappa)^{-1}$. Our $T_0$ of (4.14) may be written as $\frac{1}{2}N^2(i\kappa - p)^{-1}(i\kappa + p)^{-1}$, the residue of the pole at $i\kappa$ being $-\frac{1}{4}iN^2/\kappa$. The poles are identical for $N^2 = 4\kappa(1 - r\kappa)^{-1}$. If, on the other hand, $T_L/p^{2L}$ has a zero near threshold, the following expansion converges better:

$$pT_L = \tan \delta_L (1 - i \tan \delta_L)^{-1}, \qquad \tan \delta_L = p^{2L+1}(a_L + b_L p^2). \tag{5.7}$$

The terms of order $p^{2L+2}$ in $T_L$ are identical in (5) and (7) for

$$r_L = -2b_L/a_L^2. \tag{5.8}$$

Unitarity can be incorporated also in the case of several open 2-particle channels and in the case of unstable particles. We drop the partial-wave index $L$ or $J$ and rewrite the multichannel matrix equation (4-4.11) as

$$(T^+)^{-1} - T^{-1} = 2i\tilde{p}\theta, \qquad \tilde{p}_{ij} \equiv \delta_{ij} p_i, \qquad \theta_{ij} \equiv \delta_{ij}\theta(s - s_i), \tag{5.9}$$

which ensures unitarity of the open-channel $S$-matrix $1 + 2i\tilde{p}^{\frac{1}{2}} T \tilde{p}^{\frac{1}{2}}$, (4-8.25). As explained in appendix B-1, unitary matrices are easily constructed from Hermitean ones. We use the $K$-matrix (B-1.5) because it can incorporate poles. However, we must generalize the construction if we want to include the diagonal matrix $\tilde{p}$ which comes from the two-particle phase spaces, and also include unstable single-particle states (Branson 1965). We therefore introduce the Hermitean matrix $K$ via

$$T^{-1} = K^{-1} - 8\pi s^{\frac{1}{2}} d. \tag{5.10}$$

Insertion of (10) into (9) shows that the new matrix $d$ must satisfy

$$8\pi s^{\frac{1}{2}}(d - d^+) = 2i\tilde{p}\theta, \quad (d - d^+)_{ij} = \delta_{ij} \int d \text{ Lips } (s; j). \tag{5.11}$$

The Hermitean part of $d$ is not specified by (11) and can be chosen such that $d$ is analytic.

Since $p_j$ is purely imaginary in a piece $\bar{s}_j < s < s_j$ below threshold, the simplest choice for $s > \bar{s}_j$ is

$$d = i\tilde{p}(8\pi s^{\frac{1}{2}})^{-1}, \qquad T = K(1 - i\tilde{p}K)^{-1}. \tag{5.12}$$

In this form, $K$ is also called the Wigner $R$-matrix. It was used by Heitler (1941, 1944) in his theory of radiation damping. Its role in potential theory has been reviewed by Dalitz (1961, 1962). Comparison between (12) and (7) shows that $K$ is the multichannel generalisation of $p^{-1}$ tan $\delta$. The scattering length approximation is generalized to

$$K_L = \tilde{p}^L A_L \tilde{p}^L, \tag{5.13}$$

where $A_L$ is a constant real matrix. To see its use, consider the energy-dependence of $T$ near threshold of a new channel $c$, where several new states (in the case of spin) start to contribute to the unitarity condition. Denoting the old channels by the index "0", we have

$$T = \begin{pmatrix} T_{00} & T_{0c} \\ T_{c0} & T_{cc} \end{pmatrix}, \qquad K = \begin{pmatrix} K_{00} & K_{0c} \\ K_{c0} & K_{cc} \end{pmatrix}, \qquad \tilde{p} = \begin{pmatrix} \tilde{p}_0 & 0 \\ 0 & p_c \end{pmatrix}. \tag{5.14}$$

The 00- and 0$c$-components of $T(1 - i\tilde{p}K) = K$ are then

$$T_{00}(1 - i\tilde{p}_0 K_{00}) - T_{0c} i p_c K_{c0} = K_{00}, \tag{5.15}$$

$$T_{0c}(1 - ip_c K_{cc}) - iT_{00}\tilde{p}_0 K_{0c} = K_{0c}. \tag{5.16}$$

The second equation gives

$$T_{0c} = (K_{0c} + iT_{00}\tilde{p}_0 K_{0c})(1 - ip_c K_{cc})^{-1}, \tag{5.17}$$

and insertion of this expression into (15) gives

$$T_{00}(1 - i\tilde{p}_0 K_r) = K_r, \tag{5.18}$$

$$K_r \equiv K_{00} + K_{0c}(1 - ip_c K_{cc})^{-1} ip_c K_{c0}. \tag{5.19}$$

From the form (18) we conclude that the "reduced $K$-matrix" $K_r$ must be Hermitean when channel $c$ is closed. On the other hand, $K$ must be Hermitean when $c$ is open:

$$K_{00} = K_{00}^{+}, \qquad K_{cc} = K_{cc}^{*}, \qquad K_{0c} = K_{c0}^{+}. \tag{5.20}$$

Looking now at (19) and remembering $p_c = i|p_c|$ below threshold, we see that $K_r$ is Hermitean if $K$ remains Hermitean also below threshold. In particular, $K_{cc}$ is regular at $p_c = 0$. This follows also directly from the fact that $K^{-1}$ drops out of the unitarity condition (9). It forms the basis of the approximation (13).

Unstable one-particle states $|e\rangle$ on a continuum of 2-particle states are easily included in this formalism (Lang 1975). All one has to do is to insert the one-particle phase space (1.15) on the right-hand side of (11):

$$d_{ie} = \delta_{ie}\, d_e, \qquad \operatorname{Im} d_e = \pi\, \delta(s - m_e^2). \tag{5.21}$$

The simplest analytic extension of (21) is

$$d_e = \lim_{\varepsilon \to 0} (m_e^2 - s - i\varepsilon)^{-1}. \tag{5.22}$$

Putting the new state on place "$c$" in (14) with $K_{ee} = 0$ as the appropriate modification of (1.12), we get from (18) and (19)

$$K_r = K_{00} + 8\pi s^{\frac{1}{2}}(m_e^2 - s - i\varepsilon)^{-1} K_{0e} K_{e0} \tag{5.23}$$

$$\begin{aligned} T_{00} = {} & [K_{00}(m_e^2 - s) + 8\pi s^{\frac{1}{2}} K_{0e} K_{e0}] \\ & \times [(m_e^2 - s)(1 - i\tilde{p}_0 K_{00}) - i\tilde{p}_0 8\pi s^{\frac{1}{2}} K_{0e} K_{e0}]^{-1} \end{aligned} \tag{5.24}$$

This equation is the proper generalization of (4-8.22). For $(m_e^2 - s)K_{00} \ll 8\pi s^{\frac{1}{2}} K_{0e} K_{e0}$, we can neglect $K_{00}$ and evaluate $T_{00}$ directly from (10):

$$T_{00} = K_{0e}(1 - 8\pi s^{\frac{1}{2}}\, dK)_{e0}^{-1} = 8\pi s^{\frac{1}{2}} K_{0e}\, d_e K_{e0}(1 - 64\pi^2 s\, d_e K_{e0}\, d_0 K_{0e})^{-1} \tag{5.25}$$

The last bracket is the determinant of $1 - 8\pi s^{\frac{1}{2}}\, dK$ for this case. (25) is equivalent to (4-8.22) with

$$K_{0e} = (8\pi s^{\frac{1}{2}})^{-1} T_S(R \to ab), \qquad \Gamma_e = \sum_f \Gamma_f, \tag{5.26}$$

and $\Gamma_f$ defined as in (4-4.8). On the other hand, for $(m_e^2 - s)K_{00} \gg 8\pi s^{\frac{1}{2}} K_{0e} K_{e0}$ (24) reduces to (12) with $K = K_{00}$, which shows that $K_{00}$ is identical with the $K$-matrix for the "background amplitude" outside the "resonance." Note that $K_{0e}$ remains real also in the general case, such that (26) really refers to the Born approximation of $T_s$ as in table 4-9 (note the difference to table 4-11). Next, we consider (24) for two open channels, $0 = 1$ and 2. For simplicity we take $K_{00}$ in diagonal form. Denoting the denominator of (24) by $x$, we need

$$\begin{aligned} \det x = {} & (m_e^2 - s)[(m_e^2 - s)(1 - ip_1 K_{11})(1 - ip_2 K_{22}) \\ & - im_e \Gamma_1 (1 - ip_2 K_{22}) - im_e \Gamma_2 (1 - ip_1 K_{11})] \\ x^{-1} = {} & (\det x)^{-1} \begin{pmatrix} x_{22} & -x_{12} \\ -x_{21} & x_{11} \end{pmatrix} \\ T_{11} = {} & \left[ K_{11}(m_e^2 - s) + 8\pi s^{\frac{1}{2}} K_{1e}^2 - \frac{im_e \Gamma_2 K_{22}}{1 - ip_2 K_{22}} \right] \\ & \times \left[ (m_e^2 - s)(1 - ip_1 K_{11}) - im_e \Gamma_1 - im_e \Gamma_2 \frac{1 - ip_1 K_{11}}{1 - ip_2 K_{22}} \right]^{-1} \end{aligned} \tag{5.27}$$

Thus, even in this simple example, the numerator is complex, and the denominator contains $\Gamma_2$ multiplied by a complex coefficient. If we also put $K_{22} = 0$, we get

$$T_{11} = [K_{11}(m_e^2 - s) + 8\pi s^{\frac{1}{2}} K_{1e}^2][(m_e^2 - s - im_e \Gamma_2)(1 - ip_1 K_{11}) - im_e \Gamma_1]^{-1}. \tag{5.28}$$

## 6-6 Analyticity in Two Variables. Mandelstam Representation

So far we have investigated the scattering amplitude $T(s, t)$ as a function of complex $s$ for physical $t$ and as a function of complex $t$ for physical $s$. We now study $T$ as a function of two complex variables. We start from the Cauchy integral formula corresponding to the dispersion relation (1.21) in $s$,

$$T(s, t) = \frac{R_s}{s - m_e^2} + \frac{R_u}{u - m_u^2} + \frac{1}{2\pi i}\left[\int_{s_0}^{\infty} \frac{\Delta_s(s', t)}{s' - s}\, ds' + \int_{u_0}^{\infty} \frac{\Delta_u(u', t)}{u' - u}\, du'\right], \tag{6.1}$$

$$\begin{aligned} \Delta_s(s, t) &= T(s + i\varepsilon, t) - T(s - i\varepsilon, t), \\ \Delta_u(u, t) &= T_u(u + i\varepsilon, t) - T_u(u - i\varepsilon, t). \end{aligned} \tag{6.2}$$

$R_s = -G_{abe}\, G_{cde}$ is the residuum of the $s$-channel pole analogous to (3.2). For $t$ in the physical $s$-channel region, we have of course $\Delta_s = 2i \operatorname{Im} T(s + i\varepsilon, t)$ as in (1.21), but this relation is no longer true when the first singularity in $t$ is reached. Instead, the analytic continuation of $\Delta_s$ is defined via the unitarity relation (1.8)

$$\begin{aligned} \Delta_s(s, t) &= i \sum_k \int d \text{ Lips } (s; k) T_{ik}(s, t') T_{jk}^*(s, t'') \\ &= i \sum_k p_k s^{-\frac{1}{2}} \int \frac{d\Omega_k}{16\pi^2}\, T_{ik}(s, t') T_{jk}^*(s, t''). \end{aligned} \tag{6.3}$$

In the last expression, we have assumed that $s$ is small enough that only 2-particle channels contribute to $k$. We can now insert dispersion relations of the type (3.1) for $T_{ik}(s, t')$ and $T_{jk}^*(s, t'')$:

$$T_{ik}(s, t) = \frac{R_{ik}}{t - m_e^2} + \frac{R_{uk}}{u - m_u^2} + \frac{1}{2\pi i}\left[\int_{t_k}^{\infty} \frac{\Delta_{tk}(s, t_1)}{t_1 - t}\, dt_1 + \int_{u_k}^{\infty} \frac{\Delta_{uk}(s, u_1)\, du_1}{u_1 - u}\right] \tag{6.4}$$

Insertion into (3) gives (at least for elastic scattering, $i = j$)

$$\begin{aligned} \Delta_s(s, t) = i \sum_k p_k s^{-\frac{1}{2}} \int \frac{d\Omega_k}{16\pi^2} & \left\{\frac{R_{tk}}{t' - m_e^2} + \frac{R_{uk}}{u' - m_u^2} + \int_{t_k}^{\infty} \frac{\Delta_{tk}(s, t_1)\, dt_1}{2\pi i(t_1 - t')} + \int_{u_k}^{\infty} \frac{\Delta_{uk}(s, u_1)\, du_1}{2\pi i(u_1 - u')}\right\} \\ \times & \left\{\frac{R_{tk}}{t'' - m_e^2} + \frac{R_{uk}}{u'' - m_u^2} - \int_{t_k}^{\infty} \frac{\Delta_{tk}(s, t_2)\, dt_2}{2\pi i(t_2 - t'')} - \int_{u_k}^{\infty} \frac{\Delta_{uk}(s, u_2)\, du_2}{2\pi i(u_2 - u'')}\right\}. \end{aligned} \tag{6.5}$$

When the curly brackets are multiplied term by term, one gets a large number of terms, which are illustrated by unitarity diagrams. The combination of the poles with each other leads to "square diagrams" of the types

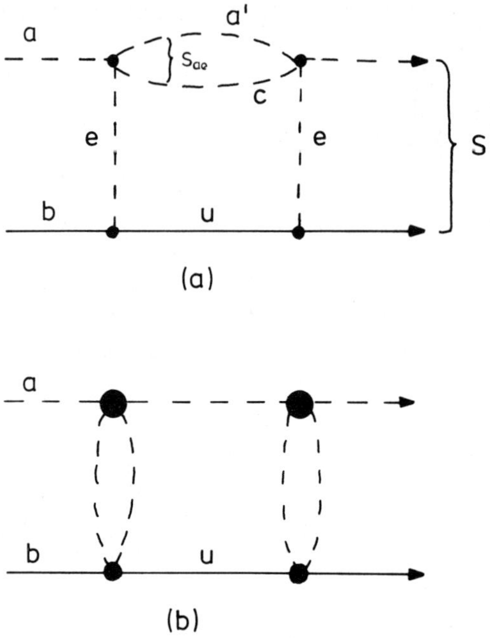


**Figure 6-6** Graphs which determine the limit (17) of the double spectral function $\rho_{st}$. For $\pi N$ scattering, the dashed line refers to pions and the full line to the nucleon.

shown in fig. 6-6. All combinations contain an integral over the solid angle $\Omega_k$ of the intermediate momentum $\mathbf{p}_k$, of the type

$$\begin{aligned} I_{ijk} &= \pi^{-1} p_k \int d\Omega_k (m_e^2 - t')^{-1} (m_e^2 - t'')^{-1} \\ &= (4\pi p_i p_j p_k)^{-1} \int d\Omega_k (z_{ik} - x')^{-1} (z_{jk} - x'')^{-1}, \end{aligned} \tag{6.6}$$

$$x' \equiv \hat{\mathbf{p}}_i \hat{\mathbf{p}}_k, \qquad x'' \equiv \hat{\mathbf{p}}_k \hat{\mathbf{p}}_j, \tag{6.7}$$

where $z_{ik}$ and $z_{jk}$ are defined as in (3.4). The integral is a bit tricky and is best evaluated by Feynman's "parametrization method"

$$a^{-1} b^{-1} = \int_0^1 dy [ay + b(1-y)]^{-2}. \tag{6.8}$$

We then have

$$\int d\Omega_k [z_{ik} y + z_{jk}(1-y) - \hat{\mathbf{p}}_k \hat{q}]^{-2} = 4\pi \{[z_{ik} y + z_{jk}(1-y)]^2 - q^2\}^{-1}; \tag{6.9}$$

$$\hat{q} \equiv y \hat{\mathbf{p}}_i + (1-y) \hat{\mathbf{p}}_j, \qquad q^2 = 2y^2 - 2y + 2xy(1-y). \tag{6.10}$$

(9) is easily derived by taking the $z$-axis along the constant vector $\hat{q}$, such that $\hat{q}\hat{\mathbf{p}}_k = q \cos\theta_k$, $d\Omega_k = d\cos\theta_k\, d\phi_k$ and $\int dx (a + qx)^{-2} = (a^2 - q^2)^{-1}$.

The curly bracket in (9) is a second order polynomial in $y$, and the integral over the auxiliary variable $y$ in (8) is easily performed:

$$\begin{aligned} p_i p_k p_j I_{ijk} &= \int_0^1 dy(cy^2 + 2\,dy + z_{jk}^2 - 1)^{-1} \\ &= \tfrac{1}{2}K^{-\frac{1}{2}} \ln \frac{cy + d - K^{\frac{1}{2}}}{cy + d + K^{\frac{1}{2}}}\bigg|_{y=0}^{y=1} \\ &= \tfrac{1}{2}K^{-\frac{1}{2}} \ln \frac{d + K^{\frac{1}{2}} + z_{jk}^2 - 1}{d - K^{\frac{1}{2}} + z_{jk}^2 - 1}, \end{aligned} \tag{6.11}$$

$$c \equiv (z_{ik} - z_{jk})^2 - 2 + 2x, \qquad d \equiv z_{ik} z_{jk} - z_{jk}^2 + 1 - x,$$

$$K \equiv d^2 - (z_{jk}^2 - 1)c = x^2 + z_{ik}^2 + z_{jk}^2 - 2x z_{ik} z_{jk} - 1.$$

We now study the singularities of (11). For that purpose, we write

$$K = (x - z_+)(x - z_-), \qquad z_\pm = z_{ik} z_{jk} \pm (z_{ik}^2 - 1)^{\frac{1}{2}}(z_{jk}^2 - 1)^{\frac{1}{2}}, \tag{6.12}$$

$$p_i p_j p_k I_{ijk} = \tfrac{1}{2}K^{-\frac{1}{2}} \ln\left[(\tfrac{1}{2}z_+ + \tfrac{1}{2}z_- - x + K^{\frac{1}{2}})/(\tfrac{1}{2}z_+ + \tfrac{1}{2}z_- - x - K^{\frac{1}{2}})\right]. \tag{6.13}$$

The factor $K^{\frac{1}{2}}$ has square root branch points at $x = z_\pm$. We put the corresponding cut from $z_-$ to $z_+$ along the real axis, with $K^{\frac{1}{2}} > 0$ for $x > z_+$ and for $x < z_-$. Due to the $K^{\frac{1}{2}}$ in the argument, the logarithm has the same branch points $z_\pm$ plus a third one at $x = \infty$ where it goes as $\ln(z^{-2})$. For $x < z_-$, the argument of the logarithm is real $> 0$ and for $x \to z_-$, the logarithm goes as $4(z_- - x)^{\frac{1}{2}}(z_+ - z_-)^{-\frac{1}{2}}$, such that the right-hand side of (13) goes in fact to $2(z_+ - z_-)^{-1}$ which is finite. Therefore, $I_{ijk}$ is regular at $x = z_-$. For $z_+ > x > z_-$, we have $K^{\frac{1}{2}} = -i\kappa^{\frac{1}{2}}$.

$$\begin{aligned} p_i p_j p_k I_{ijk} &= \frac{1}{2i\kappa^{\frac{1}{2}}} \ln \frac{1 + i\kappa^{\frac{1}{2}}(\frac{1}{2}z_+ + \frac{1}{2}z_- - x)^{-1}}{1 - i\kappa^{\frac{1}{2}}(\frac{1}{2}x_+ + \frac{1}{2}z_- - x)^{-1}} \\ &= \kappa^{-\frac{1}{2}} \arctan \frac{2\kappa^{\frac{1}{2}}}{z_+ + z_- - 2x}. \end{aligned} \tag{6.14}$$

The arc tan reaches $\pi/2$ at $x = \frac{1}{2}z_+ + \frac{1}{2}z_- = z_{ik} z_{jk}$ and $\pi$ at $x = z_+$, and (14) diverges as $\kappa^{-\frac{1}{2}}\pi$ at $x = z_+$. This is the point at which (3) gets a real part. In the following, we restrict ourselves to elastic scattering where we have $p_i = p_j \equiv p$,

$$z_{ik} = z_{jk} = [E_b E_u - \tfrac{1}{2}(m_b^2 + m_u^2 - m_e^2)]/pp_k \equiv z_k, \tag{6.15}$$

$d = 1 - x$ in (11), $z_+ = 2z_k^2 - 1$, $z_- = 1$ in (12), and with $I_{ijk} \equiv I_k$,

$$p^2 p_k I_k = \tfrac{1}{2}K^{-\frac{1}{2}} \ln \frac{z_k^2 - x + K^{\frac{1}{2}}}{z_k^2 - x - K^{\frac{1}{2}}}, \qquad K = (x - 1)(x + 1 - 2z_k^2) \tag{6.16}$$

The singularity appears in $t$ at

$$t(x = z_+) = -2p^2(1 - z_+) = 4p^2(z_k^2 - 1). \tag{6.17}$$

If the lowest state in $k$ is elastic as in $NN$ scattering, we have $z_k = 1 + m_e^2/2p^2$ as in (3.8), and

$$t(x = z_+) = 4m_e^2(1 + \tfrac{1}{4}m_e^2/p^2). \tag{6.18}$$

With $m_e - 2m_\pi$, this formula applies also to $\pi N$ scattering, since the $3\pi$-vertex is parity-forbidden. The corresponding graph is illustrated in fig. 6-6b. For ineastic states $k$ (graph ($a$) of fig. 6-6), insertion of (15) into (17) gives

$$tp_k^2 = 4[m_b^2 m_u^2 + m_b^2 p_k^2 + m_u^2 p^2 - E_b E_u(m_b^2 + m_u^2 - m_e^2)] + (m_b^2 + m_u^2 - m_e^2)^2, \tag{6.19}$$

which can be rewritten in terms of $s$ and $s_{ae}$ $(sp_k^2 = \tfrac{1}{4}\lambda(s, m_u^2, s_{ae}))$:

$$\begin{aligned}(t - 4m_e^2)sp_k^2 = s_{ae}(m_b^2 - m_e^2 - m_u^2)^2 + m_u^2(s_{ae} + m_e^2 - m_a^2)^2 \\ - (s - m_u^2 - s_{ae})(s_{ae} + m_e^2 - m_a^2)(m_b^2 - m_e^2 - m_u^2)\end{aligned} \tag{6.20}$$

For $\pi N$ scattering, the lowest value of $s_{ae}$ is $4m_\pi^2$, and of course $m_b = m_u = m_N$. For large enough $s$, the curve extends to lower $t$-values than (18). Limiting curves for crossed-channel singularities are given by Frazer and Fulco (1960).

Quite generally, it is clear that the fuction $\Delta_s(s, t)$ defined by (3) is an analytic function of $t$, with cuts along the real $t$-axis starting at a function $b(s)$ from direct channel unitarity and $b_u(s)$ from crossed-channel unitarity. Its asymptotic behaviour for $t \to \infty$ is unknown, however. If it is polynomial-bounded, we can write down dispersion relations for $\Delta_s$, and for simplicity we neglect possible subtraction constants:

$$\Delta_s(s', t) = \frac{1}{2\pi i}\left[\int_{b(s)}^{\infty} \frac{dt'}{t' - t}\, \Delta_{st}(s', t') + \int_{b_u(s)}^{\infty} \frac{du''}{u'' - u'}\, \Delta_{su}(s', u'')\right], \tag{6.21}$$

$$u' = \sum m^2 - s' - t$$

where $\Delta_{st}$ is the discontinuity of $\Delta_s$ across the $t$-channel cut. Since $\Delta_s$ is purely imaginary below the cut, $\Delta_{st}$ is purely real. It is normally denoted by $-4\rho_{st}(s, t')$, where $\rho$ is called the "double spectral function" (remember that $\Delta_{st}$ is the discontinuity in $t$ of the discontinuity in $s$). A corresponding relation is written down for $\Delta_u(u, t)$:

$$\Delta_u(u', t) = \frac{2i}{\pi}\left[\int_{c(u)}^{\infty} \frac{dt'}{t' - t}\rho_{ut}(u', t') + \int_{c_s(u)}^{\infty} \frac{ds''}{s'' - s'}\rho_{us}(u', s'')\, ds''\right]. \tag{6.22}$$

Inserting (21) and (22) into (1) and neglecting the poles, we obtain

$$\begin{aligned}T = \pi^{-2}\iint \Bigg[&\frac{ds'\,dt'}{(s' - s)(t' - t)}\rho_{st}(s't') + \frac{ds'\,du''}{(s' - s)(u'' - u')}\rho_{su}(s', u'') \\ &+ \frac{du'\,dt'}{(u' - u)(t' - t)}\rho_{ut}(u't') + \frac{du'\,ds''}{(u' - u)(s'' - s')}\rho_{us}(u', s'')\Bigg].\end{aligned} \tag{6.23}$$

Since the order of integration does not matter, we have $\rho_{us}(u', s'') = \rho_{su}(s'', u')$. The last integral can be rewritten as $ds'\, du''\rho_{su}(s', u'')(u'' - u)^{-1} \times (s' - s'')^{-1}$ with $s'' = \sum m^2 - t - u''$. Eliminating also $u'$ in favour of $s'$ in (21), we can combine the second and fourth integrals in (23):

$$(u'' - \sum m^2 + t + s')^{-1}\left(\frac{1}{s' - s} + \frac{1}{u'' - u}\right) = \frac{1}{s' - s}\frac{1}{u'' - u},$$

$$T = \pi^{-2} \iint \left[\frac{ds'\, dt'}{(s' - s)(t' - t)} \rho_{st}(s't') + \frac{dt'\, du'}{(t' - t)(u' - u)} \rho_{tu}(t', u') + \frac{ds'\, du''}{(s' - s)(u'' - u)} \rho_{su}(s', u'')\right]. \tag{6.24}$$

This form is due to Mandelstam (1958) and is called the Mandelstam representation. Its practical use depends of course on the asymptotic behavior of $\rho$ when both arguments are large. When one of the incident particles is composite in the sense of (3.21), eq. (17) still gives singular values of $t$. However, in this case $T(s, t)$ will be singular also in a region of complex $s$ and $t$. In such cases, the Mandelstam representation is not valid.

It is also interesting to find the minimum of $t(s)$ as given by (20). For that purpose, we define

$$m_a^2 - s_{ae} - m_e^2 = 2m_e s_{ae}^{\frac{1}{2}} A, \qquad m_b^2 - m_u^2 - m_e^2 = 2m_e m_u B, \tag{6.25}$$

$$s = (m_u + s_{ae}^{\frac{1}{2}})^2 + 2m_u s_{ae}^{\frac{1}{2}}\, \delta_s, \qquad sp_k^2 = 4m_u^2 s_{ae}\, \delta_s(2 + \delta_s). \tag{6.26}$$

In terms of these expressions, (20) assumes the form ($\eta \equiv 1 - t/4m_e^2$)

$$\begin{aligned} -\eta\, \delta_s(2 + \delta_s) &= (A + B)^2 + 2\, \delta_s AB, \\ \delta_s &= -1 - AB/\eta \pm \eta^{-1}[(\eta - A^2)(\eta - B^2)]^{\frac{1}{2}}. \end{aligned} \tag{6.27}$$

The minima of $t(s)$ occur at the zeros of the square root:

$$\eta_1 = A^2, \qquad \delta_{s1} = -1 - B/A; \qquad \eta_2 = B^2, \qquad \delta_{s2} = -1 - A/B. \tag{6.28}$$

The corresponding $t$-values are

$$t_1 = 4m_e^2(1 - A^2), \qquad t_2 = 4m_e^2(1 - B^2), \tag{6.29}$$

and coincide with the triangle singularity (3.25) according to (3.21), $t_2$ is on the physical sheet only for $B > 0$, and correspondingly for $t_1$. If neither condition is satisfied $t_{\min}$ is just $4m_e^2$.

## 6-7 Analytic Structure of Partial-Wave Amplitudes

So far, we have discussed resonances, the unitarity cut and possible poles of partial-wave amplitudes. Additional singularities arise from the crossing relations. For the scattering of spinless particles, the partial-wave amplitude

$T_L(s)$ is defined by the inversion of (2-5.7). Using (2-5.10),

$$T_L(s) = (16\pi s^{\frac{1}{2}})^{-1} \int_{-1}^{1} T(s, t) P_L(x)\, dx. \tag{7.1}$$

The corresponding formula for spin-$\frac{1}{2}$–spin-0 scattering is given in (2-5.34) and contains some additional "kinematical singularities" which follow from (2.5). However, in the following we consider elastic scattering, in which case (2.5) provides only a $s^{\frac{1}{2}}$-singularity which is also present in (1).

In (1), $s$ is fixed and $t = -2p^2(1-x)$ varies with $x$ between $-4p^2$ and 0. Because of $s + t + u = \sum m^2$, a pole of $T(s, t)$ at $u = m_u^2$ produces a cut in $T_L(s)$, along those values of $s$ that satisfy

$$s - 2p^2(1-x) = 2m_a^2 + 2m_b^2 - m_u^2, \qquad -1 \le x \le 1. \tag{7.2}$$

Its branchpoints are given by the extrema of (2). They occur at $x = \pm 1$ and are denoted by $s_\pm$:

$$s_+ = 2m_a^2 + 2m_b^2 - m_u^2, \qquad s_- = m_u^{-2}(m_a^2 - m_b^2)^2 \tag{7.3}$$

In the second equation we have used $s - 4p^2 = 2(m_a^2 + m_b^2) - s^{-1}(m_a^2 - m_b^2)^2$, which follows from (2-2.9). Similarly, the $u$-channel cut of $T(s, t)$ produces a left-hand cut in $T_L(s)$. To find its branchpoint, we merely replace in $s_+$, $m_u^2$ by the branchpoint $u_0$:

$$\bar{s}_+ = 2m_a^2 + 2m_b^2 - u_0. \tag{7.4}$$

In $\pi\pi$, $\pi N$ and $\bar{K}N$ scattering we have $u_0 = (m_a + m_b)^2$ and therefore

$$\bar{s}_+ = (m_a - m_b)^2 = \bar{s}_-. \tag{7.5}$$

Finally, the $t$-channel cut of $T(s, t)$ (with branchpoint $t_0$, say) produces two cuts in $T_L(s)$, characterized by

$$2p^2 < -t_0(1-x)^{-1} < -\tfrac{1}{2}t_0. \tag{7.6}$$

The values of $s$ corresponding to a given $p^2$ are

$$s = [(m_a^2 + p^2)^{\frac{1}{2}} + (m_b^2 + p^2)^{\frac{1}{2}}]^2 = m_a^2 + m_b^2 + 2p^2 + 2(m_a^2 + p^2)^{\frac{1}{2}}(m_b^2 + p^2)^{\frac{1}{2}}. \tag{7.7}$$

We assume $m_b > m_a$. Then $s$ is negative for $p^2 < -m_b^2$ and reaches zero for $p^2 \to -\infty$, i.e., $T_L(s)$ has a cut for $s < 0$. Since $t_0$ is at most $4m_a^2$, (6) is also satisfied in the interval $-m_b^2 < p^2 < -m_a^2$, in which $s$ is complex:

$$s^{\frac{1}{2}} = (m_b^2 + p^2)^{\frac{1}{2}} \pm i(-m_a^2 - p^2)^{\frac{1}{2}}, \qquad |s| = m_b^2 - m_a^2. \tag{7.8}$$

Thus $T_L(s)$ has also a circular cut in the complex $s$-plane, with radius $m_b^2 - m_a^2$ (see fig. 6-7). It is absent only for $m_a = m_b$. In the simplest models of $\pi N$ scattering, it is neglected.

We now want to construct partial-wave amplitudes which in addition to the threshold behavior (5.4), unitarity (5.9) and a possible resonance or

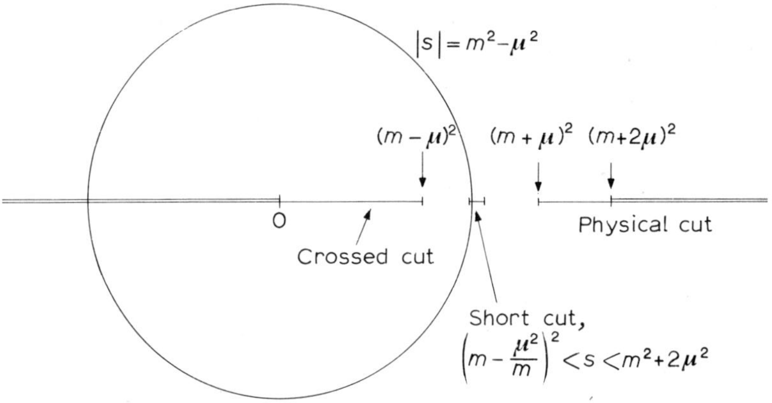


**Figure 6-7** The singularities of the partial-wave amplitudes $f_L \pm (s)$ of $\pi N$-scattering in the complex $s$-plane. $m \equiv m_N$, $\mu \equiv m_\pi$.

bound state also have a given left-hand cut. The threshold behavior is incorporated by considering instead of $T_L$ the matrix

$$F_L(s) = 8\pi s^{\frac{1}{2}} \tilde{p}^{-L} T_L \tilde{p}^{-L} = \frac{1}{\pi}\left(\int_{-\infty}^{\bar{s}_0} + \int_{s_0}^{\infty}\right) \frac{ds'}{s' - s - i\varepsilon} \operatorname{Im} F_L(s'). \quad (7.9)$$

The $8\pi s^{\frac{1}{2}}$ is included to cancel the factor in (1). (9) can be solved by the socalled $N/D$-method of Chew and Mandelstam (1960):

$$F_L = ND^{-1}, \qquad \operatorname{Im} N(s > \bar{s}_0) = \operatorname{Im} D(s < s_0) = 0, \quad (7.10)$$

where both the numerator $N$ and the denominator $D$ matrices are real-analytic functions. $N$ has no right-hand cut and $D$ has no left-hand cut ($D$ is analogous to (4.7) and is called the Jost function). From these definitions, it follows that $N$ satisfies

$$N(s) = \frac{1}{\pi} \int ds'(s' - s - i\varepsilon)^{-1} \operatorname{Im} F_L(s')\, D(s'). \quad (7.11)$$

Im $D$ follows from the unitarity equation (5.9) for the direct channel,

$$\begin{aligned} \operatorname{Im} D_{ij} &= \sum_k \operatorname{Im} (F^{-1})_{ik} N_{kj} = -(8\pi s^{\frac{1}{2}})^{-1} p_i^{2L+1} \theta_i N_{ij} \\ &= -p_i^{2L} \tfrac{1}{2}\, d \operatorname{Lips}(s; i) N_{ij}, \end{aligned} \quad (7.12)$$

and the dispersion relation for $D_{ij}$ assumes the form

$$D_{ij}(s) = \sum_p \frac{\gamma_{ijp}}{s - s_p} - \tfrac{1}{8}\pi^{-2} \int_{s_i}^{\infty} ds'(s' - s - i\varepsilon)^{-1} p_i(s')^{2L+1}(s')^{-\frac{1}{2}} N_{ij}. \quad (7.13)$$

The possible poles in (13) at $s_p < s_i$ are called CDD-poles after their inventors (Castillejo, Dalitz, Dyson 1956) and correspond to zeros of $F$, $F(s_p) = 0$. They also produce nearby zeros of $D$ and thereby poles in $F$. The connection to the $K$-matrix is best given via (5.10), $T^{-1} = K^{-1} - 8\pi s^{\frac{1}{2}} d$:

$$DN^{-1} = \tilde{p}^L (K^{-1}(8\pi s^{\frac{1}{2}})^{-1} - d)\tilde{p}^L. \quad (7.14)$$

In practice, one can only include a small number of 2-particle channels in the matrices $K$, $d$, $D$ and $N$, and unitarity will always be violated at high energies. Since the integral in (13) extends over all energies, one frequently contends oneself with introducing the $N/D$-method for the elastic channel alone

$$F_{L,\,\mathrm{el}} = N/D, \qquad \mathrm{Im}\, D = -(8\pi s^{\frac{1}{2}})^{-1} p^{2L+1} R N, \qquad R \equiv -p^{-1}\, \mathrm{Im}\, (T_{L,\,\mathrm{el}})^{-1}. \tag{7.15}$$

$R$ must lie between 1 and 2. To see this, we recall the inequalities which partial-wave unitarity (4-8.25) imposes on the elastic channel

$$T_{L,\,\mathrm{el}} = (2ip)^{-1}(S_{\mathrm{el},\,L} - 1), \qquad T_{i\neq f,\,L} = (2ip_i^{\frac{1}{2}} p_f^{\frac{1}{2}})^{-1} S_{i\neq f,\,L},$$
$$|S_{\mathrm{el},\,L}|^2 + \sum_f |S_{i\neq f,\,L}|^2 = 1. \tag{7.16}$$

Obviously, $S_{\mathrm{el},\,L}$ can be parametrized by an "inelasticity" $\eta_L$ and real $\delta_L$:

$$S_{\mathrm{el},\,L} = \eta_L e^{2i\,\delta_L}, \qquad 0 \le \eta_L \le 1, \qquad T_{\mathrm{el},\,L} = (2ip)^{-1}(\eta_L e^{2i\,\delta_L} - 1), \tag{7.17}$$

and $\sum_f |S_{i\neq f,\,L}|^2 = 1 - \eta_L^2$. The elastic and "reaction" partial-wave cross sections (2-5.15) become

$$\sigma_{\mathrm{el},\,L} = 4\pi(2L+1)|T_{\mathrm{el},\,L}|^2 = \pi p^{-2}(2L+1)|1 - \eta_L e^{2i\,\delta_L}|^2, \tag{7.18}$$
$$\sigma_{r,\,L} = 4\pi(2L+1)p^{-1} \sum_{f\neq i} p_f\, |T_{if,\,L}|^2 = \pi p^{-2}(2L+1)(1 - \eta_L^2). \tag{7.19}$$

The sum of these two gives the total cross section

$$\sigma_{\mathrm{tot},\,L} = 2\pi(2L+1)(1 - \eta_L \cos 2\,\delta_L), \tag{7.20}$$

and the partial-wave unitarity (2-5.11) from which we started can be rewritten as

$$\mathrm{Im}\, T_{\mathrm{el},\,L} = [4\pi(2L+1)]^{-1} p\, \sigma_{\mathrm{tot},\,L}. \tag{7.21}$$

All these formulas are well-known from the nonrelativistic theory. The parameter $R$ of (15) becomes

$$R = p^{-1}\, \mathrm{Im}\, T_L |T_L|^{-2} = \sigma_{\mathrm{tot},\,L}/\sigma_{\mathrm{el},\,L}, \tag{7.22}$$

and is easily seen to satisfy $1 \le R \le 2$.

For proper coupled 2-particle channels, the $K$-matrix method is easier to handle than the multichannel $N/D$-method. With (5.11) as the only restriction on the diagonal matrix $d$, one can clearly define $d$ such that it contains the whole $s$-channel unitarity cut and no crossed-channel cut (Lang 1975). The threshold behavior (5.13) is incorporated by writing (5.10) as

$$F_L \equiv 8\pi s^{\frac{1}{2}} \tilde{p}^{-L} T_L \tilde{p}^{-L} = 8\pi s^{\frac{1}{2}} A_L (1 - p^L\, d\, p^L 8\pi s^{\frac{1}{2}} A_L)^{-1}, \tag{7.23}$$

where $s^{\frac{1}{2}}A_L$ is analytic at $s = 0$. We must therefore define $d_{L_j} = p_j^{2L}\, d_j$ such that it (i) fulfills the unitarity condition (5.11) everywhere and (ii) is analytic:

$$\operatorname{Im} d_{L_j} = p_j^{2L+1}(8\pi s^{\frac{1}{2}})^{-1}\theta(s - s_j), \qquad \operatorname{Re} d_{L_j} = \frac{P}{\pi}\int_{s_j}^{\infty}\frac{ds' p_j(s')^{2L+1}}{(s'-s)8\pi s'^{\frac{1}{2}}}, \quad (7.24)$$

$$p_j(s) = \tfrac{1}{2}s^{-\frac{1}{2}}(s - s_j)(s - \bar{s}_j), \qquad s_j = (m_a + m_b)^2, \qquad \bar{s}_j = (m_a - m_b)^2. \quad (7.25)$$

The dispersion integral in (24) diverges and must be rewritten with $L + 1$ subtractions. Since the indefinite integral $\int_s^x ds' \ldots$ can be evaluated analytically, one can simply subtract from it the integral for $s = s_1$ and take $x \to \infty$ afterwards. Putting them $\operatorname{Re} d_{L_j}(s_1) = 0$, one obtains for $L = 0$

$$d_{0_j}(s) = f_j(s) - \operatorname{Re} f_j(s_1) \quad (7.26)$$

$$\begin{aligned} f_j(s_j < s) &= -(8\pi^2 s^{\frac{1}{2}})^{-1}\left[\frac{1}{2}s^{-\frac{1}{2}}(m_b^2 - m_a^2)\ln(m_b/m_a)\right.\\ &\qquad\left. + p_j \ln\frac{2p_j s^{\frac{1}{2}} + s - m_a^2 - m_b^2}{2m_a m_b} - i\pi p_j\right] \\ &= (16\pi^2 s)^{-1}\left[(m_b^2 - m_a^2)\ln(m_b/m_a)\right.\\ &\qquad\left. + \lambda^{\frac{1}{2}}\ln\left((m_a^2 + m_b^2 - s - \lambda^{\frac{1}{2}})/2m_a m_b\right)\right]. \end{aligned} \quad (7.27)$$

In the last form we have written $-i\pi$ as $\ln(-1)$ and used the triangle function $\lambda$ of (2-2.9) in order to emphasize the analyticity of $f_j$. In the following, we specialize to the equal-mass case, $m_a = m_b \equiv m$, and also drop the index $j$. The analytic continuations of (27) below threshold are

$$\begin{aligned} f(0 \le s < 4m^2) &= -(8\pi^2)^{-1}(4m^2/s - 1)^{\frac{1}{2}}\tan^{-1}\{[s(4m^2 - s)^{-1}]^{\frac{1}{2}}\},\\ f(s < 0) &= -(8\pi^2)^{-1}(1 - 4m^2/s)^{\frac{1}{2}}\ln \\ &\quad \times\{1 - 2m^2/s + (1 - 4m^2/s)^{\frac{1}{2}}(-m^2 s)^{-\frac{1}{2}}\}. \end{aligned} \quad (7.28)$$

For threshold expansions, it is convenient to take $s_1 = 4m^2$, because of $f(4m^2) = 0$. We then have

$$d(s > 4m^2) = -(4\pi^2 s^{\frac{1}{2}})^{-1}p\ln(p/m + s^{\frac{1}{2}}/2m) + ip(8\pi s^{\frac{1}{2}})^{-1}. \quad (7.29)$$

Now $s^{\frac{1}{2}}A_L$ of (23) is expanded as follows:

$$s^{\frac{1}{2}}A_L = 2m(a_L + b_L p^2), \quad (7.30)$$

$$T_L = p^{2L}A_L[1 - 16\pi\, d\, p^{2L}m(a_L + b_L p^2)]^{-1}. \quad (7.31)$$

A factor $2m$ has been included in (30) such that $a_L$ coincides with the old scattering length of (5.7) at $s^{\frac{1}{2}} = 2m$. Inserting (29) for $d$, one sees that (31) reduces to the scattering length approximation (5.4) at threshold.

For $L = 1$, a twice-subtracted dispersion relation must be used. Starting from (3-11.22), we find

$$f(z) = \frac{z - z_2}{z_1 - z_2} f(z_1) - \frac{z - z_1}{z_1 - z_2} f(z_2) + \frac{(z - z_1)(z - z_2)}{2\pi i} \oint \frac{dz' f(z')}{(z' - z)(z' - z_1)(z' - z_2)}. \tag{7.32}$$

In view of the later application to the electric form factor of the pion, it is convenient to put $z_1 = 0$, because this permits incorporation of the condition $F_\pi(0) = 1$. The second subtraction point is arbitrary. However, if there is an unstable particle state at $s = m_e^2$, one should require

$$\text{Re } d_{1j}(s = m_e^2) = 0, \tag{7.33}$$

because otherwise the position of the resonance gets shifted from $m_e^2$ to some other value $m_R^2$, according to (5.22) and (5.25). For the $\rho$-resonance in $\pi\pi$-scattering for example, we have $T_S = 3^{-\frac{1}{2}} 2Gp$ (table 4-9),

$$K_{0e} = (8\pi s^{\frac{1}{2}})^{-1} 3^{-\frac{1}{2}} 2Gp, \tag{7.34}$$

according to (5.26) and (in the absence of a background amplitude)

$$F_1 = \tfrac{4}{3} G^2 (m_e^2 - s - \tfrac{4}{3} d_1 G^2)^{-1} = \tfrac{4}{3} G^2 (m_e^2 - s - \tfrac{4}{3} G^2 \text{ Re } d_1 - im\Gamma(s))^{-1}. \tag{7.35}$$

Since the resonance position is defined by the vanishing of the real part of the denominator in this case, we would thus get $m_\rho^2 = m_e^2 - \frac{4}{3} G^2 d_1(m_\rho^2)$.

In the approximation (34) with constant $G$, (35) is equivalent to the $N/D$-method (10) with $N =$ constant and no CDD-poles. The only difference between $D$ and the denominator of (35) is that one defines $D(0) = 1$. One thus finds (Gounaris and Sakurai 1968):

$$D(s) = [m_\rho^2 - s - i m_\rho^2 \Gamma_\rho p^3 (s^{\frac{1}{2}} p_\rho^3)^{-1} \theta(s - 4m_\pi^2) + l(s) m_\rho \Gamma_\rho] [m_\rho^2 + l(0) m_\rho \Gamma_\rho]^{-1}, \tag{7.36}$$

$$\begin{aligned} l(s) &\equiv m_\rho p^2 p_\rho^{-3} \left[ h(s) - h_\rho + p_\rho^2 p^{-2} \left( \frac{dh}{ds} \right)_\rho (m_\rho^2 - s) \right], \\ h &\equiv (2\pi s^{\frac{1}{2}})^{-1} p \ln (p/m_\pi + s^{\frac{1}{2}}/2m_\pi), \\ l(0) &= \pi^{-1} m_\pi^2 p_\rho^{-2} \left( 3 \ln \frac{m_\rho + 2p_\rho}{2m_\pi} + \tfrac{1}{2} m_\rho p_\rho / m_\pi^2 - m_\rho / p_\rho \right) = 0.48, \end{aligned} \tag{7.37}$$

where all quantities with an index $\rho$ are taken at $s = m_\rho^2$. Near $m_\rho^2$, $l$ is proportional to $(m_\rho^2 - s)^2$ and therefore negligible, but near threshold it introduces important modifications.

## 6-8 Electromagnetic Form Factors of Hadrons

The principle for the calculation of electromagnetic form factors of hadrons follows from sections 3-11 and 3-12: One starts with the unitarity equation (3-11.5) for $e\bar{e} \to b\bar{b}$, separates the $e\bar{e}$- and $\gamma\gamma$-contributions (3-11.6) and (3-11.20), and arrives at a unitarity equation for only hadronic intermediate states $k$, in which both $T_{ij}$ ($i = e\bar{e}$, $j = b\bar{b}$) and $T_{ik}$ can be approximated by one-photon exchange. For example, when $b$ is a spin-$\frac{1}{2}$ nucleon or nucleus, $T_{ij}$ contains the hadron spinors $u$ and $\bar{v}$ in the combination $\bar{v}0^{\mu}u$, with $0^{\mu}$ given by (2-7.11), and $T_{kj}$ is of the more general form of (A-3.1). The resulting unitarity equation in the photon rest frame is

$$\gamma \text{ Im } G_M(t) - m_b^{-1}\mathbf{p}_b \text{ Im } F_2(t) = \frac{1}{2}\sum_k \int d \text{ Lips } (t; k)\bar{\mathbf{J}}(k)Q(k \to b\bar{b}) \quad (8.1)$$

and allows the separate evaluation of Im $G_M$ and Im $F_2$. The form factors for $t < 0$ are then obtained from dispersion relations:

$$F_2(t) = \pi^{-1}\int_{t_0}^{\infty} dt'(t'-t)^{-1} \text{ Im } F_2(t'), \quad (8.2)$$

$$F_1(t) = G_M - F_2 = Z + \frac{t}{\pi}\int_{t_0}^{\infty} dt'[t'(t'-t)]^{-1} \text{ Im } (G_M(t') - F_2(t')). \quad (8.3)$$

(3) is written with a subtraction to ensure $F_1(0) = Z$. One can also use dispersion relations for $G_M$ and $G_E$ of (2-7.27), but these functions are related by

$$G_M(4m_b^2) = G_E(4m_b^2), \quad (8.4)$$

because otherwise $F_1$ and $F_2$ would have poles at $t = 4m_b^2$ according to (2-7.28).

To simplify the unitarity relation (1), one decomposes the form factors $F_i$ of a given isomultiplet into isoscalar and isovector components $F_i^s$ and $F_i^v$. For nucleons, this is done in (4-14.6). The isoscalar part contains contributions from the $I = 0$ vector mesons $\omega$, $\phi$, $\psi$ ... which are evaluated in the stable-particle approximation (1.15). $\mathbf{J}(\omega) = f_{\omega}^{-1}m_{\omega}^2\,\boldsymbol{\varepsilon}^*(M)$ follows from (4-9.18), and the $\omega \to b\bar{b}$ coupling for a real $\omega$ has the same structure as the $\gamma \to b\bar{b}$ coupling (2-7.13). For an arbitrary vector meson $v$,

$$\begin{aligned} Q(v \to b\bar{b}) &= [G_{vb}^V\gamma^{\mu} - \tfrac{1}{2}G_{vb}^T m_b^{-1}\sigma^{\mu\nu}P_{k\nu}]\varepsilon_{\mu}(M) \\ &= -[(G_{vb}^V + G_{vb}^T)\boldsymbol{\gamma} - G_{vb}^T\mathbf{p}_b/m_b]\boldsymbol{\varepsilon}(M), \end{aligned} \quad (8.5)$$

where $G_{vb}^V$ and $G_{vb}^T$ are the "vector" and "tensor" coupling constants of the vector meson to the baryon of given charge. For the coupling to nucleons,

$$G_{\omega p} = G_{\omega n} \equiv G_{\omega}, \qquad G_{\phi p} = G_{\phi n} \equiv G_{\phi}. \quad (8.6)$$

After summation over the spin $M$ using (2-8.9), we thus find for nucleons

$$\gamma \text{ Im } G_M^s - \mathbf{p}_b \text{ Im } F_2^s/m_b$$
$$= \pi\, \delta(t - m_\omega^2) f_\omega^{-1} m_\omega^2 [(G_\omega^V + G_\omega^T)\gamma - G_\omega^T \mathbf{p}_b/m_b] + \dots \quad (8.7)$$

and insertion into (2) and (3) gives a pole at the $\omega$-mass (see fig. 6-8a)

$$F_2^s(t) = f_\omega^{-1}(1 - t/m_\omega^2)^{-1} G_\omega^T + \dots, \quad F_1^s(t) = f_\omega^{-1}(1 - t/m_\omega^2)^{-1} G_\omega^V + \dots \quad (8.8)$$

In practice, however, the analysis proceeds in the opposite direction: one first tries to determine $F_i^v(t)$ and $F_i^s(t)$ for $t < 0$ from electron scattering on protons and neutrons as explained in section 2-7, then includes poles of the type (8) in $F^s$ and determines the constants $G_\omega^V$ and $G_\omega^T$, using $f_\omega^{-1}$ from $e\bar{e} \to \omega \to \pi^+\pi^0\pi^-$ (see Höhler et al., 1976). The proton form factors are best known and reasonably reproduced by a simple "dipole fit":

$$G_{E_p} \approx (1 + \kappa_p)^{-1} G_{M_p} \approx (1 - t/0.71 \text{ GeV}^2)^{-2}. \quad (8.9)$$

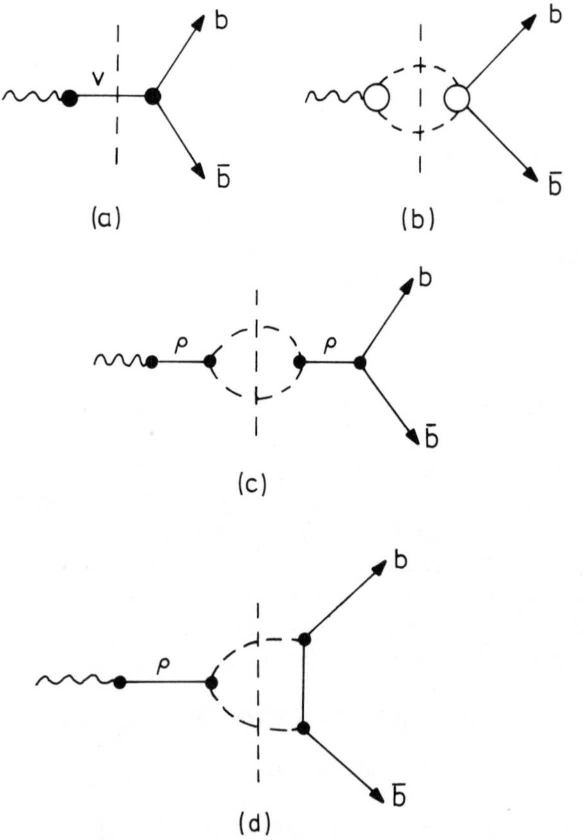


**Figure 6-8** Unitarity diagrams for $\gamma \to b\bar{b}$. (a) The stable-particle approximation for isoscalar vector mesons, (b) the $\pi\pi$-continuum in the isovector part, (c) + (d) insertion of the model amplitude (27) for the $\pi\pi \to b\bar{b}$ amplitude on the right-hand side of (b).

For the neutron, the anomalous magnetic moment $\kappa_n$ is precisely known, and from the scattering of thermal neutrons on electrons, one finds (expanding the curly bracket of (2-7.26) to first order in $t$)

$$F_{1n}(t) = 0 + F'_{1n}t + 0(t^2), \qquad F'_{1n} = (0.17 \pm 0.03) \cdot 10^{-2}\ \text{fm}^2. \quad (8.10)$$

The remaining neutron form factor data come from the analysis of elastic and inelastic $ed$-scattering and are less reliable. Roughly,

$$\kappa_n^{-1} G_{Mn}(t) \approx G_{Ep}(t). \quad (8.11)$$

Under these circumstances, the separation of $F$ into $F^s$ and $F^v$ is not unique, and the values of $G_\omega^V$ and $G_\omega^T$ depend also on the remaining contributions ($\phi$, $\psi$, continuum states) included in (8). A reasonable estimate is

$$G_\omega^{V2}/4\pi \approx 13, \qquad G_\omega^T/G_\omega^V \leq 0.2. \quad (8.12)$$

If one takes $F^s$ as a sum of two poles at the $\omega$ and $\phi$ positions plus a small correction, one gets $G_\phi^V/G_\omega^V < 0$, in order to simulate the dipole fit (9).

In the isovector form factors, the $\rho$-meson can be treated similarly, but its large width and the smallness of the pion mass introduce modifications. We therefore return to the unitarity equation for the electromagnetic form factor of an arbitrary hadron. Inserting (4-4.7) with $S = 1$ for the decay matrix element $T(\hat{\gamma}_M \to b\bar{b})$ of (3-6.13), we obtain in the $t$-channel cms

$$\mathbf{J}(b\bar{b}) = e^{-1} \sum_M \boldsymbol{\varepsilon}^*(M) 3^{\frac{1}{2}} D^1_{M\lambda}(\theta, \phi) T_1(\lambda_b, \lambda_{\bar{b}}), \qquad \lambda \equiv \lambda_b - \lambda_{\bar{b}}. \quad (8.13)$$

Having separated the narrow-resonance contributions, we assume that only 2-hadron intermediate states $k = a\bar{c}$ contribute to the unitarity equation (3-11.5). The strong amplitude $T_{b\bar{b},k}$ is decomposed into partial waves according to (4-4.9), and after the angular integration only the $J = 1$ partial wave amplitude $T_1(t, \lambda_b \lambda_{\bar{b}}, \lambda_a, \lambda_{\bar{c}})$ contributes:

$$\text{Im}\ T_1(\lambda_b, \lambda_{\bar{b}}) = \sum_k p_k \sum_{\lambda_a \lambda_{\bar{c}}} T_1(\lambda_a \lambda_{\bar{c}}) T_1^*(\lambda_b, \lambda_{\bar{b}}, \lambda_a, \lambda_{\bar{c}}). \quad (8.14)$$

The important intermediate states are (for $I = 1$) $k = \pi^+\pi^-$, $K\bar{K}$, $\pi^0\omega$ (plus of course $N\bar{N}$ etc. in the case of nuclei, due to the anomalous thresholds). In the following, we consider only the case $k = \pi^+\pi^-$. We then have $\lambda_a = \lambda_{\bar{c}} = 0$,

$$T_1(\hat{\gamma} \to \pi^+\pi^-) = 3^{-\frac{1}{2}} 2eF_\pi(t) p_\pi \quad (8.15)$$

according to (3-6.18) and table 4-9, with $F_\pi$ = pion form factor. Let us first discuss the case $b = \pi^+$: Below $t = (m_K + m_\omega)^2$,

$$\text{Im}\ F_\pi = p_\pi F_\pi T_1^*(\pi\pi) = F_\pi \exp(-i\,\delta_1) \sin \delta_1, \quad (8.16)$$

where elastic unitarity (2-5.13) has been used for the $p$-wave $\pi\pi$ amplitude $T_1(\pi\pi)$. Since the left-hand side of (16) is real, $F_\pi(t)$ must have the phase $\delta_1$ of $p$-wave $\pi\pi$-scattering. The subtracted dispersion relation for $F_\pi$ is now simply

$$F_\pi(t) = 1 + \pi^{-1} t \int_{4m_\pi^2}^{\infty} [t'(t' - t - i\varepsilon)]^{-1}\, dt'\ \text{Im}\ F_\pi(t'). \quad (8.17)$$

In practice, $F_\pi$ can be identified with $D^{-1}$ of (7.36), because it has the same value at $t = 0$, cut and phase as $D^{-1}$. To be precise, suppose that we know the phase of $F_\pi$ everywhere:

$$F_\pi(t) = |F_\pi(t)| \exp(i\,\delta(t)), \qquad \delta \equiv \delta_1 + \delta', \tag{8.18}$$

where $\delta_1$ is the $\pi\pi$-phase and $\delta'$ a remaining phase. We can define a function

$$G(z) = F_\pi(z)/P(z), \qquad P(0) = 1, \tag{8.19}$$

where $P(z)$ is a polynomial which cancels possible zeros of $F_\pi$. The function $\ln G(z)$ is also analytic in the cut $z$-plane, and its discontinuity across the cut is given by

$$(2i)^{-1}(\ln G(t + i\varepsilon) - \ln G(t - i\varepsilon)) = \arctan(\operatorname{Im} F_\pi/\operatorname{Re} F_\pi) = \delta(t). \tag{8.20}$$

A dispersion relation for $z^{-1} \ln G(z)$ leads to

$$\begin{aligned} \ln G &= \pi^{-1} t \int_{t_0}^{\infty} [t'(t' - t - i\varepsilon)]^{-1}\, dt'\, \delta(t'), \\ F_\pi &= P \exp\left\{\frac{t}{\pi} \int_{t_0}^{\infty} \frac{dt'\,\delta(t')}{t'(t' - t - i\varepsilon)}\right\}. \end{aligned} \tag{8.21}$$

We can thus construct $F_\pi$ from a knowledge of its phase and its zeros. (21) is called the "phase representation" of $F_\pi$. The splitting (18) of the phase is made explicit by using the function of Omnès (1958)

$$\begin{aligned} F_\pi(t) &= P(t)\Omega(t) \exp\left\{\frac{t}{\pi} \int_{(m_\omega + m_\pi)^2}^{\infty} \frac{dt'\,\delta(t')}{(t' - t - i\varepsilon)t'}\right\}, \\ \Omega &\equiv \exp\left\{\frac{t}{\pi} \int_{t_0}^{\infty} \frac{dt'\,\delta_1(t')}{t'(t' - t - i\varepsilon)}\right\}. \end{aligned} \tag{8.22}$$

Since $D^{-1} = F/N$, $D^{-1}$ is zero at the poles of $N$. In the Gounaris-Sakurai model, $N$ is just a constant, and the polynomial $P$ of (19) is 1, at least for $\delta' = 0$. The $\rho$-resonance now appears naturally as part of the $\pi\pi$-continuum. As a very crude approximation, one can of course also use (4-9.26), remembering $\Gamma(t < 4m_\pi^2) = 0$.

We are now prepared to tackle the isovector form factor of baryons (Frazer and Fulco 1960). As before, we only keep the $\pi^+\pi^-$-state in $k$ in (14), see fig. 6-8b:

$$\operatorname{Im} T_1(\tfrac{1}{2}, \pm\tfrac{1}{2}) = 3^{-\frac{1}{2}} 2eF_\pi(t) p_\pi^2 T_\pm^{1*}(t). \tag{8.23}$$

$T_\pm^1$ are the $J = 1$ $\pi\pi \to b\bar{b}$ partial-wave amplitudes as defined in (2.24) and (2.25). The $T_1(\lambda_b, \lambda_{\bar{b}})$ is obtained by writing the matrix element $-e\boldsymbol{\varepsilon}(M)\mathbf{J}_{b\bar{b}}$ for $\hat{\gamma}_M \to b\bar{b}$ as in (13)

$$-e\boldsymbol{\varepsilon}(M)\mathbf{J}_{b\bar{b}} = -e\bar{v}(\lambda_{\bar{b}})(\boldsymbol{\gamma} G_M - \mathbf{p}_b F_2/m_b)u(\lambda_b)\boldsymbol{\varepsilon}(M) = 3^{\frac{1}{2}}\, d_{M\lambda}^1(\theta_t) T_1(\lambda_b, \lambda_{\bar{b}}). \tag{8.24}$$

For the $\gamma \cdot \varepsilon$-part, we can adopt the result from table 4-9, omitting the minus sign from the negative electron charge. The $-\mathbf{p}_b \cdot \varepsilon$-part is evaluated as in (2.21) with $B = 0$, $A = e\mathbf{p}_b \varepsilon F_2/m_b$. From (2.23) we learn that it contributes only for $\lambda_b = \lambda_{\bar{b}}$, i.e., $\lambda = 0$. It then suffices to consider $M = 0$, with $d^1_{00} = \cos\theta_t$, $\mathbf{p}_b \cdot \varepsilon(0) = p_b \cos\theta_t$. One thus finds

$$T_1(\lambda_b \lambda_{\bar{b}}) = 3^{-\frac{1}{2}} 2e[(m_b G_M + p_b^2 F_2 m_b^{-1})\, \delta_{\lambda_b \lambda_{\bar{b}}} + 2^{-\frac{1}{2}} t^{\frac{1}{2}} G_M\, \delta_{\lambda_b, -\lambda_{\bar{b}}}]. \quad (8.25)$$

With $p_b^2 = \frac{1}{4}t - m_b^2$, the coefficient of $\delta_{\lambda_b\lambda_{\bar{b}}}$ in the square bracket is just $m_b$ times $G_E$ of (2-7.27). We can thus rewrite (23) as

$$\operatorname{Im} G_E = F_\pi p_\pi^2 m_b^{-1} T_+^{1*}, \qquad \operatorname{Im} G_M = 2^{\frac{1}{2}} F_\pi p_\pi^2 t^{-\frac{1}{2}} T_-^{1*}. \quad (8.26)$$

The imaginary parts of $F_1$ and $F_2$ can be obtained from (26). Now we come to the most difficult part, namely the $P$-wave $\pi\pi \to b\bar{b}$ amplitudes $T^1_\pm$ (Nielsen, Oades 1972). From (23) and (18), we see that $T^1_\pm$ has the phase $\delta_1$ of elastic $\pi\pi$-scattering for $t < (m_\omega + m_\pi)^2$. If $\pi\pi$-scattering is approximated by the $\rho$-pole, one may assume

$$T^1_\pm \approx T_\pm(\rho) + e^{i\,\delta_1(\rho)} T^{1(u)}_\pm \quad (8.27)$$

where $T_\pm(\rho)$ is the $\rho$-exchange contribution. From (4-8.22),

$$T_\pm(\rho) = (8\pi t^{\frac{1}{2}})^{-1}(m_\rho^2 - t - i m_\rho \Gamma(t))^{-1} T_1(\rho \to \pi\pi) T_\pm(\rho \to b\bar{b}), \quad (8.28)$$

$$T_1(\rho \to \pi\pi) = 3^{-\frac{1}{2}} 2 G_{\rho\pi\pi} p_\pi, \qquad T_+(\rho \to b\bar{b}) = 3^{-\frac{1}{2}} 2 G^E_{\rho b} m_b, \qquad T_-(\rho \to b\bar{b}) = (\tfrac{2}{3})^{\frac{1}{2}} G^M_{\rho b} t^{\frac{1}{2}}. \quad (8.29)$$

$G^E_{\rho b}$ and $G^M_{\rho b}$ are the electric and magnetic $\rho$ "couplings" and can be expressed in terms of the coupling constants $G^V$ and $G^T$ of (5) as

$$G^M = G^V + G^T, \qquad G^E = G^V + \tfrac{1}{4} t m_b^{-2} G^T \quad (8.30)$$

as in (2-7.13) and (2-7.27). $T^{1(u)}_\pm$ is the baryon octet + decuplet exchange contribution, i.e., $N$ and $\Delta$ exchange in the case of nucleons. The octet exchange contribution dominates and has been given in (3.31). It is particularly important for $b = \Sigma^\pm$, where $u = \Lambda$ has an anomalous threshold. But also in the case of nucleons, it is essential because the logarithmic singularity of $Q_0$,

$$t_{\text{tr}} = 4m_\pi^2(1 - m_\pi^2/m_N^2) = 3.91 m_\pi^2 \quad (8.31)$$

occurs very close to the threshold $4m_\pi^2$. Apart from the extra phase in front of $T^{1(u)}_\pm$, the model amplitude (27) corresponds to the unitarity diagrams (c) and (d) of fig. 6-8. Diagram (c) corresponds to diagram (a) for stable vector mesons: Inserting (28) into (23) and using (4-9.26) for $F_\pi$ and $\frac{2}{3} p_\pi^3 t^{-\frac{1}{2}} G^2_{\rho\pi\pi}/4\pi = m_\rho \Gamma(t)$ of table 4-9, we find

$$\begin{aligned} \operatorname{Im} T_1(\tfrac{1}{2}, \pm\tfrac{1}{2}) &= e f_\rho^{-1} m_\rho \Gamma m_\rho^2 [(m_\rho^2 - t)^2 + m_\rho^2 \Gamma^2]^{-1} T_\pm(\rho \to b\bar{b}) \\ &\to e f_\rho^{-1} m_\rho^2 \pi\, \delta(m_\rho^2 - t) T_\pm(\rho \to b\bar{b}) \end{aligned} \quad (8.32)$$

in the limit $m_\rho \Gamma \to 0$, which corresponds exactly to (7).

The isospin and $SU_3$-properties of $vb\bar{b}$ couplings follow from those of $pb\bar{b}$ ($p$ = pseudoscalar meson, table 4-12 and section 5-8) by replacing

$$\pi \to \rho, \qquad K \to K^*, \qquad \eta_8 \to \phi_8 . \tag{8.33}$$

For the nucleons, for example, we have

$$G_{\rho p\bar{p}}^{(V,\,T)} = -G_{\rho n\bar{n}}^{(V,\,T)} \equiv G_\rho^{(V,\,T)}. \tag{8.34}$$

Analysis of the isovector nucleon form factors then leads to

$$(G_\rho^V)^2/4\pi \approx 0.55, \qquad G_\rho^T/G_\rho^V \gg 1. \tag{8.35}$$

(Höhler and Pietarinen 1975).

## 6-9 Soft Pions

In section 5-7 we have split the matrix elements $A_{if}$ of the hyperneutral current operator $a^{(0)\mu}$ into a one-pion-exchange (= OPE) contribution $A_{if}^\mu(\pi)$ (5-7.8) and a remainder $A_{if}^{\mu'}$. The PCAC condition (5-7.14) led us to (5-7.15). In the limit $q_\mu \to 0$ (soft pion), we get

$$J_{\pi if}(P_i = P_f) = 0, \tag{9.1}$$

at least if $A_{if}$ is regular at $q_\mu = 0$. When $i$ and $f$ have equal parities, this is simply a consequence of the pseudoscalar nature of the pion. For (5-7.9), for example, $\bar{u}(P_\mu, M_d)\gamma_5 u(P_\mu, M_b) = 0$. Adler (1965) investigated the case of opposite parities, adding one real pion in the initial or final state. With $f = b'\pi$ or $i = b\pi$, $J_{\pi if}$ is the amplitude for $\pi b \to \pi b'$ with one soft pion (denoted by $\hat{\pi}$), and (1) leads to

$$T(\hat{\pi} b \to \pi b') = T(\pi b \to \hat{\pi} b') = 0. \tag{9.2}$$

This is called Adler's self-consistency condition. The kinematics is that of $b \to \pi' b'$ or $b' \to b\pi$ "decays," but since $b$ and $b'$ are stable against strong pionic decays, the decay momentum must be imaginary and the points (2) are only reached by analytic continuation. Expressed in terms of the Mandelstam variables $s$, $t$, $u$ and the squares of the pion 4-momentum $q_1^2$, $q_2^2$, we have for $m_b = m_b'$

$$T(m_b^2, m_\pi^2, m_b^2, 0, m_\pi^2) = T(m_b^2, m_\pi^2, m_b^2, m_\pi^2, 0) = 0. \tag{9.3}$$

Under $s$-$u$ crossing, $q_1^2$ and $q_2^2$ must be exchanged. The crossing-symmetric and -antisymmetric amplitudes $T^{(\pm)}$ of (1.26) are now defined as

$$T^{(\pm)} = \tfrac{1}{2}T(s, t, u, q_1^2, q_2^2) \pm \tfrac{1}{2}T(u, t, s, q_2^2, q_1^2). \tag{9.4}$$

If we understand that one pion has $q^2 = m_\pi^2$ and the other $q^2 = 0$, we can rewrite (3) as

$$T^{(\pm)}(m_b^2, m_\pi^2, m_b^2) = 0. \tag{9.5}$$

The equation $T^{(-)}(s = u) = 0$ is of course trivial according to (1.26).

When $\hat{\pi} b \to \pi b'$ has a pole at $s = m_b^2$ (which is the case for $b = N, \Sigma, \Xi$), we must use instead of (1) the form

$$\lim_{q_\mu \to 0} q_\mu A_{if}^{\mu'}(b) = f_\pi J_{\pi if}, \tag{9.6}$$

which is obtained from (5-7.15) by keeping the $b$-pole in the matrix element of the axial current operator. Specializing to $i = \pi^- p, f = p'$ for simplicity and putting $m_b = m$,

$$\begin{aligned} A_{if}^{\mu'}(b) &= -\bar{u}'\gamma_5\gamma^\mu f_A(m^2 - s)^{-1}[(P_p + P_\pi)\cdot\gamma + m]2^{\frac{1}{2}}G\gamma_5 u \\ &= -G^2 f_\pi m^{-1} F_{NN\pi}(0)\bar{u}'\gamma^\mu(m^2 - s)^{-1}[(P_p + P_\pi)\cdot\gamma - m]u. \end{aligned} \tag{9.7}$$

The coupling constant $2^{\frac{1}{2}}G$ comes from (5-8.4), and the Goldberger-Treiman relation (5-7.18) has been used in the last form. Writing $q = P_\pi + P_p - P_p'$ and using $\bar{u}'\gamma^\mu P_\mu' = \bar{u}'m$, we find

$$\begin{aligned} q_\mu A_{if}'^{\mu}(b) &= -G^2 f_\pi m^{-1} F_{NN\pi}(0)(m^2 - s)\bar{u}'[(P_\pi + P_p)\cdot\gamma - m]u \\ &= G^2 f_\pi F_{NN\pi}(0)\bar{u}'[m^{-1} + 2(m^2 - s)^{-1}((P_\pi + P_p)\cdot\gamma - m)]u. \end{aligned} \tag{9.8}$$

The singular part of this expression is identical with that of $f_\pi J_{\pi if}$ which can be seen from (2.7) with $G_{eab}^2 = 2G^2$. Let us see now what (6) implies for the invariant amplitudes $A$ and $B$ introduced in (2.1). Using $P_\pi = P_p' + q - P_p$ and the Dirac equation, we have $\frac{1}{2}(P_\pi + q)\cdot\gamma B = q\cdot\gamma B$, which shows that the regular part of $B$ becomes negligible in the limit $q_\mu \to 0$. Thus the only new consequence of (6) is

$$A(s = u = m_b^2, t = m_\pi^2) = (2m_b)^{-1}G_{bb\pi}^2 F_{bb\pi}(0). \tag{9.9}$$

For the crossing-symmetric and -antisymmetric combinations, we find

$$A^{(+)}(m_b^2, m_\pi^2, m_b^2) = (2m_b)^{-1}G_{bb\pi}^2 F_{bb\pi}(0), \qquad A^{(-)}(m_b^2, m_\pi^2, m_b^2) = 0. \tag{9.10}$$

The extrapolation from $q_\mu = 0$ to $(q_0 = m_\pi, \mathbf{q} = 0)$ at fixed $s$ and $u$ is relatively harmless and will be discussed later. The extrapolation in $s$ at fixed $q^2$ and $t$ is conveniently done in the variable $\nu = \frac{1}{4}m_b^{-1}(s\text{-}u)$ of (1.28):

$$A^{(+)}(\nu) = A^{(+)}(0) + \nu^2 \tfrac{1}{2}A^{(+)\prime\prime}(0), \qquad A^{(-)}(\nu) = \nu A^{(-)\prime}(0). \tag{9.11}$$

We thus find that to order $\nu^2$, $A^{(+)}$ is given by (10) also at the threshold of physical $\pi b$-scattering, where $\nu = m_\pi$. The amplitude $B_{\pi N}^{(+)}$ is well approximated by the pole of (2.15) at threshold, which is $-A^{(+)}m_\pi^{-1} + 0(m_\pi/m_p)$. We thus get from (2.17)

$$T^{(+)}(\lambda, \lambda', \omega = m_\pi) = 2m_b\,\delta_{\lambda\lambda'}[A^{(+)}(m_\pi) + m_\pi B^{(+)}(m_\pi)] \approx 0, \tag{9.12}$$

which will be discussed in section 7-2 below. We can also take both pions off-shell, by considering the interaction of two lepton currents with a particle $b$ (fig. 6-9). In that case we have according to (3-4.4) and (5-1.11)

$$S_{aa}^{(2)} = -\tfrac{1}{2}G_V^2 \int d^4x\, d^4y\, l^{+\mu}(x) l^\nu(y) \langle b' | T a_\mu^{(0)}(x) a_\nu^{(0)+}(y) | b \rangle. \tag{9.13}$$

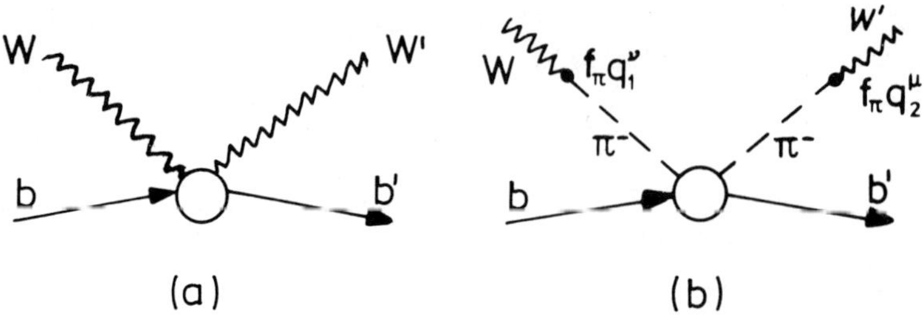


**Figure 6-9** (a) The interaction of a particle $b$ with two lepton currents; (b) its double-pion pole.

Because of the presence of the time-ordering operator $T$ in (13), $\partial^\mu a_\mu^{(0)}(x) = 0$ does not lead to (1) in the limit of one vanishing pion momentum any longer. We first perform the $y$-integration as in (3-4.13), obtaining

$$\begin{aligned} S_{if}^{(2)} &= i(2\pi)^4\, \delta_4(P_b + q_1 - P_b' - q_2)\tfrac{1}{2}G_V^2\, L_2^{*\mu} L_1^\nu T_{\mu\nu}, \\ T_{\mu\nu} &= i \int d^4x e^{iq_2\cdot x}\langle b' | T a_\mu(x) a_\nu^+(0) | b\rangle. \end{aligned} \tag{9.14}$$

In the last expression, we have dropped the "hyperneutrality index" (0) of $a_\mu^{(0)}$. The time-ordering can be rewritten as

$$T a_\mu(x) a_\nu^+(0) = \theta(x_0)[a_\mu(x), a_\nu^+(0)] + a_\nu^+(0) a_\mu(x). \tag{9.15}$$

If we now wish to get at $q_2^\mu T_{\mu\nu}$ from $\partial^\mu a_\mu$ using partial integration, we must also differentiate the $\theta$-function

$$\partial^\mu \theta(x_0) = \delta_{\mu 0}\, \delta(x_0). \tag{9.16}$$

The integral over $x_0$ is now trivial in this extra term, and we find

$$\begin{aligned} q_2^\mu T_{\mu\nu} &= U_\nu - \int d^3r e^{-iq_2 r}\langle b' | [a_0(0, \mathbf{r}), a_\nu^+(0)] | b\rangle, \\ U_\nu &\equiv \int d^4x e^{iq_2\cdot x}\langle b' | T\, \partial^\mu a_\mu(x) a_\nu^+(0) | b\rangle. \end{aligned} \tag{9.17}$$

(17) is called a generalized Ward identity. $U_\nu$ is obviously zero if $\partial^\mu a_\mu = 0$, but the other integral, which is the Fourier transform of the matrix element of the equal-time-commutator of $a_0^+$ and $a_\nu$, remains. This integral is determined by the "current algebra" of Gell-Mann (1962). The result itself is due to Adler (1965) and Weisberger (1965). Our derivation follows that given in the book of de Alfaro et al. (1973).

We are only interested in the limit $\mathbf{q}_2 = 0$, in which case the 3-dimensional Fourier transform in (17) gives just the "axial charge" operator

$$Q_+^5 \equiv \int d^3r a_0(0, \mathbf{r}) = Q_1^5 + iQ_2^5, \tag{9.18}$$

which is analogous to the weak charge operator $Q_+$ of (5-6.3). By CVC, $Q_+$

is identical with the operator $I_+$ which generates isospin rotations. Starting from $[I_+, I_-] = 2I_3$, one finds

$$[Q_+, v_\nu^+(x)] = 2j_\nu^v(x) \tag{9.19}$$

for the commutator between the weak charge and the vector current. ($j_\nu^v$ is the isovector component of the electromagnetic current). Inspired by a quark model in which the hyperneutral current operator is taken as $\bar{\Psi}_u \gamma^\mu(1-\gamma_5)\Psi_d$ in analogy with the lepton current, Gell-Mann postulated

$$[Q_+^5, a_\nu^+(x)] = 2j_\nu^v(x) \tag{9.20}$$

in addition to (19). Inserting this into (17) and taking also $q_1 \to 0$, one obtains

$$\begin{aligned} q_2^\mu T_{\mu\nu} &= -2J_{bb'\nu}^v = -2I_{b3}(P_b + P_b')_\mu\, \delta_{\lambda_b\lambda_{b'}}, \\ q_2^\mu q_1^\nu T_{\mu\nu} &= -2I_{b3}\, 2m_b \nu\, \delta_{\lambda_b\lambda_{b'}}, \qquad 2m_b\nu = 2q_1 \cdot P_b = s - m_b^2. \end{aligned} \tag{9.21}$$

It now remains to evaluate the singular part of $T_{\mu\nu}$ on the left-hand side of (21). The double pion pole of fig. 6-9 contributes

$$q_2^\mu q_1^\nu T_{\mu\nu}(\pi_1\pi_2) = -f_\pi^2 T(\pi_1^- b \to \pi_2^- b'). \tag{9.22}$$

Although we have omitted isospin-indices on $T_{\mu\nu}$, it follows from the fact that $a_\mu(x)$ is charge-raising that our pion carries negative charge. For most baryons, there are also singular contributions from the baryon poles in the direct or crossed channels, depending on the charge of $b$. However, these contributions are identical for $\pi^+$ and $\pi^-$ elastic scattering and therefore contribute only to the charge combination $T^{(+)}$ which was evaluated already in (12). If we therefore restrict ourselves to the antisymmetric charge combination

$$T_b^{(-)} = \tfrac{1}{2}T(\pi^- b) - \tfrac{1}{2}T(\pi^+ b) = 8\pi s^{\frac{1}{2}}\, \delta_{\lambda_b\lambda_{b'}} F^{(-)}, \tag{9.23}$$

we can neglect the baryon poles. The amplitude $F^{(-)}$ was introduced in (2.20). For $\pi p$-scattering at threshold, we obtain

$$F^{(-)}(\nu = m_\pi) = \tfrac{1}{3}(a_1 - a_3), \tag{9.24}$$

where $a_1$ and $a_3$ are the $I = \frac{1}{2}$ and $\frac{3}{2}$ scattering lengths, the particular combination following from (4-2.7) and (4-2.10).

Assuming now that the difference between $T_b^{(-)}$ at threshold ($\nu = m_\pi$) and $T_b^{(-)}$ for massless pions ($\nu = 0$) is given by the square of the pion form factor $F_{\pi NN}(0)$ (one factor for each virtual pion as in the Goldberger-Treiman relation), we find from (21)

$$4\pi f_\pi^2 \tfrac{1}{3}(1 + m_\pi/m_p)(a_1 - a_3)F_{\pi NN}^2(0) = m_\pi. \tag{9.25}$$

This "Tomozawa-Weinberg relation" is the low-energy version of the original Adler-Weisberger relation which involves a dispersion integral. Its numerical success will be discussed in section 7-2. A more precise derivation is given by Brown et al. (1971). See also the reviews by Reya (1974) and Pagels (1975).

Let us now look at the structure of the "chiral symmetry" implied by current algebra. With $i = 1, 2, 3$ as isovector index, we define the following charges and axial charges:

$$Q_i(t) = \int d^3x v_{0i}^{(0)}(t, \mathbf{x}), \qquad Q_i^5(t) = \int d^3x a_{0i}^{(0)}(t, \mathbf{x}). \tag{9.26}$$

From CVC alone, it follows that the $Q_i$ have the following commutation relations among themselves and with the $Q_i^5$:

$$[Q_i, Q_j] = i\varepsilon_{ijk} Q_k, \qquad [Q_i, Q_j^5] = i\varepsilon_{ijk} Q_k^5. \tag{9.27}$$

These relations must be taken at equal times, but if the currents were strictly conserved, $Q_i$ and $Q_i^5$ would of course be time-independent. The commutator of $Q_i^5$ and $Q_j^5$ would also be conserved and thus refer to the charge of another conserved current, which must again be 4-vectorial (since $\gamma_5^2 = 1$) and isovector (since it is antisymmetric in $i$ and $j$). Knowing that there are no other conserved currents, we must in fact have

$$[Q_i^5, Q_j^5] = ia\varepsilon_{ijk} Q_k, \tag{9.28}$$

where $a$ is a universal constant which determines the scale between the vector and axial currents. Gell-Mann (1962) postulated $a = 1$. Introducing the "chiral charges"

$$Q_i^{\pm} = \tfrac{1}{2}(Q_i \pm Q_i^5), \tag{9.29}$$

one can rewrite (27) and (28) as

$$[Q_i^{\pm}, Q_j^{\pm}] = i\varepsilon_{ijk} Q_k^{\pm}, \qquad [Q_i^{+}, Q_j^{-}] = 0. \tag{9.30}$$

The algebra thus corresponds to two separate $SU_2$-groups, one for each chirality (righthanded and lefthanded currents). The symmetry is $SU_2 \times SU_2$ and is trivially extended to $SU_3 \times SU_3$, thus providing a natural explanation for the equality of the vectorial and axial Cabibbo angles (5-1.23). However, the large value of $m_K/m_\pi$ indicates that $SU_3 \times SU_3$ is less accurate than $SU_2 \times SU_2$.

## 6-10 High Energy Behavior. Regge Pole Model

The high-energy behavior of a scattering amplitude $T(s, t)$ is important not only for high-energy collisions, but also influences $T(s, t)$ at low energies, via the dispersion integral. In particular, if $T$ goes as $s^{n-\alpha}$, $0 < \alpha < 1$ for $s \to \infty$ at fixed $t$, the dispersion relation must be $n$ times subtracted, as discussed in section 3-11. Froissart (1961) has shown that for forward elastic scattering

$$T(s \to \infty, t = 0) < \text{const} \cdot s \cdot \ln^2 s, \tag{10.1}$$

such that two subtractions are sufficient in this case. From the optical theorem (2-5.4), Froissart then concludes

$$\sigma_{\text{tot}}(s \to \infty) < \text{const} \ln^2 s. \tag{10.2}$$

Various experimental total cross sections are shown in fig. 6-10.1. We first present Froissart's proof of (1), then show that a $t$-channel Born term for the exchange of a particle of spin $S$ behaves as $s^S$ for $s \to \infty$, and then present the Regge poles which are well-behaved at high energies. A suitable book on these topics is that of Collins (1977).

Froissart's proof uses the partial-wave amplitudes $T_L(s)$ of (3.13). The lower partial waves are majorized using (7.17)

$$|T_L| \leq p^{-1}, \tag{10.3}$$

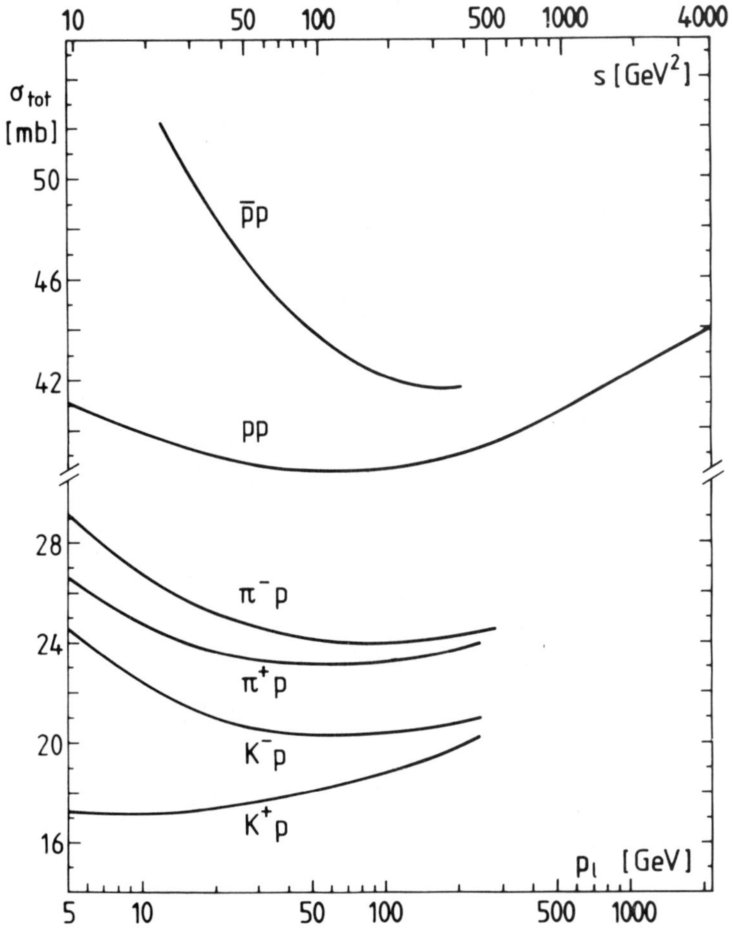


**Figure 6-10.1** (a) Total cross sections of $\pi^{\pm}$, $K^{\pm}$, $p$ and $\bar{p}$ on protons at high energies from Giacomelli 1976).

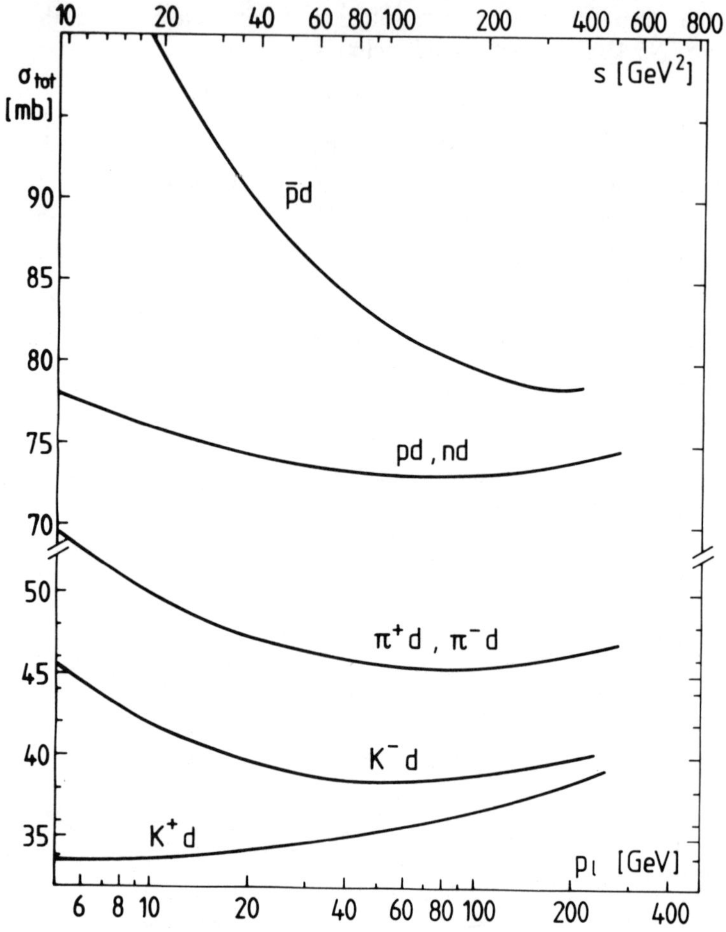


**Figure 6-10.1** (b) Total cross sections of $\pi^\pm$, $K^\pm$, $p$, $\bar{p}$ on deuterons at high energies (from Giacomelli 1976).

and for the higher ones, one can use

$$\lim_{L\to\infty} Q_L(z) = (\pi/L)^{\frac{1}{2}}(1 - e^{-2\xi})^{-\frac{1}{2}} \exp\{-(L + \tfrac{1}{2})\xi\}, \qquad z \equiv \cosh \xi. \quad (10.4)$$

It is true that from the assumption $T(s, t) \sim s^{n-\alpha}$, (3.13) requires $n$ subtractions, but one can convince oneself that the partial waves with $L > n$ are not affected by the subtractions. For $L \to \infty$, $T_L$ is dominated by the minimum $|z|$, which we called $z_n$ in section 6-3. For definiteness, we take $z_n$ from (3.8), obtaining

$$\xi_n(s \to \infty) = \ln(z_n + (z_n^2 - 1)^{\frac{1}{2}}) = m_e/p. \quad (10.5)$$

Since by assumption $|\text{Im}\, T_t(s \to \infty, t')| < s^n$, we have for $s \to \infty$, $L \to \infty$

$$T_L \sim \exp\{-Lm_e/p + n' \ln s\}, \qquad p = \tfrac{1}{2}s^{\frac{1}{2}}, \quad (10.6)$$

where $n'$ is another constant. The partial-wave expansion can now be truncated at

$$L_{\max} = cs^{\frac{1}{2}} \ln s, \tag{10.7}$$

where $c = 2n'/m_e$ is some constant. Using now (3) up to $L_{\max}$ and inserting $P_L(x = 1) = 1$, we find the desired $s$-bound

$$T(s \to \infty, x = 1) < 8\pi s^{\frac{1}{2}} \sum_0^{L_{\max}} (2L + 1)2s^{-\frac{1}{2}} \sim 16\pi L_{\max}^2 = 16\pi c^2 s \ln^2 s. \tag{10.8}$$

For non-forward scattering ($x < 1$), we can use the stronger limit

$$\lim_{L \to \infty} P_L(x < 1) = g(x)L^{-\frac{1}{2}}, \qquad |g| < 1, \tag{10.9}$$

and the summation (8) leads to

$$T(s \to \infty, x < 1) < 16\pi \sum_0^{L_{\max}} (2L + 1)L^{-\frac{1}{2}} = \text{const} \cdot s^{\frac{3}{4}} \ln^{\frac{3}{2}} s. \tag{10.10}$$

Next, we investigate the $s$-behavior of the Born term for exchange of a meson of spin $S$ in the $t$-channel. For $S = 0$, the full $t$-channel Born term is $s$-independent, even when the external particles carry arbitrary spins. For $S = 1$, on the other hand, the meson propagator contains a $g_{\mu\nu}$ from the spin summation (2-8.9); in its contraction with the 4-vectors available at each vertex, it produces a term $P_a^\mu g_{\mu\nu} P_b^\nu \sim \frac{1}{2}s$. For this reason, the Born terms of electron scattering (section 2-7) are all linear in $s$. For arbitrary $S$ and external spins, the propagator for a particle of spin $S$ must have $S$ free 4-vector indices for each of the two vertices which it connects, which produces a polynomial of degree $S$ in $s$ in the Born term (when the external particles are spinless, this is essentially the Legendre polynomial $P_L$ $(\cos \theta_t)$, where $\cos \theta_t$ is linear in $s$ (see 22 below)). This polynomial enters $R_t$ in the dispersion relation (3.1) at fixed $s$. Clearly, at large $s$ there must be large cancellations between the pole and the dispersion integral, at least for $S > 1$, in order to satisfy (8).

Regge (1959) found a modification of the Born terms which does not violate (9) or (10),

$$T_R(s \to \infty, t) = \beta(t)s^{\alpha(t)}, \qquad \alpha(m_e^2) = S_e, \qquad \alpha(t < 0) = \text{real}, < 1. \tag{10.11}$$

The function $\alpha(t)$ is called a "Regge trajectory," and a plot of Re $\alpha(t)$ versus $t$ is a "Chew-Frautschi-plot" (1961, fig. 6-10.2). While Re $\alpha(t)$ can (at most) be measured for $t < 0$, it is tempting to speculate that mesons with equal hypercharge, isospin and parity all belong to one common Regge trajectory. The resulting Chew-Frautschi plot for $t > 0$ turns out to be a straight line, at least for $I = 1$. The function $\alpha(t)$ thus interpolates between the spins of the various mesons. The derivation of (11) starts from an analytic continuation of the partial-wave amplitude $T_L(s)$ in $L$ in the $t$-channel. In the following, we first derive $s$-channel Regge poles and then use crossing symmetry. The

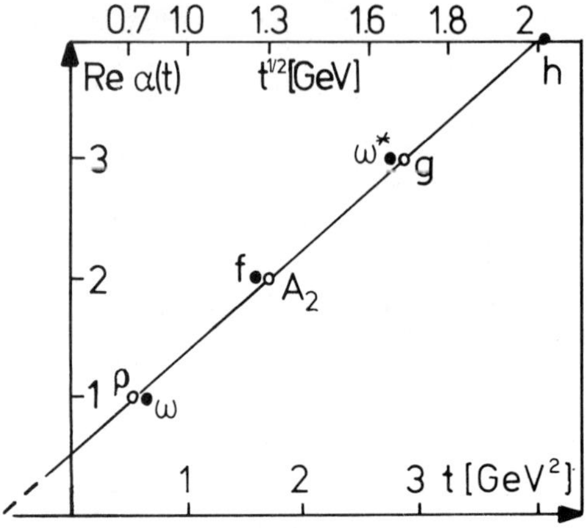


**Figure 6-10.2** Chew-Frautschi plot of the hyperneutral mesons having parity $(-1)^L$. The straight line is a fit to the $I = 1$ mesons (open circles (courtesy of H. Müller, Karlsruhe 1976)).

Legendre functions $P_\nu(x)$ and $Q_\nu(x)$ for complex $\nu$ and $x$ are defined in terms of the hypergeometric function

$$F(a, b; c; z) = 1 + zab/c + (2!)^{-1}z^2a(a+1)b(b+1)/c(c+1) + \dots \quad (10.12)$$

$$P_\nu(x) = F(-\nu, \nu+1; 1; \tfrac{1}{2}(1-x)),$$

$$Q_\nu(x) = F(\nu+1, \nu+1; 2\nu+2; 2(1-x)^{-1}) \cdot \frac{1}{2}\frac{\Gamma^2(\nu+1)}{\Gamma(2\nu+1)}[\tfrac{1}{2}(x^2-1)]^{-\nu-1} \quad (10.13)$$

$F(z)$ is analytic in $z$ except for a cut from $+1$ to $\infty$. For $P_\nu$, this implies a cut in $x$ from $-\infty$ to $-1$. For integer $\nu = L$, the product $(-L)(-L+1)\dots$ can at most have $L$ factors, thus $P_L$ is in fact a polynomial. Viewed as a function of $\nu$, $P_\nu(x)$ is an entire function with an essential singularity at $\infty$, while $Q_\nu$ has additional poles at $-1, -2, -3, \dots$ In principle, we could now define $T_\nu(s)$ from (7.1) by taking $L \to \nu$. However, the resulting amplitude diverges for $|\nu| \to \infty$

$$\lim_{|\nu|\to\infty} P_\nu(x) = (\pi\nu)^{-\frac{1}{2}}e^{-\xi/2}(1-e^{-2\xi})^{-\frac{1}{2}}[e^{(\nu+\frac{1}{2})\xi} + i\,\text{sign}\,(\text{Im}\,x)e^{-(\nu+\frac{1}{2})\xi}] \quad (10.14)$$

where $\xi$ is defined as in (4). The asymptotic behavior of $Q_\nu$ for $|\nu| \to \infty$, on the other hand, is still given by (4), with $L \to \nu$, and converges exponentially for Re $\nu > -\frac{1}{2}$. Therefore, if we want to use dispersion integrals in $\nu$, we must use the Froissart-Gribov projection (3.13) for $T_\nu$ (a theorem by Carlson states that any other definition diverges for $|\nu| \to \infty$ at least as $\exp(\pi|\nu|)$, for Re $\nu >$ const).

A complication arises from the crossed cut: neglecting the poles but keeping both cuts, we may rewrite (3.13) as

$$T_L(s) = (8\pi^2 s^{\frac{1}{2}})^{-1} P\left[\int_{z_n}^{\infty} dz Q_L(z) \text{ Im } T_t(s, z) - \int_{-\infty}^{z_L} dz Q_L(z) \text{ Im } T_u(s, z)\right]. \tag{10.15}$$

From $P_L(-x) = (-1)^L P_L(x)$ and (3.7), we find $-Q_L(-z) = (-1)^L Q_L(z)$. The analytic continuation of the innocent factor $(-1)^L$ is $\exp(i\pi\nu)$ and diverges exponentially as $\text{Im } \nu \to -\infty$. Therefore, (15) cannot be used either. Instead, one must separate the even and odd $L$-values of $T_L$ before making the continuation in $L$. This is done by taking the combinations

$$T_\nu^{\pm}(s) = (8\pi^2 s^{\frac{1}{2}})^{-1} P \int_{z_n}^{\infty} dz\, (\text{Im } T_t(s, z) \pm \text{Im } T_u(s, z)) Q_\nu(z). \tag{10.16}$$

Thus the $T_L$ have in fact two independent analytic continuations, $T_\nu^+$ for even $L$ and $T_\nu^-$ for odd $L$.

The partial-wave expansion (2-5.7) itself can be rewritten in terms of a Cauchy integral

$$T(s, t) = 4\pi s^{\frac{1}{2}} i \int_{C_0} d\nu\, (\sin \pi\nu)^{-1} (2\nu + 1) T_\nu(s) P_\nu(-x), \tag{10.17}$$

where the path $C_0$ encloses all nonnegative integers. The integrand has poles at all integers $L$ of $\nu$. Near the poles, $\sin \pi\nu$ can be expanded as $(-1)^L \times (\nu - L)\pi$. Observing the Cauchy integral formula (2-6.9), and $P_L(-x) = (-1)^L P_L(x)$, we recover in fact (2-5.7) from (17). For the combinations $T_\nu^{\pm}$ of (16), the path of integration can now be distorted into $C_1$ as shown in fig. 6-10.3:

$$T(s, t) = T_b + T_{\text{cut}} + T_R; \tag{10.18}$$

$$T_b = 2\pi s^{\frac{1}{2}} i \int_{-\frac{1}{2} - i\infty}^{-\frac{1}{2} + i\infty} (\sin \pi\nu)^{-1} (2\nu + 1)[T_\nu^+ P_\nu^+(x) + T_\nu^- P_\nu^-(x)]\, d\nu, \tag{10.19}$$

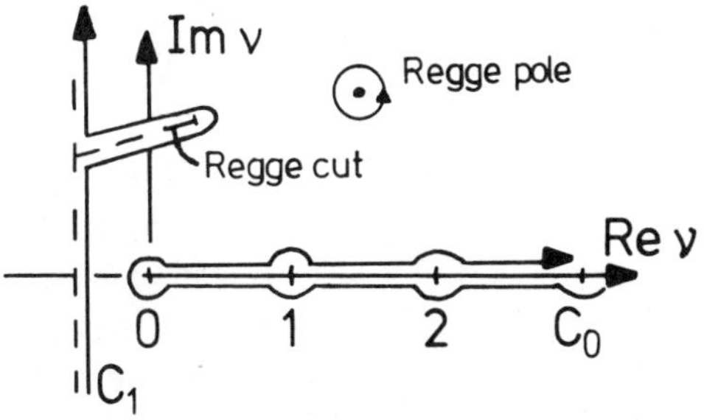


**Figure 6-10.3** The contour $C_0$ in the Sommerfeld–Watson transform (10.18) is transformed into $C_1$ + Regge pole contributions.

$$T_{\text{cut}} = 2\pi s^{\frac{1}{2}} i \int_{\text{Regge, cuts}} (\sin \pi\nu)^{-1}(2\nu + 1)[\text{disc } T_\nu^+ P_\nu^+ + \text{disc } T_\nu^- P_\nu^-]\, d\nu,$$

$$T_R = -4\pi^2 s^{\frac{1}{2}} \sum_{i \text{ poles}} (\sin \pi\alpha_i^\pm)^{-1}(2\alpha_i^\pm(s) + 1)\beta_i^\pm(s) P_{\alpha i}^\pm(x); \tag{10.20}$$

$$P_\nu^\pm(x) = P_\nu(-x) \pm P_\nu(x). \tag{10.21}$$

Eqs. (18)–(20) constitute the relativistic Sommerfeld-Watson transform (Sommerfeld 1949). It is assumed that the $T_\nu^\pm$ are analytic in $\nu$ at least down to Re $\nu = -\frac{1}{2}$ except for some cuts (Regge cuts) and poles (Regge poles). The poles appear here in the variable $\nu$. Viewed as functions of $s$, the $\alpha_i^\pm(s)$ are Regge trajectories, and the $\sin \pi\alpha_i^\pm(s)$ in the denominators of $T_R$ produce poles (= Regge poles) at Re $\alpha_i^+ = 0, 2, 4 \ldots$ and Re $\alpha_i^- = 1, 3, 5 \ldots$, the corresponding $s^{\frac{1}{2}}$ being the masses of the stable particles (if Im $\alpha = 0$) or resonances.

To derive (11), we must exchange the $s$- and $t$-channels. The variable $x$ in $P_{\alpha i}^\pm(x)$ in (20) is then $\cos \theta_t$, and is obtained from (2-4.7) by exchanging $s$ and $t$ according to (1.1). The result for elastic $ab$-scattering has been given in (3.15):

$$\cos \theta_t = \tfrac{1}{2}\left(\frac{t}{4} - m_a^2\right)^{-\frac{1}{2}}\left(\frac{t}{4} - m_b^2\right)^{-\frac{1}{2}}(s + \tfrac{1}{2}t + m_a^2 + m_b^2), \tag{10.22}$$

and shows that for fixed $t$ and $s \to \infty$, $\cos \theta_t$ becomes $\infty$. One can then use the asymptotic form of $P_\alpha$ ($\cos \theta_t$),

$$\pi^{\frac{1}{2}} P_\alpha(x \to \infty) = (\alpha - \tfrac{1}{2})!\,(\alpha!)^{-1}(2x)^\alpha + (-\alpha - \tfrac{3}{2})!\,((-\alpha - 1)!)^{-1}(2x)^{-\alpha-1}. \tag{10.23}$$

For $\alpha > -\frac{1}{2}$ the first term dominates and leads to the desired $s^\alpha$ in (11). Equation (23) also shows that the "background integral" $T_b$ in (19) can be neglected for $s \to \infty$; it vanishes as $s^{-\frac{1}{2}}$ in that limit.

For some time, it appeared that total cross sections would tend to constants at high energies. It was then shown by Pomeranchuk (1958) that these constants must be equal for particles and antiparticles on the same target, for example $\sigma_{\text{tot}}(K^+p) = \sigma_{\text{tot}}(K^-p)$. In a naive Regge pole model, a particular Regge pole (the "Pomeron" $P$) was introduced, with $\alpha_P(0) = 1$ to account for the constancy of total cross sections. Today we know that cuts are necessary in a parametrization of forward elastic amplitudes at high energies.

We now return to a discussion of $s$-channel Regge poles. The $\pm$-sign of the Regge trajectory is called its "signature." For mesons of given isospin we need at least two trajectories. For $I = 1$, we can call these $\alpha_1^-$ (for $\rho$ and $g$) and $\alpha_1^+$ (for $A_2$). The fact that these trajectories are practically identical (fig. 6-10.2) is called "exchange degeneracy," and is hardly understood. A similar thing happens for the $I = 0$ mesons $\omega, f, \omega^*, h$. The linearity of $\alpha_1^-(s)$,

$$\text{Re } \alpha_1^-(s) = c_1 + d_1 s, \qquad c_1 = 0.5, \qquad d_1 = 0.9 \text{ GeV}^{-2} \tag{10.24}$$

is not understood either. Near an integer value $L$, one can put

$$\text{Re}\ \alpha = L + d(s - m_e^2), \qquad \text{Im}\ \alpha = dm\Gamma, \tag{10.25}$$

$$\sin \pi\alpha = \pi(-1)^L(\alpha - L) = -\pi(-1)^2\, d(m_e^2 - s - im\Gamma), \tag{10.26}$$

which shows that Im $\alpha$ is proportional to the width $\Gamma$.

## 6-11 Effective Range Expansion in the Presence of Coulomb Forces: Hadronic Atoms

We saw in section 1-8 that the partial wave amplitude of Coulomb scattering has infinitely many poles, corresponding to the atomic bound states $(nL)$. Therefore the effective range expansion (5.5) cannot apply to the scattering of two charged particles. A modified expansion has been given by Bethe (1949). We derive it here by the method of Lambert (1969).

We start from the Schrödinger equation in the form (4.1), with

$$U = 2p\eta/r + V_h(r), \tag{11.1}$$

where $V_h$ comprises the hadronic potential as well as deviations from the point Coulomb potential. For $r$ larger than a certain distance $R$, $V_h$ is negligible with respect to $U$, such that the solution of (4.1) can be written as a linear combination of the Coulomb wave functions $F_L$ and $G_L$ of A-2:

$$\phi_L(r) = C(p)(F_L \cot \delta_{Lh} - G_L) \tag{11.2}$$

From the asymptotic formulas for $F_L$ and $G_L$ we see that $\delta_{Lh}$ is the extra phase shift due to $V_h$, $\delta_{Lh} = \delta_L - \sigma_L$. According to (A-2.2) the combination $C_L^{-1}p^{-L-1}F_L$ is an entire function of $r$ and $p^2$. The trick is now to find a second linearly independent solution $E_L$ of the pure Coulomb problem which is also entire in $p$. Then we can express $\phi_L$ as

$$\phi_L = BC_L^{-1}p^{-L-1}F_L - AE_L, \tag{11.3}$$

where $A$ and $B$ are entire functions of $p^2$. The function $C_L p^L G_L$ still contains the singularities of the function $h(\eta)$ of (A-2.13), but the combination

$$E_L = C_L p^L(G_L + 2\eta C_0^{-2}hF_L) \tag{11.4}$$

is again entire, which is easily checked for $L = 0$. Comparison with (2) gives

$$B/A = p^{2L+1}(C_L^2 \cot \delta_{Lh} + 2\eta h(\eta)C_L^2/C_0^2). \tag{11.5}$$

In the limit $\eta \to 0$, we have $C_L^2 = ((2L + 1)!!)^{-2}$ according to (A-2.5) and $B/A$ reduces to $((2L + 1)!!)^{-2}p^{2L+1} \cot \delta_L$, where $\delta_L$ is the phase shift in absence of the Coulomb potential. Since $B/A$ is free of Coulomb poles, it stays close to its value at $\eta = 0$. Inserting the effective range expansion (5.5), we find

$$((2L + 1)!!)^2 p^{2L+1} C_L^2 (\cot \delta_{Lh} + 2\eta h C_0^{-2}) = a_L^{-1} + \tfrac{1}{2} r_L p_L^2. \tag{11.6}$$

This is the desired effective range expansion. A more powerful formula, particularly for coupled channels, is obtained by inserting $\cot \delta_L = i - p_c^{-1} T_{cc}^{-1}$, where $T_{cc}$ and $p_c$ refer to the "charged channel" under consideration:

$$((2L+1)!!)^2 C_L^2 (\cot \delta_{Lh} + 2\eta h C_0^{-2}) = i + p_c^{-1} T_{cc}^{-1}, \tag{11.7}$$

where $T_{cc}$ may be parametrized by any of the methods of sections 6-5 and 6-7. However, if the hadronic force has a large range or if the position of a nearby singularity (a potential pole, for example) is changed by the Coulomb potential, this treatment can be insufficient. For nearby poles, one may use the method indicated in section 4-9. The poles in $pp$-scattering and $\alpha\alpha$-scattering are of this type.

The atomic bound states occur at $\cot \delta_{Lh} = i$, which follows from (2) and the discussion of the analogous problem in section 1-8. If $T_{cc}$ is not too large, the binding energy $B$ will differ from the unperturbed Coulomb binding energy $B_n$ only by a small amount $\varepsilon_h$, and $\eta$ will be close to $\eta_B$ of (1-8.18):

$$B = B_n - \varepsilon_h, \qquad \eta = \eta_B + \varepsilon_h \eta_B B_n^{-1}. \tag{11.8}$$

To lowest order in $\alpha$, relativistic corrections can be ignored, such that $\eta_B = inB_n = \frac{1}{2} Z^2 \alpha^2 \mu n^{-2}$, $p_c = (-2\mu B_n)^{\frac{1}{2}} = iZ\alpha\mu/n$ (with $Z_a Z_b = -Z$). We then have $C_0^2 = 4B_n/\varepsilon_h$. The function $h(\eta)$ is also proportional to $B_n/\varepsilon_h$ (due to the term $s = n$ in the sum of A-2.13), but this factor is cancelled by the $C_0^{-2}$ in (6) and (7). We thus obtain in the scattering length approximation ($r_L = 0$ in 6) using (A-2.5)

$$\begin{aligned} \varepsilon_h &= -(-1)^L \prod_{s=1}^{L} (1 - n^2/s^2) 4B_n (Z\alpha\mu/n)^{2L+1} a_L \\ &= -\prod_{s=1}^{L} (s^{-2} - n^{-2}) 2n^{-3} \mu^{-1} (Z\alpha\mu)^{2L+3} a_L . \end{aligned} \tag{11.9}$$

This result is more easily derived by bound state perturbation theory. For $L = 0$ the perturbing potential $V_h$ is just $a_0\, \delta(\mathbf{r})$. ($V_h$ is the Fourier transform of $T_{cc}$). When other channels are open below threshold, $a_L$ and therefore $\varepsilon_h$ are complex,

$$\varepsilon_h = E_h - \tfrac{1}{2} i \Gamma_h, \tag{11.10}$$

and $\Gamma_h$ as obtained from (9) agrees with (2-3.21) if one uses the unitarity condition (2-5.12) or better (7.21).

CHAPTER SEVEN

# Particular Hadronic Processes

## 7-1 $\pi\pi$-Scattering

Due to the low threshold of the $\pi\pi$ system ($s_0 = 4m_\pi^2$), knowledge of low-energy scattering is essential in the analysis of other processes such as $\pi N$ scattering or $NN$ scattering. Unfortunately, the instability of pions has excluded direct collision experiments so far. The many indirect methods are described in a book by Martin et al. (1976).

After the discovery of the $\rho$ resonance in $p$-wave $\pi\pi$-scattering in the reaction $\pi N \to \pi\pi N$ (see fig. 4-6.1), the first evidence for an $s$-wave $\pi\pi$-amplitude came from forward-backward asymmetry in the $\pi^+\pi^-$ angular distribution in $\pi^- p \to \pi^+\pi^- n$ in the $\pi\pi$-cms (see the discussion following 4-13.8). The analysis is complicated by possible final state interactions between the outgoing pion and nucleon (Estabrooks and Martin 1974). The resulting $s$-wave phase shift in the $I = 0$-channel is shown in fig. 7-1. Near threshold, its behaviour is determined by the scattering length $a_0^0$ which can be extracted from an analysis of the final state interaction in the decay $K \to e\nu\pi\pi$ (Rosselet et al. 1977). The rapid rise of $\delta_0^0$ above 700 MeV can be parametrized by two resonances, $S^*$ and $\varepsilon$:

$$\begin{aligned} m(S^*) &\sim 993 \text{ MeV}, \qquad \Gamma(S^*) \sim 40 \text{ MeV}, \\ m_\varepsilon &\sim 1200 \text{ MeV}, \qquad \Gamma_\varepsilon \sim 600 \text{ MeV}. \end{aligned} \tag{1.1}$$

However, cot $\delta_0^0$ is not just the superposition of two Breit-Wigner resonances. The $K\bar{K}$ channel which opens around 990 MeV has a very large $\pi\pi \to K\bar{K}$ scattering length, and the $S^*$ is needed to describe this effect.

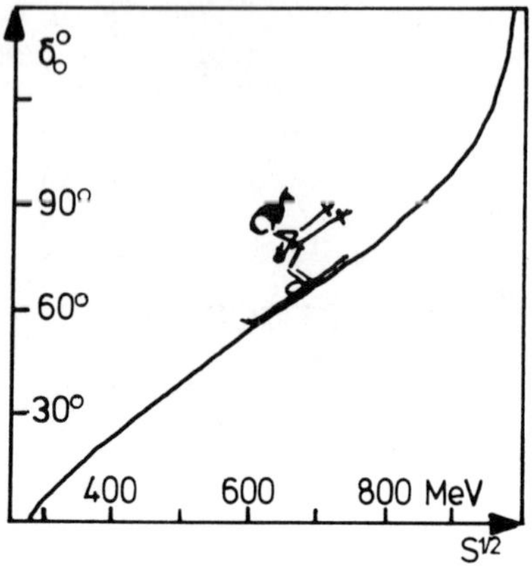


**Figure 7-1** The $I = L = 0$ phase shift up to the $K\bar{K}$ threshold, assuming $a_0^0 = 0.3m_\pi^{-1}$.

From table 4-2.3, we find the isospin decomposition of $\pi\pi$ scattering (remembering the orthogonality of the transformation)

$$|\pi^\pm\pi^\pm\rangle = |2, \pm 2\rangle, \qquad |\pi^\pm\pi^0\rangle = 2^{-\frac{1}{2}}|2, \pm 1\rangle \pm 2^{-\frac{1}{2}}|1, \pm 1\rangle,$$
$$|\pi^\pm\pi^\mp\rangle = 3^{-\frac{1}{2}}|0, 0\rangle \pm 2^{-\frac{1}{2}}|1, 0\rangle + 6^{-\frac{1}{2}}|2, 0\rangle, \tag{1.2}$$
$$|\pi^0\pi^0\rangle = (\tfrac{2}{3})^{\frac{1}{2}}|2, 0\rangle - 3^{-\frac{1}{2}}|0, 0\rangle.$$

Application of (4-2.6) leads to

$$T(\pi^\pm\pi^\pm) = T^2, \qquad T(\pi^0\pi^\pm) = \tfrac{1}{2}(T^1 + T^2).$$
$$T(\pi^+\pi^-) = \tfrac{1}{3}T^0 + \tfrac{1}{2}T^1 + \tfrac{1}{6}T^2, \qquad T(\pi^0\pi^0) = \tfrac{1}{3}T^0 + \tfrac{2}{3}T^2, \tag{1.3}$$
$$T(\pi^+\pi^- \to \pi^0\pi^0) = -\tfrac{1}{3}T^0 + \tfrac{1}{3}T^2.$$

For elastic pion scattering on any target $b$, the $s$- and $u$-channels are identical except possibly for their isospin content. When the target has zero isospin, $I_b = 0$, the $s$-$u$-crossing of (6-1.2) imposes the "crossing symmetry"

$$T(u, t, s) = T(s, t, u). \tag{1.4}$$

For $I_b > 0$, (4) applies only for elastic $\pi^0$ scattering. However, the isospin amplitudes $T^I(u, t)$ must be linear combinations of the $T^I(s, t)$

$$T^I(u, t, s) = \sum_{I'} C_{II'} T^{I'}(s, t, u). \tag{1.5}$$

In particular, for $I = I_b + 1$ and $I_{b3} = I_b$, the left-hand side of (5) is just the $\pi^+b$ amplitude, and the right-hand side must be the $\pi^-b$-amplitude, such that the $C_{I_b+1, I'}$ are the coefficients of the isospin decomposition of $T(\pi^-b, I_{b3} = I_b)$, which are the squares of the appropriate Clebsch-Gordan coefficients:

$$C_{I_b+1, I'} = (C^{1,\ I_b,\ I'}_{-1,\ I_b,\ I_b-1})^2 \tag{1.6}$$

For $I_b = 1$, the last row of the "crossing matrix" thus consists of the elements $(\frac{1}{3}, \frac{1}{2}, \frac{1}{6})$ according to (3). Finally, since "crossing twice" is the identity operator, we must have

$$\sum_{I'} C_{II'} C_{I'I''} = \delta_{II''}. \tag{1.7}$$

For $I_b = 1$, (4) applies to the combination $T^1 + T^2$ according to (3). Using also (5) and (7), we find

$$C_{II'} = \begin{pmatrix} \frac{1}{3} & -1 & \frac{5}{3} \\ -\frac{1}{3} & \frac{1}{2} & \frac{5}{6} \\ \frac{1}{3} & \frac{1}{2} & \frac{1}{6} \end{pmatrix} \tag{1.8}$$

So much for the general isospin formalism of $s$-$u$-crossing in pion scattering. For $\pi\pi$-scattering, we have in addition to (5) also

$$T^I(s, u, t) = (-1)^I T^I(s, t, u), \tag{1.9}$$

which expresses the generalized Bose-Einstein principle. It allows only even partial waves for $T^0$ and $T^2$ and odd partial waves for $T^1$. Low-energy $\pi\pi$ scattering is therefore described by the three partial-wave amplitudes $T_0^0$, $T_0^2$ and $T_1^1$, which are parametrized by (6-7.31). To order $p^2$, one can neglect $b_1^1$, such that the scattering is described by the five real parameters $a_0^0, b_0^0, a_0^2, b_0^2$ and $a_1^1$ (table 7-1). The crossing relations (5) impose further restrictions on these. Since $T_L(s)$ as given by (6-7.31) is not a polynomial in $s$, crossing symmetry requires in fact an infinite number of partial waves. However, for $m_\pi a_L^I \ll 1$ we can neglect unitarity effects, thus getting approximately

$$\begin{aligned} T^I &\approx 8\pi s^{\frac{1}{2}} T_0^I \approx 16\pi m_\pi (a_0^I + p^2 b_0^I) \qquad \text{for} \quad I = 0 \quad \text{and} \quad 2, \\ T^1 &\approx 8\pi s^{\frac{1}{2}} 3 T_1^1 \cos\theta \approx 48\pi m_\pi a_1^1 p^2 \cos\theta. \end{aligned} \tag{1.10}$$

If we now substitute

$$p^2 = s/4 - m^2, \qquad p^2 \cos\theta = s/4 - m^2 + t/2 = m^2 - s/4 - u/2, \tag{1.11}$$

we see that (10) is in fact linear in $s$ and $t$. Using thus (10) for the discussion of crossing symmetry, we find that (5) is satisfied if

$$3a_1^1 = -b_0^2 = \tfrac{1}{2} b_0^0 = \tfrac{1}{6}(2a_0^0 - 5a_0^2) \equiv L. \tag{1.12}$$

**Table 7-1** Approximate values of the scattering lengths $a_L^I$ and parameters $b_L^I$ defined in (6-7.30) for $\pi\pi$ scattering (in pion mass units).

| | $a_L^I$ | $b_L^I$ |
|---|---|---|
| $I = 0$ | $0.30 \pm 0.15$ | |
| $I = 2$ | $-0.02 \pm 0.03$ | $-0.086 \pm .016$ |
| $I = 1$ | $0.04 \pm 0.01$ | |

Current algebra requires

$$L = (4\pi f_\pi^2)^{-1} m_\pi \approx 0.09 m_\pi^{-1}. \tag{1.13}$$

The derivation goes as in (6-9.21), with $2m_b \nu = 2m_\pi^2$ at threshold, but with an extra factor 2 on the right-hand side of (6-9.22) because the initial state contains two identical particles (remember the discussion in section 3-4). Then (13) is most easily derived at threshold where $T^1 = 0$, $T^I = 16\pi m_\pi a_0^I$, for $I = 0$ and 2, such that

$$T^{(-)} = \tfrac{1}{2}T(\pi^-\pi^+) - \tfrac{1}{2}T(\pi^+\pi^+) = 16\pi m_\pi \tfrac{1}{2}(\tfrac{1}{3}a_0^0 - \tfrac{5}{6}a_0^2) \tag{1.14}$$

according to (3). It must be remembered that (13) applies to massless pions, and the combination $\frac{1}{3}a_0^0 - \frac{5}{6}a_0^2$ should be somewhat larger due to form factors as in (6-9.25).

Adler's condition (6-9.5) requires

$$T_b^{(+)}(s = t = u = m_\pi^2) = 0, \qquad T_b^{(+)} \equiv \tfrac{1}{2}T(\pi^- b) + \tfrac{1}{2}T(\pi^+ b), \qquad b = (+, 0) \tag{1.15}$$

and guarantees that all $\pi\pi$ amplitudes are of the order of $m_\pi^2$ at threshold. One can also investigate the way in which chiral symmetry of real pions is broken, by including in the Hamiltonian density $H(x)$ a chiral-symmetry breaking piece $H_1$:

$$H(x) = H_0(x) + \varepsilon H_1(x). \tag{1.16}$$

One can then use in (6-9.17)

$$\partial^\mu a_\mu(x) = i\varepsilon[H_1(x), Q_+^5(x_0)] \tag{1.17}$$

as the local and covariant generalization of the equation $\dot{Q}_+^5 = i\int d^3x[H(x), Q_+^5(x_0)]$. It turns out that the symmetry breaking is given by the commutator of (17) with $Q_-^5(x_0)$, which is the so-called $\sigma$-commutator. It is symmetric in the charge indices and vanishes automatically for $T^{(-)}$. Assuming moreover that it has no $I = 2$ piece, one finds that (15) must be valid also for physical pions at threshold ($s = u = 4m^2$, $t = 0$) for $\pi^b = \pi^+$. Inserting again the isospin decomposition (3), we find

$$2a_0^0 + 7a_0^2 = 0. \tag{1.18}$$

If this is true, then $|a_0^2|$ must be quite small according to table 7-1.

## 7-2 $\pi N$-Scattering

Total and elastic $\pi^\pm p$ cross sections below 900 MeV have been shown in fig. 4-2, where it was also mentioned that inelastic channels are negligible in $\pi^+ p$ scattering below 500 MeV. The same remark applies to $\pi^- p \to \pi^- p$ and $\pi^- p \to \pi^0 n$ to the extent that these two amplitudes are isospin invariant, i.e., to the extent that (4-2.10) and (4-2.11) are true. The isospin partial-wave amplitudes $f^I_{L\pm}$ can be parametrized as in (6-7.17)

$$f^I_{L\pm} = (2ip)^{-1}(\eta^I_{L\pm} \exp(2i\,\delta^I_{L\pm}) - 1), \qquad 0 \le \eta^I_{L\pm} \le 1, \tag{2.1}$$

where $\eta$ is the "inelasticity." Below 500 MeV we thus have $\eta = 1$, and (1) reduces to the simpler expression (2-5.28). The first partial wave which becomes inelastic is that of $I = J = \frac{1}{2}$ and total parity $+1$, namely $P_{11}$ in the notation $L_{2I,\,2J}$ (see fig. 7-2.1). The inelasticity is due to the $\pi\pi N$ final states, and the quantum numbers are understood as follows: At the $\pi\pi N$ threshold, all orbital angular momenta must be zero in the final state, $L_{\pi\pi} = L_{N(\pi\pi)} = 0$ and therefore $J = \frac{1}{2}$. Since the extra pion in the final state requires positive total parity, the initial state must be in a $p$-wave. The $I = \frac{1}{2}$ finally follows mainly from the large $I = 0$ $s$-wave $\pi\pi$ scattering length, $a^0_0 \gg a^2_0$ (see table 7-1).

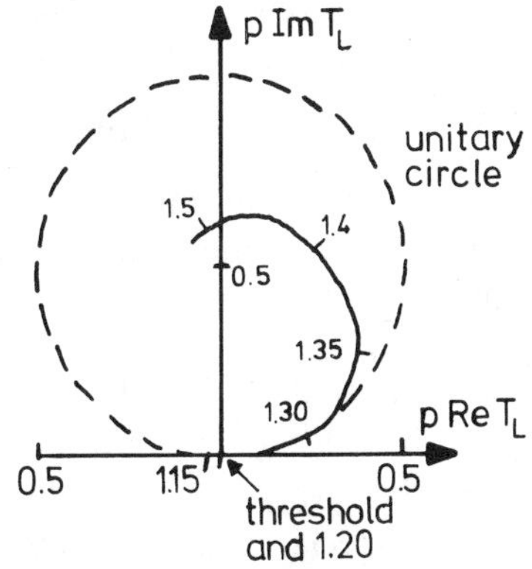


**Figure 7-2.1** "Argand" plot of the dimensionless quantity $pT_L$, $T_L$ being the elastic partial wave amplitude (6-7.17). The curve refers to the $P_{11}$ $\pi N$ scattering amplitude $f^{\frac{1}{2}}_{1+}$ (Höhler et al. 1978). The numbers along the curve give the cms energy in GeV. The inelasticity $\eta_L$ (defined in 6-7.17) is twice the distance from the centre of the "unitary circle." Note that Im $T_L$ is negative between threshold and 1.20 GeV, with a minimum at 1.15 GeV.

At very low energies, the reaction $\pi^- p \to \gamma n$ competes with $\pi^- p \to \pi^- p$ and $\pi^- p \to \pi^0 n$. In particular, the "Panofsky ratio" is defined as the ratio of $\pi^0 n$ and $\gamma n$ final states. For pions captured from the $S$-states of $\pi^- p$ atomic bound states,

$$R = \Gamma(\pi^- p \to \pi^0 n)/\Gamma(\pi^- p \to \gamma n) = 1.53. \tag{2.2}$$

One must then treat the $\pi^- p$, $\pi^0 n$ and $\gamma n$ channels jointly by $K$-matrices (see Rasche and Woolcock 1976, and section 6-11). Of all the isospin-breaking effects, we include here only the Coulomb amplitudes which were discussed in sections 1-10 and 2-9. At least the relativistic point Coulomb amplitudes must be treated separately even for $\pi^+ p$ scattering, because otherwise the partial-wave expansion diverges. We first express the differential cross section (2-5.23) in terms of $f_1$ and $f_2$ of (2-5.29):

$$d\sigma/d\Omega = \cos^2\tfrac{1}{2}\theta\,|f_1 + f_2|^2 + \sin^2\tfrac{1}{2}\theta\,|f_1 - f_2|^2. \tag{2.3}$$

Restricting ourselves to $J < \frac{5}{2}$ in the hadronic amplitudes, we obtain from (2-5.32) and (2-5.33)

$$f_1 \pm f_2 = f_C^{\frac{1}{2}0}\begin{pmatrix}1\\R\end{pmatrix} + f'_{0+} - f'_{2-} \pm (f'_{1-} - f'_{1+}) + 3\cos\theta(f'_{1+} \pm f'_{2-}), \tag{2.4}$$

$$f'_{L\pm} \equiv \exp(2i\sigma_{L\pm})f_{L\pm}. \tag{2.5}$$

Here $f_C^{\frac{1}{2}0}$ and $f_C^{\frac{1}{2}0}R$ are the helicity-nonflip and helicity-flip Coulomb amplitudes; they differ from their Born approximations (2-9.15) and (2-9.17) by the Coulomb phase (2-6.38) which may show up at small angles. The $\sigma_{L\pm}$ of (5) are the partial-wave projections of the Coulomb amplitudes; their appearance has these same reason as in (1-10.4). The polarization (1-7.30) is proportional to Im $fg^*$. According to (2-5.35) and (2-5.36), we have

$$2\,\mathrm{Im}\,fg^* = -2\sin\theta\,\mathrm{Im}\,f_1 f_2^* = \sin\theta\,\mathrm{Im}\,(f_1 + f_2)(f_1 - f_2)^*. \tag{2.6}$$

Present-day determinations of $\pi N$ scattering lengths rely largely on experiments above threshold and use crossing symmetry (1.5) as an additional constraint. From (1.6) and (1.7), we find

$$C_{II'} = \begin{pmatrix}\frac{1}{3} & -\frac{4}{3}\\ -\frac{2}{3} & -\frac{1}{3}\end{pmatrix}. \tag{2.7}$$

The $f_{L\pm}$ are obtained from the invariant amplitudes $A$ and $B$ of (6-2.1) by inserting (6-2.5) into (2-5.34). Specializing the formulas to elastic scattering, we find

$$f_{L\pm} = (16\pi s^{\frac{1}{2}})^{-1}\int dx\,\{(E+m)[A + B(s^{\frac{1}{2}} - m)]P_L + (E-m)[-A + B(s^{\frac{1}{2}} + m)]P_{L\pm 1}\}. \tag{2.8}$$

We can now show that the $f_L$ have the usual threshold behaviour $p^{2L}$. From (6-5.3) we remember that this is essentially due to the $P_L$. The $P_{L-1}$ which appears in (8) for $J = L - \frac{1}{2}$ gives only a factor $p^{2L-2}$, but it is multiplied by $E - m = p^2(E+m)^{-1} \to p^2/2m$. Therefore, we can use our threshold parametrizations of section 6-5, and in particular (6-5.7) in the form

$$\begin{aligned} pf^I_{L\pm} &= \tan\delta^I_{L\pm}(1 - i\tan\delta^I_{L\pm})^{-1},\\ \tan\delta^I_{L\pm} &= p^{2L+1}(a^I_{L\pm} + b^I_{L\pm}p^2). \end{aligned} \tag{2.9}$$

Approximate numerical values are collected in table 7-2.1. The quantities which are directly determined are the combinations

$$\begin{aligned} 3a^{(-)} &\equiv a^{\frac{1}{2}}_{0+} - a^{\frac{3}{2}}_{0+} = 0.26 \pm 0.001,\\ 3a^{(+)} &\equiv a^{\frac{1}{2}}_{0+} + 2a^{\frac{3}{2}}_{0+} = -0.02 \pm 0.01. \end{aligned} \tag{2.10}$$

**Table 7-2.1** Approximate values of the threshold parameters $a^I_{L\pm}$ and $b^r_{0+}$ of $\pi N$ scattering, in pion mass units.

| | $a_{0+}$ | $b_{0+}$ | $a_{1-}$ | $a_{1+}$ |
|---|---|---|---|---|
| $I = \frac{1}{2}$ | 0.165 | −0.024 | −0.09 | −0.04 |
| $I = \frac{3}{2}$ | −0.096 | −0.041 | −0.04 | 0.2 |

$a^{(+)}$ is almost zero, which is a consequence of Adler's consistency condition (6-9.12). The $pp\pi^0$ coupling constant $G$ is conveniently determined from the dispersion relation (6-2.20), with

$$\omega_e = \tfrac{1}{2}m_\pi^2 m_N^{-1}, \qquad R_e = -2G^2 m_\pi^2(16\pi m_N^2)^{-1} = -2f^2 \tag{2.11}$$

according to (6-2.19), (6-2.10) and (5-8.4). There is no unphysical region contribution in this case, $\omega_{\min} = m_\pi$. One finds

$$f^2 = 0.0794, \qquad G^2/4\pi = 14.4, \qquad G = 13.45. \tag{2.12}$$

With the $n$ and $\pi$ decay constants of (5-1.13), we can use the Goldberger-Treiman relation (5-7.18) to find the strong $NN\pi$ form factor at zero momentum transfer

$$F_{NN\pi}(0) = 2^{\frac{1}{2}} 6.722 m_\pi 1.26 (0.960 m_\pi 13.45)^{-1} = 0.93, \tag{2.13}$$

which in turn can be used to check the Tomozawa-Weinberg relation (6-9.25). With $f_\pi F_{NN\pi}(0) = 0.893 m_\pi$ and (10), the relation is satisfied to within 2%.

Now back to $\pi N$ scattering. The contributions of the $s$-channel nucleon pole to $f_{L\pm}$ are obtained from (8) by inserting for $\pi^\pm p$ scattering $A_\pm = 0$, $B_+ = 0$, $B_- = 2G^2(m_N^2 - s)^{-1}$ according to (6-2.8). Since $B_-$ is independent of $x$, we use $\int dx P_L = 2\,\delta_{L0}$ and find

$$f_{0+}^{\frac{1}{2}(e)} = -(8\pi s^{\frac{1}{2}})^{-1}(E + m_N)3G^2(s^{\frac{1}{2}} + m_N)^{-1}, \qquad f_{L\pm}^{\frac{3}{2}(e)} = 0 \tag{2.14}$$

$$f_{1-}^{\frac{1}{2}(e)} = -(8\pi s^{\frac{1}{2}})^{-1}(E - m_N)3G^2(s^{\frac{1}{2}} - m_N)^{-1}. \tag{2.15}$$

The pole contribution to $a_{0+}^{\frac{1}{2}}$ is obtained from (14) by putting $E = m_N$, $s^{\frac{1}{2}} = m_N + m_\pi$:

$$a_{0+}^{\frac{1}{2}(e)} = -\tfrac{3}{2}m_N^{-1}G^2/4\pi \approx -3.2 m_\pi^{-1}, \tag{2.16}$$

which is much larger than $a_{0+}$ of table 7-2.1 and of opposite sign. It demonstrates more quantitatively the large cancellation between pole and nonpole contributions required by Adler. In a way, this was anticipated by the static model of Chew and Low (1956) which neglects nuclear recoil: the nucleon only has a spin degree of freedom, and the negative intrinsic parity of the pion allows absorption and re-emission of pions only in $P$-states.

The Chew-Low model has been taken as an explanation of the $\Delta$ resonance in the $P_{33}$ partial wave. We present here an equivalent $N/D$ calculation of the $P$-wave amplitudes (Hamilton 1967). The short cut in fig. 6-7 which arises from the $u$-channel pole is approximated by a pole at $s = m^2$. Of the other partial-wave singularities, only the physical cut is included. As far as the poles are concerned, this treatment is crossing symmetric. Under $s \leftrightarrow u$ crossing, we have $t_u = t$ or

$$p^2(1 - x) = p_u^2(1 - x_u), \qquad p_u^2 = \tfrac{1}{4}u^{-1}\lambda(u, m_N^2, m_\pi^2). \tag{2.17}$$

In general, (17) leads to a nonlinear relation between $x$ and $x_u$, such that a partial wave in the direct channel contributes to all partial waves in the crossed channel. In the static approximation, however, the substitution $P^\mu_{\pi u} \to -P^\mu_\pi$ implies $E_{\pi u} \to -E_\pi$ and therefore $p_u^2 = p^2$. Then (17) requires $x_u = x$, i.e., the cms scattering angle remains invariant under $s$-$u$ crossing: $S$-waves cross into $S$-waves, $P$-waves into $P$-waves and so on. Since spin and isospin appear symmetrically in the static model for the $P$-waves (the pion has $I = L = 1$), the crossing coefficients from the $J = \frac{1}{2}$ $u$-channel are identical with those of the first column of (7). An additional minus sign comes from (6-2.2), such that we have

$$f_J^{I(u)}(s) = -C_{I\frac{1}{2}}C_{J\frac{1}{2}}\, f_{\frac{1}{2}}^{\frac{1}{2}(e)}(s), \tag{2.18}$$

$$f_{\frac{3}{2}}^{\frac{3}{2}(u)}(s) = -\tfrac{4}{9} f_{\frac{1}{2}}^{\frac{1}{2}(e)}(s) \approx \tfrac{2}{3}p^2 s^{-\frac{1}{2}}(s - m_N^2)^{-1}G^2/4\pi. \tag{2.19}$$

The last approximation has been made only to obtain a simple pole in $s$ for the amplitude $F$ defined in (6-7.9):

$$F \equiv 8\pi s^{\frac{1}{2}} f_{\frac{3}{2}}^{\frac{3}{2}} p^{-2} \equiv N/D, \qquad N = \tfrac{4}{3}G^2(s - m_N^2)^{-1}. \tag{2.20}$$

Using $D(m_N^2) = 1$, we can instead of (6-7.13) write a once-subtracted dispersion relation for $D$,

$$\operatorname{Re} D(s) = 1 - \frac{2}{3}\frac{G^2}{4\pi}(s - m_N^2)\frac{P}{\pi}\int_{s_0}^{\infty} ds' \frac{p'^3 R(s')}{(s' - s)(s' - m_N^2)^2}\, s'^{-\frac{1}{2}}, \tag{2.21}$$

with $R$ given by (6-7.15) and (6-7.22). The solution of (21) has been studied by Hamilton and Tromborg (1972, 1976. For $R = 1$, $F$ is an elliptic function, and an algebraic approximation is given). The resonance energy $m_\Delta$ is determined by $\operatorname{Re} D(m_\Delta^2) = 0$. It turns out that (21) does have a zero, but it occurs considerably above $(1.232\ \text{GeV})^2$. Additional interactions are required to reproduce position $m_\Delta$ and width $\Gamma(m_\Delta)$ of the resonance. On the other hand, the interaction range $r_0$ in eq. (4-11.13) for $m_\Delta \Gamma(s)$ can be calculated from (20): we use (2-5.13) with $T_L$ replaced by $f_J^I$, which leads to

$$\cot \delta_J^I = i + (pf_J^I)^{-1} = i + p^{-3} 8\pi s^{\frac{1}{2}} D/N. \tag{2.22}$$

In the $P_{\frac{3}{2}}^{\frac{3}{2}}$ partial wave, we must have

$$\cot \delta_{\frac{3}{2}}^{\frac{3}{2}} = (m_\Delta \Gamma(s))^{-1}(m_\Delta^2 - s) \tag{2.23}$$

in a region around the resonance. Inserting here (4-11.13) for $m_\Delta \Gamma(s)$ and comparing the expression with (22), we see that the factor $p^{-3}s^{\frac{1}{2}}$ is common to both expressions. Next, we note that $\operatorname{Re} D(s)$ is linear in $s$ over a large region including $\operatorname{Re} D(s) = 0$ (see fig. 7-2.2). Then we must have

$$(1 + r_0^2 p^2)/(1 + r_0^2 p_\Delta^2) \approx (s - m_N^2)/(m_\Delta^2 - m_N^2) \tag{2.24}$$

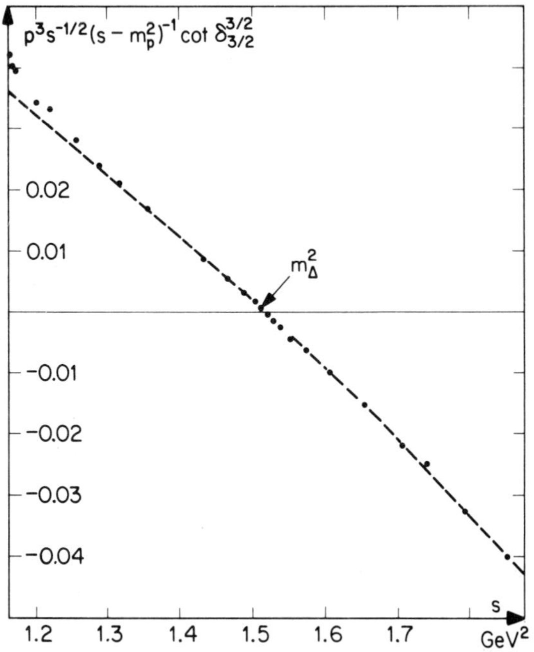


**Figure 7-2.2** "Relativistic Chew-Low plot". The points refer to the (Coulomb corrected) $\pi^+ p$ $P_{33}$ phase shifts of Koch and Pietarinen (1979) and agree within errors with those of Höhler et al. (1979). The plot is used to determine position, width and shape (2.26) of the $\Delta$ resonance (dotted line). The pole at $s = m_p^2$ is an effective one (compare fig. 6-7), its residue ($\approx 0.07$) is 30% below the value required by (20).

in that region. This near equality is an identity at $s = m_\Delta^2$. Postulating it also to hold at threshold $s = s_0 = (m_N + m_\pi)^2$, we find

$$r_0^2 = p_\Delta^{-2}(s_0 - m_N^2)^{-1}(m_\Delta^2 - s_0) = 24.6 \text{ GeV}^{-2}, \qquad r_0 = 0.98 \text{ fm} \quad (2.25)$$

in good agreement with the best fit values of Koch and Pietarinen (1979):

$$r_{0\Delta^{++}} = (1.03 \pm 0.02) \text{ fm}, \qquad m_{\Delta^{++}} = (1231.1 \pm 0.3) \text{ MeV}, \qquad \Gamma_{\Delta^{++}} = (112 \pm 1) \text{ MeV}. \quad (2.26)$$

There exist many more resonances in $\pi N$ scattering (see table 7-2.2), all of which couple also to inelastic channels. In addition, they normally require a "background" $K$-matrix $K_{00}$, which means that they must be parametrized by expressions that are at least as complicated as (6-5.28).

At very high energies, the parametrization in terms of Regge poles becomes appropriate. For intermediate energies, one might be tempted to simply add Regge poles to the existing resonances, $T = T_{\text{Regge}} + T_{\text{res}}$. However, as the Sommerfeld-Watson transform (6-10.18) is an alternative of the partial-wave expansion and not just a supplement, this procedure is probably incorrect. It has been suggested that one should use

$$T = T_{\text{Regge}} + T_{\text{res}} - \langle T_{\text{res}} \rangle, \quad (2.27)$$

**Table 7-2.2** Resonances in $\pi N$ scattering below 2000 MeV (for $\Delta$ (1232) see table C-5). The symbols $N$ and $\Delta$ stand for $I = \frac{1}{2}$ and $\frac{3}{2}$, respectively. The numbers are approximate and are taken from the Particle Data Group (1978).

| Symbol (mass [MeV]) | Partial Wave ($J^{\mathscr{P}}$) | $\Gamma$[MeV] | Main decay channels and fractions (%) $N\pi$ | $N\pi\pi$ | $N\eta$ |
|---|---|---|---|---|---|
| $N$ (1470) | $P_{11}$ ($\frac{1}{2}^+$) | 200 | 60 | 25 | 18 |
| $N$ (1520) | $D_{13}$ ($\frac{3}{2}^-$) | 125 | 55 | 45 | < 1 |
| $N$ (1530) | $S_{11}$ ($\frac{1}{2}^-$) | 100 | 30 | 5 | 65 |
| $\Delta$ (1650) | $S_{31}$ ($\frac{1}{2}^-$) | 140 | 35 | 65 | — |
| $N$ (1670) | $D_{15}$ ($\frac{5}{2}^-$) | 155 | 45 | 55 | < 0.5 |
| $\Delta$ (1670) | $D_{33}$ ($\frac{3}{2}^-$) | 200 | 15 | 85 | — |
| $N$ (1688) | $F_{15}$ ($\frac{5}{2}^+$) | 140 | 60 | 40 | < 0.3 |
| $N$ (1690) | $S_{11}$ ($\frac{1}{2}^-$) | 150 | 55 | 30 | |
| $N$ (1780) | $P_{11}$ ($\frac{1}{2}^+$) | 200 | ~ 20 | > 40 | ~ 10 |
| $N$ (1810) | $P_{13}$ ($\frac{3}{2}^+$) | 200 | ~ 20 | ~ 70 | < 5 |
| $\Delta$ (1890) | $F_{35}$ ($\frac{5}{2}^+$) | 250 | ~ 15 | ~ 80 | — |
| $\Delta$ (1910) | $P_{31}$ ($\frac{1}{2}^+$) | 200 | ~ 20 | ? | — |
| $\Delta$ (1940) | $F_{37}$ ($\frac{7}{2}^+$) | 220 | ~ 40 | > 25 | — |

where $\langle T_{\text{res}} \rangle$ is an average over several neighboring resonances. Then $T_{\text{Regge}}$ would be "dual" to $T_{\text{res}}$ in the sense that it describes the same gross features. See Fukugita and Igi (1977) for references.

In the triangle limited by $t < 4m_\pi^2$, $s < (m_\pi + m_N)^2$, $u < (m_\pi + m_N)^2$,

$$|\nu| < m_\pi + \tfrac{1}{4} m_N^{-1} t, \tag{2.28}$$

according to (6-1.28), the $\pi N$ amplitudes are real. The invariant amplitudes

$$A^{(\pm)} = \tfrac{1}{2}(A_- \pm A_+), \qquad \tilde{B}^{(\pm)} \equiv \tfrac{1}{2}(B_- \pm B_+ - B_-(\text{pole})) \tag{2.29}$$

have no singularities in this region and according to (6-2.2) or (6-2.14), the combinations $A^{(+)}$, $\nu^{-1}A^{(-)}$, $\nu^{-1}\tilde{B}^{(+)}$ and $\tilde{B}^{(-)}$ are all even functions of $\nu$. They can be expanded in the form (see Nagels et al., 1979)

$$X^i(\nu, t) = \sum_{m,\, n=0}^{\infty} x_{mn}^i \nu^{2m} t^n. \tag{2.30}$$

## 7-3 Recoupling of Two-Particle Helicity States

In the next two sections, we shall encounter relativistic 2-particle states where both particles $a$ and $b$ carry spin. The helicity states of section 4-4 are convenient for the angular dependence (4-4.9) of scattering amplitudes, but the functions $T_J(s; \lambda_a \cdots \lambda_d)$ are normally inconvenient. (4-4.25) shows that they do not refer to parity eigenstates. For that reason, we switched to the functions $f_{L\pm}$ of (2-5.26) for $\pi N$-scattering. We shall first discuss general parity eigenstates and then introduce states whose matrix elements also have simple threshold behaviour.

According to (4-4.21), the parity operator is $\exp\{i\pi J_y\} \cdot Y$. Using the matrix elements (B-3.11) we get

$$\mathscr{P}|\lambda_a, \lambda_b, JM\rangle = \eta(-1)^{S_a+S_b-J}|-\lambda_a, -\lambda_b, JM\rangle. \tag{3.1}$$

The general parity eigenstates are

$$\begin{aligned} |\lambda_a, \lambda_b, JM\rangle_\pm &= 2^{-\frac{1}{2}}|\lambda_a, \lambda_b, JM\rangle \pm 2^{-\frac{1}{2}}|-\lambda_a, -\lambda_b, JM\rangle, \\ \mathscr{P}|\lambda_a, \lambda_b, JM\rangle_\pm &= \pm\eta(-1)^{S_a+S_b-J}|\lambda_a, \lambda_b, JM\rangle_\pm . \end{aligned} \tag{3.2}$$

For the $\pi N$-system, these states are automatically eigenstates of the orbital angular momentum $L$ which dictates the threshold behaviour $p^L$. When both particles carry spin, this is normally not the case. The final choice of states is then a question of physical approximations. If particle $b$ is much heavier than particle $a$ (as in electron scattering or photoproduction on nucleons), it is convenient to use states $|\lambda_a J_a, JM\rangle$ which are eigenstates of $\mathbf{J}_a^2$ with $\mathbf{J}_a = \mathbf{S}_a + \mathbf{L}$, where $\mathbf{L}$ is the orbital angular momentum of the 2-body system. They are constructed from the states $|\lambda_a J_a J_{az}\rangle$ and the target spin states $|M_b\rangle$ by means of Clebsch-Gordans:

$$|\lambda_a J_a, JM\rangle = \sum_{M_b} C^{J_a\ S_b\ J}_{J_{az}\ M_b\ M}|\lambda_a J_a, J_{az}\rangle|M_b\rangle. \tag{3.3}$$

Here particle $a$ is still in the helicity basis while particle $b$ is in the "old spin basis." To find the connection between the states $|\lambda_a\lambda_b JM\rangle$ and (4), we need the inverse of (4-4.5). For this purpose, we observe that (B-3.3) and (B-3.2) imply

$$\int d\phi\, d\cos\theta\, D^J_{M\lambda}(\phi, \theta, -\phi)\, D^{j*}_{m\lambda}(\phi, \theta, -\phi) = 4\pi(2J+1)^{-1}\,\delta_{Jj}\,\delta_{Mm}. \tag{3.4}$$

The normalization coefficient $C_J$ in (4-4.5) is conveniently chosen such that it also normalizes the inverse transformation. We then find from (4)

$$|\lambda_a\lambda_b JM\rangle = C_J \int d\Omega\, D^{J*}_{M\lambda}(\phi, \theta, -\phi)|\lambda_a\lambda_b\hat{\mathbf{p}}\rangle, \qquad C_J^2 = (2J+1)/4\pi. \tag{3.5}$$

One-particle helicity states are now defined as in (4-4.5)

$$|\lambda_a\hat{\mathbf{p}}\rangle = \sum_{J_a, J_{az}} \left(\frac{2J_a+1}{4\pi}\right)^{\frac{1}{2}} D^{J_a}_{J_{az}, \lambda_a}(\phi, \theta, -\phi)|\lambda_a J_a, J_{az}\rangle, \tag{3.6}$$

and the 2-particle helicity states are

$$\begin{aligned} |\lambda_a\lambda_b\hat{\mathbf{p}}\rangle = |\lambda_a\hat{\mathbf{p}}\rangle \sum_{M_b} D_{M_b, -\lambda_b}|M_b\rangle = \sum_{J_a J_{az} M_b J} \left(\frac{2J_a+1}{4\pi}\right)^{\frac{1}{2}} \\ \times C^{J_a\ S_b\ J}_{J_{az}\ M_b\ M} C^{J_a S_b\ \ J}_{\lambda_a\ -\lambda_b\ \lambda} D^J_{M\lambda}|\lambda_a J_a, J_{az}\rangle|M_b\rangle. \end{aligned} \tag{3.7}$$

In the last expression, we have inserted (6) and used the Clebsch-Gordan series (B-3.6). We have now succeeded in expressing the two-particle helicity states in the same basis as (3), and even the first Clebsch-Gordan is the same. The angular integration in (5) is easily performed, and comparison with (3) shows that

$$|\lambda_a\lambda_b JM\rangle = \sum_{J_a}\left(\frac{2J_a+1}{2J+1}\right)^{\frac{1}{2}} C^{J_a\,S_b\;J}_{\lambda_a\,-\lambda_b\,\lambda}\,|\lambda_a J_a, JM\rangle. \tag{3.8}$$

If on the other hand both particles have similar masses, a symmetric treatment of $a$ and $b$ is appropriate. One then uses states $|LSJM\rangle$ which are eigenstates of $\mathbf{L}^2$ and $\mathbf{S}^2$ with $\mathbf{S} = \mathbf{S}_a + \mathbf{S}_b$:

$$|LSJM\rangle = \sum_{M_aM_b} C^{S_a\,S_b\,S}_{M_a\,M_b\,M}\, C^{L\;S\;J}_{m\,M_S\,M}\,|M_a\rangle|M_b\rangle|Lm\rangle. \tag{3.9}$$

The two-particle helicity states (7) are now decomposed as follows:

$$|\lambda_a\lambda_b\hat{\mathbf{p}}\rangle = \sum_{M_aM_b} D^{S_a}_{M_a\lambda_a} D^{S_b}_{M_b-\lambda_b}\,|M_a\rangle|M_b\rangle \sum_{Lm}\left(\frac{2L+1}{4\pi}\right)^{\frac{1}{2}} D^{L}_{m0}|Lm\rangle. \tag{3.10}$$

The triple product of $D$-functions is then reduced to a double sum over functions $D^J_{M\lambda}$ by applying the Clebsch-Gordan series twice, the integral (5) goes as before, and comparison with (9) gives

$$|\lambda_a\lambda_b JM\rangle = \sum_{L,S}\left(\frac{2L+1}{2J+1}\right)^{\frac{1}{2}} C^{S_a\,S_b\;S}_{\lambda_a\,-\lambda_b\,\lambda}\, C^{L\,S\,J}_{0\,\lambda\,\lambda}\,|LSJM\rangle. \tag{3.11}$$

Since the states $|LSJM\rangle$ are $\mathbf{L}^2$-eigenstates, they are also parity eigenstates. We can also transform (8) into an equation between the parity eigenstates $|\lambda_a\lambda_b JM\rangle_\pm$ and $|\lambda_a J_a, JM\rangle_\pm$, by means of the symmetry of Clebsch-Gordans

$$C^{J_a\;S_b\;\;J}_{-\lambda_a\,\lambda_b\,-\lambda} = (-1)^{J_a+S_b-J} C^{J_a\,S_b\;J}_{\lambda_a\,-\lambda_b\,\lambda}. \tag{3.12}$$

For $S_b = \frac{1}{2}$ and $\lambda_b = -\frac{1}{2}$, (8) becomes

$$\begin{aligned}|\lambda_a - \tfrac{1}{2}JM\rangle_\pm &= \left(\frac{2J}{2J+1}\right)^{\frac{1}{2}} C^{J-\frac{1}{2}\,\frac{1}{2}\,J}_{\lambda_a\;\frac{1}{2}\,\lambda}\,|\lambda_a J_a, JM\rangle_\pm \\ &\quad + \left(\frac{2J+2}{2J+1}\right)^{\frac{1}{2}} C^{J+\frac{1}{2}\,\frac{1}{2}\,J}_{\lambda_a\;\frac{1}{2}\,\lambda}\,|\lambda_a J_a JM\rangle_\mp \\ &= \left(\frac{J+\lambda}{2J+1}\right)^{\frac{1}{2}} |\lambda_a J_a JM\rangle_\pm - \left(\frac{J-\lambda+1}{2J+1}\right)^{\frac{1}{2}} |\lambda_a J_a JM\rangle_\mp,\end{aligned} \tag{3.13}$$

using (B-4.2). The state with $\lambda_b = +\frac{1}{2}$ is identical with (13), apart from a phase. For electrons and real photons, the index $\lambda_a$ is also superfluous. For photons, $J_a$ is the multipolarity and normally denoted by $L$. The case $\lambda_\gamma = 0$ (longitudinal multipoles) occurs only for virtual photons. The antisymmetric combination $|0LJM\rangle_-$ does not exist in this case. The parity of the one-photon states $|\lambda_\gamma LJM\rangle_\pm$ is identical to that of $|\lambda_a LJ_{az}\rangle_\pm$, since the spin state $|M_b\rangle$ is unaffected by parity. Analogous to (1), we have with $\eta_\gamma = -1$, $S_\gamma = 1$:

$$\mathscr{P}|\lambda_\gamma LJM\rangle_\pm = \pm(-1)^L|\lambda_\gamma LJM\rangle_\pm, \tag{3.14}$$

which shows that $|1LJM\rangle_+$ is the electric $|EL\rangle$ and $|1LJM\rangle_-$ the magnetic $|ML\rangle$ multipole.

## 7-4 Pion Photo- and Electroproduction

The reactions $\gamma N \to \pi N$ and $eN \to e'\pi N$ are closely related to elastic $\pi N$ scattering, particularly at low energies. The isospin decomposition of these reactions has already been given in (4-14.4) and (4-14.5). The partial-wave decomposition has the general form (4-4.9). As both the photon and the pion have negative intrinsic parities, $\eta = \eta' = -1$ must be used in the parity invariance equation (4-4.25):

$$T_J(-\lambda_\gamma, -\lambda_b, -\lambda_d) = -T_J(\lambda_\gamma, \lambda_b, \lambda_d). \tag{4.1}$$

For the construction of parity-conserving matrix elements, it suffices to use parity eigenstates either in the initial or in the final state. We adapt the parity notation to the final $\pi N$ state. With $J \equiv l + \frac{1}{2}$ as in (2-5.26), the notation is (Walker 1969):

$$A_{l+} = {}_+\langle\tfrac{1}{2}|T_J|1\tfrac{1}{2}\rangle, \qquad A_{(l+1)-} = {}_-\langle\tfrac{1}{2}|T_J|1\tfrac{1}{2}\rangle,$$

$$B_{l+} = 2[l(l+2)]^{-\frac{1}{2}}{}_+\langle\tfrac{1}{2}|T_J|1-\tfrac{1}{2}\rangle, \quad B_{(l+1)-} = 2[l(l+2)]^{-\frac{1}{2}}{}_-\langle\tfrac{1}{2}|T_J|1-\tfrac{1}{2}\rangle. \tag{4.2}$$

The amplitudes $A$ and $B$ have initial helicities $\lambda = \frac{1}{2}$ and $\frac{3}{2}$, respectively. Their multipole decomposition is

$$A_{l+} = \tfrac{1}{2}lM_{l+} + (\tfrac{1}{2}l+1)E_{l+}, \qquad A_{(l+1)-} = (\tfrac{1}{2}l+1)M_{(l+1)-} - \tfrac{1}{2}lE_{(l+1)-},$$

$$B_{l+} = E_{l+} - M_{l+}, \qquad B_{(l+1)-} = -E_{(l+1)-} - M_{(l+1)-}. \tag{4.3}$$

The $M_{l\pm}$ have "multipolarity" $l$, whereas $E_{l\pm}$ have multipolarity $l \pm 1$. In particular, $E_{0+}$ and $E_{2-}$ are both $E1$-amplitudes. The value of the index in front of the $\pm$ in (3) is the $N\pi$ orbital angular momentum.

A more practical form of (4-4.9) is

$$T(\lambda_\gamma\lambda_b\lambda_d, s, \Omega) = 8\pi s^{\frac{1}{2}}\chi_0'^+(\lambda_d)F\chi_0(\lambda_b) \tag{4.4}$$

$$F = \boldsymbol{\sigma\varepsilon}F_1 + \boldsymbol{\sigma}\hat{\mathbf{q}}(-i\boldsymbol{\sigma}(\hat{\mathbf{k}} \times \boldsymbol{\varepsilon})F_2 + \hat{\mathbf{q}}\boldsymbol{\varepsilon}F_4 + \hat{\mathbf{k}}\boldsymbol{\varepsilon}F_6 - \varepsilon^0F_7)$$
$$+ \boldsymbol{\sigma}\hat{\mathbf{k}}(\hat{\mathbf{q}}\boldsymbol{\varepsilon}F_3 + \hat{\mathbf{k}}\boldsymbol{\varepsilon}F_5 - \varepsilon^0F_8) \tag{4.5}$$

In this section, $\mathbf{k}$ and $\mathbf{q}$ denote the cms momenta of photon and pion, respectively.

For real photons, we have $\varepsilon^0 = 0$, $\hat{\mathbf{k}} \cdot \boldsymbol{\varepsilon} = 0$, such that only $F_1 \cdots F_4$ contribute. For electroproduction,

$$\varepsilon^\mu = -et^{-1}\bar{u}_e'\gamma^\mu u_e \tag{4.6}$$

according to (2-7.6) and (2-6.27) with $Z_e = -1$. Gauge invariance permits one to eliminate two of the amplitudes $F_5 \cdots F_8$

$$F_1 + \hat{\mathbf{k}}\hat{\mathbf{q}}F_3 + F_5 = F_8 K^0/k, \qquad \hat{\mathbf{k}}\hat{\mathbf{q}}F_4 + F_6 = F_7 K^0/k. \tag{4.7}$$

Following Behrends, Donnachie and Weaver (1967), we eliminate $F_6$ and $F_7$, which amounts to replacing $\varepsilon^\mu$ in (5) by

$$b^\mu = \varepsilon^\mu - (\boldsymbol{\varepsilon}\hat{\mathbf{k}})K^\mu/k, \qquad \mathbf{b} \cdot \hat{\mathbf{k}} = 0. \tag{4.8}$$

In this form, the longitudinal photons are eliminated in favour of the scalar ones. Using now table B-5 to express the $d$-functions in terms of the derivatives of Legendre polynomials, one finds

$$\tilde{F} = \sum_{l=0}^{\infty} G_l \tilde{M}_l, \qquad \tilde{S} = \sum_{l=0}^{\infty} H_l \tilde{S}_l$$

$$\tilde{F} \equiv \begin{bmatrix} F_1 \\ F_2 \\ F_3 \\ F_4 \end{bmatrix}, \qquad \tilde{M}_l \equiv \begin{bmatrix} E_{l+} \\ E_{l-} \\ M_{l+} \\ M_{l-} \end{bmatrix} \qquad \tilde{S} \equiv \begin{bmatrix} F_7 \\ F_8 \end{bmatrix}, \qquad \tilde{S}_l \equiv \begin{bmatrix} S_{l+} \\ S_{l-} \end{bmatrix}$$

$$H_l = \begin{bmatrix} -(l+1)P'_l & lP'_l \\ (l+1)P'_{l+1} & -lP'_{l-1} \end{bmatrix}, \tag{4.9}$$

$$G_l = \begin{bmatrix} P'_{l+1} & P'_{l-1} & lP'_{l+1} & (l+1)P'_{l-1} \\ 0 & 0 & (l+1)P'_l & lP'_l \\ P''_{l+1} & P''_{l-1} & -P''_{l+1} & P''_{l-1} \\ -P''_l & -P''_l & P''_l & -P''_l \end{bmatrix}.$$

The differential cross section for unpolarized photons and nucleons is

$$d\sigma/d\Omega = \frac{1}{2}\frac{q}{k} \sum_{\lambda_b \lambda_d} |\chi_b'^{+}(\lambda_d) F(\lambda_\gamma = 1)\chi_0(\lambda_b)|^2, \tag{4.10}$$

where averaging over the photon helicity $\lambda_\gamma$ has been omitted due to parity conservation (1). Near threshold ($q \to 0$), $E_{0+}$ dominates in $\pi^\pm$ production, in which case

$$F(q \to 0) = i\boldsymbol{\varepsilon\sigma}E_{0+}, \qquad d\sigma/d\Omega \to \frac{q}{k}|E_{0+}|^2. \tag{4.11}$$

Above threshold, the production of $\Delta^+$ and $\Delta^0$ grows rapidly, and since the $E_{0+}$-amplitude is very small in $\pi^0 n$ and $\pi^0 p$ final states, one sees the $\Delta$ practically free of background in these states. The $\Delta$ resonance can be excited both by $E_{1+}$ and $M_{1+}$ amplitudes. A multipole analysis shows that $E_{1+}$ is much smaller than $M_{1+}$ and moreover that $E_{1+}$ is mostly nonresonant (Behrends and Donnachie 1975 and fig. 7-4.1). Another resonance which is strongly excited in pion photoproduction is the $D_{13}(\frac{3}{2}^-)$ at 1520 MeV (see

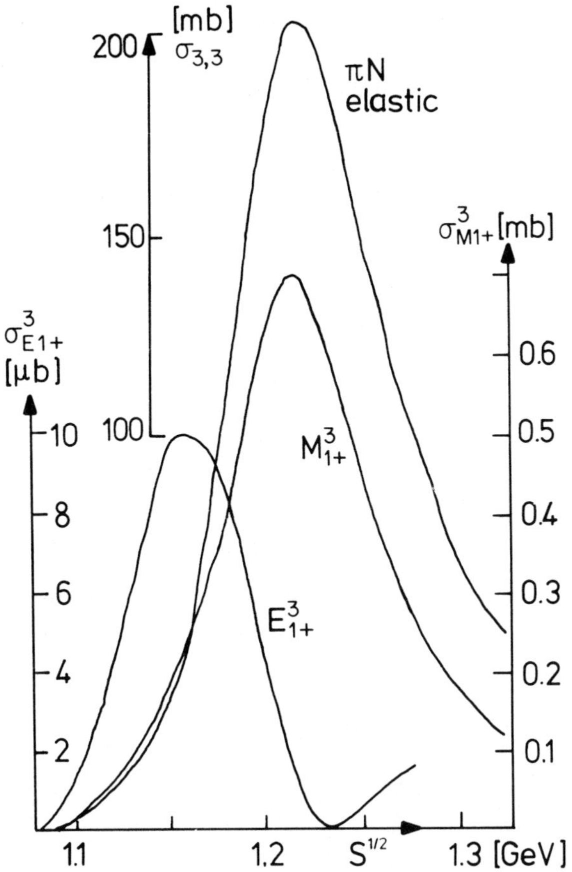


**Figure 7-4.1** The low-energy $J = I = \frac{3}{2}$ $P$-wave cross sections (after Olsson 1976). Near the $\Delta$ resonance, $\sigma(\gamma p \to \pi^0 p) \approx \frac{4}{9}\sigma^3_{M1+}$.

table 7-2.2). $M_{2-}$ is much smaller than $E_{2-}$ for that resonance. Consequently only $E_{0+}$, $E_{1\pm}$, $M_{1\pm}$ and $E_{2-}$ are appreciable up to the $D_{13}$ resonance energy. The differential cross section for unpolarized photons and nucleons is then

$$
\begin{aligned}
d\sigma/d\Omega &= \frac{q}{k}(A + B\cos\theta + C\cos^2\theta + 18\,\mathrm{Re}\,E^*_{1+}E_{2-}\cos^3\theta),\\
A &= |E_{0+}|^2 + |M_{1-}|^2 + \tfrac{9}{2}|E_{1+}|^2 + \tfrac{5}{2}|M_{1+}|^2 + \tfrac{5}{2}|E_{2-}|^2\\
&\quad + \mathrm{Re}\,(M^*_{1+}M_{1-} - E^*_{0+}E_{2-} - 3M^*_{1+}E_{1+} + 3M^*_{1-}E_{1+}),\\
B &= 2\,\mathrm{Re}\,(E^*_{0+}M_{1+} - M^*_{1}E_{2-} - E^*_{0+}M_{1-}\\
&\qquad + M^*_{1-}E_{2-} + 3E^*_{0+}E_{1+} - E^*_{1+}E_{2-}),\\
C &= \tfrac{3}{2}(3|E_{1+}|^2 - |M_{1+}|^2 - |E_{2-}|^2)\\
&\quad + 3\,\mathrm{Re}\,(E^*_{0+}E_{2-} - M^*_{1-}M_{1+} + 3E^*_{1+}M_{1+} - 3M^*_{1-}E_{1+}).
\end{aligned}
\tag{4.12}
$$

Once the resonating multipoles $E^R_{l\pm}$ and $M^R_{l\pm}$ are known, one can compute the radiative partial width $\Gamma(R \to N\gamma)$ from the residue factorization. At $s^{\frac{1}{2}} = m_R$, insertion of (4-8.23) into (4-8.24) gives

$$\Gamma(R \to ab)\Gamma(R \to cd) = \Gamma^2 pp' \sum_{\lambda_a \cdots \lambda_d} |T^R_J(ab \to cd)|^2, \tag{4.13}$$

where $J$ is the total angular momentum of the resonance. According to (1), (2) and (3), the sum in (13) is

$$\begin{aligned} 2|A^R_{l+}|^2 + \tfrac{1}{2}l(l+2)|B^R_{l+}|^2 &= (l+1)[l|M^R_{l+}|^2 + (l+2)|E^R_{l+}|^2] \\ &\qquad \text{for} \quad J = l + \tfrac{1}{2} \\ 2|A^R_{l-}|^2 + \tfrac{1}{2}l(l+2)|B^R_{l-}|^2 &= l(l-1)|E^R_{l-}|^2 + l(l+1)|M^R_{l-}|^2 \\ &\qquad \text{for} \quad J = l - \tfrac{1}{2}. \end{aligned} \tag{4.14}$$

These formulas are independent of isospin invariance. For the $\Delta$ resonance, we may put $E^R_{1+} \approx 0$, $\Gamma(\Delta \to N\pi^0) \approx \frac{2}{3}\Gamma$, getting

$$\Gamma(\Delta \to N\gamma) \approx 3\Gamma kq |M^R_{1+}(\gamma N \to N\pi^0)|^2 = \tfrac{4}{3}|M^3_{1+}|^2. \tag{4.15}$$

At energies where the inelastic partial-wave cross section is negligible, $\sigma^I_{L\pm}(\pi N \to \pi\pi N) \ll \sigma^I_{L\pm}(\pi N)$, the isovector and isoscalar photoproduction amplitudes $E^{vI}_{L\pm}$, $M^{vI}_{L\pm}$ and $E^s_{L\pm}$, $M^s_{L\pm}$ have the same phases as the corresponding elastic scattering amplitudes $f^I_{L\pm}$. This is called Watson's (1954) theorem. Its mathematical basis is analogous to that of (4-8.29): To first order in the electric charge $e$, the matrix $T_J$ of (4-4.11) can be identified with a submatrix $T^I_{L\pm}$ of given isospin, total angular momentum $J = L \pm 1$ and parity $-(-1)^L$ of the hadronic system. It is symmetric by time-reversal invariance (4-4.19). For $I = \frac{3}{2}$, it is of the form

$$T^{\frac{3}{2}}_{L\pm} = \begin{bmatrix} f^{\frac{3}{2}}_{L\pm} & E^{v3}_{L\pm} & M^{v3}_{L\pm} & f^{\frac{3}{2}}_{L\pm}(\pi N \to \pi\pi N) \\ E^{v3}_{L\pm} & 0 & 0 & E^{v3}_{L\pm}(\pi\pi N) \\ M^{v3}_{L\pm} & 0 & 0 & M^{v3}_{L\pm}(\pi\pi N) \\ f^{\frac{3}{2}}_{L\pm}(\pi N \to \pi\pi N) & E^{v3}_{L\pm}(\pi\pi N) & M^{v3}_{L\pm}(\pi\pi N) & f^{\frac{3}{2}}_{L\pm}(\pi\pi N) \end{bmatrix} \tag{4.16}$$

The Compton scattering amplitudes $T(\gamma N \to \gamma' N')$ which occupy the center of this matrix vanish to this order in $e$, but the amplitudes for $\pi N \to \pi\pi N$, $\pi\pi N \to \pi\pi N$ or $\gamma N \to \pi\pi N$ need not vanish. The 3-particle state $\pi\pi N$ is neglected among the intermediate states $k$ on the right-hand side of the unitarity equation (4-4.10), because its phase space is small or even zero. The state $k = \gamma N$ is also negligible because it contributes to the unitarity condition only to order $e^2$ (in $\pi N$ elastic scattering) or $e^3$ (in photoproduction). Thus the only important state is $k = \pi N$, and (4-4.10) becomes

$$\text{Im } T^I_{if,\,L\pm} = p_{\pi N} T^I(i \to \pi N, L\pm) T^{I*}(f \to \pi N, L\pm) \tag{4.17}$$

for *any* pair of channels $(i, f)$ of the matrix (16). Specifying now $f = \pi N$ and $f^I_{L\pm} \sim \sin \delta^I_{L\pm} \exp(i\,\delta^I_{L\pm})$, we obtain Watson's theorem from (14). According to our discussion of section 7-2, the first breakdown of the theorem is expected in the $P_{11}$ partial wave, but the effect appears to be small up to 450 MeV photon lab energy.

By Low's theorem (3-13.18), terms of order $k^{-1}$ and $k^0$ in photoproduction are given by the Born approximation. Neglecting the nucleon recoil, we have $k \approx m_\pi$ at threshold. Therefore, terms of order $m_\pi^{-1}$ and $m_\pi^0$ at threshold are given by the Born terms (Kroll-Ruderman theorem, 1954). The corresponding graphs are shown in fig. 7-4.2. The $t$-channel Born term contributes only to $\pi^\pm$-production:

$$T^{(t)} = -2^{\frac{1}{2}} G \bar{u}_f \gamma_5 u_i 2eP_\pi \cdot \varepsilon \, (m_\pi^2 - t)^{-1}, \qquad t = (P_\gamma - P_\pi)^2. \tag{4.18}$$

For $\pi^+$ production, the minus sign comes from (5-8.4), while for $\pi^-$ production, it comes from $Z(\pi^-) = -1$. The $s$-channel baryon exchange Born term for $\pi^\pm$ production is obtained from (2-6.28) and (5-8.4)

$$T^{(e)} = -Z_\pi 2^{\frac{1}{2}} G \bar{u}_p \gamma_5 \frac{(P_i + K)\cdot\gamma + m}{m^2 - s} e\left(Z_i \gamma \cdot \varepsilon - \frac{\kappa_i}{2m} \varepsilon_\mu \sigma^{\mu\nu} K_\nu\right) u_i, \tag{4.19}$$

where $Z_i$ and $\kappa_i$ are the charge and anomalous magnetic moment of the initial baryon. Similarly, the $u$-channel Born term is

$$T^{(u)} = -Z_\pi 2^{\frac{1}{2}} G \bar{u}_f e\left(Z_f \gamma \cdot \varepsilon - \frac{\kappa_f}{2m} \varepsilon_\mu \sigma^{\mu\nu} K_\nu\right) \frac{(P_f - K)\gamma + m}{m^2 - u} \gamma_5 u_i. \tag{4.20}$$

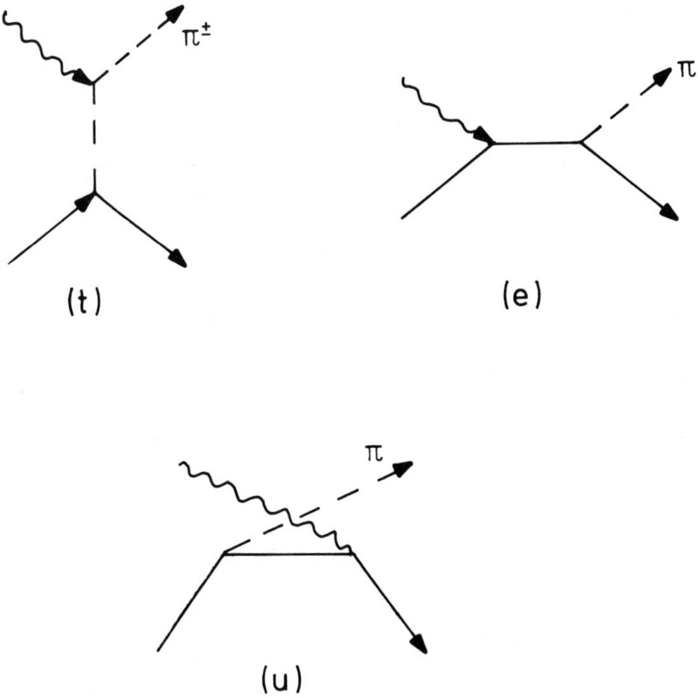


**Figure 7-4.2** The Born terms of pion photoproduction.

For $\pi^0$ production, the factor $-Z_\pi 2^{\frac{1}{2}} G$ is replaced by $G$ in $\gamma p \to \pi^0 p$ and by $-G$ in $\gamma n \to \pi^0 n$, according to (5-8.4).

The gauge invariance of the sum of the Born terms (18) and (20) for $\pi^-$ production was already demonstrated following (5-6.8), at least for the charge coupling ($Z_f = 1$). The anomalous magnetic moment coupling is separately gauge invariant, due to $K_\mu \sigma^{\mu\nu} K_\nu = 0$. Similarly, for $\pi^+$ production, the sum of (18) and (19) is gauge invariant.

So far our Born terms apply both to photo- and electroproduction. For real photons, we have $\varepsilon^0 = 0$, $P_\pi \cdot \varepsilon = -\mathbf{p}_\pi \boldsymbol{\varepsilon}$, $\gamma \cdot \varepsilon = -\boldsymbol{\gamma\varepsilon}$ etc. Thus (18) vanishes at threshold. Its $P$-wave contribution is not particularly large either, because of (A-3.10), $\bar{u}_f \gamma_5 u_i \sim (-t)^{\frac{1}{2}} \approx m_\pi$. The dominant contribution at threshold comes from the large component of $\bar{u}_f \gamma_5 \gamma \cdot \varepsilon u_i$ in (19) and (20). To evaluate it, we replace $m$ by $-P_j \cdot \gamma$ in (19) and by $-P_i \cdot \gamma$ in (20), obtaining

$$(P_i + K - P_f)\gamma = P_\pi \cdot \gamma \approx m_\pi \gamma^0 \tag{4.21}$$

in (19) and $-m_\pi \gamma^0$ in (20). This expression anticommutes both with $\gamma_5$ and with $\varepsilon \cdot \gamma = -\boldsymbol{\varepsilon\gamma}$, and $\bar{u}_f \gamma^0 = u_f$ for $\mathbf{p}_f = 0$. With $\bar{u}_f \gamma_5 \boldsymbol{\gamma\varepsilon} u = 2m\chi'^+ \boldsymbol{\sigma\varepsilon}\chi$ according to (A-3.12), we find

$$T^{(e)}(s^{\frac{1}{2}} = m_\pi + m) = -Z_\pi 2^{\frac{1}{2}} G \frac{m_\pi e Z_i - 2m}{m^2 - s} \chi'^+ \boldsymbol{\sigma\varepsilon}\chi \approx 2^{\frac{1}{2}} Ge\left(1 - \frac{m_\pi}{2m}\right)\chi'^+ \boldsymbol{\sigma\varepsilon}\chi \tag{4.22}$$

for $Z_\pi = Z_i = 1$. Comparison with (4) and (12) shows that

$$E_{0+}(\pi^+) \approx 2^{-\frac{1}{2}}(4\pi)^{-1} Gem^{-1}(1 - \tfrac{3}{2}m_\pi/m) = 0.190/\text{GeV}. \tag{4.23}$$

The charge coupling of (20) does not contribute to (23) because of $Z_f = 0$. The anomalous magnetic moments of (19) and (20) contribute only in their isoscalar combination $\kappa^s = \frac{1}{2}\kappa_p + \frac{1}{2}\kappa_n$ which is negligible. Similarly, for $E_{0+}(\pi^-)$ at threshold it suffices to evaluate the charge coupling of (20). The calculation goes as above, and with $\boldsymbol{\gamma\varepsilon}\gamma_5 = -\gamma_5 \boldsymbol{\gamma\varepsilon}$, also the sign of the numerator remains the same. The denominator is different, however. With $t \approx m_\pi^2 + 2kq \cos\theta$ and $s + t + u = 2m^2 + 2m_\pi^2$, we have

$$2mm_\pi(m^2 - u)^{-1} = 2mm_\pi(2mm_\pi + t)^{-1} \approx 1 + \tfrac{1}{2}m_\pi/m - kq(m_\pi m)^{-1}\cos\theta, \tag{4.24}$$

such that the final result for $q = 0$ is

$$E_{0+}(\pi^-) \approx -2^{-\frac{1}{2}}(4\pi)^{-1} Gem^{-1}(1 - m_\pi/2m) = -0.226/\text{GeV}. \tag{4.25}$$

The agreement between (23) and (25) and the experimental values of $E_{0+}(\pi^\pm)$ at threshold is better than one could expect from this derivation, which is only valid to order $m_\pi^0$. With current algebra constraints one gets the $m_\pi/m$-corrections unambiguously, but the agreement with experiment is not improved (de Baenst 1970). It is also interesting to note that in the full one-pion exchange matrix element, the pion propagator modification cancels against its electric form factor by means of the Ward identity (Fischer and Minkowski 1972). A similar cancellation occurs in the nucleon charge form factor.

For $\gamma p \to \pi^0 p$, one must add (23) and (25) and multiply the result by $-2^{-\frac{1}{2}}$:

$$E_{0+}(p\pi^0) \approx -\tfrac{1}{2}(4\pi)^{-1}Gem_\pi/m^2 = -0.026/\text{GeV}. \tag{4.26}$$

It is also of some interest to compute the Born term contributions to the multipoles $M_{1+}^{v3}$ and $E_{1+}^{v3}$ at threshold, because of the $\Delta$ resonance in these multipoles. According to (10), $M_{1+}$ enters $F_1$ with a factor $P_2' = 3\cos\theta$. $T^{(t)}$ and $T^{(e)}$ contain no term $\boldsymbol{\sigma\varepsilon}\cos\theta$ and thus do not contribute to $M_{1+}$ (as in elastic $\pi N$-scattering, $T^{(e)}$ contributes only to the $J = I = \frac{1}{2}$ state). Writing the e.m. current matrix element in the form $(Z_f + \kappa_f)\gamma^\mu - \frac{1}{2}\kappa_f m^{-1}(P_i + P_f)^\mu$, as in (2-6.30), we get the whole $\boldsymbol{\sigma\varepsilon}\cos\theta$-contribution from the first piece which is treated as in the case of $E_{0+}$. The coefficient of $\cos\theta$ follows from (24) and gives for $\gamma n \to p\pi^-$

$$M_{1+}^{(u)}(\pi^-) \approx 2^{-\frac{1}{2}}(12\pi m_\pi m^2)^{-1}Gekq(1 + \kappa_p), \tag{4.27}$$

while for $\gamma p \to n\pi^+$, $1 + \kappa_p$ is replaced by $-\kappa_n$. For $I = \frac{3}{2}$, we need the combination

$$T^{v3} = \tfrac{1}{2}[2^{-\frac{1}{2}}T(\pi^+) + 2^{-\frac{1}{2}}T(\pi^-) + T(p\pi^0) + T(n\pi^0)] \tag{4.28}$$

according to (4-14.4). The $\pi^0$-production amplitudes contribute as much as the $\pi^\pm$-production amplitudes to (28), such that we obtain

$$M_{1+}^{(u)v3} = \tfrac{1}{2}(12\pi m_\pi m^2)^{-1}Gekq(1 + 2\kappa^v), \qquad \kappa^v = \tfrac{1}{2}(\kappa_p - \kappa_n). \tag{4.29}$$

The amplitude $E_{1+}^{(u)v3}$ is one order $m_\pi$ smaller than (29). If the Born term had nothing to do with the $\Delta$ resonance, one would simply add it as nonresonant contribution. However, in the simple dispersion relation (2.21), the elastic Born term is the "driving force" of the resonance. The corresponding model for pion photoproduction shows that the magnetic and electric decay constants of the $\Delta$ are proportional to $M_{1+}^{(u)v3}$ and $E_{1+}^{(u)v3}$. This gives a qualitative understanding of the dominance of the $M_{1+}$-amplitude in $\Delta \to N\gamma$ decays. Moreover, $\Gamma(\Delta \to N\gamma)/\Gamma$ is determined by the ratios of the corresponding $u$-channel Born terms as follows: we assume that the resonant amplitude factorizes as in (4-8.22) and that the only energy dependence of $r = T_S(\Delta \to N\gamma)/T_S(\Delta \to N\pi)$ is the $P$-wave threshold factor $k/q$. Apart from this factor, $r$ can then be evaluated at threshold, where it is given by the ratio of the corresponding Born terms. With (2.19), we find

$$(gM_{1+}^{v3}/kf_{1+}^{\frac{3}{2}})_{s=m_\Delta^2} = \tfrac{1}{2}G^{-1}e(1 + \kappa_p - \kappa_n). \tag{4.30}$$

However, the argument is not compelling. In particular, the power series expansion which forms the basis of Low's theorem (3-13.18) converges poorly near a sharp resonance. An alternative description of pion resonance production is provided by the quark model (see Moorhouse et al. 1974).

## 7-5 *NN*-Scattering

Ever since the prediction of a light boson as the carrier of nuclear forces (Yukawa 1935), the nucleon-nucleon interaction has played a special role in particle theory. The subject is reviewed in most textbooks on nuclear theory. For an elaborate presentation of the theoretical framework, see Brown and Jackson (1976). The phenomenological analysis has been presented by Breit and Harasz (1967).

We begin with the partial-wave decomposition (4-4.9) for an arbitrary reaction $ab \to cd$ between spin-$\frac{1}{2}$ particles of equal parities (for example $pp \to pp$ or $\Sigma^- p \to \Lambda n$). According to (4-4.25), parity conservation requires

$$T_J(-\lambda_a \cdots -\lambda_d) = T_J(\lambda_a \cdots \lambda_d). \tag{5.1}$$

Parity eigenstates are given by (3.2), and the partial-wave matrix elements for these states are

$$\begin{gathered} T^{J\pm}_{\sigma\sigma'} = {}_\pm\langle\sigma'|T^J|\sigma\rangle_\pm = T_J(\lambda_a \cdots \lambda_d) \pm T_J(\lambda_a, \lambda_b, -\lambda_c, -\lambda_d), \\ \sigma = 4\lambda_a\lambda_b, \qquad \sigma' = 4\lambda_c\lambda_d, \qquad \lambda_a \equiv +\tfrac{1}{2}. \end{gathered} \tag{5.2}$$

At low energies, the states $|LSJM\rangle$ of (3.9) are convenient because of their threshold behaviour

$$\lim \begin{pmatrix} p \to 0 \\ p' \to 0 \end{pmatrix} \langle L'S'JM|T|LSJM\rangle = p^L(p')^{L'}. \tag{5.3}$$

In general neither $L$ nor $S$ need be conserved. For $NN$-scattering, isospin conservation entails conservation of $S$, as we shall see. With $S_a = S_b = \frac{1}{2}$, $S$ can only be 0 or 1, and the second Clebsch-Gordan of (3.9) requires $L = J$ or $L = J \pm 1$. The states $|\sigma JM\rangle_\pm$ have parity $\pm(-1)^{1-J} = \mp(-1)^J$ according to (3.2), while the $|LSJM\rangle$ have parity $(-1)^L$. Thus parity conservation requires $L = J \pm 1$ for $|\sigma JM\rangle_+$, and $L = J$ for $|\sigma JM\rangle_-$. The second case is particularly simple, since

$$C^{J\,0\,J}_{0\,0\,0} = 1, \qquad C^{J\,1\,J}_{0\,0\,0} = 0, \qquad C^{J\,1\,J}_{0\,1\,1} = -2^{-\frac{1}{2}}. \tag{5.4}$$

The spin-addition coefficient of (3.11) is $2^{-\frac{1}{2}}$ except for $\lambda = 1$, when it is unity. Finally, we have a factor $2^{-\frac{1}{2}}$ from (3.2) and a factor 2 from the second term of (3.2) (using the symmetry 3.12), such that the net result is

$$|1JM\rangle_- = |L0JM\rangle, \qquad |-1JM\rangle_- = -|L1JM\rangle, \qquad L = J. \tag{5.5}$$

The first state has spin $S = 0$ which is antisymmetric under exchange of the two particles, while the second has $S = 1$ which is symmetric. For $pp$ scattering, the Pauli principle requires even $L$ for $S = 0$ and odd $L$ for $S = 1$, such that no transition between the two types of states (5) occurs. For $pn$ scattering, the isospin decomposition (4-2.4) leads to the generalized Pauli principle, and to the extent that transitions from $I = 0$ to $I = 1$ are negligible, transitions between the two types of states are again forbidden. Since the remaining states $|\sigma JM\rangle_+$ have $L = J \pm 1$, they must have $S = 1$. Thus $S$ is conserved in nucleon-nucleon scattering. For the states $|\sigma JM\rangle_+$, we find from (3.11)

$$\begin{matrix} |1, JM\rangle_+ \\ |-1, JM\rangle_+ \end{matrix} = \left(\frac{J}{2J+1}\right)^{\frac{1}{2}} |J \mp 1, 1, JM\rangle \mp \left(\frac{J+1}{2J+1}\right)^{\frac{1}{2}} |J \mp 1, 1, JM\rangle. \quad (5.6)$$

Explicit expressions for $J < 4$ are collected in table 7-5.1. Transitions from $L = J - 1$ to $L = J + 1$ are allowed. For $J = 1$, we find from table 7-5.1

$$T\begin{pmatrix} S \to S \\ D \to D \end{pmatrix} = \tfrac{1}{3} T^{1+}_{\pm 1, \pm 1} + \tfrac{2}{3} T^{1+}_{\mp 1, \mp 1} \pm \tfrac{1}{3} 2^{\frac{1}{2}} (T^{1+}_{-1, 1} + T^{1+}_{1, -1}),$$
$$T\begin{pmatrix} S \to D \\ D \to S \end{pmatrix} = \tfrac{1}{3} 2^{\frac{1}{2}} (T^{1+}_{-1, -1} - T^{1+}_{1, 1}) - \tfrac{2}{3} T^{1+}_{\pm 1, \mp 1} + \tfrac{1}{3} T^{1+}_{\mp 1, \pm 1}. \quad (5.7)$$

**Table 7-5.1** Expansion of $|\sigma JM\rangle_\pm$ states in terms of $|^{2S+1}L_J, M\rangle$ states for $J < 4$.

| $J$ | $\|+1, JM\rangle_+$ | $\|-1, JM\rangle_+$ | $\|+1, JM\rangle_-$ | $\|-1, JM\rangle_-$ |
|---|---|---|---|---|
| 0 | $-{}^3P_0 (I = 1)$ | | ${}^1S_0 \; (I = 1)$ | |
| 1 | $3^{-\frac{1}{2}}\, {}^3S_1 - (\frac{2}{3})^{\frac{1}{2}}\, {}^3D_1$ | $(\frac{2}{3})^{\frac{1}{2}}\, {}^3S_1 + 3^{-\frac{1}{2}}\, {}^3D_1$ | ${}^1P_1 \; (I = 0)$ | $-{}^3P_1 \; (I = 1)$ |
| 2 | $(\frac{2}{5})^{\frac{1}{2}}\, {}^3P_2 - (\frac{3}{5})^{\frac{1}{2}}\, {}^3F_2$ | $(\frac{3}{5})^{\frac{1}{2}}\, {}^3P_2 + (\frac{2}{5})^{\frac{1}{2}}\, {}^3F_2$ | ${}^1D_2 \; (I = 1)$ | $-{}^3D_2 \; (I = 0)$ |
| 3 | $(\frac{3}{7})^{\frac{1}{2}}\, {}^3D_3 - (\frac{4}{7})^{\frac{1}{2}}\, {}^3G_3$ | $(\frac{4}{7})^{\frac{1}{2}}\, {}^3D_3 + (\frac{3}{7})^{\frac{1}{2}}\, {}^3G_3$ | ${}^1F_3 \; (I = 0)$ | $-{}^3F_3 \; (I = 1)$ |

For elastic scattering, time-reversal invariance (4-4.19) requires $T^{J\pm}_{1,-1} = T^{J\pm}_{-1,1}$ and so $T(S \to D) = T(D \to S)$. The deuteron pole in *pn*-scattering for example has $J = 1$ and positive parity, i.e., $L = 0$ and 2. Its wave function has the form

$$\psi_M = r^{-1}[u(r) + 8^{-\frac{1}{2}} w(r) S_{12}] \chi^1_M, \qquad \chi^1_{\pm 1} = |\pm\tfrac{1}{2}\rangle |\pm\tfrac{1}{2}\rangle, \quad (5.8)$$

$$S_{12} = 3(\boldsymbol{\sigma}_n \hat{\mathbf{r}})(\boldsymbol{\sigma}_p \hat{\mathbf{r}}) - \boldsymbol{\sigma}_n \boldsymbol{\sigma}_p, \qquad \chi^1_0 = 2^{-\frac{1}{2}}(|\tfrac{1}{2}\rangle|-\tfrac{1}{2}\rangle + |-\tfrac{1}{2}, \tfrac{1}{2}\rangle). \quad (5.9)$$

Here $S_{12}$ is the "tensor operator" which transforms the $S$-state $|0, 1, 1, M\rangle$ into the $D$-state $|2, 1, 1, M\rangle$. The $8^{-\frac{1}{2}}$ in (8) is a normalization constant: when acting on $\chi^1_M$, $\boldsymbol{\sigma}_n \boldsymbol{\sigma}_p$ can be replaced by 1, and $S^2_{12}$ becomes

$$S^2_{12} = [3\boldsymbol{\sigma}_n \hat{\mathbf{r}} \boldsymbol{\sigma}_p \hat{\mathbf{r}} - 1]^2 = 9 - 6\boldsymbol{\sigma}_n \hat{\mathbf{r}} \boldsymbol{\sigma}_p \hat{\mathbf{r}} + 1 = 8 - 2S_{12}. \quad (5.10)$$

Since $\chi^{1+}_{M'} S_{12} \chi^1_M = 0$, only the 8 of (10) contributes to the wave function normalization:

$$4\pi \int (u^2 + w^2)\, dr = 1, \qquad 4\pi \int w^2\, dr \equiv c^2 \approx 0.07. \quad (5.11)$$

Quite generally, baryon-baryon scattering has $2^4 = 16$ independent scattering amplitudes. Parity conservation reduces this number to 8. For elastic

**Table 7-5.2** The parameters of the $S$-wave effective range expansion (6-5.5) for nucleon-nucleon scattering (the sign convention for the scattering length is opposite to that of nuclear physics). For $pp$ scattering, (6-11.6) is used. When a potential model is used and the Coulomb potential is removed, one finds $a_{pp} = 17.1$ fm (see Brown and Jackson 1976).

| | $a$ [fm] | | $r$ [fm] | |
|---|---|---|---|---|
| | $S = 0$ | $S = 1$ | $S = 0$ | $S = 1$ |
| $pp$ | 7.83 | | 2.84(3) | |
| $np$ | 23.75 | −5.42 | 2.75(5) | 1.76 |
| $nn$ | 16.4(1) | | 2.86(3) | |

scattering, time-reversal invariance eliminates two of the remaining amplitudes. The corresponding potential contains 6 different operators

$$V_{ab} = V_c + V_S \boldsymbol{\sigma}_a \boldsymbol{\sigma}_b + V_T S_{12} + V_{LS} \mathbf{LS} + V_5 Q_{ab} + V_6(\mathbf{S}_a - \mathbf{S}_b)\mathbf{L}, \quad (5.12)$$

$$Q_{ab} = \tfrac{1}{2}[(\boldsymbol{\sigma}_a \mathbf{L})(\boldsymbol{\sigma}_b \mathbf{L}) + (\boldsymbol{\sigma}_b \mathbf{L})(\sigma_a \mathbf{L})]. \quad (5.13)$$

For identical particles, the requirement $V_{ab} = V_{ba}$ eliminates $V_6$. The operators **LS** and $Q_{ab}$ operate only between $L \neq 0$ components and are normally neglected for the deuteron. With $\boldsymbol{\sigma}_a \boldsymbol{\sigma}_b = 1$ and $V = V_c + V_S$, the Schrödinger equation for the deuteron assumes the form ($\mu$ = reduced mass)

$$\begin{aligned} &\frac{d^2u}{dr^2} + 2\mu(E - V(r))u + 8^{\frac{1}{2}} V_T w = 0, \\ &\frac{d^2w}{dr^2} + \left[2\mu(E - V(r)) - \frac{6}{r^2}\right] w + 8^{\frac{1}{2}} V_T u = 0. \end{aligned} \quad (5.14)$$

Parameters of the effective range expansion of $NN$ scattering are collected in table 7-5.2. Their different values for the $pp$, $np$ and $nn$ states are partly due to Coulomb effects and partly to isospin breaking in the hadronic amplitude proper.

As mentioned in section 6-3, the nuclear potential is constructed from the nearest $t$-channel singularities. The poles and parts of the cuts are represented by the one-boson-exchange potential (OBEP), and the rest comes in a two-boson-exchange potential (2BEP) (see fig. 7-5.1). The OBEP-potential

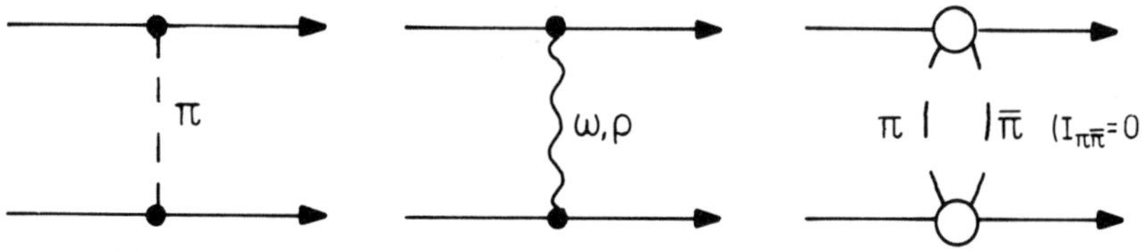


**Figure 7-5.1** The most important graphs for the construction of the $NN$-potential.

is the Fourier transform of the Born approximation as in (2-10.2). The necessary integrals are derived from (3-11.25) and (2-10.12), (2-10.13):

$$V_c = (2\pi)^{-3} \int (m_e^2 + q^2)^{-1} F(q^2) e^{i\mathbf{qr}}\, d^3q \to (4\pi r)^{-1} e^{-m_e r}, \tag{5.15}$$

$$-i\, \partial_j V_c \to i r_j (4\pi r^2)^{-1} (m_e + r^{-1}) e^{-m_e r}, \tag{5.16}$$

$$\begin{aligned} -\partial_i\, \partial_j V_c &= (\delta_{ij}\, \partial_i^2 - \partial_i\, \partial_j) V_c - \delta_{ij}\, \partial_i^2 V_c \\ &= \frac{1}{4\pi r} (\tfrac{1}{3}\, \delta_{ij} - \hat{\mathbf{r}}_i \hat{\mathbf{r}}_j) \left[ m_e^2 + m_e \frac{3}{r} + \frac{3}{r^2} \right] e^{-m_e r} \\ &\quad - \tfrac{1}{3}\, \delta_{ij} \left[ \frac{1}{4\pi r} m_e^2 e^{-m_e r} - \delta_3(\mathbf{r}) \right]. \end{aligned} \tag{5.17}$$

In the simplest version, the two-pion-exchange potential (2PEP) is approximated by the exchange of $\rho$ and a scalar, isoscalar meson which we call $\varepsilon$* but whose mass may be quite different from (1.1). Using the nonrelativistic reduction (2-10.8), we find

$$V_\varepsilon \approx -g_\varepsilon^2 (4\pi r)^{-1} e^{-m_\varepsilon r} [1 + (2m_a^2)^{-1} (m_\varepsilon r^{-1} + r^{-2}) \mathbf{LS}]. \tag{5.18}$$

Here we have neglected a relativistic correction to the central part of the potential. Note that $V_\varepsilon$ is attractive for equal signs of the coupling constants ($g_{\varepsilon a} = g_{\varepsilon b} \equiv g_\varepsilon$), whereas the Coulomb potential is repulsive for $eZ_a = eZ_b$. The minus sign in the case of (18) comes from (2-10.2), whereas the corresponding minus sign for vector exchange is cancelled by the sign of $-g^{\mu\nu}$ in the spin summation (2-8.9). The VEP is obtained from (2-10.15) to (2-10.17) and the above substitutions of the radial integrals. We replace $Ze \to G_\omega^V$, $\kappa e \to G_\omega^T$ according to (6-8.5), put $m_a = m_b$, and neglect relativistic corrections in the central part of the potential:

$$V_\omega = \frac{1}{4\pi r} e^{-m_\omega r} \left\{ G_\omega^{V2} - (2m_a^2)^{-1} G_\omega^V (\tfrac{3}{2} G_\omega^V + 2G_\omega^T) \frac{1}{r} \left( m_\omega + \frac{1}{r} \right) \mathbf{LS} \right\} + V_{SS\omega}. \tag{5.19}$$

One hopes that the central part of $V_\omega$, namely $(4\pi r)^{-1} G_\omega^{V2} \exp(-m_\omega r)$, is so strongly repulsive ("repulsive core") that $\psi(r = 0) \approx 0$, in which case one can neglect the $\delta$-function in $V_{SS}$ (2-10.16):

$$V_{SS} = (G^V + G^T)^2 (16\pi r m_a^2)^{-1} m^2 \left[ \tfrac{2}{3} \boldsymbol{\sigma}_a \boldsymbol{\sigma}_b - \left( \frac{1}{3} + \frac{1}{rm} + \frac{1}{r^2 m^2} \right) S_{12} \right] e^{-mr}. \tag{5.20}$$

(This term comes from $\bar{u}'_a \gamma u_a \bar{u}'_b \gamma u_b$, the spin-dependent part of which gives $(\boldsymbol{\sigma}_a \times \mathbf{q})(\boldsymbol{\sigma}_b \times \mathbf{q}) = \boldsymbol{\sigma}_a \boldsymbol{\sigma}_b q^2 - (\boldsymbol{\sigma}_a \mathbf{q})(\boldsymbol{\sigma}_b \mathbf{q}) = q^2 \sigma_a^\perp \sigma_b^\perp$, in analogy with the last term of 2-10.9). Corresponding expressions describe $\rho$-exchange in *pp*- and *nn*-scattering. To include *np* scattering where also $\rho^\pm$ is exchanged, one must pay attention to the Clebsch-Gordan (5-8.4). This is normally done by introducing an isospinor **N** and writing the coupling of isovector bosons ($\pi$ or $\rho$)

* It is also called $\sigma$.

as $\mathbf{N} + \tau\mathbf{N}$ ($\tau$ is given in 4-2.3). From invariance under isospin rotations, it is clear that inclusion of charged meson exchange produces an extra factor

$$\tau_a \tau_b = \{1 \quad \text{for} \quad I_{NN} = 1, \qquad -3 \quad \text{for} \quad I_{NN} = 0\}. \tag{5.21}$$

The exchange of pseudoscalar mesons is best calculated from the pseudovector coupling $(4\pi)^{\frac{1}{2}} m_\pi^{-1} f \bar{u}'_a \gamma_5 (-\boldsymbol{\gamma}\mathbf{q}) u_a$ of (6-2.9), the large components of which have $\boldsymbol{\sigma}$ instead of $\boldsymbol{\gamma}$, according to (A-3.12). The spin-dependence of pseudoscalar exchange is thus $(\boldsymbol{\sigma}_a \mathbf{q})(\boldsymbol{\sigma}_b \mathbf{q})$ in this approximation. Using again (17) and neglecting the $\delta_3(\mathbf{r})$, we find the one-pion-exchange potential (OPEP) as

$$V_\pi = \tau_a \tau_b f^2 \frac{1}{r} \left[ \tfrac{1}{3} \boldsymbol{\sigma}_a \boldsymbol{\sigma}_b + \left( \frac{1}{3} + \frac{1}{r m_\pi} + \frac{1}{r^2 m_\pi^2} \right) S_{12} \right] e^{-m_\pi r}. \tag{5.22}$$

Although the numerical value (2.12) of $f^2$ is small, the potential dominates at large distances, because of the smallness of $m_\pi$ (see fig. 7-5.2). The $\pi^+ - \pi^0$ mass difference can be approximated by expanding $\Pi(t)$ of (4-15.19) around $t = 0$ (the $m_\pi^{-1}$ in the definition of the pseudovector coupling must be kept fixed, however).

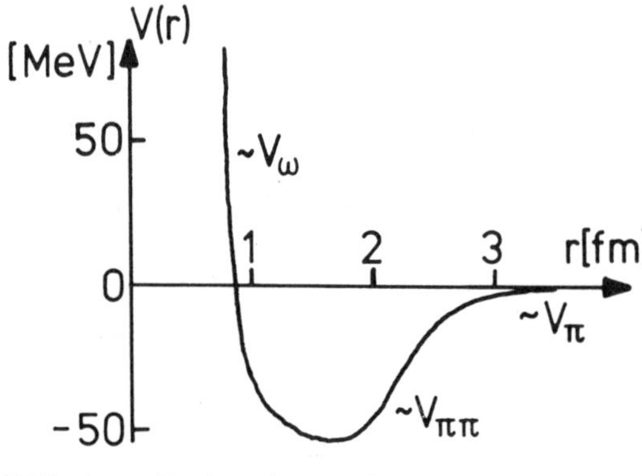


**Figure 7-5.2** A qualitative picture of a nucleon-nucleon potential.

Unfortunately, the theoretical basis of the potential treatment of $NN$-scattering is much weaker than in the case of Coulomb scattering, due to the lack of gauge invariance (see the discussion in section 3-14). This is also evident from the fact that the OPEP (22) diverges as $r^{-3}$ for $r \to 0$. It necessitates the inclusion of form factors, and inhibits a reliable calculation of the 2PEP (Gross 1969). For VEP, a part of the form factor is hidden in the use of "effective coupling constants." In the case of $\omega$-exchange, for example, the value 13 of (6-8.12) is reduced to $\sim 10$.

For the long-range OPEP, $F_{NN\pi}(t)$ should be restricted by (2.13). Since only the square of this function is needed in $NN$-scattering, a convenient parametrization is

$$F_{NN\pi} = (\Lambda_\pi^2 + q^2)^{-\frac{1}{2}} (\Lambda_\pi^2 - m_\pi^2)^{\frac{1}{2}}, \tag{5.23}$$

the square of which is a simple pole in $q^2$. Moreover, one can decompose

$$F_{NN\pi}^2(m_\pi^2 + q^2)^{-1} = (m_\pi^2 + q^2)^{-1} - (\Lambda_\pi^2 + q^2)^{-1}, \qquad (5.24)$$

which shows that the modified OPEP is just (22) minus the corresponding expression which has $m_\pi$ replaced by $\Lambda_\pi$ (according to the model (4-9.16), $\Lambda_\pi^{-1} = r_0$ is the "radius of a nucleon" as seen by a pion).

The calculation of a 2PEP along the lines indicated in 3-14 (i.e., from the sum of the 4th order graphs, $T^{(4)}$) has been carried out by Brueckner and Watson (1953) and Taketani et al. (1952), among others. Since it diverges for pseudovector coupling, one can only use pseudoscalar coupling. The result is incorrect in any case, because perturbation theory is inappropriate for strong interactions. This should not come as a surprise. When the two pions in fig. 7-5.1b are put on their mass shell, the upper and lower halves of that graph contain the analytic continuation of the $\pi N$ scattering amplitude, which is known to be different both from the *ps* and from the *pv* Born terms. There is only one exception: because of the current algebra constraint (6-9.12), the *s*-wave $\pi N$ amplitude near threshold is much closer to the pseudovector Born term (6-2.12) than to the pseudoscalar one (6-2.7). Thus, if one calculates $T^{(4)}$ from *ps*-coupling and then throws away those terms which obviously violate current algebra, the result is quite acceptable for large $r$. This method is called "pair suppression." Of course, it omits both the $\Delta$ resonance and $\pi\pi$-scattering.

A better method is to first consider the amplitudes $T^{J\pm}_{(t)\sigma\sigma'}$ which refer to $N\bar{N} \to N\bar{N}$ transitions in states of definite angular momentum, parity and helicity products. Two-pion intermediate states ($N\bar{N} \to \pi\bar{\pi} \to N\bar{N}$) occur only in $T^{J+}_{(t)\sigma\sigma'}$ because of parity conservation (remember (4-5.16) and our discussion of the states $|\sigma JM\rangle_+$ earlier in this section. The negative "intrinsic parity" of $N\bar{N}$ scattering does not modify our selection rules for elastic scattering, of course). We can now define the $2\pi$-exchange amplitudes $T^{\pi\pi}_{\sigma\sigma'}$ as the analytic continuation of amplitudes whose partial-wave projections $T^{\pi\pi J}_{\sigma\sigma'}$ satisfy, for $4m_\pi^2 < t < 4m_N^2$, the extended unitarity equation (6-3.17):

$$(2i)^{-1}(T^{\pi\pi J}_{\sigma\sigma'} - (-1)^J T^{\pi\pi J}_{\sigma'\sigma}) = (-1)^J p_{\pi\pi} T^J_\sigma T^{J*}_{\sigma'}, \qquad (5.25)$$

where the $T^J_\sigma$ are the $N\bar{N} \to \pi\bar{\pi}$ helicity amplitudes (6-2.24) and (6-2.25). The full amplitude $T^{J+}_{(t)\sigma\sigma'}$ will also contain a 4-pion-cut, starting at $t = 16m_\pi^2$, but this gives rise to a 4PEP which is hopefully unimportant (it could be, however, that the $\omega\pi$-cut produces a potential which imitates the form factor in the OPEP).

For the analytic continuation, one must use "invariant amplitudes" which are defined without the $u$'s and $\bar{u}$'s and $\gamma$-matrices, and which depend on the Mandelstam variables $s$, $t$, $u$ only:

$$T(\lambda_a \ldots \lambda_d, s, t) = \sum_{i=1}^{6} \bar{u}_d \bar{u}_c P_i u_a u_b A_i(s, t). \qquad (5.26)$$

Various sets of amplitudes are in use here and can be transformed into each other. The "perturbative amplitudes" go with the operators

$$P_1 = 1, \qquad P_2 = \tfrac{1}{2}\gamma_a(P_b + P_d) + \tfrac{1}{2}\gamma_b(P_a + P_c),$$
$$P_3 = \tfrac{1}{4}\gamma_a(P_b + P_d)\gamma_b(P_a + P_c), \qquad P_4 = \gamma_a\gamma_b, \qquad P_5 = \gamma_a^5\gamma_b^5. \tag{5.27}$$

The operator $P_6$ has the same structure as $P_2$ but a minus sign between the two terms. It is not needed for identical particles. For the application of dispersion relations, $P_2$ and $P_3$ must be replaced by momentum-free operators, in order to avoid kinematical singularities (Goldberger et al. 1960, Scadron and Jones 1968). In addition to $P_1$, $P_4$ and $P_5$, only

$$A = -\gamma_a^5\gamma_a\gamma_b\gamma_b^5, \qquad T = \tfrac{1}{2}\sigma_a^{\mu\nu}\sigma_{b\mu\nu} \tag{5.28}$$

are of that type. One finds

$$mP_2 = \frac{t}{4} + \frac{u-s}{4}(P_1 + P_5), \qquad 4P_3 = 4m^2P_5 - tA - (u-s)P_4. \tag{5.29}$$

The contributions of (25) to the perturbative amplitudes are elaborated in appendix B to chapter 8 in the book of Brown and Jackson. One finds $A_3 = A_5 = 0$. The rest is then similar to the treatment of the isovector nucleon form factors. In particular, the $I = 1$ $\pi\pi$-continuum can be approximated by $\rho$-exchange and $N$ and $\Delta$ Born terms. The $\rho\,NN$ coupling constant $G_\rho^M$ of (6-8.30) is determined from $t = 0$ $NN$ dispersion relations

$$(G_\rho^M)^2/4\pi \approx 20, \tag{5.30}$$

which in connection with (6-8.35) gives

$$G_\rho^T/G_\rho^V \approx 5, \qquad (G_\rho^T)^2/4\pi \approx 14. \tag{5.31}$$

The $I = 0$ $\pi\pi$ continuum contains only even partial waves. The $J = 2$ partial wave contributes the $f^0$-meson, $J \geq 4$ partial waves are negligible, so the only nontrivial partial wave is that of $J = 0$. It has $\sigma = +1$, since $\sigma = -1$ corresponds to helicity $|\lambda| = |\lambda_a - \lambda_b| = 1$ which is forbidden for $J = 0$. The amplitude $T_+^0(t > 4m_\pi^2)$ of (26) is constructed from $T(\pi N \to \pi N, t < 0)$ and the $\pi\pi$-phase $\delta_0^0$ (fig. 7-1) by the method of section 6-8.

To obtain the 2PEP $V_{\pi\pi}$, one must subtract from $T_{\sigma\sigma'}^{\pi\pi}$ the first iteration of $V_\pi$ (i.e., the second Born approximation to the potential scattering induced by $V_\pi$), and then Fourier transform the result.

## 7-6 $\bar{K}N$ Interactions and $SU(3)$-Symmetry for Pseudoscalar Meson Couplings

Already at threshold ($s_0^{\frac{1}{2}} = m_p + m_K = 1436$ MeV), $K^-p$ interactions show appreciable inelasticity, due to the open $\Lambda\pi$ and $\Sigma\pi$ channels (fig. 7-6.). Below threshold, the $P$-wave resonance $\Sigma(1385)$ occurs in the $I = 1$ state (see table C-5), and an $S$-wave resonance $\Lambda(1405)$ occurs in the $I = 0$ state (see

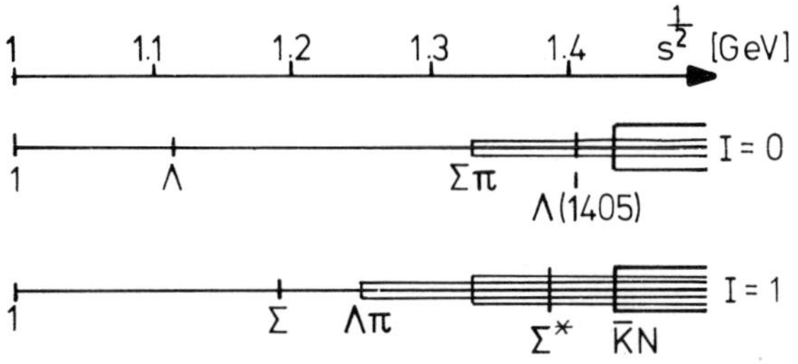


**Figure 7-6** The poles, cuts and resonances of amplitudes connecting meson-baryon states of zero hypercharge.

table 7-6.1). The isospin decomposition of $pK^+$, $pK^0$, $nK^0$ and $nK^+$ goes as in (4-2.4), and since we use identical phases for antiparticle multiplets, the same decomposition applies also to the isospin doublet $(\bar{K}^0, K^-)$:

$$|p\bar{K}^0\rangle = |1, 1\rangle, \qquad |pK^-\rangle = 2^{-\frac{1}{2}}(|1, 0\rangle + |0, 0\rangle),$$
$$|nK^-\rangle = |1, -1\rangle, \qquad |n\bar{K}^0\rangle = 2^{-\frac{1}{2}}(|1, 0\rangle - |0, 0\rangle). \tag{6.1}$$

We thus obtain

$$T(pK^-) = T(n\bar{K}^0) = \tfrac{1}{2}(T_N^1 + T_N^0), \qquad T(pK^- \to n\bar{K}^0) = \tfrac{1}{2}(T_N^1 - T_N^0). \tag{6.2}$$

The $\Lambda\pi$ states are pure $I = 1$ states, while the isospin decomposition of $\Sigma\pi$ states is obtained from (1.2) by replacing the first pion by the corresponding $\Sigma$. This leads to

$$T(pK^- \to \Sigma^\pm\pi^\mp) = 6^{-\frac{1}{2}}T_{N\Sigma}^0 \pm \tfrac{1}{2}T_{N\Sigma}^1,$$
$$T(n\bar{K}^0 \to \Sigma^\pm\pi^\mp) = -6^{-\frac{1}{2}}T_{N\Sigma}^0 \pm \tfrac{1}{2}T_{N\Sigma}^1, \tag{6.3}$$
$$T(pK^- \to \Sigma^0\pi^0) = -T(n\bar{K}^0 \to \Sigma^0\pi^0) = -6^{-\frac{1}{2}}T_{N\Sigma}^0$$

In practice, $\bar{K}^0$ collisions are part of the $K_L^0$ collisions, according to (4-1.4).

**Table 7-6.1** Baryon resonances of zero hypercharge below 1850 MeV (for Σ (1385) see table C-5). The numbers are approximate and are taken from the Particle Data Group (1978).

| Symbol (mass [MeV]) | Partial wave ($J^{\mathscr{P}}$) | Γ [MeV] | Main decay channels and fractions [%] $N\bar{K}$ | $\Lambda\pi$ | $\Sigma\pi$ | $\Lambda\pi\pi$ | $\Sigma\pi\pi$ | $\Lambda\eta$ | $\Sigma\eta$ | other |
|---|---|---|---|---|---|---|---|---|---|---|
| Λ (1405) | $S_{01}$ $(\frac{1}{2}^-)$ | 40 | — | — | 100 | — | — | — | — | — |
| Λ (1519) | $D_{03}$ $(\frac{3}{2}^-)$ | 15 | 46 | — | 42 | 10 | 0.9 | — | — | — |
| Λ (1670) | $S_{01}$ $(\frac{1}{2}^-)$ | 40 | 30 | — | 40 | | | 30 | — | |
| Λ (1690) | $D_{03}$ $(\frac{3}{2}^-)$ | 60 | 25 | — | 30 | 25 | 20 | | — | |
| Λ (1820) | $F_{05}$ $(\frac{5}{2}^+)$ | 85 | 60 | — | 12 | | | | | 20[a] |
| Λ (1830) | $D_{05}$ $(\frac{5}{2}^-)$ | 95 | | — | ~ 60 | | | | | |
| Σ (1670) | $D_{13}$ $(\frac{3}{2}^-)$ | 50 | 20 | | 40 | | | | | |
| Σ (1750) | $S_{11}$ $(\frac{1}{2}^-)$ | 75 | 30 | 15 | | | | | ~ 40 | |
| Σ (1773) | $D_{15}$ $(\frac{5}{2}^-)$ | 130 | 41 | 14 | 1 | | | | 16[b] | 10[a] |

(a) Σ (1385)π (b) Λ (1520)π

Partial-wave decomposition and dispersion relations for $KN$ and $\bar{K}N$ amplitudes have already been discussed in section 6-2, and the relevant coupling constants $G_{\Lambda N\bar{K}}$ and $G_{\Sigma N\bar{K}}$ have been defined in (5-8.8) and (5-8.9). Analysis of (6-2.20) for $K^{\pm}n$-scattering gives

$$G_{\Sigma N\bar{K}}^2/4\pi \leq 3. \tag{6.4}$$

For $K^{\pm}p$ scattering, only a combination of $\Lambda$ and $\Sigma$ poles can be determined, due to the fact that the two poles are close to each other and far from the physical regions. Since one expects $G_{\Sigma N\bar{K}}^2 < G_{\Lambda N\bar{K}}^2$ one assumes an effective pole at the $\Lambda$-pole position, with effective coupling constant

$$G_Y^2 = G_{\Lambda N\bar{K}}^2(1 + R_\Sigma/R_\Lambda) = G_{\Lambda N\bar{K}}^2 + 0.84G_{\Sigma N\bar{K}}^2, \qquad G_Y^2/4\pi \approx 17. \tag{6.5}$$

The complex scattering lengths are

$$A_N^0 = (-1.66 + i0.75) \text{ fm}, \qquad A_N^1 = (0.35 + i0.66) \text{ fm}. \tag{6.6}$$

Unitarity can only be exploited if the $N\bar{K}$, $\Lambda\pi$ and $\Sigma\pi$ channels are treated simultaneously, for each isospin and partial wave. This is achieved by the $K$-matrix method of section 6-5. For $I = 1$ and 0, we have explicitly

$$K^1 = \begin{pmatrix} K_N^1 & K_{N\Lambda}^1 & K_{N\Sigma}^1 \\ K_{N\Lambda}^1 & K_\Lambda^1 & K_{\Lambda\Sigma}^1 \\ K_{N\Sigma}^1 & K_{\Lambda\Sigma}^1 & K_\Sigma^1 \end{pmatrix}, \qquad K^0 = \begin{pmatrix} K_N^0 & K_{N\Sigma}^0 \\ K_{N\Sigma}^0 & K_\Sigma^0 \end{pmatrix}. \tag{6.7}$$

With $0 = N$ in (6-5.18), we see that the scattering lengths (6) are the $N\bar{K}$-channel reduced $K$-matrices at the $N\bar{K}$ threshold. In particular for $I = 0$, (6-5.19) gives

$$A_N^0 = K_N^0 + ip_\Sigma K_{N\Sigma}^{0\,2}(1 - ip_\Sigma K_\Sigma^0)^{-1}. \tag{6.8}$$

For the $I = 1$ channels, a constant $K$-matrix can be used at low energies. For the $I = 0$ channels, a simultaneous description of the $\Lambda(1405)$ resonance requires some energy dependence of $K^0$ (Baillon et al. 1976). Nevertheless, it is instructive to see that a constant $K$-matrix can produce a $\Sigma\pi$ resonance: we neglect the $ip_\Sigma K_\Sigma^0$ in the denominator of (8) below threshold, such that we have $K_N^0 \approx \text{Re } A_N^0$, $p_\Sigma K_{N\Sigma}^{0\,2} \approx \text{Im } A_N^0$. The reduced $K$-matrix with respect to the $\Sigma\pi$ channel is

$$(K_r^0)_\Sigma = K_\Sigma^0 + ip_N K_{N\Sigma}^{0\,2}(1 - ip_N K_N^0)^{-1}. \tag{6.9}$$

It is real below $s^{\frac{1}{2}} = m_N + m_K$ because $p_N$ is imaginary there. Since $ip_N$ is negative, the denominator of (9) vanishes at $ip_N = (K_N^0)^{-1}$, and the $I = 0$ $\Sigma\pi$ scattering amplitude has a pole near this position. From the point of view of the $N\bar{K}$ system, it corresponds to a bound state with binding energy $-p_N^2/2\mu = (2\mu K_N^{0\,2})^{-1} \approx 22$ MeV ("virtual bound state").

The values (4) and (5) may be compared with the $SU(3)$-predictions for these quantities. However, due to the baryon mass splitting in the relation (6-2.10), $G = (4\pi)^{\frac{1}{2}} f(m_a + m_c)m_\pi^{-1}$, the relation between $G$ and $f$ is somewhat different for each baryon combination, even if the meson mass in this relation is kept fixed at $m_{\pi^+}$ (it was included in (6-2.9) for dimensional reasons, not for dynamical ones). Therefore, the $SU(3)$-predictions depend on the choice of coupling. In particular, the parameter $\alpha$ which enters the predictions in the last column of table 4-12 can be obtained from PCAC in $\Sigma \to \Lambda e\nu$ decays. It comes out as 0.64 for pseudovector coupling (see table 5-7) and 0.78 for pseudoscalar coupling, due to the extra factor $(m_\Sigma + m_\Lambda)/2m_N \approx 1.23$ in the latter case. The success of the Cabibbo theory (compare 5-8.12) gives a first indication that pv coupling is more appropriate. The resulting predictions for $f^2$ and $G^2/4\pi$ are collected in table 7-6.2. For comparison, the value $\alpha_{ps} = 0.78$ would give $G^2_{\Sigma N\bar{K}}/4\pi > 4$, in contradiction with (4).

**Table 7-6.2** Pseudovector $SU(3)$ predictions for the pseudoscalar meson-baryon coupling constants, using $f^2_{NN\pi} = 0.0794$, $\alpha = 0.639$ and the relations of table 4-12 for the $f$'s. The values for the $G$'s are then obtained from (6-2.10).

| | $NN\pi$ | $\Lambda\Sigma\pi$ | $\Sigma\Sigma\pi$ | $\Xi\Xi\pi$ | $N\Lambda K$ | $N\Sigma K$ | $NN\eta_8$ | $\Lambda\Lambda\eta_8$ | $\Sigma\Sigma\eta_8$ |
|---|---|---|---|---|---|---|---|---|---|
| $1000\,f^2$ | 79.4 | 4.3 | 41 | 6.1 | 78 | 6.1 | 5.2 | 43 | 43 |
| $G^2/4\pi$ | 14.4 | 11.8 | 12 | 2.2 | 17.0 | 1.4 | 0.9 | 11.0 | 12.6 |

A value for $G_{\Sigma\Sigma\pi}$ can be obtained from the double charge exchange reaction $\pi^-\Sigma^+ \to \pi^+\Sigma^-$. This reaction has not yet been detected, but it has the unique property that only $I = 2$ states can be exchanged in the $t$-channel. Since no mesons with these quantum numbers exist, the scattering amplitude should tend to zero quite rapidly at high energies.

The Regge behavior for $s \to \infty$ (6-10.11) gives for the invariant amplitudes $A$ and $B$ of (6-2.1)

$$A \sim s^{\alpha(t)}, \qquad B \sim s^{\alpha(t)-1}. \tag{6.10}$$

One expects $\alpha(0) < 0$ and therefore $sB \to 0$ for $s \to \infty$. One can thus multiply the dispersion relation (6-2.14) for $B^+$ by $s$ and take the limit $s \to \infty$. This gives a "superconvergence relation," which for the above process reads

$$G^2_{\Sigma\Sigma\pi} - G^2_{\Lambda\Sigma\pi} = \pi^{-1} \int_{s_0}^{\infty} \text{Im}\, B^+(\pi^-\Sigma^+ \to \pi^+\Sigma^-) \tag{6.11}$$

(Babu et al. 1967). Taking now $t = 0$ and inserting for $B$ the known contributions from the $\Lambda$ and $\Sigma$ resonances (table 7-6.1), one finds that $G^2_{\Sigma\Sigma\pi}$ must be at least as large as $G^2_{\Lambda\Sigma\pi}$. Again, this is compatible with $SU(3)$-invariance only for pseudovector coupling.

## 7-7 Hadron Production in Hadron Collisions

In most proton accelerators, secondary hadrons are produced in collisions with light nuclei such as $^9$Be. These hadrons are collected into secondary beams, and hadron production has been mainly studied in the collisions of these beams with hydrogen or deuterium.

Total and elastic $pp$ cross sections at intermediate energies are shown in fig. 7-7.1. The total cross section reaches a maximum of 47.5 mb at $p_{\text{lab}} = 1.7$ GeV ($s^{\frac{1}{2}} = 2.3$ GeV) and then drops slowly towards a broad minimum at 50 GeV (fig. 6-10.1). The maximum is largely due to the reaction $pp \to pn\pi^+$ which reaches 19 mb at the same energy and falls rapidly afterwards (10 mb at $p_{\text{lab}} = 4$ GeV). The reaction $pp \to pp\pi^0$ reaches 4 mb at the maximum and then falls, too. The first inelastic reaction is of course $pp \to \pi^+ d$. It has a peak value of 3 mb at $p_{\text{lab}} = 1.2$ GeV and then drops to zero.

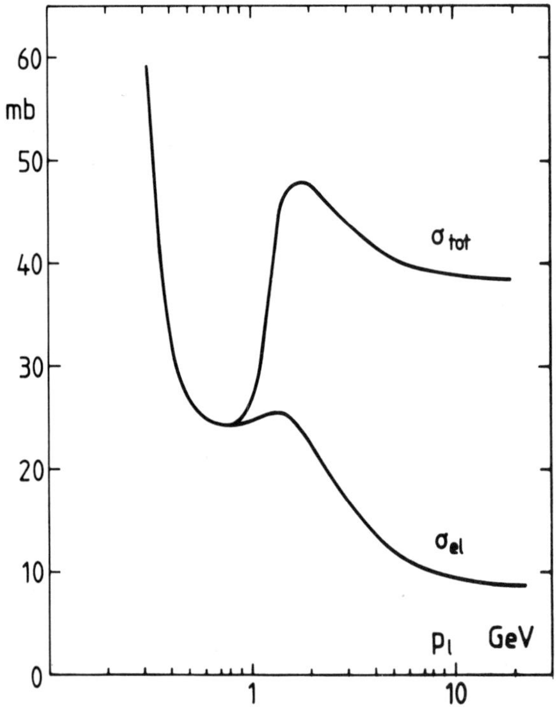


**Figure 7-7.1** Total and elastic $pp$ cross sections at intermediate energies (after Giacomelli 1970).

Single pion production in $NN$ collisions proceeds largely via the $\Delta$ resonance, $NN \to N\Delta \to NN\pi$, also at higher energies. In fact, when the Dalitz plot becomes considerably larger than the $\Delta$ width, the $\Delta$ peak is clearly seen. (In this respect, the situation is similar to the reaction $\pi^+ p \to \pi^0 \pi^+ p$, a Dalitz plot of which has been shown in fig. 4-6.1). As the energy increases, the production of more pions and other particles increases. Average multiplicities of $\pi^\pm$, $K^\pm$, $p$ and $\bar{p}$ in $pp$ collisions are shown in fig. 7-7.2a. Similar curves apply to $\pi p$ collisions.

The momenta of the outgoing particles parallel and perpendicular to the incident momentum are called $p_l$ and $p_t$. The average $\langle p_t \rangle$ remains small even at high energies (fig. 7-7.2b). Differential one-particle spectra without regard of possible additional hadrons are called "inclusive" spectra. At high energies, they obey the empirical scaling law (Feynman 1969):

$$E_i\, d^3\sigma(\mathbf{p}_i)/dp_l\, dp_t^2 = f(p_t, x), \qquad x \equiv 2s^{-\frac{1}{2}} p_l \approx p_l/p_{l\,\max}. \tag{7.1}$$

After integration over $p_l$, one finds approximately

$$d^2\sigma/dp_t^2 = c \exp\{-\lambda(p_t^2 + m^2)^{\frac{1}{2}}\}, \qquad \lambda \approx 6/\text{GeV}. \tag{7.2}$$

It must be remembered that many of the produced hadrons originate from the decays of resonances. For example, the overall reaction $\pi N \to \pi\pi\pi\pi N$ gets contributions from $\pi N \to A_2 N'$, $\omega N'$, $\rho N'$ ... where $N'$ (or $\Delta$) is one of the $\pi N$ resonances of table 7-2.2. A theory of particle production at a few GeV is therefore largely a theory of quasi-two-particle reactions $ab \to cd$. The fact that the inclusive cross section falls rapidly with $p_t$ is then to some extent a combined effect of a rapid fall of $d\sigma(ab \to cd)/dt$ and small resonance decay momenta. The rapid fall of $d\sigma(ab \to cd)/dt$ in turn is ascribed to the residue functions ($\beta(t)$ in 6-10.11) of Regge poles, but also to "absorption corrections," which represent the competition of other reactions in the lower partial waves (see for example Irving and Warden (1977) for a recent review). Because all Regge trajectories except the "Pomeron" have

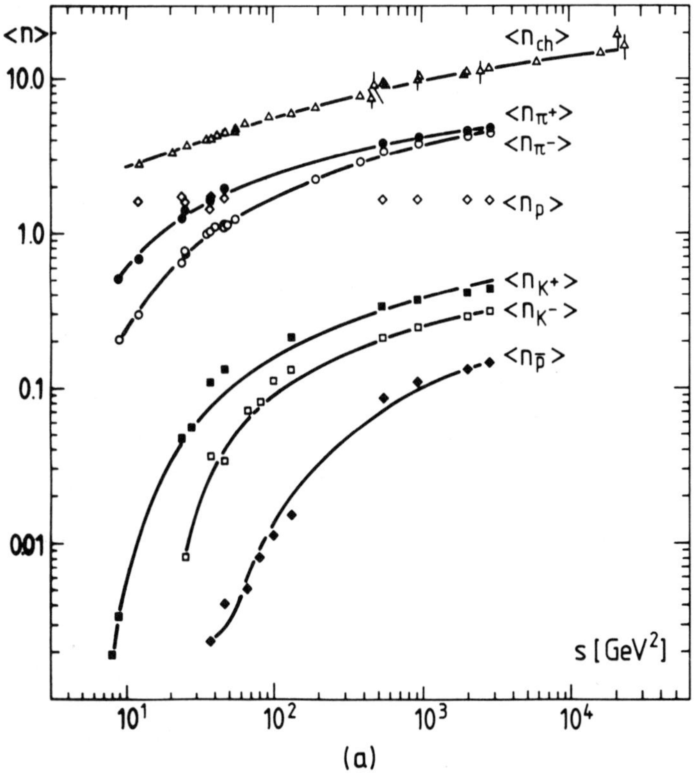


**Figure 7-7.2** (a) Average multiplicity and transverse momentum

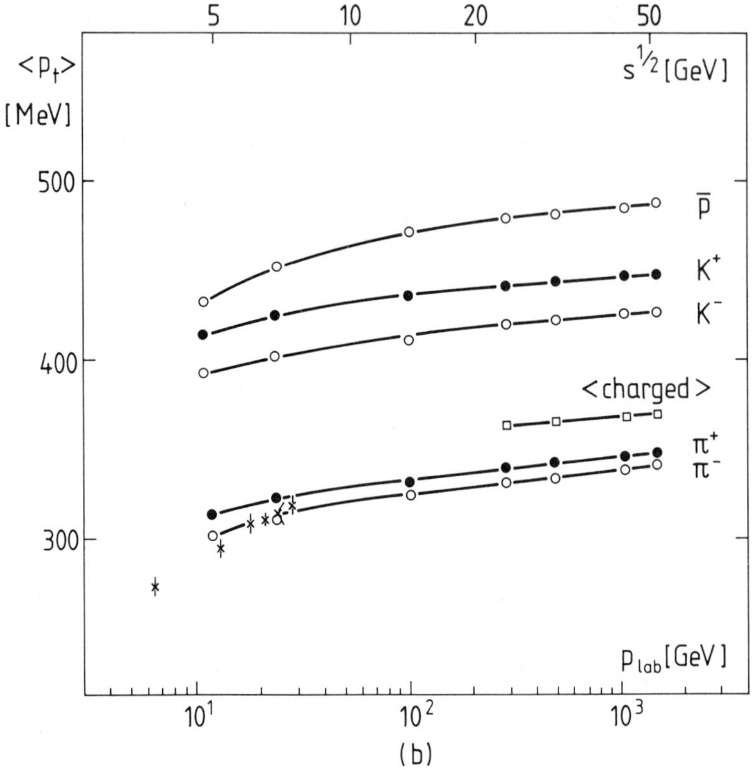


**Figure 7-7.2** (b) of charged particles in high-energy *pp*-collisions (Rossi et al. 1975).

$\alpha(t < 0) < 1$, all reactions $ab \to cd$ eventually tend to zero for $s \to \infty$, with the exception of Pomeron exchange reactions. In these latter reactions, discrete quantum numbers such as isospin and hypercharge of the outgoing particles $c$ and $d$ are identical with those of the incoming ones. The spin may change (if it changes by one unit, the parity must also change).

Another name for such reactions is "diffraction dissociation," which will be justified in section 7-9. Frequently, one of the incident particles remains in its ground state. An important kinematical quantity is the maximum value of $t$. For $m_b = m_d = m$, one can rewrite $t_{\max}$ of (2-4.11) as follows

$$\begin{aligned} \lambda\lambda' &= (\lambda + \lambda')^2 - 4d^2(s + m^2 - \tfrac{1}{2}m_a^2 - \tfrac{1}{2}m_c^2), \qquad d \equiv m_c^2 - m_a^2, \\ t_{\max} &\approx -2\, d^2 m^2 (\lambda + \lambda')^{-1}[1 + d^2(sm^2/\lambda + \tfrac{1}{2})/\lambda] \approx -(d/2p)^2, \end{aligned} \tag{7.3}$$

which shows that $t_{\max}$ tends to zero as $s^{-2}$ in such cases. The most important "diffraction dissociation" product of the nucleon appears to be the $N(1470)$ of table 7-2.2, but the peak of the $N\pi$ effective mass distribution occurs already at 1410 MeV. The pion can only diffraction dissociate into an odd number of pions, and in fact none of the resonances discussed so far can contribute. Instead, there appears a broad peak around $s_{\pi\pi\pi} = 1.2\ \text{GeV}^2$ in the $J^{\mathscr{P}} = 1^+$ partial wave. This is frequently called the "$A_1$ resonance," although it is unlikely that the $\pi\pi\pi \to \pi\pi\pi$ partial wave amplitude traces a resonance circle.

A more general hypothesis is the "limiting fragmentation" (Benecke et al. 1969), which states that for $s \to \infty$, the distribution of the slow particles in the lab system depends on the quantum numbers of the target only (i.e., it is independent of the energy and type of projectile) and the distribution of the fast particles depends on the quantum numbers of the projectile only and is energy-independent in the projectile rest frame. It is then useful to consider the "rapidity" of a particle

$$y = \sinh^{-1}[p_e(m^2 + p_t^2)^{-\frac{1}{2}}] = \tfrac{1}{2} \ln [(E + p_l)/(E - p_l)], \tag{7.4}$$

because this quantity is additive under Lorentz transformations (compare section 1-2). The limiting fragmentation hypothesis says that although the rapidity distribution $d\sigma/dy$ gets larger as $s$ increases, it does not change its form near the ends. One can thus distinguish between a projectile fragmentation region (large $y$), a target fragmentation region (small $y$) and a "central region."

A baryon normally has a baryon among its fragments. In fact, in high-energy $NN$ collisions, the fast outgoing baryon carries about 50% of the incident energy. It is then natural to describe these collisions by "double-peripheral" graphs (fig. 7-7.3), where Regge poles of zero baryon number are exchanged.

For the relatively small group of collisions in which particles with large $p_t$ are observed, it is normally possible to find a new axis along which the transverse momenta are again small (the "jet axis"). Such collisions are described in the quark model as being caused by the large-angle scattering of two quarks, one from the projectile and one from the target. See Sivers et al. (1976) and Ellis and Stroynowski (1977).

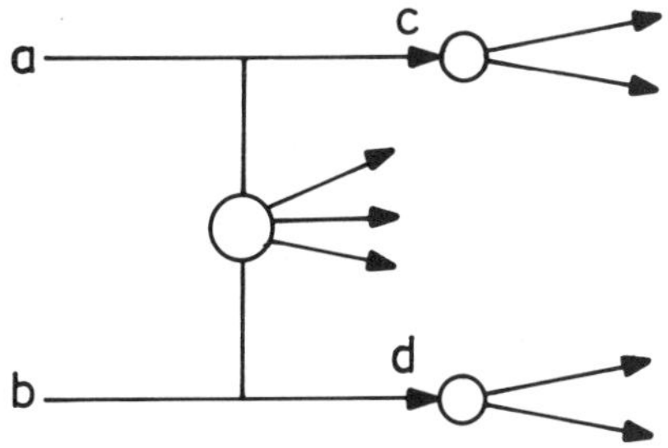


**Figure 7-7.3** A double-peripheral graph leading to projectile and target fragmentation and additional particles in the central $y$-region.

## 7-8 More about Triangle Singularities and Impulse Approximations

In section 6-4 we defined the impulse approximation (6-4.18). The distance of the pole in $p_k^2$ from the physical region $p_k^2 > 0$ is $\kappa^2 = 2\mu b$, which is 0.002089 GeV$^2$ for the deuteron, using table 4-2.1. But even in this favourable case, the vertex function $G(P_e^2)$ cannot be approximated by a constant. Its nearest singularity is of the triangle type (fig. 7-8.1a) and has an anomalous threshold (Blankenbecler and Cook 1960). According to (6-3.23), its position is at $z = -1$, where $z$ is given by the first expression of (6-3.19) (the second one applies only to $c\bar{c}$ intermediate and $b\bar{b}$ final states). For arbitrary masses as indicated in fig. 7-8.1b, one finds $z^2 = 1$ at

$$m_1^2 m_2^2 m_3^2 - m_1^2(P_2 P_3)^2 - m_2^2(P_3 P_1)^2 - m_3^2(P_1 P_2)^2 + 2(P_1 P_2)(P_2 P_3)(P_3 P_1) = 0, \quad (8.1)$$

$$P_i P_j = -\tfrac{1}{2}(s_k - m_i^2 - m_j^2), \qquad k \neq i, j.$$

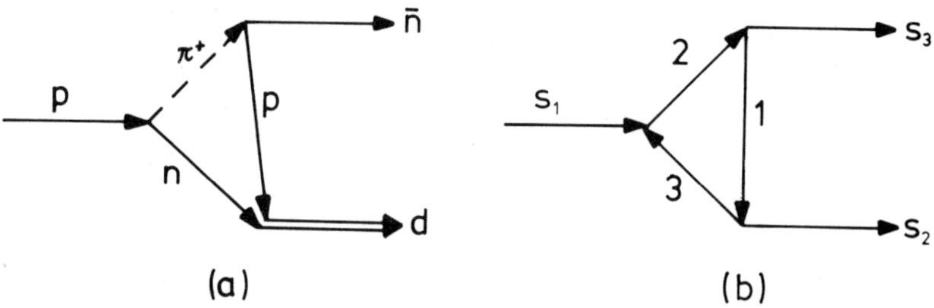


**Figure 7-8.1** The triangle singularity (a) in the deuteron vertex function as a function of $P_p^2$, (b) for arbitrary masses in a symmetric notation. The uniform sense of the loop momenta is essential in (b).

Thus although the derivation from extended unitarity is asymmetric in the indices 1, 2, 3, the result is symmetric. The corresponding derivation from Feynman integrals is more elegant in this respect (see Eden et al. 1966). Here the singularity is given by the Landau-Cutkosky rules:

$$\sum_{j\,(\text{loop})} \alpha_j P_j = 0, \qquad \begin{cases} P_i^2 = m_i^2, \ \alpha_i > 0 \\ \text{or } \alpha_i = 0 \end{cases} \quad (8.2)$$

These rules apply to arbitrary Feynman graphs with arbitrary numbers of loops. The first rule puts all particles on the mass shell, (compare 6-1.31) and connects the momenta in each loop (the sense of the loop momenta must be uniform). It can also be derived from the requirement that the particles involved are real and interact over macroscopic space-time intervals (Coleman and Norton 1965, Pham 1967, Iagolnitzer 1967). For example, if the proton disintegrates into $\pi^+$ and $n$ at $\mathbf{x} = t = 0$, these two particles will be at $x_\pi^\mu = P_\pi^\mu \tau_\pi / m_\pi$ and $x_n^\mu = P_n^\mu \tau_n / m_n$, respectively, where $\tau_\pi$ and $\tau_n$ are the corresponding proper times. Somewhere along $x_\pi^\mu$, a proton is emitted. The classical condition that it hits the neutron at some later time is $x_p^\mu = x_n^\mu$ or

$$m_\pi^{-1} \tau_\pi P_\pi^\mu + m_p^{-1} \tau_p P_p^\mu = m_n^{-1} \tau_n P_n^\mu, \quad (8.3)$$

which coincides with (2) for $\alpha_i = \tau_i / m_i$ and $P_1 = -P_n$.

Returning to our one-loop diagram, we can multiply the loop equation by $P_i$, getting

$$\sum_{j\,(\text{loop})} \alpha_j P_j P_i = 0, \tag{8.4}$$

which has a solution $\alpha_j > 0$ only if

$$\begin{vmatrix} P_1^2 & P_1 P_2 & P_1 P_3 \\ P_2 P_1 & P_2^2 & P_2 P_3 \\ P_3 P_1 & P_3 P_2 & P_3^2 \end{vmatrix} = 0. \tag{8.5}$$

This equation is identical with (1). Solving it with respect to $P_2 P_3$ we find for $m_1^2 = s_3$, $P_1 P_2 = \frac{1}{2} m_2^2$:

$$s_1 - m_3^2 = \tfrac{1}{2}(s_2 + m_1^2 - m_3^2) m_2^2 m_1^{-2} + 2 s_2^{\frac{1}{2}} \kappa (1 - m_2^2/4m_1^2)^{\frac{1}{2}} m_2/m_1, \tag{8.6}$$

where $\kappa$ is the imaginary decay momentum of $d$ as in (6-4.9). For $s_2 \approx (m_1 + m_3)^2 - 2bs_2^{\frac{1}{2}}$ and $m_2 = m_\pi \ll 2m_1$,

$$s_1 - m_3^2 \approx s_2^{\frac{1}{2}}(m_\pi + 2\kappa) m_\pi/m_1, \qquad \kappa^2 \approx 2bm_1 m_3 s_2^{-\frac{1}{2}}. \tag{8.7}$$

Two-pion exchange produces an anomalous threshold at $s_1 - m_3^2 \approx 4m_\pi^2 s_2^{\frac{1}{2}}/m_1$, and the "normal" threshold starts at $s_1 - m_3^2 = 2m_3 m_\pi + m_\pi^2$ which is still larger. For the deuteron, the resulting singularities are shown in fig. 7-8.2. In practice, the whole vertex function is frequently approximated by one pole, at a position $-p^2 = \alpha^2 \gg \kappa^2$. The corresponding $S$-wave function $u$ of (5.8) is that of Hulthén:

$$u = C(e^{-\kappa r} - e^{-\alpha r}), \qquad C \approx (2\kappa)^{\frac{1}{2}}(1 + 2\kappa/\alpha) \tag{8.8}$$

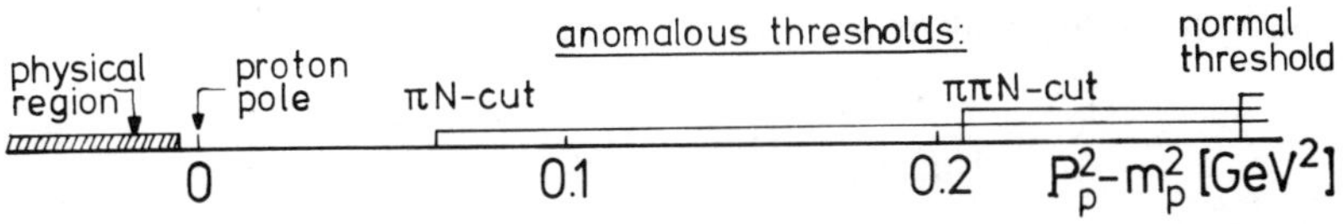


**Figure 7-8.2** Singularities of the Fourier transform of the deuteron wave function.

Graphs of the type shown in fig. 7-8.3 in which the exchanged nucleon $e$ recombines with the recoiling nucleus $u$ into a nuclear state $d$ are referred to as "impulse approximation." Here we treat all particles as spinless. According to the Feynman rules,

$$T = -i \int \frac{d^4 P_u}{(2\pi)^4} G_{beu}(P_e^2) G_{de'u}(P_e'^2) T(ae \to ce') \\ [(m_u^2 - P_u^2)(m_e^2 - P_e^2)(m_e'^2 - P_e'^2)]^{-1}. \tag{8.9}$$

In principle, the vertex functions $G$ are also functions of $P_u^2$, but we shall immediately put the "subnucleus" $u$ on its mass shell. As in the

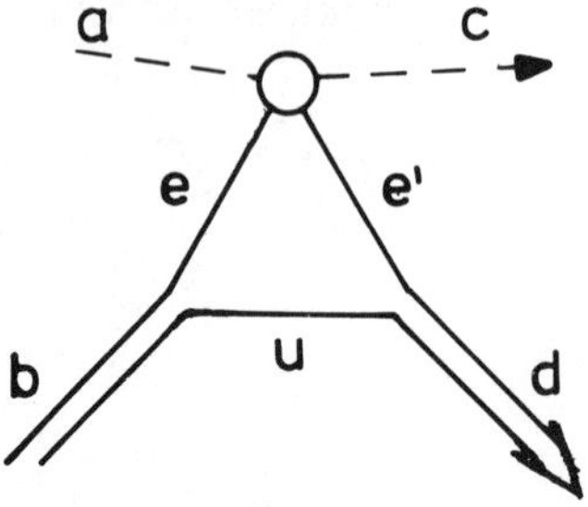


**Figure 7-8.3** The impulse approximation for the reaction $ab \to cd$. $b$ and $d$ are nuclei, $e$ and $e'$ are nucleons.

Bethe-Salpeter equation, the $u$-propagator is approximated by $2\pi i(2E_u)^{-1}\,\delta(E_u - P_u^0)$, and the antiparticle poles of the $e$ and $e'$ propagators as well as all $P^0$-singularities of the vertex functions are neglected. The integration over $P_u^0$ then reduces (9) to

$$T = (2\pi)^{-3} \int \frac{d^3 p_u}{8E_u E_e E'_e} G_{beu}(p_e^2) G_{de'u}(p_e'^2) T(ae \to ce')(E_e - P_e^0)^{-1}(E'_e - P_e^{0\prime})^{-1}, \tag{8.10}$$

$$E_e = (m_e^2 + p_e^2)^{\frac{1}{2}}, \qquad P_e^0 = E_b - E_u, \qquad P_e'^0 = E_d - E_u.$$

This expression exhibits the energy denominators of noncovariant perturbation theory. Here we only wish to show that $T$ has (among others) the singularities discussed in section 6-6. For that purpose, we approximate the functions $G$ by the coupling constants (6-4.23):

$$G_{beu} G_{de'u} \approx 4\pi N_{beu} N_{de'u}. \tag{8.11}$$

To begin with, we consider elastic scattering $c = a$, $d = b$, $e' = e$. We work in the cms where $E'_b = E_b$ and thus $P_e'^0 = P_e^0$ (but $E'_e \pm E_e$). We also put $\mathbf{p}_u = \mathbf{p}_k$ in this system, to conform with the notation of section 6-6. This gives us

$$T = \pi^{-2} N^2 \int \frac{d^3 p_u}{16 E_u E_e E'_e} T(ae \to a'e')(E_e - P_e^0)^{-1}(E'_e - P_e^0)^{-1}. \tag{8.12}$$

The particles $u$, $e$ and $e'$ will be nonrelativistic. $T(ae \to a'e')$ may depend on $s_{ae}^{\frac{1}{2}}$ which is the total energy of the $ae$-system in its own cms. In pion-nucleus scattering for example, $T(ae) = T(\pi N)$ goes as $G_\Delta^2(m_\Delta^2 - s_{ae})^{-1}$ near the $\Delta$ resonance. One can then eliminate $p_k^2$ in favour of $s_{ae}$. When the $ae$-system moves nonrelativistically in the overall cms, one has

$$s^{\frac{1}{2}} - m_u - s_{ae}^{\frac{1}{2}} = \tfrac{1}{2} p_k^2 (m_u^{-1} + s_{ae}^{-\frac{1}{2}}). \tag{8.13}$$

For $m_a \ll m_e$, particle $a$ may of course still be relativistic. In any case, we can put $d^3 p_k = p_k^2\, dp_k\, d\Omega_k$ and do the angular integration at fixed $p_k$, i.e., at fixed $s_{ae}$. With $m_b = m_e + m_u - b$ and $p_a^2 = p_a'^2 = p^2$, we have

$$P_e^0 = E_b - E_u = m_e - b + p^2/2m_b - p_k^2/2m_u,$$

$$E_e - P_e^0 = b + \tfrac{1}{2} p^2 (m_e^{-1} - m_b^{-1}) + \tfrac{1}{2} p_k^2 (m_e^{-1} + m_u^{-1}) - p p_k x'/m_e, \tag{8.14}$$

$$x' \equiv \hat{\mathbf{p}}_b \hat{\mathbf{p}}_k.$$

where we have used $E_e = m_e + (\mathbf{p}_b - \mathbf{p}_u)^2/2m_e$. The corresponding expression for $E'_e - P^0_e$ is obtained from (14) by substituting $x'' = \hat{\mathbf{p}}_{b'}\,\hat{\mathbf{p}}_k$ for $x'$. We may now write the angular integration of (12) in the form of (6-6.6) by putting

$$\begin{gathered} E_e - P^0_e = (Z_k - x')pp_k/m_e, \\ Z_k = (bm_e + p^2m_u/2m_b + p_k^2m_b/2m_u)/pp_k, \qquad Z_k > 1. \end{gathered} \tag{8.15}$$

This shows that the position of the singularity is again given by (6-6.17). Note also that the $m_e$ of (15) cancels the $E_e$ in the denominator of (12). For more information on the use of analyticity in nuclear physics, see Locher and Mizutani (1978).

## 7-9 Eikonal Approximation and Glauber Theory

In the collisions of fast particles with composite targets such as atoms or nuclei, the incident particle frequently emerges at a relatively small scattering angle. For such cases the eikonal approximation is useful. It was first developed by Molière (1947) for the electric scattering of charged particles on atoms*. Glauber (1959) put it into the form of a multiple-scattering theory, which allows one to calculate the scattering amplitude on a nucleus from the scattering amplitudes on its nucleons without using a potential. For a general review, see Joachain and Quigg (1974). We start from the Klein-Gordon equation in the form of (1-1.22) with $\mathbf{A} = 0$ and assume $\psi$ in the form of a modulated plane wave, with $\mathbf{k}$ along the $z$-axis:

$$\psi(\mathbf{r}) = e^{ikz}\varphi(\mathbf{r}). \tag{9.1}$$

For large enough $k$, one can neglect $\nabla^2\varphi$ relative to $ik\,\partial_z\varphi$ and also approximate $U = 2EV$. Insertion of (1) into (1-1.22) gives $ik\,\partial_z\varphi = EV\varphi$ and thus

$$\psi = \exp\left\{ikz - iv^{-1}\int_{-\infty}^{z} dz'V(\mathbf{r}')\right\}, \qquad v \equiv k/E. \tag{9.2}$$

To find the scattering amplitude $f(\mathbf{k}, \mathbf{k}')$, we first transform (1-1.22) into the Lippmann-Schwinger equation:

$$\psi = e^{ikz} - \int d^3r' e^{ik|\mathbf{r}-\mathbf{r}'|}(4\pi|\mathbf{r}-\mathbf{r}'|)^{-1}U(\mathbf{r}')\psi(\mathbf{r}'). \tag{9.3}$$

For $\mathbf{r} \to \infty$, $|\mathbf{r} - \mathbf{r}'| \to r - \mathbf{r}\mathbf{r}'/r$, and comparison with (1-3.7) gives

$$f = -(4\pi)^{-1}\int e^{-i\mathbf{k}'\mathbf{r}}U(\mathbf{r})\psi(\mathbf{r})\,d^3r, \qquad d\sigma/d\Omega = |f|^2. \tag{9.4}$$

Note that for $\varphi = 1$, $f$ reduces to the Born approximation (2-6.32) with $\mathbf{q} = \mathbf{k} - \mathbf{k}'$. When $\psi$ is given by (2), cylindrical coordinates $(z, \mathbf{b})$ are appro-

* See also Scott (1963).

priate, where $\mathbf{b}$ are the components of $\mathbf{r}$ perpendicular to $z$. For elastic scattering with $q \ll k$, $\mathbf{q}$ is also perpendicular to $z$, such that (4) is of the form

$$f = -(4\pi)^{-1} \int d^2 b e^{i\mathbf{qb}} I(\mathbf{b}),$$
$$I = \int_{-\infty}^{\infty} dz U(\mathbf{b}, z) \exp\left\{-\frac{i}{v}\int_{-\infty}^{z} V(\mathbf{b}, z')\, dz'\right\}. \tag{9.5}$$

We can write $U = 2EV = 2ik(-iv^{-1}V)$, the bracket being the derivative of the function in the exponent with respect to $z$,

$$I = 2ik \int_{-\infty}^{\infty} dz \exp\{\ \}\frac{d}{dz}\{\ \} = 2ik \exp\{\ \}_{z=-\infty}^{z=\infty} = 2ik(e^{\chi(\mathbf{b})} - 1),$$
$$\chi(b) = -\frac{i}{v}\int_{-\infty}^{\infty} dz' V(\mathbf{b}, z'). \tag{9.6}$$

This is already the desired result. From its derivation, one would conclude that it is only valid up to terms of the order $\mathbf{q} \cdot \mathbf{k}$. However, these terms are actually included in (5) and (6). To understand this, we note that time-reversal invariance requires $f(\mathbf{k}, \mathbf{k}') = f(-\mathbf{k}', -\mathbf{k})$, and that $f$ as given by (5) satisfies this requirement. There exists thus a better derivation, in which the $z$-axis is taken along $\mathbf{k} + \mathbf{k}'$, such that $\mathbf{q} \cdot (\mathbf{k} + \mathbf{k}') = 0$ implies $\mathbf{q} \cdot \mathbf{r} = \mathbf{q} \cdot \mathbf{b}$ (Breit frame). Classically, $\mathbf{b}$ is the average impact parameter of initial and final states, but we shall simply call it "impact parameter."

Consider now the scattering on a nucleus composed of $A$ nucleons (for the moment, we need not distinguish between protons and neutrons). If the incident particle moves fast enough, it will see the nucleons standing still at positions $\mathbf{r}_1 \ldots \mathbf{r}$ (see fig. 7-9.1), with probability amplitudes $\langle \mathbf{r}_1 \ldots \mathbf{r}_A | a\rangle = \psi_a(\mathbf{r}_1 \ldots \mathbf{r}_A)$. Denoting the elastic scattering amplitude on the whole nucleus by $F$, we have

$$F = \frac{ik}{2\pi}\int d^2 b e^{i\mathbf{qb}} \langle a | 1 - \exp\{\chi_t(\mathbf{b}, \mathbf{r}_1 \ldots \mathbf{r}_A)\} | a\rangle. \tag{9.7}$$

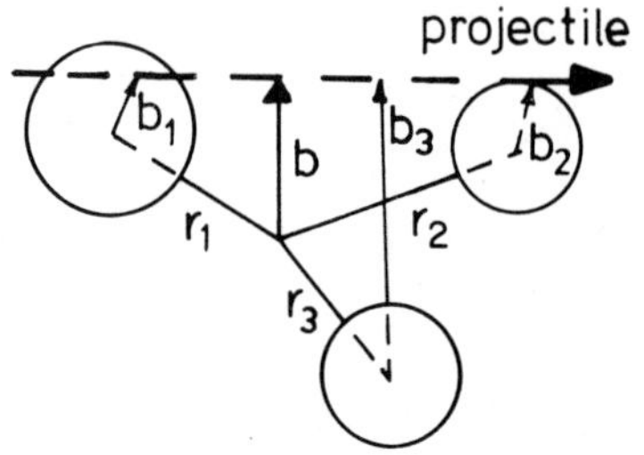


**Figure 7-9.1** A projectile trajectory through an $A = 3$ nucleus. The nucleons are at positions $\mathbf{r}_1, \mathbf{r}_2, \mathbf{r}_3$, at distances $\mathbf{b}_1, \mathbf{b}_2, \mathbf{b}_3$ from the trajectory.

We now assume that the potential $V(\mathbf{r}, \mathbf{r}_1 \ldots \mathbf{r}_A)$ which determines $\chi_t$ is the sum of the potentials of the single nucleons,

$$V(\mathbf{r}, \mathbf{r}_1 \ldots \mathbf{r}_A) = \sum_{i=1}^{A} V(\mathbf{r} - \mathbf{r}_i). \tag{9.8}$$

This leads to

$$\chi_t(\mathbf{b}, \mathbf{r}_1 \ldots \mathbf{r}_A) = \sum_{i=1}^{A} \chi(\mathbf{b} - \mathbf{s}_i), \tag{9.9}$$

where $\mathbf{s}_i$ are the components of $\mathbf{r}_i$ perpendicular to $z$. Each term on the right-hand side of (9) can be expressed in terms of the scattering amplitude $f(\mathbf{q})$ which now refers to the scattering on a single nucleon. Calling $\mathbf{b} - \mathbf{s}_i \equiv \mathbf{b}_i$, we can invert (5) to give

$$1 - e^{i\chi(\mathbf{b}_i)} \equiv \Gamma(\mathbf{b}_i) = (2\pi ik)^{-1} \int d^2q e^{-i\mathbf{q}\mathbf{b}_i} f(\mathbf{q}). \tag{9.10}$$

We can thus rewrite the function in (7) as

$$\begin{aligned}\Gamma(\mathbf{b}, \mathbf{b}_1 \ldots \mathbf{b}_A) &= 1 - \exp\left\{\sum_i \chi(\mathbf{b}_i)\right\} = 1 - \prod_{i=1}^{A} (1 - \Gamma(\mathbf{b}_i)) \\ &= \sum_{i=1}^{A} \Gamma(\mathbf{b}_i) - \sum_{i<j} \Gamma(\mathbf{b}_i)\Gamma(\mathbf{b}_j) + \sum_{i<j<k} \cdots\end{aligned} \tag{9.11}$$

This is the desired multiple scattering expansion. The first sum gives just the impulse approximation. Let us see in detail how this works for the simplest composite nucleus, namely the deuteron.

After separation of the nuclear cms motion (which gives momentum conservation $\delta_3(\mathbf{k} + \mathbf{p}_d - \mathbf{k}' - \mathbf{p}_d')$, compare (5-2.2)), we can put the deuteron cms at the origin. With $\mathbf{r} = \mathbf{r}_p - \mathbf{r}_n$, the proton and neutron positions are $\mathbf{r}/2$ and $-\mathbf{r}/2$, respectively. Denoting the components of $\mathbf{r}$ perpendicular to $z$ by $\mathbf{s}$, we have $\mathbf{s}_p = \mathbf{s}/2$, $\mathbf{s}_n = -\mathbf{s}/2$ in (9). Thus (7) becomes (for spinless nucleons)

$$\begin{aligned}F(\mathbf{q}) &= \frac{ik}{2\pi} \int d^2b e^{i\mathbf{q}\mathbf{b}} \int d^3r \psi_d^2(\mathbf{r})(1 - \exp\{\chi_p(\mathbf{b} - \mathbf{s}/2) + \chi_n(\mathbf{b} + \mathbf{s}/2)\} \\ &= \frac{ik}{2\pi} \int d^2b e^{i\mathbf{q}\mathbf{b}} \int d^3r \psi_d^2(\mathbf{r}) \\ &\quad \times [\Gamma_p(\mathbf{b} - \tfrac{1}{2}\mathbf{s}) + \Gamma_n(\mathbf{b} + \tfrac{1}{2}\mathbf{s}) - \Gamma_p(b - \tfrac{1}{2}\mathbf{s})\Gamma_n(b + \tfrac{1}{2}\mathbf{s})].\end{aligned} \tag{9.12}$$

We first consider the integral over $\Gamma_p$. For that purpose, we replace the integration variables $\mathbf{b}$ by $\mathbf{b} - \mathbf{s}/2$ and write $\exp i\mathbf{q}\mathbf{b}$ as $\exp i\mathbf{q}(\mathbf{b} - \mathbf{s}/2) \exp(i\mathbf{q}\mathbf{s}/2)$. From (5) or (11), we then find

$$F_p(\mathbf{q}) = f_p(\mathbf{q}) \int d^3r \psi_d^2(\mathbf{r}) e^{i\mathbf{q}\mathbf{s}/2} = f_p(\mathbf{q}) S(\mathbf{q}/2), \tag{9.13}$$

$$S(\mathbf{q}) = \int d^3r \psi_d^2(\mathbf{r}) e^{i\mathbf{q}\mathbf{r}}. \tag{9.14}$$

$S(\mathbf{q})$ is the deuteron form factor. The neutron contribution $F_n$ is analogous to (13). To evaluate the "double-scattering" contribution $F_{pn}$, we must use (10) both for $\Gamma_p$ and $\Gamma_n$:

$$F_{pn} = -\frac{ik}{2\pi}\int d^2b e^{i\mathbf{qb}} \int d^3r\psi_d^2(\mathbf{r})(2\pi ik)^{-2} \int d^2q_p\, d^2q_n\, f_p(\mathbf{q}_p) f_n(\mathbf{q}_n) \times e^{-i\mathbf{q}_p(\mathbf{b}-\mathbf{s}/2)} e^{-i\mathbf{q}_n(\mathbf{b}+\mathbf{s}/2)}. \tag{9.15}$$

Next, we replace $\mathbf{q}_p$ and $\mathbf{q}_n$ by new integration variables $\mathbf{q}'$ and $\mathbf{q}''$:

$$\mathbf{q}_p = \mathbf{q}' + \tfrac{1}{2}\mathbf{q}'', \qquad \mathbf{q}_n = -\mathbf{q}' + \tfrac{1}{2}\mathbf{q}'', \tag{9.16}$$

because the whole $\mathbf{b}$-dependence is then simply $\exp i\mathbf{b}(\mathbf{q} - \mathbf{q}'')$, which is integrated over $\mathbf{b}$ to give $(2\pi)^2\, \delta_2(\mathbf{q} - \mathbf{q}'')$. We thus obtain

$$F_{pn} = \frac{i}{2\pi k}\int d^2q'\, d^3r\psi_d^2(\mathbf{r})e^{i\mathbf{q}'\mathbf{s}} f_p(\mathbf{q}' + \tfrac{1}{2}\mathbf{q}) f_n(\mathbf{q}' - \tfrac{1}{2}\mathbf{q}). \tag{9.17}$$

The integral over $\mathbf{r}$ produces the deuteron form factor $S(\mathbf{q}')$. Collecting our pieces, we have

$$F(\mathbf{q}) = [f_p(\mathbf{q}) + f_n(\mathbf{q})]S(\tfrac{1}{2}\mathbf{q}) + \frac{i}{2\pi k}\int d^2q' S(\mathbf{q}') f_p(\mathbf{q}' + \tfrac{1}{2}\mathbf{q}) f_n(\mathbf{q}' - \tfrac{1}{2}\mathbf{q}). \tag{9.18}$$

The total cross section on deuterons is given by the optical theorem:

$$\sigma_{\text{tot}} = k^{-1}4\pi \text{ Im } F(0). \tag{9.19}$$

(Compare 2-5.4 with $T = 8\pi s^{\frac{1}{2}} F$). Since the form factor (14) is unity at zero momentum transfer,

$$\begin{aligned} \sigma_{\text{tot}} &= \sigma_{\text{tot}}(p) + \sigma_{\text{tot}}(n) - \delta\sigma, \\ \delta\sigma &= -2k^{-2} \text{ Re} \int d^2\mathbf{q}' S(\mathbf{q}') f_p(\mathbf{q}') f_n(\mathbf{q}'). \end{aligned} \tag{9.20}$$

The double scattering correction $\delta\sigma$ is negative when $f_p$ and $f_n$ are dominantly imaginary at small momentum transfers. For a qualitative estimate, one may say that the deuteron is so loosely bound that $S(\mathbf{q}')$ is proportional to $\delta_2(\mathbf{q}')$. This gives

$$\delta\sigma \approx -2k^{-2} \text{ Re } [f_p(0) f_n(0)] \int d^2q S(q). \tag{9.21}$$

The last integral is $2\pi\langle r^{-2}\rangle_d$. When Re $f_p$ and Re $f_n$ are neglected, one has $-\text{Re } (f_p f_n) = \text{Im } (f_p) \text{ Im } (f_n)$, and according to the optical theorem

$$\delta\sigma \approx \sigma_{\text{tot}}(p)\sigma_{\text{tot}}(n)(4\pi)^{-1}\langle r^{-2}\rangle_d. \tag{9.22}$$

This expression can be understood classically as a "screening effect" in the scattering on two extended nucleons.

The single-scattering terms in (18) dominate at small $q$, but fall rapidly due to the factor $S(\mathbf{q}/2)$. This factor is not present in the double-scattering term, which involves $S(\mathbf{q}')$. In fact, when $S(\mathbf{q}')$ is approximated by a $\delta$-function, the double-scattering term becomes proportional to $f_p(\mathbf{q}/2)f_n(\mathbf{q}/2)$. If $f_p(\mathbf{q})$ and $f_n(\mathbf{q})$ are dominantly imaginary, the interference between single- and double-scattering terms remains destructive as $q$ increases, and the differential cross section has a minimum where the two terms are equal (this is in the case of only one scattering amplitude. Since the deuteron has spin 1, the minimum is filled by the $D$-state portion of the deuteron wave function, see fig. 7-9.2). Beyond the minimum, double scattering dominates over single-scattering. Similar statements hold for heavier nuclei. The quantitative success of Glauber's theory for the scattering on such nuclei is at least also partly due to the fact that the theory contains the limit of black sphere scattering. This becomes obvious if all target nucleons are put into identical one-nucleon wave functions $\varphi(\mathbf{r}_i)$ with $\varphi^2(\mathbf{r}_i) = \rho(\mathbf{r}_i) =$ 1-nucleon-density. Insertion of (11) into (7) then gives

$$\begin{aligned} F &= \frac{ik}{2\pi}\int d^2b e^{i\mathbf{q}\mathbf{b}} \prod_{i=1}^{A} d^3r_i \rho(\mathbf{r}_i)[1 - \prod(1 - \Gamma(\mathbf{b}_i))] \\ &= \frac{ik}{2\pi}\int d^2b e^{i\mathbf{q}\mathbf{b}} \left\{1 - \left[1 - \int d^3r_1 \Gamma(\mathbf{b} - \mathbf{s}_1)\rho(\mathbf{r}_1)\right]^A\right\}. \end{aligned} \tag{9.23}$$

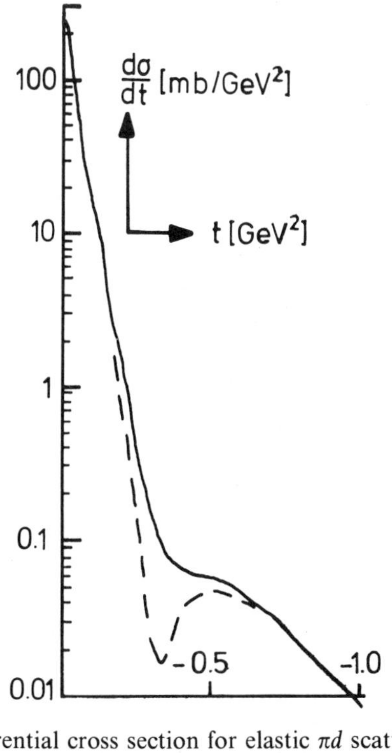


**Figure 7-9.2** The differential cross section for elastic $\pi d$ scattering at high energies. The dashed curve neglects the $D$-state in the deuteron. Its minimum separates the impulse approximation region from the double scattering region.

For large nuclei, $\rho(\mathbf{r}_1)$ varies much less with $\mathbf{s}_1$ than $\Gamma(\mathbf{b}-\mathbf{s}_1)$, which contributes only for $\mathbf{s}_1 \approx \mathbf{b}$. Thus the integral in (23) can be approximated by the two factors

$$\int d^2b\Gamma(\mathbf{b}) = \frac{2\pi}{ik} f(0) \approx \tfrac{1}{2}\sigma_{\text{tot}}(p), \qquad \int dz_1 \rho(b, z_1) \equiv A^{-1}T(\mathbf{b}). \quad (9.24)$$

For large nucleon number $A$ and nearly imaginary $f(0)$, (23) is thus approximately

$$F \approx \frac{ik}{2\pi}\int d^2be^{i\mathbf{qb}}\{1 - \exp\left[-\tfrac{1}{2}\sigma T(\mathbf{b})\right]\}. \quad (9.25)$$

The "target thickness" $T(\mathbf{b})$ is the two-dimensional matter density at impact parameter $\mathbf{b}$. The nucleus is "black" at all $\mathbf{b}$ for which the exponent in the curly bracket of (25) is negligible. When $\rho(\mathbf{r}_1)$ is spherically symmetric, one can put

$$\int d\varphi \exp\,(iqb\cos\varphi) = 2\pi J_0(qb). \quad (9.26)$$

Comparing this with (6-7.16)

$$F = \sum_L (2L+1)P_L(\cos\theta)\frac{i}{2k}(1 - S_L) = ik\int bdbJ_0(qb)(1 - e^{\chi}) \quad (9.27)$$

and remembering

$$P_L\,(\cos\theta) \approx J_0((2L+1)\sin\tfrac{1}{2}\theta) = J_0(qb), \qquad b = (L+\tfrac{1}{2})/k, \quad (9.28)$$

we see that for large $L$, $e^{\chi}$ is approximately $S_L$. The case $S_L = 0$ corresponds to complete absorption ($\eta_L = 0$ in 6-7.17). In the "black sphere" limit, one approximates $S_L$ by $e^{\chi} = \theta(b - R)$. Integration of (27) gives then

$$F(\text{black sphere}) = ikq^{-2}\int_0^{qR} x\,dxJ_0(x) = iR\left(2\sin\frac{\theta}{2}\right)^{-1} J_1(qR). \quad (9.29)$$

The formalism can also be applied to nuclear excitation to final states $|e\rangle$. For these cases, (7) is replaced by

$$F_{ae} = \frac{ik}{2\pi}\int d^2be^{i\mathbf{qb}}\langle e|\Gamma(b, \mathbf{r}_1 \ldots \mathbf{r}_A)|a\rangle. \quad (9.30)$$

This is most reliable in cases where the matrix elements (11) of $\Gamma$ are large, such as the excitation of $2^+$ rotational levels. Also nucleus-nucleus scattering can be treated by Glauber theory. For example, in deuteron-nucleus collisions, the deuteron wave function (of relative proton-neutron motion) is distroted by absorption, such that it develops components in the continuum of $np$-states. This is called "diffraction dissociation." Even if the incident particle is "elementary" but is coupled to a few excited states $f$, with amplitudes

$$f_{if} = \frac{ik}{2\pi}\int d^2be^{i\mathbf{qb}}\Gamma_{if}(\mathbf{b}) \quad (9.31)$$

on a single nucleon, one can calculate the corresponding production amplitude on a nucleus. To first order in $\Gamma_{if}$,

$$F_{if} = \frac{ik}{2\pi} \int d^2b e^{i\mathbf{qb}} \sum_k \Gamma_{if}(\mathbf{b}_k) \prod_{j \neq k} [1 - \theta(z_i - z_j)\Gamma_{ij}(\mathbf{b}_j) - \theta(z_j - z_i)\Gamma_{fj}(\mathbf{b}_j)], \tag{9.32}$$

where $\Gamma_{ij}$ and $\Gamma_{fj}$ are the profile functions for the elastic scattering of particles $i$ and $f$ on nucleon $j$. The transformation from $i$ to $f$ occurs on the $k$th nucleon at position $\mathbf{r}_k = (\mathbf{b}_k, z_k)$, and the $\theta$-functions in (32) tell whether the particle is in state $i$ or $f$ when it scatters on the other nucleons. When no distinction is made between $p$ and $n$ in the target, the summation over $k$ in (32) gives $A$ times the term for $k = 1$. Using again the simple product wave function which led us to (23) and making approximations (24), we arrive at the following integral

$$\int_{-\infty}^{\infty} dz A\rho(\mathbf{b}, z) \exp\left\{-\tfrac{1}{2}\sigma_i A \int_{-\infty}^{z} dz' \rho(\mathbf{b}, z') - \tfrac{1}{2}\sigma_f A \int_{z}^{\infty} dz' \rho(\mathbf{b}, z')\right\}$$
$$= 2(\sigma_f - \sigma_i)^{-1}[\exp\{-\tfrac{1}{2}\sigma_i T(\mathbf{b})\} - \exp\{-\tfrac{1}{2}\sigma_f T(\mathbf{b})\}]. \tag{9.33}$$

With these approximations, we obtain

$$F_{if} \approx f_{if}(0) 2(\sigma_f - \sigma_i)^{-1} \int d^2b e^{i\mathbf{qb}}[\exp\{-\tfrac{1}{2}\sigma_i T(\mathbf{b})\} - \exp\{-\tfrac{1}{2}\sigma_f T(\mathbf{b})\}]. \tag{9.34}$$

An important application of this formula is the coherent photoproduction of $\rho$-mesons on nuclei, $\gamma A \to \rho A$, because it provides a means to measure the total $\rho N$ cross section $\sigma_{\rho N}$, at least at high energies. Since the total $\gamma N$ cross section is negligible, (34) simplifies to

$$F_{\gamma\rho}(\mathbf{q}) \approx 2 f_{\gamma\rho}(0) \sigma_{\rho N}^{-1} \int d^2b e^{i\mathbf{qb}}[1 - \exp\{-\tfrac{1}{2}\sigma_{\rho N} T(\mathbf{b})\}]. \tag{9.35}$$

The resulting $\sigma_{\rho N}$ in the 10 GeV region is about 27 mb. For further details, see the review by Yennie (1971).

CHAPTER EIGHT

# Particular Electromagnetic Processes in Collisions with Atoms and Nuclei

## 8-1 Ionization by Fast Particles

When a charged particle $a$ enters a large target, it continually loses energy to the atomic electrons, either by atomic excitation or by ionization (electron emission). Typical energy losses in single collisions

$$\Delta E_n = E_{aL} - E_{aL}(n) \tag{1.1}$$

($n$ characterizes the final state of the excited target) are much smaller than the kinetic energy of particle $a$. The particle then moves approximately on a straight line, with an energy loss per target density

$$\varepsilon\,(\equiv -N^{-1}\,dE/dx) = \sum_n \Delta E_n \int d\sigma_n = \sum_n \int \Delta E_n\, dt\, d\sigma_n/dt. \tag{1.2}$$

The quantity $\varepsilon$ is called the "stopping cross section." As long as the velocity $\beta = p_{aL}/E_{aL}$ is much larger than the electron velocity ($\sim \alpha$) in the target, $\varepsilon$ can be evaluated in the Born approximation. The derivation is due to Bethe and Bloch (see Livingstone and Bethe 1937). One first considers all those final states $n$ in which the target electrons remain nonrelativistic. The corresponding cross section is given by (2-7.17), with $\mathbf{J}_{bd}$ approximated by the nonrelativistic electron current contribution:

$$\mathbf{J}_{bd}(\mathbf{q}) = 2m_b(2m_e i)^{-1} \prod_{i=1}^{Z_b} \int d^3r_i \sum_{j=1}^{Z_b} e^{-i\mathbf{q}\mathbf{r}_j}(\psi_b \nabla_j \psi_n^* - \psi_n^* \nabla_j \psi_b). \tag{1.3}$$

Gauge invariance $q_\mu J^\mu_{bd} = 0$ with $\mathbf{q} = \mathbf{p}_a - \mathbf{p}'_a$, $q_0 = \Delta E_n$ is used to find $J^0_{bd}$:

$$J^0_{bd} = q_0^{-1}\mathbf{q}\mathbf{J}_{bd}(\mathbf{q}). \tag{1.4}$$

As $q^\mu$ remains small for these states, $J^\mu_{aa'}$ is approximated by $2Z_a P^\mu_a$, independent of the spin of the incident particle $a$ (for example, using $S^{\mu\nu} \approx 2P^\mu_a P'^\nu_a + 2P^\nu_a P'^\mu_a$ in (2-7.20) gives the same result as (2-7.18)). We may thus use in (2-7.17)

$$J_{aa'\mu} J^\mu_{bd} = -2(\mathbf{p}_a - \mathbf{q}E_{aL}/\Delta E_n)\mathbf{J}_{bd}, \tag{1.5}$$

and neglect both the projectile spin and the electron spin. Near $t = t_{\max}$, we may use

$$p'_{aL} = [(E_{aL} - q_0)^2 - m_a^2]^{\frac{1}{2}} \sim p_{aL} - E_{aL} q_0/p_{aL}, \tag{1.6}$$

$$\begin{aligned} -t &= -q_0^2 + (\mathbf{p}_{aL} - \mathbf{p}'_{aL})^2 \approx -q_0^2 + (p_{aL} - p'_{aL})^2 + p^2_{aL}\vartheta_L^2 \\ &= q_0^2 m_a^2 p_{aL}^{-2} - E_{aL} q_0/p_{aL}, \end{aligned} \tag{1.7}$$

where $\vartheta_L$ is the lab scattering angle. Thus, the integration over $-t$ in (2) starts at $\Delta E_n^2 m_a^2 p_{aL}^{-2}$. The remaining calculation for these states is entirely nonrelativistic (see Landau and Lifshitz, Vol. 3 (1971) § 146 and Vol. 4 (1975) § 82a), the result being the approximate analytic expression

$$\varepsilon_{nr} \approx 2\pi Z_b Z_a^2 \alpha^2 m_e^{-1}[\beta^{-2} \ln(-t_{\text{cut}} p^2_{aL}/m_a^2 I^2) - 1]. \tag{1.8}$$

Here $I$ is an average ionization potential ($I \sim Z_b I_0$, $I_0 \sim 12$ eV, see Whaling 1958, Fano 1963), and $-t_{\text{cut}}$ is a maximum value of $-t$ such that the final electrons are still nonrelativistic.

To the nonrelativistic contribution $\varepsilon_{nr}$ we must now add $\varepsilon_r$, which is the contribution from final states with free electrons having $t < t_{\text{cut}}$. There exists a large region of possible choices of $t_{\text{cut}}$ such that the electrons can be approximated by $Z_b$ free electrons at rest for this category of states. Thus, we find

$$t = 2m_e^2 - 2m_e E'_{eL} = -2m_e\,\Delta E, \tag{1.9}$$

$$\varepsilon_r = \tfrac{1}{2} m_e^{-1} Z_b \int_{t_{\min}}^{t_{\text{cut}}} t\, dt\, d\sigma(ae \to a'e')/dt. \tag{1.10}$$

When the projectile is not an electron or positron, we can still approximate it by a spinless point particle, i.e., $d\sigma(ae \to a'e')$ is obtained from (7.21) by exchanging $a$ and $b$:

$$d\sigma(ae \to a'e')/dt = \pi p_{aL}^{-2} m_e^{-2} Z_a^2 \alpha^2 [m_a^2 + 4E^2_{aL} m_e^2/t + 2E_{aL} m_e]. \tag{1.11}$$

Here we have used $P_a P_b = E_{aL} m_e$ and (2-7.22) to eliminate $P'_a P_b$. The integration in (10) is now easily performed. Using $t_{\min} = -4p^2 = -4p^2_{aL} m_e^2/s_{ae}$ from (2-4.10) and (2-4.14), and $s_{ae} \approx m_a^2 + 2E_{aL} m_e$, we find

$$\varepsilon_r = 2\pi Z_b Z_a^2 \alpha^2 m_e^{-1}[\beta^{-2} \ln(-4p^2_{aL} m_e^2 t_{\text{cut}}^{-1} s_{ae}^{-1}) - 1]. \tag{1.12}$$

Adding this to (8), we see that the unphysical $t_{\text{cut}}$ disappears:

$$\varepsilon = \varepsilon_{nr} + \varepsilon_r = 4\pi Z_b Z_a^2 m_e^{-1}[\beta^{-2} \ln (2p_{aL}^2 m_e m_a^{-1} I^{-1} s_{ae}^{-\frac{1}{2}}) - 1]. \quad (1.13)$$

This is the famous Bethe-Bloch formula. One normally assumes $2E_{aL} m_e \ll m_a^2$, in which case the argument of the logarithm can be rewritten as $2m_e \beta^2 (1 - \beta^2)^{-1} I^{-1}$. See Northcliffe and Schilling (1976) and Inokuti et al. (1978) for corrections.

## 8-2 Coherent Photon Exchange Reactions on Nuclei

By coherent reactions on nuclei we mean reactions of the type $aA \to cA$, where the projectile is excited, but the target nucleus remains in its ground state. Such reactions are particularly important for photon exchange, because of the long range of the interaction and because the coupling constant $Ze$ enters quadratically (the sum over all excited nuclear states brings only a factor $Ze^2$). Examples of such reactions are bremsstrahlung $eA \to e'\gamma A$ and pair production $\gamma A \to e\bar{e}A$, but also the "Primakoff effect" $\gamma A \to \pi^0 A$ (fig. 8-2) at small angles, or Coulomb excitation of composite projectiles. The application of the eikonal approximation to these reactions dates back to times when a straightforward calculation of the cross section in the Born approximation (the matrix element being given by 3-5.8 in the case of bremsstrahlung) was not available. The approximation is known as the "Weizsäcker-Williams-method" or "method of virtual photons" and is normally derived from classical considerations (see Jackson (1975), chapter 15). Later, it was understood that it can include effects which go beyond the Born approximation. It then becomes a particular type of distorted-wave Born approximation (Fäldt 1972).

One normally uses the black sphere model, which does not resolve the nucleus into protons and neutrons but works directly with the profile function $\Gamma_{if}(b)$ for the whole nucleus. For $b > R$ ($R$ being the radius of the black sphere), $\Gamma$ is simply given by the Born approximation (multiplied by Coulomb phases if $a$ and $c$ are charged), while for $b < R$ one puts $\Gamma_{if} = 0$.

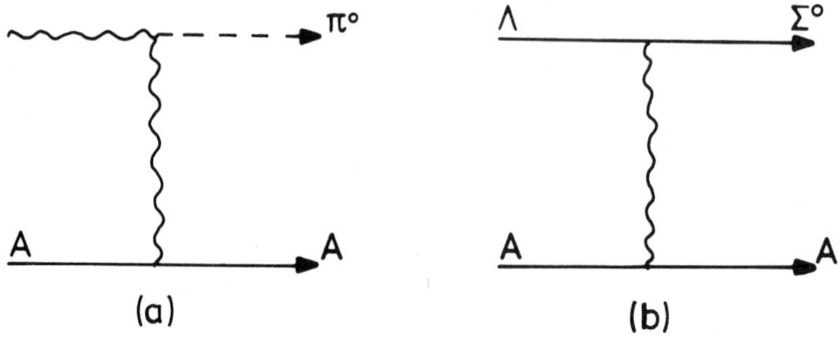


**Figure 8-2** The "Primakoff" graphs, from which the lifetimes of $\pi^0$ (graph a) and $\Sigma^0$ (graph b) are determined.

One then has in the lab system

$$\Gamma_{if}(b, q_l) = (2\pi ik)^{-1}\theta(b - R)\int d^2q_t e^{-i\mathbf{q}_t\mathbf{b}} T_{if}^{\text{Born}}(\mathbf{q}_t, q_l). \tag{2.1}$$

Here $q_t$ and $q_l$ are the components of $\mathbf{q}$ transverse and parallel to $\mathbf{k}$, respectively. One cannot simply put $q_l = 0$ because of the photon propagator $t^{-1}$. In the lab system one has $q_0 = 0$ (the recoil energy is negligible) and therefore

$$t = -q_t^2 - q_l^2, \qquad q_l = (-t_{\max})^{\frac{1}{2}} \approx d/2k, \qquad d = m_c^2 - m_a^2. \tag{2.2}$$

as in (7-7.3). In the case of bremsstrahlung, $c$ refers to a 2-particle continuum and $m_c^2$ is replaced by $s_c$. In any case, $q_l$ is independent of $q_t$ within the applicability of the model. The production amplitude is now given by (7-9.30):

$$T_{if} = ik(2\pi)^{-1}\int d^2b e^{i\mathbf{q}_t\mathbf{b}}\Gamma_{if}(\mathbf{b}). \tag{2.3}$$

It differs from $T_{if}^{\text{Born}}$ because of the $\theta$-function in (1) and a possible Coulomb phase. The general form of $T_{if}^{\text{Born}}$ has been given in (2-7.16). Target spin effects can be neglected, such that $J_{aa'}$ is always given by (2-7.9)

$$T_{if}^{\text{Born}} = -Z_b e^2 F_b(t)[q^{-2}J_{ac}^0(E_b + E_b') + t^{-1}J_{ac}^{\perp}(\mathbf{p}_b + \mathbf{p}_b')^{\perp}]. \tag{2.4}$$

Note that we have switched from the $F_{if}$ of the previous section back to our "invariant" $T_{if}$. The angular-integrated cross section is, from (2-4.9)

$$\sigma(q_l) = \lambda^{-1}\int d^2q_t\overline{|T_{if}/4\pi|^2} = \tfrac{1}{4}m_b^{-2}\int_R^\infty d^2b\overline{|\Gamma_{if}(\mathbf{b}, q_l)/4\pi|^2}, \tag{2.5}$$

where the bar has the meaning of (2-3.6), since the target spin is completely decoupled.

The $q$-dependence of (4) is best evaluated by multipole expansions of $J^0$ and $J^{\perp}$ in the rest system of $c$ (Jäckle and Pilkuhn 1975). Here we consider only the limit $k \to \infty$, $d$ fixed. In this case, the average value of $t$ tends also to zero, such that the exchanged photon in (4) is almost real. With $q^2 = q_0^2 - t \approx q_0^2$, $q_0 = (2m_c)^{-1}d$, we see that $q$ remains finite in this frame. Thus the first term in (4) (the "Coulomb excitation" proper, which dominates at low energies) becomes relatively unimportant for $k \to \infty$. The remainder is due to the exchange of transverse photons. The length of $(p_b + p_b')^{\perp} \approx 2p_b^{\perp}$ is $2p(c)\sin\vartheta$, with $\sin\vartheta = q_t/q$ and $p(c) = m_A k/m_c$ according to (2-4.14). We therefore have

$$J_{ac}^{\perp}(p_b + p_b')^{\perp} = \mathbf{J}_{ac}(p_b + p_b')^{\perp} = 2J_{ac}q_t m_b k(qm_c)^{-1}\cos\varphi. \tag{2.6}$$

Inserting this into (4) and (1), neglecting $F_b(t)$ and using (7-9.26), we obtain

$$\Gamma_{if}(\mathbf{b}) = 2Z_b e^2(qm_c)^{-1}m_b J_{ac} b\cos\varphi q_l K_1(q_l b), \tag{2.7}$$

where $K_1$ is the modified Bessel function of order 1. The integrals (3) and (5) can also be solved analytically. For (5), the result is

$$\begin{aligned}\sigma(q_l) &= \pi Z_b^2 \alpha^2 \overline{|J_{ac}|^2} (qm_c)^{-2} \tfrac{1}{2} (q_l R)^2 [K_0(q_l R) K_2(q_l R) - K_1^2(q_l R)] \\ &= \pi Z_b^2 \alpha^2 \overline{|J_{ac}|^2} 4\, d^{-2} [\ln(2/q_l R) - C - \tfrac{1}{2}], \qquad C = 0.5772.\end{aligned} \tag{2.8}$$

The produced particle $c$ is necessarily unstable against the decay $c \to a\gamma$. The corresponding partial width can be written as

$$\Gamma_{ca} = \tfrac{1}{2} \alpha m_c^{-2} q (2S_c + 1)^{-1} \sum_{M_c, M_a} |J(M_c, M_a)|^2, \tag{2.9}$$

where the decay momentum has been denoted by $q$. This gives us

$$\overline{|J_{ac}|^2} = (2S_c + 1)(2S_a + 1)^{-1} 2\Gamma_{ca} m_c^2 q^{-1} \alpha^{-1}. \tag{2.10}$$

This formula can be inserted into (8), which can then be used for determining $\Gamma_{ca}$. A good example is the measurement of $\Gamma(\Sigma^0 \to \Lambda\gamma)$ (Dydak 1977). In other cases, a coherent $\omega$-exchange amplitude must also be admitted. When the incident particle is a photon as in the case of the $\pi^0$ and $\eta$ lifetime measurements, (9) gets multiplied by $\frac{1}{2}$, and the factor $(2S_a + 1)$ must be omitted in (10) (compare 4-7.10). The value $\Gamma(\pi^0 \to \gamma\gamma) = 7.86$ eV of table C-2 corresponds to $g_{\pi\gamma\gamma} = 0.277/\text{GeV}$.

When $c$ represents a system of particles, the differential cross section with respect to $s_c$ is according to (4-4.6)

$$d\sigma/ds_c = (2\pi)^{-1} \int d \text{ Lips } (s_c; \ldots) \sigma(q_l) \tag{2.11}$$

Moreover, we have from (2-3.16)

$$\sigma_\gamma = \tfrac{1}{8} e^2 (qm_c)^{-1} \int \overline{|J_{ac}|^2}\, d \text{ Lips } (s_c; \ldots), \tag{2.12}$$

where an extra factor $\frac{1}{2}$ accounts for averaging over the photon spins. We can thus write (9) as

$$d\sigma/ds_c = Z_b^2 \alpha (\pi q s_c^{\frac{1}{2}})^{-1} \sigma_\gamma [\ln(2/q_l R) - 1.0772]. \tag{2.13}$$

This formula is normally rewritten in terms of the photon energy $\omega$ in the rest frame of the incident particle, $\omega = d/2m_a$, $d\omega = ds_c/2m_a$. Using also $qs_c^{\frac{1}{2}} = \omega m_a$, we obtain from (13)

$$d\sigma/d\omega = \sigma_\gamma(\omega) n(\omega), \qquad n = 2\alpha Z_b^2 (\pi\omega)^{-1} [\ln(2/q_l R) - 1.0772], \tag{2.14}$$

$n$ being the "number of equivalent photons." This number diverges for $k \to \infty$, unless the electron screening of the nuclear charge is included. In that case (4) must be multiplied by $1 - F_e$, where $F_e$ is the Fourier transform of the electron charge distribution. We take $F_e$ as a simple pole

$$F_e = a^{-2}(a^{-2} - t)^{-1}, \qquad a \approx 111 Z_b^{-1/3} m_e^{-1}, \tag{2.15}$$

$$-t^{-1}(1 - F) = (a^{-2} - t)^{-1} = q_t^2 + q_l^2 + a^{-2} \equiv q_t^2 + \tilde{q}_l^2. \tag{2.16}$$

Since this is the only place where $q_l$ occurs, we can simply replace $q_l \to \tilde{q}_l$ in our previous formulas. Better form factors have been discussed by Tsai (1974).

## 8-3 Bremsstrahlung and Pair Creation. Radiation Length. Furry-Sommerfeld-Maue Wave Functions

The matrix element for bremsstrahlung has been written down in section 3-5 in the Born approximation. There remains a long way to the experimentally interesting quantities such as the energy loss of relativistic electrons. The important results are presented in various reviews (Koch and Motz, 1959, Motz, Olsen and Koch 1964, 1969, Tsai, 1974). We shall only derive the easiest part, namely bremsstrahlung at very large energies where the typical longitudinal momentum transfer $q_l$ is much smaller than $a^{-1}$, $a$ being the screening radius (2.15). In this limit we can use the "equivalent photon number" (2.14) in the rest frame of the incident electron, with $q_l$ replaced by $\tilde{q}_l \approx a^{-1}$ according to (2.16):

$$d\sigma/d\omega' = \int n(\omega)\, d\omega\, d\sigma_c(\omega, \omega')/d\omega', \qquad n \approx 2\alpha Z_b^2(\pi\omega)^{-1}\left(\ln \frac{2a}{R} - 1.0772\right) \tag{3.1}$$

where $d\sigma_c/d\omega'$ is the cross section (3-7.22) for Compton scattering. Unfortunately one cannot use $R$ = nuclear radius when the incident or outgoing particles are electrons: The eikonal approximation breaks down for impact parameters $b < m^{-1}$, as can be seen from the uncertainty principle: the reaction at impact parameter $b$ takes a time $b/c$, during which the impact parameter changes by an amount $bv^{\perp}/c$. This amount must be small compared to $b$, i.e., the transverse motion may only be mildly relativistic. By the uncertainty principal $\Delta_p^{\perp}\, \Delta b \sim 1$, we find $\Delta b > m^{-1}$. We therefore substitute $R^{-1} \to m$ in (1). For a more detailed discussion, see Tsai (1974).

In practice, one is interested in $d\sigma/d\omega'_L$, where $\omega'_L$ is the energy of the bremsstrahlung photon in the nuclear rest frame, which is the true lab frame. In (1), on the other hand, $\omega'$ is the same quantity in the initial electron rest frame. Denoting by $E_L$ the initial lab energy of the electron, we need a boost with $\cosh \eta = E_L/m$ (1-6.8) in the direction opposite to the virtual photon momentum. Before the boost, the final photon has a momentum component $-\omega' \cos \vartheta$ in this direction, such that the Lorentz transformation (1-2.8) gives

$$\omega'_L = \omega' E_L/m - \omega' \cos \vartheta p_L/m \approx \omega'(1 - \cos \vartheta)E_L/m, \tag{3.2}$$

for electrons with $p_L \approx E_L$. The factor $1 - \cos \vartheta$ is taken from (3-7.19):

$$\omega'_L = (1 - \omega'/\omega)E_L, \qquad \omega'/\omega = 1 - \omega'_L/E_L = E'_L/E_L, \qquad d\omega'/\omega = -d\omega'_L/E_L, \tag{3.3}$$

$$d\sigma = d\omega'_L E_L^{-1} mC \int_{\omega_{\min}}^{\infty} d\omega\omega^{-2}$$

$$\times \left[ E'_L/E_L + E_L/E'_L - 1 + \left(1 + \frac{m}{\omega} - \frac{m}{\omega} E_L/E'_L\right)^2 \right] \tag{3.4}$$

$$= d\omega'_L \omega_L'^{-1} C\, 2\,[1 - \tfrac{2}{3}E'_L/E_L + (E'_L/E_L)^2],$$

$$C = 2\alpha Z_b^2[\ln{(2am)} - 1.0772]r_e^2. \tag{3.5}$$

The lower integration limit in (3) is obtained by putting $\cos\vartheta = -1$ in (2), $\omega'_{\min} = \omega'_L m/2E_L$ and $\omega_{\min} = \omega'_{\min} E_L/E'_L$. The total bremsstrahlung cross section has the normal soft photon divergence $\int d\omega'_L/\omega'_L$. More important is the energy loss

$$-dE_L/dR = \int_0^{E_L - m} \omega'_L \rho\, d\sigma$$

$$\approx 2C\rho \int_0^{E_L} d\omega'_L[\tfrac{4}{3} - \tfrac{4}{3}\omega'_L/E_L + (\omega'_L/E_L)^2] = 2C\rho E_L. \tag{3.6}$$

Also in refined calculations, the energy loss of extremely relativistic electrons remains proportional to the energy (which is due to the fact that the electron mass is neglected). One therefore defines

$$\text{(a)}\ -dE/dR \equiv \Lambda_R^{-1}E, \qquad \text{or} \quad \text{(b)}\ -dE/dx = \Lambda_R^{-1}E, \tag{3.7}$$

where $\Lambda_R$ is the "radiation length." It differs from the definition of the stopping cross section (1.2) only by the target density $\rho$ in version (a) and $\rho/\rho_m = N_A/A$ in version (b) (see the discussion at the end of section 2-3).

For practical applications, (4) is insufficient. For atoms heavier than He, one may use in the complete screening limit

$$d\sigma = \omega_L'^{-1}\, d\omega'_L 4Z^2\alpha r_e^2\{(\tfrac{4}{3} - \tfrac{4}{3}y + y^2)$$

$$\times [Z^2(\ln{(184.15z^{-1/3})} - f) + Z\ln{(1194z^{-2/3})}] \tag{3.8}$$

$$+ \tfrac{1}{9}(1 - y)(Z^2 + Z)\}, \qquad y \equiv \omega'_L/E_L, \qquad Z \equiv Z_b.$$

Except for $f$ which will be discussed below, the $Z^2$-terms give the exact Bethe-Heitler (1934) expression for one-photon exchange with the screened Coulomb field of the nucleus, with $\ln 184.15 \approx \ln 111 + \frac{1}{2}$. The $Z$-terms give the one-photon exchange with the electrons' Coulomb field (screened by the nucleus). The last term $\frac{1}{9}(1 - y)(Z^2 + Z)$ is always small ($\leq 2.5\%$) and is omitted in the definition of the "standard radiation length" $X_0$:

$$X_0^{-1} = 4\alpha r_e^2 N_A A^{-1}[Z^2(\ln{(184Z^{-\frac{1}{3}})} - f) + Z\ln{(1194Z^{-\frac{2}{3}})}]. \tag{3.9}$$

The cross section for pair production $\gamma A \to e^+e^-A$ in the complete screening limit ($k_L \to \infty$) is

$$\sigma(\gamma A \to e^+e^-A) = \tfrac{7}{9}AN_A^{-1}X_0^{-1} + \tfrac{2}{27}\alpha r_e^2(Z^2 + Z). \tag{3.10}$$

The second term here is always less than 1% of the first one, which shows that $X_0$ is really the appropriate unit of length (or target thickness) for the description of electromagnetic cascades at high energies. Some typical values of $X_0$ as given by Tsai (1974) have been included in table 2-3.

The function $f$ in (9) is a Coulomb correction to the one-photon exchange (Bethe and Maximon 1954). With $Z_b^2\alpha^2 = z$,

$$f = z\sum_{n=1}^{\infty} n^{-1}(n^2 + z)^{-1} \approx 1.202z - 1.037z^2 + z^3(1 + z)^{-1}, \qquad (3.11)$$

It arises from the use of Coulomb distorted waves instead of plane waves in the electron field operator. Instead of the exact operator (3-3.18), one may use approximate wave functions in parabolic coordinates which are due to Furry, Sommerfeld and Maue. In the nonrelativistic limit, these functions are exact solutions of the Schrödinger equation. For spinless particles, the Klein-Gordon equation (1-1.22) is identical to the Schrödinger equation if one neglects $V^2$ and replaces $E \to m$ in $U(r)$. Therefore, the Klein-Gordon solution to the point Coulomb potential can be written in a form analogous to the nonrelativistic case. With the ansatz $\psi = F \exp(i\mathbf{kr})$, the equation becomes

$$(\nabla^2 + 2i\mathbf{k}\,\nabla - U)F = 0, \qquad U = 2EV. \qquad (3.12)$$

The normalized solution is

$$\psi = e^{Z\alpha\pi/2}\Gamma(1 - iZ\alpha)e^{i\mathbf{kr}}\,{}_1F_1(iZ\alpha E/k,\, 1,\, i(kr - \mathbf{kr})). \qquad (3.13)$$

Its partial-wave projection differs from the exact solution (1-8.4) only by terms of order $Z^2\alpha^2L^{-1}$. For electrons, one may start from (1-5.4) inserting $\boldsymbol{\pi} = -i\nabla$ and $\pi^0 = i\,\partial_t + V = -E + V$:

$$[(E - V)^2 + \nabla^2 - i\boldsymbol{\sigma}\text{ grad } V]\psi_r = m^2\psi_r, \qquad (3.14)$$

which differs from the corresponding Klein-Gordon equation by the term $-i\boldsymbol{\sigma}$ grad $V$. We now try

$$\psi_r = \exp(i\mathbf{kr})[Fu_r(\mathbf{k}) + \varphi] \qquad (3.15)$$

where $F$ is the same function as in (12) and $u_r$ is the right-handed free-particle spin function. Then

$$\begin{aligned}&(\nabla^2 + k^2 - U - i(2E)^{-1}\boldsymbol{\sigma}\text{ grad } U)\psi_r = 0\\ &\to (\nabla^2 + 2i\mathbf{k}\,\nabla - U)\varphi \approx i(2E)^{-1}F\,\nabla U\boldsymbol{\sigma}u_r(\mathbf{k}),\end{aligned} \qquad (3.16)$$

which is solved by $\varphi = i(2E)^{-1}\boldsymbol{\sigma}\,\nabla F u_r(\mathbf{k})$. Thus the complete Furry-Sommerfeld-Maue wave function is

$$\psi_r = e^{Z\alpha\pi/2}\Gamma(1 - iZ\alpha)e^{i\mathbf{kr}}(1 + i\boldsymbol{\sigma}\,\nabla/2E)\,{}_1F_1\,u_r(\mathbf{k}), \qquad (3.17)$$

and correspondingly for the left-handed spinor $\psi_l$. The accuracy of this solution is the same as in the spinless case. For the brems-strahlung of very hard photons, it becomes insufficient.

APPENDIX A

# Some Formulas for Partial Waves and Fermions

## A-1 Spherical Bessel, Neuman and Hankel Functions

These functions satisfy the differential equation (1-3.10). The solution which is regular at the origin is $j_L(\rho) = (\pi/2\rho)^{\frac{1}{2}} J_{L+\frac{1}{2}}(\rho)$. Its behavior for $\rho \to 0$ and $\rho \to \infty$ is

$$\begin{aligned} j_L(\rho \to 0) &= [(2L+1)!!]^{-1} \rho^L [1 - \tfrac{1}{2}\rho^2 (2L+3)^{-1}], \\ j_L(\rho \to \infty) &= \rho^{-1} \sin(\rho - \tfrac{1}{2}L\pi). \end{aligned} \tag{1.1}$$

The spherical Neuman function, $n_L = (-1)^{L+1}(\pi/2\rho)^{\frac{1}{2}} J_{-L-\frac{1}{2}}$ is $\frac{3}{2}\pi$ out of phase at $\rho \to \infty$:

$$\begin{aligned} n_L(\rho \to \infty) &= -\rho^{-1} \cos(\rho - \tfrac{1}{2}L\pi), \\ n_L(\rho \to 0) &= -(2L-1)!!\,\rho^{-L-1}[1 + \tfrac{1}{2}\rho^2(2L-1)^{-1}]. \end{aligned} \tag{1.2}$$

The spherical Hankel function is

$$h_L = j_L + i n_L, \qquad h_L(\rho \to \infty) = -i\rho^{-1} \exp(i\rho - \tfrac{1}{2}i\pi L). \tag{1.3}$$

The following recurrence relation is valid for all 3 types of functions:

$$j_{L+1} = \rho^{-1}(2L+1) j_L - j_{L-1}. \tag{1.4}$$

The functions for $L = 0$, 1 and 2 are collected in table A1, together with the penetration factors

$$v_L = \rho^{-2} |h_L|^{-2}. \tag{1.5}$$

**Table A1** The spherical Bessel, Neumann, and Hankel functions and the penetration factors for $L = 0, 1, 2$.

| | $L = 0$ | $L = 1$ | $L = 2$ |
|---|---|---|---|
| $j_L$ | $\rho^{-1} \sin \rho$ | $\rho^{-2} \sin \rho - \rho^{-1} \cos \rho$ | $(3\rho^{-3} - \rho^{-1}) \sin \rho - 3\rho^{-2} \cos \rho$ |
| $n_L$ | $-\rho^{-1} \cos \rho$ | $-\rho^{-2} \cos \rho - \rho^{-1} \sin \rho$ | $-(3\rho^{-3} - \rho^{-1}) \cos \rho - 3\rho^{-2} \sin \rho$ |
| $h_L$ | $-i\rho^{-1} e^{i\rho}$ | $-(\rho^{-1} + i\rho^{-2}) e^{i\rho}$ | $-(3\rho^{-2} + i3\rho^{-3} - i\rho^{-1}) e^{i\rho}$ |
| $v_L$ | 1 | $\rho^2(1 + \rho^2)^{-1}$ | $\rho^4(9 + 3\rho^2 + \rho^4)^{-1}$ |
| $h_L(i\rho)$ | $-\rho^{-1} e^{-\rho}$ | $i(\rho^{-1} + \rho^{-2}) e^{-\rho}$ | $(3\rho^{-2} + 3\rho^{-3} + \rho^{-1}) e^{-\rho}$ |

The variable $\rho$ may be complex. We also recall

$$j_L(-\rho) = (-1)^L j_L(\rho), \qquad n_L(-\rho) = -(-1)^L n_L(\rho),$$
$$h_L(-\rho) = (-1)^L h_L^*(\rho^*), \tag{1.6}$$

$$\rho j_L\, \partial_\rho(\rho n_L) - \rho n_L\, \partial_\rho(\rho j_L) = 1 \qquad \text{(Wronski determinant)}. \tag{1.7}$$

## A-2 Coulomb Wave Functions

These functions satisfy the differential equation

$$(\rho^{-2}\, \partial_\rho \rho^2\, \partial_\rho + 1 - 2\eta/\rho - \rho^{-2} L(L + 1)) R_L = 0, \qquad \eta = Z_a Z_b \alpha/v. \tag{2.1}$$

The solution which is regular at the origin is

$$\rho^{-1} F_L = C_L \rho^L e^{-i\rho} F(L + 1 - i\eta,\, 2L + 2,\, 2i\rho), \tag{2.2}$$

$$F(a, b, x) = 1 + \frac{a}{b} x + \frac{a(a + 1)}{b(b + 1)} \frac{1}{2!} x^2 + \cdots = e^x F(b - a, b, -x), \tag{2.3}$$

$$C_L = 2^L \exp(-\tfrac{1}{2}\eta\pi) |\Gamma(L + 1 + i\eta)| / (2L + 1)!, \tag{2.4}$$

$$C_0^2 = 2\pi\eta (e^{2\pi\eta} - 1)^{-1}, \qquad C_L = (1 + \eta^2/L^2)^{\frac{1}{2}} (2L + 1)^{-1} C_{L-1}. \tag{2.5}$$

Its asymptotic form is

$$\rho^{-1} F_L(\rho \to \infty) = \rho^{-1} \sin(\rho - \eta \ln 2\rho - \tfrac{1}{2} L\pi + \sigma_L),$$
$$\sigma_L \equiv \arg \Gamma(L + 1 + i\eta), \tag{2.6}$$

$$\arg \Gamma(a + i\eta) = \eta\psi(a) + \sum_{n=0}^{\infty} \left( \frac{\eta}{a + n} - \arctan \frac{\eta}{a + n} \right), \qquad \psi \equiv \Gamma'/\Gamma. \tag{2.7}$$

Insertion of $a = L + 1$ gives

$$\sigma_L = \sigma_0 + \sum_{n=1}^{L} \arctan \frac{\eta}{n},$$
$$\sigma_0 = -C\eta + \sum_{n=1}^{\infty} (\eta/n - \arctan(\eta/n)), \qquad C = 0.5772. \tag{2.8}$$

The irregular solution is defined to be $3\pi/2$ out of phase asymptotically,

$$\rho^{-1}G_L(\rho \to \infty) = -\rho^{-1}\cos(\rho - \eta \ln 2\rho - \tfrac{1}{2}L\pi + \sigma_L). \tag{2.9}$$

These definitions are analogous to those of $j_L$ and $n_L$.

For $\eta = 0$ we have $\rho^{-1}F_L = j_L$, $\rho^{-1}G_L = n_L$. For small $\rho$, we write

$$\rho^{-1}G_L = -C_L\rho^L e^{-i\rho}G(L + 1 + i\eta, 2L + 2, 2i\rho), \tag{2.10}$$

$$\begin{aligned} G(a, b, z) = (2\pi i)^{-1}(e^{2i\pi a} - 1) & \\ \times \Bigg\{ (2 \ln z + \pi \cot \pi a - i\pi)F(a, b, z) & \\ - 2e^{i\pi a}\frac{\Gamma(b)}{\Gamma(a)}\sum_{r=1}^{b-1}\frac{\Gamma(r)}{\Gamma(b-r)\Gamma(r-a+1)}z^{-r} & \qquad (2.11) \\ -2\sum_{m=0}^{\infty}[\psi(m+1) + \psi(b+m) - \psi(a+m)] \times \frac{\Gamma(a+m)\Gamma(b)}{m!\,\Gamma(a)\Gamma(b+m)}z^m \Bigg\}. & \end{aligned}$$

See Abramowitz and Stegun (1965), chapter 6.3 for $\psi$ and chapter 14 for Coulomb wave functions. For $L = 0$ we have

$$G_0 = -C_0^{-1}\{1 + 2\eta\rho[\ln 2\eta\rho + 2C + h(\eta)]\}, \tag{2.12}$$

$$h(\eta) = \mathrm{Re}\,\psi(1 + i\eta) - \tfrac{1}{2}\ln \eta^2 = \eta^2 \sum_{S=1}^{\infty} S^{-1}(S^2 + \eta^2)^{-1} - C - \tfrac{1}{2}\ln \eta^2, \tag{2.13}$$

and for $L > 0$

$$G_L = -[(2L + 1)C_L\rho^L]^{-1}[1 + \mathcal{O}(\eta\rho/L)]. \tag{2.14}$$

## A-3 Reduction of Dirac Matrices

Lorentz invariance requires the number of fermions to be conserved modulo 2. It is sufficient to consider the matrix element $T$ for a reaction containing one spin-$\frac{1}{2}$ particle of momentum $P_\mu$ and magnetic quantum number $M$ in the initial state and another one of quantum numbers $P'_\mu$, $M'$ in the final state. If these particles are represented by Dirac spinors, $T$ has the form

$$T = \bar{u}(P'_\mu, M')Qu(P_\mu, M), \tag{3.1}$$

where $Q$ is a $4 \times 4$ matrix. It can be written as a superposition of 16 linearly independent matrices: $\gamma^\mu$, 1, $\gamma_5$, $\gamma^\mu\gamma_5$, $\sigma^{\mu\nu}$. Products of the form $\gamma^\mu\gamma^\nu\gamma_5$ can be rewritten (by using $\gamma_5 = i\gamma^0\gamma^1\gamma^2\gamma^3$) in terms of the antisymmetric tensor $\sigma^{\mu\nu}$

(1-5.13) and $g^{\mu\nu} \cdot 1$. We now express $T$ in terms of the Pauli spinors $\chi_0$ for particles at rest. We do this first in the "low-energy" representation (1-6.12)

$$u = \begin{pmatrix} u_g \\ u_k \end{pmatrix} = \begin{pmatrix} (E+m)^{\frac{1}{2}} & \chi_0 \\ (E+m)^{-\frac{1}{2}} & \boldsymbol{\sigma}\mathbf{p}\chi_0 \end{pmatrix}, \qquad Q = \begin{pmatrix} Q_{gg} & Q_{gk} \\ Q_{kg} & Q_{kk} \end{pmatrix} \tag{3.2}$$

$$\bar{u} = (\chi_g^+, -\chi_k^+) = ((E+m)^{\frac{1}{2}}\chi_0^+, -(E+m)^{-\frac{1}{2}}\chi_0^+\boldsymbol{\sigma}\mathbf{p}) \tag{3.3}$$

Insertion of (2) into (1) gives

$$T = \chi_0'^+(M')R\chi_0(M), \tag{3.4}$$

$$R = (E+m)^{\frac{1}{2}}(E'+m')^{\frac{1}{2}} \times \left[Q_{gg} - \frac{\boldsymbol{\sigma}\mathbf{p}'}{E'+m'}Q_{kk}\frac{\boldsymbol{\sigma}\mathbf{p}}{E+m} + Q_{gk}\frac{\boldsymbol{\sigma}\mathbf{p}}{E+m} - \frac{\boldsymbol{\sigma}\mathbf{p}'}{E'+m'}Q_{kg}\right]. \tag{3.5}$$

The Dirac matrices are in this representation given by (1-5.11), (1-5.13) and (1-6.18)

$$\gamma^0 = \begin{pmatrix} \sigma_0 & 0 \\ 0 & -\sigma_0 \end{pmatrix}, \quad \gamma^i = \begin{pmatrix} 0 & \sigma^i \\ -\sigma^i & 0 \end{pmatrix}, \quad \gamma_5 = \begin{pmatrix} 0 & \sigma_0 \\ \sigma_0 & 0 \end{pmatrix},$$
$$\sigma^{0i} = \begin{pmatrix} 0 & \sigma^i \\ \sigma^i & 0 \end{pmatrix}, \quad \sigma^{ij} = -i\varepsilon_{ijk}\begin{pmatrix} \sigma^k & 0 \\ 0 & \sigma^k \end{pmatrix}. \tag{3.6}$$

$R$ simplifies in the rest frame of one of the particles, say the initial particle (lab system):

$$R_L = k_+\left[Q_{gg} - \frac{\boldsymbol{\sigma}\mathbf{p}'}{E'+m'}Q_{kg}\right] = k_+Q_{gg} - k_-\boldsymbol{\sigma}\hat{\mathbf{p}}'Q_{kg}, \qquad \hat{\mathbf{p}}' \equiv \mathbf{p}'/p', \tag{3.7}$$

$$k_\pm \equiv [2m(E' \pm m')]^{\frac{1}{2}} = [(m \pm m')^2 - t]^{\frac{1}{2}}. \tag{3.8}$$

The last expression is obtained from (2-4.5). The following special cases of $Q$ are needed:

$$Q = 1 \quad \text{or} \quad \gamma_0\colon \quad R_L = k_+ \tag{3.9}$$

$$Q = \gamma_5 \quad \text{or} \quad \gamma_5\gamma_0\colon \quad R_L = -k_-\boldsymbol{\sigma}\hat{\mathbf{p}}' \tag{3.10}$$

$$Q = \boldsymbol{\gamma}\mathbf{a} \quad \text{or} \quad \boldsymbol{\gamma}\gamma_0\,\mathbf{a}\colon \quad R_L = k_-\boldsymbol{\sigma}\hat{\mathbf{p}}'(\boldsymbol{\sigma}\mathbf{a}) = k_-[\hat{\mathbf{p}}'\mathbf{a} + i\boldsymbol{\sigma}(\hat{\mathbf{p}}' \times \mathbf{a})], \tag{3.11}$$

$$Q = \gamma_5\boldsymbol{\gamma}\mathbf{a} \quad \text{or} \quad \gamma_5\boldsymbol{\gamma}\mathbf{a}\gamma_0\colon \quad R_L = k_+\boldsymbol{\sigma}\mathbf{a}, \tag{3.12}$$

$$Q = \gamma^i\gamma^j\colon \quad R_L = -k_+\sigma^i\sigma^j = -k_+(\delta_{ij} + i\varepsilon_{ijk}\sigma^k). \tag{3.13}$$

If the spinor $\bar{u}$ in (1) represents an antiparticle, it is sufficient to replace $M'$ by $-\bar{M}$ and $P'_\mu$ by $-\bar{P}_\mu$ in the relevant formulas. Instead of (4), (5) and (7), we thus have

$$T = \chi_0'^+(-\bar{M})\bar{R}\chi_0(M),$$
$$\bar{R} = (E+m)^{\frac{1}{2}}(-\bar{E}+m')^{\frac{1}{2}} \times \left[\bar{Q}_{gg} - \frac{\boldsymbol{\sigma}\bar{\mathbf{p}}}{\bar{E}-m'}\bar{Q}_{kk}\frac{\boldsymbol{\sigma}\mathbf{p}}{E+m} + \bar{Q}_{gk}\frac{\boldsymbol{\sigma}\mathbf{p}}{E+m} - \frac{\boldsymbol{\sigma}\bar{\mathbf{p}}}{\bar{E}-m'}\bar{Q}_{kg}\right], \tag{3.14}$$

$$\bar{R}_L = ik_-\left[\bar{Q}_{gg} - \frac{\sigma\bar{\mathbf{p}}}{\bar{E} - m'}\bar{Q}_{kg}\right] = ik_-\bar{Q}_{gg} - ik_+\sigma\hat{\mathbf{p}}\bar{Q}_{kg}. \tag{3.15}$$

Thus $\bar{R}$ is obtained from $R$ by substituting $Q \to \bar{Q}$ and

$$k_+ \to ik_-, \qquad \mathbf{p}' \to -\bar{\mathbf{p}}, \qquad k_- \to -ik_+. \tag{3.16}$$

There is thus a relative change of sign between $k_+$ and $k_-$. If one uses $\bar{v} = -i\bar{u}$ (1-6.26), the factor $i$ is omitted in (14)-(16). In the cms, (5) is particularly simple for helicity states:

$$R = (E + m)^{\frac{1}{2}}(E' + m')^{\frac{1}{2}} \times \left[Q_{gg} - \frac{4\lambda\lambda' pp'}{(E + m)(E' + m')}Q_{kk} + \frac{2\lambda p}{E + m}Q_{gk} - \frac{2\lambda' p'}{E' + m'}Q_{kg}\right]. \tag{3.17}$$

For elastic scattering, we have

$$R = (E + m)Q_{gg} - 4\lambda\lambda'(E - m)Q_{kk} + 2\lambda p Q_{gk} - 2\lambda' p' Q_{kg}. \tag{3.18}$$

In the case of the unit matrix, $Q_{gg} = Q_{kk} = 1$, $Q_{gk} = Q_{kg} = 0$ we obtain

$$\begin{aligned} R(Q = 1) &= E(1 - 4\lambda\lambda') + m(1 + 4\lambda\lambda') = 2m\,\delta_{\lambda\lambda'} + 2E\,\delta_{\lambda,-\lambda'}, \\ \bar{u}'(\lambda')u(\lambda) &= 2\,d^{\frac{1}{2}}_{\lambda\lambda'}(\theta)(m\,\delta_{\lambda\lambda'} + E\,\delta_{\lambda,-\lambda'}), \end{aligned} \tag{3.19}$$

using (1-4.21). The matrix elements of the electromagnetic and weak currents are more directly derived from the high-energy representation (1-5.17), with $\psi_r$ and $\psi_l$ given by (1-6.11):

$$\bar{u}'\gamma^\mu u = J_r^\mu + J_l^\mu, \tag{3.20}$$

$$J_r^\mu = \tfrac{1}{2}\bar{u}'\gamma^\mu(1 + \gamma_5)u = \chi_0'^+(E' + \mathbf{p}'\sigma)^{\frac{1}{2}}\sigma^\mu(E + \mathbf{p}\sigma)^{\frac{1}{2}}\chi_0 \tag{3.21}$$

$$= (E' + 2\lambda' p')^{\frac{1}{2}}(E + 2\lambda p)^{\frac{1}{2}}\chi_0'^+\sigma^\mu\chi_0, \tag{3.22}$$

$$J_l^\mu = \tfrac{1}{2}\bar{u}'\gamma^\mu(1 - \gamma_5)u = \chi_0'^+(E' - \mathbf{p}'\sigma)^{\frac{1}{2}}\sigma_l^\mu(E - \mathbf{p}\sigma)^{\frac{1}{2}}\chi_0 \tag{3.23}$$

$$= (E' - 2\lambda' p')^{\frac{1}{2}}(E - 2\lambda p)^{\frac{1}{2}}\chi_0'^+\sigma_l^\mu\chi_0. \tag{3.24}$$

Expressions (22) and (24) are of course only valid for helicity states. For elastic scattering in the cms,

$$\bar{u}'\gamma^0 u = 2\chi_0'^+\chi_0(E\,\delta_{\lambda\lambda'} + m\,\delta_{\lambda,-\lambda'}), \tag{3.25}$$

$$\bar{u}'\boldsymbol{\gamma} u = 4\lambda p\chi_0'^+\sigma\chi_0\,\delta_{\lambda\lambda'}, \tag{3.26}$$

or using eqs. (1-4.21) for $\chi_0'^+\chi_0$ and $\chi_0'^+\sigma\chi_0$ and the coordinate system of fig. 2-4 ($\varphi = 0$)

$$\bar{u}'\gamma^0 u = 2\,d^{\frac{1}{2}}_{\lambda\lambda'}(E\,\delta_{\lambda\lambda'} + m\,\delta_{\lambda,-\lambda'}), \qquad \bar{u}'\gamma_z u = 2p\,d^{\frac{1}{2}}_{\lambda\lambda'}\delta_{\lambda\lambda'}, \tag{3.27}$$

$$\bar{u}'\gamma_x u = 2p\sin\frac{\theta}{2}\,\delta_{\lambda\lambda'}, \qquad i\bar{u}'\gamma_y u = -4\lambda p\sin\frac{\theta}{2}\,\delta_{\lambda\lambda'}. \tag{3.28}$$

As explained below (3-6.21), a particle-antiparticle helicity state contains the helicities $\lambda$ and $\bar{\lambda} = -\bar{M}$ in spin states $\chi_0(\lambda)$ and $\chi_0(\bar{\lambda})$ which both refer to the same direction. We thus get from (14), omitting the factor $i$:

$$T = \chi_0^+(\bar{\lambda})\bar{R}\chi_0(\lambda),$$

$$\bar{R} = (E+m)^{\frac{1}{2}}(\bar{E}-m')^{\frac{1}{2}}\left[Q_{gg} + \frac{4\lambda\bar{\lambda}p^2 Q_{kk}}{(E+m)(\bar{E}-m')} + \frac{2\lambda p Q_{gk}}{E+m} + \frac{2\bar{\lambda}pQ_{kg}}{\bar{E}-m'}\right]. \quad (3.29)$$

## A-4 Fermion Spin Summation

Once the relativistic spinors are reduced to Pauli spinors, the spin algebra proceeds as in the nonrelativistic case. The matrix element $R$ of (3.5) is decomposed as

$$R = R_0 + \mathbf{R}\boldsymbol{\sigma}, \quad (4.1)$$

and the transition probability is proportional to (2-3.6):

$$\overline{|T|^2} = \sum_{MM'} \rho_{MM'} \sum_{M_f} T(M, M_f)T^*(M', M_f) = \sum_{M_f} \chi_0'^+(M_f)Y\chi_0'(M_f) \quad (4.2)$$

$$Y = (R_0 + \mathbf{R}\boldsymbol{\sigma})\tfrac{1}{2}(1 + \mathbf{P}\cdot\boldsymbol{\sigma})(R_0^* + \mathbf{R}^*\cdot\boldsymbol{\sigma})$$

Here we have used $\chi_m(M) = \delta_{mM}$ from (1-4.1) and $\rho$ from (2-3.8). Summation over $M_f$ leads to

$$\overline{|T|^2} = \sum_{M_f i, j} \chi_{0i}'^+(M_f)Y_{ij}\chi_{0j}'(M_f) = \text{trace } Y, \quad (4.3)$$

using the completeness of the spin functions $\chi_0'$. After a cyclic permutation of the factors in trace $Y$, we find

$$\begin{aligned}\overline{|T|^2} &= \tfrac{1}{2}\,\text{trace }(R_0^* + \mathbf{R}^*\boldsymbol{\sigma})(R_0 + \mathbf{R}\boldsymbol{\sigma})(1 + P\cdot\boldsymbol{\sigma}) \\ &= R_0R_0^* + \mathbf{R}\mathbf{R}^* + \mathbf{P}(2\,\text{Re } R_0^*\mathbf{R} + i\mathbf{R}^* \times \mathbf{R}).\end{aligned} \quad (4.4)$$

Matrix elements involving virtual fermions always contain a summation over the spin of that fermion. If only "undotted" spinors $\chi_r$ occur, we have

$$\sum_M \chi_r(\mathbf{p}, M)\chi_r^+(\mathbf{p}, M) = \sum_M A(\mathbf{p})\chi_0(M)\chi_0^+(M)A(\mathbf{p}) = A^2(\mathbf{p}) = m^{-1}(P_0 + \mathbf{p}\boldsymbol{\sigma}), \quad (4.5)$$

using (1-6.9) for $A$. If only "dotted" spinors $\chi_l$ occur, we get $A^{-2} = m^{-1}(P_0 - \mathbf{p}\boldsymbol{\sigma})$ instead. For parity-conserving interactions, Dirac spinors are more convenient:

$$\sum_M u(M)\bar{u}(M) = P\cdot\gamma + m, \qquad \sum_M v(M)\bar{v}(M) = P\cdot\gamma - m, \quad (4.6)$$

The first relation follows for example from (1-6.11)

$$u = m^{\frac{1}{2}}\begin{pmatrix} A \\ A^{-1}\end{pmatrix}\chi_0, \qquad \sum_M u\bar{u} = m\begin{pmatrix} A \\ A^{-1}\end{pmatrix}(A^{-1}, A) = m\begin{pmatrix} 1 & A^2 \\ A^{-2} & 1\end{pmatrix}, \quad (4.7)$$

and the relation for $v$ follows from (1-6.26). The $\gamma^0$ which is hidden in $\bar{u}$ introduces a minor complication: with $T$ given by (3.1), we get for unpolarized particles:

$$\frac{1}{2}\sum_M |T|^2 = \frac{1}{2}\sum_M \bar{u}'Qu(M)u^+(M)Q^+\bar{u}'^+ = \frac{1}{2}\sum_M \bar{u}'Qu(M)\bar{u}(M)\tilde{Q}u'$$
$$= \bar{u}'Q(P\cdot\gamma + m)\tilde{Q}u', \qquad \tilde{Q} \equiv \gamma_0 Q^+ \gamma_0 . \tag{4.8}$$

Using the following property of the $\gamma$-matrices

$$\gamma_0 \gamma_\mu^+ \gamma_0 = \gamma_\mu, \qquad \gamma_0 \gamma_5^+ \gamma_0 = -\gamma_5, \tag{4.9}$$

we see that $\tilde{Q}$ is obtained from $Q$ simply by inverting the order of $\gamma$-matrices, complex conjugating their coefficients and giving an extra minus sign for each $\gamma_5$.

Summation over the magnetic quantum number $M_f$ of $u'$ leads to

$$\frac{1}{2}\sum_{M,M_f} |T|^2 = \frac{1}{2}\sum_{ij} [Q(P\cdot\gamma + m)\tilde{Q}]_{ij}(P'\cdot\gamma + m')_{ji}$$
$$= \tfrac{1}{2}\,\text{trace}\; Q(P\cdot\gamma + m)\tilde{Q}(P'\cdot\gamma + m'). \tag{4.10}$$

Of the 16 linearly independent $\gamma$-matrices enumerated in (A-3), only the scalar matrix 1 has a nonvanishing trace, trace $1 = 4$. The tracelessness of $\gamma_\mu$ and $\gamma_5$ is evident from (3.6). Also, as $\gamma_5$ anticommutes with $\gamma_\mu$ and trace $(ABC) =$ trace $(CAB)$, the trace of any odd number of $\gamma$-matrices vanishes, for example

$$\text{trace}\;\gamma_\mu = \text{trace}\;\gamma_\mu\gamma_5^2 = \text{trace}\;\gamma_5\gamma_\mu\gamma_5 = -\text{trace}\;\gamma_\mu\gamma_5^2 . \tag{4.11}$$

The trace of $\sigma_{\mu\nu}$ vanishes because of its antisymmetry. According to (1-5.12), we can decompose

$$\gamma_\mu\gamma_\nu = 1\cdot g_{\mu\nu} + \sigma_{\mu\nu}, \qquad \text{trace}\;\gamma_\mu\gamma_\nu = 4g_{\mu\nu}. \tag{4.12}$$

It also follows that

$$P_1\cdot\gamma\, P_2\cdot\gamma = 2P_1\cdot P_2 - P_2\cdot\gamma\, P_1\cdot\gamma. \tag{4.13}$$

Similarly, by "pulling $P_4 \cdot \gamma$ through to the left," one finds

$$\tfrac{1}{4}\,\text{trace}\; P_1\cdot\gamma\, P_2\cdot\gamma\, P_3\cdot\gamma\, P_4\cdot\gamma$$
$$= P_1\cdot P_2 P_3\cdot P_4 + P_1\cdot P_4 P_2\cdot P_3 - P_1\cdot P_3 P_2\cdot P_4 . \tag{4.14}$$

Finally, we compute the trace of $\gamma_5\gamma_\alpha\gamma_\beta\gamma_\gamma\gamma_\delta$. We first observe

$$\gamma_5\sigma_{\mu\nu} = i\varepsilon_{\mu\nu\alpha\beta}\sigma^{\alpha\beta}, \tag{4.15}$$

which implies

$$\text{trace}\;\gamma_5\gamma_\mu\gamma_\nu = g_{\mu\nu}\,\text{trace}\;\gamma_5 + \text{trace}\;\gamma_5\sigma_{\mu\nu} = 0. \tag{4.16}$$

Next, we observe that trace $\gamma_5 \gamma_\alpha \gamma_\beta \gamma_\gamma \gamma_\delta$ is antisymmetric in all four indices. This follows for example from splitting $\gamma_\alpha \gamma_\beta$ as in (12) and using $g_{\alpha\beta}$ trace $\gamma_5 \gamma_\gamma \gamma_\delta = 0$ according to (16). Since the only available antisymmetric tensor with 4 indices is $\varepsilon_{\alpha\beta\gamma\delta}$ of (4-7.3), it is sufficient to verify for $\alpha = 0$, $\beta = 1$, $\gamma = 2$, $\delta = 3$ that

$$\text{trace } \gamma_5 \gamma_\alpha \gamma_\beta \gamma_\gamma \gamma_\delta = 4i\varepsilon_{\alpha\beta\gamma\delta} . \tag{4.17}$$

APPENDIX B

# The Groups $SU_2$ and $SU_3$

## B-1 Unitarity Matrices and Their Representations

Let $M^+$, $M^*$ and $M_{\text{tr}}$ denote the Hermitean conjugate, complex conjugate and transpose of a matrix $M$, respectively. A square matrix $S$ is said to be unitary if

$$S^+ \equiv S^*_{\text{tr}} = S^{-1} \quad \text{or} \quad S^+S = SS^+ = 1. \tag{1.1}$$

The matrix equation $SS^+ = 1$ has the components

$$\sum_m S_{im}(S^+)_{mf} = \sum_m S_{im} S^*_{fm} = \delta_{if}. \tag{1.2}$$

Any unitary matrix may be written as $e^{iH}$, where $H$ is Hermitean:

$$S = e^{iH} \equiv 1 + iH - \tfrac{1}{2}H^2 + \cdots, \tag{1.3}$$

$$H^+ = H, \qquad S^+ = e^{-iH^+} = e^{-iH} = S^{-1}. \tag{1.4}$$

The advantage of representation (3) is that a general Hermitean matrix may be constructed as a superposition of fixed, linearly independent Hermitean matrices. The construction of unitary matrices is more complicated. Instead of (3), one may use the algebraic expression

$$S = (1 + iK)(1 - iK)^{-1}, \qquad 2K = \text{tg}\, H = H + \tfrac{1}{3}H^3 + \tfrac{2}{15}H^5 + \cdots \tag{1.5}$$

Obviously $K$ is also Hermitean. Unitary matrices are important in quantum mechanics, because the scalar product of two quantum mechanical state vectors

$$\langle b \,|\, a\rangle = \sum_i a_i b_i^* \tag{1.6}$$

is invariant under such transformations:

$$\langle bS \,|\, Sa\rangle = \sum_{ijk} S_{ij} a_j (S_{ik} b_k)^* = \sum_{ijk} (S^+)_{ki} S_{ij} a_j b_k^* = \sum_{jk} \delta_{kj} a_j b_k^* . \tag{1.7}$$

The product of two unitary matrices is also unitary, and a unitary matrix is diagonalized by another unitary matrix $U$,

$$(U^+ SU)_{mm'} = e^{i\phi m}\, \delta_{mm'}. \tag{1.8}$$

In connection with time-reversal invariant interactions, symmetric matrices $S$ are needed, $S = S_{\text{tr}}$. The corresponding matrices $H$ and $K$ are also symmetric by (3) and (5), and since they are Hermitean, they are in fact real, $H = H^*$, $K = K^*$. The matrix $U$ of (8) which diagonalizes $S$ is then also real. This follows from the fact that $U$ diagonalizes not only $S$, but also $H$ and $K$, and that real symmetric matrices are diagonalized by real matrices.

A unitary matrix $U$ has $|\det U| = 1$ and can therefore be decomposed into

$$U = e^{i\alpha} SU, \qquad \det SU = 1. \tag{1.9}$$

Matrices of unit determinant are called "special" or "unimodular" ($SU$ denotes one matrix, not a matrix product !). The phase $e^{i\alpha}$ commutes with all matrices and can therefore be omitted in the study of group properties. When $SU$ is written in the form $e^{iH}$, $H$ must have zero trace, which is obvious after diagonalisation of $SU$. Such an n × n matrix $H$ can be expressed in terms of $n^2 - 1$ fixed and linearly independent matrices $\lambda^k$ with real coefficients $\omega^k$

$$SU_n(\omega^k) = \exp\left\{-i \sum_{k=1}^{n^2-1} \omega^k \lambda^k\right\}, \qquad \lambda^k = \lambda^{k+}, \qquad \text{trace } \lambda^k = 0. \tag{1.10}$$

The application of group theory to physics consists of studying the irreducible representations of groups. A representation $D_k$ of the group $SU_n$ is again a group of unitary matrices of dimension $k$ (which may be different from $n$), which has the same number of parameters as $SU_n$. Moreover, the assignment of parameters to the $D$ matrices preserves the multiplication properties of the $SU_n$ matrices: if $SU_n(\alpha^k) = SU_n(\beta^k) \cdot SU_n(\gamma^k)$, then $D(\alpha^k) = D(\beta^k)D(\gamma^k)$. For this property, it is both necessary and sufficient that $D$ may be written in the form

$$D(\alpha_1 \cdots \alpha_{n^2-1}) = \exp\left[i \sum_{k=1}^{n^2-1} \omega^k F_k\right], \tag{1.11}$$

where the generators $F_k$ have the same commutation relations as the $\lambda^k$ of (10). Thus the "structure constants" $f_{ijk}$ in the commutation relations

$$[F_i, F_j] = 2if_{ijk} F_k \tag{1.12}$$

should be identical with the constants in the commutation relations of the corresponding $\lambda^k$. This is the basic property of Lie groups. For the group $SU_2$ we have $F_k = 2J^k$, where $J^k$ are the angular momentum operators.

Let $|\xi\rangle$ with components $\xi_\alpha$, $\alpha = 1 \cdots n$ denote a vector on which the matrices $SU_n$ act. The direct product of $r$ such vectors $\xi^a \cdots \xi^r$ transforms under $U_n$ like a tensor of rank $r$. Consider for example the transformation law of the second-rank tensor $T_{\alpha\beta}$,

$$T_{\alpha\beta} \equiv \xi_\alpha^a \xi_\beta^b, \qquad (V \cdot T)_{\alpha\beta} = U_{\alpha\alpha'} U_{\beta\beta'} T_{\alpha'\beta'}. \tag{1.13}$$

The matrix $V$ is the "direct product" $U \times U$. It is a $n^2 \times n^2$ unitary matrix, the first (second) index of which is given by the index pair $\alpha\beta$ ($\alpha'\beta'$). The group of $V$ matrices which are of the form $U_n \otimes U_n$ form a representation of $U_n$. For $n > 1$, this representation is reducible, which means that $T$ may be split into two parts such that $V$ never mixes components of different parts. The reduction is accomplished by splitting $T$ into its symmetric and antisymmetric parts

$$T_{\alpha\beta} = T_{\{\alpha\beta\}} + T_{[\alpha\beta]}, \quad T_{\{\alpha\beta\}} = \tfrac{1}{2}(\xi_\alpha^a \xi_\beta^b + \xi_\beta^a \xi_\alpha^b), \quad T_{[\alpha\beta]} = \tfrac{1}{2}(\xi_\alpha^a \xi_\beta^b - \xi_\beta^a \xi_\alpha^b), \tag{1.14}$$

which have $\frac{1}{2}n(n+1)$ and $\frac{1}{2}n(n-1)$ nonzero components, respectively. Since the symmetric tensor remains symmetric under the transformation law (13), while the antisymmetric tensor remains antisymmetric, it follows that a division of each index pair into symmetric and antisymmetric form brings $V$ into the reduced form

$$V = \begin{pmatrix} V_{\{\alpha\beta\},\{\alpha'\beta'\}} & 0 \\ 0 & V_{[\alpha\beta],[\alpha'\beta']} \end{pmatrix} \begin{matrix} \leftarrow \frac{1}{2}n(n+1) \text{ comp} \\ \leftarrow \frac{1}{2}n(n-1) \text{ comp} \end{matrix} \Big\} n^2 \text{ comp.} \tag{1.15}$$

Symbolically, one writes $2 \times 2 = 3 \times 1$ in the case of $SU_2$, $3 \times 3 = 6 + \bar{3}$ in the case of $SU_3$, etc. (The symbol "$\bar{3}$" is used to distinguish the transformation of the antisymmetric tensor from that of the 3-vector. In $SU_3$, $\bar{V}$ denotes the complex conjugate representation of $V$, see below.)

Similarly, the tensor product $\xi_\alpha^a \xi_\beta^b \xi_\gamma^c$ of rank three may be decomposed into one totally symmetric part, one totally antisymmetric part and two parts of mixed symmetry. The totally symmetric and antisymmetric parts are of dimensions $\frac{1}{6}n(n+1)(n+2)$ and $\frac{1}{6}n(n-1)(n-2)$, respectively, whereas the remaining parts are of dimension $\frac{1}{3}n(n^2-1)$ each. By forming tensors of arbitrary rank and reducing them with respect to their permutation properties, all irreducible representations of $SU_n$ are found. This method will be used for $SU_3$. For $SU_2$, the more familiar approach is to study the commutation relations (12).

## B-2 Schur's Lemma and Orthogonality Relations

If $D^1$ and $D^2$ are two irreducible sets of $n_1 \times n_1$ and $n_2 \times n_2$ matrices, then any fixed $n_1 \times n_2$ matrix $A$ for which

$$D^1(\alpha^i) \cdot A = A \cdot D^2(\alpha^i) \tag{2.1}$$

for all values of the parameters $\alpha^i$, is either zero or nonsingular. This statement is the first part of Schur's lemma. The term "nonsingular" means $\det A \neq 0$. Since only square matrices can be nonsingular, the latter possibility implies $n_1 = n_2$.

Without loss of generality, we take $n_1 \geq n_2$. The matrix multiplication on the left-hand side of (1), $\sum_{k=1}^{n_1} D_{ik} A_{kl}$, may be viewed as the transformation of $n_2$ different vectors $A_l$ of $n_1$ components $A_{kl}$. The right-hand side of (1) states that the transformed vectors are linear combinations of the old vectors $A_l$. If this is true for all matrices $D^1$, then these matrices leave the space spanned by the $A_l$ invariant. For $n_1 > n_2$ this would be a proper subspace, which is inconsistent with our assumption that the $D^1$ are irreducible, unless of course

$$A_l = 0 \quad \text{for} \quad l = 1 \cdots n_2, \qquad n_2 > n_1. \tag{2.2}$$

For $n_1 = n_2$, the space spanned by the $A_l$ is the full space, provided that the $A_l$ are linearly independent, i.e., $\det A \neq 0$. This proves the first part of Schur's lemma.

The second part investigates in more detail the case $n_1 = n_2$. We exclude the trivial possibility $A = 0$, in which case (1) may be written as

$$D^2(a^i) = A^{-1} \cdot D^1(\alpha^i) \cdot A \quad \text{for all} \quad \alpha^i. \tag{2.3}$$

This is a similarity transformation which does not affect the matrix elements between state vectors. Representations $D^2$ and $D^1$ which are connected by (3) are called equivalent. The second part of Schur's lemma states that for $D^2 = D^1$, $A$ must be a multiple of the unit matrix. This is proven as follows: In the relation $AD = DA$, we may add to $A$ any $\mu \cdot 1$ multiple of the unit matrix 1, since this matrix certainly commutes with any $D$. Specifically we now choose $\mu$ such that

$$\det\,(A - \mu \cdot 1) = 0. \tag{2.4}$$

If we allow complex values for $\mu$, this equation has always a solution. By the first part of Schur's lemma, we find

$$A - \mu \cdot 1 = 0, \tag{2.5}$$

since the other possibility, $\det\,(A - \mu \cdot 1) \neq 0$ is barred by construction.

As an application of Schur's lemma, we derive the orthogonality relations of the matrix elements of irreducible representations. For this purpose, we choose an arbitrary fixed $n_1 \times n_2$ matrix $B$ and construct

$$A = \int d\Omega_i\, D^1(\alpha^i) B D^{2+}(\alpha^i), \tag{2.6}$$

where the integration covers the whole parameter space $\Omega_i$. Since $D(\alpha^{i'})\, D(\alpha^i)$ gives another matrix $D(\alpha^{i''})$ of the same representation, we have

$$D^1(\alpha^{i'}) A D^{2+}(\alpha^{i'}) = \int d\Omega_i''\, D^1(\alpha^{i''}) \cdot B \cdot D^{2+}(\alpha^{i''}) = A. \tag{2.7}$$

Actually we have assumed that the integral (6) is finite (compact groups), and that the Jacobian determinant of the transformation of variables $\alpha^i \to \alpha^{i''}$ is one. The latter requirement restricts the form of the differential $d\Omega_i$. For $SU_1$ we have $d\Omega_i = d\alpha^1$. For $SU_2$, the differential is given in the next section.

From (7), we conclude that $A$ has the property (1). Now Schur's lemma requires $A = 0$ for inequivalent irreducible representations. Since the matrix $B$ was completely arbitrary, we conclude that

$$\int d\Omega_i\; D^1_{ik}(\alpha^i)\; D^{2*}_{lm}(\alpha^i) = 0. \tag{2.8}$$

On the other hand, for $D^1 = D^2$, $A$ is a multiple of the unit matrix, which implies

$$\int d\Omega_i\; D_{ik}(\alpha^i)\; D^*_{lm}(\alpha^i) = \delta_{il}\; d_{km}. \tag{2.9}$$

Setting $i = l$ and summing, we get from the unitarity of the $D$-matrices

$$\int d\Omega_i\; D^a_{ik}(\alpha^i)\; D^{a*}_{im}(\alpha^i) = \delta_{km} \int d\Omega_i = n_a\; d_{km}, \tag{2.10}$$

where $n_a$ is the dimension of the $D$-matrices. Equations (8), (9) and (10) may be combined into one single equation:

$$\int d\Omega_i\; D^a_{ik}(\alpha^i)\; D^{b*}_{lm}(\alpha^i) = \delta_{ab}\; \delta_{il}\; \delta_{km} n_a^{-1} \int d\Omega_i. \tag{2.11}$$

## B-3 The Functions $D$ and $d$ of $SU_2$. Wigner–Eckart Theorem

The parametrization (1.11) is conceptually simple, but in practice it is simpler to decompose $D$ into a product of 3 unitary matrices, two of which are in diagonal form, and the third matrix being real. With the matrix $J_z$ in diagonal form and $iJ_y$ real, this decomposition is

$$D^j(\alpha\beta\gamma) = \exp\,(-i\alpha J^j_z)\,\exp\,(-i\beta J^j_y)\,\exp\,(-i\gamma J^j_z), \tag{3.1}$$

It generalizes (1-4.16) to arbitrary dimensions $n = 2j + 1$. The corresponding matrix elements are

$$\begin{gathered}(e^{-i\alpha J_z j})_{mm'} = e^{-i\alpha m}\,\delta_{mm'}, \qquad (e^{-i\beta J_y j})_{mm'} = d^j_{mm'}(\beta), \\ D^j_{mm'} = e^{-i\alpha m - i\gamma m'}\, d^j_{mm}.\end{gathered} \tag{3.2}$$

The differential $d\Omega_i$ of the last section is $d\alpha\; d\cos\beta\; d\gamma$, which follows from the geometrical interpretation of the Euler angles $\alpha, \beta, \gamma$. Thus the orthogonality relation (2.11) reads

$$\int d\alpha\; d\cos\beta\; d\gamma\; D^{j*}_{\lambda\mu}(\alpha\beta\gamma)\; D^{j'}_{\lambda'\mu'}(\alpha\beta\gamma) = 8\pi^2(2j+1)^{-1}\,\delta_{jj'}\,\delta_{\lambda\lambda'}\,\delta_{\mu\mu'}. \tag{3.3}$$

Let $|jm\rangle$ be the eigenstates of $J_z$ with eigenvalue $m$, and let $|\lambda, \alpha\beta\gamma\rangle$ be the eigenstates of $J'_z$, where $z'$ is the $z$-axis in a coordinate system rotated by the angles $\alpha$, $\beta$, $\gamma$. Then we have

$$|j\lambda, \alpha\beta\gamma\rangle = \sum_m D^j_{m\lambda}(\alpha\beta\gamma)|jm\rangle. \tag{3.4}$$

The Clebsch-Gordan coefficients $C$ are defined by the expansion of product states $|j_1 m_1\rangle|j_2 m_2\rangle$ in terms of the basic states of irreducible representations:

$$|j_1 m_1\rangle|j_2 m_2\rangle = \sum_j C^{j_1\ j_2\ j}_{m_1\, m_2\, m}|jm\rangle, \qquad m = m_1 + m_2. \tag{3.5}$$

Inserting now a rotation according to (4), we obtain the famous Clebsch-Gordan series:

$$D^{j_1}_{m_1 m_1'} D^{j_2}_{m_2 m_2'} = \sum_j C^{j_1\ j_2\ j}_{m_1\, m_2\, m} C^{j_1\ \ j_2\ \ j}_{m_1'\, m_2'\, m'} D^j_{mm'}. \tag{3.6}$$

From the knowledge of $d^{\frac{1}{2}}$ (1-4.15) and eqs. (3) and (6), one can calculate $d$, $D$ and $C$ recurrently for all $j$. In most textbooks, one calculates instead the matrix elements of $J_\pm = J_x \pm iJ_y$ directly for arbitrary $j$,

$$\begin{aligned} (J_\pm)_{mm'} &= \langle m'|J\pm|m\rangle = \delta_{m',\, m\pm 1} C_\pm, \\ C_\pm &= [j(j+1) - m'(m' \mp 1)]^{\frac{1}{2}}. \end{aligned} \tag{3.7}$$

Before computing the $d$'s, we discuss their symmetry properties. By definition of the phases, they are elements of real unitary matrices (= orthogonal matrices). This implies

$$d_{mm'}(-\theta) = d_{m'm}(\theta). \tag{3.8}$$

Moreover, writing $J_y$ as $-\frac{1}{2}i(J_+ - J_-)$, we see from (7) that the matrix elements of $J_y$ are invariant under the exchange $m \leftrightarrow -m'$. Consequently,

$$d_{mm'} = d_{-m',\, -m}. \tag{3.9}$$

Next, the fact that the matrix $J_+ - J_-$ is antisymmetric and has nonzero matrix elements only if $m$ and $m'$ differ by one, leads to the following property: an even (odd) power of $J_y$ is symmetric (antisymmetric) and has nonzero matrix elements only if $m$ and $m'$ differ by an even (odd) number. This can be summarized as

$$d_{mm'} = (-1)^{m-m'} d_{m'm}. \tag{3.10}$$

Finally, the relation $d(2\pi) = (-1)^{2j} d(0)$ which expresses the double-valuedness of half-integer $j$ representations is sufficient to determine $d(\pi)$ as

$$d^j_{mm'}(\pi) = (-1)^{j+m} \delta_{m,\, -m'} = (-1)^{j-m'} \delta_{m,\, -m'}. \tag{3.11}$$

Combination of (9) and (10) yields

$$d^j_{mm'} = (-1)^{m-m'} d_{-m,\, -m'}. \tag{3.12}$$

In the text, we only need rotations with rotation axis in the $xy$-plane. The angle between the rotation axis and the $y$-axis is called $\varphi$ or $\phi$, and the angle of rotation itself is called $\vartheta$ or $\theta$. This corresponds to the Euler angles $\beta = \theta$, $\alpha = -\gamma = \phi$:

$$D^j_{mm'}(\phi, \theta, -\phi) = e^{i(m'-m)\phi}\, d_{mm'}(\theta). \tag{3.13}$$

Equation (12) now implies

$$D^j_{mm'}(\phi, \theta, -\phi) = (-1)^{m-m'}\, D^{j*}_{-m,\,-m'}(\phi, \theta, -\phi). \tag{3.14}$$

An irreducible tensor operator of rank $r$ consists of $2r + 1$ operators $T^r_\nu$ ($\nu = -r \cdots +r$) which transform into each other under unitary transformations:

$$T^r_\nu(\Omega) = D(\Omega)T^r_\nu D^{-1}(\Omega) = \sum_{\nu'} T^r_{\nu'}\, D^r_{\nu'\nu}(\Omega). \tag{3.15}$$

The matrix elements of such operators between states $|j\lambda\Omega\rangle$ and $|j'\lambda'\Omega\rangle$ are obviously independent of the choice of $\Omega$; the Wigner-Eckart theorem states that they depend on $\lambda$ and $\lambda'$ only via a Clebsch-Gordan coefficient:

$$\langle j'\lambda' | T^r_\nu | j\lambda\rangle = C^{jrj'}_{\lambda\nu\lambda'}\langle j' \| T^r \| j\rangle. \tag{3.16}$$

The last quantity is called a "reduced matrix element." The theorem is easily derived from the $CG$-series (6) and the orthogonality relations (3) of the $D$-functions. We insert the transformation properties (15) and (4)

$$\langle j'\lambda'\Omega | T^r_\nu(\Omega) | j\lambda\Omega\rangle = \sum_{mm'\nu'} D^j_{m\lambda}\, D^r_{\nu'\nu}\, D^{j'*}_{m'\lambda'}\langle j'm' | T^r_{\nu'} | jm\rangle, \tag{3.17}$$

reduce the product of $D^j$ and $D^r$ according to (6)

$$D^j_{m\lambda}\, D^r_{\nu'\nu} = \sum_k C^{j\ r\ k}_{m\,\nu'\,\mu} C^{j\ r\ k}_{\lambda\,\nu\,\sigma}\, D^k_{\mu\sigma}, \tag{3.18}$$

and integrate (17) over all $\Omega$. On the left-hand side, this produces merely a factor $8\pi^2$, and on the right-hand side, the product $D^k\, D^{j'*}$ is replaced by $8\pi^2(2k + 1)^{-1}\, \delta_{kj'}\, \delta_{\mu m'}\, \delta_{\sigma\lambda'}$ according to (3). We thus get

$$\langle j'\lambda' | T^r_\nu | j\lambda\rangle = \left(\sum_{mm'\nu'} (2j' + 1)^{-1} C^{j\ r\ j'}_{m\,\nu'\,m'}\langle j'm' | T^r_{\nu'} | jm\rangle\right) C^{j\ r\ j'}_{\lambda\,\nu\,\lambda'}. \tag{3.19}$$

The quantity in brackets is independent of $\lambda$ and $\lambda'$ and represents the reduced matrix element of (16).

## B-4 Calculation of the $d$-Functions

The explicit calculation of $d$-functions proceeds conveniently recurrently, using the Clebsch-Gordan series (3.6). By means of the orthogonality relations of the Clebsch-Gordan coefficients, (3.6) can be transformed into

$$C^{j_1'\ j_2\ j}_{m_1'\,m_2'\,m'} \cdot D^j_{mm'} = \sum_{m_1} C^{j_1\ j_2\ j}_{m_1\,m_2\,m}\, D^{j_1}_{m_1 m_1'}\, D^{j_2}_{m_2 m_2'}. \tag{4.1}$$

All these formulae are valid for the $d$-functions as well. We start by calculating $d^{\frac{1}{2}}$ directly from its definition $\exp(-\frac{1}{2}i\theta\sigma_y)$. The result is found in table B-4.

**Table B-4** The functions $d^S_{MM'}(\theta)$ for $S = \frac{1}{2} \cdots \frac{5}{2}$. The sign convention of Rose (1957) is chosen, and $d^S_{MM'} = d^S_{-M'-M} = (-1)^{M-M'} d_{M'M}$

| M \ M′ | $+\frac{1}{2}$ | $-\frac{1}{2}$ |
|---|---|---|
| $+\frac{1}{2}$ | $\cos \frac{1}{2}\theta$ | $-\sin \frac{1}{2}\theta$ |
| $-\frac{1}{2}$ | $\sin \frac{1}{2}\theta$ | $\cos \frac{1}{2}\theta$ |

| M \ M′ | $+1$ | $0$ | $-1$ |
|---|---|---|---|
| $+1$ | $\cos^2 \frac{1}{2}\theta$ | $-2^{-\frac{1}{2}} \sin \theta$ | $\sin^2 \frac{1}{2}\theta$ |
| $0$ | $2^{-\frac{1}{2}} \sin \theta$ | $\cos \theta$ | $-2^{-\frac{1}{2}} \sin \theta$ |
| $-1$ | $\sin^2 \frac{1}{2}\theta$ | $2^{-\frac{1}{2}} \sin \theta$ | $\cos^2 \frac{1}{2}\theta$ |

| M \ M′ | $+\frac{3}{2}$ | $+\frac{1}{2}$ | $-\frac{1}{2}$ | $-\frac{3}{2}$ |
|---|---|---|---|---|
| $+\frac{3}{2}$ | $\cos^3 \frac{1}{2}\theta$ | $-3^{\frac{1}{2}} \cos^2 \frac{1}{2}\theta \sin \frac{1}{2}\theta$ | $3^{\frac{1}{2}} \cos \frac{1}{2}\theta \sin^2 \frac{1}{2}\theta$ | $-\sin^3 \frac{1}{2}\theta$ |
| $+\frac{1}{2}$ | $3^{\frac{1}{2}} \cos^2 \frac{1}{2}\theta \sin \frac{1}{2}\theta$ | $\cos \frac{1}{2}\theta(1 - 3 \sin^2 \frac{1}{2}\theta)$ | $\sin \frac{1}{2}\theta(1 - 3 \cos^2 \frac{1}{2}\theta)$ | $3^{\frac{1}{2}} \cos \frac{1}{2}\theta \sin^2 \frac{1}{2}\theta$ |
| $-\frac{1}{2}$ | $3^{\frac{1}{2}} \cos \frac{1}{2}\theta \sin^2 \frac{1}{2}\theta$ | $\sin \frac{1}{2}\theta(3 \cos^2 \frac{1}{2}\theta - 1)$ | $\cos \frac{1}{2}\theta(1 - 3 \sin^2 \frac{1}{2}\theta)$ | $-3^{\frac{1}{2}} \cos^2 \frac{1}{2}\theta \sin \frac{1}{2}\theta$ |
| $-\frac{3}{2}$ | $+\sin^3 \frac{1}{2}\theta$ | $3^{\frac{1}{2}} \cos \frac{1}{2}\theta \sin^2 \frac{1}{2}\theta$ | $3^{\frac{1}{2}} \cos^2 \frac{1}{2}\theta \sin \frac{1}{2}\theta$ | $\cos^3 \frac{1}{2}\theta$ |

| M \ M′ | $2$ | $1$ | $0$ | $-1$ | $-2$ |
|---|---|---|---|---|---|
| $2$ | $\cos^4 \frac{1}{2}\theta$ | $-2 \cos^3 \frac{1}{2}\theta \sin \frac{1}{2}\theta$ | $6^{\frac{1}{2}} \cos^2 \frac{1}{2}\theta \sin^2 \frac{1}{2}\theta$ | $-2 \cos \frac{1}{2}\theta \sin^3 \frac{1}{2}\theta$ | $\sin^4 \frac{1}{2}\theta$ |
| $1$ | $2 \cos^3 \frac{1}{2}\theta \sin \frac{1}{2}\theta$ | $\cos^2 \frac{1}{2}\theta(\cos^2 \frac{1}{2}\theta - 3 \sin^2 \frac{1}{2}\theta)$ | $-\frac{3^{\frac{1}{2}}}{2} \sin \theta \cos \theta$ | $\sin^2 \frac{1}{2}\theta(3 \cos^2 \frac{1}{2}\theta - \sin^2 \frac{1}{2}\theta)$ | |
| $0$ | $6^{\frac{1}{2}} \cos^2 \frac{1}{2}\theta \sin^2 \frac{1}{2}\theta$ | $\frac{3^{\frac{1}{2}}}{2} \sin \theta \cos \theta$ | $\frac{1}{2}(3 \cos^2 \theta - 1)$ | $-\frac{3^{\frac{1}{2}}}{2} \sin \theta \cos \theta$ | |

| M \ M′ | $\frac{5}{2}$ | $\frac{3}{2}$ | $\frac{1}{2}$ | $-\frac{1}{2}$ | $-\frac{3}{2}$ | $-\frac{5}{2}$ |
|---|---|---|---|---|---|---|
| $\frac{5}{2}$ | $c^5$ | $-5^{\frac{1}{2}}c^4s$ | $10^{\frac{1}{2}}c^3s^2$ | $-10^{\frac{1}{2}}c^2s^3$ | $5^{\frac{1}{2}}cs^4$ | $-s^5$ |
| $\frac{3}{2}$ | $5^{\frac{1}{2}}c^4s$ | $c^3(c^2 - 4s^2)$ | $2^{\frac{1}{2}}sc^2(3s^3 - 2c^2)$ | $2^{\frac{1}{2}}s^2c(3c^2 - 2s^2)$ | $s^3(s^2 - 4c^2)$ | $5^{\frac{1}{2}}cs^4$ |
| $\frac{1}{2}$ | $10^{\frac{1}{2}}c^3s^2$ | $2^{\frac{1}{2}}sc^2(2c^2 - 3s^2)$ | $c(c^4 + 3s^4 - 6c^2s^2)$ | $s(6s^2c^2 - s^4 - 3c^4)$ | $c \equiv \cos \frac{1}{2}\theta,\ s \equiv \sin \frac{1}{2}\theta$ | |

Next, we put $j_2 = \frac{1}{2}$ and $j_1 = j - \frac{1}{2}$ in (1). The Clebsch-Gordans for $j_1 = j \pm \frac{1}{2}$ are

$$C_{m_1\ \ m_2\ m}^{j-\frac{1}{2},\frac{1}{2},\ j} = \left(\frac{j + 2m_2 m}{2j_1 + 1}\right)^{\frac{1}{2}}, \qquad C_{m_1\ \ m_2\ m}^{j+\frac{1}{2},\frac{1}{2},\ j} = -2m_2\left(\frac{j - 2m_2 m + 1}{2j_1 + 1}\right)^{\frac{1}{2}}. \quad (4.2)$$

We set $m_2' = +\frac{1}{2}$ in (1) and obtain

$$(j + m')^{\frac{1}{2}}\, d_{mm'}^{j} = (j + m)^{\frac{1}{2}}\, d_{m-\frac{1}{2},\, m'-\frac{1}{2}}^{j-\frac{1}{2}} \cos\frac{\theta}{2} + (j - m)^{\frac{1}{2}}\, d_{m+\frac{1}{2},\, m'-\frac{1}{2}}^{j-\frac{1}{2}} \sin\frac{\theta}{2}. \quad (4.3)$$

If we apply this formula twice, we obtain

$$\begin{aligned} 2(j + m')^{\frac{1}{2}}(j + m' - 1)^{\frac{1}{2}}\, d_{mm'}^{j} &= (j + m)^{\frac{1}{2}}(j + m - 1)^{\frac{1}{2}}(1 + \cos\theta)\, d_{m-1,\, m'-1}^{j-1} \\ &\quad + 2(j^2 - m^2)^{\frac{1}{2}} \sin\theta\, d_{m,\, m'-1}^{j-1} \\ &\quad + (j - m)^{\frac{1}{2}}(j - m - 1)^{\frac{1}{2}}(1 - \cos\theta)\, d_{m+1,\, m'-1}^{j-1}. \end{aligned} \quad (4.4)$$

In the theory of scattering of particles with spin, one needs the $d$-functions for arbitrary values of $j$ but for small values of $m$ and $m'$. For that case, a convenient recurrence relation in $m$ has been given by Jacob and Wick (1959):

$$d_{m',\, m\pm 1}^{j} = (j \pm m + 1)^{-\frac{1}{2}}(j \mp m)^{-\frac{1}{2}}(-m'/\sin\theta + m\cot\theta \mp \partial_\theta)\, d_{m'm}^{j}. \quad (4.5)$$

The proof of this formula uses the relation

$$e^{-i\theta J_y} J_z e^{i\theta J_y} = \cos\theta J_z + \sin\theta J_x, \quad (4.6)$$

which is evident from the geometrical interpretation in terms of rotations. It allows one to express $J_x \exp(-i\theta J y)$ in terms of $J_z$ and $\exp(-i\theta J_y)$. Using the relation

$$-iJ_y e^{-i\theta J_y} = \partial_\theta e^{-i\theta J_y}, \quad (4.7)$$

we arrive at

$$(J_x \pm iJ_y)e^{-i\theta J_y} = (\sin\theta)^{-1} e^{-i\theta J_y} J_z - \cot\theta J_z e^{-i\theta J_y} \mp \partial_\theta e^{-i\theta J_y}. \quad (4.8)$$

Now we obtain (5) by inserting the matrix elements (3.7) into this last equation.

## B-5 *D*-Functions for Integer *j* and Legendre Polynomials

The functions $d_{00}^{L}(\theta)$ are identical with the Legendre polynomials $P_L$ (cos $\theta$). To show this, we first notice from (4.4) that $d_{mm'}^{L}$ is a polynomial of degree $L$ in cos $\theta$ and sin $\theta$. For $m = m' = 0$, the symmetry (3.8) shows that $d_{00}^{L}$ is a polynomial in cos $\theta$ only. The orthogonality relations for $d$-functions follow from (3.3) and the decomposition (3.1):

$$\int_{-1}^{1} d(\cos\theta)\, d^j_{mm'}\, d^{j'}_{mm'} = 2(2j+1)^{-1}\,\delta_{jj'}, \tag{5.1}$$

which is just the orthogonality relation (2-5.10) for $P_L$, with $L = j$.

Next, the spherical harmonics are defined by

$$\begin{aligned} Y^L_M(\theta, \phi) &= (4\pi)^{-\frac{1}{2}}(2L+1)^{\frac{1}{2}}\, D^{L*}_{M0}(\phi, \theta, -\phi) \\ &= (4\pi)^{-\frac{1}{2}}(2L+1)^{\frac{1}{2}}\, d^L_{M0}(\theta)e^{iM\phi}. \end{aligned} \tag{5.2}$$

From (1), it is obvious that these functions are orthonormal on a unit sphere:

$$\int d\Omega Y^{L*}_M(\Omega) Y^{L'}_{M'}(\Omega) = \delta_{MM'}\,\delta_{LL'}. \tag{5.3}$$

The spherical harmonics addition theorem

$$\begin{aligned} P_L(\cos\theta'') &= 4\pi(2L+1)^{-1} \sum_M Y^{L*}_M(\theta, \phi) Y^L_M(\theta', \phi'), \\ \cos\theta'' &\equiv \cos\theta\cos\theta' + \sin\theta\sin\theta'\cos(\phi - \phi'), \end{aligned} \tag{5.4}$$

is a consequence of the group property of the $D$-functions,

$$D_{m'm''}(\alpha''\beta''\gamma'') = \sum_m D_{m'm}(\alpha\beta\gamma)\, D_{mm''}(\alpha'\beta'\gamma') \tag{5.5}$$

for $m' = m'' = 0$. Finally, we compute the functions $d^L_{M0} = d^L_{0,-M}$ by means of the recurrence relation (4.5) for $m' = 0$:

$$d^L_{0,M-1} = [(L-M+1)(L-M)]^{-\frac{1}{2}}(M\cot\theta + \partial_\theta)\, d^L_{0M}. \tag{5.6}$$

Starting with $M = 0$, the $d_{0M}$ may be expressed in terms of $P_L$ and its derivatives. The first two cases are

$$d^L_{0,-1} = d^L_{10} = -L^{-\frac{1}{2}}(L+1)^{-\frac{1}{2}} \sin\theta\, P'_L(x), \tag{5.7}$$

$$d^L_{20} = [L(L^2-1)(L+2)]^{-\frac{1}{2}}[2P'_{L-1}(x) - L(L-1)P_L(x)]. \tag{5.8}$$

The prime means derivation with respect to $x = \cos\theta$. In deriving (8) from (6), we have applied eqs. (11) below to eliminate the second derivative.

From (7) and (8) we can now construct $d^j_{MM'}$ for small values of $M$ and $M'$. The values given in table B-5 are taken from the recurrence relation (4.3). The Legendre polynomials are given by $P_0 = 1$, $P_1 = x$, and for $L > 1$ by

$$P_L(x) = x(2 - L^{-1})P_{L-1}(x) - (1 - L^{-1})P_{L-2}(x). \tag{5.9}$$

There exist various recurrence relations involving the derivatives of $P_L$. The basic relation of this type is

$$P'_{L+1} - xP'_L = (L+1)P_L. \tag{5.10}$$

**Table B-5** Some $d$-functions for half-integer $j$.

$$(j+\tfrac{1}{2})\, d^j_{\frac{1}{2}\frac{1}{2}} = \cos\frac{\theta}{2}\,(P'_{j+\frac{1}{2}} - P'_{j-\frac{1}{2}}) = (j+\tfrac{1}{2})\, d^j_{-\frac{1}{2}-\frac{1}{2}}$$

$$(j+\tfrac{1}{2})\, d^j_{-\frac{1}{2}\frac{1}{2}} = \sin\frac{\theta}{2}\,(P'_{j+\frac{1}{2}} + P'_{j-\frac{1}{2}}) = -(j+\tfrac{1}{2})\, d^j_{\frac{1}{2}-\frac{1}{2}}$$

$$(j+\tfrac{1}{2})\, d^j_{\frac{1}{2}\frac{3}{2}} = \sin\frac{\theta}{2}\,(j-\tfrac{1}{2})^{-\frac{1}{2}}(j+\tfrac{3}{2})^{-\frac{1}{2}}[(j-\tfrac{1}{2})P'_{j+\frac{1}{2}} + (j+\tfrac{3}{2})P'_{j-\frac{1}{2}}]$$

$$(j+\tfrac{1}{2})\, d^j_{-\frac{1}{2}\frac{3}{2}} = \cos\frac{\theta}{2}\,(j-\tfrac{1}{2})^{-\frac{1}{2}}(j+\tfrac{3}{2})^{-\frac{1}{2}}[-(j-\tfrac{1}{2})P'_{j+\frac{1}{2}} + (j+\tfrac{3}{2})P'_{j-\frac{1}{2}}]$$

It is proven by integration by parts. Other relations are obtained by differentiating (9) and combining it with (10), for example

$$(1-x^2)P'_L = LP_{L-1} - LxP_L. \tag{5.11}$$

From these equations, one derives the orthogonality relation

$$\frac{1}{2}\int_{-1}^{1} dx\; P'_{L'}(x)(P_{L-1}(x) - P_{L+1}(x)) = \delta_{LL'}. \tag{5.12}$$

## B-6 $SU_3$-Transformations

The $3 \times 3$ unitary matrices of determinant 1 form the group $SU_3$. This group has $3^2 - 1 = 8$ Hermitean and traceless matrices as generators (compare 1.10). It contains $SU_2$ as a subgroup, and it is therefore convenient to take the first 3 generators as trivial extensions of the Pauli matrices $\sigma^i$ of (1-4.2):

$$SU_3 = \exp\left\{-i\sum_{k=1}^{8}\omega^k\lambda^k\right\}, \qquad \lambda^k = \begin{pmatrix}\sigma^k & 0\\ 0 & 0\end{pmatrix} \qquad \text{for } k = 1, 2, 3. \tag{6.1}$$

The eigenstates of $\lambda^3$ with eigenvalues $+1$, $-1$ and 0 are called $u$ (for "isospin up"), $d$ (for "isospin down") and $s$ ("strange"), respectively:

$$u = \begin{pmatrix}1\\0\\0\end{pmatrix}, \qquad d = \begin{pmatrix}0\\1\\0\end{pmatrix}, \qquad s = \begin{pmatrix}0\\0\\1\end{pmatrix}. \tag{6.2}$$

In this case the subgroup $SU_2$ with generators $\lambda^1$, $\lambda^2$, $\lambda^3$ leaves the state $s$ invariant. Generally, $SU_2$ is contained in $SU_3$ in 3 different ways, namely (i) as isospin transformations (which leave $s$ invariant), (ii) $V$-spin transformations (which leave $d$ invariant) and (iii) $U$-spin-transformations (which leave

$u$ invariant). Correspondingly we can take the generators $\lambda^4, \lambda^5$ ($\lambda^6, \lambda^7$) as $\sigma^1$ and $\sigma^2$ in the first and third (second and third) rows and columns:

$$\lambda^4 = \begin{pmatrix} 0 & 0 & 1 \\ 0 & 0 & 0 \\ 1 & 0 & 0 \end{pmatrix}, \qquad \lambda^5 = \begin{pmatrix} 0 & 0 & -i \\ 0 & 0 & 0 \\ i & 0 & 0 \end{pmatrix},$$
$$\lambda^6 = \begin{pmatrix} 0 & 0 & 0 \\ 0 & 0 & 1 \\ 0 & 1 & 0 \end{pmatrix}, \qquad \lambda^7 = \begin{pmatrix} 0 & 0 & 0 \\ 0 & 0 & -i \\ 0 & i & 0 \end{pmatrix}, \tag{6.3}$$

There is now one linearly independent generator left, which could be taken as $\sigma^3$ either in the first and third or second and third rows and columns. Because of the importance of the isospin subgroup, one chooses that combination of the second alternative and $\lambda^3$ which commutes with all 3 isospin generators (1):

$$\lambda^8 = 2 \cdot 3^{-\frac{1}{2}} \begin{pmatrix} 0 & 0 \\ 0 & \sigma_3 \end{pmatrix} + 3^{-\frac{1}{2}} \begin{pmatrix} \sigma^3 & 0 \\ 0 & 0 \end{pmatrix} = 3^{-\frac{1}{2}} \begin{pmatrix} 1 & 0 & 0 \\ 0 & 1 & 0 \\ 0 & 0 & -2 \end{pmatrix}. \tag{6.4}$$

$SU_3$ is a group of "rank 2," which means that the basis states of its irreducible representations carry two additive quantum numbers, namely the eigenvalues of the operators $F_3$ and $F_8$ which represent $\lambda^3$ and $\lambda^8$. Together with the unit matrix, the 8 matrices $\lambda^k$ form a complete set of Hermitean $3 \times 3$ matrices. The "structure constants" of (1.12) are defined by their commutation relations,

$$[\lambda^i, \lambda^i] = 2if_{ijk}\lambda^k. \tag{6.5}$$

As all $\lambda^i$ are traceless, the trace of $\lambda^i\lambda^j$ must be entirely due to a possible component of the unit matrix in this product. We have normalized our $\lambda^i$ in the same way as the Pauli matrices, therefore we find, using (1-4.3)

$$\text{trace } (\lambda^i\lambda^j) = 2\,\delta_{ij}. \tag{6.6}$$

It is now easy to see that the structure functions are not only antisymmetric in $i$ and $j$ (which follows directly from their definitions (5)), but also in $i$ and $k$. Inserting (5) into (6), we find

$$\begin{aligned} 4if_{ijk} &= \text{trace } \lambda^k[\lambda^i, \lambda^j] = \text{trace } \lambda^k(\lambda^i\lambda^j - \lambda^j\lambda^i) \\ &= \text{trace } (\lambda^k\lambda^i\lambda^j - \lambda^i\lambda^k\lambda^j) = -\text{trace } [\lambda^i, \lambda^k]\lambda^j. \end{aligned} \tag{6.7}$$

In other words, $f_{ijk}$ is antisymmetric in all three indices. Its only nonzero components are given in table B-6. They follow most easily from $[\sigma^i, \sigma^j] = 2i\varepsilon_{ijk}\sigma^k$ and from (4).

Contrary to $SU_2$, there exists no unitary matrix which transforms all matrices of $SU_3$ into their complex conjugates, i.e., $SU_3$ and $SU_3^*$ are "inequivalent." When the representations are simply denoted by their dimensions, an additional symbol is needed for the complex conjugate

**Table B-6** The nonzero components of $f_{ijk} = -f_{jik} = -j_{kij} = f_{jki}$.

| $ijk$ | 123 | 147 | 156 | 246 | 257 | 345 | 367 | 458 | 678 |
|---|---|---|---|---|---|---|---|---|---|
| $f_{ijk}$ | 1 | $\frac{1}{2}$ | $-\frac{1}{2}$ | $\frac{1}{2}$ | $\frac{1}{2}$ | $\frac{1}{2}$ | $-\frac{1}{2}$ | $(\frac{3}{4})^{\frac{1}{2}}$ | $(\frac{3}{4})^{\frac{1}{2}}$ |

representation. One writes "3" for $SU_3$ and "$\bar{3}$" for $SU_3^*$. Representations of higher dimensions are obtained from the direct product (1.13). According to the dimensional argument following (1.14), the antisymmetric product $T_{[\alpha\beta]} = \frac{1}{2}(\xi_\alpha^a \xi_\beta^b - \xi_\beta^a \xi_\alpha^b)$ must be either the representation 3 or $\bar{3}$. To decide between these two possibilities, we consider the 3-vector with an upper index

$$\eta^\alpha = \varepsilon_{\alpha\beta\gamma} T_{[\beta\gamma]} = \varepsilon_{\alpha\beta\gamma} \xi_\beta^a \xi_\gamma^b. \tag{6.8}$$

Under (1.13), $\eta^\alpha$ transforms into

$$\eta'^\alpha = \varepsilon_{\alpha\beta\gamma} U_{\beta\beta'} U_{\gamma\gamma'} T_{[\beta'\gamma']}. \tag{6.9}$$

We insert a factor $\delta_{\alpha\lambda} = U^*_{\alpha\alpha'} U_{\lambda\alpha'}$ into this equation, obtaining

$$\begin{aligned} \eta'^\alpha &= \delta_{\alpha\lambda} \varepsilon_{\lambda\beta\gamma} U_{\beta\beta'} U_{\gamma\gamma'} T_{[\beta'\gamma']} = U^*_{\alpha\alpha'} \varepsilon_{\lambda\beta\gamma} U_{\lambda\alpha'} U_{\beta\beta'} U_{\gamma\gamma'} T_{[\beta'\gamma']} \\ &= \det (U) U^*_{\alpha\alpha'} \varepsilon_{\alpha'\beta'\gamma'} T_{[\beta'\gamma']} = U^*_{\alpha\alpha'} \eta^{\alpha'}, \end{aligned} \tag{6.10}$$

since $\det (U) = 1$. This shows that

$$3 \otimes 3 = \bar{3} \oplus 6. \tag{6.11}$$

The transformation law (10) can be summarized as $\eta' = \eta U^+ = \eta U^{-1}$. Therefore, the combination $\eta^\alpha \xi_\alpha$ is $SU_3$-invariant. The remaining part,

$$T_\alpha^\beta = \eta^\alpha \xi_\beta - \tfrac{1}{3}\, \delta_{\alpha\beta} \eta^\gamma \xi_\gamma \tag{6.12}$$

is irreducible and gives a unitary octet. We thus have

$$3 \otimes \bar{3} = 1 \oplus 8. \tag{6.13}$$

It is clear from the construction that these two representations have in fact $1 = \bar{1}$, $8 = \bar{8}$. The reduction of the product $6 \times 3$ proceeds along similar lines. The totally symmetric part of

$$T_{\{\alpha\beta\}\gamma} \equiv T_{\{\alpha\beta\}} \xi_\gamma \tag{6.14}$$

may be written down directly; it is $\frac{1}{6}$ of the sum of all permutations of the three indices. The part which is antisymmetric in $\beta$ and $\gamma$ is

$$T_\alpha^\delta = T_{\{\alpha\beta\}\gamma} \cdot \varepsilon_{\beta\gamma\delta}. \tag{6.15}$$

The trace of this mixed tensor is again zero, which shows that it belongs to the representation 8.

Instead of (15), we now consider the combinations

$$T_i = \tfrac{1}{2} \operatorname{trace} (T \cdot \lambda^i) = \tfrac{1}{2} T_\alpha^\beta \lambda_{\alpha\beta}^i = \tfrac{1}{2} \eta^\alpha \lambda_{\alpha\beta}^i \xi_\beta. \tag{6.16}$$

Under infinitesimal unitary transformations, $\xi$ and $\eta$ are multiplied by $1 + i\omega^k\lambda^k$ and $1 - i\omega^k\lambda^k$, respectively. The effect on the combination (16) is

$$\begin{aligned} T_i' &= \eta^{\alpha'}(\delta_{\alpha\alpha'} - i\omega^k\lambda^k_{\alpha\alpha'})\tfrac{1}{2}\lambda^i_{\alpha\beta}(\delta_{\beta\beta'} + i\omega^k\lambda^k_{\beta\beta'})\xi_{\beta'} \\ &= \eta^\alpha \cdot \tfrac{1}{2}\lambda^i_{\alpha\beta}\xi_\beta - \tfrac{1}{2}i\omega^k\eta^\alpha[\lambda^k, \lambda^i]_{\alpha\beta}\xi_\beta = T_i + 2\omega^k f_{kij} T_j. \end{aligned} \tag{6.17}$$

Comparing this with (1.11), we find the result that in the form (16) of the 8-dimensional representation, the generators are related to the structure constants:

$$(T_k)_{ij} = -if_{kij}. \tag{6.18}$$

APPENDIX C

# Units and Particle Tables

## C-1 Units and Fundamental Constants, Leptons

The units $1 = c = \hbar$ are used in this book:

$$\begin{aligned} 1 &= 2.997925 \times 10^{10}\ \text{cm/sec} = 6.58218 \times 10^{-22}\ \text{MeV sec} \\ &= 1.054592 \times 10^{-27}\ \text{erg sec}. \end{aligned} \tag{1.1}$$

There is then only one dimension left. This is conveniently taken as the MeV or GeV,

$$1\ \text{GeV} = 10^3\ \text{MeV} = 10^9\ \text{eV} \tag{1.2}$$

For example, if the lifetime $\Gamma^{-1}$ of an unstable particle is expressed in sec, the corresponding width in MeV is $6.582 \times 10^{-22}/\Gamma^{-1}$. By multiplying $c$ and $\hbar$ in (1), we get

$$1 = 1.973289 \times 10^{-11}\ \text{MeV cm} = 197.329\ \text{MeV fm} \tag{1.3}$$

$$1\ \text{fm} = 10^{-13}\ \text{cm} = 10^{-5}\ \text{A} = 5.0687\ \text{GeV}^{-1} \tag{1.4}$$

For cross sections one uses the millibarn

$$\begin{aligned} \text{GeV}^{-2} &= 0.38935\ \text{mb}, \\ 1\ \text{mb} &= 10^{-27}\ \text{cm}^2 = 0.1\ \text{fm}^2 = 10^3\ \mu\text{b} = 10^6\ \text{nb} \end{aligned} \tag{1.5}$$

For electromagnetic quantities we use cgs units. The elementary charge $e > 0$ is measured by the Coulomb potential $Z_a Z_b \alpha/r$

$$e = 0.30282, \qquad \alpha = \frac{e^2}{4\pi} = 137.036^{-1} \text{ (Taylor 1969)} \tag{1.6}$$

For lengths and cross sections in quantum electrodynamics, one uses the electron radius,

$$r_e = \alpha m_e^{-1}, \qquad r_e^2 = (2.81794 \text{ fm})^2 = 79.4078 \text{ mb}, \tag{1.7}$$

and for extreme relativistic energies one uses radiation lengths (see 8-3). Occasionally the pion Compton wave length is used in hadronic scattering:

$$(m_{\pi^+})^{-1} = 1.414 \text{ fm} \approx 2^{\frac{1}{2}} \text{ fm}. \tag{1.8}$$

In thermonuclear reactions, one needs Boltzmann's constant

$$k_B = 1.38062 \times 10^{-16} \text{ erg/°K} = 0.86171 \times 10^{-4} \text{ eV/°K} \tag{1.9}$$

which follows from the last two numbers in (1).

**Table C-1** Masses, lifetimes and magnetic moments of the charged leptons. The bracketed digit indicates the experimental uncertainty in the last digit of the preceding number, e.g., $m(e\mp) = (0.511003 \pm 0.000001)$ MeV. In this and the following tables, experimental numbers are taken from the Particle Data Group (1978), unless stated differently. Theoretical values are denoted by (th). In the charged weak currents, each charged lepton has its own neutral partner lepton (neutrino: $\nu_e$, $\nu_\mu$, $\nu_\tau$), which is massless within experimental errors ($m(\nu_e) < 60$ eV, $m(\nu_\mu) < 0.57$ MeV, $m(\nu_\tau) < 300$ MeV).

| Symbol | m[GeV] | m²[GeV²] | $\Gamma^{-1}[10^{-6}$ sec] | Magn. moment |
|---|---|---|---|---|
| $e^\mp$ | $0.511003(1) \times 10^{-3}$ | $2.61124 \times 10^{-7}$ | Stable | 1.0011596524(2) |
| $\mu^\mp$ | 0.1056595(2) | 0.01116393 | 2.19713(8) | 1.00116592(1) |
| $\tau^\mp$ | 1.782(7)[1]) | 3.176 | $2.8 \times 10^{-7}$ (th)[2]) | 1.0012 (th) |

1. From Bacino et al. (1978).

2. Flügge (1979) quotes the following decay channels and fractions: $\nu_\tau e\bar{\nu}$ 0.17, $\nu_\tau \mu\bar{\nu}_\mu$ 0.16, $\nu_\tau \pi$ 0.09(2), $\nu_\tau \rho$ 0.24(9), $\nu_\tau A_1$ 0.10(2), other 3 + more meson decays 0.28(4).

**Table C-2** The pseudoscalar mesons (see text to table C-1)

| Name | $m$ [GeV] / $m^2$ [GeV$^2$] | $\Gamma^{-1}$ [sec] / $\Gamma$ [eV] | Decay channel | Fraction [percent] | $\Gamma_{\text{partial}}$ [eV] | $p$ or $p_{\max}$ [MeV] |
|---|---|---|---|---|---|---|
| $\pi^\pm$ | 0.139567(1) | $2.603(2) \cdot 10^{-8}$ | $\mu\nu$ | 99.975 | $2.528 \times 10^{-8}$ | 29.79 |
| | 0.0194789 | $2.529 \cdot 10^{-8}$ | $e\nu$ | 0.0127(7) | $3.21 \times 10^{-12}$ | 69.78 |
| | | | $\mu\nu\gamma$ | 0.0124(25) | $3.14 \times 10^{-12}$ | 29.79 |
| $m_{\pi^\pm} - m_{\pi^0} = 4.604(4)$ MeV | | | $\pi^0 e\nu$ | $1.02(7) \cdot 10^{-6}$ | $2.58 \times 10^{-16}$ | 4.58 |
| $\pi^0$ | 0.134963 | $0.828 \cdot 10^{-16}$ | $\gamma\gamma$ | 98.83 | 7.86 | 67.48 |
| | 0.0182149 | 7.95(55) | $\gamma e\bar{e}$ | 1.17 | 0.093 | 67.48 |
| $\theta_{\pi\eta} = 0.0107$, using (4-15.10) | | | $e\bar{e}e\bar{e}$ | 0.0033 | 0.00026 | 67.4 |
| $K^\pm$ | 0.243707 | $1.237(3) \cdot 10^{-8}$ | $\mu\nu$ | 63.61(16) | $3.385 \cdot 10^{-8}$ | 235.5 |
| | | $5.321 \cdot 10^{-8}$ | $\pi^\pm\pi^0$ | 21.05(14) | $1.120 \cdot 10^{-8}$ | 205 |
| | | | $\pi^\pm\pi^\pm\pi^\mp$ | 5.59(3) | $2.974 \cdot 10^{-9}$ | 125 |
| | | | $\pi^\pm\pi^0\pi^0$ | 1.73(5) | $0.921 \cdot 10^{-9}$ | 133 |
| | | | $\pi^0 e\nu$ | 4.82(5) | $2.56 \cdot 10^{-9}$ | 228 |
| | | | $\pi^0\mu\nu$ | 3.20(9) | $1.70 \cdot 10^{-9}$ | 215 |
| | | | $\pi^0 e\nu\gamma$ | 0.037(14) | $1.97 \cdot 10^{-11}$ | 228 |
| $m_{K^+} - m_{K^0} = -4.0(1)$ MeV | | | $\pi^\pm\pi^0\gamma$ | 0.027(2) | $1.44 \cdot 10^{-11}$ | 205 |
| | | | $e\nu$ | 0.00154(9) | $8.19 \cdot 10^{-13}$ | 247 |
| $K_S$ | 0.4977(1) | $0.893(2) \cdot 10^{-10}$ | $\pi^+\pi^-$ | 68.67(25) | $5.06 \cdot 10^{-6}$ | 206 |
| | 0.24768 | $7.37 \cdot 10^{-6}$ | $\pi^0\pi^0$ | 31.33(25) | $2.31 \cdot 10^{-6}$ | 209 |
| | | | $\pi^+\pi^-\gamma$ | 0.20(4) | $1.5 \cdot 10^{-8}$ | 206 |

| | | | | | | |
|---|---|---|---|---|---|---|
| $K_L$ | 0.4977(1) | $5.18(4) \cdot 10^{-8}$ | $\pi e \nu$ | 39.00(50) | $4.95 \cdot 10^{-9}$ | 229 |
| | 0.24768 | $1.270 \cdot 10^{-8}$ | $\pi \mu \nu$ | 27.00(50) | $3.43 \cdot 10^{-9}$ | 216 |
| | | | $\pi^0\pi^0\pi^0$ | 21.40(70) | $2.72 \cdot 10^{-9}$ | 139 |
| | | | $\pi^+\pi^-\pi^0$ | 12.25(18) | $1.56 \cdot 10^{-9}$ | 133 |
| | | | $\pi e \nu \gamma$ | 1.30(80) | $1.6 \cdot 10^{-10}$ | 229 |
| | | | $\pi^+\pi^-$ | 0.201(6) | $2.55 \cdot 10^{-11}$ | 206 |
| | | | $\pi^0\pi^0$ | 0.094(19) | $1.19 \cdot 10^{-11}$ | 209 |
| | | | $\gamma\gamma$ | 0.049(5) | $6.2 \cdot 10^{-12}$ | 249 |
| $m_L - m_S = 3.52(1) \cdot 10^{-6}$ eV | | | $\pi^+\pi^-\gamma$ | 0.006(2) | $7. \cdot 10^{-14}$ | 206 |
| $\eta$ | 0.5488(6) | $7.7 \cdot 10^{-19}$ | $\gamma\gamma$ | 38.0(10) | 323 | 274 |
| | 0.30118 | 850(120) | $\pi^0\pi^0\pi^0$ | 29.9(11) | 254 | 180 |
| | | | $\pi^0\gamma\gamma$ | 3.1(11) | 26 | 258 |
| | | charged channels 29.0(7)% { | $\pi^+\pi^-\pi^0$ | 23.6(6) | 201 | 175 |
| | | | $\pi^+\pi^-\gamma$ | 4.89(13) | 42 | 236 |
| | | | $e\bar{e}\gamma$ | 0.61 | 5.2 | 274 |
| | | | $\pi^+\pi^- e\bar{e}$ | 0.03 | 0.26 | 236 |
| | | | $\mu\bar{\mu}$ | 0.0022(8) | 0.02 | 253 |
| $\eta'$ or $X$ | 0.9576(3) | — | $\eta\pi^+\pi^-$ | 43(1) | | 231 |
| | 0.9170 | ~0.5 MeV | $\eta\pi^0\pi^0$ | 23(1) | | 238 |
| | | | $\rho^0\gamma$ | 30(2) | | 166 |
| | | | $\omega\gamma$ | 2.1(4) | | 159 |
| | | | $\gamma\gamma$ | 2.0(3) | | 479 |

**Table C-3** The vector mesons. (See text to table C-1.) The decay constants $f_V$ and $G$ are explained in section 4-9, finite-width corrections to $f_V$ are given in (4-14.22).

| Name | $m$ [GeV]<br>$m^2$ [GeV$^2$] | $\Gamma$ [MeV]<br>$m\Gamma$ [eV$^2$] | Decay channel | Fraction [percent] | $\Gamma_{part}$ [MeV] | $G^2/4\pi$<br>$f_V^2/4\pi$ | $p$ or $p_{max}$ [MeV] |
|---|---|---|---|---|---|---|---|
| $\rho^\pm$ | 0.766(3) | 148(4) | $\pi^\pm\pi^0$ | 99.98 | 148 | (2.84) | 357 |
| | 0.587 | 0.113 | $\pi^\pm\gamma$ | 0.024(7) | 0.036 | | 370 |
| $\rho^0$ | 0.774(3) | 150(3) | $\pi^+\pi^-$ | 99.98 | 150 | 2.84 | 361 |
| | 0.599 | 0.116 | $\pi^0\gamma$ | 0.02 | 0.030 | | 375 |
| | | | $e^+e^-$ [1] | 0.0043(5) | 0.0065 | 2.26 | 387 |
| $\omega$ | 0.7827(3) | 10.0(4) | $\pi^+\pi^0\pi^-$ | 89.8(6) | 7.03 | | 327 |
| | 0.6126 | 0.00783 | $\pi^0\gamma$ | 8.8(5) | 0.69 | 0.66 GeV$^{-2}$ | 380 |
| | | | $\pi^+\pi^-$ | 1.3(3) | 0.13 | 0.00244 | 365.6 |
| | | | $\pi^0 e\bar{e}$ | 0.075 | 0.0059 | | 380 |
| | | | $e\bar{e}$ [1] | 0.0076(17) | 0.0006 | 18.4 | 391 |
| $\phi$ | 1.0197(3) | 4.1(2) | $K^+K^-$ | 48.6(12) | 1.99 | (1.50) | 128 |
| | 1.0398 | 0.00418 | $K_L K_S$ | 35.1(12) | 1.44 | 1.50 | 111 |
| | | | $\pi^+\pi^-\pi^0(\rho\pi)$ | 14.7(4) | 0.60 | | 462 |
| | | | $\eta\gamma$ | 1.6(2) | 0.066 | | 362 |
| | | | $\pi^0\gamma$ | 0.14(5) | 0.0057 | | 501 |
| | | | $e^+e^-$ [1] | 0.032(2) | 0.0013 | 14.3 | 510 |
| | | | $\mu^+\mu^-$ | 0.025(3) | 0.0010 | | 499 |
| $K^{*\pm}$ | 0.8922(5) | 49.4(18) | $K^0\pi^+$, $\bar{K}^0\pi^-$ | 66 | 32.6 | $2\times 0.84$ | 286 |
| | 0.7960 | 0.0441 | $K^\pm\pi^0$ | 34 | 16.7 | 0.84 | 290 |
| | | | $K^\pm\gamma$ | 0.15(7) | 0.07 | | 310 |
| $K^{*0}$ | 0.8963(6) | 50.4 | $K^\pm\pi^\mp$ | 67 | 33.8 | $2\times 0.84$ | 291 |
| $\bar{K}^{*0}$ | 0.80335 | 0.0452 | $K^0\pi^0$, $\bar{K}^0\pi^0$ | 33 | 16.6 | 0.84 | 290 |
| $m(K^{*0}) - m(K^{*\pm}) = 4.1(6)$ MeV | | | $K^0\gamma$, $\bar{K}^0\gamma$ | | | | 310 |

[1]) From $e\bar{e}$-collisions. The decays $V\to\mu\bar{\mu}$ are less precisely measured. Theoretically, $\Gamma(V\to\mu\bar{\mu}) = 1.00\Gamma(V\to e\bar{e})$ (see table 4-9).

**Table C-4** The baryon octet (see text to table C-1).

| Name | $m$ [GeV]<br>$m^2$ [GeV$^2$] | $\Gamma^{-1}$ [sec]<br>$\Gamma$ [eV] | Decay channel | Fraction [percent] | $\Gamma_{partial}$ [eV] | $\mu$ [e/2$m_p$] | $p$ or $p_{max}$ [MeV] |
|---|---|---|---|---|---|---|---|
| $p$ | 0.938280(3) | $> 2 \cdot 10^{30} y$ | | | | | |
| | 0.880369 | | | | | 2.792846(1) | |
| $n$ | 0.939573(3) | 918(14) | $pe\bar{\nu}$ | 100 | $7.17 \cdot 10^{-19}$ | | 1.19 |
| | 0.882797 | $7.17 \cdot 10^{-19}$ | | | | −1.91315(7) | |
| $m_n - m_p = 1.29343(4)$ MeV | | | | | | | |
| $\Lambda$ | 1.11560(5) | $2.63(2) \cdot 10^{-10}$ | $p\pi^-$ | 64.2 | $1.61 \cdot 10^{-6}$ | | 100.5 |
| | 1.24456 | $2.50 \cdot 10^{-6}$ | $n\pi^0$ | 35.8 | $0.90 \cdot 10^{-6}$ | −0.614(5)[1] | 103.9 |
| | | | $pe\bar{\nu}$ | 0.081(3) | $2.0 \cdot 10^{-9}$ | | 163.2 |
| $\theta_{\Lambda\Sigma} = 0.0113$, using (4-15.9) | | | $p\mu\bar{\nu}$ | 0.016(4) | $3.9 \cdot 10^{-10}$ | | 130.9 |
| $\Sigma^+$ | 1.18937(6) | $0.802(5) \cdot 10^{-10}$ | $p\pi^0$ | 51.6 | $4.25 \cdot 10^{-6}$ | | 189 |
| | 1.40115 | $8.21 \cdot 10^{-6}$ | $n\pi^+$ | 48.4 | $3.98 \cdot 10^{-6}$ | | 185 |
| | | | $p\gamma$ | 0.12(2) | $0.99 \cdot 10^{-8}$ | 2.6(4) | 225 |
| $m_0 - m_+ = 3.11$ MeV | | | $n\pi^+\gamma$ | 0.09(1) | $7.7 \cdot 10^{-9}$ | | 185 |
| $\Sigma^0$ | 1.19247(8) | $0.58(13) \cdot 10^{-19}$ | $\Lambda\gamma$ | 99.5 | 11300 | $\mu_{\Lambda\Sigma} =$ | 74.4 |
| | 1.42198 | 11000 | $\Lambda e\bar{e}$ | 0.545 | 62 | 1.90(26) | 74.4 |
| $\Sigma^-$ | 1.19735(6) | $1.48(2) \cdot 10^{-10}$ | $n\pi^-$ | 99.8 | $4.43 \cdot 10^{-6}$ | | 193 |
| | 1.43365 | $4.44 \cdot 10^{-6}$ | $ne\bar{\nu}$ | 0.108(4) | $4.80 \cdot 10^{-9}$ | | 230 |
| | | | $n\mu\bar{\nu}$ | 0.045(4) | $2.0 \cdot 10^{-9}$ | −1.4(4) | 210 |
| $m_- - m_0 = 4.88$ MeV | | | $n\pi^-\gamma$ | 0.046(6) | $2.0 \cdot 10^{-9}$ | | 193 |
| $\Xi^0$ | 1.3149(6) | $3.0(1) \cdot 10^{-10}$ | $\Lambda\pi^0$ | 100 | $2.22 \cdot 10^{-6}$ | • | 135 |
| | 1.7290 | $2.22 \cdot 10^{-6}$ | $\Lambda\gamma$ | 0.5(5) | $1.1 \cdot 10^{-8}$ | — | 184 |
| $\Xi^-$ | 1.3213(1) | $1.65(2) \cdot 10^{-10}$ | $\Lambda\pi^-$ | 100 | $3.98 \cdot 10^{-6}$ | | 139 |
| | 1.7459 | $3.98 \cdot 10^{-6}$ | $\Lambda e\bar{\nu}$ | 0.07(2) | $2.8 \cdot 10^{-9}$ | −.9(8) | 190 |
| $m_- - m_0 = 6.4(6)$ MeV | | | $\Lambda\mu\bar{\nu}$ | 0.03(3) | $1.2 \cdot 10^{-9}$ | | 163 |

1. From Schachinger et al. (1978).

**Table C-5** The baryon decuplet. Only numbers with bracketed errors are taken from experiment (Particle Data Group 1978). All other numbers are either unknown or measured less precisely. They are calculated from broken $SU(2)$-symmetry (section 4-15 and eqs. 4-9.17, 4-11.13, see also Myhrer and Pilkuhn 1976). $\Gamma(\Delta \to N\gamma)$ is calculated from (7-4.15) and from fig. 7-4.1, the photonic decay rates of $\Sigma^*$ and $\Xi^*$ use $SU(3)$-invariance.

| Name | $m$ [GeV]<br>$m^2$ [GeV$^2$] | $\Gamma$ [MeV]<br>$m\Gamma$ [GeV$^2$] | Decay channel | Fraction [percent] | $\Gamma_{part}$ [MeV] | $p$ [MeV] |
|---|---|---|---|---|---|---|
| $\Delta^{++}$ | 1.2311(4)<br>1.5156 | 111.5(4)<br>0.137 | $p\pi^+$ | 100 | 111.5 | 226 |
| $\Delta^+$ | 1.2303<br>1.5136 | 113.5<br>0.140 | $p\pi^0$<br>$n\pi^+$<br>$p\gamma$ | 67<br>32<br>0.63(8) | 76<br>37<br>0.7 | 228<br>225<br>257 |
| $\Delta^0$ | 1.2325(6)<br>1.5191 | 115.7<br>0.142 | $p\pi^-$<br>$n\pi^0$<br>$n\gamma$ | 33<br>66<br>0.6 | 38<br>77<br>0.7 | 228<br>229 |
| $\Delta^-$ | 1.2359<br>1.5274 | 116.0<br>0.143 | $n\pi^-$ | 100 | 116 | 230 |
| $\Sigma^{*+}$ | 1.3830(5)<br>1.9127 | 35(2)<br>0.0485 | $\Lambda\pi^+$<br>$\Sigma^+\pi^0$<br>$\Sigma^0\pi^+$<br>$\Sigma^+\gamma$ | 87<br>6.6<br>5.6<br>0.6 | 31<br>2.3<br>2.0<br>0.2 | 205.6<br>128.9<br>120.5<br>180 |
| $\Sigma^{*0}$ | 1.385<br>1.9182 | 37<br>0.0512 | $\Lambda\pi^0$<br>$\Sigma^+\pi^-$<br>$\Sigma^-\pi^+$<br>$\Lambda\gamma$<br>$\Sigma^0\gamma$ | 88<br>6.0<br>4.8<br>0.7<br>0.1 | 33<br>2.2<br>1.8<br>0.3<br>0.04 | 210<br>127<br>117<br>243<br>179 |
| $\Sigma^{*-}$<br>$m(\Sigma^{*-}) - m(\Sigma^{*+}) = 6$ MeV | 1.3890(6)<br>1.9293 | 38<br>0.0528 | $\Lambda\pi^-$<br>$\Sigma^-\pi^0$<br>$\Sigma^0\pi^-$ | 89<br>5.3<br>5.5 | 34<br>2.0<br>2.1 | 211<br>126<br>128 |
| $\Xi^{*0}$ | 1.5318(3)<br>2.3464 | 9.1(5)<br>0.0139 | $\Xi^0\pi^0$<br>$\Xi^-\pi^+$<br>$\Xi^0\gamma$ | 35<br>62<br>2.7 | 3.2<br>5.7<br>0.24 | 158<br>146<br>202 |
| $\Xi^{*-}$ | 1.5350(6)<br>2.3562 | 9.9<br>0.0151 | $\Xi^-\pi^0$<br>$\Xi^0\pi^-$ | 33<br>67 | 3.25<br>6.65 | 154<br>158 |
| $\Omega^-$<br>$\Gamma^{-1} = 1.0(2) \times 10^{-10}$ sec[1] | 1.6722(4)<br>2.7963 | $6.6 \times 10^{-12}$ | $\Lambda K^-$ [2]<br>$\Xi^0\pi^-$ [2]<br>$\Xi^-\pi^0$ [2] | ~72<br>~17<br>~ 8 | 5<br>1.1<br>0.5 | 211<br>294<br>290 |

[1]) Including Hemingway et al. (1978).
[2]) Assuming $\Delta I = \frac{1}{2}$.

**Table C-6** Masses, total and partial decay widths of the tensor mesons. (See text to table C-1.)

| Name | $m$ [GeV]<br>$m^2$ [GeV$^2$] | $\Gamma$ [MeV]<br>$m\Gamma$ [GeV$^2$] | Decay channel | | Fraction [percent] | $\Gamma_{part}$ [MeV] | $p$ or $p_{max}$ [MeV] |
|---|---|---|---|---|---|---|---|
| $A_2^{\pm}$ | 1.312(5) | 102(5) | $\rho^{\pm}\pi^0$ | | 35.1 | 35.8 | 411 |
| | 1.721 | 0.134 | $\rho^0\pi^{\pm}$ | | 35.1 | 35.8 | 411 |
| | | | $\eta\pi^{\pm}$ | | 14.4 | 14.7 | 531 |
| | | | $\omega\pi^0\pi^{\pm}$ | | 10.6 | 10.8 | 356 |
| | | | $KK^{\pm}$ | | 4.7 | 4.8 | 430 |
| | | | $\pi^{\pm}\gamma$ | | 0.5 | 0.5 | 649 |
| $A_2^0$ | 1.312(5) | 102(5) | $\rho^+\pi^-$ | | 35.1 | 35.8 | 411 |
| | 1.721 | 0.134 | $\rho^-\pi^+$ | | 35.1 | 35.8 | 411 |
| | | | $\eta\pi^0$ | | 14.4 | 14.7 | 531 |
| | | | $\omega\pi^+\pi^-$ | | 10.6 | 10.8 | 356 |
| | | | $K^+K^-$ | | 2.4 | 2.4 | 430 |
| | | | $K_SK_S + K_LK_L$ | | 2.3 | 2.4 | 430 |
| $f$ | 1.271(5) | 180(20) | $\pi^+\pi^-$ | | 53.5 | 96.3 | 620 |
| | 1.615 | 0.23 | $\pi^0\pi^0$ | | 26.8 | 48.2 | 621 |
| | | | $K^+K^-$ | | 1.6 | 2.9 | 400 |
| | | | $K_SK_S + K_LK_L$ | | 1.5 | 2.7 | 395 |
| | | | $\pi^+\pi^+\pi^-\pi^-$ | | 2.8 | 5.0 | 557 |
| | | | $\pi^+\pi^-\pi^0\pi^0$ | | seen | | 560 |
| $f'$ | 1.516(10) | 65(10) | $K^+K^-$ | | ~30 | | 575 |
| | 2.298 | 0.10 | $K_SK_S + K_LK_L$ | | ~30 | | 572 |
| | | | $\eta\eta$ | | | | 523 |
| | | | $\pi\pi$ | | 1.2(4) | | 745 |
| | | | $\underline{K_N^+}$ | $\underline{K_N^0}$ | | | |
| $K_N^{\pm}$ or $K^{*\pm}$ (1430) | | | $K^0\pi^+$ | $K^+\pi^-$ | 32.7 | 32.7 | 623 |
| | 1.430(5) | 100(10) | $K^+\pi^0$ | $K^0\pi^0$ | 16.4 | 16.4 | 623 |
| | 2.045 | 0.14 | $K^{*0}\pi^+$ | $K^{*+}\pi^-$ | 18.0 | 18.0 | 424 |
| | | | $K^{*+}\pi^0$ | $K^{*0}\pi^0$ | 9.0 | 9.0 | 424 |
| $K_N^0$, $\bar{K}_N^0$ or $K^{*0}$ (1430), | | | $K^0\rho^+$ | $K^+\rho^-$ | 4.4 | 4.4 | 327 |
| $\bar{K}^{*0}$ (1430) | | | $K^+\rho^0$ | $K^0\rho^0$ | 2.2 | 2.2 | 327 |
| | | | $K^*\pi\pi$ | $K^*\pi\pi$ | 11.2 | 11.2 | 374 |
| | | | $K^+\omega$ | $K^0\omega$ | 3.7 | 3.7 | 320 |
| | | | $K^+\eta$ | $K^0\eta$ | 2.5 | 2.5 | 492 |

**Table C-7** The new bosons (see text to table C-1). $D^0$, $\bar{D}^0$, $D^\pm$ and $F^\pm$ decay weakly; $J/\psi$, $\chi(0^+, 1^+, 2^+)$, $\psi'$ and $\Upsilon$, $\Upsilon'$, $\Upsilon''$ are narrow resonances. Most of these states have many appreciable channels, of which only the first few are included.

| Name ($J^{\mathscr{P}}$) | $m$ [GeV]<br>$m^2$ [GeV$^2$] | $\Gamma$ [MeV] | Decay channel | Fraction [percent] | Decay channel | Fraction [percent] |
|---|---|---|---|---|---|---|
| $D^0$, $\bar{D}^0$<br>$(0^-)$ | 1.8633(9)<br>3.472 | | $D^0$: $K^-\pi^+\pi^0$<br>$K^0\pi^+\pi^-$ | 12(6)<br>4(1) | $e^\pm \ldots$<br>$\mu^\pm \ldots$ | 10(1)<br>$\sim e^\pm$ (th) |
| $D^\pm$<br>$(0^-)$ | 1.8683(9)<br>3.491 | $\sim 3 \times 10^{-9}$ | $D^+$: $K^-\pi^+\pi^+$<br>$\bar{K}^0\pi^+$ | 4(1)<br>2(1) | $e^\pm \ldots$<br>$\mu^\pm \ldots$ | 10(1)<br>$\sim e^\pm$ (th) |
| $D^{*0}$, $\bar{D}^{*0}$<br>$(1^-)$ | 2.006(2)<br>4.024 | | $D^{*0}$: $D^0\pi^0$<br>$D^0\gamma$ | 55(15)<br>45(15) | | |
| $D^{*\pm}$<br>$(1^-)$ | 2.0086(15)<br>4.034 | | $D^{*+}$: $D^+\pi^0$<br>$D^+\gamma$ } | 40(15) | $D^0\pi^+$ | 60(15) |
| $F^\pm$<br>$(0^-)$ | 2.0395(10)<br>4.160 | | $\eta\pi^\pm$ [1]) | | | |
| $F^{*\pm}$<br>$(1^-)$ | 2.15(5)<br>4.62 | | $F^\pm\gamma$ [1]) | | | |
| $J/\psi$<br>$(1^-)$ | 3.097(2)<br>9.591 | 0.067<br>(12) | $\pi^+\pi^+\pi^-\pi^-\pi^0$<br>$(\pi^+\pi^-)^3\pi^0$ | 3.7(5)<br>2.9(7) | hadrons, total<br>$e\bar{e}$, $\mu\bar{\mu}$ | 86(2)<br>7(1) each |

Table C-7 (*continued*)

| | | | | | | |
|---|---|---|---|---|---|---|
| $\chi$<br>$(0^+)$ | 3.413<br>11.65 | | $\gamma J/\psi$<br>$(\pi^+\pi^-)^3$ | 3(1)<br>2.0(7) | $\pi^+\pi^+\pi^-\pi^-$<br>$\pi^+\pi^-K^+K^-$ | 4.7(9)<br>4(1) |
| $P_c$ or $\chi$<br>$(1^+)$ | 3.508<br>12.31 | | $\gamma J/\psi$<br>$(\pi^+\pi^-)^3$ | 23.4(8)<br>2.4(8) | $\pi^+\pi^+\pi^-\pi^-$<br>$\pi^+\pi^-K^+K^-$ | 1.5(6)<br>0.9(4) |
| $\chi$<br>$(2^+)$ | 3.554<br>12.63 | | $\gamma J/\psi$<br>$(\pi^+\pi^-)^3$ | 16(3)<br>1.1(7) | $\pi^+\pi^+\pi^-\pi^-$<br>$\pi^+\pi^-K^+K^-$ | 2.3(6)<br>2.0(6) |
| $\psi'$<br>$(1^-)$ | 3.686(3)<br>13.59 | 0.23(6) | $\pi^+\pi^-J/\psi$<br>$\pi^0\pi^0J/\psi$ | 33(3)<br>17(2) | hadrons, total<br>$e\bar{e}$, $\mu\bar{\mu}$ | 98.1(3)<br>0.9(1) each |
| $\psi''$<br>$(1^-)$ | 3.771(4)<br>14.22 | 26(5) | $D^+D^-$<br>$D^0\bar{D}^0$ | ~50<br>~50 | $e\bar{e}$ | 0.0010(2) |
| $(1^-)$ | $\psi$ 4.040<br>$\psi$ 4.156<br>$\psi$ 4.414 | 55<br>50<br>33 | $D\bar{D}$, $D\bar{D}^* + D^*\bar{D}$, $D^*\bar{D}^*$ | | $e\bar{e}$<br>$e\bar{e}$<br>$e\bar{e}$ | 0.0014(4)<br>0.0009(3)<br>0.0010(3) |
| $(1^-)$ | $\Upsilon$ 9.46(1)<br>$\Upsilon'$ 10.01(1)<br>$\Upsilon''$ 10.41(5) | 0.05<br><br>[2] | | | $\Gamma(e\bar{e}) =$ | 1.3(4) keV<br>0.3(1) " |

[1]) Brandelik et al. (1977, 1979). [2]) Ueno et al. (1979).

# References*

Aachen-Berlin-Birmingham-Bonn-Hamburg-London (IC)-München collaboration (1964): *Nuovo Cim.* **34**, 495 (160)
E. S. Abers and B. W. Lee, 1973, *Phys. Reports* **9**, 1 (258)
M. Abramowitz and I. A. Stegun, 1965, *Handbook of Mathem. Functions*, Dover (377)
S. L. Adler, 1965, *Phys. Rev. Letters* **14**, 1051 (311, 312)
M. Ademollo and R. Gatto, 1964, *Phys. Letters* **13**, 264 (247)
A. I. Akhiezer and V. B. Berestetskii, 1965, "*Quantum Electrodynamics*", Interscience (ix, 117)
M. Alston et al., 1961, *Phys. Rev. Letters* **6**, 300 (169)
L. W. Alvarez, 1938, *Phys. Rev.* **53**, 606 and **54**, 486 (209)
C. D. Anderson, 1933, *Phys. Rev.* **43**, 491 (103, 136)
H. L. Anderson, T. Fujii, R. H. Miller, and L. Tau, (1959): *Phys. Rev. Letters* **2**, 53 (214)
T. Appelquist, R. M. Barnett, and K. Lane, 1978, *Ann, Rev. Nucl. Part. Sci.* **28**, 387 (194)
J. J. Aubert et al., 1974, *Phys. Rev. Letters* **33**, 1404 (205)
J.-E. Augustin et al., 1974, *Phys. Rev. Letters* **33**, 1406 (205)
P. R. Auvil and J. J. Brehm, *Phys. Rev.* **145**, 1152 (67)
P. Babu, F. T. Gilman, and M. Suzuki, 1967, *Phys. Letters* 24B, 65 (351)
W. Bacino et al., 1978, *Phys. Rev. Letters* **41**, 13 (399)
H. Bacry, J. Nuyts, and L. van Hove, 1965, *Nuovo Cim.* **35**, 510 (178)
W. L. Bade and H. Jehle, 1953, *Rev. Mod. Phys.* **25**, 714 (18)
A. Badertscher et al., 1978, *Phys. Letters* **79** B, 371 (233)
J. N. Bahcall and R. M. May, 1969, *Astrophys. J.* **155**, 501 (41)

* The numbers in parentheses appearing at the end of the reference indicate the pages of this book.

J. Bailey et al., 1972, *Nuovo Cim.* A9, 369 (4)

D. Bailin, 1971, *Reports on Progr. in Physics* **34**, 491 (245); 1977, *Weak Interactions*, Sussex Univ. Press (216)

P. Baillon et al., 1976, *Nucl. Phys.* **B105**, 365 (350)

W. Bambynek et al., 1977, *Rev. Mod. Phys.* **49**, 77 (216)

V. Bargmann, L. Michel, and V. Telegdi, 1959, *Phys. Rev. Letters* **2**, 435 (37)

A. D. Barut, 1967, *The Theory of the Scattering Matrix*, Macmillan Comp. (269)

M. A. Bég and A. Sirlin, 1974, *Ann. Rev. Nucl. Sci.* **24**, 379 (258)

F. A. Behrends, A. Donnachie, and D. L. Weaver, 1967, *Nucl. Phys.* B**4**, 1 and 54 (336)

F. A. Behrends and A. Donnachie, 1975, *Nucl. Phys.* B**84**, 342 (336)

H. Behrens and W. Bühring, 1971, *Nucl. Phys.* A **162**, 111 (229, 230)

H. Behrens and J. Jänecke, 1969, *Numerical Tables for Beta Decay and Electron Capture*, Landolt-Börnstein new series Vol. I/4 (Springer, Berlin) (211)

H. Behrens and L. Szybisz, 1974, Nucl. Phys. A **223**, 268 (220)

J. Benecke, T. T. Chou, C. N. Yang, and E. Yen, 1969, *Phys. Rev.* **188**, 2159 (355)

D. Berger, 1972, Diplomarbeit (Karlsruhe, unpublished) (144)

K. E. Bergkvist, 1972, *Nucl. Phys.* B **39**, 317, 371 (210)

L. Bertanza et al., 1962, *Phys. Rev. Letters* **9**, 180 (169)

H. A. Bethe, 1949, *Phys. Rev.* **76**, 38 (321)

H. A. Bethe and W. Heitler, 1934, *Proc. Roy. Soc.* (*London*) A **146**, 83 (372)

H. A. Bethe and E. E. Salpeter, 1951, *Phys. Rev.* **84**, 1232 (132)

H. A. Bethe and L. C. Maximon, 1954, *Phys. Rev.* **93**, 768 (373)

C. P. Bhalla and M. E. Rose, 1961, Oak Ridge report ORNL-2954 (231)

J. D. Bjorken, 1969, *Phys. Rev.* **179**, 1547 (267)

J. D. Bjorken and S. D. Drell, 1965, *Relativistic Quantum Mechanics, Relativistic Quantum Fields*, McGraw Hill (ix, 117)

R. Blankenbecler and L. F. Cook, 1960, *Phys. Rev.* **119**, 1745 (356)

R. J. Blin-Stoyle, 1973, *Fundamental Interactions and the Nucleus* (North-Holland/American Elsevier) (140, 216, 238, 240, 241, 252)

N. N. Bogoliubov, A. A. Logunov, and I. T. Todorov, 1975, *Introduction to Aximatic Quantum Field Theory*, Benjamin (92)

W. Bozzoli et al., 1978, *Nucl. Phys.* B **144**, 317 (136)

R. Brandelik et al., 1977, *Phys. Lett.* 70 B, 132; 1979, *Phys. Lett.* 80 B, 412 (208, 407)

D. Branson, 1965, *Ann. of Phys.* **35**, 351 (292)

G. Breit and R. D. Harasz, 1967, in *High Energy Physics*, Ed. E. H. S. Burhop (Academic Press) (342)

G. Breit, 1929, *Phys. Rev.* **34**, 553 (73)

S. J. Brodsky, 1971, in *Atomic Physics and Astrophysics*, Vol. 1, Ed. M. Chretien (77, 134)

S. J. Brodsky and J. R. Primack, 1968, *Phys. Rev.* **174**, 2071; 1969, *Ann. of Phys.* **52**, 315 (77)

G. E. Brown and A. D. Jackson, 1976, *The Nucleon-Nucleon Interaction* (North Holland/American Elsevier) (342, 344)

L. S. Brown, W. J. Pardee, and R. D. Peccei, 1971, *Phys. Rev.* D4, 2801 (313)

K. M. Brueckner, 1952, *Phys. Rev.* **86**, 106 (185)

K. M. Brueckner, K. M. Watson, 1953, *Phys. Rev.* **92**, 1023 (347)

D. Bryman and C. Picciotto, 1978, *Rev. Mod. Phys.* **50**, 11 (232)

W. Bühring, 1965, *Nucl. Phys.* **61**, 110 (231); 1966, *Z. Physik* **192**, 13 (35)

N. Cabibbo, 1963, *Phys. Rev. Letters* **10**, 531 (245)
C. G. Callan, Jr., and D. J. Gross, 1968, *Phys. Rev. Letters* **22**, 156 (267)
A. A. Carter et al., 1971, *Nucl. Phys.* B **26**, 445 (144)
P. Carruthers, 1966, *Introduction to Unitary Symmetry*, Interscience (179)
L. Castillijo, R. H. Dalitz, and F. J. Dyson, 1956, *Phys. Rev.* **101**, 453 (301)
W. E. Caswell and G. P. Lepage, 1978, *Phys. Rev.* A**18**, 810 (133)
J. Chadwick, 1932, *Proc. Roy. Soc.* A **136**, 692 (136)
O. Chamberlain, E. Segré, C. Wiegand, and T. Ypsilantis, 1955, *Phys. Rev.* **100**, 947 (136)
G. F. Chew and S. C. Frautschi, 1961, *Phys. Rev. Letters* **8**, 41 (317)
G. F. Chew and F. E. Low, 1956, *Phys. Rev.* **113**, 1640 (329)
G. F. Chew, S. Mandelstam, 1960, *Phys. Rev.* **119**, 464 (301)
W. Chinowsky and J. Steinberger, 1954, *Phys. Rev.* **95**, 1561 (152)
H.-Y. Chiu, 1968, *Stellar Physics*, Vol. 1, Blaisdell (41)
L. M. Chounet, J.-M. Gaillard, and M. K. Gaillard, 1972, *Phys. Reports* 4C, 201 (245)
J. M. Christenson, J. W. Cronin, V. L. Fitch, and R. Turlay, 1964, *Phys. Rev. Letters* **13**, 138 (139)
S. Coleman and S. L. Glashow, 1961, *Phys. Rev. Letters* **6**, 423 (202)
S. Coleman and R. E. Norton, 1965, *Nuovo Cim.* **38**, 438 (356)
P. D. B. Collins, 1977, *An Introduction to Regge Theory and High Energy Physics*, Cambridge Univ. Press (315)
E. D. Commins, 1973, *Weak Interactions*, McGraw Hill (216)
E. U. Condon and G. H. Shortley, 1935, *The Theory of Atomic Spectra*, Cambridge Univ. Press (143)
I. Curie and F. Joliot, 1934, *Compt. Rend.* **198**, 254 (209)
R. E. Cutkosky, 1960, *J. Math. Phys.* **1**, 429 (269)
R. H. Dalitz, 1953, *Phil. Mag.* **44,** 1068 (159); 1957, *Rep. Progr. Phys.* **20,** 163 (139); 1961, *Rev. Mod. Phys.* **33,** 471; 1962, *Strange Particles and Strong Interactions*, Oxford Univ. Press (293)
G. Danby et al., 1962, *Phys. Rev. Letters* **9**, 36 (137)
M. Daum et al., 1978, *Phys. Letters* **74B**, 126 (4)
R. Davies, Jr., D. S. Harmer, and K. C. Hoffman, 1968, *Phys. Rev. Lett.* **20**, 1205 (232)
V. de Alfaro, S. Fubini, G. Furlan, and C. Rossetti, 1973, *Currents in Hadron Physics*, North-Holland (312)
P. de Baenst, 1970, *Nucl. Phys.* B 24, 633 (340)
C. W. De Jager, H. De Vries, and C. De Vries, 1974, *Atomic Data and Nucl. Tables* **14**, 479 (141)
J. J. de Swart, 1963, *Rev. Mod. Phys.* **35**, 916 (179)
E. De Vries and J. E. Jonker, 1968, *Nucl. Phys.* B **6**, 213 (25)
A. M. Diamant-Berger et al., 1976, *Phys. Letters* **62B**, 485 (233)
P. M. A. Dirac, 1928, *Proc. Roy. Soc.* A **117**, 610 (18)
G. Donaldson et al., 1974, *Phys. Rev.* D**9**, 2960 (248, 249)
L. Durand, III, 1964, *Phys. Rev.* **135**, B 310 (231)
F. Dydak et al., 1977, *Nucl. Phys.* B **118**, 1 (370)
R. J. Eden and J. Goldstone, 1963, *Nucl. Phys.* **49**, 33
R. J. Eden, P. V. Landshoff, D. I. Olive, and J. C. Polkinghorne, 1966, *The Analytic S-Matrix*, Cambridge University Press (269, 278)
T. Eichten et al., 1973, *Phys. Letters* **B46**, 274 (255)

S. D. Ellis and R. Stroynowski, 1977, *Rev. Mod. Phys.* **49**, 753 (266, 355)
M. Ericson, A. Figureau and C. Thévenet, 1973, *Phys. Letters* **45B**, 19 (245)
T. E. O. Ericson and M. Locher, 1970, *Nucl. Phys.* A **148**, 1 (289)
A. Erwin, R. March, W. Walker, and E. West, 1961, *Phys. Rev. Letters* **6**, 628 (169)
P. Estabrooks and A. D. Martin, 1974, *Nucl. Phys.* B**79**, 301 (323)
E. Fabri, 1954, *Nuovo Cim.* **11**, 479 (159)
G. Fäldt, 1972, *Nucl. Phys.* B **43**, 591 (368)
U. Fano, 1963, *Ann. Rev. Nucl. Sci.* **13**, 1 (367)
H. W. Fearing, E. Fischbach, and J. Smith, 1970, *Phys. Rev.* D **2**, 542 (249)
B. T. Feld, 1969, Models of Elementary Particles, Blaisdell Publ. Co. (192)
G. Feinberg and J. Sucher, 1970, *Phys. Rev.* **A2,** 2395 (128)
E. Fermi, 1934, *Nuovo Cim.* **11**, 1 and *Z. Phys.* **88**, 161 (210)
E. Fermi and C. N. Yang, 1949, *Phys. Rev.* **76**, 1739 (192, 211)
H. Feshbach and F. Villars, 1958, *Rev. Mod. Phys.* **30**, 24 (8)
R. P. Feynman, 1961, *Quantum Electrodynamics*, Benjamin (18); 1969, *Phys. Rev. Letters* **23,** 1415 (353)
R. P. Feynman and M. Gell-Mann, 1958, *Phys. Rev.* **109**, 193 (18, 214, 215, 236)
W. E. Fischer and P. Minkowski, 1972, *Nucl. Phys.* B**36**, 519 (341)
S. M. Flatté et al., 1966, *Phys. Rev.* **145**, 1050 (176)
D. Flothmann, W. Wiesner, R. Löhken, and H. Rebel, 1969, *Z. Phys.* **225**, 164 (220)
G. Flügge, 1979, Z. Physik C 1, 121 (399)
L. L. Foldy and S. A. Wouthuysen, 1950, *Phys. Rev.* **78**, 29 (25)
H. Frauenfelder et al., 1957, *Phys. Rev.* **106**, 386 (212)
W. R. Frazer and J. R. Fulco, 1960, *Phys. Rev.* **117**, 1603, 1609 (298, 308)
J. L. Friar, 1976, *Ann. Phys.* **98**, 490 (71, 74)
M. Froissart, 1961, *Phys. Rev.* **123**, 1053 (314)
M. Fukugita and K. Igi, 1977, *Phys. Reports* **31C**, 237 (332)
Gargamelle collaboration, 1976, *Nucl. Phys.* B**114**, 189 (257)
R. Garwin, L. Lederman, and M. Weinrich, 1957, *Phys. Rev.* **105**, 1415 (212)
S. Gasiorowicz, 1966, *Elementary Particle Physics*, Wiley (79, 179, 188)
M. Gell-Mann, 1956, *Nuovo Cimento* **4,** Suppl. 2, 848 (147); 1961, *Report CTSL-20* (reprinted in the book of Gell-Mann and Neéman, 1964) (177, 179); 1962, *Phys. Rev.* **125,** 1067 (312, 314); 1964, *Phys. Rev. Letters* **12,** 155 (147, 188)
M. Gell-Mann and M. Levy, 1960, *Nuovo Cim.* **16**, 705 (244)
M. Gell-Mann and Y. Ne'eman, 1964, *The Eightfold Way*, Benjamin
M. Gell-Mann, and D. Sharp, and W. G. Wagner, 1962, *Phys. Rev. Letters* **8**, 261 (176)
M. Gell-Mann and K. M. Watson, 1954, *Ann. Rev. Nucl. Sci.* **4**, 219 (183)
S. S. Gerstein and J. B. Zeldovitch, 1955, *Z. Eksperim. Teor. Fiz.* **29,** 698 (215, 236)
G. Giacomelli, 1970, *Progr. in Nucl. Phys.* **12**, 77 (352); 1976, *Phys. Reports* **23C**, 124 (315, 316)
V. Glaser and B. Jaksić, 1957, *Nuovo Cim.* **5**, 1197 (76)
S. L. Glashow, 1961, *Nucl. Phys.* **22**, 579 (258)
S. L. Glashow, J. Iliopoulos, and L. Maiani, 1970, *Phys. Rev.* D**2**, 1285 (206, 258, 263)
R. J. Glauber, 1959, in *Lectures in Theoretical Physics*, ed. W. C. Brittin et al., Vol. 1, Interscience (359)
R. L. Glückstern and S.-R. Lin, 1964, *J. Math. Phys.* **5**, 1549 (35)
M. Goldberg et al., 1964, *Phys. Rev. Letters* **12**, 546 (146)
M. L. Goldberger, M. T. Grisaru, S. W. MacDowell, and D. Y. Wong, 1960, *Phys. Rev.* **120**, 2250 (348)

M. L. Goldberger and S. B. Treiman, 1958, *Phys. Rev.* **110**, 1178, 1478; **111**, 354 (244)
M. L. Goldberger and K. M. Watson, 1964, *Collision Theory*, Wiley
G. Goldhaber et al., 1976, *Phys. Rev. Letters* **37**, 255 (207)
G. J. Gounaris and J. J. Sakurai, 1968, *Phys. Rev. Letters* **21**, 244 (304)
M. Gourdin, 1963, *Nuovo Cim.* **28**, 533; 1966, *Diffusion des Electrons de Haute Energie*, Masson & Cie., Paris (70); 1967, *Unitary Symmetries*, North-Holland (179, 188); 1971, in *Lectures in Theoretical Physics*, Vol. XII-B, Gordon and Breach; 1974, *Physics Reports* **11C,** 30 (173, 175, 257)
G. Grammer and D. R. Yennie, 1973, *Phys. Rev.* **D8,** 4332 (130)
O. W. Greenberg, 1964, *Phys. Rev. Letters* **13,** 598 (194); 1978, *Ann. Rev. Nucl. Part. Sci.*, **28,** 327 (194, 267)
O. W. Greenberg and C. A. Nelson, 1977, *Phys. Reports* **32C**, 71 (194)
F. Gross, 1969, *Phys. Rev.* **186**, 1448 (133, 346)
H. Grotch and D. R. Yennie, 1969, *Rev. Mod. Phys.* **41**, 350 (134)
J. Hamilton, 1967, in *High Energy Physics*, Vol. I (ed. E. H. S. Burhop), Academic Press (329)
J. Hamilton and B. Tromborg, 1972, *Partial Wave Amplitudes and Resonance Poles*, Clarendon Press; 1976, *Nuovo Cim.* **33A**, 605 (330)
H. Harari, 1978, *Phys. Reports* **42C**, 235 (266)
F. Hasert et al., 1973, *Phys. Letters* **46B**, 121 (257)
W. Heitler, 1941, *Proc. Cambridge Phil. Soc.* **37**, 291; 1944, *The Quantum Theory of Radiation, Clarendon* (293)
R. J. Hemingway et al., 1978, *Nucl. Phys.* B **142**, 205 (404)
S. W. Herb et al., 1977, *Phys. Rev. Letters* **39**, 252 (208)
P. W. Higgs, 1964, *Phys. Rev. Letters* **12,** 132 (260)
G. Höhler, F. Kaiser, R. Koch, and E. Pietarinen, 1979, *Handbook of Pion-Nucleon Scattering*, Physik-Daten 12-1, Fachinformationszentrum Karlsruhe, D7514 Eggenstein-Leopoldshafen 2 (144, 326, 331)
H. Höhler and E. Pietarinen, 1975, *Nucl. Phys.* B **95**, 210 (310)
G. Höhler et al., 1976, *Nucl. Phys.* B **114**, 505 (141, 306)
M. Holder et al., 1978, *Phys. Letters* **74B**, 277 (232)
S. Holmgren et al., 1977, *Nucl. Phys.* B **119**, 261 (186)
B. R. Holstein, 1974, *Rev. Mod. Phys.* **46**, 789 (227)
G. t'Hooft, 1971, *Nucl. Phys.* B **33**, 173; B **35**, 167 (260)
V. W. Hughes and C. S. Wu, 1975, *Muon Physics*, Academic Press (73)
B. D. Hyams et al., 1968, *Nucl. Phys.* B **7**, 1 (195)
D. Iagolnitzer, 1965, *J. Math. Phys.* **6**, 1576 (356)
J. Iizuka, 1966, *Progr. Theor. Phys. Suppl.* **37-38**, 21 (193)
M. Inokuti, Y. Itakawa, and J. E. Turner, 1978, *Rev. Mod. Phys.* **50**, 23 (368)
A. C. Irving and C. Michael, 1974, *Nucl. Phys.* B **82**, 282 (169)
A. C. Irving and R. P. Warden, 1977, *Phys. Reports* **34C**, 118 (353)
J. D. Jackson, 1975, *Classical Electrodynamics*, Wiley (ix, 25, 368); 1976, *Rev. Mod. Phys.* **48,** 417 (37)
M. Jacob and G. C. Wick, 1959, *Ann. Phys.* **7**, 404 (150, 391)
R. Jäckle and H. Pilkuhn, 1975, *Nucl.* Phys. A **247**, 521 (369)
J. M. Jauch and F. Rohrlich, 1976, *The Theory of Photons and Electrons*, Springer (ix, 131, 204)
C. J. Joachain and C. Quigg, 1974, *Rev. Mod. Phys.* **46**, 279 (359)
H. Joos, 1962, *Fortschr. Physik* **10**, 65 (67)
R. Jost, 1965, "The General Theory of Quantized Fields", American Math. Soc. Providence (252)

P. K. Kabir, 1963, *Development of Weak Interaction Theory*, Gordon and Breach (216)
G. Källén, 1958 in *Encyclopedia of Physics V 1*, Springer (ix, 92, 117); 1964, *Elementary Particle Physics*, Addison-Wesley (269)
R. Karplus, C. M. Sommerfield, and E. H. Wichman, 1958, *Phys. Rev.* **111**, 1187 (286)
T. W. B. Kibble, 1960, *Phys.Rev.* **117**, 1159 (270)
K. Kleinknecht, 1976, *Ann. Rev. Nucl. Sci.* **26**, 1 (254)
M. Kobayashi and K. Maskawa, 1973, *Progr. Theor. Phys.* **49**, 652 (265)
H. W. Koch and J. W. Motz, 1959, *Rev. Mod. Phys.* **31**, 920 (371)
R. Koch and E. Pietarinen, 1979, *Nucl. Phys.* A (331)
J. J. Kokkedee, 1969, *The Quark Model*, Benjamin (192)
E. J. Konopinski, 1966, *The Theory of Beta Radio-Activity*, Clarendon (216, 226)
E. J. Konopinski and H. M. Mahmoud, 1953, *Phys. Rev.* **92**, 1045 (232)
R. A. Krajcik and L. L. Foldy, 1970, *Phys. Rev. Letters* **24**, 545 (77)
N. M. Kroll and M. A. Ruderman, 1954, *Phys. Rev.* **93**, 233 (399)
E. Lambert, 1969, *Helv. Phys. Acta* **42**, 667 (321)
L. D. Landau, 1957, *Nucl. Phys.* **3,** 127 (233); 1959, *Nucl. Phys.* **13,** 181 (269)
L. D. Landau and E. M. Lifshitz, 1971, *Course of Theoretical Physics 4*, Pergamon (ix, 86, 117, 367); 1975, *Lehrbuch der theoretischen Physik 4*, Akademie-Verlag (ix, 126, 277, 367)
C. B. Lang, 1975, *Nucl. Phys.* B **93**, 415 (293, 302)
C. M. G. Lattes, G. P. S. Occhialini, and C. F. Powell, 1947, *Nature*, London **160**, 453 (137)
B. E. Lautrup, A. Peterman, and E. de Rafael, 1972, *Phys. Reports* **3C,** 193 (128)
T. D. Lee, R. Oehme, C. N. Yang, 1957, *Phys. Rev.* **106,** 340 (252)
T. D. Lee and C. N. Yang, 1956, *Phys. Rev.* **104,** 254 (211); 1957, *Phys. Rev.* **105,** 1671 (233); 1960, *Phys. Rev.* **119,** 1410 (242)
Y. K. Lee, L. W. Mo, and C. S. Wu, 1963, *Phys. Rev. Letters* **10,** 253 (238)
O. I. Leipunskii, B. N. Novozhilov, and V. N. Sakharov, 1965, *The Propagation of Gamma Quanta in Matter*, Pergamon (114)
D. B. Lichtenberg, 1970, *Unitary Symmetry and Elementary Particles*, Academic Press (179)
H. J. Lipkin, 1965, *Lie Groups for Pedestrians*, North-Holland (180); 1973, *Phys. Reports* **8C,** 174 (192)
M. S. Livingstone and H. A. Bethe, 1937, *Rev. Mod. Phys.* **9**, 245 (366)
M. Locher and T. Mizutani, 1978, *Phys. Reports* **46C**, 44 (359)
F. E. Low, 1958, Phys. Rev. **110**, 974 (131)
B. C. Maglic, L. W. Alvarez, A. H. Rosenfeld, and M. L. Stevenson, 1961, *Phys. Rev. Letters* **7**, 178 (169)
S. Mandelstam, 1958, *Phys. Rev.* **112,** 1344 (299); 1960, *Phys. Rev. Lett.* **4,** 84 (286)
J. B. Mann and W. R. Johnson, 1971, *Phys. Rev.* A**4**, 41 (74)
W. Marciano and H. Pagels, 1978, *Phys. Reports* **36C**, 138 (194)
R. E. Marshak, Riazuddin, and C. P. Ryan, 1968, *Theory of Weak Interactions in Particle Physics*, Wiley-Interscience (216, 245)
A. D. Martin and T. D. Spearman, 1970, *Elementary Particle Theory*, North-Holland (150)
B. R. Martin, D. Morgan, G. Shaw, 1976, "Pion-Pion Interaction in Particle Physics", Academic Press (323)
T. S. Mast et al., 1969, *Phys. Rev.* **183**, 1200 (161, 251)

W. H. McMaster, 1961, *Rev. Mod. Phys.* **33**, 8 (110)
E. Merzbacher, 1970, "Quantum Mechanics", Wiley (84)
A. Messiah, 1961, *Quantum Mechanics*, North-Holland (36)
T. C. Mo and C. H. Papas, 1971, *Phys. Rev.* D**4**, 3566, D**6**, 2296 (36)
G. Molière, 1947, *Z. Naturforschung* **2A**, 133 (359)
R. G. Moorhouse, H. Oberlack, and A. H. Rosenfeld, 1974, *Phys. Rev.* D **9**, 1 (341)
G. Mopurgo, 1970, *Ann, Rev. Nucl. Science* **20**, 105 (192)
M. Morita, 1973, *Beta Decay and Muon Capture*, Benjamin (216)
N. F. Mott and H. S. W. Massay, 1965, *The Theory of Atomic Collisions*, Clarendon (28)
J. W. Motz, H. A. Olsen, and H. W. Koch, 1964, *Rev. Mod. Phys.* **36**, 881; **41** (1969), 581 (371)
N. Mukhopadhyay, 1977, *Phys. Reports* **30C**, 1 (216)
P. Musset and J.-P. Vialle, 1978, *Phys. Reports* **39C**, 1 (257)
F. Myhrer and H. Pilkuhn, 1976, *Z. Physik* A **276**, 29 (404)
M. M. Nagels et al., 1976, *Nucl. Phys.* B **109**, 1 (186); 1979, *Nucl. Phys.* B **147**, 189 (213, 247, 332)
Y. Nambu, 1960, *Phys. Rev. Letters* **4**, 380 (244)
N. F. Nasrallah and K. Schilcher, 1978, *Phys. Letters* **78B**, 84 (204)
Y. Ne'eman, 1961, *Nucl. Phys.* **26**, 222 (177)
H. Nielsen and G. C. Oades, 1972, *Nucl. Phys.* B **49**, 586 (309)
K. Nishijima, 1955, *Progr. Theor. Phys.* **13**, 285 (147)
L. C. Northcliffe and R. F. Schilling, 1976, *Nuclear Data Tables* A **7**, 233 (368)
S. Okubo, 1962, *Progr. Theor. Phys.* **27**, 949 (179); 1964, *Phys. Letters* **8**, 362 (188)
H. A. Olsen, 1968, *Springer Tracts in Modern Phys.* **44**, 83 (110, 113)
M. G. Olsson, 1976, *Phys. Rev.* D **13**, 2502 (337)
R. Omnès, 1958, *Nuovo Cim.* **8**, 316 (308)
H. Pagels, 1975, *Phys. Reports* **16C**, 219 (313)
Particle Data Group, 1976, *Rev. Mod. Phys.* 48 (251); 1978, *Phys. Letters* **75B**, 1 (47, 187, 332, 349, 399–407)
W. Pauli, 1958, *Handbuch der Physik 5/1* (S. Flügge ed.), Springer (25)
Y. Perez-y-Jorba and F. Renard, 1977, *Phys. Reports* **31C**, 1 (205)
M. L. Perl et al., 1975, *Phys. Rev. Letters* **35**, 1489 (137)
A. Pevsner et al., 1961, *Phys. Rev. Letters* **7**, 421 (146)
F. Pham, 1967, *Ann. de L'Inst. H. Poincaré VI*, 89 (356)
H. Pilkuhn, 1967, "The Interactions of Hadrons", North-Holland (197); 1972, Landolt-Börnstein, Group I, Vol. 6 (H. Schopper ed.) (195); 1974, *Nucl. Phys.* B **82**, 365 (172)
I. Pomeranchuk, 1958, *JETP* **34**, 725 (320)
B. Pontecorvo, 1947, *Phys. Rev.* **72**, 246 (211)
C. F. Powell, 1950, *Reports on Progr. in Phys.* **13**, 350 (137)
C. Y. Prescott et al., 1978, *Phys. Letters* **77B**, 347 (258)
S. Raman and N. B. Gove, 1973, *Phys. Rev.* C **7**, 1995 (228, 240)
S. Raman, T. A. Walkiewicz, and H. Behrens, 1975, *Atomic Data and Nuclear Data Tables* **16**, 451 (241)
W. Rarita and J. Schwinger, 1941, *Phys. Rev.* **60**, 61 (68)
G. Rasche and W. S. Woolcock, 1976, *Helv. Phys. Acta* **49**, 557 (327)
T. Regge, 1959, *Nuovo Cim.* **14**, 951 (317)
F. Reines and C. L. Cowan, 1959, *Phys. Rev.* **113**, 273 (209, 232)

F. Reines and M. F. Crouch, 1974, *Phys. Rev. Letters* **32**, 493 (140)
F. M. Renard, 1970, *Nucl. Phys.* B **15,** 267 (201)
B. Renner, 1968, *Current Algebras and their Applications*, Pergamon
E. Reya, 1974, *Rev. Mod. Phys.* **46**, 545 (313)
Riazuddin, 1959, *Phys. Rev.* **114,** 1184 (204)
F. Rohrlich, 1965, *Classical Charged Particles*, Addison-Wesley (36)
M. E. Rose, 1957, *Elementary Theory of Angular Momentum*, Wiley (390); 1961, *Relativistic Electron Theory*, Wiley (30)
M. N. Rosenbluth, 1950, *Phys. Rev.* **79**, 615 (65)
L. Rosselet et al., 1977, *Phys. Rev.* D **15**, 574 (323)
A. M. Rossi et al., 1975, *Nucl. Phys.* B **84**, 269 (354)
S. Sakata, 1956, *Progr. Theor. Phys.* **16**, 686 (177)
J. J. Sakurai, 1967, *Advanced Quantum Mechanics*, Addison-Wesley (86)
A. Salam, 1957, *Nuovo Cim.* **5**, 299 (233)
A. Salam and J. C. Ward, 1964, *Phys. Letters* **13**, 168 (258)
M. D. Scadron and H. F. Jones, 1968, *Phys. Rev.* **173**, 1734 (348)
L. Schachinger et al., 1978, *Phys. Rev. Letters* **41**, 1348 (403)
H. F. Schopper, 1966, *Weak Interactions and Nuclear Beta Decay*, North-Holland (216, 229)
L. Schülke, 1964, *Z. Physik* **179**, 331 (229)
S. S. Schweber, 1961, *An Introduction to Relativistic Quantum Field Theory*, Row, Peterson and Co. (79)
W. T. Scott, 1963, *Rev. Mod. Phys.* **35**, 231 (359)
D. Shay and R. H. Good, 1969, *Phys. Rev.* **179**, 1410 (70)
D. Sivers, S. J. Brodsky and R. Blankenbecler, 1976, *Phys. Reports* **23C**, 2 (355)
A. Sommerfeld, 1949, *Partial Differential Equations in Physics*, Academic Press (320)
P. P. Srivastava and G. Sudarshan, 1958, *Phys. Rev.* **110**, 765 (158)
B. Stech and L. Schülke, 1964, *Z. Physik* **179**, 314 (216)
H. Stöckel, 1976, *Fortschritte der Physik* **24**, 417 (36)
R. F. Streater and A. S. Wightman, 1964, *PCT, Spin and Statistics and all that*, Benjamin (89, 252)
M. A. Stroscio, 1975, *Physics Reports* **22C**, 216 (116)
E. C. G. Sudarshan and R. E. Marshak, *Phys. Rev.* **109**, 1860 (214)
H. Taketani, A. Machida, and H. Ohnuma, 1952, *Prog. Theor. Phys.* **7,** 45 (347)
B. N. Taylor, W. H. Parker, and D. N. Langenberg, 1969, *Rev. Mod. Phys.* **41**, 375 (398)
J. C. Taylor, 1976, *Gauge Theories of Weak Interactions*, Cambridge Univ. Press (258)
J. R. Taylor, 1972, *Scattering Theory*, Wiley (12, 288)
H. A. Tolhoek, 1956, *Rev. Mod. Phys.* **28**, 277 (110)
Yung-Su Tsai, 1974, *Rev. Mod. Phys.* **46**, 815 (371, 373)
H. Überall, 1971, *Electron Scattering from Complex Nuclei*, Academic Press (35, 59)
K. Ueno et al., 1979, *Phys. Rev. Letters* **42**, 486 (407)
B. L. Van der Waerden, 1929, *Göttinger Nachrichten*, 100 (18)
A. Vitale, A. Berlin, G. Carboni, 1975, *Phys. Rev.* D **11,** 2441 (244)
S. Waldenström and H. Olsen, 1971, *Nuovo Cim.* **3A**, 491 (65)
R. L. Walker, 1969, *Phys. Rev.* **182**, 1729 (335)
A. H. Wapstra and N. B. Gove, 1971, *Nuclear Data Tables*, A**9**, 265 (141)
K. M. Watson, 1954, *Phys. Rev.* **95**, 228 (338)
H. A. Weidenmüller, 1961, *Rev. Mod. Phys.* **33**, 574 (227)

S. Weinberg, 1964, *Phys. Rev.* **133,** B 1318; **135,**B 1049 (67); 1967, *Phys. Rev. Letters* **19,** 1264 (258)

W. I. Weisberger, 1965, *Phys. Rev. Letters* **14,** 1047 (312)

W. Whaling, 1958, *Handbuch der Physik* **34,** Springer (367)

W. Wiesner, D. Flothmann, and H. J. Gils, 1973, *Nucl. Instr. and Methods* **112,** 449 (220)

D. H. Wilkinson, 1969, *Isospin in Nuclear Physics,* North-Holland (140)

L. Wolfenstein, 1964, *Phys. Rev. Letters* **13,** 562 (139, 255)

B. Wiik, G. Wolf, 1979, *Springer Tracts in Modern Physics* (205)

C. S. Wu et al., 1957, *Phys. Rev.* **105,** 1413 (212)

C. S. Wu and S. A. Moszkowski, 1966, *Beta Decay,* Interscience (216)

C. N. Yang and R. Mills, 1954, *Phys. Rev.* **96,** 191 (260)

D. R. Yennie, 1971, in *Hadronic Interactions of Electrons and Photons* (J. Cumming and H. Osborn eds.), Academic Press (365)

H. Yukawa, 1935, *Proc. Phys. Math. Soc. Japan* **17,** 48 (137, 342)

A. Zee, 1972, *Phys. Reports* **3C,** 127 (204)

G. Zweig, 1964, *CERN* preprint TH 412 (unpublished) (193)

# Index